Techniques in Sedimentology

Techniques in Sedimentology

Edited by
MAURICE TUCKER
BSc, PhD
Department of Geological Sciences
University of Durham
Durham DH1 3LE
UK

Blackwell Scientific Publications
OXFORD LONDON EDINBURGH BOSTON PALO ALTO MELBOURNE

Editorial offices:
Osney Mead, Oxford OX2 0EL
(*Orders:* Tel. 0865 240201)
8 John Street, London WC1N 2ES
23 Ainslie Place, Edinburgh EH3 6AJ
3 Cambridge Center, Suite 208
Cambridge, Massachusetts 02142, USA
667 Lytton Avenue, Palo Alto
California 94301, USA
107 Barry Street, Carlton
Victoria 3053, Australia

First published 1988

Set by Setrite Typesetters Ltd, Hong Kong
Printed and bound in Great Britain by
William Clowes, Beccles, Suffolk

DISTRIBUTORS

USA
Blackwell Scientific Publications Inc
PO Box 50009, Palo Alto
California 94303
(*Orders:* Tel. (415) 965-4081)

Canada
Oxford University Press
70 Wynford Drive
Don Mills
Ontario M3C 1J9
(*Orders*: Tel: (416) 441–2941)

Australia
Blackwell Scientific Publications
(Australia) Pty Ltd
107 Barry Street
Carlton, Victoria 3053
(*Orders:* Tel. (03) 347 0300)

British Library
Cataloguing in Publication Data

Techniques in sedimentology.
1. Sedimentology — Technique
I. Tucker, Maurice E.
551.3′04′028 QE471

ISBN 0-632-01361-3
ISBN 0-632-01372-9 Pbk

Library of Congress
Cataloging-in-Publication Data

Techniques in sedimentology.

Bibliography: p.
Includes index.
1. Sedimentology. I. Tucker, Maurice E.
471.T37 1988 552′.5 87-34120
ISBN 0-632-01361-3
ISBN 0-632-01372-9 (pbk.)

Contents

List of contributors

IAN FAIRCHILD
Department of Geological Sciences, The University, Birmingham B15 2TT.

JOHN GRAHAM
Department of Geology, Trinity College, Dublin, Eire.

RON HARDY
Department of Geologial Sciences, The University, Durham DH1 3LE.

GILL HARWOOD
Department of Geology, The University, Newcastle upon Tyne NE1 7RU.

GRAHAM HENDRY
Department of Geological Sciences, The University, Birmingham B15 2TT.

JOHN MCMANUS
Department of Geology, The University, Dundee DD1 4HN.

JOHN MILLER
Grant Institute of Geology, The University, Edinburgh EH9 3JW.

MARTIN QUEST
Department of Geological Sciences, The University, Birmingham B15 2TT.
Present address: Core Laboratories, Isleworth, Middlesex TW7 5AB.

NIGEL TREWIN
Department of Geology and Mineralogy, The University, Aberdeen AB9 1AS.

MAURICE TUCKER
Department of Geological Sciences, The University, Durham DH1 3LE.

Preface

Sedimentologists are keen to discover the processes, conditions and environments of deposition and diagenesis of their rocks. They currently use a whole range of quite sophisticated instruments and machines, in addition to routine fieldwork and microscopic study. Although geologists have been making field observations for nearly 200 years, in the last two decades there have been many new approaches to the collection and processing of field data. New sedimentary structures and relationships are still being found in 'classic', well-studied rocks. The microscopic examination of sediments is an essential tool in the description and interpretation of sediments, but there are ways to maximize the information a thin slice of rock will yield. Chemical analyses are being increasingly used to prise the stories out of sedimentary minerals and rocks and many of the analytical procedures have come from other branches of the earth sciences or are more frequently used by other geologists.

This book aims to cover all the various techniques used in the study of sedimentary rocks. It aims to provide instructions and advice on the various approaches and to give examples of the information obtained and interpretations possible.

One chapter is concerned with the collection of field data, with the emphasis on how these data can be analysed and presented. The following chapter looks at grain size analyses, and grain size parameters and their interpretation. Two chapters deal with microscopic studies: one being concerned with the production of thin sections, peels and slices, and the other with the description and interpretation of sedimentary minerals and depositional and diagenetic textures. The now popular technique of cathodoluminiscence, which can reveal hidden structures, follows. The X-ray diffraction of mudrocks, carbonates and cherts, providing information on mineralogy and composition, is treated in some depth. A chapter on Scanning Electron Microscopy explains how the machine works and how samples are best prepared and viewed, and then gives examples of SEM uses in soft-rock geology. A final chapter reviews the principles behind the chemistry of sedimentary rocks and discusses the collection and preparation of samples, followed by the techniques of electron beam microanalysis, XRF, AAS, ICP, INAA and stable isotope MS. Sections are included on analytical quality and the reporting of results and the chapter concludes with examples of the application of chemcial analysis to sedimentary problems. This book is a multi-authored volume and so naturally there are different levels of treatment and emphasis throughout the text.

Techniques in Sedimentology is written for final year undergraduates and postgraduates, to give them information and ideas on how to deal with their rocks in the field and samples in the laboratory during dissertation and thesis research. Much useful background material is also provided which will be relevant to lecture courses. The book will also be invaluable to professional sedimentologists, in industry and academia alike, and to other earth scientists, as a source book for the various techniques covered and for tips and recipes on extracting information from sedimentary rocks.

Maurice Tucker
Durham, March 1988

1 Introduction

MAURICE TUCKER

1.1 INTRODUCTION AND RATIONALE

The study of sediments and sedimentary rocks has come a long way from the early days of field observations followed by a cursory examination of samples in the laboratory. Now many sophisticated techniques are applied to data collected in the field and to specimens back in the laboratory. Some of these techniques have been brought in from other branches of the earth sciences, while some have been specifically developed by sedimentologists.

Research on sediments and sedimentary rocks is usually a progressive gathering of information. First, there is the fieldwork, an essential part of any sedimentological project, from which data relating to the conditions and environments of deposition are obtained. With modern sediments, measurements can be made of the various environmental parameters such as salinity, current velocity and suspended sediment content, and the sediments themselves can be subject to close scrutiny and sampling. With ancient sediments, the identification of facies types and facies associations follows from detailed examination of sedimentary structures, lithologies, fossil content etc., and subsequent laboratory work on representative rocks. After consideration of depositional environment, the larger scale context of the sequence in its sedimentary basin may be sought, necessitating information on the broad palaeogeographical setting, the tectonics of the region, both in terms of synsedimentary and post-sedimentary movements, and the subsurface structure, perhaps with input from seismic sections. With an understanding of a sedimentary rock's deposition and tectonic history, leading to an appreciation of the rock's burial history, the diagenetic changes can be studied to throw light on the patterns of cementation and alteration of the original sediment, and on the nature of pore fluids which have moved through the sedimentary sequence. Although much information on the diagenesis can be obtained from petrograpic microscopic examination of thin sections of the rock, sophisticated instruments are increasingly used to analyse the rocks and their components for mineralogical composition, major, minor and trace elements, isotopic signatures and organic content. The data obtained provide much useful information on the diagenesis, and also on the original depositional conditions.

Which techniques to use in sedimentological research depend of course on the questions being asked. The aims of a project should be reasonably clear before work is commenced; knowing what answers are being sought makes it much easier to select the appropriate technique. There is usually not much point in hitting a rock with all the sophisticated techniques going, in the hope that something meaningful will come out of all the data. It may be that the problem would be solved with a few simple field measurements or five minutes with the microscope on a thin section, rather than a detailed geochemical analysis giving hundreds of impressive numbers which add little to one's understanding of the rock.

The techniques available to sedimentologists, and those covered in this book, often cannot be used for all sedimentary rock types. It is necessary to be aware of what all the various instruments available in a well-found earth sciences department can do and how they can be used with sedimentary rocks. Many such instruments are more frequently used and operated by hard rocks petrologists and geochemists, but they can be used with great success on sedimentary rocks, as long as one is still seeking an answer to a particular question rather than just another analysis. Certain techniques are best suited to specific sedimentary rock types and cannot be used generally to analyse any rock. In the next sections of this introduction (1.2 to 1.9), the techniques covered in this book, in Chapters 2 to 9 are briefly reviewed. Not all the possible techniques available are included in this book and Section 1.10 notes what is omitted and where to find details. It also indicates where new techniques in soft-rock geology are frequently published, so that the keen student can keep up to date with developments in this field.

1.2 COLLECTION AND ANALYSIS OF FIELD DATA: CHAPTER 2

In Chapter 2, John Graham examines the rationale behind fieldwork, and the various ways in which field data can be collected and presented in graphic form are shown. The various sedimentary structures and the identification of lithologies and fossils are not described in detail since there are textbooks on these topics (e.g. Collinson & Thompson, 1982; Tucker, 1982), but the problems of recognizing certain structures are aired. The collection and analysis of palaeocurrent data (described in Section 2.3) are important in facies analysis and palaeogeographical reconstruction and statistical treatments are available to make the data more meaningful. There are many ways in which to examine a sedimentary sequence for rhythms and cycles (Section 2.4) and again the field data can be manipulated by statistical analysis to reveal trends. This chapter also shows the many ways in which information from the field can be presented for publication.

1.3 GRAIN SIZE DETERMINATION AND INTERPRETATION: CHAPTER 3

It is very important to know quite precisely what the grain size distribution is in a sediment sample and the procedures here are described by John McManus. Sample preparation varies from the unnecessary to having to break up the rock into its constituent grains, dissolve out the cement in acid, or make a thin section of the sample. Sieving, sedimentation methods and Coulter counter analysis can be used for unconsolidated or disaggregated samples, but microscopic measurements are required for fully lithified sandstones and most limestones. With a grain size analysis at hand, various statistical parameters are calculated. From these, with care, it is possible to make deductions on the sediment's conditions and environment of deposition.

1.4 MICROSCOPIC TECHNIQUES I: SLICES, SLIDES, STAINS AND PEELS: CHAPTER 4

The rock thin section is the basis of much routine description and interpretation but all too often the production of the slide is not given thought. John Miller explains how the best can be achieved by double-polished thin sections and describes the various techniques of impregnating, staining and etching to encourage the slide to give up more of its hidden secrets. Acetate peels are frequently made of limestones and the manufacture of these is also discussed.

1.5 MICROSCOPIC TECHNIQUES II: PRINCIPLES OF SEDIMENTARY PETROGRAPHY: CHAPTER 5

This chapter is by Gill Harwood and follows on from the previous one by explaining how the various minerals and textures in sedimentary rocks can be recognized and interpreted. The chapter is written in such a way so that it is applicable to all sedimentary rock types, rather than discussing each separately, as is frequently the case in sedimentary petrography texts. There is a huge textbook literature on 'sed. pet.' and many of the books will be readily available in a university or institute library (see, e.g. Folk, 1966; Scholle, 1978, 1979; Tucker, 1981; Blatt, 1982). Thus, in depositional fabrics (Section 5.2), grain identification, modal composition, point counting techniques, grain morphology, size and orientation, and provenance studies are briefly treated with pertinent literature references and many diagrams, photomicrographs and tables. In diagenetic fabrics (Section 5.3), again a topic with a voluminous literature, the various diagenetic environments and porosity types are noted, and then compaction-related fabrics are described and illustrated. Here, compaction is divided into that resulting from mechanical processes, from chemical (solution) processes between grains, and from chemical processes in lithified sediments. Cementation is a major factor in a rock's diagenesis and Section 5.3.3 demonstrates the variety of cements in sandstones and limestones, their precipitational environments and how the timing of cementation can be deduced. Typical fabrics of dissolution, alteration and replacement are described and illustrated, with emphasis on how these can be distinguished from other diagenetic fabrics. This overview of microscopic fabrics shows what can be seen, how they can be described, and their significance in terms of depositional and diagenetic processes. Sound microscope work is a fundamental prerequisite for geochemical analyses, and of course it provides much basic information on the nature, origin and history of a sedimentary rock.

1.6 CATHODOLUMINESCENCE: CHAPTER 6

This chapter is presented by John Miller and describes a technique which has been very popular amongst carbonate sedimentologists for the last few years. Very pretty colour photographs can be obtained with CL and these have enhanced many a lecture and published paper. The specimen in a vacuum chamber is bombarded with electrons and light is emitted if activator elements are present. An explanation of the luminescence is given, with a consideration of excitation factors and luminescence centres, and then a discussion of equipment needs and operation. Sample preparation is relatively easy; polished thin sections (or slabs) are used. General principles of the description and interpretation of CL results are given, along with applications to sedimentology. It is with carbonate rocks that CL is most used and here it is particularly useful for recognizing different cement generations and for distinguishing replacements from cements. In sandstones it can differentiate between different types of quartz grain, help to spot small feldspar crystals, and reveal overgrowths on detrital grains. Good photography is important in CL studies and hints are provided on how the quality of photomicrographs can be improved. These days, a study of carbonate diagenesis is not complete without consideration of cathodoluminescence and the textures it reveals.

1.7 X-RAY DIFFRACTION OF SEDIMENTARY ROCKS: CHAPTER 7

X-ray diffraction is a routine technique in the study of mudrocks and is frequently used with carbonate rocks too, and cherts. Ron Hardy and Maurice Tucker provide a brief general introduction to XRD, the theory and the instrument. XRD is the standard technique for determining clay mineralogy and various procedures are adopted to separate the different clay minerals. Examples are given of how XRD data from muds can be used to infer palaeoclimate, transport direction, conditions of deposition, and the pattern of diagenesis. With carbonates, XRD is mostly used to study the composition of modern sediments, the Mg content of calcite, and the stoichiometry and ordering of dolomites. The procedure is relatively straightforward and the precision is good, and much useful information is provided. Fine-grained siliceous rocks are often difficult to describe petrographically, but XRD enables the minerals present, opal A, opal C-T or quartz, to be determined readily. It has been especially useful in documenting the diagenesis of deep sea siliceous oozes through to radiolarian and diatom cherts.

1.8 SCANNING ELECTRON MICROSCOPY IN SEDIMENTOLOGY: CHAPTER 8

The SEM has become popular for studying fine-grained sedimentary rocks and for examining the ultrastructure of grains, fossils and cements. Nigel Trewin briefly describes the microscope and provides an account of how sedimentary materials are prepared for the machine. The SEM is a delicate machine and often the picture on the screen or the photographs may not be as good as expected. Comments are given on how such difficulties can be overcome or minimized. The SEM also has the facility for attachments providing analysis, EDS and EDAX, and these can be most useful when the elemental composition of the specimen is not known. An SEM can also be adjusted to give a back-scattered electron image and with mudrocks this can reveal the nature of the clay minerals themselves. The SEM has been applied to many branches of sedimentology, particularly the study of the surface textures of grains, both carbonate and clastic. In diagenetic studies, the SEM is extensively used with sandstones, to look at the nature of clay cements, evidence of grain dissolution and quartz overgrowths. In carbonates too, the fine structure of ooids and cements is only seen with SEM examination.

1.9 CHEMICAL ANALYSIS OF SEDIMENTARY ROCKS: CHAPTER 9

In many branches of sedimentology, chemical analyses are made to determine major, minor and trace element concentrations and stable isotope signatures, to give information on the conditions of deposition and diagenesis, and on long- and short-term variations in seawater chemistry and elemental cycling. In this chapter, largely written by Ian Fairchild, with contributions from his colleagues Graham Hendry and Martin Quest, and Maurice Tucker, a quite detailed background is given on some of the important principles of sedimentary geochemistry: concentrations and activities, equilibrium, adsorption, incorporation of trace elements and partition coefficients, and stable isotope fractionation This chapter should help the reader appreciate some of the problems in interpreting geochemical data from rocks where, commonly, inferences are being made about the nature of fluids from which precipitation took place. In sedimentary geochemistry much emphasis is now placed on the sample itself since there is a great awareness of the chemical inhomogeneities in a coarse-grained, well-cemented rock. Individual grains or growth zones in a cement are now analysed where possible, rather than the bulk analyses of whole rocks.

The techniques covered in this chapter are X-ray fluorescence, atomic absorption spectrometry, inductively-coupled plasma optical emission and mass spectrometry, electron microbeam analysis, neutron activation analysis and stable isotope (C.O.S) analysis. With the treatment of most of these techniques, the accent is not on the instrument operation, or theory — since there are many textbooks covering these aspects (e.g. Potts, 1987) — but on how sedimentary rocks can be analysed by these methods and the sorts of data that are obtained. A further section discusses precision and accuracy, the use of standards and how data can be presented.

To illustrate the use of geochemical data from sedimentary rocks, applications are described to the study of provenance and weathering, the deduction of environmental parameters, diagenesis and pore fluid chemistry, and elemental cycling.

1.10 TECHNIQUES NOT INCLUDED

This book describes most of the techniques currently employed by sedimentologists in their research into facies and diagenesis. It does not cover techniques more in the field of basin analysis, such as seismic stratigraphic interpretation, and decompaction, backstripping and geohistory analysis. A recent book on this which includes wire-line log interpretation and the tectonic analysis of basins is published by the Open University (1987). The measurement of porosity-permeability is also not covered.

This book does not discuss the techniques for collecting modern sediments through shallow coring, including vibracoring. The latter is described by Lanesky *et al.* (1979). Smith (1984) and others. There are many papers describing very simple inexpensive coring devices for marsh, tidal flat and shallow subtidal sediments (see, e.g. Perillo *et al.*, 1984). Also with modern deposits (and some older unconsolidated sands), large peels can be taken to demonstrate the sedimentary structures. Cloth is put against a smoothed, usually vertical surface of damp sand and a low viscosity epoxy resin sprayed or painted on to and through the cloth to the sand. On drying and removal, the sedimentary structures are neatly and conveniently preserved on the cloth. This technique is fully described by Bouma (1969).

The techniques used by sedimentologists are constantly being improved and new ones developed. Many sedimentological journals publish the occasional accounts of a new techique or method, and in many research papers there is often a methods section, which may reveal a slightly different, perhaps better, way of doing something. The *Journal of Sedimentary Petrology* publishes many 'research-methods papers', all collected together into one particular issue of the year. It is useful to keep an eye out for this section for the latest developments in techniques in sedimentology.

2 Collection and analysis of field data

JOHN GRAHAM

2.1 INTRODUCTION

Much of the information preserved in sedimentary rocks can be observed and recorded in the field. The amount of detail which is recorded will vary with the purpose of the study and the amount of time and money available. This chapter is primarily concerned with those studies involving sedimentological aspects of sedimentary rocks rather than structural or other aspects. Common aims of such studies are the interpretation of depositional environments and stratigraphic correlation. Direction is towards ancient sedimentary rocks rather than modern sediments, since techniques for studying the latter are often different and specialized, and are admirably covered in other texts such as Bouma (1969). In fieldwork the tools and aids commonly used are relatively simple, and include maps and aerial photographs, hammer and chisels, dilute acid, hand lens, penknife, tape, camera, binoculars and compass-clinometer.

During fieldwork, information is recorded at selected locations within sedimentary formations. This selection is often determined naturally such that all available exposures are examined. In other cases, e.g. in glaciated terrains, exposure may be sufficiently abundant that deliberate sampling is possible. The generation of natural exposures may well include a bias towards particular lithologies, e.g. sandstones tend to be exposed preferentially to mudrocks. These limitations must be considered if statements regarding bulk properties of rock units are to be made. For many purposes, vertical profiles of sedimentary strata are most useful. In order to construct these, continuous exposures perpendicular to dip and strike are preferred. With such continuous exposures, often chosen where access is easy, one must always be cautious of a possible bias because of an underlying lithological control.

The main aspects of sedimentary rocks which are likely to be recorded in the field are:

Lithology:	mineralogy/composition and colour of the rock.
Texture:	grain size, grain shape, sorting and fabric.
Beds:	designation of beds and bedding planes, bed thickness, bed geometry, contacts between beds.
Sedimentary structures:	internal structures of beds, structures on bedding surfaces and larger scale structures involving several beds.
Fossil content:	type, mode of occurrence and preservation of both body fossils and trace fossils.
Palaeocurrent data:	orientation of palaeocurrent indicators and other essential structural information.

In some successions there will be an abundance of information which must be recorded concisely and objectively. Records are normally produced in three complementary forms and may be augmented by data from samples collected for further laboratory work. These are:

(i) Field notes: These are written descriptions of observed features which will also include precise details of location. Guidance on the production of an accurate, concise and neat notebook is given in Barnes (1981), Moseley (1981) and Tucker (1982).

(ii) Drawings and photographs: Many features are best described by means of carefully labelled field sketches, supplemented where possible by photographs. All photographs must be cross referenced to field notes or logs and it is important to include a scale on each photograph and sketch.

(iii) Graphic logs: These are diagrams of measured vertical sections through sedimentary rock units. There are a variety of formats which are discussed below (Section 2.2.9). Although many logs are constructed on pre-printed forms, additional field notes accompany them in most cases.

2.2 RECORDING IN THE FIELD

2.2.1 Lithology identification and description

The ability to recognize different sedimentary rock types is embodied in most geology courses and is amply covered in texts such as Tucker (1981) and Blatt (1982). Such identification is generally quicker and more reliable with increased experience in the field, acquired initially under controlled conditions, i.e. with supervision and laboratory back up. Although there is a huge range of sedimentary rock types, by far the majority of successions contain only mudrocks, sandstones, conglomerates, limestones and dolomites, evaporites, and their admixtures. Thus some comments are made here on the recording of these major rock types.

MUDROCKS

Mudrocks can be subdivided in the field according to a simple objective scheme such as the widely accepted one shown in Table 2.1 (Ingram, 1953). It involves only the approximate determination of grain size and fissility. Colour, which is also particularly useful in mudrocks, is generally employed as a prefix. Application of more sophisticated laboratory techniques is necessary to obtain compositional information (Chapters 7, 8 and 9).

SANDSTONES

The lithology of sandstones, in terms of the grains/matrix ratio, the main detrital constituents, and the type of cement, can commonly be identified in the field, although detailed description and classification require thin section analysis (Chapters 4 and 5). The problem of matrix percentage and origin is difficult, even in thin section (Blatt, 1982), but it is often possible in a crude way to distinguish matrix-rich (wackes) from matrix-poor (arenites) sandstones in the field. This is most difficult when lithic grains are dominant and the sandstones are dark coloured and slightly metamorphosed and/or deformed.

Table 2.1. Scheme for nomenclature of fine-grained clastic sedimentary rocks

		Breaking characteristic	
Grain size	General terms	Non-fissile	Fissile
Silt + clay	Mudrock	Mudstone	Shale
Silt ≫ clay	Siltrock	Siltstone	Silt shale
Clay ≫ silt	Clayrock	Claystone	Clay shale

CONGLOMERATES

Conglomerates contrast with other rock types in that most of the measurement, description and classification is undertaken in the field, and laboratory study often takes a secondary role. A full description will involve measurement of size, determination of clast or matrix support, description of internal fabric and structures and data on composition (Fig. 2.1). Some commonly used descriptive terms for these coarse grained sedimentary rocks are:

Diamictite: a non-genetic term referring to any poorly sorted, terrigenous, generally non-calcareous, clast-sand-mud admixture regardless of depositional environment.

Breccia: a term used when the majority of the clasts are angular (in the sense of Section 2.2.2).

Extraformational: a term to describe clasts from source rocks outside the basin of deposition.

Intraformational: a term to describe clasts from fragmentation processes that take place within the basin of deposition and that are contemporaneous with sedimentation.

Oligomict: a term to describe conglomerates where one clast type, usually of stable, resistant material, is dominant.

Polymict (petromict): a term to describe conglomerates where several clast types are present.

Description can be enhanced by using the dominant clast size and clast type as prefixes, e.g. granite boulder conglomerate. A special series of terms is used where volcanic processes are involved in conglomerate formation (Lajoie, 1984).

Further information on the sedimentary structures present in conglomerates can be conveyed by use of the concise lithofacies codes as developed by Miall (1977, 1978), Rust (1978) and Eyles, Eyles & Miall (1983) (Table 2.2). Although these have been developed specifically for alluvial fan, fluvial and glacial lithofacies, there is every likelihood that they will and can be used for all conglomerates. These

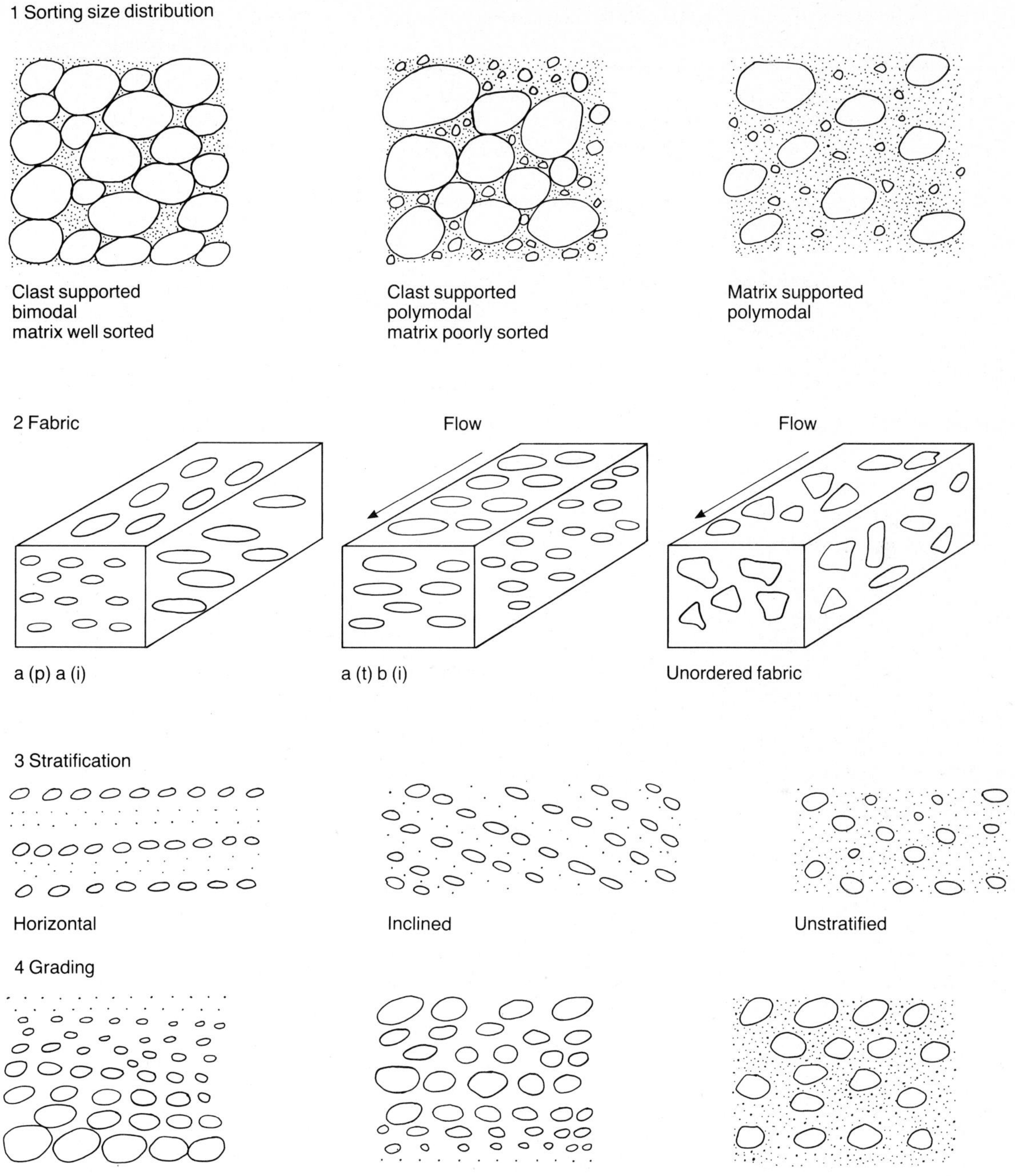

1 Sorting size distribution

Clast supported
bimodal
matrix well sorted

Clast supported
polymodal
matrix poorly sorted

Matrix supported
polymodal

2 Fabric

Flow

Flow

a (p) a (i)

a (t) b (i)

Unordered fabric

3 Stratification

Horizontal

Inclined

Unstratified

4 Grading

Normal

Inverse

Ungraded

Fig. 2.1. Features used in a textural and structural classification of conglomerate (from Harms, Southard & Walker, 1982). Under fabric, codes a and b refer to long and intermediate axes respectively; p = parallel to flow, t = transverse to flow, i = imbricate. (Reproduced by permission of SEPM.)

Table 2.2. Use of concise lithofacies codes in field description

(a) Criteria used for low sinuosity and glaciofluvial stream deposits (modified from Miall, 1977)

Code	Lithofacies	Sedimentary structures
Gms	Massive, matrix-supported gravel	None
Gm	Massive or crudely bedded gravel	Horizontal bedding, imbrication
Gt	Gravel, stratified	Trough crossbeds
Gp	Gravel, stratified	Planar crossbeds
St	Sand, medium to coarse, may be pebbly	Solitary (theta) or grouped (pi) trough crossbeds
Sp	Sand, medium to coarse, may be pebbly	Solitary (alpha) or grouped (omicron) planar crossbeds
Sr	Sand, very fine to coarse	Ripple marks of all types
Sh	Sand, very fine to very coarse, may be pebbly	Horizontal lamination, parting or streaming lineation
Sl	Sand, fine	Low angle (10°) crossbeds
Se	Erosional scours with intraclasts	Crude crossbedding
Ss	Sand, fine to coarse, may be pebbly	Broad, shallow scours including eta cross-stratification
Sse, She, Spe	Sand	Analogous to Ss, Sh, Sp
Fl	Sand, silt, mud	Fine lamination, very small ripples
Fsc	Silt, mud	Laminated to massive
Fcf	Mud	Massive with freshwater molluscs
Fm	Mud, silt	Massive, dessication cracks
Fr	Silt, mud	Rootlet traces
C	Coal, Carbonaceous mud	Plants, mud films
P	Carbonate	Pedogenic features

(b) Diagnostic criteria for recognition of common matrix-supported diamict lithofacies (from Eyles *et al.*, 1983)

Code	Lithofacies	Description
Dmm	Matrix-supported, massive	Structureless mud/sand/pebble admixture
Dmm(r)	Dmm with evidence of resedimentation	Initially appears structureless but careful cleaning, macro-sectioning, or X-ray photography reveals subtle textural variability and fine structure (e.g. silt or clay stringers with small flow noses). Stratification less than 10% of unit thickness

Code	Lithofacies	Sedimentary structures
Dmm(c)	Dmm with evidence of current reworking	Initially appears structureless but careful cleaning, macro-sectioning, or textural analysis reveals fine structures and textural variability produced traction current activity (e.g. isolated ripples or ripple trains). Stratification less than 10% of unit thickness
Dmm(s)	Matrix-rupported, massive, sheared	Dense, matrix supported diamict with locally high clast concentrations. Presence of distinctively shaped flat-iron clasts oriented parallel to flow direction, sheared
Dms	Matrix-supported, stratified diamict	Obvious textural differentiation or structure within diamict. Stratification more than 10% of unit thickness
Dms(r)	Dms with evidence of resedimentation	Flow noses frequently present; diamict may contain rafts of deformed silt/clay laminae and abundant silt/stringers and rip-up clasts. May show slight grading. Dms(r) units often have higher clast content than massive units; clast clusters common. Clast fabric random or parallel to bedding. Erosion and incorporation of underlying material may be evident
Dms(c)	Dms with evidence of current reworking	Diamict often coarse (winnowed) interbedded with sandy, silty and gravelly beds showing evidence of traction current activity (e.g. ripples, trough or planar cross-bedding). May be recorded as Dmm, St, Dms, Sr etc. according to scale of logging. Abundant sandy stringers in diamict. Units may have channelized bases
Dmg	Matrix-supported, graded	Diamict exhibits variable vertical grading in either matrix or clast content; may grade into Dcg
Dmg(r)	Dmg — with evidence of resedimentation	Clast imbrication common

schemes are still being refined and modified (cf. Eyles *et al.*, 1983 with McCabe, Dardis & Hanvey, 1984 and Shultz, 1984) and the overlap between the D (diamictite) and G (gravel) codes needs further clarification. The codes should not be regarded as all that is needed for environmental interpretation

(Dreimanis, 1984; Kemmis & Hallberg, 1984) but simply a concise and convenient shorthand description of some of the main observable features.

In addition to information on depositional processes and environments, polymict conglomerates can yield some information on the relative contribution of various source lithologies. However, there are many factors which affect the presence and size of clasts in conglomerates. The initial size of fragments released from the source area varies with lithology, being related to features such as bed thickness, joint spacing and resistance to weathering. In addition clasts have varying resistances to size reduction during transport. To avoid spurious size-related effects, compositional data can be compared at constant size. This can be achieved either by counting the clast assemblage for a given size class or by a more detailed analysis in which the proportion of clast types is examined over a spectrum of size classes at a single site.

To determine the distribution of clast types by size, an area of several square metres should be chosen on a clean exposure surface on which all clasts can be identified easily. Areas of strong shape selection, common in some proximal fluvial and beach environments, should be avoided since this can introduce a bias towards anisotropic clast lithologies. Preliminary observations and the nature of the study will determine the number of lithological types to which clasts are assigned. Often crude discriminants, e.g. granite porphyry, etc., will suffice. More detailed studies require more subtle subdivision and may involve thin section checks on field identification.

Clast counting should proceed from the finest clast size interval. It is important that all clasts are counted to eliminate bias; repetition may be avoided by using chalk to mark counted clasts. As coarser size clasts are counted, the area over which clasts are counted may be increased as a representative clast population (>100) is sampled. These data can be summarized by constructing a plot of percentage clast types against grain size. The proportion of any clast type at a given grain size can easily be read from the plot, allowing direct comparison from locality to locality (Fig. 2.2). To present data from many localities, stratigraphic columns can be used to show changes in clast composition, or a map can be devised to show the regional distribution of clast types (Figs 2.3 and 2.4).

Many published studies of clast composition are of limited value, either due to lack of specification of clast size or to problems of how the clasts to be measured were selected. Techniques of random selection of clasts involve the placing of a sampling grid, for example chalked squares or a piece of fish net, over the exposure and measuring either at grid intersections or within small grid squares. The former may be difficult to apply if only certain sizes are accepted; the latter may introduce bias if only a limited number of clasts per square are to be measured. It is important that the method of data collection is clearly stated so that its limitations can be assessed by later workers.

LIMESTONES AND DOLOMITES

Field distinction of carbonate rocks is possible, but detailed description is best performed in the laboratory using thin sections and acetate peels (Chapter 4) although under favourable conditions the latter may be made in the field. Dilute 10% HCl is a standard field aid. Whilst limestones will react vigorously, most dolomites will show little or no reaction unless they are powdered. The addition of Alizarin red S in HCl will stain limestones but not dolomite (Chapter 4) and this can be used in the field. In addition many dolomites are yellow or brown weathering, harder than limestones and may show poor fossil preservation. In sequences of alternating

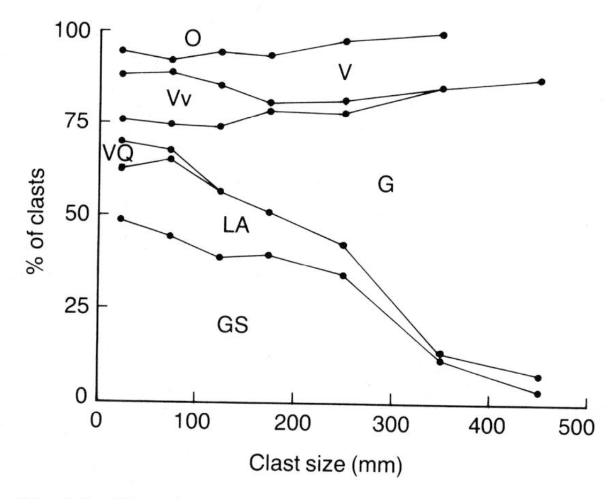

Fig. 2.2. Plot of percentage clast types versus grain size for one locality for a Lower Devonian fluvial conglomerate (diagram kindly supplied by Peter Haughton, University of Glasgow). Key to clast types: GS = greenschist, LA = lithic arenite, VQ = vein quartz, G = granite, Vv = vesicular volcanics, V = other volcanics, O = other.

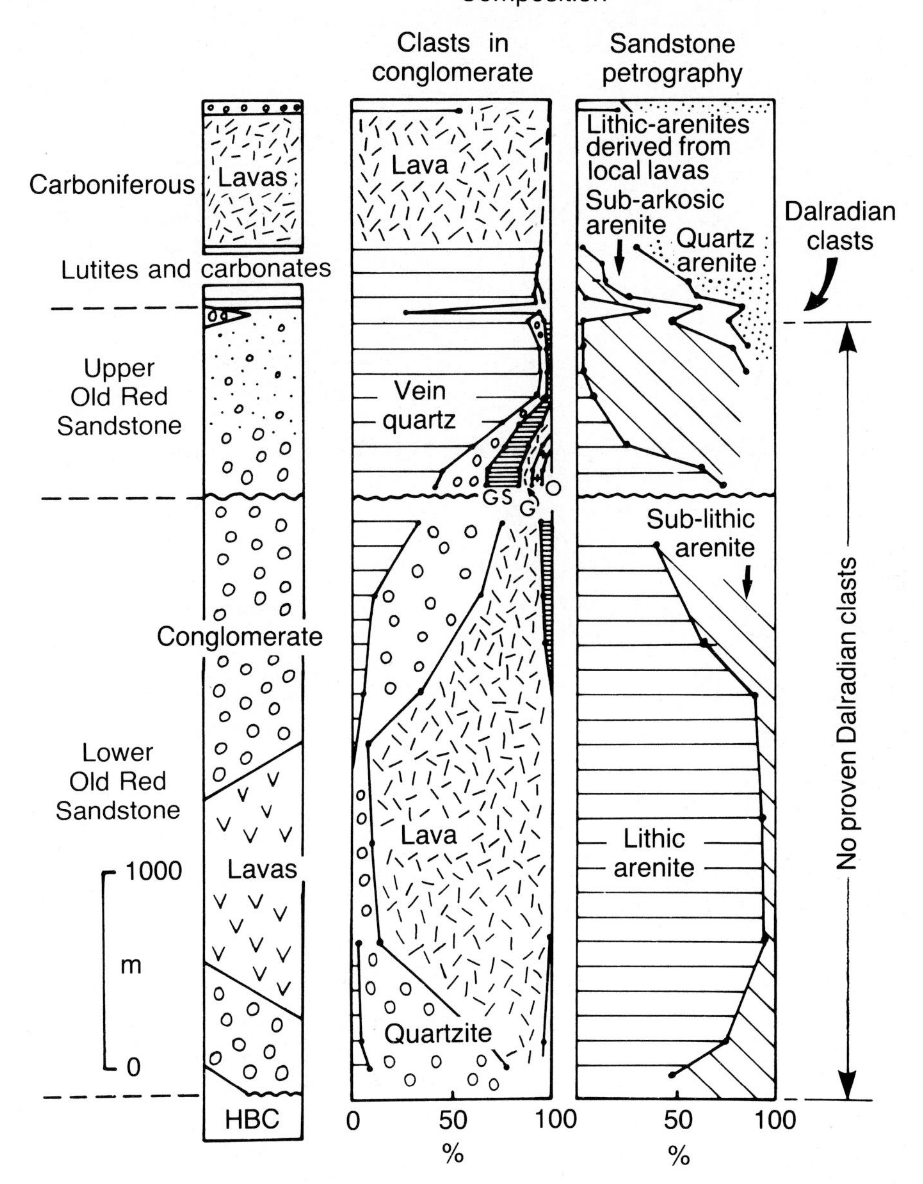

Fig. 2.3. Diagram to show changes in clast composition with time for the Midland Valley of Scotland (from Bluck, 1984). GS = greenschist, G = granite, O = other. (Reproduced by permission of the Royal Society, Edinburgh.)

limestones and dolomites, the beds of dolomite are commonly more intensely jointed and fractured than the beds of limestone.

Limestones are classified according to two well tried and tested schemes which, although designed for microscope work, can often be applied in the field. It is more likely that a field identification can be made using the classification of Dunham (1962), while later laboratory work can identify the constituent components through which the Folk (1962) classification may be applied. Even if a full identification is not possible, any recognizable allochem

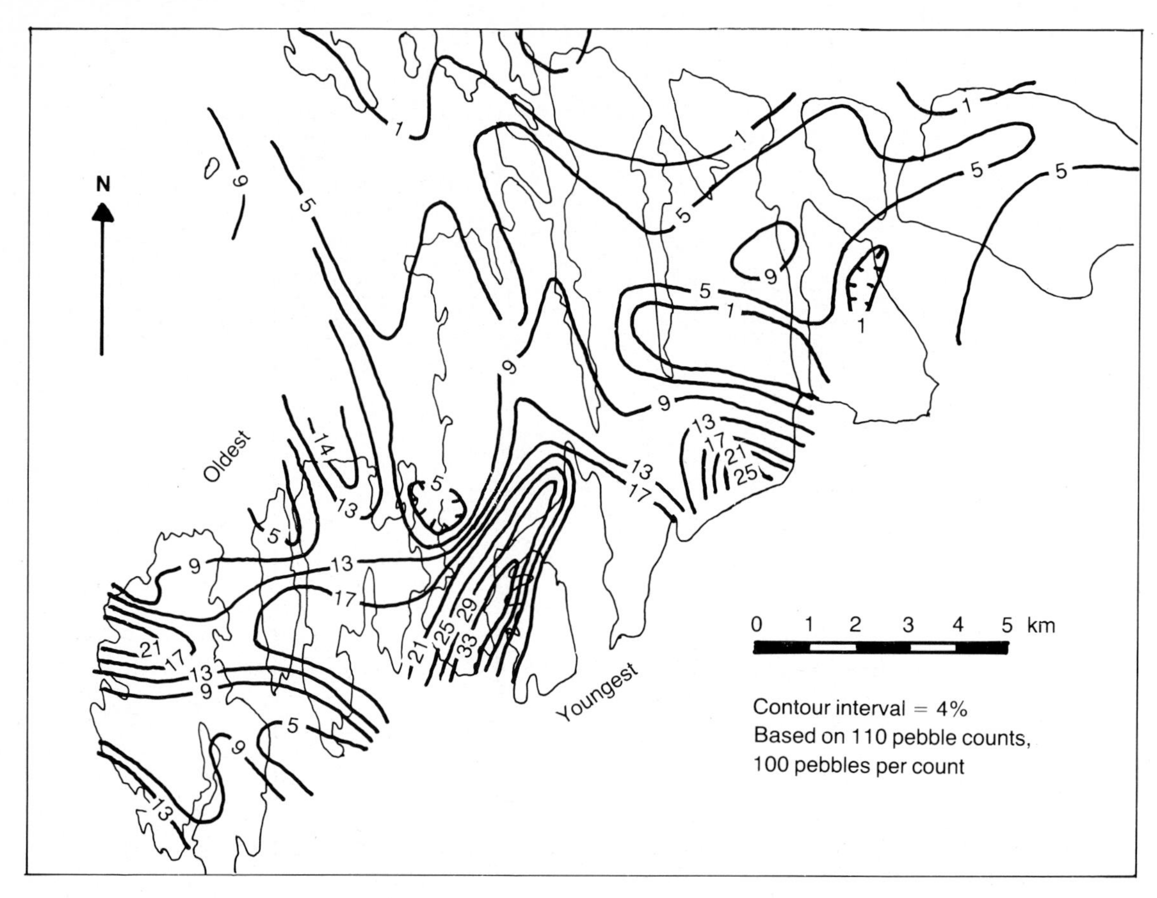

Fig. 2.4. Regional distribution of gabbro pebbles in the Solund Conglomerate, Devonian, Norway (from Nilsen, 1968). (Reproduced by permission of *Norges Geologiske Undersokelse*.)

type or fossil component should be recorded. If the degree of weathering is suitable, these can usually be determined with a hand lens.

EVAPORITES

Evaporites are chemical sediments that have been precipitated from water following the evaporative concentration of dissolved salts. Where these rocks are preserved at outcrop there is little difficulty in field description and preliminary identification (see Kendall, 1984 for a recent review). However, because of their soluble nature, evaporites are frequently dissolved or replaced in the subsurface. It is thus important to be aware of the typical pseudomorphs of evaporites which can be recognized, e.g. cubic halite crystals. Replacement is commonly by calcite, quartz or dolomite. Collapse features are also common where dissolution has occurred and a careful check should be made where disrupted or brecciated horizons are found.

MIXED LITHOLOGIES

Lithologies which are essentially mixtures of other rock types are frequent in some successions. At present there seems to be little agreement on the nomenclature of admixtures. An objective and consistent procedure of labelling these rocks whilst identifying the main and admixing lithologies is necessary. The commonest admixtures are of siliciclastic and carbonate sediments and a textural and compositional classification for these has recently been proposed by Mount (1985).

2.2.2 Texture

GRAIN SIZE

Because the determination of grain size is basic to sedimentological fieldwork, it is fortunate that the grade scale proposed by Wentworth (1922) is now internationally accepted as a standard (see Chapter 3 and Table 3.4). However, the choice of what to measure can be somewhat different for different parts of the grade scale. For conglomerates it is commonly maximum clast size that is measured, although it is advisable to estimate modal size(s) as well. Techniques for measuring maximum clast size have varied considerably among different workers, as shown in Table 2.3. In the absence of an accepted standard it is thus necessary to state how the measurement was made. The maximum size of clasts is used as a parameter partly because the measurement is easily made but also because of its known relationship to flow competence.

For sandstone it is usually modal size which is estimated by means of visual comparison to a reference card or block. These reference sets are constructed either by glueing sieved sand of each size

Table 2.3. Differing techniques for measuring maximum clast size in conglomerates

Author	Measurement of maximum clast size
Bluck (1967)	Average of 10 largest clasts in 0.5 × 0.5 m square
Steel *et al.* (1977)	Average of 10 largest clasts (after omission of 'outsized' clasts) from an area over several metres on either side of the point of bed thickness measurement
Heward (1978)	Average of longest axes of 25 largest clasts 1–5 m each side of section line at intervals 1 m
Surlyk (1978)	Measure 12 largest clasts from each bed; omit two largest then average
Allen (1981)	Average of 10 largest clasts in rectangular sampling area, height 0.5 m, width determined by size of largest clast such that $l = 0.69\ r^2$ where l = lateral sampling distance (cm), r = long axis/2 of largest clast (cm)

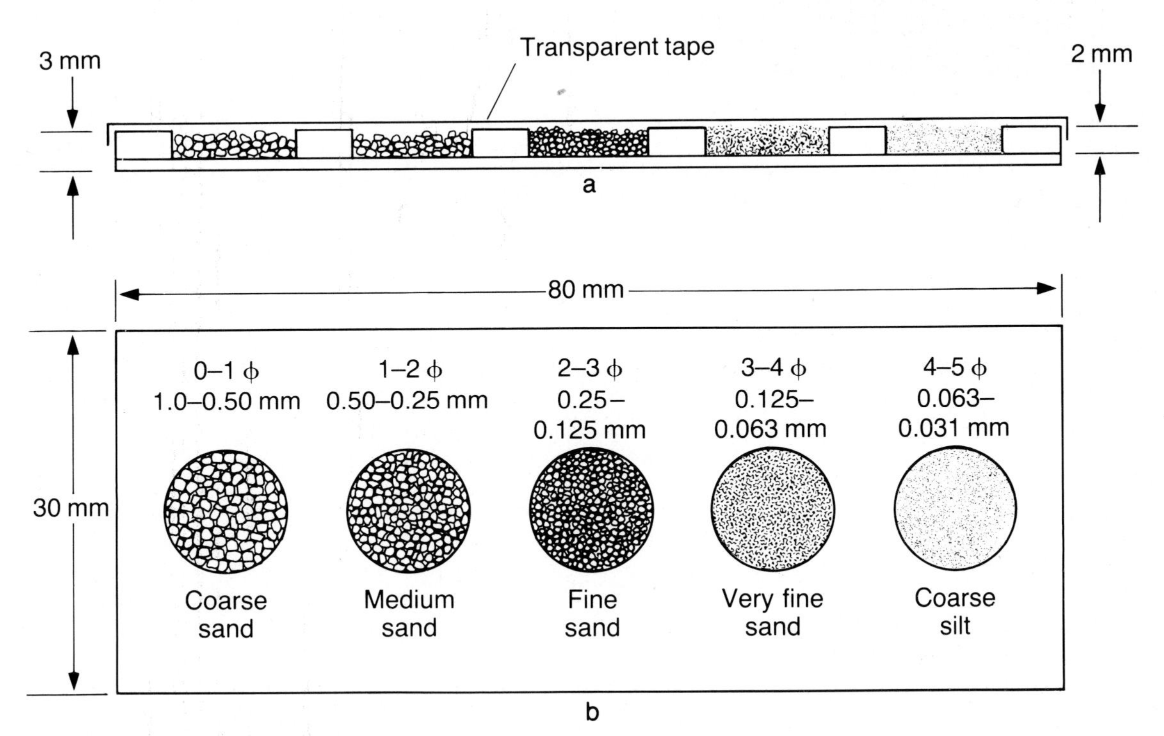

Fig. 2.5. Construction of simple grain size comparator for field use in (a) sectional and (b) plan views (after Blatt, 1982). (Reproduced by permission of Freeman.)

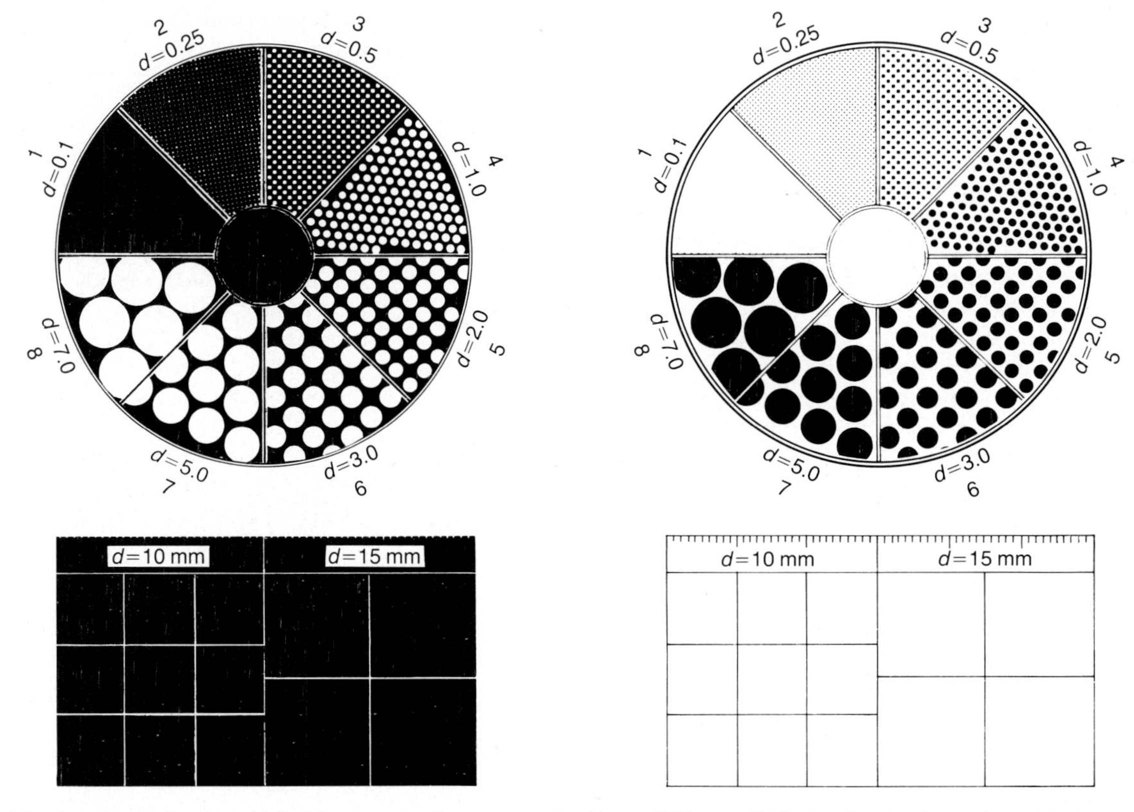

Fig. 2.6. Graph for determining the size of sedimentary grains (from Chilingar, 1982). Sand grains (or rock particles) are placed in the central part of the circle — light particles in the left chart and dark particles in the right chart — and compared with those on the graph using a hand lens. The numbers are used for notebook reading. (Reproduced by permission of the American Geological Institute.)

fraction on to an annotated card (e.g. Friend *et al.*, 1976) or by placing sieved sand within a series of depressions in a plastic ruler (e.g. Blatt, 1982 and Fig. 2.5). These reference sets must be small and durable. A similar approach for most sand sizes and finer gravels has long been used by Soviet geologists (Chilingar, 1982; Fig. 2.6). These charts can be easily fixed into the field notebook.

Grain size of mudrocks is much more difficult to estimate in the field and it is probably unrealistic to attempt more than the determination of whether silt exceeds clay or vice versa. This can be accomplished by using a combination of hand lens and 'feel'. Nibbling a small piece of mudrock with the teeth can be a useful test. Lack of abrasion plus a generally greasy or soapy feel will suggest dominance of clay; a gritty abrasive feel indicates the presence of silt grade quartz, of which some estimate may be possible with a powerful hand lens. In mudrocks where there is a regionally developed cleavage, the spacing of cleavage planes may locally be a useful guide.

GRAIN SHAPE

Shape of sedimentary particles is a complex property such that there are even differences of opinion as to what constitutes shape. It is taken here in the sense of Barrett (1980) as comprising form, roundness and surface texture. Only aspects of the first two are normally observable in the field.

Roundness is generally described by means of comparison to standard images which can be readily taped into the field note book. It has been shown that there is considerable operator variation with these

(Rosenfield & Griffiths, 1953; Folk, 1955) but nevertheless their use may serve to highlight problems requiring more rigorous laboratory investigation.

Similarly form (overall shape) is normally described by assigning grains to one of the four major classes erected by Zingg (1935) (Fig. 2.7). The parameter of form includes, but is not completely defined by, sphericity, the measure of approach of a particle to a sphere. It is also possible to use a visual comparison chart which combines the measures of roundness and sphericity (Powers, 1982; Fig. 2.8). Such information may help to indicate features such as lithological control of particle shape, shape sorting and depositional fabric.

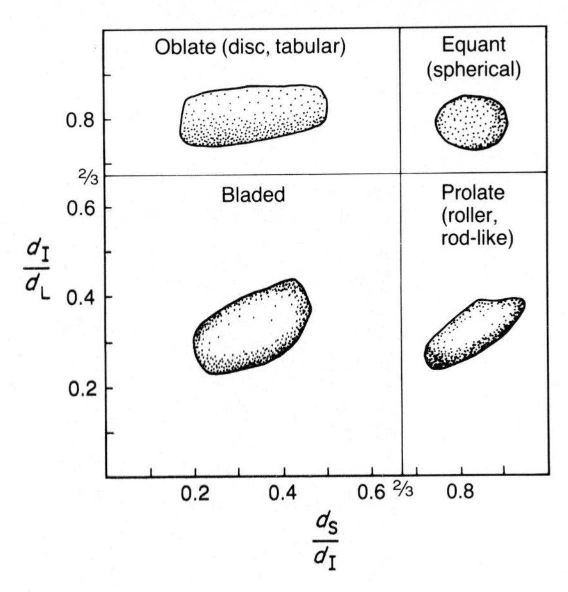

Fig. 2.7. Major shape classes of sedimentary particles (after Zingg, 1935). d_S, d_I, d_L = shortest, intermediate and longest diameters respectively.

SORTING

In the field degrees of sorting are most commonly determined for sandstones, usually by visual comparison with a number of standard images which can be taped into the field note book (Fig. 2.9). Field estimation of sorting in mudrocks is generally not possible. When studying limestones and conglomerates the presence of clast or matrix support is important in description and classification, e.g. in the application of the Dunham nomenclature.

FABRIC

Fabric refers to the mutual arrangement of grains within a rock, i.e. their orientation and packing. Fabrics may be produced during sedimentation or by later tectonic processes. Careful measurement will generally allow this distinction. The commonest fabrics are those produced by the orientation of elongate particles such as fossils in limestones, discs, blades and rods in conglomerates. This alignment may define bedding or may provide evidence of palaeocurrent direction as in imbrication (Section 2.3.2; Potter & Pettijohn, 1977). Fabrics also occur in finer-grained sedimentary rocks but are not always observable in the field.

POROSITY AND PERMEABILITY

It may be possible to make rough estimates of the amount and origin of porosity and permeability in the field even though accurate measurement is not possible. For example, one may note partially filled openings in fossils, or irregular channels formed by fracturing, leaching or alteration. However, there are many problems such as lateral variability of porosity over small distances and the fact that surface samples may not be representative of the rock's average porosity.

2.2.3 Colour

This can an important attribute in the description of many sedimentary rocks. There are basically two methods of description. By far the most rapid and simple is the subjective impression of the geologist in the field. Whilst this may be adequate for some general surveys it does lack objectivity, for colour perception is known to vary considerably among workers. More objective is the use of a standard colour chart such as that published by the Geological Society of America (Goddard *et al.*, 1975) based on the Munsell Colour System. The form and arrangement of the colour system are shown in Fig. 2.10. The first step is to determine the hue of the rock. There are 10 major hues (Fig. 2.10), each one divided into 10 divisions. Thus 5P would refer to the mid-point of the purple hue, 10P to a hue mid-way between purple and purple-blue and 7.5P to a hue mid-way between these two. Numbering is clockwise as shown in Fig. 2.10. After this a value is selected from 1 to 9 with 1 being the darkest and 9 the

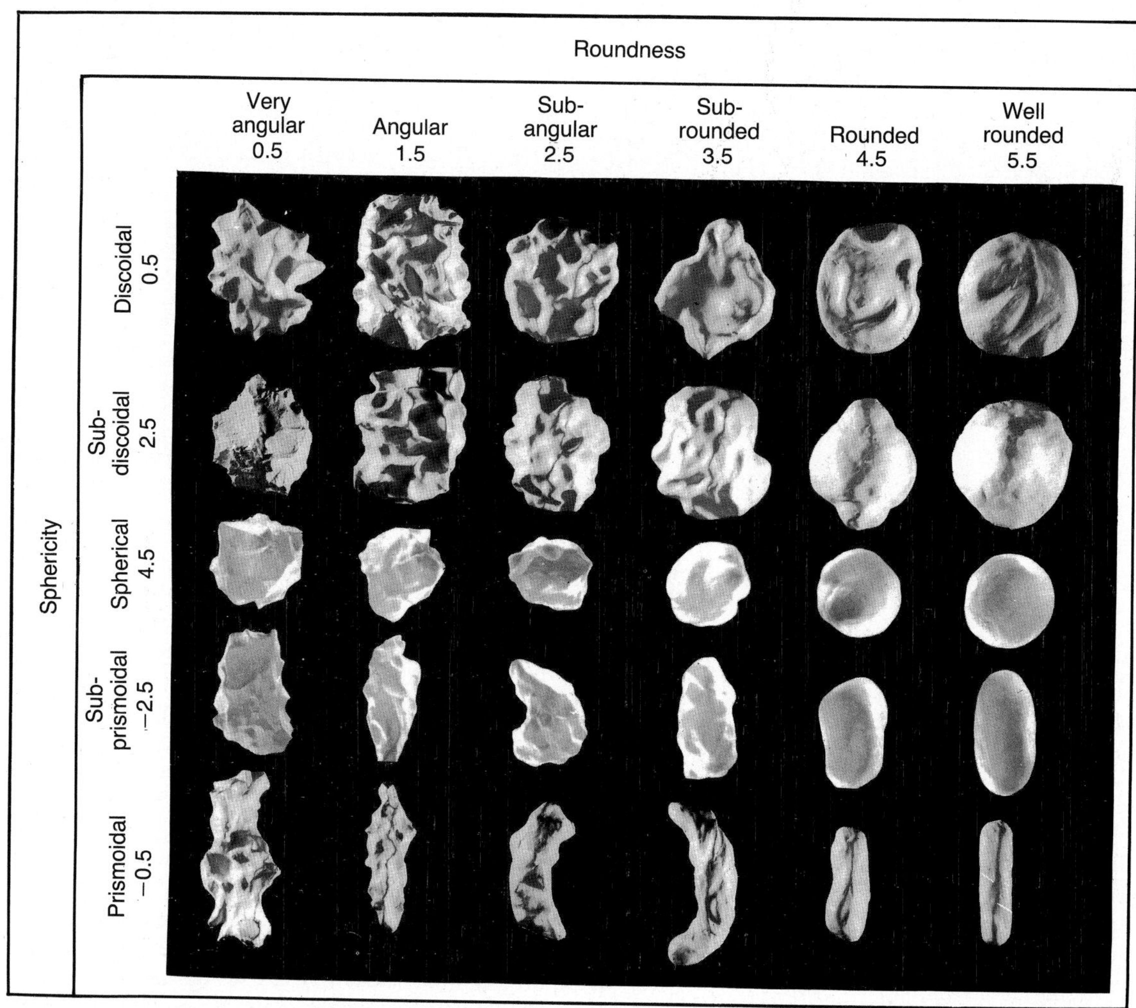

Fig. 2.8. Visual comparison chart for estimating roundness and sphericity (from Powers, 1982). (Reproduced by permission of the American Geological Institute.)

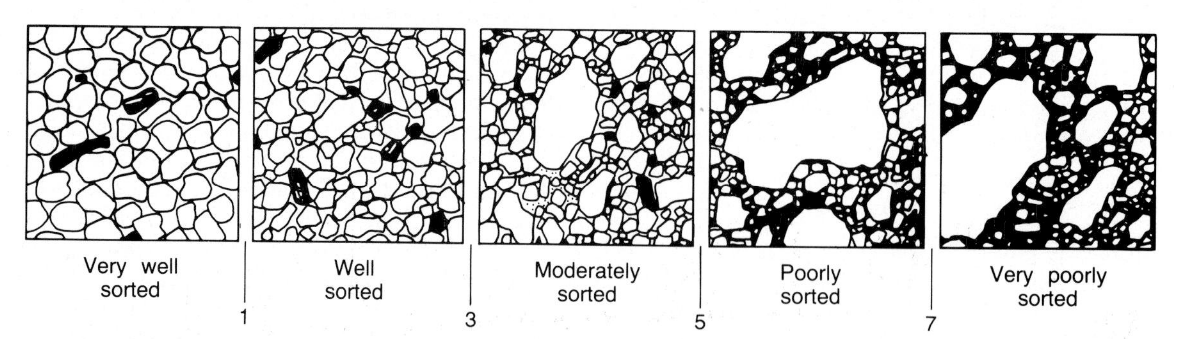

Fig. 2.9. Sorting images and standard terms (from Compton, 1962). The numbers indicate the number of size classes included by *c.* 80% of the material. The drawings represent sandstones seen with a hand lens. (Reproduced by permission of Wiley.)

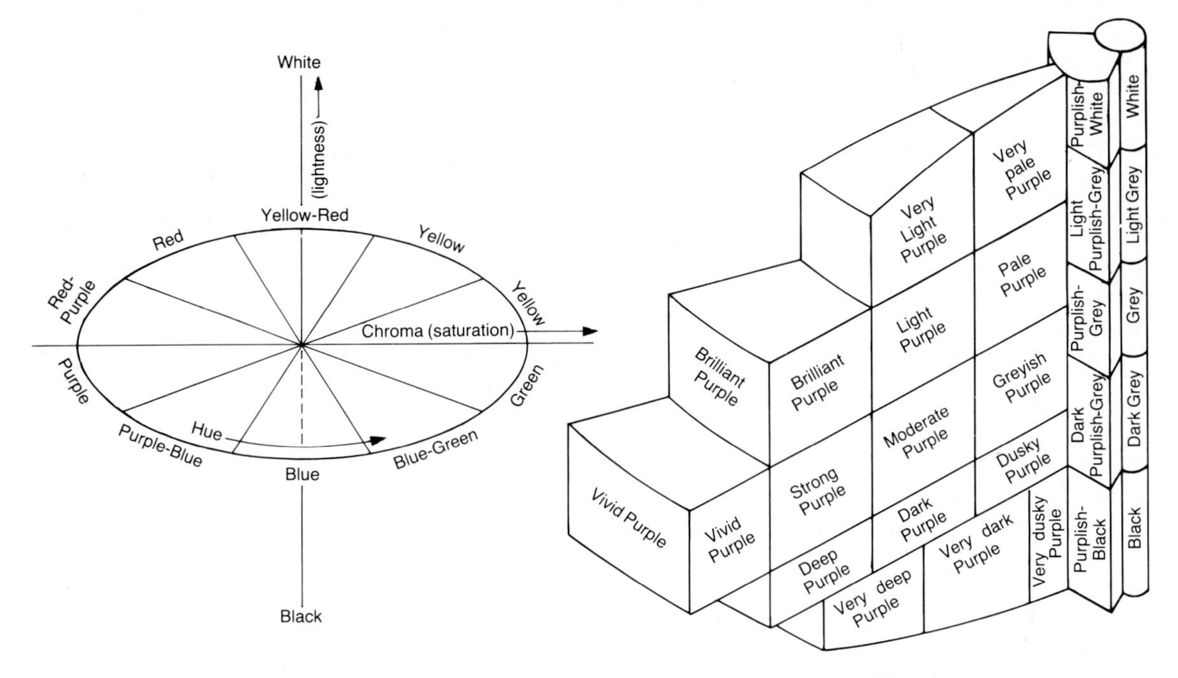

Fig. 2.10. The form and arrangement of the Munsell Colour System (from Goddard *et al.*, 1975). See text for explanation. (Reproduced by permission of the Geological Society of America.)

lightest (black to white when there is no hue). Also selected is a chroma which is a degree of colour saturation. These are given values from O (no colour saturation) to 6 for the most vivid colours. Thus a colour can be represented by a code such as 5P4/2 as well as a name — greyish-purple. Several standard colours are arranged on sheets in a small booklet which is taken into the field and then visually compared with the rocks. Intermediates are estimated.

2.2.4 **Induration and degree of weathering**

Although induration (hardness) cannot be readily quantified, it may be useful to record the variation in this property using the qualitative scheme below:

Unconsolidated:	Loose.
Very friable:	Crumbles easily between fingers.
Friable:	Rubbing with fingers frees numerous grains. Gentle blow with hammer disintegrates sample.
Hard:	Grains can be separated from sample with a steel probe. Breaks easily when hit with hammer.
Very hard:	Grains are difficult to separate with a steel probe. Difficult to break with a hammer.
Extremely hard:	Sharp hard hammer blow required. Sample breaks across most grains.

The degree of weathering will be a complex function of lithology, topography, climate and vegetation. In particular in humid tropical regions there may be an extremely thick regolith and little available 'fresh' bedrock. Some general comments on the nature and degree of weathering are appropriate, especially where these may influence or restrict interpretation.

2.2.5 **Bedding**

In descriptive terms, a bed is a layer that is sufficiently distinct from adjoining layers. In genetic terms, a bed represents a depositional episode during which conditions were *relatively* uniform. The main problem is that these two definitions do not always coincide, leading to some contention over exactly what constitutes a bed.

Campbell (1967) stated that distinction of beds depends on recognition of bedding surfaces produced during periods of non-deposition or by an abrupt change of conditions. This is the concept of the 'sedimentation unit' (Otto, 1938) which has been adopted by many later workers (e.g. Reineck & Singh, 1975; Collinson & Thompson, 1982). Other workers follow the convention that beds are recognizably distinct strata that have some lithologic or structural unity (Pettijohn, 1975, p. 102). Campbell acknowledged that bedding surfaces (using his definition) might not be readily recognized or may appear to be discontinuous. Thus there can be different judgements of the positions, thicknesses and features of beds depending on which concept is followed. The problem can be demonstrated by exposures of turbidites in the Carboniferous of Morocco (Figs 2.11 and 2.12). Following Otto (1938) or Campbell (1967), a 'bed' in Fig. 2.11 would be (1) + (2), a turbidite, or (3), a hemipelagite; the bedding surfaces would be between (3) and (1) and between (2) and (3). However, in many examples the boundary between units (2) and (3) is not recognizable due either to low quality exposure or simply to a lack of lithological contrast between units (2) and (3). Here the 'beds' recognized may well be (1) and (2) + (3). Such ambiguity of description is clearly undesirable but at times difficult to avoid. In general the problems will be least where exposure is best and these are the areas in which detailed measurement is most likely to be made. However, the problem illustrates that there should be a consistent single definition of a bed.

The definition of a bed as a sedimentation unit, after Campbell (1967) and Reineck & Singh (1975) should be retained, despite its problems. Assignment of a particular thickness limitation is not necessary; thicknesses may range from a few millimetres to several metres and can be described using the terms in Fig. 2.13. The main use of bed thickness measurements has been in helping to estimate competence and capacity of the transporting medium, particularly in terrestrial aqueous systems. For example, Bluck (1967), Steel (1974) and Nemec & Steel (1984) presented many plots of maximum particle size versus bed thickness and show that the relationship varies for different depositional processes (Fig. 2.14). Potter & Scheidegger (1965) suggested a correlation between maximum particle size and bed thickness in turbidites and between median clast diameter and bed thickness in ash falls. In all these cases interpretation is clearly dependent upon the sedimentation unit concept for bed thickness. Thus if there is any uncertainty in the definition of

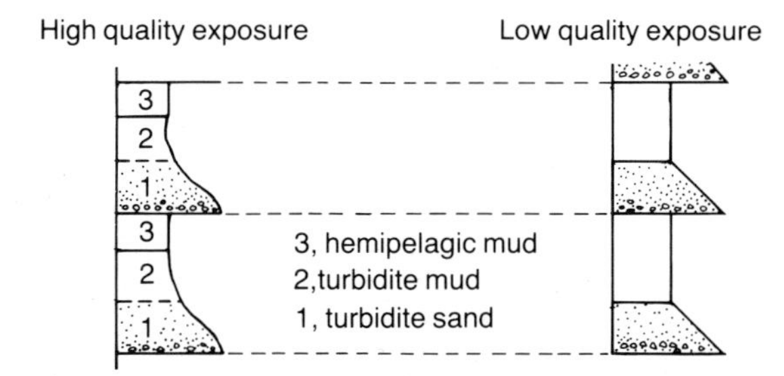

Fig. 2.11. Recognition of beds with varying quality of exposure, an example from interbedded turbidites and hemipelagites.

Fig. 2.12. Interbedded turbidites and hemipelagites from the Lower Carboniferous of Morocco showing rapid variation in exposure quality over short distances. Only the bedding surfaces at the base of turbidite sands (1) can be followed with any certainty.

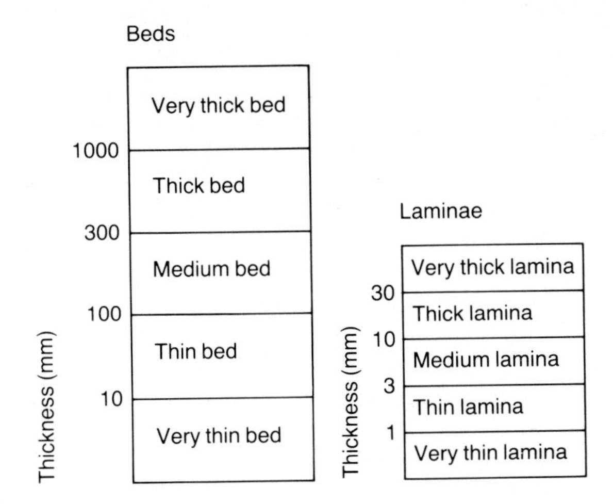

Fig. 2.13. Terminology for thickness of beds and laminae (modified after Ingram, 1954, Campbell, 1967 and Reineck & Singh, 1975). (Reproduced by permission of Springer.)

bedding surfaces, this must be made very clear in the field notes. This problem may be particularly acute in many limestone successions where the bedding surfaces may show extreme diagenetic modification, usually accompanied by pressure solution and the formation of stylolites. Recognition is complicated because bedding parallel stylolitic seams may form at levels unrelated to true bedding surfaces (Simpson, 1985). A summary of the complexity observed in a series of tabular limestone beds from the Lower Carboniferous of South Wales is shown in Fig. 2.15 (after Simpson, 1985).

A single bed may be internally homogeneous, show continuous gradational variation, or be internally layered, these smaller layers being termed laminae. Laminae are generally a few millimetres thick but may exceptionally reach a few centimetres (Fig. 2.13). Beds may also show a variety of internal structures, such as types of cross stratification. It is common to find two superimposed beds that are

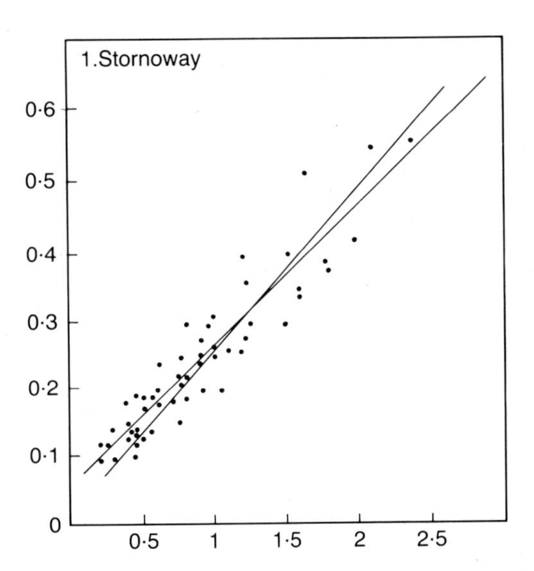

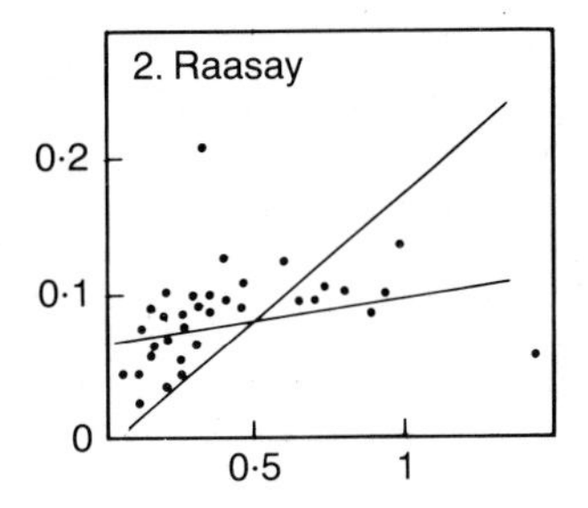

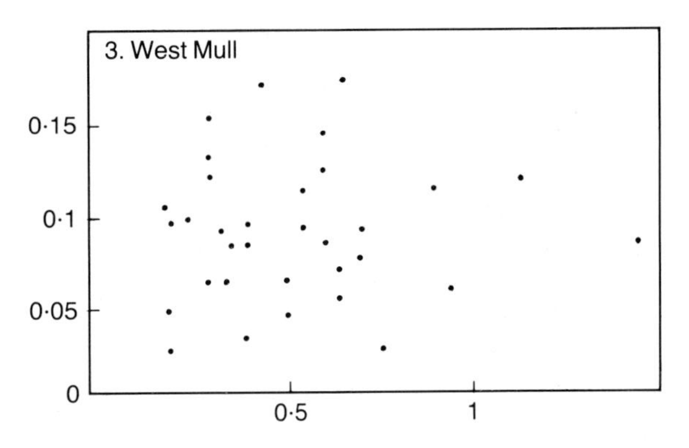

Fig. 2.14. Plots of maximum particle size (vertical axis) against bed thickness (horizontal axis) showing differences between different processes (from Steel, 1974) in the New Red Sandstone of western Scotland. Locality: 1—mudflow units, 2—stream flood conglomerates, 3—braided stream conglomerates. All measurements in metres. (Reproduced by permission of SEPM.)

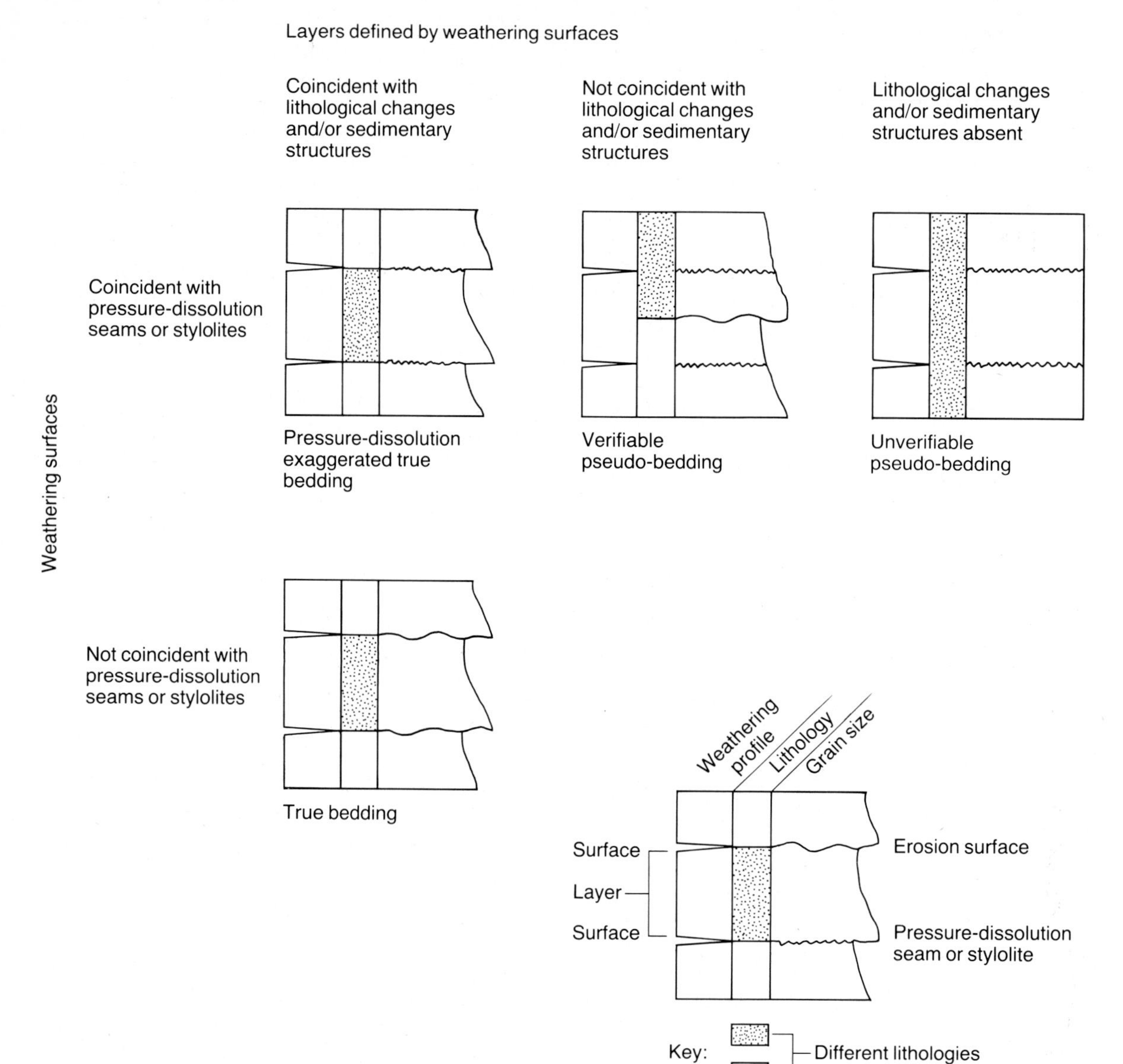

Fig. 2.15. Complexity of depositional units (beds) and bedding parallel weathering surfaces in limestones from the Lower Carboniferous of South Wales (from Simpson, 1985).

related genetically and these are termed bedsets. They are called simple where the superimposed beds are similar and composite where they are different in appearance.

The primary nature of contacts between beds and bedsets may be most important in environmental analysis. Any information which can be observed relating to the time span of non-deposition, e.g. burrowed surfaces, hardgrounds, soil formation, palaeokarsts, or evidence of erosional contacts should be carefully recorded.

BED GEOMETRY

Most description of bedding concentrates on the vertical changes in layered successions. Sedimentary layers usually do have a large lateral extent relative to their thickness, but it is, of course, finite. The three-dimensional geometry of beds and bedsets can yield very valuable information where it can be observed. Beds terminate laterally by: (i) convergence and intersection of bedding surfaces; (ii) lateral gradation of material comprising the bed into another in which bedding surfaces become indistinguishable; (iii) abutting against a fault, unconformity or cross cutting feature such as a channel.

The way in which beds thicken and thin is important. For example channelling is manifested by change in the position of the lower bedding surface whereas preserved bedforms such as ripples, dunes and hummocks are characterized by changing position of the upper bedding surface. It is important to measure the spacing between such thickness changes to see if there is any pattern such as rhythmic spacing. The relationship of internal structures to the external geometry is also important in interpretation.

Some beds can be delineated within a single exposure whereas others may extend for tens or even hundreds of kilometres. The lateral extent can be useful in environmental distinction of superficially similar deposits, e.g. thin outer fan turbidites versus thin interchannel turbidites of the upper fan, sheetflood versus fluvial channel deposits.

2.2.6 Sedimentary structures

The accurate recognition and recording of sedimentary structures is vital to most attempts at environmental reconstruction. Description and interpretation of such structures are the subject of several texts at various levels, e.g. Allen (1982), Collinson & Thompson (1982), Conybeare & Crook (1968), Pettijohn & Potter (1964), Tucker (1982, chapter 5). These are not duplicated here but some comments on methods of recording and recognition are appropriate. A list of the common structures and groups of structures classified predominantly by position (after Pettijohn & Potter, 1964) is given in Table 2.4.

Almost all sedimentary structures are three-dimensional, although recognition of many is based on observation of two-dimensional sections. The value of commenting on the three-dimensional geometry of any available structures cannot be overstated. It is always important to examine any exposed bedding surfaces as well as vertical sections. In recording sedimentary structures reliance is placed both on objective observation and measurement and also on existing theoretical framework. Every field record of sedimentary structures will be a compromise between measuring all basic and indisputable parameters on the one hand and using available classes of structures on the other hand.

For example, many schemes for recording cross-stratification are so constricted that they imply that there are two basic types of cross-bedding — tabular and trough. If such a scheme was used to record cross stratification in a section containing hummocky cross stratification (HCS) (Fig. 2.16) then these would probably be recorded as trough cross strata, possibly with the lower angle of cross stratal dip noted. In this case the use of a limited theoretical framework leads to the amalgamation of HCS with cross strata of rather different origins and to the loss of valuable information. If data on angle of cross stratal dip and three-dimensional geometry of

Table 2.4. Sedimentary structures

Sedimentary structures
Observed primarily as internal structures of beds in section
Cross stratification
Lamination
Grading
Soft sediment deformation
Bioturbation and trace fossils, stromatolites
Pedogenic horizons, hardgrounds
Cavities (mainly in limestones)
Concretions
Stylolites
Observed primarily on bedding surfaces
(i) Bottom surfaces:
Flute marks
Tool marks
Load casts
Geometry due to scour or topographic fill
(ii) Top surfaces:
Surface topography
Bedforms, e.g. ripples, dunes, hummocks
Primary current lineation
Shrinkage cracks
Trace fossils
Sand volcanoes
Raindrop impressions

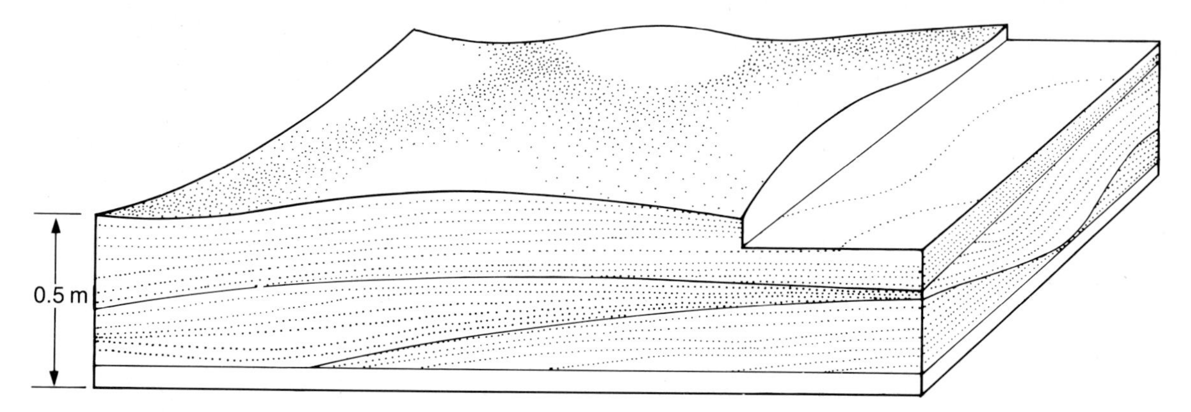

Fig. 2.16. Block diagram showing the main features of hummocky cross stratification (HCS) (from Harms *et al.*, 1982). Current directions unknown.

the structures had been recorded, later interpretation of the structures as HCS would remain possible. How many examples of HCS can be confidently identified from pre-1965 publications? In spite of this the recording of all parameters of cross strata in every case would be too time consuming.

In most cases the available theoretical framework is used much more than is realized. Thus some general guidelines are available for field recording.

(a) Be as familiar as possible with useful distinctions between the various sedimentary structures. If differences from a typical example of a particular structure are suggested it is better to over-divide than the reverse.
(b) Remember that our knowledge is never complete. If the observed structures do not neatly fit any of our present theoretical pigeonholes, there is a need for objective and descriptive measurement.
(c) Have a clear knowledge, preferably as a written record, of the range of variation accepted within any particular category of structure that is used.
(d) If a suite of structures is relatively unfamiliar to you, although well known in general, carry summary diagrams or photographs in the field until familiarity is achieved.

2.2.7 **Fossil and trace fossil content**

Fossils are important components of many sedimentary rocks and, even where present in small numbers, they can provide useful, often critical, information. Extracting the maximum information from fossils will usually require the services of a specialist palaeontologist. Nevertheless, important observations, particularly those useful for environmental rather than stratigraphic use, can and should be made by any investigator of sedimentary rocks. A useful checklist is given by Tucker (1982) (see also Table 2.5).

Similarly, trace fossils and other biogenic structures provide valuable information. Biogenic structures vary from trace fossils possessing a definite form that can be described and named, to rather vague disruptions or even complete homogenization of stratification. The latter can often only be described as bioturbation. Trace fossils can provide information on both palaeoecology and environment and may be especially valuable where body fossils are limited or absent. Description and classification of trace fossils utilize three main approaches:

(a) Taxonomic: traces can be assigned to morphological ichnogenera. Details of this approach can be found in Crimes & Harper (1970), Frey (1975), Hantzschel (1975), Basan (1978) and Ekdale, Bromley & Pemberton (1984). A concise summary of a procedure to follow is given in Collinson & Thompson (1982, chapter 9) (see also Table 2.6).
(b) Tophonomic: description is based on mode of preservation of the trace fossils (Fig. 2.17).
(c) Behavioural (ethological): subdivision is made on the interpretation of behavioural pattern represented by the trace fossil, i.e. crawling, grazing, resting, dwelling, feeding or escaping.

Table 2.5. Checklist for the examination of fossils in the field

Distribution of fossils in sediment
(i) Fossils largely in growth position
(a) Do they constitute a reef? — characterized by colonial organisms; interaction between organisms, e.g. encrusting growth; presence of original cavities (infilled with sediment and/or cement); unbedded appearance:
Describe growth forms of colonial organisms; do these change up through reef? Are some skeletons providing a framework?
(b) If non-reef, are fossils epifaunal or infaunal?; if infaunal, how have fossils been preserved?
(c) Do epifaunal fossils have a preferred orientation, if so, measure
(d) Are fossils encrusting substrate, i.e. is it a hardground surface?
(e) Are the plant remains rootlets?
(ii) Fossils not in growth position
(a) Are they concentrated into pockets, lenses or laterally persistent beds or are they evenly distributed throughout the sediment?
(b) Do fossils occur in a particular lithofacies; are there differences in the faunal content of different lithofacies?
(c) If fossil concentrations occur, what proportion of fossils are broken and disarticulated? Are delicate skeletal structures preserved? Check for sorting, degree of rounding; look for sedimentary structures
(d) Do fossils show a preferred orientation?, if so, measure
(e) Have fossils been bored or encrusted?
(f) Note the degree of bioturbation and any trace fossils present

Fossil assemblages and diversity
(i) Determine the composition of the fossil assemblages by estimating the relative abundance of the different fossil groups.
(ii) Are fossil assemblages different for different lithofacies?
(iii) Consider the composition of the fossil assemblage. Is it dominated by only a few species, are they euryhaline or stenohaline? Are certain groups notably absent? Do all fossil groups present have a similar mode of life? Do pelagic forms dominate? Are infaunal organisms absent?

Diagnesis of fossil skeletons
(i) Is the original mineralogy preserved or have the skeletons been replaced, if so, by what?
(ii) Have the fossils been dissolved out to leave moulds?
(iii) Do the fossils preferentially occur in nodules?

Table 2.6. Trace fossils: how to describe them and what to look for

(1) Sketch (and/or photograph) the structures; measure length, width, diameter etc.
(2) For trails and tracks: (seen on bedding surfaces)
(a) Note whether regular or irregular pattern, whether trial is straight, sinuous, curved, coiled, meandering or radial
(b) Has the trail a continuous ridge or furrow?; is there any central division or ornamentation?; if there are appendage marks or footprints measure the size and spacing of the impressions; look for tail marks
(3) For burrows (best seen within beds but also on bedding surfaces):
(a) Describe shape and orientation to bedding, i.e. horizontal, oblique, vertical; simple straight tube, simple curved or irregularly disposed tube. If branching note if regular or irregular branching pattern and any changes in burrow diameter
(b) Examine burrow wall; is the burrow lined with mud or pellets?; are there scratch marks?; are laminae in adjacent sediment deflected by the burrow?
(c) Examine the burrow fill; is it different from adjacent sediment?; e.g. coarser or finer, richer or poorer in skeletal debris?; is the fill pelleted?; are there curved backfill laminae within the burrow fill sediments?

The last two approaches can usually be performed in the field without specialist knowledge. Taxonomic identification of common trace fossils is often possible but is best aided by accurate drawings and photographs and by consultation with specialists.

2.2.8 Measurement of stratigraphic sections

The methods employed in measuring sections in sedimentary rocks will be determined by the degree of detail required, by the physical nature of the terrain and the exposures, and by the time, funds, equipment and personnel available. Details of the various procedures available are given in Compton (1962), Krumbein & Sloss (1963) and Kottlowski (1965). In cases of suitable exposure measurement can be made simply by direct contact of a tape or ruler held normal to both dip and strike directions.

Where this is feasible the most common method for measuring vertical sections is probably the use of

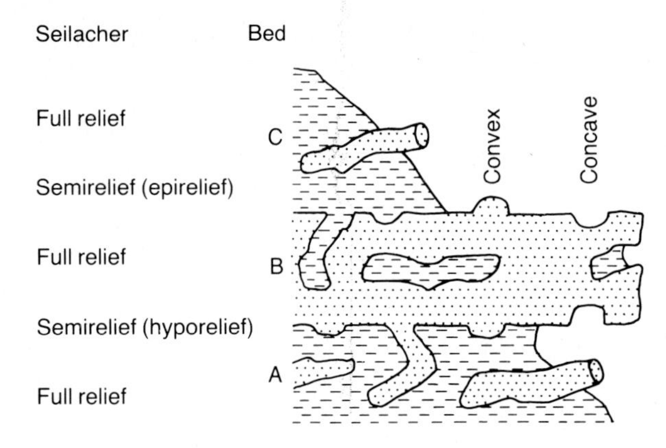

Martinsson

Exichnia

Epichnia
(top of bed B)

Endichnia
(interior of bed B)

Hypichnia
(bottom of bed B)

Exichnia
(outside bed B)

Fig. 2.17. A toponomic classification of trace fossils for use in the field (after Ekdale *et al.*, 1984). (Reproduced by permission of SEPM.)

a compass and tape measure. This is relatively quick and with care is sufficiently precise for most purposes. It is most difficult to be accurate where the dip of the strata is at a low angle to the section to be measured. The procedure is easier with two persons but can be performed by individuals. Important points to note where the line of section is not at 90° to the dip of the beds are: (a) measure carefully the slope of the surface along with the strike and dip of the beds; (b) read the apparent thickness of beds, bedsets or facies units from the stretched tape (Fig. 2.18); (c) correct for both slope angles and oblique sections (Fig. 2.19); (d) keep readings separate between changes of slope.

In many cases where the beds are very steeply inclined relative to the surface of measurement, the tape can be used as a small Jacob staff (see below) with reasonable accuracy, rendering the laborious corrections for slope unnecessary.

An alternative to the use of a tape, where terrain permits, is to use a graduated pole termed a Jacob staff (Fig. 2.20). The pole provides a steady support for a level or clinometer, an Abney hand level being the most accurate. The dip of the units to be measured is determined and then this value is set on the

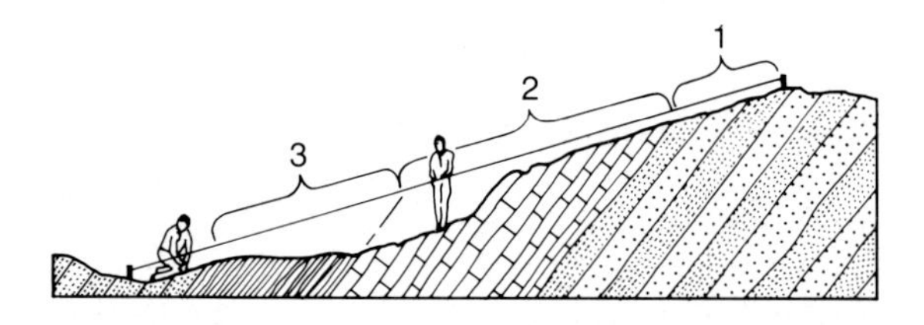

Fig. 2.18. Measurement of strata on a slope by reference to a stretched tape. Note the projection of contact between units (2) and (3) to the tape (from Compton, 1962). (Reproduced by permission of Wiley.)

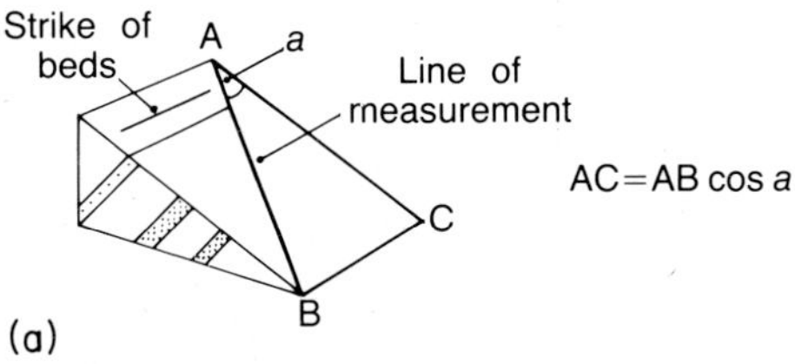

a. Slope and dip are opposed and angle of slope (y) plus angle of dip (x) is $< 90°$

$BC = AB \sin (x+y)$

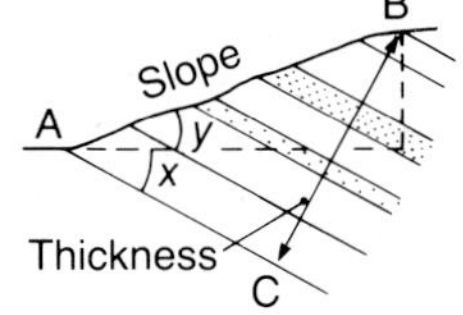

b. Slope and dip are opposed and angle of slope plus angle of dip is $> 90°$

$BC = AB \cos (x+y-90°)$
or $BC = AB \sin [180° - (x+y)]$

c. Slope and dip are in the same direction, with dip $>$ slope

$AC = AB \sin (x-y)$

d. Slope and dip are in the same direction, with dip $<$ slope

$BC = AB \sin (y-x)$

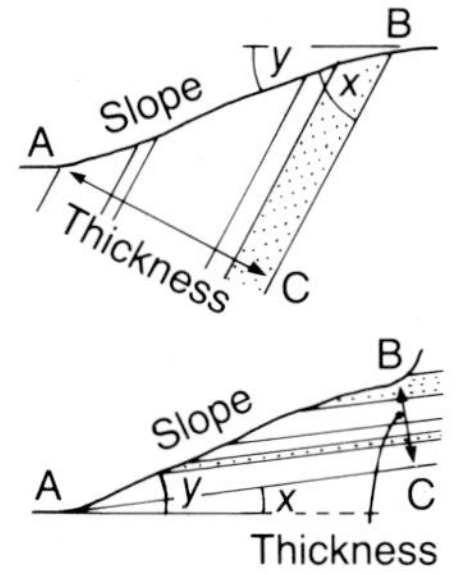

(b)

Fig. 2.19. Corrections for slope angles and oblique sections (from Compton, 1962).
(a) Correction of slope distance that was measured oblique to the dip of the beds.
(b) Formulae used for various combinations of direction and amount of ground slope and dip of beds. (Reproduced by permission of Wiley.)

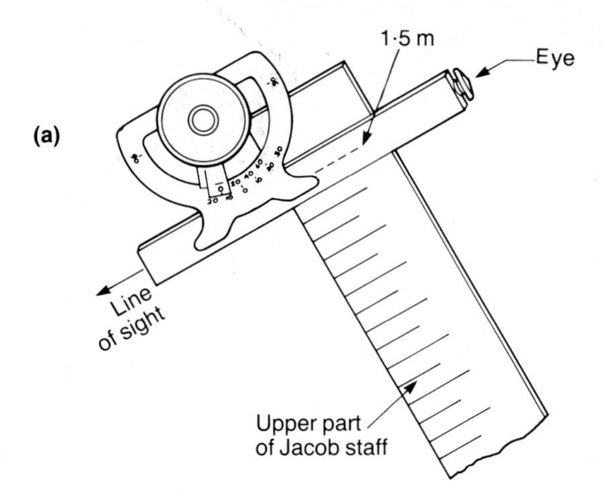

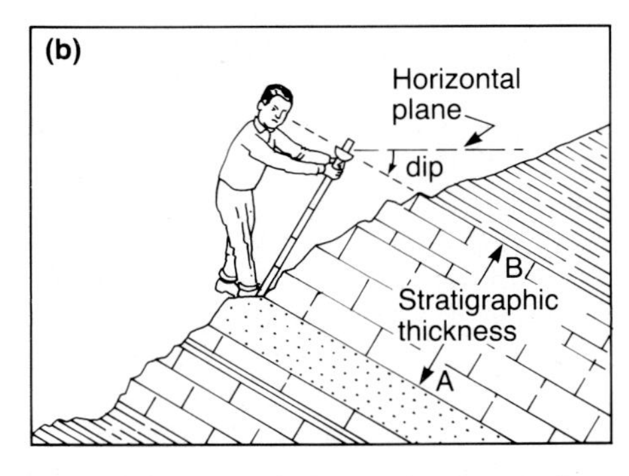

Fig. 2.20. Use of the Jacob staff for measuring sections (from Kottlowski, 1965).
(a) Setting dip on clinometer of Abney hand level used with a Jacob staff.
(b) Measuring stratigraphic thickness AB.
(c) Measuring a unit with thickness less than the length of a Jacob staff.
(Reproduced by permission of CBS.)

clinometer (Fig. 2.20). The base of the staff is placed at the base of the unit to be measured and the staff tilted forward until the level bubble of the clinometer is centred. The line of sight is then in a direction parallel to the dip of the beds and the Abney level can be slid up and down the staff as required (Hansen, 1960) until the top surface of the unit is in the line of sight (Fig. 2.20). If the staff is oriented within 10° of normal to the dip, then the error is likely to be no more than 2% (Compton, 1962). Outcrops are rarely exposed in a straight upslope line but a stepwise course of movements can easily be made.

2.2.9 Graphic logs

The prime moving force behind the use of graphic logs in the field is the need to record very large numbers of observations, often in a routine repetitive manner. The first major formulation of such a scheme was presented by Bouma (1962) who demonstrated a successful application to a succession of turbidites. He used in the field a series of preprinted recording sheets of the format shown in Fig. 2.21. These were later supplemented by an accompanying sheet based on laboratory investigations. However, the field record remained as the final display, with the added laboratory data shown alongside. Bouma's recording format was accompanied by a complex series of codes and notation symbols which permitted an unambiguous and detailed record.

Broadly similar techniques have since been employed by many other workers investigating a variety of sedimentary successions. Some examples of

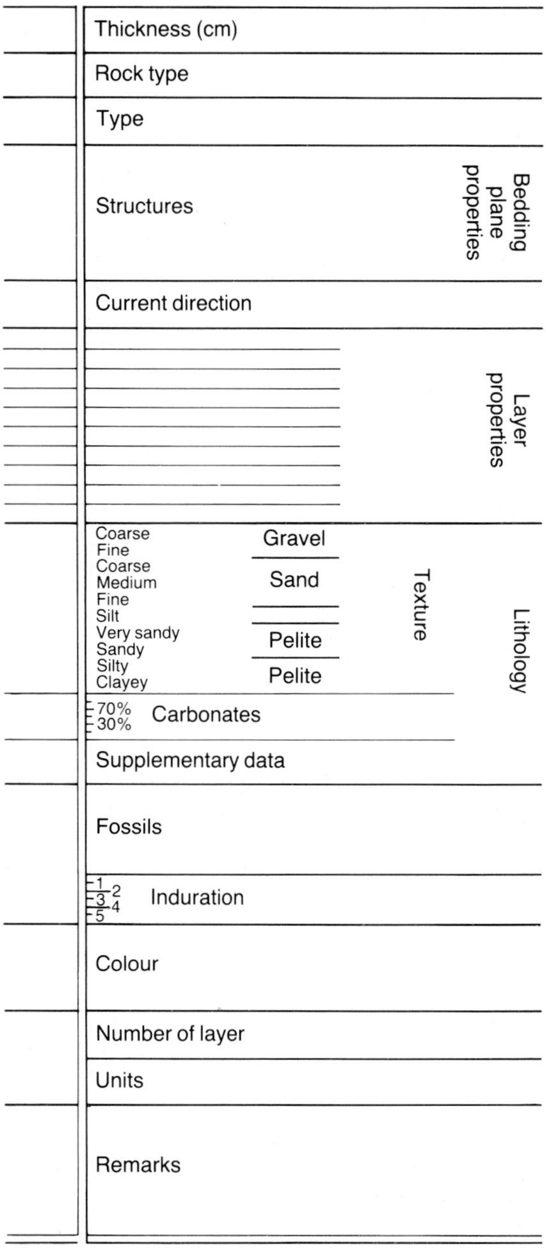

Thickness (cm)
Rock type
Type
Structures
Bedding plane properties
Current direction
Layer properties
Coarse / Fine — Gravel
Coarse / Medium / Fine — Sand
Silt
Very sandy / Sandy — Pelite
Silty / Clayey — Pelite
Texture
Lithology
70% / 30% Carbonates
Supplementary data
Fossils
1 2 3 4 5 Induration
Colour
Number of layer
Units
Remarks

Fig. 2.21. Format for field recording sheets designed by Bouma (Bouma, 1962). Field records are drawn to scale (chosen according to the nature of the succession) and the sheets are accompanied by a complex set of notation symbols. (Reproduced by permission of Elsevier.)

different recording sheets are shown in Fig. 2.22. What are the reasons for this multiplicity of recording sheets?

(1) The features commonly recorded vary considerably with the type of succession, i.e. turbidites, fluvials, carbonates etc. If the broad nature of the succession is known in advance, there is an advantage in adapting the recording form to the particular data to be recorded. This is usually the case for advanced, detailed studies that follow a reconnaissance examination. An example of a complex but informative scheme is that of the Utrecht school used in their studies of ancient marginal-marine sediments in southwest Ireland (Kuijpers, 1972; Van Gelder, 1974; De Raaf, Boersma & Van Gelder, 1977; Fig. 2.22a).

(2) One can maintain at a higher level and for a longer period a discipline in the routine listing of features than one can in a free-format notebook. Conversely, there may be some loss of flexibility, leading to a tendency for the field geologist to neglect features other than those on the check list.

(3) Subsequent data retrieval is generally easier and more rapid than from written notebook records.

(4) The use of a standard format can give greater consistency of recording/observation where different operators are employed.

(5) If the logs are drawn to scale in the field, the investigator has a useful visual display which may assist thought processes and the formulation of hypotheses.

(6) There is often a tendency to think that the construction of a graphic log provides a complete description of a succession. In fact, vertical changes in the succession become over-emphasized at the expense of lateral relations.

(7) There is a strong similarity between most field logs and the final published logs used in formation description and interpretation.

(8) Some sort of remarks column is invariably necessary to accommodate features which are rare or were neglected during design of the form. Even with this, a field notebook remains a necessary aid for sketches etc. which cannot be accommodated on the recording forms.

A natural extension of graphic logs as the main recording technique was the attempt to design and use machine readable forms for field records (Alexander-Marrack, Friend & Yeats, 1970; Piper,

Harland & Cutbill, 1970; Friend *et al.*, 1976). The main advantages over and above the use of graphic logs were speed, rapid data retrieval and processing, and easier data presentation, e.g. the data for computer-drawn stratigraphic sections are already stored in a retrievable form. The main disadvantage is the difficulty of building flexibility into the form whilst maintaining a manageable size and simplicity. Friend *et al.* (1976) demonstrated the value of this approach for recording thick, relatively monotonous successions under difficult and expensive field conditions (Fig. 2.23).

PRESENTATION OF GRAPHIC LOGS

Graphic logs are used in a large proportion of publications dealing with sedimentary rock successions but for appropriate reasons there is no standardization. Information must be presented on a variety of scales, and for a wide variety of facies associations. Even the basic unit of the logs may vary from being the bed or bedset to 'facies' which are defined either on the basis of a detailed field log or during a pilot study. An attempt to provide a comprehensive scheme of notation and symbols to cover all eventualities may be either too limiting for very detailed studies or too complex. Nevertheless there are many conventions which make comparison of different logs easier.

Thickness is almost always the vertical scale and the horizontal scale is most frequently grain size. Other features which can be readily presented are lithology, sedimentary structures and nature of contacts between units. Palaeocurrent data (Section 2.3) and some other data on vertical sequences (Section 2.4) can be conveniently located at the side of these logs. Figure 2.24 shows a variety of typical examples.

An overall aim in the presentation of logs should be that the reader has a good basis for comparison with the literature and the data on which to build preliminary interpretations of depositional environments and time-dependent changes.

2.2.10 Recording lateral relations

Lateral variability of rock units is often obvious where there are large natural or man-made exposures. In humid, temperate regions these are mainly cliff faces, road cuts and quarry sections in shallowly dipping strata. Extensive strike sections also occur in steeply dipping strata on many foreshore sections and in recently glaciated areas. In arid areas with shallowly dipping strata, large lateral exposures are common. Description of lateral changes generally relies on the use of photo-mosaics supplemented by field sketches, use of binoculars, examination of accessible portions and, where possible, measuring spaced vertical logs. If the exposure is accessible, plane table mapping can also be useful, e.g. Bluck (1981) (Fig. 2.25).

For reports and publications it is more common to demonstrate lateral relations with line diagrams than with photographs although in some cases both approaches can be useful (e.g. Allen & Matter, 1982, fig. 7). This is often due to the difficulty of presenting photographs at a scale where they are clear but can still be accommodated within the report format. This problem can also apply to line drawings and in many journals pull outs are only rarely tolerated due to the expense of production. Some examples of the presentation of lateral variability are shown in Figs 2.25 to 2.28.

2.3 PALAEOCURRENT DATA

2.3.1 Introduction

Palaeocurrent data are those which provide information on the direction(s) of sediment transport in the past. As such they are an important part of sedimentological analysis on a variety of scales from individual bedforms to the whole sedimentary basin. The main categories of interpretation which may result from analysis of palaeocurrent data are direction of palaeoslope, patterns of sediment dispersal, relationship of palaeocurrent direction to lithosome (Krumbein & Sloss, 1963, p. 300) geometry and location of source area. Such interpretations may have economic as well as academic applications when dealing with phenomena such as washouts in coalfields or placer deposits.

The subject of palaeocurrents and basin analysis has been exhaustively treated by Potter & Pettijohn (1977). They emphasized that a variety of observations can contribute palaeocurrent data and these may be grouped into two classes:

(a) Properties which acquire directional significance only when mapped regionally. These consist of:

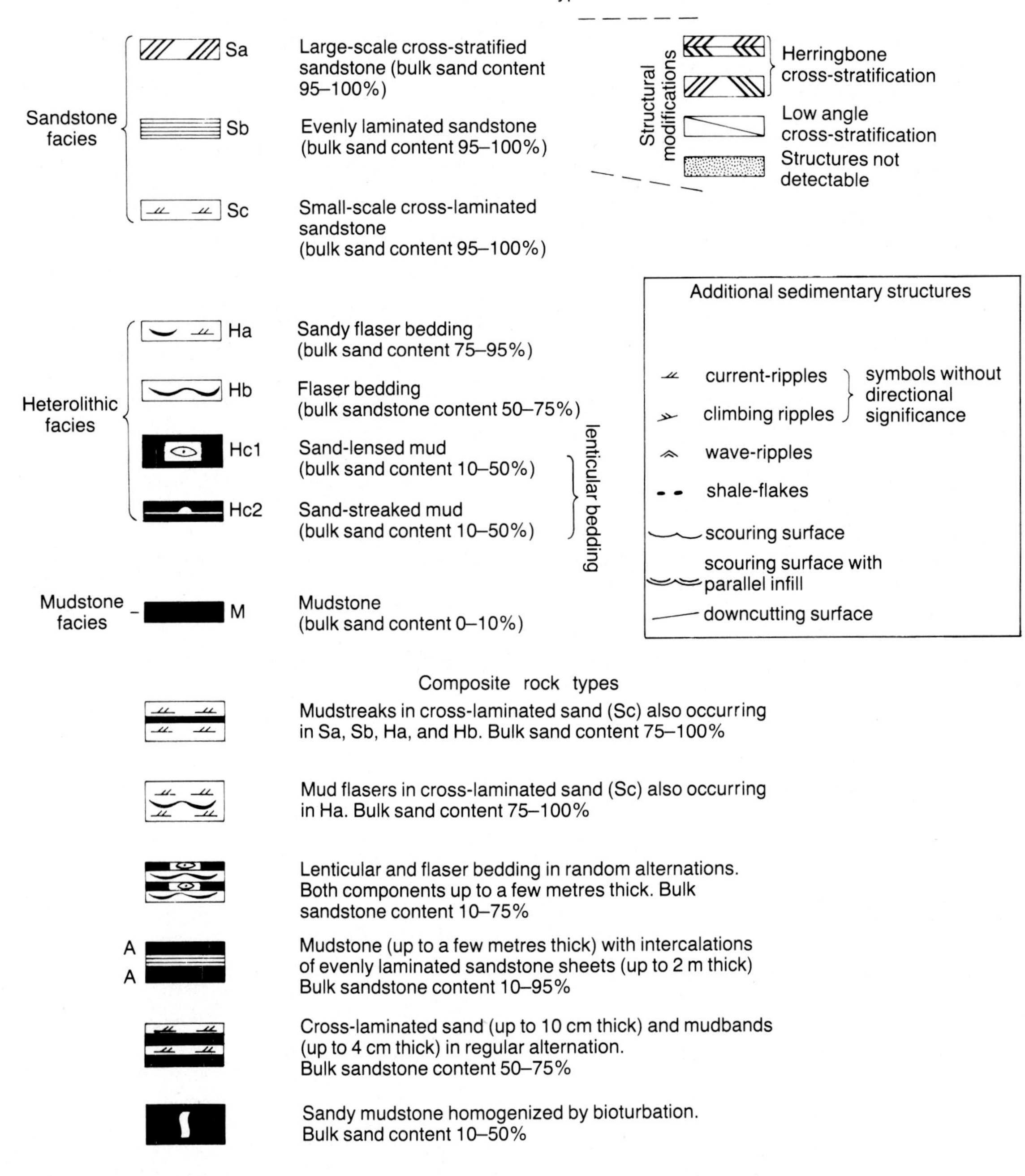
Basic rock types
Sandstone facies
Sa
Large-scale cross-stratified sandstone (bulk sand content 95–100%)
Sb
Evenly laminated sandstone (bulk sand content 95–100%)
Sc
Small-scale cross-laminated sandstone (bulk sand content 95–100%)
Structural modifications
Herringbone cross-stratification
Low angle cross-stratification
Structures not detectable
Heterolithic facies
Ha
Sandy flaser bedding (bulk sand content 75–95%)
Hb
Flaser bedding (bulk sandstone content 50–75%)
Hc1
Sand-lensed mud (bulk sand content 10–50%)
Hc2
Sand-streaked mud (bulk sand content 10–50%)
lenticular bedding
Mudstone facies
M
Mudstone (bulk sand content 0–10%)
Additional sedimentary structures
current-ripples
climbing ripples
symbols without directional significance
wave-ripples
shale-flakes
scouring surface
scouring surface with parallel infill
downcutting surface
Composite rock types
Mudstreaks in cross-laminated sand (Sc) also occurring in Sa, Sb, Ha, and Hb. Bulk sand content 75–100%
Mud flasers in cross-laminated sand (Sc) also occurring in Ha. Bulk sand content 75–100%
Lenticular and flaser bedding in random alternations. Both components up to a few metres thick. Bulk sandstone content 10–75%
A
A
Mudstone (up to a few metres thick) with intercalations of evenly laminated sandstone sheets (up to 2 m thick) Bulk sandstone content 10–95%
Cross-laminated sand (up to 10 cm thick) and mudbands (up to 4 cm thick) in regular alternation. Bulk sandstone content 50–75%
Sandy mudstone homogenized by bioturbation. Bulk sand content 10–50%

(a)

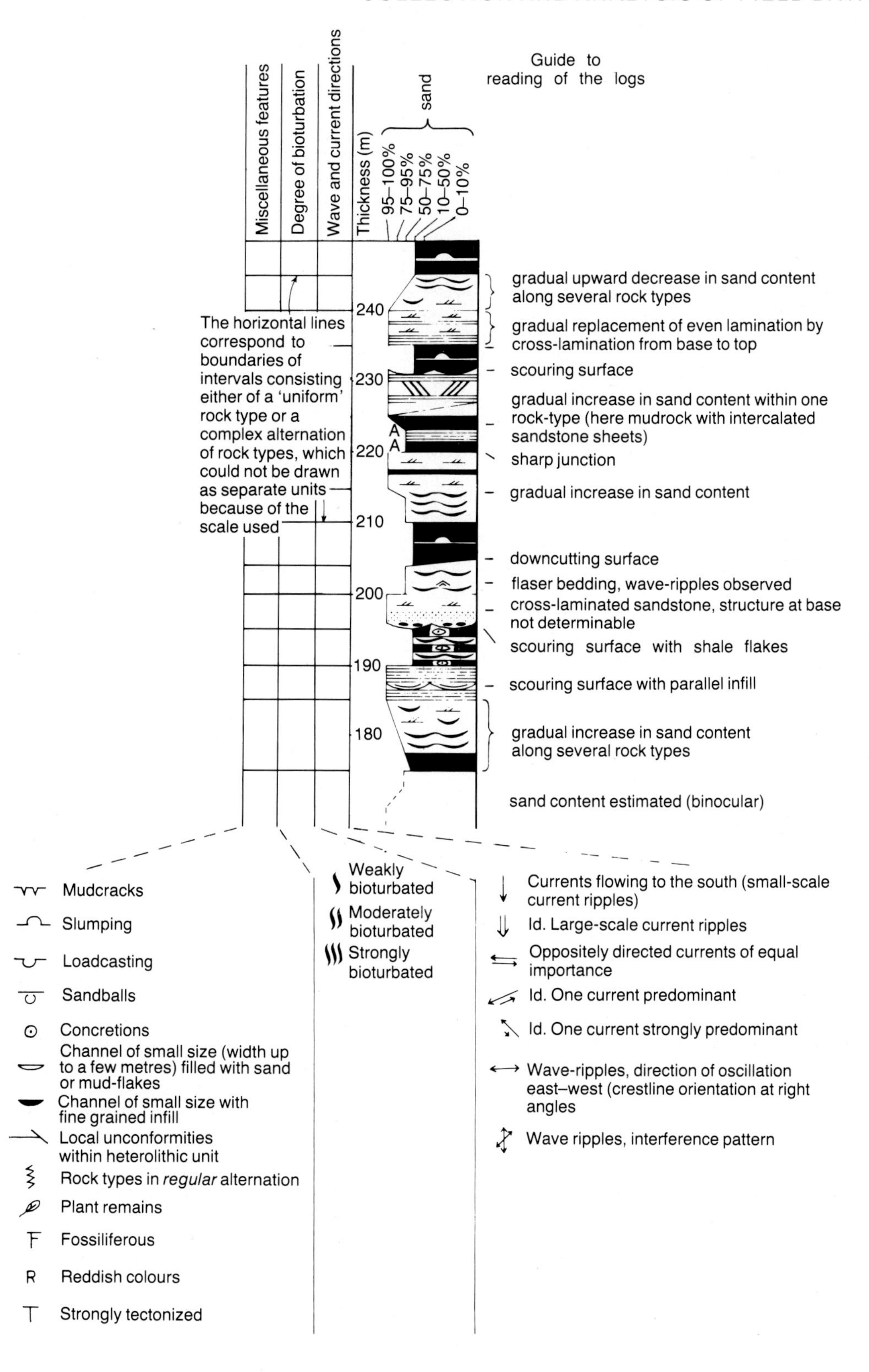

Guide to reading of the logs
Miscellaneous features
Degree of bioturbation
Wave and current directions
Thickness (m)
sand
95–100%
75–95%
50–75%
10–50%
0–10%
240
230
220
210
200
190
180
A
A
The horizontal lines correspond to boundaries of intervals consisting either of a 'uniform' rock type or a complex alternation of rock types, which could not be drawn as separate units because of the scale used
gradual upward decrease in sand content along several rock types
gradual replacement of even lamination by cross-lamination from base to top
scouring surface
gradual increase in sand content within one rock-type (here mudrock with intercalated sandstone sheets)
sharp junction
gradual increase in sand content
downcutting surface
flaser bedding, wave-ripples observed
cross-laminated sandstone, structure at base not determinable
scouring surface with shale flakes
scouring surface with parallel infill
gradual increase in sand content along several rock types
sand content estimated (binocular)
Mudcracks
Slumping
Loadcasting
Sandballs
Concretions
Channel of small size (width up to a few metres) filled with sand or mud-flakes
Channel of small size with fine grained infill
Local unconformities within heterolithic unit
Rock types in regular alternation
Plant remains
F Fossiliferous
R Reddish colours
T Strongly tectonized
Weakly bioturbated
Moderately bioturbated
Strongly bioturbated
Currents flowing to the south (small-scale current ripples)
Id. Large-scale current ripples
Oppositely directed currents of equal importance
Id. One current predominant
Id. One current strongly predominant
Wave-ripples, direction of oscillation east–west (crestline orientation at right angles
Wave ripples, interference pattern

Section log

Sheet no.	in sequence of

Scale

Geologist	
Date(s)	

Locality		
Grid co-ordinates	from	
	to	
Altitude	from	to
Map sheet no.		
Aerial photo no.		
Locality no.		

Age	Fm.	Mbr.	Cum-thick	Lithology	Stuctures / S: cl, sl, f, m, c, gr	Orient.	Fossils, trace fossils	Sample no.	Photo no.	Remarks

(b)

(i) attributes — presence/absence of some distinctive feature such as boulder type or mineral assemblage; (ii) scalars — magnitude of some property such as grain size, roundness or unit thickness.

In these cases there is little extra recording of field data but merely a selection of specific properties from available records. Accordingly this class is given little treatment here apart from some examples.

(b) Properties which provide some directional information at the point of observation, although more information will be conveyed by presenting many observations in map form. These directional properties can indicate either: (i) a line of flow, e.g. primary current lineation, groove marks, symmetrical ripple crests or (ii) a unique direction of flow, e.g. imbrication, cross stratification, flute marks.

Directional properties form the basis of most palaeocurrent analysis because of the greater amount of information they contain.

(c)

Location ....

Sheets ....

Continues from ....

— = Massive bedding
═ = Flat bedding
≡ = Laminated bedding
- - - = Parting lineation

= X-bedded
= X-laminated
TR = Trough
TAB = Tabular

= Mud cracks
= Convoluted bedding
= Channel
//// = Imbrication

Erosive
Sharp, non-erosive —
Gradational - - -
Deformed (load casts etc.)
Abrupt (not known if erosional) — ?
Undulating
Not seen *

G = good M = moderate P = poor

B = black Gr = green Lt = light
G = grey P = purple Dk = dark D = dull

Cum thick	Unit thick	Dom. grain size							Rock type	Structures	L.B.S.	Sorting/ round	Colour	Uncorrected palaeo- current	Local bedding	Remarks (+ comp. of coarse clasts)	PSL number
				F	M	C											
		C	S	S	S	S	G	C									

Fig. 2.22. Examples of different recording sheets.
(a) A detailed and complex scheme devised specifically for marginal marine sediments by De Raaf, Boersma, Leflef & Kuijpers (from Kuijpers, 1972).
(b) A more general scheme for recording vertical sequences in sedimentary successions designed for the Greenland Geological Survey by L.B. Clemmensen and F. Surlyk.
(c) A scheme with a limited number of symbols and columns designed specifically for recording large thicknesses of fluvial sediments (from Russell, 1984).

2.3.2 **Measurement of directional structures**

For each measurement it is important to record the exact location, type and scale of structure, as well as the lithofacies in which it occurs. For example, it may be useful to investigate the relationship of cross bedding direction to set thickness, foreset dip angle or foreset shape, as environmental interpretation becomes more refined.

LINEAR STRUCTURES

In this category are structures in which it is not possible to distinguish between one 'end' of a structure and the other, e.g. primary current lineation, striae, groove marks, gutter casts, channels, some oriented elongate fossils and symmetrical ripple crests. The orientation of the lineation is recorded as a pitch on inclined strata or in horizontal beds as a direction. Techniques for unambiguous recording of

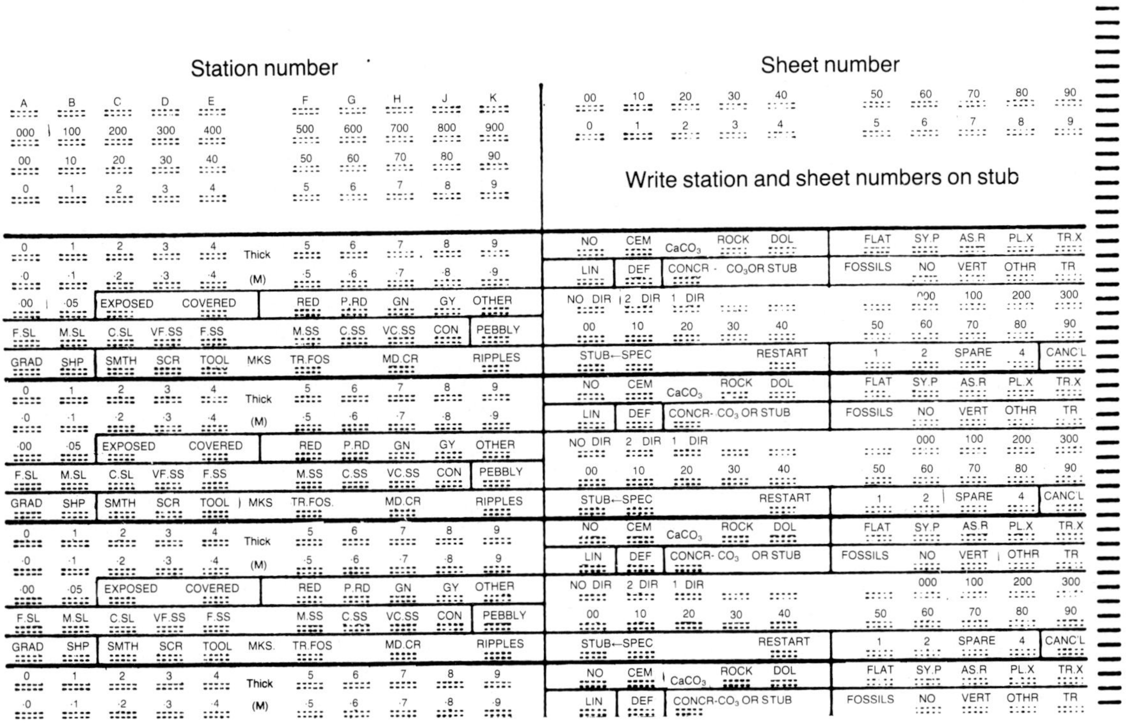

Station number

A B C D E F G H J K
000 100 200 300 400 500 600 700 800 900
00 10 20 30 40 50 60 70 80 90
0 1 2 3 4 5 6 7 8 9

Sheet number

00 10 20 30 40 50 60 70 80 90
0 1 2 3 4 5 6 7 8 9

Write station and sheet numbers on stub

0 1 2 3 4 Thick 5 6 7 8 9 | NO CEM $CaCO_3$ ROCK DOL | FLAT SY.P AS.R PL.X TR.X
·0 ·1 ·2 ·3 ·4 (M) ·5 ·6 ·7 ·8 ·9 | LIN DEF CONCR-CO_3 OR STUB | FOSSILS NO VERT OTHR TR
·00 ·05 EXPOSED COVERED RED P.RD GN GY OTHER | NO DIR 2 DIR 1 DIR | 000 100 200 300
F.SL M.SL C.SL VF.SS F.SS M.SS C.SS VC.SS CON PEBBLY | 00 10 20 30 40 | 50 60 70 80 90
GRAD SHP SMTH SCR TOOL MKS TR.FOS MD.CR RIPPLES | STUB←SPEC RESTART | 1 2 SPARE 4 CANC'L

0 1 2 3 4 Thick 5 6 7 8 9 | NO CEM $CaCO_3$ ROCK DOL | FLAT SY.P AS.R PL.X TR.X
·0 ·1 ·2 ·3 ·4 (M) ·5 ·6 ·7 ·8 ·9 | LIN DEF CONCR-CO_3 OR STUB | FOSSILS NO VERT OTHR TR
·00 ·05 EXPOSED COVERED RED P.RD GN GY OTHER | NO DIR 2 DIR 1 DIR | 000 100 200 300
F.SL M.SL C.SL VF.SS F.SS M.SS C.SS VC.SS CON PEBBLY | 00 10 20 30 40 | 50 60 70 80 90
GRAD SHP SMTH SCR TOOL MKS TR.FOS MD.CR RIPPLES | STUB←SPEC RESTART | 1 2 SPARE 4 CANC'L

0 1 2 3 4 Thick 5 6 7 8 9 | NO CEM $CaCO_3$ ROCK DOL | FLAT SY.P AS.R PL.X TR.X
·0 ·1 ·2 ·3 ·4 (M) ·5 ·6 ·7 ·8 ·9 | LIN DEF CONCR-CO_3 OR STUB | FOSSILS NO VERT OTHR TR
·00 ·05 EXPOSED COVERED RED P.RD GN GY OTHER | NO DIR 2 DIR 1 DIR | 000 100 200 300
F.SL M.SL C.SL VF.SS F.SS M.SS C.SS VC.SS CON PEBBLY | 00 10 20 30 40 | 50 60 70 80 90
GRAD SHP SMTH SCR TOOL MKS. TR.FOS MD.CR RIPPLES | STUB←SPEC RESTART | 1 2 SPARE 4 CANC'L

0 1 2 3 4 Thick 5 6 7 8 9 | NO CEM $CaCO_3$ ROCK DOL | FLAT SY.P AS.R PL.X TR.X
·0 ·1 ·2 ·3 ·4 (M) ·5 ·6 ·7 ·8 ·9 | LIN DEF CONCR-CO_3 OR STUB | FOSSILS NO VERT OTHR TR

Fig. 2.23. Example of a machine readable field recording form (from Alexander-Marrack *et al.*, 1970). This form was used for recording thick, relatively monotonous fluvial successions in East Greenland (Friend *et al.*, 1976) where fieldwork was expensive, time very limited and compatability among records from different investigators was important.

Marks are made on the form with soft pencil and each form has an accompanying stub on which additional notes can be made. (Reproduced by permission of Academic Press.)

lines and planes are given in texts such as Ragan (1973) and Compton (1962).

CROSS-STRATIFICATION

The most commonly used sedimentary structure for palaeocurrent analysis is cross stratification. The foresets represent the former slip faces of bedforms that migrated in the direction of foreset dip. The average direction of foreset dip is a measure of the average local flow direction. The geometrical variability of cross stratification mirrors the natural variability of bedforms. Although most workers categorize cross stratification in the two very broad subdivisions of planar and trough varieties (McKee & Weir, 1953; Section 2.2.4) there is, in nature, much more of a form continuum between these two (Meckel, 1967; Allen, 1982). The dimensions of the foresets will clearly be related to the size of the bedform, but this relationship is not simple. Commonly each foreset is truncated by an erosion surface, and thus the set thickness can only give a minimum estimate of bedform height. Nevertheless set thickness is valuable information, important in the interpretation of palaeoenvironmental conditions and must always be recorded. Similarly the angle of inclination of the foresets is the result of a complex balance of factors such as grain size, current velocity, sediment load etc., but it can still yield useful information on both the bedforms and transport processes. Thus the dip and shape of the foresets should also be recorded.

The measurement of cross stratification is most simply accomplished where there is a reasonable degree of three-dimensional exposure. In these cases

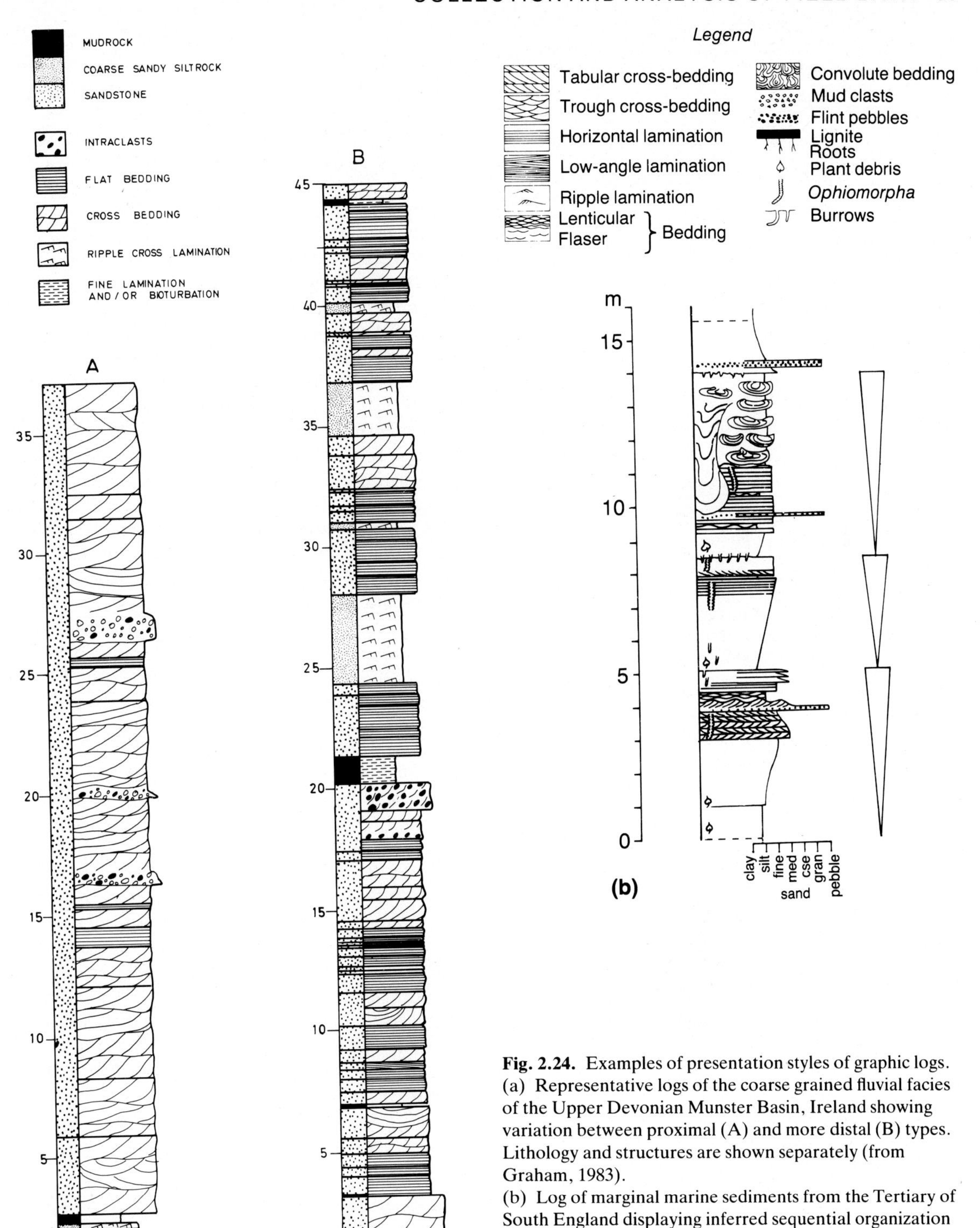

Fig. 2.24. Examples of presentation styles of graphic logs. (a) Representative logs of the coarse grained fluvial facies of the Upper Devonian Munster Basin, Ireland showing variation between proximal (A) and more distal (B) types. Lithology and structures are shown separately (from Graham, 1983).
(b) Log of marginal marine sediments from the Tertiary of South England displaying inferred sequential organization as well as data on structures and trace fossils (from Plint, 1983).

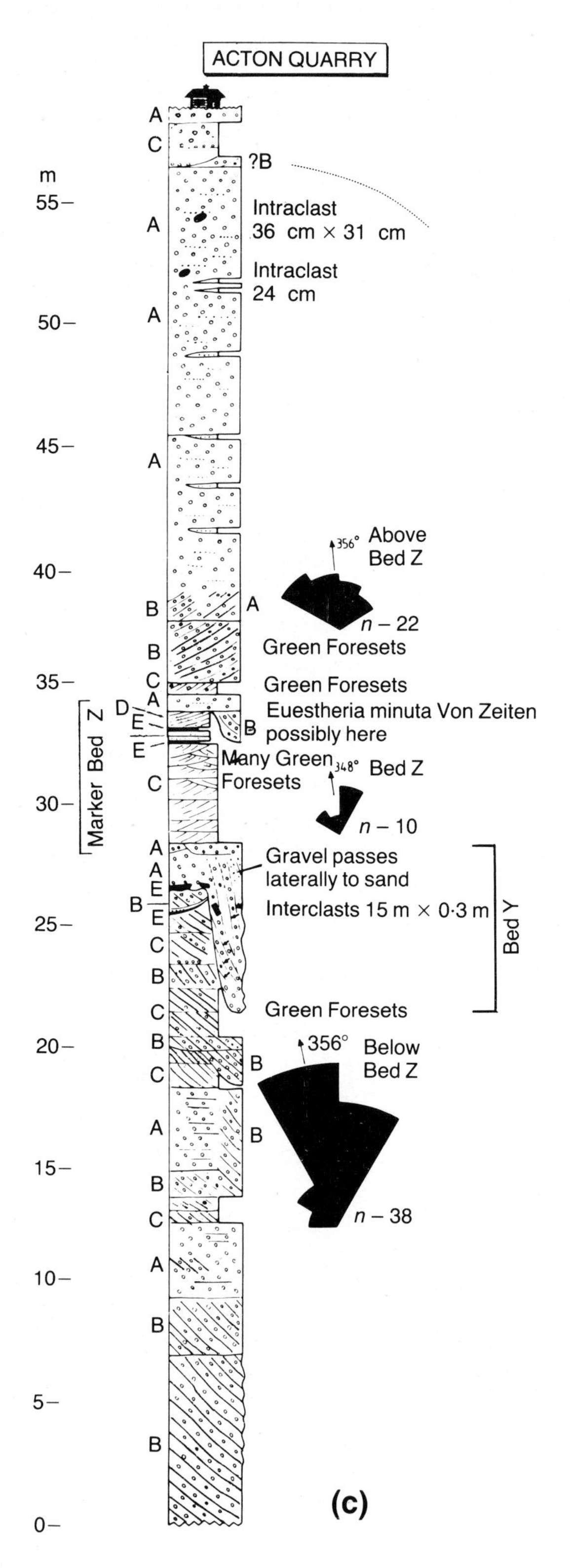

the orientation of a foreset plane is measured directly with a compass/clinometer (along with any additional structural information as noted in Section 2.3.3). Where such suitable exposure is lacking, the true (maximum) foreset dip can be calculated by measuring two apparent dips and solving by means of a stereographic projection.

Measurement problems generally centre around the ability to distinguish the three-dimensional geometry of the bedforms at outcrop. These problems may be compounded when the filling of troughs is asymmetrical. Thus it has been noted by several workers that trough axis orientations are much less variable than cross strata orientations (Meckel, 1967; Dott, 1973; Michelson & Dott, 1973; Slingerland & Williams, 1979) and are to be preferred for measurement. Unfortunately suitable exposures are only uncommonly available and less accurate approximations must generally be made. Careful recording of the geometry of the structures on the measured faces is most important for subsequent interpretation.

Techniques for deriving palaeocurrent data from common exposure types have been suggested by Slingerland & Williams (1979) and De Celles, Longford & Schwartz (1983). One method suggested by the latter is to attempt to measure both opposing trough limbs in relatively equal numbers on normal oblique exposures. Plotting these data on a stereographic net should yield two clusters of opposing limb sets from which the trough axis orientation can be estimated (Fig. 2.29). This technique may be useful for relatively flat lying strata but would be particularly susceptible to tectonic modification since the plunge of undisturbed trough axes is commonly less than 10°.

De Celles *et al.* (1983) also suggested a semi-quantitative technique which relies on assessing the overall geometry of the troughs by inspection of the foreset geometry relative to that of the basal scour surface, claiming accuracy to within 25° of the true palaeocurrent direction. The closer that a section approximates a true longitudinal section, the greater the apparent widening of the trough and the number of foresets that are truncated.

(c) Log of Triassic fluvial sediments from central England which shows palaeocurrent data and comments on specific features of interest. Assignment of particular beds and bedsets to defined lithofacies is shown to the left of the log (from Steel & Thompson, 1983).

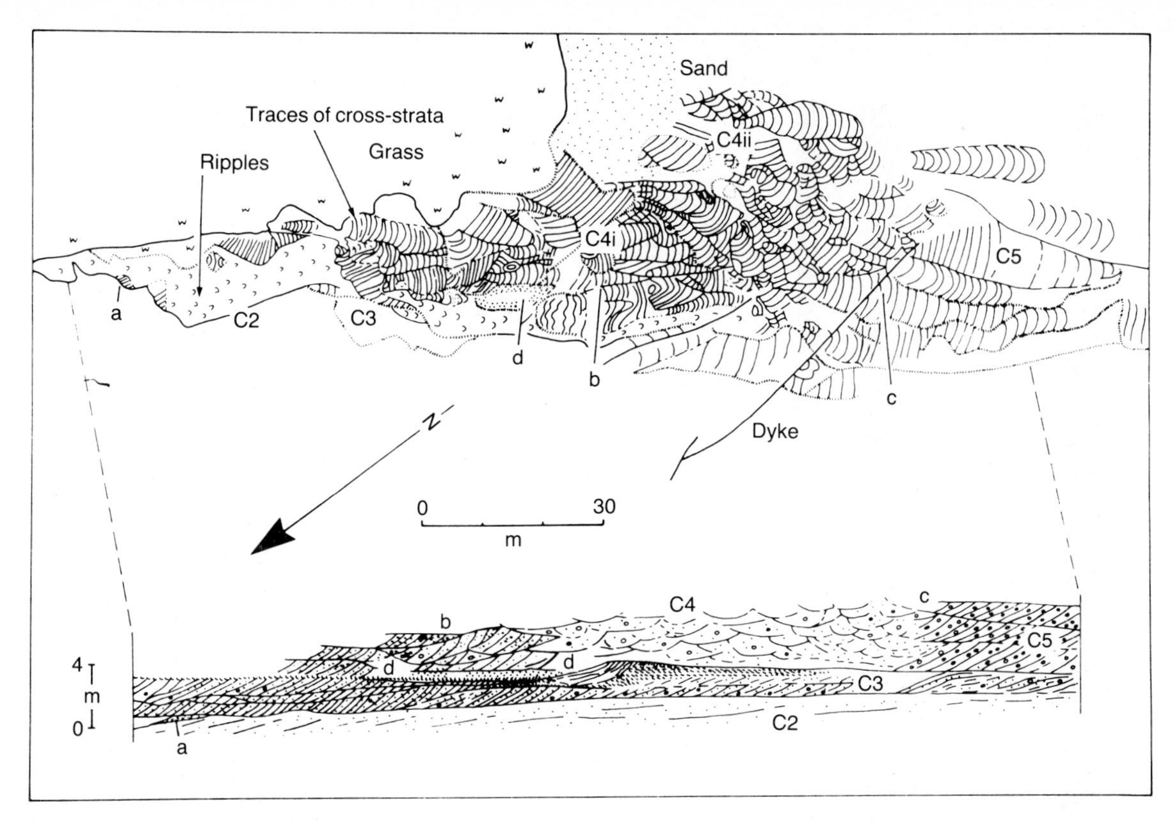

Fig. 2.25. A plane table map of an Upper Old Red Sandstone sand body from central Scotland and accompanying section. Both lateral and vertical relations of bedforms are well demonstrated. C1, C2 etc. refer to lithofacies described in text. a, b, c refer to zones of facies interfingering and d to an area of soft sediment deformation (from Bluck, 1981). (Reproduced by permission of IAS.)

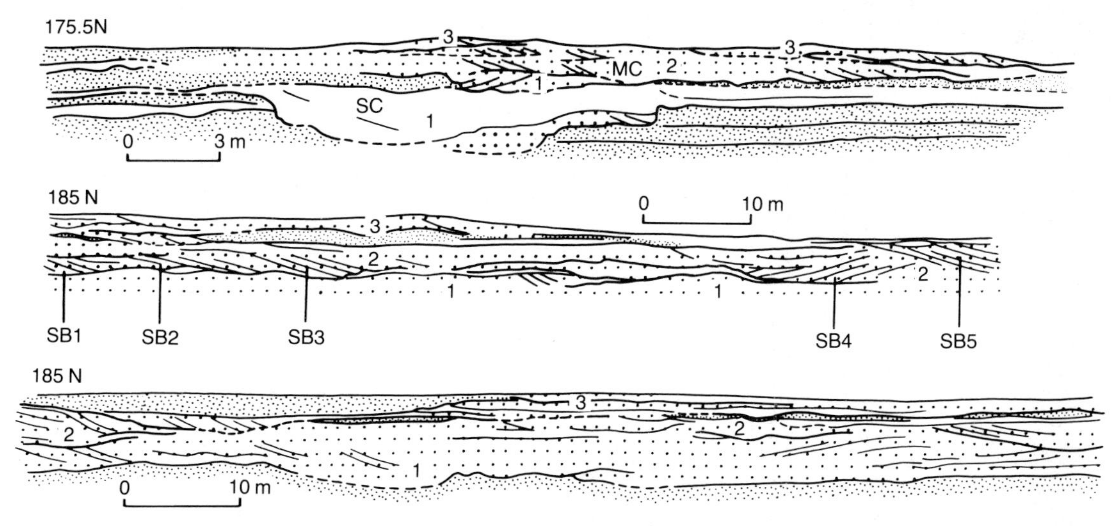

Fig. 2.26. Field sketches of extensive motorway cuttings in shallowly dipping strata which demonstrate lateral relationships within multistorey conglomerate channel bodies. Numbers refer to storeys: dense stipple — overbank sandstones and mudstones; blank — channel sandstone and wings; diffuse stipple — conglomerate; SB1 to 5 — sidebar units (from Allen *et al.*, 1983). (Reproduced by permission of the Geological Society of London.)

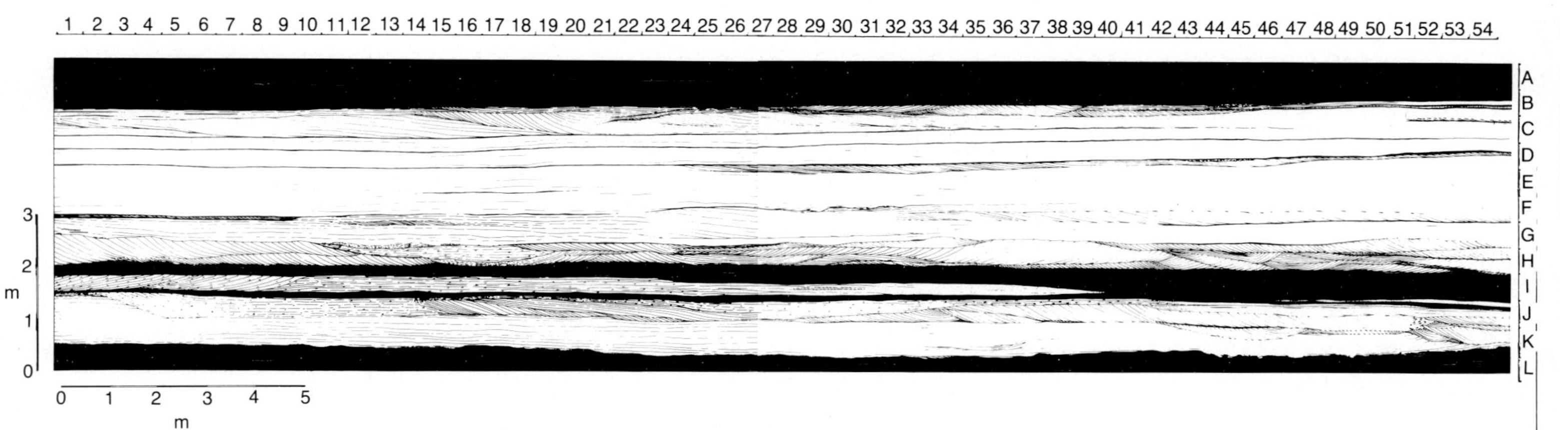

Fig. 2.27. Diagram to show lateral and vertical relationships in well exposed sandstone bodies in the Upper Devonian Old Red Sandstone of southern Ireland. To construct this a grid of 2.5 × 1.5 m graticules was laid down oriented parallel to a base line in underlying, even bedded siltrock. Internal structures in each graticule were drawn to scale on graph paper and the sheets were later redrawn to a reduced scale as a large composite diagram which was then photo-reduced for presentation (from Russell, 1984).

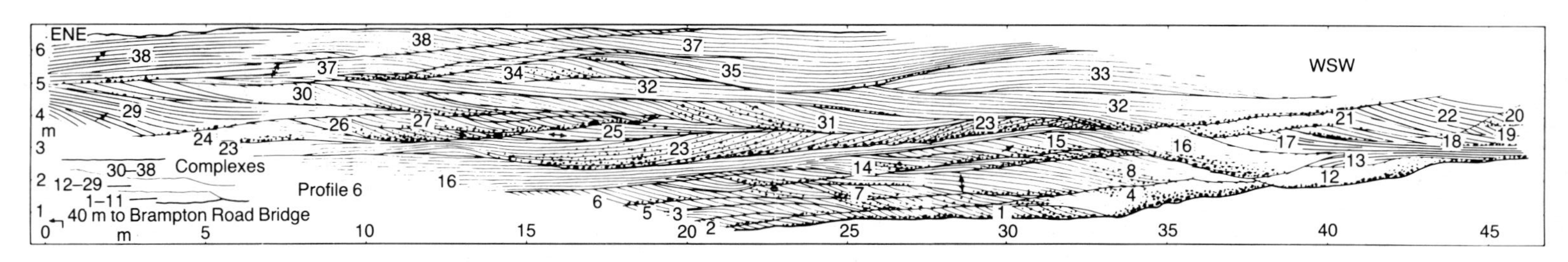

Fig. 2.28. Diagram to show lateral and vertical relationships in fluvial sandstones exposed in extensive road cuttings from the Lower Devonian of the Welsh Borders. The diagram was prepared by using enlarged prints of photographs as 'base maps' to which were attached overlays of acetate film on which information from binoculars and limited direct observation was recorded (from Allen, 1983). (Reproduced by permission of Elsevier.)

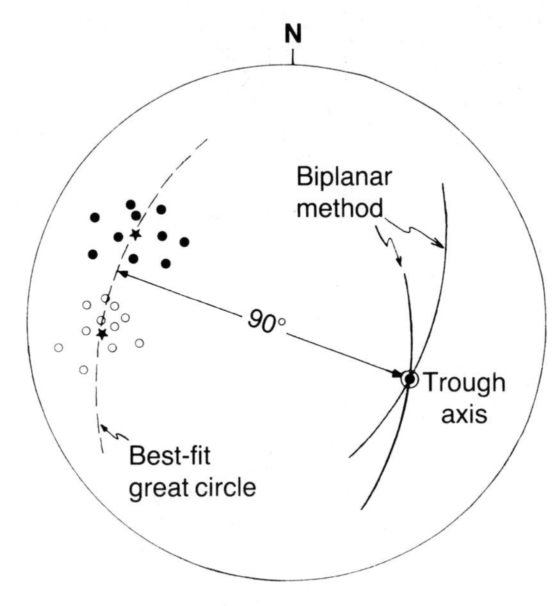

Fig. 2.29. A method of estimating trough axes from measurements on trough limbs using a stereo net. Trough limb data show clusters of right hand (open circles) and left hand (closed circles) poles; average poles are indicated by stars. The intersection of the two corresponding great circles (biplanar method) and the pole to the best fit great circle both give the trough axis (from De Celles *et al.*, 1873). (Reproduced by permission of SEPM.)

IMBRICATION AND CLAST ORIENTATION

Ellipsoidal clasts frequently show a preferred alignment, particularly in coarse grained sedimentary rocks. This alignment is visible in both plan view, here termed orientation, and in vertical sections, here termed imbrication. In some cases this alignment is due to the position of the clasts within structures such as foresets of cross strata. Here they show a similar downstream dip direction to the foresets but commonly a shallower inclination (Johansson, 1965). In contrast the dip of the clasts not contained in large foresets is preferentially upstream. Clearly the presence or absence of foresets in these coarse grained rocks must be determined at the outset. Several factors appear to control the range of inclination including clast size, clast sphericity, degree of clast contact, and palaeohydraulic conditions (Johansson, 1965; Rust, 1975; Koster, Rust & Gendzwill, 1980). It has also been shown that the orientation of the longest *a*-axis (where $a>b>c$) tends to be either normal to or parallel to the flow direction, the difference depending in part on the nature of the flow. The former is the characteristic result of clasts rolling on the bed (Johansson, 1965; Rust, 1972) whereas the latter appears to be common only in conglomerates associated with sediment gravity flows (Walker, 1975), although the precise mechanism of formation is less clear.

In sedimentary rocks the best measure of palaeocurrent direction is to obtain the mean vector of the *ab* plane. The need to measure rod- or disc-shaped clasts involves some selection, usually by means of ignoring clasts more equant than say 2:2:1 (Rust, 1975). The dip and strike of the *ab* plane is estimated by eye and measured from a notebook held in that plane. If the rocks are poorly indurated the clasts may be extracted after measurement and the orientation of the *a*-axis determined. Application of this technique clearly requires good quality exposures. Random selection of clasts can be achieved by placing a sampling grid, e.g. a piece of fish net or chalk lines, over the bedding surface. The stronger alignment of larger clasts which has commonly been observed (Johansson, 1965; Rust, 1972, 1975) may suggest that some selection in terms of size as well as shape is useful. The number of clasts to be measured should be such that a clear non-random distribution is achieved (Koster *et al.*, 1980). Experience suggests that about 40 clasts is generally sufficient.

In many cases the quality of exposure does not permit this technique of measurement. Where there is extensive bedding plane exposure but estimation of the dip of *ab* plane is difficult, a technique for measurement described by Nilsen (1968) can be employed. The elongation directions of 50 randomly selected clasts with axial ratios 1.5:1 are measured. This technique can also be used with a photograph and overlay if preferred. Such orientation readings may not provide an unambiguous palaeocurrent direction as long axis orientation modes tend to be either current normal or current parallel. Thus the same technique must also be applied to measure imbrication on vertical sections. Where possible vertical sections either parallel to the long axis orientation or perpendicular to it or both have commonly been chosen (Nilsen, 1968; Davies & Walker, 1974; Fig. 2.30). Davies & Walker (1974), in a study of the deep water Cap Enragé Formation of Quebec, noted that most of the available sections were fortuitously oriented parallel to modal clast orientation and that orthogonal faces had an incliniation not significantly different from horizontal compared to a mean of 12°

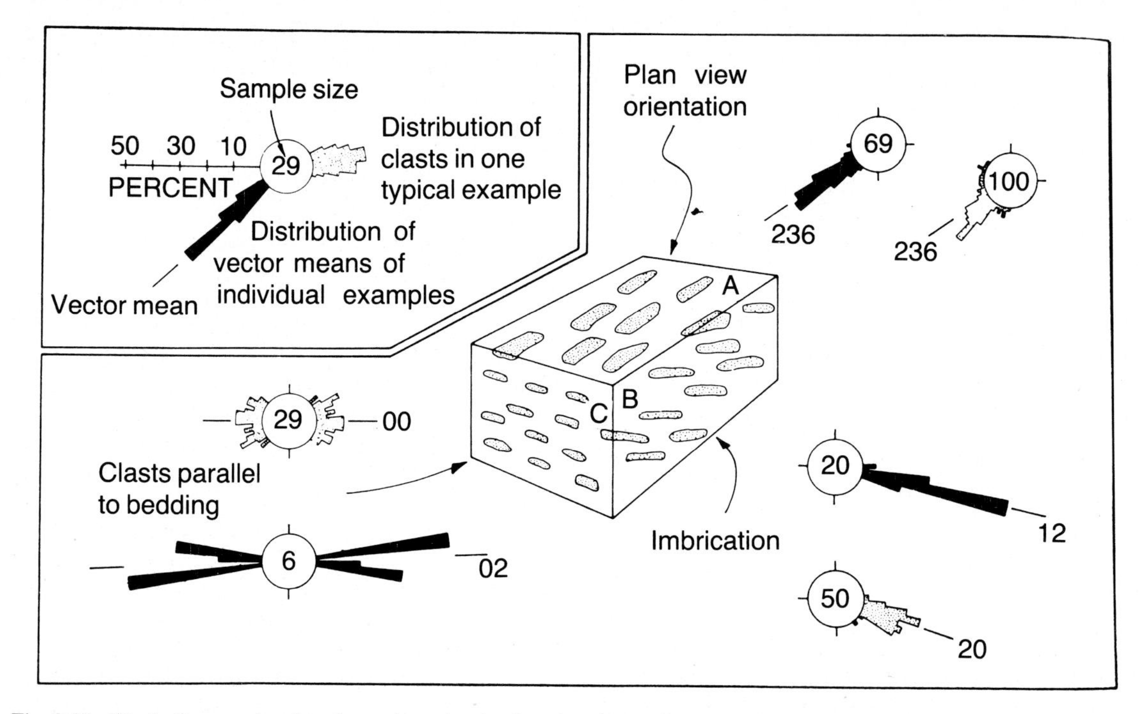

Fig. 2.30. Block diagram showing clast orientation in plan view (face A), imbrication (face B) and clasts parallel to bedding (face C). Solid rose diagrams show the distribution of vector means of individual examples and stippled rose diagrams show distribution of individual clast orientations in a typical example. Data are from the Cap Enrage Formation (deep water Cambro–Ordovician) of Quebec (from Davies & Walker, 1974). (Reproduced by permission of SEPM.)

for the current parallel — in this case long axis parallel — sections. Thus they were able to obtain a statistical three-dimensional alignment.

However, in many cases the available exposures may not be parallel to the palaeocurrent, and bedding plane exposures from which orientation can be estimated may also not be available. In these cases measurement is more difficult and less precise. It is necessary to measure imbrication on two faces, preferably <90° apart. That showing the greater dip would be closer to the true palaeocurrent direction, but there is no quantitative way of solving this apparent dip problem as there is for foresets of cross strata. Thus accuracy in many cases will be limited to an octant of the compass for each reading and this should be clearly stated.

FLUTE MARKS

Flute marks are found predominantly but not exclusively in turbidite successions. The origin of these structures and their relationship to fluid flow has been considered at length by Allen (1971), 1982). They are usually seen on the bases of sandstone beds, commonly in association with other sole marks such as groove casts. Whereas most sole marks are linear structures, flute marks uniquely specify a direction with the deeper, bulbous end indicating upcurrent. Flutes generally occur in large numbers on a particular bedding plane and a mean direction should be specified, although variation is typically very small. Measurement is as for linear structures although the end of the structure which is bulbous must also be recorded.

SLUMP FOLDS

Slump folds are produced when unconsolidated sediments resting on a slope become unstable and move downslope under the influence of gravity. Thus if the direction of that movement could be estimated a means of determining palaeoslope is obtained. It is clearly necessary as a first stage to distinguish slump folds from (a) those produced by vertical soft sediment movement; these generally produce little or no preferred orientation of axes, and (b) tectonic folds. The latter distinction is frequently more difficult. Criteria are summarized by Woodcock (1976a) who suggested that the more objective and reliable ones are: (i) angular discordance at some upper slump sheet contacts; (ii) miscellaneous but important features such as burrowed folds, dewatering structures, sand volcanoes, undeformed clasts and fossils; (iii) absence of tension cracks and veins in folds; (iv) absence of geometrically related macroscopic folds; (v) arrangement of slump sheets in a sequence in which unit thicknesses have strongly skewed distributions.

In many other respects slump folds and tectonic folds may be geometrically indistinguishable, e.g. similar range of layer and surface shape, similar lineation and cleavage patterns and similar spatial attitudes.

If slump folds can be confidently recognized, then a method of determining mean downslope direction is needed. Variability occurs mainly because slump sheets are three-dimensional structures and commonly lobate in plan view. Detailed procedures are given by Woodcock (1979) and only a brief summary is given below. The two methods used are known as the *mean axis method* which estimates downslope direction as perpendicular to the means slump fold axis, and the *separation arc method* which uses the bisector of a planar separation angle between groups of folds with opposite downplunge asymmetry (Fig. 2.31).

The simpler mean axis method is preferred as it is amenable to the derivation of confidence limits, it is applicable where asymmetry data are unavailable, and it relies on average properties of the data. The average trend is taken either as the mode or, in most quantitative studies, as the mean of the axes. However, the separation arc method, which relies more on the extreme properties of the data, has advantages with strongly skewed distributions. Most studies have used data specified in two dimensions only but the technique is equally applicable to three-dimensional data. The method specifies a strike of palaeoslope and leaves two alternatives for dip. Solution generally comes from considering either the vergence or facing of the slump folds, or from regional considerations (Fig. 2.32).

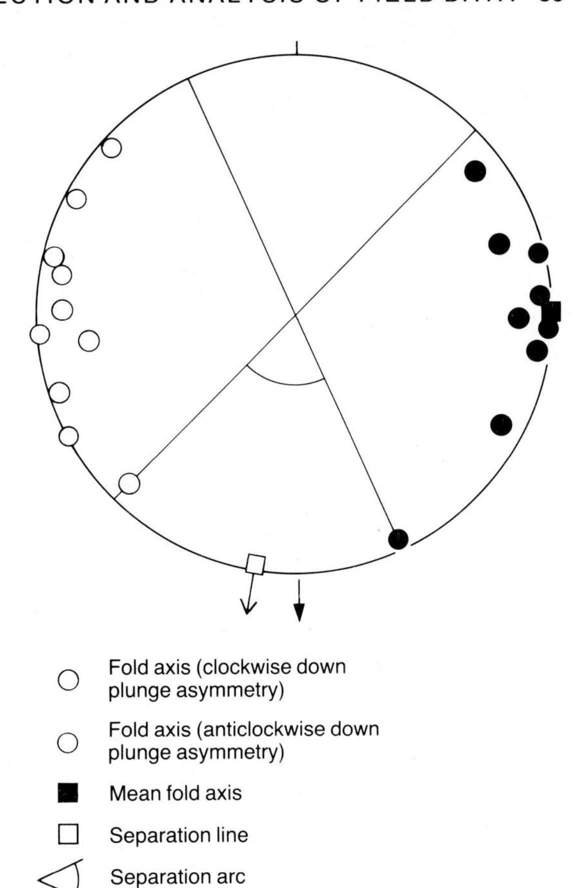

Fig. 2.31. An equal area stereo plot of hypothetical fold axis distribution to illustrate the mean axis and separation arc methods (see text). The palaeoslope dips due south (from Woodcock, 1979).

SAMPLING — HOW MANY MEASUREMENTS?

In palaeocurrent analysis, population parameters can almost never be evaluated and it is only possible to make estimates based on samples. A close identity between the orientation of directional structures at outcrop and those not available for measurement is assumed. In cases of poor exposure or restricted development of particular units there may be an

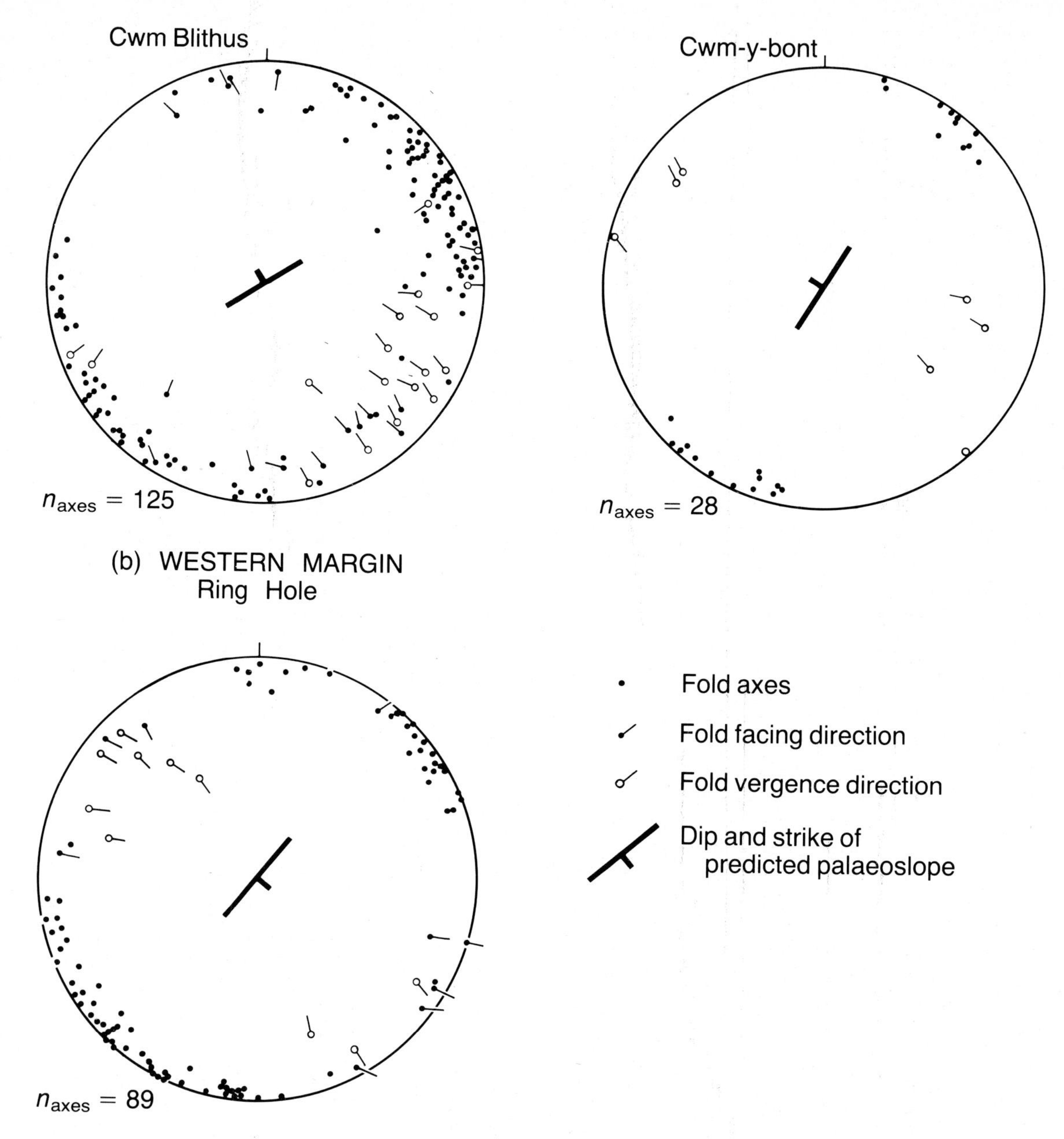

Fig. 2.32. Use of vergence and facing to determine palaeoslope direction in slump folds from Silurian strata in Wales. The facing direction is a line within the axial surface, locally perpendicular to the fold hinge. The tick indicates the facing sense. A similar convention is used to indicate fold vergence (from Woodcock, 1976b). (Reproduced by permission of the Geologists' Association.)

enforced sample such that every available measurement must be taken. Designed sampling can only occur where there is a surfeit of data and may be determined by cost (time + access) or for purely analytical reasons.

Over larger areas sampling usually aims at equalizing area coverage and uses some sort of pre-devised grid system. Selection of sample points within grid squares should be random but this is only rarely possible. In practice, outcrops are commonly sampled as encountered and some effort is made to distribute sample points evenly within a given area. The number of measurements will depend largely on the objectives of the study which must be defined before any specifications can be given. There has perhaps been a tendency to concentrate on estimation of mean values at the expense of investigating variability. Both are very important parameters in environmental interpretation. Despite the obvious advantages there have been few attempts at 'nested' or hierarchical sampling where an attempt is made to analyse different components of the total variance. An interesting exception is the work of Kelling (1969) which is discussed in Section 2.3.5.

2.3.3 Removal of tectonic effects

As it is the original current direction at the time of deposition that is required for interpretation, any subsequent reorientation or deformation of the primary structures being used should be estimated and removed. The former of these operations is relatively simple but the latter may be difficult and in some cases impossible.

Neglect of tectonic modification will clearly introduce error into the resultant observation. For example, the error that is introduced by failing to untilt beds has been compiled graphically by Ramsay (1961, fig. 2). The error will depend on the angle of dip and the angle between the sedimentary structure (expressed as a lineation) and the fold axis. For tilts less than 30° there is little introduction of error, and the original orientation of directional structures as measured in the field could be employed. For tilts in excess of this it is necessary to make a correction. Removal of tilt can be accomplished by rotating the bedding about an axis parallel to its strike until it comes to lie in a horizontal position. The method is applicable for simple flexural folds having horizontal axes and is most conveniently accomplished using a stereographic net.

If the fold axis is inclined (i.e. a plunging fold), the strike of a bed at any position on the fold can never be parallel to it. This fact invalidates the method of unfolding about the strike which would introduce an amount of error varying with both the dip of the beds and the plunge of the fold. This error was presented graphically by Ramsay (1961, fig. 5) who showed that for steeply dipping or overturned beds it can be very large. To prevent this error the structure must first be unfolded by rotation about the fold axis to a position of lowest dip (equal to the plunge of the fold axis) and then the final tilt removed by rotation about the strike. Two worked examples for the common cases of cross bedding and sole marks are shown in Figs 2.33 and 2.34.

Correct for tectonic effects is much more problematical when flexural folding has been accompanied by significant compression (flattening) and when similar folding is involved. In these cases lines and planes suffer significant distortion which must be removed in the following order: (i) removal of compression effects, (ii) removal of shear folding, (iii) removal of plunge and tilt.

For the removal of compression it is necessary to know both the orientation of the tectonic axes and the amount of compression. Although techniques for this exist (Ramsay, 1961, 1967), they require a considerable amount of detailed structural information which in some cases may be unobtainable. The errors which may be introduced by neglecting these effects may be large (see graphs in Ramsay, 1961).

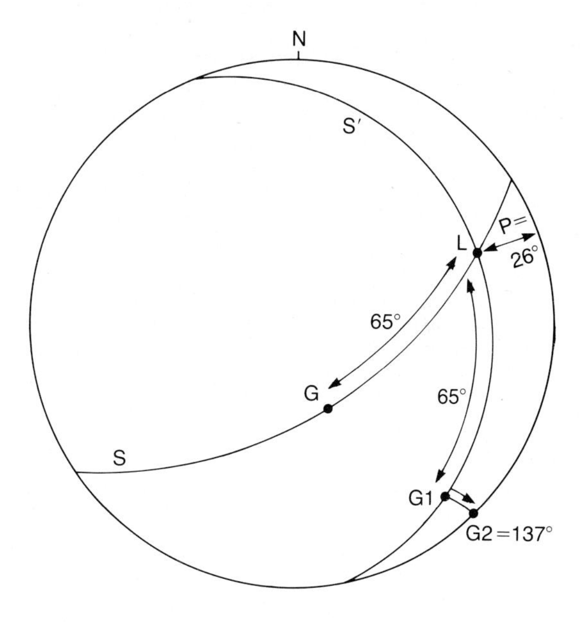

Fig. 2.33. A worked example of re-orientation of groove marks.
Field record: Bedding (S) 055/60SE. Intersection lineation (L) pitches 30°NE on bedding. Groove marks (G) ptich 85°SW on bedding.
(1) Rotate bedding (S) about the fold axis (L) to position where dip = plunge (P). G moves to G1 such that L ∠ G = 65° = L ∠ G1.
(2) Rotate the bedding (S') to horizontal. G1 moves along a small circle to give the original direction of the lineation, G2 = 137°.

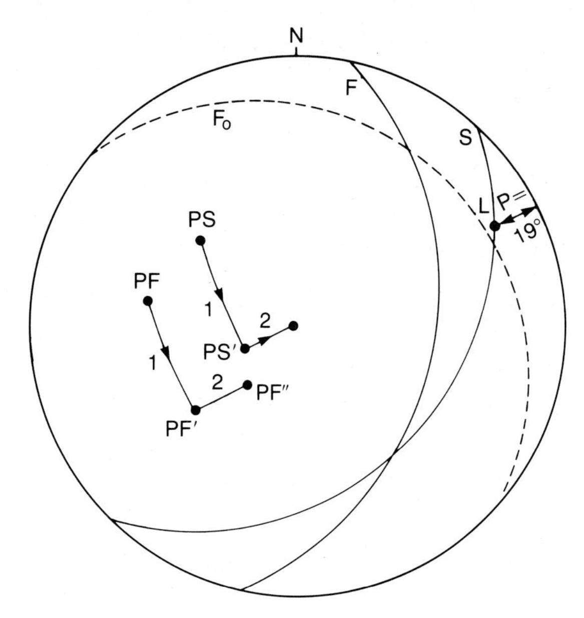

Fig. 2.34. A worked example of reorientation of cross beds.
Field record: Bedding (S) 040/40SE. Foreset (F) 010/48SE. Intersection lineation pitches 30°NE on bedding. Plot the poles to the bedding (PS) and foreset (PF) as well as the bedding (S) and foreset (F) planes. Plot the intersection lineation (L).
(1) Unfold by rotation about the fold axis (L) to position where dip = plunge (p). PS and PF move along great circles to PS′ and PF′ as S rotates about L. Position PS′ determined by L. PF ∠ PF′ = PS ∠ PS′.
(2) Rotate the bedding to the horizontal. Return PS′ to the centre along small circle and move PF′ the same angle along a small circle to PF″. This is the pole of the original foreset orientation (F_0). Read off the direction and amount of dip (25° to 040).

Even for those sedimentologists who are clearly aware of structural limitations (e.g. Dott, 1974; McDonald & Tanner, 1983), there is seldom evidence that regional compression and fold style have been adequately considered. Unless considerable effort is spent in collecting the necessary structural data, palaeocurrent analysis in rocks which have undergone significant shear and compression can at best be considered as approximate and at worst may be misleading.

Several published palaeocurrent studies showed that reoriented foresets commonly showed inclinations much higher than the stable angle of repose of sand (Pelletier, 1958; Cassyhap, 1968). This is to be expected if the mechanisms of folding are considered (Ramsay, 1961, 1967). In flexural folding there is generally a systematic increase in angle between the cross bedding and the regional bedding on one fold limb and a decrease in angle on the other limb. Where flexural folding is accompanied by compression normal to the axial plane the angle of inclination of the cross bedding is further modifed. Abnormally high and low cross bedding inclinations are produced on both folded limbs (Fig. 2.35). The magnitude of the modifications is dependent on the radius of curvature of the base of the bed, the thickness of the cross bedded unit, the amount of dip of the bedding and the position of the foreset/regional bedding intersection before folding. Most complex modifications result from the combination of shear folding and compression. This can result in abnormally high or low inclinations on both fold limbs and, in the extreme case of the limbs of isoclinal folds, the structure may be completely unrecognizable.

It is clear that the removal of tectonic effects from palaeocurrent readings can be laborious and time-consuming. Such a repetitive process is amenable to the application of computer techniques and some authors have advocated doing this in the field with pocket calculators (Freeman & Pierce, 1979). A program that adequately considers both plunge and tilt removal is presented by Cooper & Marshall

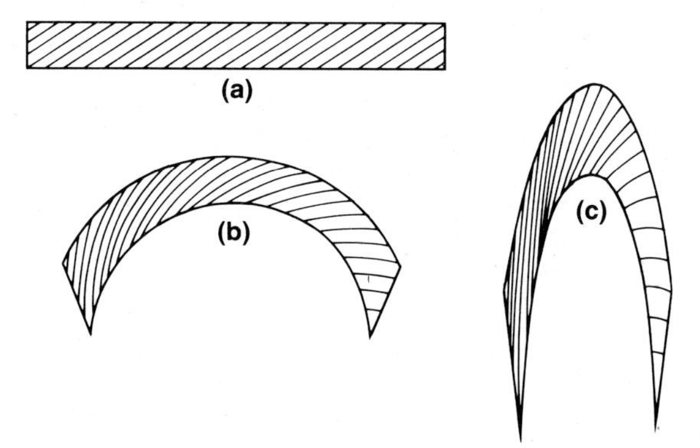

Fig. 2.35. Demonstration of the distortion of angles with flexural folding and compression.
(a) Original orientation of bedding with foresets inclined at 30°.
(b) The same bed subjected to flexural folding.
(c) The same bed subjected to flexural folding and 50% compression.
Note how abnormally high and low foreset dips can be generated depending on position within the fold structure. (Reproduced by permission of McGraw-Hill.)

(1981), who reviewed most earlier attempts. Such programs also frequently present the data in some of the standard formats outlined in the next section. These packages are now widely available and will probably be used increasingly in the future. Despite this some workers in structurally complex areas prefer to perform these operations by hand so that a careful check can be kept on the precision of each recording (McDonald & Tanner, 1983). It is important to state clearly any manipulations that were performed and to note any likely limitations, such as not taking account of tectonic deformation.

2.3.4 Presentation of data

In a few cases where data are scarce but geologically significant, e.g. major channel orientations or giant cross beds (Collinson, 1968), each palaeocurrent measurement may be shown individually. More normally, measurements are grouped into classes of 10°, 15°, 20° or 30°, the choice of interval depending on both the number of readings and directional variability. In general, the lower the variability the smaller the class interval although too small an interval can lead to irregular class frequencies. These are presented in the form of a histogram converted to a circular distribution and termed a current rose (see Fig. 2.36). These histograms (roses) can plot either total number of observations per class or percentages, the latter often being useful for comparative purposes. In the geological literature the current rose conventionally indicates the direction toward which the current moved.

Although many structures used to infer palaeocurrents are vectors, i.e. possessing magnitude as well as direction, they are all usually considered as having unit magnitude. This relates to the geological uncertainty of objectively assigning a magnitude component. However, there are many cases where presentation of only standard current roses may lead to loss of valuable data. Thus amongst workers in aeolian sediments it is common practice to present data on a polar stereographic projection (Reiche, 1938; Kiersch, 1950; McKee, 1966; Carruthers, 1985; Fig. 2.37). Not only does this allow all readings to be shown individually but it also places visual emphasis on the larger dip angles as indicators of palaeowind directions. In addition Glennie (1970) has suggested the possibility that the type of dune structure can be recognized from the distribution of poles of dip planes.

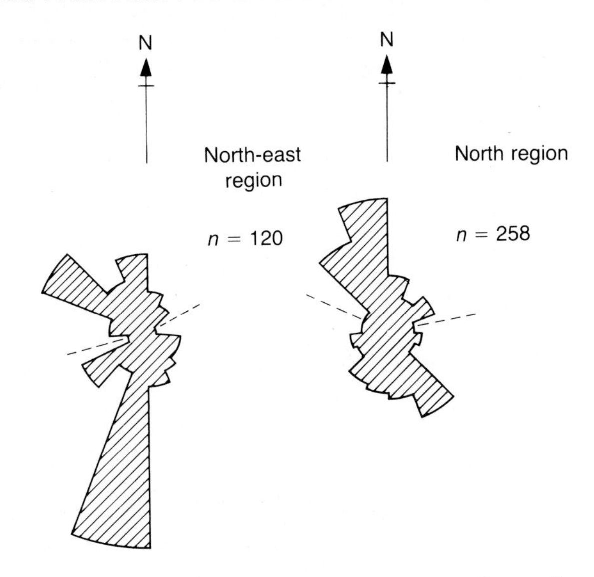

Fig. 2.36. Rose diagrams based on semi-octant classes of cross-stratal azimuths. Dotted lines represent the partitioning into two distributions for statistical analysis (from Kelling, 1969). (Reproduced by permission of SEPM.)

Most palaeocurrent studies are concerned with the preferred orientation direction, if one exists, and the degree of spread about that orientation. These are measures which are analogous to the means and variance of linear distributions but must be calculated differently to account for the circular nature of the data. The simplest form of calculation is:

$V = \Sigma_1^n \cos \theta$,
$W = \Sigma_1^n \sin \theta$,
$\bar{x} = \arctan W/V$,
$R = (V^2 + W^2)^{1/2}$,
$L = (R/n) . 100$,
θ = individual azimuth,
n = number of readings,
$\bar{x}$ = vector mean,
R = magnitude of resultant vector,
L = magnitude of resultant vector (%).

The values of L and n can be used for the Rayleigh test of significance outlined below.

For two-dimensional orientation data such as primary current lineation or goove casts, all orientations can be expressed within a 180° range. In these cases the calculations become:

$V = \Sigma_1^n \cos 2\theta$,
$W = \Sigma_1^n \sin 2\theta$,

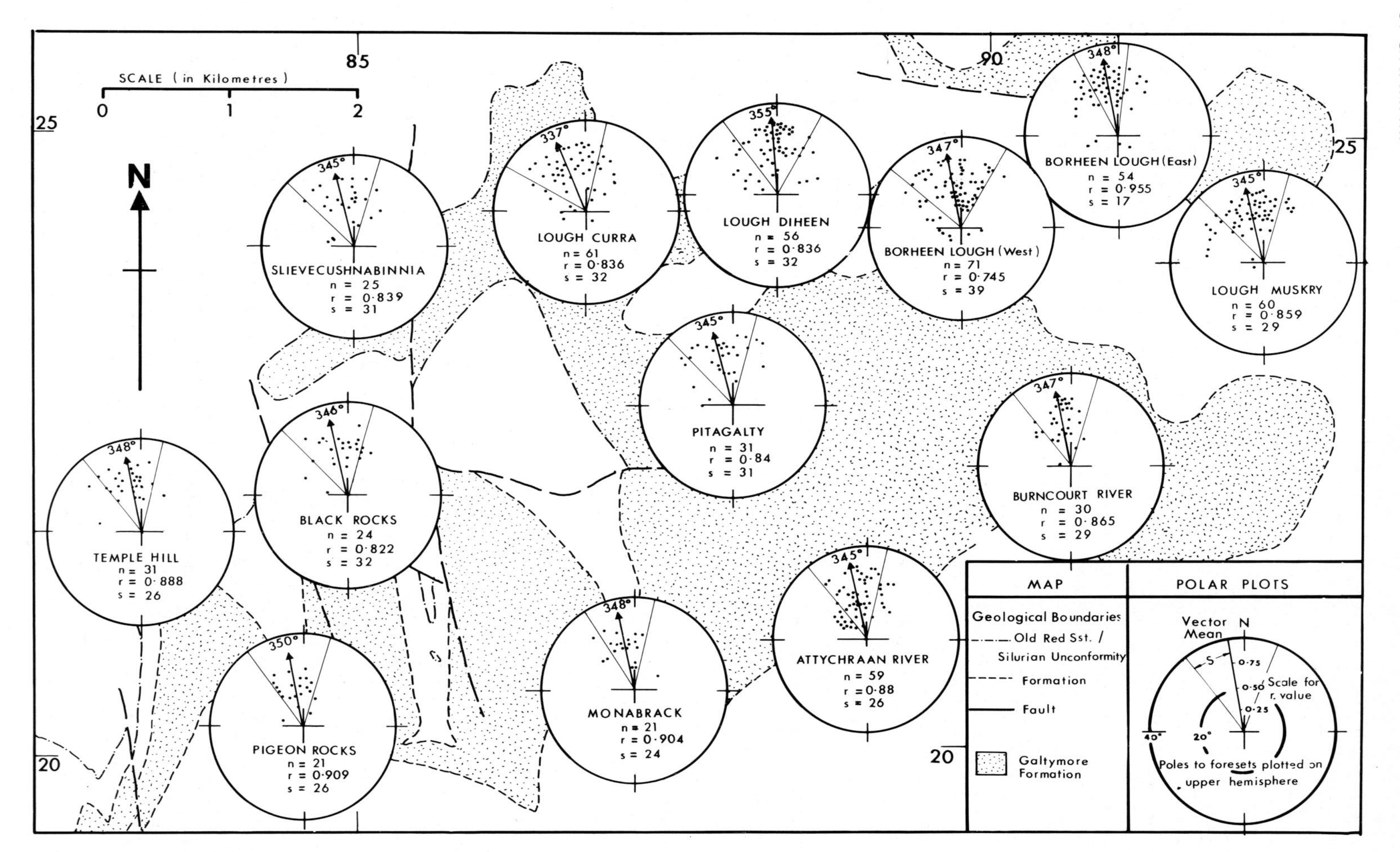

Fig. 2.37. A polar stereographic projection of aeolian foresets from the Galtymore Formation of the Old Red Sandstone, Southern Ireland. Dip angles of foresets as well as directions can be clearly seen (from Carruthers, 1985).

$\bar{x}$ = 1/2 arctan W/V,
$R = (W^2 - V^2)^{1/2}$,
$L = (R/n) . 100$.

The vector mean ($\bar{x}$) is an expression of preferred orientation whilst the vector magnitude (%) (L) is a sensitive measure of dispersion.

It is necessary to test whether a set of palaeocurrent data possesses a distribution of orientation which is significantly different from random. Several ways of performing this test are reviewed by Potter & Pettijohn (1977, pp. 377–380). The simplest is probably the Rayleigh test as outlined by Curray (1956), for which a graph can be used (Fig. 2.38) to read the probability that the measurements are uniformly distributed in the interval 0–360°. A desired level of significance, usually $p < 0.05$, is chosen at which the null hypothesis of a uniform distribution is rejected.

The use of vector statistics in a comparable manner to the use of moments of a normal distribution implies the presence of a von Mises (circular normal) distribution of data. The density function of the von Mises distribution may be written as $f(x) = \frac{1}{2}\pi I_o(K)$ $(e^{k \cos (x-\theta)})$ where K is a parameter that expresses the concentration of the mass and θ is the angle where the function takes its maximum value. I_o (K) is a Bessel function whose values are tabulated. Such distributions are not always present as is often assumed, e.g. Sengupta & Rao (1966), although this can be readily checked by the method outlined by Harvey & Ferguson (1976).

It is generally vector means which are presented in published data either on maps or accompanying stratigraphic columns. The number of readings and value of L (or R) should also be readily accessible. It is often useful to present some visual impression of the amount of confidence attached to a vector mean by showing a confidence interval (of, say, 90 or 95%), e.g. Kelling (1969). This can be most readily produced by using a Batschelet chart as shown in Fig. 2.39. The techniques for calculating these confidence intervals numerically from the raw data are given by Mardia (1972, pp. 144–145).

It is also necessary to quantify the spread of values around the vector mean. For the von Mises distribution, R is an estimate of the spread of the angular values around a unit circle. The mean angular deviation, s, can be calculated by $s = \sqrt{(2(1 - R))}$ where s is measured in radians. Although s behaves mostly

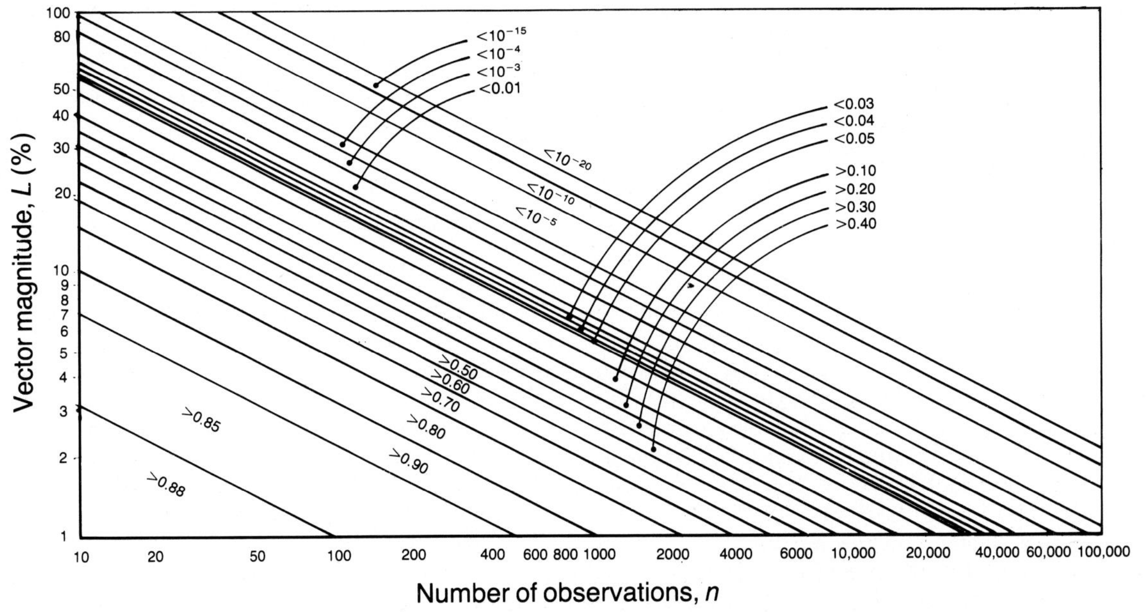

Fig. 2.38. Rayleigh test of significance. The graph plots the probability of a given vector magnitude (L) occurring by chance for varying numbers of observations. A significance level is chosen, commonly $p < 0.05$, at which the hypothesis of a random distribution of orientations is rejected (from Curray, 1956). (Reproduced by permission of University of Chicago Press.)

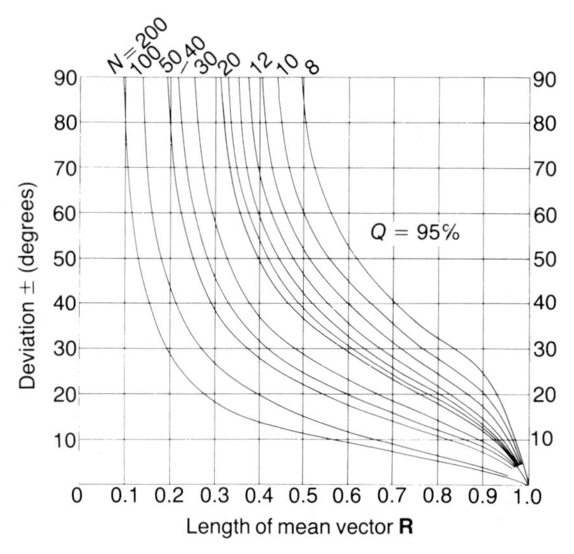

Fig. 2.39. Batschelet chart for obtaining a 95% confidence interval for the mean direction (from Batschelet, 1981). (Reproduced by permission of Academic Press.)

in the same way as the standard deviation of a normal distribution, it is not its circular analogue which is K. Tables are available for converting R to K and indeed K is used in many further tests (Mardia, 1972).

Having obtained reliable estimates for preferred orientation and concentration of values some further tests may be desirable or possible depending on the nature of the problem. Perhaps the commonest problem is that of comparing preferred current orientations for significant differences. Before advancing to testing equality of mean directions it is necessary to test that the samples are drawn from populations with equal concentration (Mardia, 1972, pp. 158–62; Cheeny, 1983, pp. 102–103). The test for equality of preferred direction is somewhat complex as it depends in part on the ratio of the two sample sizes and also on the strength of the concentration parameter, K, of the samples combined. Details and worked examples are given by Reyment (1971, pp. 48–50), Mardia (1972, pp. 152–155) and Cheeney (1983, pp. 101–106).

Although vector statistics are both precise and powerful, it is clear that the underlying populations of directional data must be thoroughly understood and defined. For example, sole marks and ripple marks from the bases and tops of beds could, in the case of storm-modified turbidites, represent different current systems operating at different times. Grouping of data and assumption of one population may be statistically possible but would not be geologically reasonable. Moreover the application of vector statistics implies a unimodal distribution and they are not directly applicable when the data are bimodal or polymodal. For this reason in particular current roses should always be drawn and the data inspected qualitatively.

Some methods have been proposed for further treatment of bimodal and polymodal distributions. For example the problem of bimodality may be resolved by calculating separate vector means for two circular frequency distributions, the overlap being separated by partition about the midpoint of the two shared classes containing the lowest frequencies (Fig. 2.36). This is possible where the two modes are almost opposite in direction but difficult to apply where they are closer together.

For polymodal distributions where vector statistics are inappropriate, simpler, semi-quantitative techniques can be used to express preferred direction (Tanner, 1959; Picard & High, 1968; Fig. 2.40). A compass is divided into 12 × 30° intervals and the mean number of occurrences per interval and the standard deviation are calculated. Those intervals that contain palaeocurrent directions in excess of one standard deviation above or below the mean represent prominent modes or nodes respectively. Intervals within one standard deviation are considered to be qualitatively indistinguishable from random distributions.

2.3.5 Interpretation of results

BEDFORM HIERARCHIES

It has long been realized that sedimentary structures show an hierarchical arrangement such that no one sedimentary structure fully specifies a complex flow system. Such a specification can only be made after sampling all types of structure generated within the limits of preservation. The concept of bedform hierarchies was formalized by Allen (1966), Bluck (1974) and Miall (1974) by whom parts of a fluvial system were given orders or ranks (Figs 2.41 and 2.42). In a general sense, the smaller structures in the hierarchy tend to be more variable because the currents forming them are controlled both by the overall transport direction and also by local deviations caused by larger bedforms. This phenomenon is particularly well known from fluvial systems (Collinson, 1968,

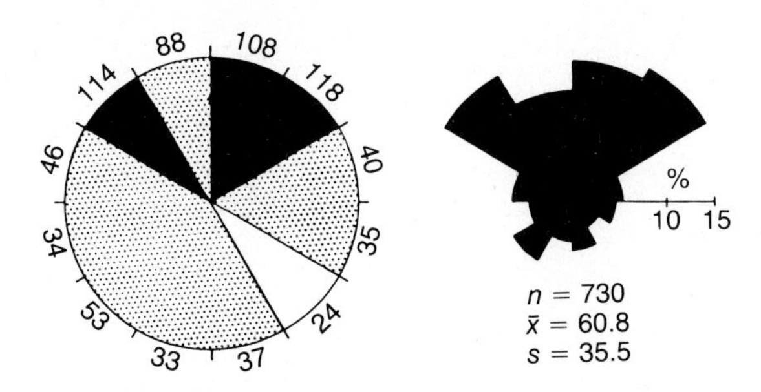

Fig. 2.40. Recognition of modes and nodes in polymodal distributions (see text for explanation) (from Picard & High, 1968). (Reproduced by permission of SEPM.)

1971; Bluck, 1974; Miall, 1974) but also applies in varying degrees to other environments (Field *et al.*, 1981; Link, Squires & Colburn, 1984). Thus one would be likely to regard the orientation of a single channel form as being in some way more significant than the orientation of a single set of trough cross beds. There have been only a few attempts to quantify this difference in scale by introducing a 'weighting factor'. In many ways this is due to the difficulty of defining an objective and operational methodology. Miall (1974) suggested using the cube of set thickness as it may crudely reflect the volume of sediment displaced by the current. Erosional loss of varying proportions of the sets clearly presents problems as do assumptions of geometrical similarity. Although this method has been little used it does qualitatively emphasize the differing significance that might be attached to different bedforms.

It is sometimes possible to analyse or even sample palaeocurrent data with this concept of hierarchical organization in mind (Olson & Potter, 1954; Potter & Olson, 1954; Kelling, 1969). However, an assumption that chosen sampling levels (e.g. outcrops) will match any natural hierarchical level needs to be justified. An example of this is given by Kelling (1969) who carefully sampled and analysed the palaeocurrents of some Upper Carboniferous fluvial

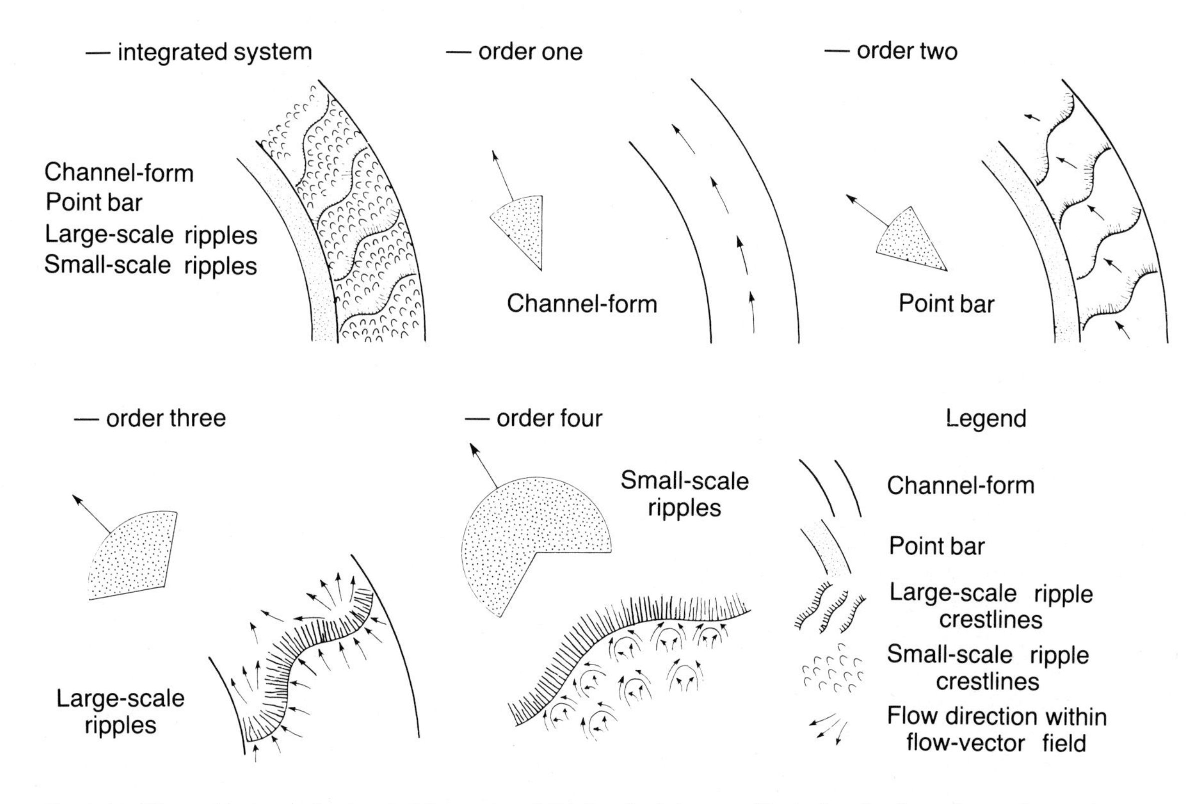

Fig. 2.41. Hierarchical organizational of flow-vector fields in a fluvial system illustrating the dependence of current-directional data on rank of bedform (from Allen, 1966). (Reproduced by permission of Elsevier.)

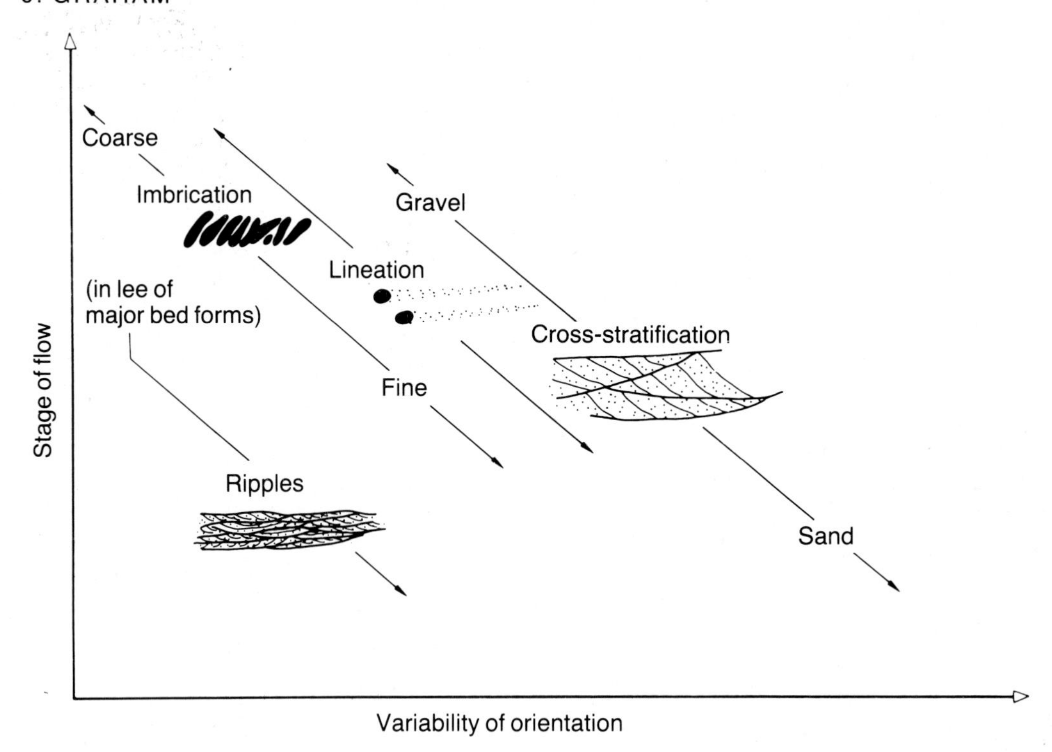

Fig. 2.42. Illustration of the possible relationships between flow stage, variability in orientation of structures and the kind of sedimentary structures in a fluvial system. Ripples are often an exception because they tend to be caught in major channel troughs and orientated along them (from Bluck, 1974).

sediments in South Wales. Sampling design followed a pilot study in a well-exposed area which gave some indication of variability and minimum number of measurements. The resultant data, summarized in Fig. 2.43, show clearly the mean direction with confidence limits for each sample sector. An hierarchical analysis of variance at four different levels was also attempted and related to natural levels of variation in a river system, i.e.

(i) different parts of drainage net or different sources;
(ii) different streams or portions of streams;
(iii) deviations of laterally or vertically adjacent sand bodies within major stream courses;
(iv) divergence of superposed bedforms.

More importantly, complementary data on the nature of the sedimentary structures (e.g. cross stratification type, set thickness, foreset dip angle) allow an objective assessment of the data by the reader.

USE OF VARIABILITY IN ENVIRONMENTAL INTERPRETATION

It can be seen from the above section that the amount of variability, even in a predominantly unidirectional transport system, will depend in part on the level of bedform hierarchy which is sampled. Much of our knowledge of palaeocurrent variability comes from specific modern environments (Allen, 1967; Miall, 1974). Attempts to distinguish environments by means of the variance of their palaeocurrent pattern have been made by Potter & Pettijohn (1977) and Long & Young (1978) who tabulated large amounts of data from case histories. However, it should be remembered that modern environments record variability over a very short time period whereas most geological studies have data bases that are fundamentally different because the strata accumulated with an appreciable time dimension such that the grouping of data over a longer time interval might be expected to increase variance. More im-

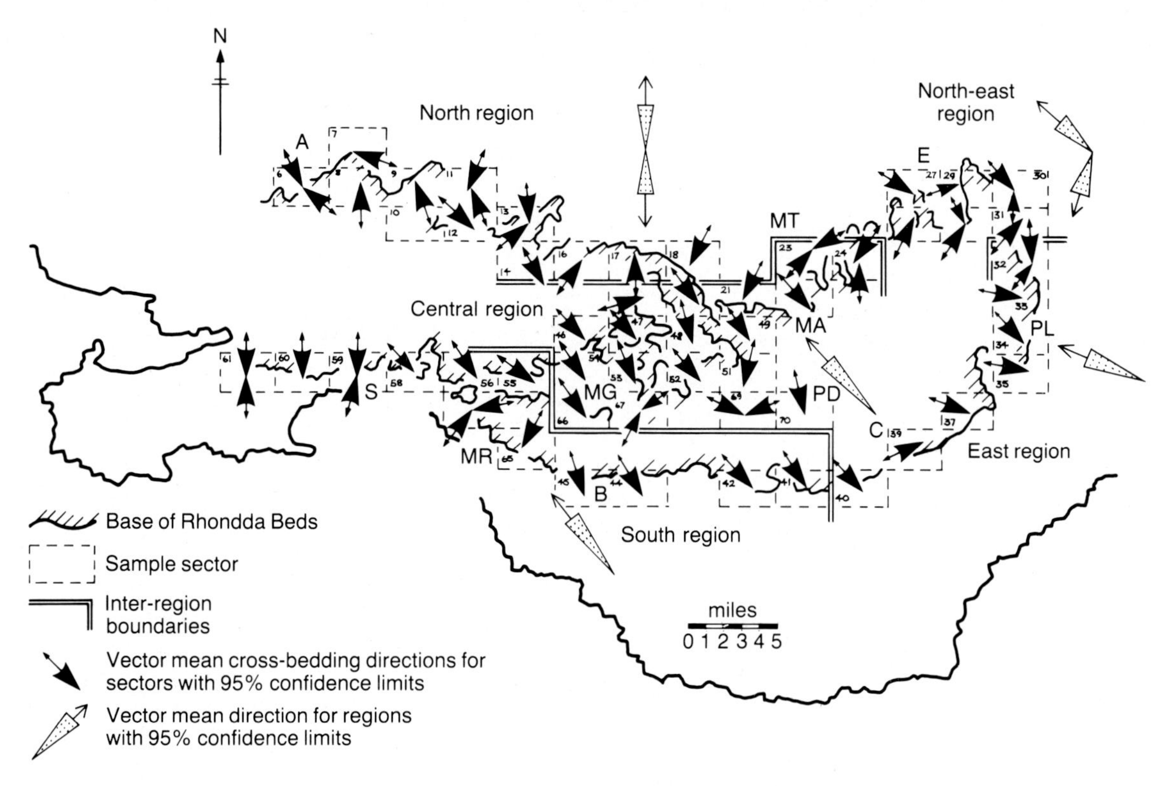

Fig. 2.43. An example of hierarchical sampling of palaeocurrent data from the Upper Carboniferous of South Wales. The map shows vector mean cross-stratal azimuths and 95% confidence limits for sectors and regions within the study area (from Kelling, 1969). (Reproduced by permission of SEPM.)

portantly, the inequalities and uncertainties of time-dependent changes and the difficulty of separating them from spatial variation make overall variance at best a very crude criterion.

In specific studies it is possible and often desirable to examine variability of palaeocurrents in both time and space (e.g. Chisholm & Dean, 1974; Miall, 1976; Pickering, 1981; Gray & Benton, 1982). Variation in time generally involves splitting vertical sections into meaningful subdivisions which can then be compared, e.g. by testing whether their vector means are significantly different. Clearly this technique will be limited by the accuracy with which such subdivisions can be correlated.

RELATIONSHIP OF SEDIMENTARY STRUCTURES TO ENVIRONMENT

Although palaeocurrent data themselves can be important environmental discriminants, their full significance usually requires independent interpretation of lithofacies and structures. It is particularly important in this respect that the structures that are measured in palaeocurrent analysis are described as thoroughly as possible. For example Bourgeois (1980) (Fig. 2.44) compared the dip amounts, as well as direction of, from HCS and trough cross beds and showed marked differences. In many earlier studies, written before the common recognition of HCS, such measurements would probably have been grouped as trough cross stratification. However if sufficient data were given on foreset inclination the possible presence of HCS might be suggested and re-investigation stimulated.

PALAEOCURRENT PATTERNS

It has been noted on a more general scale that palaeocurrent data tend to form a set pattern when seen over appreciable areas. Potter & Pettijohn

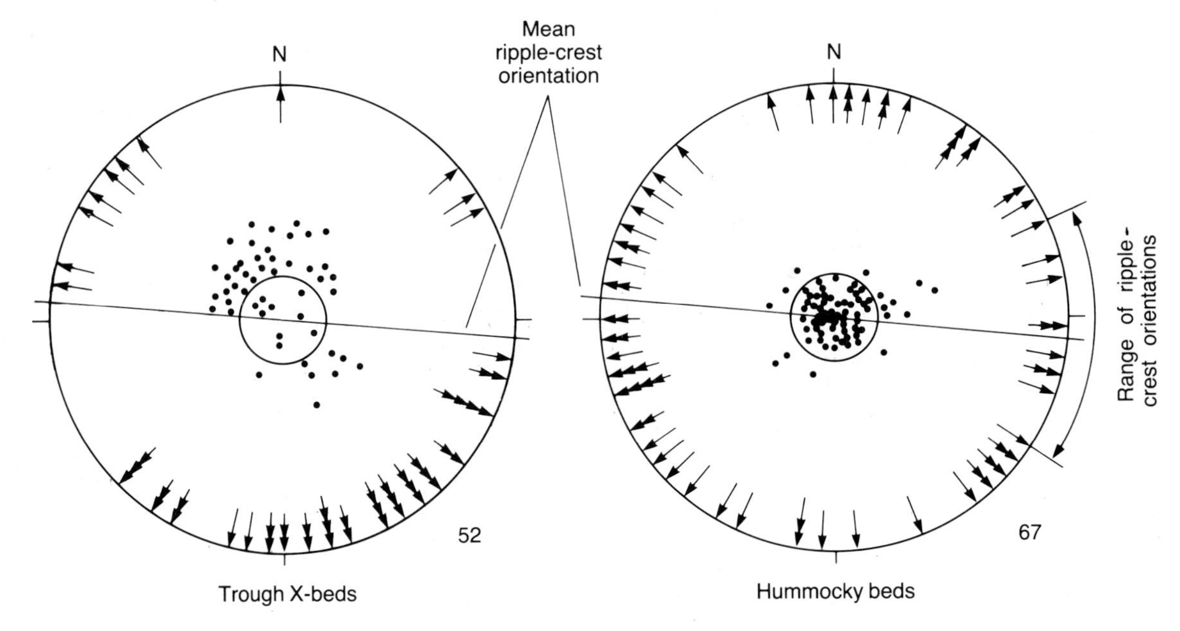

Fig. 2.44. Comparison of dip direction (arrows) and dip angle of trough cross-beds and hummocky beds in the Cape Sebastian Sandstone, Cretaceous, Oregon. Dots are poles to cross bedding; dots within the small circle represent dips less than 15° (from Bourgeois, 1980). (Reproduced by permission of SEPM.)

(1977) suggested that, based purely on geometry, there are seven basic patterns (Fig. 2.45). Attempts have also been made to produce environmental classifications of palaeocurrent patterns, (e.g. by Selley, 1968) and also to relate these to geotectonic setting (e.g. by Potter & Pettijohn 1977). At a smaller scale palaeocurrent data can support environmental discrimination suggested by other criteria. Thus Laming (1966) showed that palaeocurrents derived from strata interpreted as aeolian have a totally different pattern from those inferred to be fluvial. Chisholm & Dean (1974) used palaeocurrent data as the main evidence that parts of a largely fluvial Carboniferous formation in Scotland were deposited under the influence of tidal currents.

Fluvial palaeocurrents, because of their basic control by gravity, tend to be unidirectional although with considerable spread over small areas (e.g. Leeder, 1973). Over larger areas it has been shown that major drainage often occurs parallel to tectonic strike (Van Houten, 1974; Steel *et al.*, 1977; Bluck, 1978). Similar patterns have been shown to exist for deeper water marine sediments, particularly where there are submarine fan systems present (Jipa, 1966; Ricci Lucchi, 1975b, 1981; McDonald & Tanner, 1983).

Aeolian current directions are typically independent of other terrestrial transport directions (e.g. Hubert & Mertz, 1984) and in favourable cases it may be possible to relate patterns to trade wind circulation (e.g. Fryberger & Dean, 1979; Kocurek, 1981). Shallow marine patterns are typically the most complex (Klein, 1967) and difficult to interpret. In the cases where the currents are of different types and with different controls it is particularly important to attempt to relate palaeocurrent data to the palaeoenvironmental interpretation of the sedimentary structures from which they are derived.

2.4 SEDIMENTARY FACIES AND SEQUENCE ANALYSIS

2.4.1 Erection and use of facies

Much of the field recording of sedimentary rocks is aimed at classifying the strata being investigated into recurrent units and attempting to detect an order in the vertical and lateral arrangement of those units.

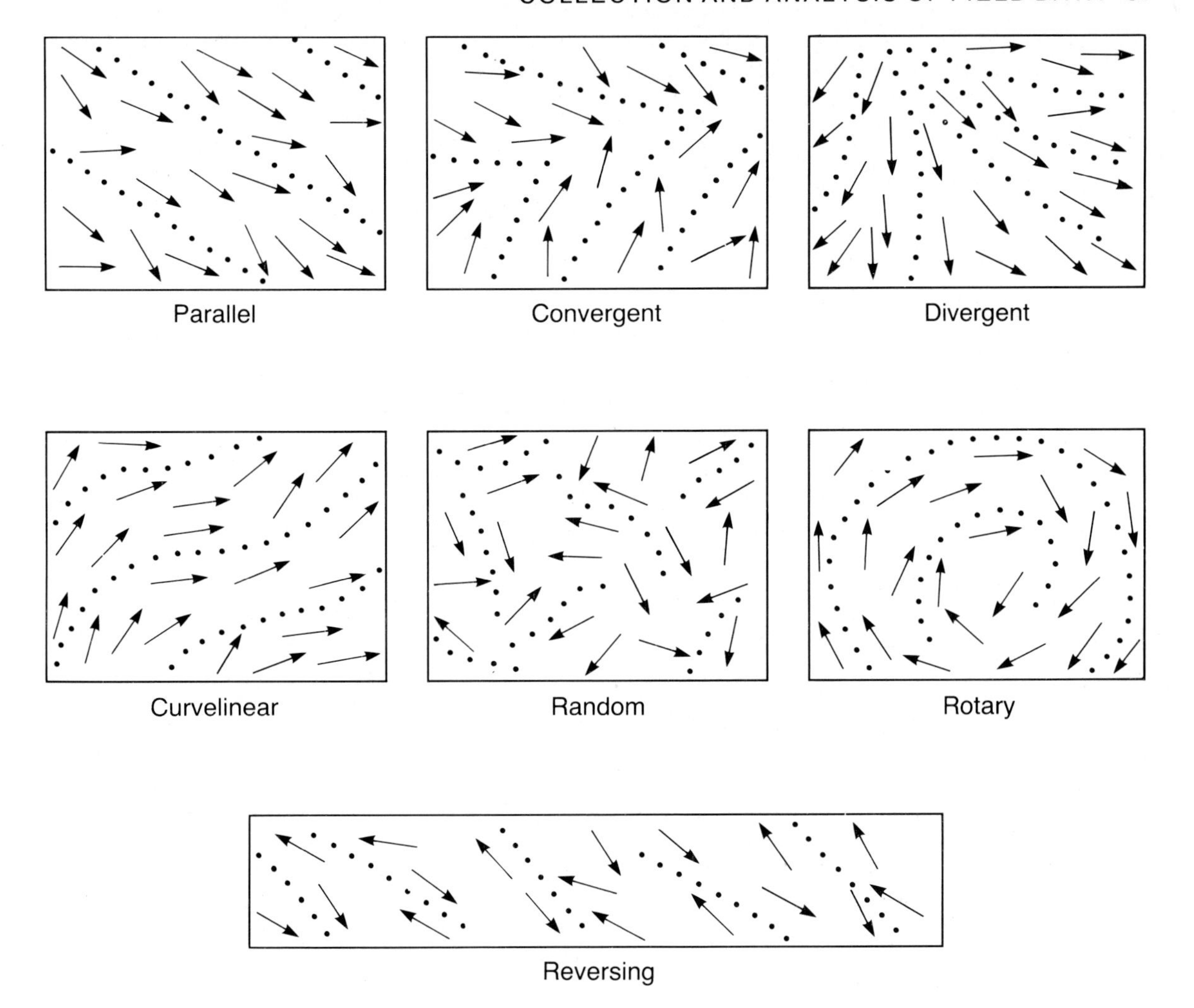

Fig. 2.45. Seven basic palaeocurrent patterns based on geometry (from Potter & Pettijohn, 1977). (Reproduced by permission of Springer.)

The term applied to these units is facies which are defined on a combination of lithological, structural and organic aspects in the field. The facies may be given informal designations, e.g. 'facies A' or brief descriptive designations, e.g. 'pebbly sandstone facies'. The main criteria used in delimiting facies will vary with the strata being investigated. A facies unit may in some cases be an individual bed or bedset but may also comprise several beds or bedsets. The origin and modern usage of the term facies has been reviewed by Reading (1978b) and Walker (1984).

The scale of facies subdivision is in part dependent on the variety of physical and biological structures in the rocks but also partly on the time and money available for the study and the scale and purpose of the investigation. Standardization becomes important particularly when several different investigators are describing the same succession, e.g. in many industrial studies, but must not be allowed to 'straightjacket' new observations. Objectivity of definition is most important. Unless the general nature of a succession is relatively well known at the outset, a reconnaissance study will generally precede facies subdivision. It is difficult to give general rules on how finely to subdivide in the field but it is better to over-subdivide than the reverse. Subdivisions can always be combined during later analysis but splitting without a field check is much more hazardous. Classic examples of the objective definition of facies are given by De Raaf, Reading & Walker (1965), Johnson (1975) and De Raaf *et al.* (1977).

2.4.2 Facies relationships

Individual facies vary in interpretative value but only a very few facies in isolation allow unambiguous interpretations. The key to interpretation is to study facies in association, in particular their relative vertical and lateral positions. The nature of the contact between two facies is fundamental in assessing their original depositional proximity.

LATERAL RELATIONSHIPS

Lateral relationships are in some cases directly observable and do not rely on an inferential step from the application of Walther's Law (Middleton, 1973) to a vertical sequence. However, the limited physical size of most exposures means that such situations are uncommon and require favourable circumstances. Such exposures are valuable in the description of the larger scale geometry of facies which is also seldom possible with limited vertical sections. For example Horne *et al.* (1978) gave some excellent examples of interpretative block diagrams constructed from laterally extensive road cuts through deltaic successions where exposure levels are between 60 and 95% over 102–103 m.

Spectacular examples of lateral facies changes on a large scale are given by Bosellini (1984) for Triassic carbonate platform margins in the Alps, by Hurst & Surlyk (1984) for Silurian carbonate platform margins in North Greenland, by Newell *et al.* (1953) for the Permian reef complex of Texas and New Mexico, by Playford (1980) for the Devonian reef complex of the Canning Basin, Western Australia, and by Ricci Lucchi (1975b) for the Miocene turbidites of the Apennines. In these and other cases unusually large natural exposures allow lateral relationships of mega-units such as carbonate platforms and slope complexes to be observed. On a smaller scale lateral tracing can, for example, illustrate the position of levee deposits in both submarine fan valleys (Ricci Lucchi, 1981) and in river systems (Horne *et al.*, 1978, fig. 7; Stewart, 1981), the geometry of fluvial channel sandstones (Friend, Slater & Williams, 1979), and the lateral pinch out of delta front sands (Ryer, 1981).

VERTICAL RELATIONSHIPS

Very extensive studies have been made of vertical sequences of facies from almost every environment. This is largely because sections with appreciable vertical thickness but limited lateral extent are the most common data sources. There is probably also some element of tradition in this emphasis in that the simplest pictorial representation of such data resembles a stratigraphic column. It is often possible from such diagrams to see repeated patterns of facies or changes with time by simple inspection. However, not all observers will have the same impression of a succession, and thus it is necessary to present and analyse the data as objectively as possible. The recognition of repeated patterns of facies (facies sequences, cycles) has proven to be a most powerful tool in environmental interpretation (Reading, 1978a).

For any given succession there are two simple techniques which will normally form the first stage of any analysis. The first is to present a graphic log of the succession (see Section 2.2.7), and the second is to produce a facies relationship diagram (FRD) to summarize the observed vertical sequence of facies (Fig. 2.46a). In Fig. 2.46(a) the basic data are still accessible but have been arranged in an order suggested by visual inspection of the graphic logs; the accompanying pictorial representation (Fig. 2.46b) helps to summarize the information.

2.4.3 Markov Chains

Few successions have as clear a vertical order as the Upper Carboniferous deltaic sediments studied by De Raaf *et al.* (1965) (Fig. 2.46). Some simple statistical techniques have been employed to investigate the possible presence of vertical order in sedimentary successions. Almost all use probability matrices and employ the idea of Markov Chains. The underlying questions can be generalized as: (i) is the observed vertical sequence of lithologies random or ordered?; (ii) in which way is it ordered?

Early work in this field was presented by Potter & Blakey (1968) and Gingerich (1969) with similar techniques proposed and used by Read (1969), Schwarzacher (1969), Selley (1969), Lumsden (1971), Doveton (1971), Miall (1973), Turner (1974),

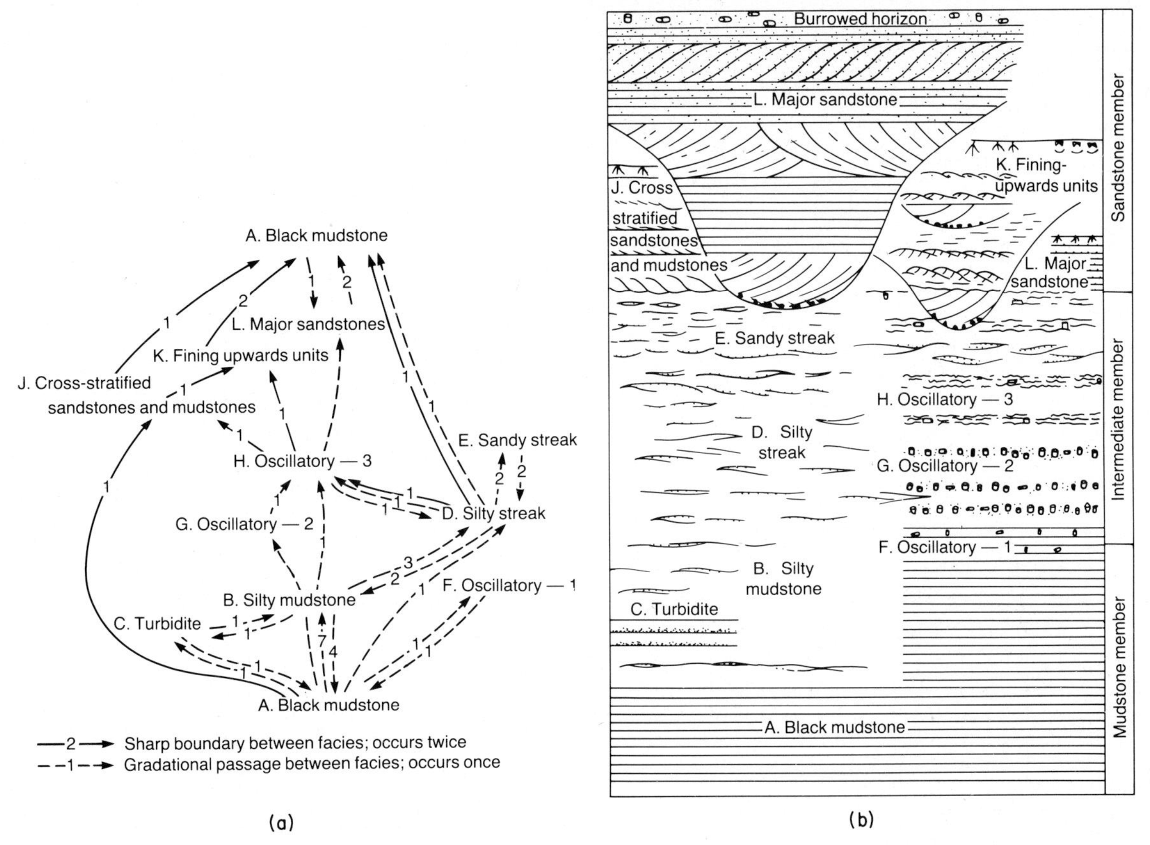

Fig. 2.46. (a) A facies relationship diagram for the Upper Carboniferous Abbotsham Formation of SW England showing both number and type of facies boundaries.
(b) Pictorial representation of (a) and suggested sedimentary 'cycle'. (From Reading, 1978b after De Raaf *et al.*, 1965.)

Ethier (1975), Hattori (1976) and Cant & Walker (1976).

A first order Markov process is a stochastic process in which the state of the system at time t_n is influenced by or dependent on the state of the system at time t_{n-1} but not the previous history that led to the state at time t_{n-1}. The presence of a Markov process implies a degree of order in a system but this must be extended to a number of states to imply cyclicity. In most studies the raw data consist of an observed number of upward transitions which are plotted in matrix form. The often quoted data of Gingerich (1969) are used here for illustration (Table 2.7a, b). This transition probability matrix is then compared to one generated by random methods (independent trials matrix — Table 2.7c, d). The comparison is by means of a χ^2 test which attempts to answer question (i) above, i.e. is the observed sequence random or ordered?

However, subsequently it has been shown that this technique is not entirely justifiable statistically and must be modified (Carr, 1982; Powers & Easterling, 1982). The matrices employed in Table 2.6 are termed *embedded matrices* because they contain structural zeros, i.e. zeros implicit in the technique and not due to sampling. These occur because transitions from a bed of one lithology or facies to another case of the same are held to be objectively non-recordable and thus the diagonal frequencies are forced to be zero. It was pointed out by Schwarzacher (1975) that the presence of previously defined zeros was a major obstacle to rigorous analysis. It becomes impossible to generate an 'independent trials matrix' by a simple independent random process.

Table 2.7. Matrices for Markov analysis (data from Gingerich, 1969)

(a) Transition count matrix

	SS	MS	LIG	LS	Total
SS	—	37	3	2	42
MS	21	—	41	14	76
LIG	20	25	—	0	45
LS	1	14	1	—	16
Total	42	76	45	16	179

(b) Transition probability matrix

	SS	MS	LIG	LS
SS	—	0.88	0.07	0.05
MS	0.28	—	0.54	0.18
LIG	0.44	0.56	—	0.00
LS	0.06	0.88	0.06	—

(c) Independent trials probability matrix

	SS	MS	LIG	LS
SS	—	0.55	0.33	0.12
MS	0.41	—	0.43	0.16
LIG	0.31	0.57	—	0.12
LS	0.26	0.47	0.27	—

(d) Independent trials matrix (data calculated by Gingerich method)

	SS	MS	LIG	LS	Total
SS	—	23.3	13.8	4.9	42.0
MS	31.0	—	33.2	11.8	76.0
LIG	14.1	25.5	—	5.4	45.0
LS	4.1	7.5	4.4	—	16.0
Total	49.2	56.3	51.4	22.0	

(e) Difference matrix (after Gingerich)

	SS	MS	LIG	LS
SS	–	+0.33	−0.26	−0.07
MS	−0.13	–	+0.11	+0.02
LIG	+0.13	−0.01	–	−0.12
LS	−0.20	+0.41	−0.21	–

(f) Independent trials matrix (after Carr, 1982 and Powers & Easterling, 1982)

	SS	MS	LIG	LS	Total
SS	—	31.38	8.17	2.56	42.01
MS	31.28	—	34.05	10.66	75.99
LIG	8.17	34.06	—	2.78	45.01
LS	2.56	10.06	2.78	—	16.00
	42.01	76.00	45.00	16.00	

(g) Difference matrix (after Carr, 1982 and Powers & Easterling, 1982)

	SS	MS	LIG	LS
SS	—	+0.14	−0.13	−0.01
MS	−0.13	—	+0.09	+0.04
LIG	+0.26	−0.20	—	−0.06
LS	−0.10	+0.21	−0.11	—

Both Carr (1982) and Powers & Easterling (1982) independently suggested a method (effectively the same one) of overcoming this problem using a technique initially developed by Goodman (1968). The model proposed by Goodman (1968) is termed quasi-independence which is similar to independence but is applied to a subset of a contingency table. This model can be applied to embedded matrices and allows preservation of both row and column totals which was not possible in earlier methods (cf. Table 2.7d, f).

The expected transition frequencies under a model of quasi-independence are generated by an iterative method illustrated in Table 2.8 using Gingerich's data as calculated by Powers & Easterling (1982). This model of quasi-independence can be tested using a chi-square statistic which has $(m - 1)^2 - m$ degrees of freedom. A comparison of the expected transition frequencies derived by quasi-independence and by the Gingerich method is given by Table 2.7(d, f). Testing for goodness of fit of the quasi-independence model yields $\chi^2 = 35.7$ (cf. 42.3 for the Gingerich method), which rejects the assumption of quasi-independence. In this case, the earlier method of Gingerich would have given a similar conclusion but the strata were deliberately chosen to demonstrate order and in other cases the conclusions would be different.

An alternative method has been proposed, although not widely used, to avoid problems caused by embedded matrices. This is to sample at regular intervals such that there are no structural zeros (e.g. Ethier, 1975). However, this also creates many

Table 2.8. Calculation of expected frequencies under a model of quasi-independence after Goodman (1968) and Powers & Easterling (1982)

The model termed quasi-independence is

$$E(n_{ij}) = a_i b_j, \quad i \neq j$$
$$= 0, \quad i = j$$

The iterative procedure for calculating a_i and b_j under the model is as follows:
First iteration

$$a_i^1 = n_{i+}/(m-1) \qquad i = 1, 2, \ldots\ldots\ldots m$$
$$b_j^1 = n_{+j}/\sum_{i \neq j} a_i^1 \qquad j = 1, 2, \ldots\ldots\ldots m$$

*I*th iteration

$$a_i^I = n_{i+}/\sum_{j \neq i} b_j^{(I-1)} \qquad i = 1, 2, \ldots\ldots\ldots m$$
$$b_j^I = n_{+j}/\sum_{i \neq j} a_i^I \qquad j = 1, 2, \ldots\ldots\ldots m$$

Beginning with a_i^1 as the average of the ith row entries provides a useful starting point. Iteration is continued until:

$$|a_i^I - a_i(^{I-1})| < 0.01\, a_i^I \quad \text{for} \quad i = 1, \ldots\ldots m$$
$$\text{and} \quad |b_j^I - b_j^{(I-1)}| < 0.01\, b_j^I \quad \text{for} \quad j = 1, \ldots\ldots m$$

When $\hat{a}_i$ and $\hat{b}_j$ are the final values of a_i^I and b_j^I then the estimated expected frequencies are given by $E_{ij} = \hat{a}_i \hat{b}_j$ for $i \neq j$. For example consider the data from Table 2.7(a). On the first iteration:

$$a_1^1 = n_{1+}/(m-1) = 42/3 = 14.0 \qquad a_2^1 = n_{2+}/(m-1) = 76/3 = 25.33$$
$$a_3^1 = n_{3+}/(m-1) = 45/3 = 15.0 \qquad a_4^1 = n_{4+}/(m-1) = 16/3 = 5.33$$
$$b_1^1 = n_{1+}/a_i^1 = 42/(25.33 + 15.0 + 5.33) = 0.92$$
$$b_2^1 = 76/(14.0 + 15.0 + 5.33) = 2.21$$
$$b_3^1 = 45/(14.0 + 25.33 + 5.33) = 1.01$$
$$b_4^1 = 16/(14.0 + 25.33 + 15.0) = 0.29$$

On the second iteration:

$$a_1^2 = 42/(2.21 + 1.01 + 0.29) = 11.97$$
$$a_2^2 = 34.23 \qquad a_3^2 = 13.16 \qquad a_4^2 = 3.86$$
$$\text{Further} \quad b_1^2 = 42/(34.23 + 13.16 + 3.86) = 0.82$$
$$b_2^2 = 2.62 \qquad b_3^2 = 0.90 \qquad b_4^2 = 0.27$$

Convergence to achieve 1% change requires 13 iterations at which point the estimates are:

$$\hat{a}_1 = 10.22 \qquad \hat{a}_2 = 42.61 \qquad \hat{a}_3 = 11.12 \qquad \hat{a}_4 = 3.48$$
$$\hat{b}_1 = 0.73 \qquad \hat{b}_2 = 3.06 \qquad \hat{b}_3 = 0.80 \qquad \hat{b}_4 = 0.25$$

Thus a value of $E_{34} = \hat{a}_3 \hat{b}_4 = 11.12 \times 0.25 = 2.78$
These values are tabulated in Table 2.7(f)

problems as the choice of sample interval is very important. The sample interval must be sufficiently small to catch important transitions and this usually leads to large diagonal frequencies with which testing the full matrix for independence can be misleading. This can be overcome by testing off-diagonal frequencies against the model of quasi-independence. In general, there seems to be little advantage in following this more complex method.

Having established that a succession contains some order, the next question concerns the nature of that order and an identification of preferred transitions (question (ii) above). Gingerich (1969) attempted to express that order by constructing a difference matrix (Table 2.7e) produced by subtracting the independent trials matrix (the 'expected' results) from the transition probability matrix (the 'observed' results). An assumption is made that the positive elements in this matrix have a higher than random chance of occurring. All positive differences are used to indicate the path of the sedimentary 'cycle'. With minor alterations most later workers have employed a similar technique. However, small positive differences can arise from a random process and failure to evaluate the significance of positive differences can lead to the identification of too many preferred transitions (Harper, 1984). Similarly, negative departures are ignored by this technique even though they contribute to the test statistic and to the rejection of the model of independence.

The currently used techniques for identifying sources of non-randomness are relatively crude and demonstrate the need for further research. Such techniques involve the identification of 'extreme' cells, i.e. those which show the largest differences between observed and expected transition frequencies. A simple method (Turk, 1979; Powers & Easterling, 1982) is to tabulate

$$Z_{ij} = (O_{ij} - E_{ij})/E_{ij}^{1/2}.$$

The Z_{ij}'s are squared and summed to yield χ^2 and they have approximately the standard normal distribution. As 95% of the standard normal distribution falls between ± 2.0, so a Z_{ij} value outside this is fairly unusual. Analysis of the Gingerich data (Table 2.9) shows that the primary contributor to the large χ^2 is the large number of transitions from lignite to sandstone.

Table 2.9. Normalized differences $(O_{ij} - E_{ij})/E_{ij}^{1/2}$

	SS	MS	LIG	LS
SS	–	+1.02	−1.81	−0.35
MS	−1.84	–	+1.19	+1.02
LIG	+4.14	−1.55	–	−1.67
LS	−0.98	+1.02	−1.07	–

Where O_{ij} is shown in Table 2.7(a), E_{ij} is shown in Table 2.7(f).

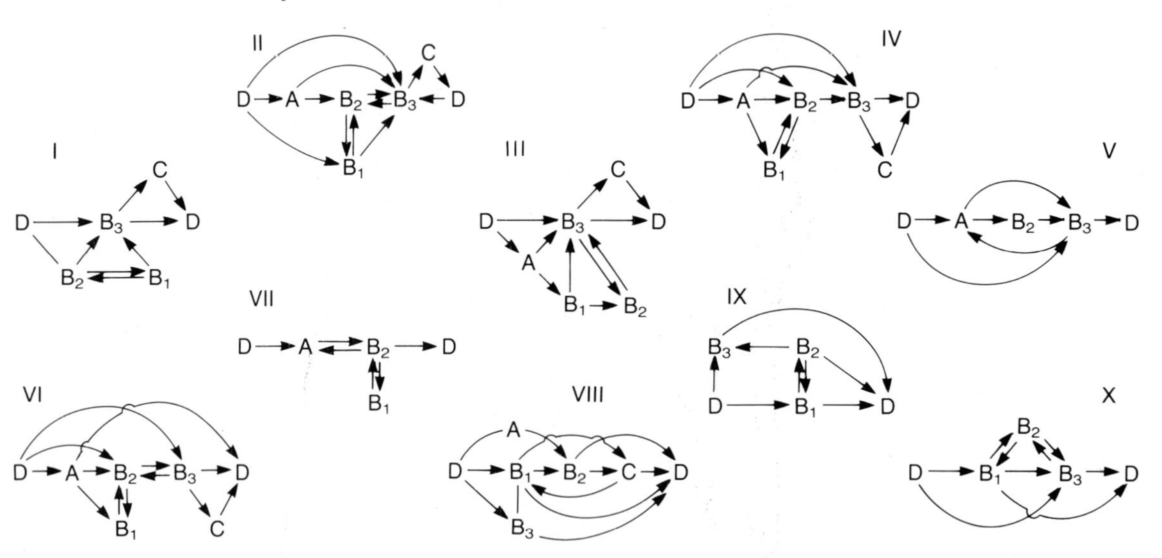

Fig. 2.47. Tree diagrams for Devonian fluvial sediments which show upward facies transition probabilities >0.15. Arbitrary selection of this value is used to imply order of sequence (from Allen, 1970). (Reproduced by permission of SEPM.)

An alternative method is described by Brown (1974) and Carr (1982). The χ^2_{ij} is calculated for each cell and that whose elimination produces the smallest χ^2 is deleted, i.e. treated as a structural zero. The remaining cells are then refitted by a quasi-independent model and the χ^2 test for quasi-independence is recalculated. This stepwise procedure is continued until a chosen level of significance. Correct identification of a single cell when it is the only cell that does not fit a quasi-independent model is a far simpler problem than identifying a subset of cells, all of which deviate from a model fitted to the other cells. An optimal solution to this latter problem has yet to be found but all extreme cells are likely to be identified. In the case of the Gingerich data, a chosen level of significance of $\alpha = 0.1$ was exceeded after selection of the first cell (sandstone over lignite). This shows good agreement with the normalized differences approach above.

DATA PRESENTATION AND INTERPRETATION

Vertical arrangement of different facies can be represented by a variety of facies relationship diagrams. Their aim is to present either the raw data or statistics based on those data in a format which aids interpretation. Figure 2.46(a) above was derived from the raw data and retained information on the nature of facies boundaries as well as number of transitions. Other workers have attempted to select from the raw data the more commonly occurring transitions (e.g. Allen, 1970 and Fig. 2.47), in some cases implying preferred sequence.

Preferred sequence has been suggested mainly following statistical analysis of data, usually by plotting the positive elements of the difference matrix (Gingerich, 1969; Miall, 1973; Hubert & Hyde, 1982) (Figs 2.48, 2.49 and 2.50). The data of Gingerich

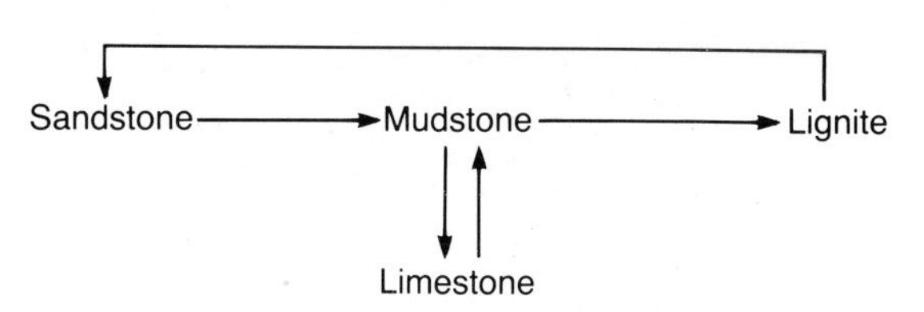

Fig. 2.48. Tree diagram constructed from the positive elements of the differences matrix (Table 2.8d) (after Gingerich, 1969). (Reproduced by permission of SEPM.)

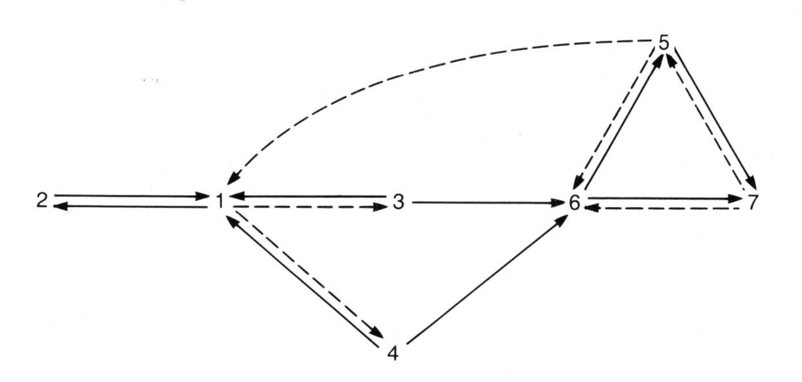

Difference matrix, Lower Peel Sound Formation

		1	2	3	4	5	6	7	
Conglomerate	1	0.00	0.25	0.02	0.02	−0.06	−0.09	−0.14	
Red sandstone	2	0.33	0.00	−0.03	−0.01	−0.20	−0.06	−0.02	
Red siltstone	3	0.19	−0.23	0.00	−0.01	0.01	0.16	−0.13	
Red limestone	4	0.20	−0.23	−0.02	0.00	−0.23	0.41	−0.13	$= d_{ij}$
Grey sandstone	5	0.03	−0.15	0.00	−0.01	0.00	0.07	0.07	
Grey siltstone	6	−0.20	−0.12	−0.02	−0.01	0.25	0.00	0.11	
Grey limestone	7	−0.21	−0.17	−0.02	−0.01	0.39	0.03	0.00	

Fig. 2.49. Difference matrix constructed for Devonian fluvial sediments, Arctic Canada, following the methodology of Gingerich (1969). The accompanying tree diagram distinguishes between 'major' transition paths (solid lines) and 'minor' transition paths (dashed lines) on the basis of values in the difference matrix (from Miall, 1973).

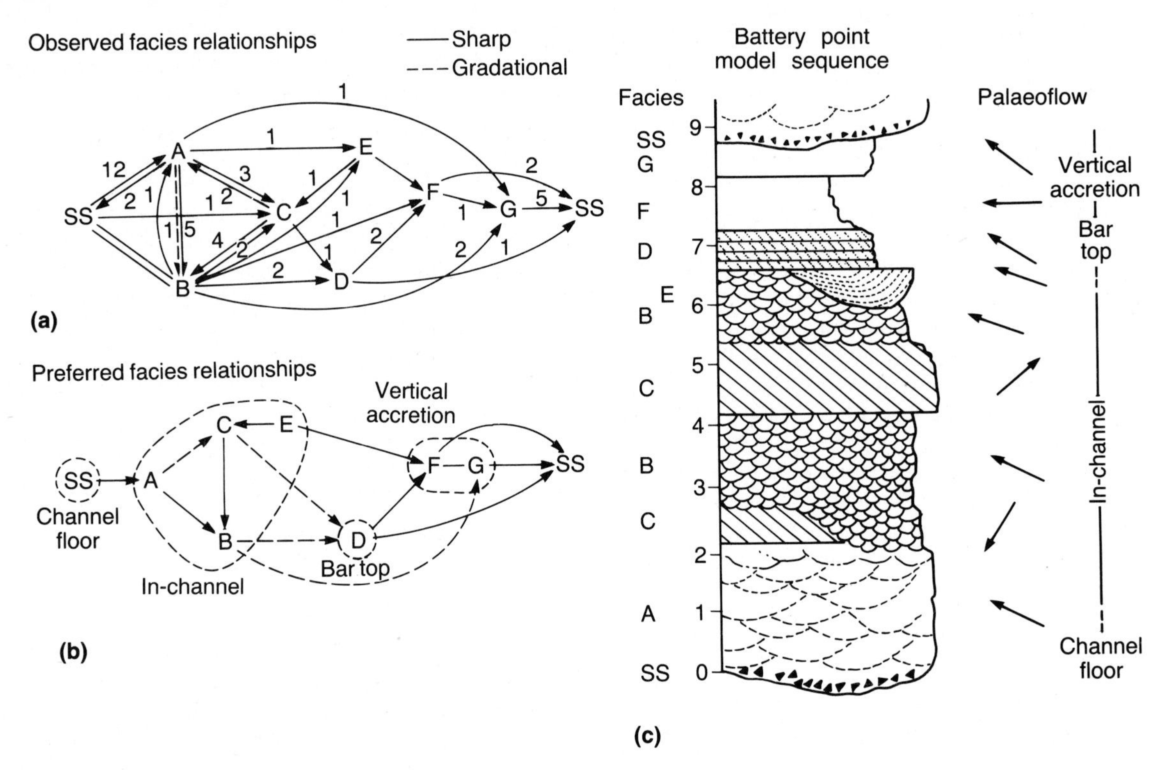

Fig. 2.50. Summary of vertical facies relationships for the Battery Point Formation, Devonian, Canada (after Cant & Walker, 1976).
(a) Facies relationship diagram which displays the basic field data (cf. Fig. 2.46a)
(b) Preferred facies relationships based on data from a difference matrix. Higher values in the difference matrix are given greater emphasis; thus, heavy solid arrows — >0.30, light solid arrows — 0.10–0.30, dotted arrows — 0.05–0.10. A generalized environmental interpretation has been added.
(c) A representation of (b) in the form of a vertical log with average facies thicknesses and palaeocurrents taken from the field data. (Reproduced by permission of *Canadian Journal of Earth Sciences*.)

(1969) can again be used to demonstrate how differences in interpretation might result. Gingerich inferred a fully developed cycle (represented by Fig. 2.48) about which random 'deviations' might occur. Thus mudstone would be expected to follow sandstone and lignite to follow mudstone. In contrast, reanalysis of these data by Carr (1982) showed that most of the non-randomness can be explained by the upward transition from lignite to sandstone. Thus lignite would be the terminal state of random succession of fluvial facies after which the system displayed a preferred transition to sandstone (?rapid channel migration). Knowledge of the nature of the facies transitions might increase confidence in the correctness of this interpretation. The original interpretation of Gingerich clearly suggests more organization than is justifiable from the data.

Other attempts to construct inferred facies relationship diagrams have also used difference matrices but have made subjective judgements on the statistical data (e.g. Figs 2.49 and 2.50). The assignment of environmental interpretations to the different facies in such diagrams is clearly useful but there is little statistical justification for the separation of values from the difference matrix into groups of differing significance. For example Fig. 2.50 was used to generate expected sequences of facies to which further data such as thickness of facies were added. Such an approach is potentially useful and care must be taken that the upward transitions used are statistically meaningful. It is particularly valuable to accompany presentation of the raw data with the number of transitions and the nature of boundaries (see Cant & Walker, 1976 and Fig. 2.50a).

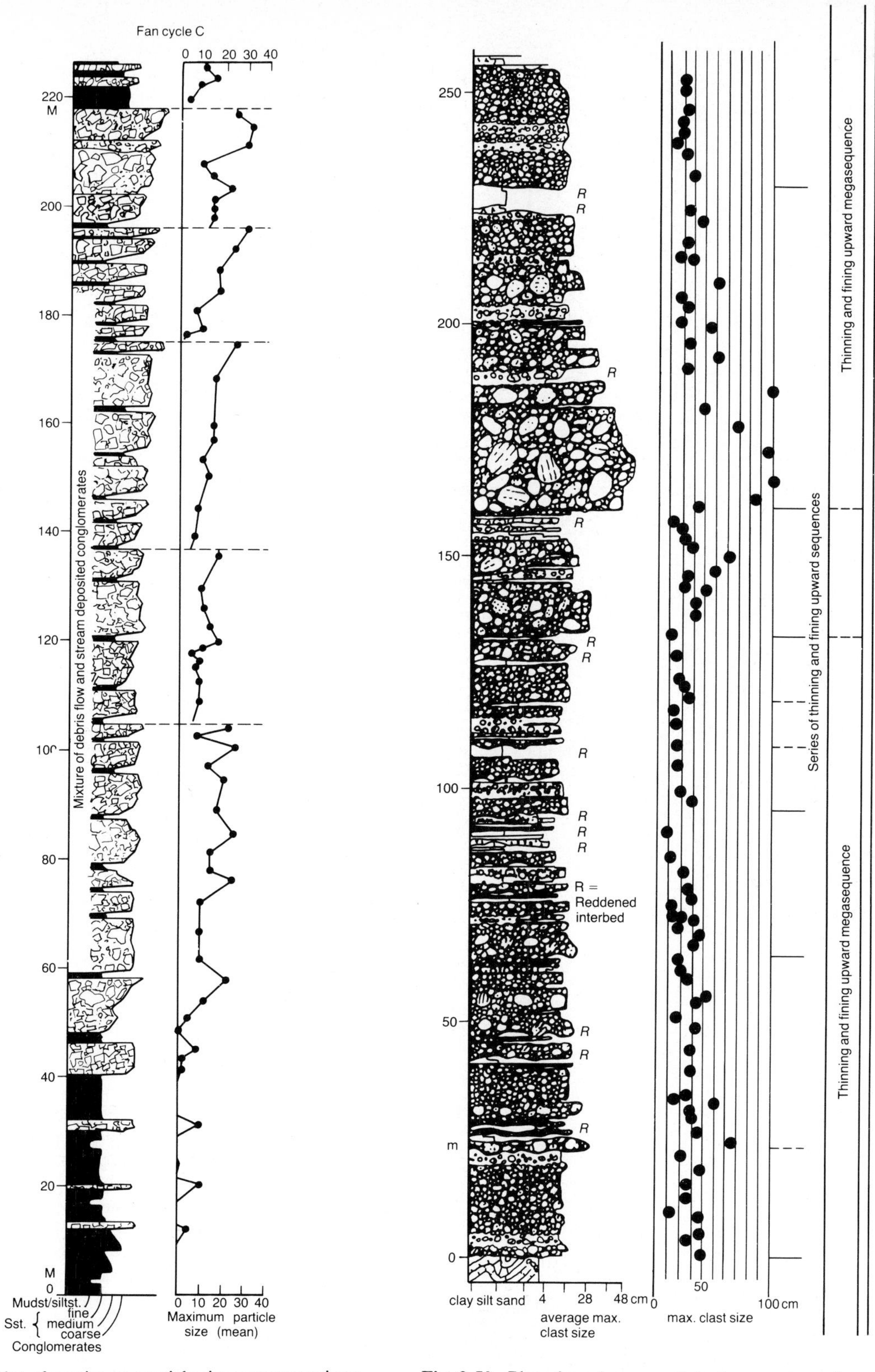

Fig. 2.51. Plot of maximum particle size accompanying a vertical log for Devonian alluvial fan conglomerates, Hornelen Basin, Norway. Sequential organization is inferred from visual inspection of the data (from Steel *et al.*, 1977). (Reproduced by permission of the Geological Society of America.)

Fig. 2.52. Plot of maximum particle size accompanying graphic logs for some Upper Carboniferous alluvial fan conglomerates, northern Spain. Sequences and megasequences are inferred from visual inspection of the data (from Heward, 1978).

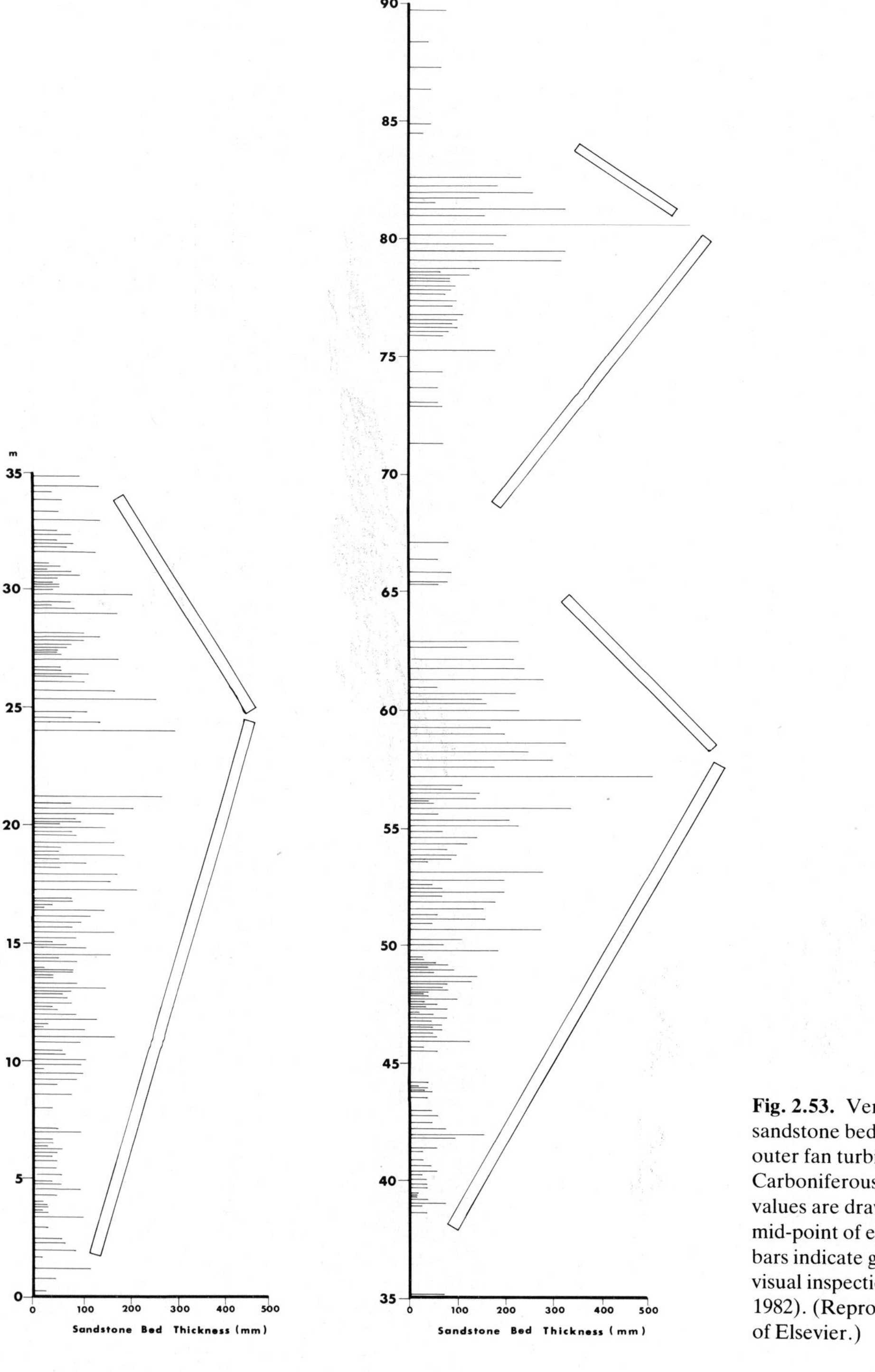

Fig. 2.53. Vertical profile of sandstone bed thickness for possible outer fan turbidites in the Lower Carboniferous of Morocco. Thickness values are drawn outwards from the mid-point of each sandstone bed. The bars indicate generalized trends from visual inspection (after Graham, 1982). (Reproduced by permission of Elsevier.)

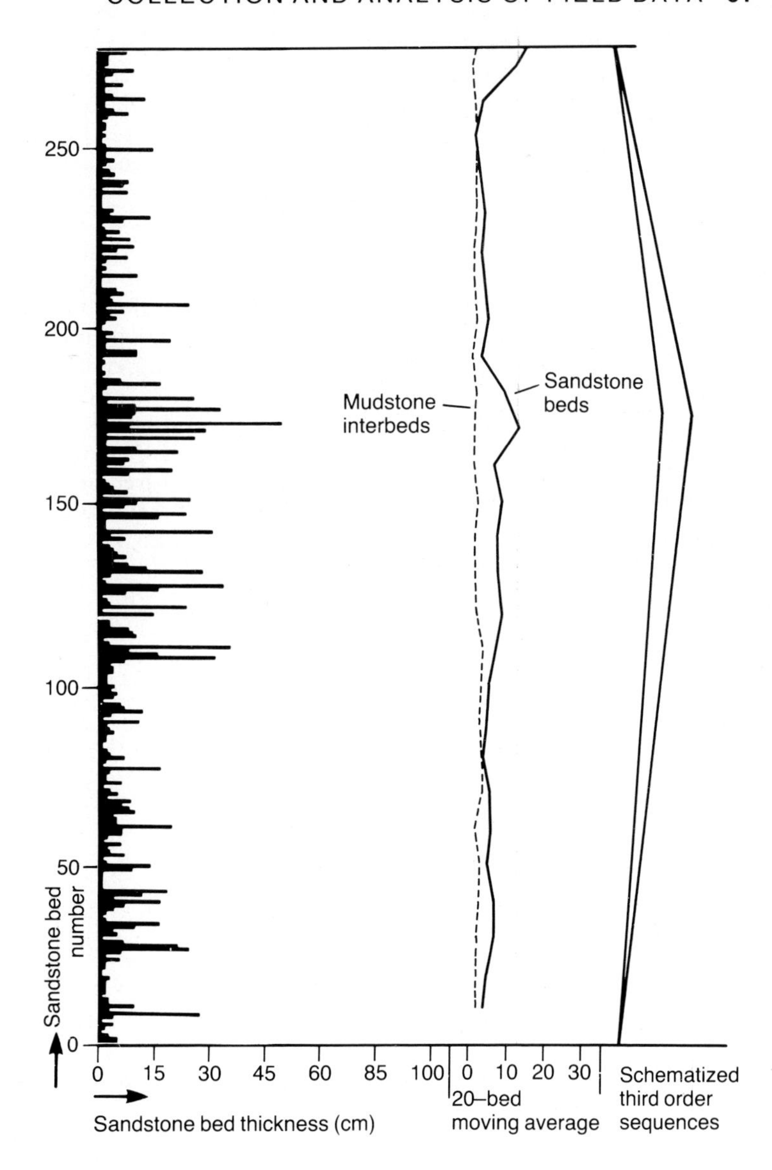

Fig. 2.54. Vertical variations of sandstone bed thickness from deep sea fan sediments in the upper Carboniferous of Cantabrian Mountains, Spain. Vertical scale is sandstone bed number and not stratigraphic thickness. Moving means are used to show some general trends but 'third-order sequences' are schematic (from Rupke, 1977). (Reproduced by permission of the University of Chicago Press.)

Environmental interpretation commonly includes statements on the lateral arrangement of facies as deduced from vertical sequences. This involves application of Walther's Law of Facies, as succinctly summarized by Reading (1978a). Inferring lateral juxtaposition often necessitates the presence of 'gradational' contacts in vertical sequence. The techniques for analysing vertical sequences do not normally consider the nature of contacts between facies. This must be borne in mind during interpretation and preferably presented on facies relationship diagrams such as that in Fig. 2.46.

Thus Markov analysis provides a powerful tool to test for order in vertical sequences. Techniques for investigating the nature of that order, where it exists, are less well developed but are available and allow for statistically definitive statements on vertical sequence.

2.4.4 Non-Markov techniques for sequence analysis

Presentation and analysis of vertical sequences of rock is not limited to Markov analysis. Several other, often simpler, techniques have been employed with considerable success. Perhaps the most common are those that consider vertical changes in grain size or bed thickness. The former have commonly been employed in the study of conglomerates (Figs 2.24c, 2.51 and 2.52) (Bluck, 1967; Steel *et al.*, 1977; Heward, 1978; Surlyk, 1978; Allen, 1981). The aim has largely been the detection of sequences showing a unidirectional variation of maximum clast size, i.e. coarsening up (CU) or fining up (FU) sequences. The detection of such sequences has often been by simple visual inspection of the data but they are also susceptible to techniques such as moving means.

Vertical changes in layer thickness have also been useful, particularly in turbidite successions (Ricci Lucchi, 1975b; Rupke, 1977; Shanmugam, 1980; Graham, 1982). Techniques of presentation show considerable variation, e.g. the vertical scale may be stratigraphic thickness as in Fig. 2.53 or sandstone number as in Fig. 2.54. A criticism of many of these presentations is that 'trends' are designated from visual inspection of the data (Fig. 2.53, Fig. 2.54 — third order sequences) without any statistical justification. Techniques such as moving means (Fig. 2.54) can be useful here although these should always be presented along with the original data as averaging will lead to displacement of the peak values that are commonly used to delimit packets of sediment (rhythms or cycles). One must take care that compared units of thickness have similar definition and that their significance is understood (see Section 2.2.3).

3 Grain size determination and interpretation

JOHN MCMANUS

3.1 INTRODUCTION

The size of the component particles is one of the fundamental textural characteristics of all fragmentary deposits and their lithified equivalents. From the earliest days of observation and recording of geological features, terms such as 'coarse', 'medium' or 'fine' grained have been applied to unconsolidated deposits. Although considered a useful lithological discriminator for tracing individual horizons during geological mapping, little systematic study of grain size characterization and measurement occurred before the end of the nineteenth century. By that time there was already an appreciation that very substantial currents of water were necessary to transport coarse sediments but, without systematic means of establishing precisely the sizes of particles and the range of particle sizes within a deposit, little scientific advance was made.

In general terms, 'size' of particles is not readily defined as it is a measure of the dimensions which best describe a specific group of particles. At one extreme the particle dimensions may exceed one metre, but at the other are the very fine particles less than one micrometre in diameter. A cube may be best described by its edge length, a sphere by its diameter, a rod perhaps by its intermediate axis and a flake by its projection area. Thus 'size' is partly dependent on shape, a fact which was gradually recognized during the early years of the present century.

The methods used in determination of size vary widely from calipers on the coarsest fragments, through sieving and techniques dependent upon settling velocity, to those detecting changes in electrical resistance as particles are passed through small electrolyte-filled orifices. No single technique of assessing particle size is applicable throughout the entire size spectrum, and each is suited to specific size ranges.

Although when measuring pebbles to determine axial lengths individually, relatively small numbers of grains are considered, in most methods large numbers of grains are examined so that a statistically meaningful set of numbers is obtained to characterize the sample.

The techniques of processing the data on grain size were largely explored during the 1920s and 1930s when a series textural classifications was introduced. In their simplest forms these classifications indicate average grain size, and the degree and form of the spread around that average. Statistical analysis is aided by means of a simple transformation of size from a metric scale to a logarithmic one which enables application of both graphic and moment statistical techniques. Explored initially for sand-dominated assembiages, the methods were refined during the 1950s and applied to thin section analysis of lithified sediments during the early 1960s. Fresh techniques of exploring the information on grain populations with a view to interpreting depositional environment or transportation experienced by the deposits continue to be introduced with varying degrees of success. Many have appeared, been explored and found applicable only under certain restricted conditions.

Even the most sanguine grain size enthusiast would agree that no universal solution exists as yet to enable the sedimentologist unequivocally to distinguish depositional environments in ancient deposits on grain size criteria alone. Nevertheless, an enormous amount of information about a sedimentary deposit may be obtained from systematic examination of the multiple sub-populations of which most are formed. Specific sub-populations may be source- or process-diagnostic, a factor which frequently emerges from regional surveys.

From the outset it is most important that the investigator should decide precisely what information is required from the sediments to be analysed (McCave, 1979a). Analysis solely to obtain numerical assessment of the texture of a deposit is unlikely to satisfy the sedimentologist but this information may be all that is required to demonstrate to a fisheries expert the potential value of a site for the introduction of an economically important species such as *Nephrops* (Scampi). In assessment of the suitability of a deposit for supplying specific coarse concrete aggregates or finer mortar sands, very precise requirements are laid down by the construction industry. These often demand restricted ranges of particle sizes in fixed proportions within the gravel. In tracing characteristic source-specific sub-populations of sediment passing through a river system the closest possible spacing of sieves may be required. In estimating the potential 'activity' of a soil it may be necessary to determine the maximum projection area of the particle surfaces of various sizes to assist in the assessment of potential cationic exchanges; scanning electron microscopy may be necessary.

In this chapter attention is devoted to the techniques used commonly by sedimentologists to determine sizes of particles, i.e. sieving, pipetting, settling and Coulter counting, coupled with optical microscopy. Where other appropriate methods exist reference will be made to useful sources of information.

3.2 SAMPLE PREPARATION

The selection of materials to be processed is made in the field and usually collections comprising many samples arrive in the laboratory at any one time. The initial procedure in preparing for analysis is the systematic checking of the materials to ensure that identification numbers and labels are all present and legible. This is most readily achieved by laying out cores, sample bags, bottles or blocks of material sequentially. Any internal numbering, unless engraved on indestructible material, may rapidly become obliterated, and relabelling with external labels is essential. Gaps in numerical sequences may be identified and samples lacking clear labels may at this stage be returned to their correct sequential position. If identification of misplaced samples cannot be completed with certainty at this stage the samples have little value for subsequent analysis and should be discarded. Better to have no data than unreliable data.

Further preparation of all sediments relates to the form of size determination to be carried out. Lithified materials may require little more than decision on the orientations of thin sections needed for later examination under the optical microscope, but special slide preparation and cementing media may be required, particularly if soluble minerals are thought to be present (see Chapter 4).

Prior to examination of large particles some cleaning may be necessary. In unconsolidated materials washing in running tap water may be sufficient to remove unwanted muddy coatings (assuming data from the muds are not required). Lithified coarse sediments frequently release large fragments once cements are weakened in warm dilute hydrochloric acid, but care must be taken that the clasts themselves are not susceptible to attack.

Sands derived from beaches, estuaries or the sea bed may contain quantities of salt within their interstitial waters. Unless removed by washing the salts become deposited during drying before sieving, cementing adjacent particles together, to produce misleading analyses. If no very fine grains are present in the sediment the sands may be freely washed in well-agitated water to remove the salts. Normally three washings, each in one litre of distilled or deionized water, are needed to remove the salt from about 200 g of sediment with thorough stirring (Buller & McManus, 1979). It is, of course, important that no particles are lost during each process of decanting the water after washing.

When fine materials are present it is necessary to separate them by wet sieving so that sands and the silt/clay fraction may be examined separately using different size determination techniques. It is most important that separation of the fines be carried out before drying of the sediment for, on heating, silts and clays produce crusts or durable pellets. Although these may be later broken down physically to release individual particles there is no guarantee that individual flakes of clay would become separated by purely physical means. Division of the sediment into two fractions at an early stage is therefore recommended. Each fraction must be labelled to permit later recombination of partial analytical results.

Silt and clay sized particles dominate many marine, estuarine and lacustrine deposits. Frequently the behaviour of the finest size fractions is determined by the concentrations, identities and valencies of salts in the interstitial waters. Whether the clays remain as discrete particles or cluster together to form flocs or aggregates with diameters much larger than the individual component particles is often a function of these waters. Modification of the fluids through washing thus not only changes the salt concentrations, but may also induce separation or aggregation of the particles, so that subsequent size analysis may not indicate the natural interrelationship between the particles during deposition or achieved since. Interpretations of settling characteristics based on the most minute elemental particles, of which many thousands may combine naturally to produce flocs or aggregates, may give a very misleading guide to the energy levels in the depositional environment, for flocs settle at much greater rates than their individual component particles.

Size analysis of fine particles therefore presents problems, for the pre-analytical preparations may largely determine the ultimate results obtained. Nevertheless, most investigators remove the salts before subjecting the samples to pipette, hydrometer or Coulter analysis, in all of which some dispersing agent is introduced to ensure separation of individual particles.

Removal of water salts from fine sediments is best achieved using dialysis. The sample, placed in a dialyser bag, is suspended in distilled or deionized water, circulation of which encourages osmotic exchange of the salts over a period of several days. Ideally many samples are dialysed simultaneously.

Once sample preparation has been completed the simplest analyses may be performed on the largest particles.

3.3 DIRECT MEASUREMENT OF GRAIN SIZE

The analysis of particle sizes of coarse unconsolidated sediments may be achieved through direct measurement of individual pebbles. The lengths of representative diameters or axes are determined with the aid of vernier calipers (Briggs, 1977). For every pebble several possible diameters may be recognized along the three principal axes of the pebble. A nominal diameter may also be derived from the volume of the pebble.

The particle is placed on a flat surface and the length of the intermediate axis, I, is determined as the shortest visible diameter. The length of the largest axis, L, at right angles to I is next measured, and rotation of the particle by 90° about that axis reveals the shortest axis, S which may then be directly measured. It should be noted that L is not the greatest possible dimension of the pebble but is quite strictly defined in geometric terms. Thus three mutually perpendicular axes may be used to characterize the pebble (Fig. 3.1). Addition of the three lengths, I, L and S, and division by three yields a mean diameter D_M for the particle. A second mean diameter is derived by immersing the pebble in water to determine the volume of water displaced, from which the volumetric diameter, D_v is calc' 'ated using the expression

$$\text{volume} = \frac{\pi}{6} D_v^3.$$

Such measurements should be repeated for 100–400 pebbles at a site. Thereafter some workers analyse the results according to the numbers of pebbles falling into specific categories. However, because it is the weight distribution rather than the frequency distribution which determines sediment behaviour during transport and deposition the weight of the pebbles falling into each size class should be used for the purposes of comparison. In most cases the length of I provides a useful assessment of the mean diameter, and if determinations are to be made in the field this diameter provides a useful measure of average size.

3.4 SIEVING

Although it may be advantageous for specific purposes to examine individual pebbles, under most circumstances the sedimentologist seeks to establish the character of an entire deposit, not simply the coarser fractions. For such purposes a more general form of mechanical analysis to determine grain size, the sieving method, is used (Krumbein & Pettijohn,

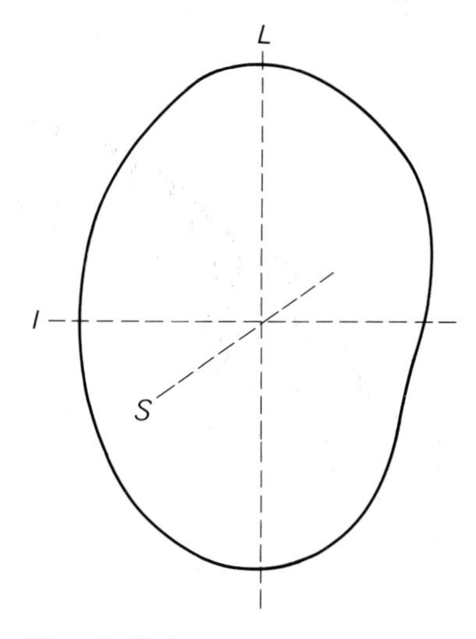

Fig. 3.1. The principal axes measured in characterizing pebble size.

1961, Folk, 1974b, Buller & McManus, 1979). With appropriate sieves, particles varying between 0.002 and 250 mm may be separated into regular size class intervals. Most commonly sieving is used for size determination in the pebble and sand ranges, i.e. particles coarser than 0.063 mm.

For the coarser particles sieve screens are formed of plates or strong wire of stainless steel or brass, and finer wire meshes are used for smaller particles. Between the wires are square openings, the numbers of which per unit length give the mesh number and the diagonal the nominal size.

The coarsest sieve required is placed at the top of a nest of sieves in which the screen openings become progressively smaller downwards. A pan is placed beneath the lowest sieve to retain the 'fines' which pass through the entire column. For the most detailed analysis the meshes are arranged at the closest possible intervals, but greater spacings by size may be used if less detail is required.

The sample for testing is placed on the uppermost sieve after weighing, the retaining lid closed and the nest of sieves arranged in a mechanical shaker (Fig. 3.2). The nest is agitated by the shaker for a predetermined time interval, usually 15–20 minutes. It is most important that the base of the sieve nest is arranged exactly parallel to the shaker base and is firmly held in position during agitation (Metz, 1985). Not uncommonly only half of the sieves required may be mounted on the shaker at a time, so the pan residue from the coarser fraction is released into the upper sieve of the finer part of the nest in order to complete the analysis. This nest also requires a terminal pan.

The material retained in each successive sieve is emptied in turn on to a sheet of glazed paper, the sieve tapped gently in a direction diagonal to the meshes and swept with appropriate sieve brushes (camel hair for fine apertures, brass for coarse screens) to dislodge particles which have become firmly held. Each fraction of sediment obtained is weighed to 0.01 g. The material retained on each sieve screen is weighed in turn and the weight of any residue in the final pan is also determined.

Most workers find it convenient to record their sieving analysis data systematically on forms designed for the purpose. The sieve mesh sizes, raw weights, weight percentages and cumulative percentages, finer or coarser than the specific sieve, may be displayed. Likewise, further groupings such as percentage of coarse sand or of very fine sand may also be calculated. An example of such a record sheet is provided in Table 3.1.

Fig. 3.2. Sieve nest and shaker (photograph — Fritsch).

3.4.1 Dry sieving

Perhaps the most commonly used method of determining particle sizes, dry sieving, is subject to potential errors from many sources.

(a) During the drying process *particle aggregation* may occur and it is most important that sieve residues should be checked, using a hand lens where necessary, to detect the presence of aggregates which form clusters considerably larger than the original single component grains.

(b) *Thorough cleaning* of the sieve meshes is important as retained particles severely restrict the aperture spaces available for grains to pass through.

(c) Overloading of the sieves restricts the opportunity for particles to progress down the nest to an appropriate mesh. The greatest load on a sieve should not exceed 4–6 grain diameters (McManus,

Table 3.1. Typical sieve analysis record sheet

Sieve number	φ values	Approx. value (mm)	Approx. value (μm)	Weight retained (g)[a]	Weight/Wentworth classification	Weight (%)	Weight %/Wentworth classification	Cumulative (%)
				BS sieve analysis				
8	−1.04	2	2000		PG		PG	
10	−0.75				VCS		VCS	
12	−0.49						Sum	
14	−0.27							
16	−0.00	1	1000					
18	0.23				CS		CS	
22	0.52						Sum	
25	0.74							
30	1.00	$\frac{1}{2}$	500					
			Tay Estuary sediments programme					
36	1.24				MS		MS	
44	1.50						Sum	
52	1.76							
60	1.99	$\frac{1}{4}$	250					
72	2.24				FS		FS	
85	2.49						Sum	
100	2.72							
120	3.01	$\frac{1}{8}$	125					
150	3.26				VFS		VFS	
170	3.49						Sum	
200	3.71							
240	3.91	$\frac{1}{16}$	62					
Pan (loss)					St/Cl		St/Cl	
Total								

Original weight . . . .

Sieving loss . . . .

[a] Recombine fractions to full φ units and store

1965). Overloading may also cause mesh distortion, a factor which is most evident on coarse screens. Distortion may be avoided by following the maximum sieve loadings recommended in BS 1377:1975 (Table 3.2). Examination of the condition of the meshes in the sieve nest should always be carried out before undertaking a major sieving programme. For most sands about 100 g of material is adequate for sieving analysis but larger weights are required for coarse gravelly deposits.

(d) The ideal duration of agitation by mechanical shaker (Ro-tap or Frisch) is still a matter of discussion. In most laboratories standardization is to 15 or 20 minutes but some researchers recommend only 10–15 minutes (Friedman & Johnson, 1982, Lewis, 1984) and still others suggest that significant changes in the final analytical figures obtained may be detected after 35 minutes of shaking (Mizutani, 1963). For most purposes acceptable reproducibility of analyses is obtained after 20 minutes.

(e) The sieving technique is strongly dependent upon sieve mesh shape. Platey particles such as mica flakes may readily pass diagonally through meshes which more nearly spherical grains of identical intermediate diameters cannot pass as a result of the length of their short diameters. Thus size determination using sieves is partly controlled by the particle shape (Rittenhouse, 1943). Substantial deviation of particle shapes from spheres may lead to underestimation of the characteristic intermediate diameter according to Ludwick & Henderson (1968). However, it takes longer for irregularly shaped particles to pass down a sieve column than grains which are smooth surfaced or equant (Kennedy, Meloy & Durney, 1985). Visual inspection of sieve residues should be undertaken to confirm the presence or absence of substantial proportions of strongly non-spherical particles in the sediment.

In a better attempt to define the intermediate diameter (I) of particles retained on sieves of specific nominal diameters (D_{sv}) Komar & Cui (1984) demonstrated that the relationship

$$I = 1.32\ D_{sv}$$

provided a means of obtaining a more closely representative grain diameter than currently given by the sieve mesh itself. This correction factor may be applied to metric measures of diameter.

Table 3.2. List of maximum permissible sieve loadings, after BS 1377:1975

BS sieve mesh (mm)	Maximum weight (kg)	Sieve diameter (mm)
20	2.0	300
14	1.5	300
10	1.0	300
6.3	0.75	300
5	0.5	300
3.35	0.3	200
2.0	0.200	200
1.18	0.100	200
0.600	0.075	200
0.425	0.075	200
0.300	0.050	200
0.212	0.050	200
0.150	0.040	200
0.063	0.025	200

3.4.2 Wet sieving

Since many sediments contain mixtures of gravels, sands and finer particles the dry sieving technique may be appropriate for examination of only part of the whole assemblage of particles. The finer particles require other methods of analysis. Separation of the coarse from the fine fractions is customarily made at 63 μm, which is the most commonly used lower limit of dry sieving. However, it is not possible to analyse satisfactorily both the coarse and the fine fractions fully once the split has been achieved.

Some workers suggest that the entire sediment sample should be oven dried at 110°C and weighed to 0.1% before being immersed in water containing a dispersant such as sodium hexametaphosphate. The sample is periodically agitated in the water for over an hour before being washed through 2 mm and 0.063 mm stainless steel meshes using wash bottles containing dispersant solution. The fines continue to be washed through the sieve until the water runs clear. The residue (not more than 105 g) on the 0.063 mm sieve is dried at 110°C and sieved as above. The total fines content is determined as the difference between the initial and the retained material weights. However, many clay sized particles, whether organic or inorganic, become structurally altered at temperatures approaching 100°C so if any further study is to be undertaken this method is not recommended. The satisfactory dispersion of initially oven dried sediments as recommended in BS 1377 is not guaranteed by this method.

If further size analysis is to be undertaken the best procedure is to use two identical sub-samples. One is wet sieved as above, and both the mud and sand fractions dried and weighed to establish their relative proportions. The second sub-sample is wet sieved with sand retention for drying and weighing while the mud fraction, whose proportional contribution to the whole sediment is now known, remains in a receptacle for further analysis.

3.5 SIZE ANALYSIS BY SEDIMENTATION METHODS

Grain size analysis of fine sediments depends not on direct measurement of the particles themselves but rather upon indirect computations of diameters based on observation of the grain behaviour in fluids or the response of the fluids to displacement by the grains. The principal methods based on the speed with which particles settle through fluids yield settling velocities from which equivalent grain diameters are computed. Characterization of particles in terms of their dynamic behaviour is thought by many to be more environmentally significant than direct measurement of particles with calipers or sieves. In consequence the use of sedimentation columns or tubes to determine the fall velocities of sediment particles has become very popular. The technique is relatively quick for coarse sediments and may also be relevant for silt sized particles, although time for completion of analyses increases greatly within the silt size range.

Computation of the diameters of grains from a knowledge of their settling velocities is dependent upon the exploitation of Stokes' Law of settling. When a particle in static water settles at a constant velocity the gravitational force exerted on the particle is balanced by fluid resistances represented by viscosity and particle drag coefficient. The balance is normally represented within the equation

$$V_s = \frac{d^2(\rho_s - \rho)g}{18\,\mu}$$

where V_s is the settling velocity, d the particle diameter, ρ_s and ρ the densities of the grain and water respectively, g is the gravitational acceleration and μ is the dynamic viscosity of the fluid.

Use of the equation is based on the precept that particles are dominantly spheres and are of identical densities. The assumption that most are of quartz permits computation of equivalent grain diameters. In practice natural sediments are rarely composed of spherical grains, and most contain assemblages of many shapes. Particles of identical composition and diameter settle at greatly differing velocities if their shapes or surface textures differ, for the drag coefficients resisting passage through the water vary with both of these properties.

3.5.1 Pipette method

Inexpensive size analysis of naturally occurring fine sediments or of material obtained from wet sieving of a sediment containing significant fines is most satisfactorily achieved by the use of the pipette method (Krumbein & Pettijohn, 1961; Galehouse, 1971b; Folk, 1974b). In essence this technique relies on the fact that in a dilute suspension, particles settle through a column of water at velocities which are dependent upon their size. If a material behaves according to Stokes' Law then, by repeatedly sampling at a fixed depth below the surface progressively finer and finer sediments are present at the sampling depth. Temporal variations of solid concentrations at that level indicate the relative abundance of particles whose diameters may be calculated.

As particles decrease in diameter through the silt and clay sizes they become increasingly cohesive as surface ionic charges grow in relative significance. In natural environments such as rivers and estuaries, these materials form aggregates or flocs within which interparticulate cohesion may be strong. The size characteristics of the flocs may be examined by permitting settling in natural waters in which sediment and ionic concentrations may be high (Peirce & Williams, 1966). Likewise where organic particles play potentially important roles in sedimentation, settling in low sediment and ionic concentrations may be useful (Duck, 1983). However, if the size of the smallest component particles in a flocculated sediment is to be determined, it is necessary to introduce a dispersing or peptising agent. Under most circumstances this is one of the first stages in the preparation of material for pipette analysis.

Sodium hexametaphosphate solution is perhaps the commonest dispersing agent used sedimentologically. It is prepared by dissolving 33 g of sodium hexametaphosphate and 7 g of sodium carbonate in distilled water to give one litre of solution. Although many variations of concentration have been tried that found most useful is 50 ml of the solution to 1000 ml of distilled water. Alternative dispersants used in some laboratories are Calgon (2 g per 1000 ml water) and Teepol (4 drops per 1000 ml of water).

A sample of 20 g of fine sediment is suspended in a 1000 ml measuring cylinder charged with water containing dispersant, and thoroughly mixed through end over end rotation or by means of a manual stirer which travels from the base to the top of the fluid column. As the suspension begins standing a timing device is started.

A pipette is inserted gradually into the fluid so that its inlet is 20 cm below the surface and a volume of 20 ml is withdrawn 58 s from the start of settling. The fluid is released into a numbered 50 ml beaker and any particles retained within the pipette washed into the beaker using distilled water.

Successive 20 ml aliquots are extracted from the column at time intervals which calculations have shown reveal particles of known diameter (Table 3.3 after Krumbein & Pettijohn, 1961). The aliquots are withdrawn from specified depths at the time intervals shown for a full analysis of the fine particle content.

Table 3.3. Table of settling times, after Krumbein & Pettijohn (1961)

Phi	mm	μm		Depth (cm)	*h*	min	s
4.0	0.063	63		20			58
4.5	—	—		20		1	56
5.0	0.0312	31.2		10		1	56
5.5	—	—		10		3	52
6.0	0.0156	15.6	silt	10		7	42
6.5	—	—		10		15	
7.0	0.0078	7.8		10		31	
7.5	—	—		10	1	1	
8.0	0.0039	3.9		10	2	3	
8.5	—	—		10	4	5	
9.0	0.00195	1.95		10	8	10	
9.5	—	—	clay	10	16	21	
10.0	0.00098	0.98		10	32	42	
10.5	—	—		5	32	42	
11.0	0.00049	0.49		5	65	25	

If no more than the contents of medium silt, fine silt or clay are required, aliquots may be withdrawn after the appropriate time intervals and the relative abundance calculated from differences in weights of suspension in each. In establishing sand : silt : clay ratios of sediments it may be necessary to withdraw samples no more often than after 58 s at 20 cm (0.063 mm) and 123 min at 10 cm (0.004 mm).

In each case the pipette should be gently inserted and withdrawn from the fluid in order to minimize disturbance. The lowering and extraction phases should each take about 10 s. Fluid withdrawal is most readily achieved with the aid of a manually operated vacuum device (Fig. 3.3). Beakers containing the fluids are placed in a ventilated oven and dried at 100°C for at least 24 hr, cooled in a desiccator, and weighed to 0.001 g.

In calculating the weight of sediment retained in each withdrawn aliquot, allowance must be made for the dispersant. The simplest computation is achieved by comparing the weights of successive

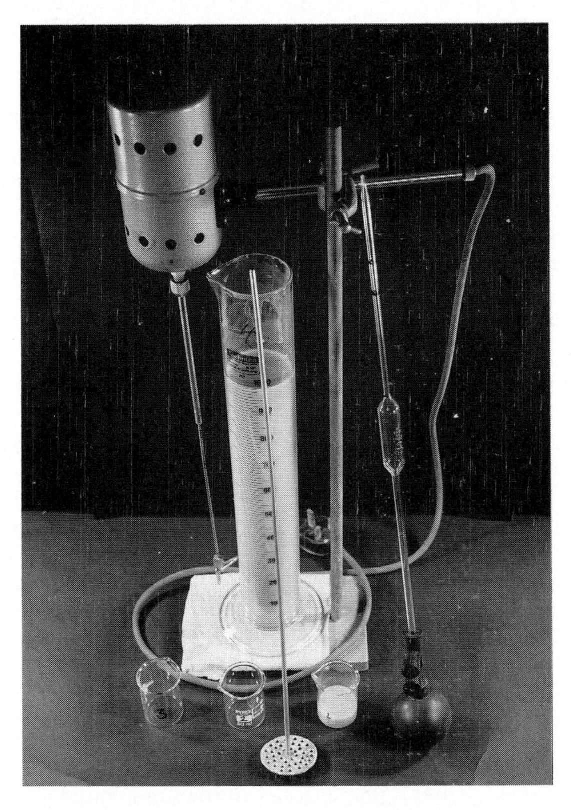

Fig. 3.3. Settling column, pipettes and beakers for estimation of particle size by withdrawal.

withdrawals. As both contain the same quantity of dispersant the difference represents the weight within the size interval concerned. Since each aliquot forms 2% of the original sample the full size distribution may be calculated through continued withdrawals to the finest required size. The total weight of fine sediment undergoing analysis is fifty times the sediment extracted in the initial (58 s) sample.

Because the method depends on Stokes' Law the dynamic viscosity of the water is important. Viscosity varies greatly with temperature and the table of timings provided is calculated for analysis at the standardized temperature of 20°C. As full analysis may take several days to complete it is very important that it be carried out in a thermally regulated laboratory. In many countries stabilization may be possible but not to the relatively warm 20°C. Modified tables may be established for lower temperatures, the greater viscosities requiring extension of the sampling times as settling velocities decrease as the temperature falls.

The quantities of initial sample used in different centres varies from 4 to 35 g per 1000 ml but in practice the most satisfactory concentrations from the viewpoint of reproducibility have been found to be between 10 and 20 g per 1000 ml. The higher the concentrations the greater the possibility of entering into the realms of hindered settling and settling convection in which upward motion of escaping waters impedes settling (Kuenen, 1968).

This time consuming analytical procedure has the advantage of simplicity and requires little specialist equipment. It also provides sufficient time to permit several analyses to be carried out simultaneously. Most workers find that four to six sets of analyses can be dealt with even if a very detailed withdrawal pattern is required. If for any reason a withdrawal time is missed the entire pattern of settling may be restarted by restirring and allowing the material to recommence settling until the required time interval has elapsed. Resuspension may also permit the analyst to leave the laboratory for sleep if the timings are carefully pre-arranged.

3.5.2 Sedimentation tube

The second widely used technique based on settling velocity is that of the sedimentation tube. However, unlike the pipette technique, this method is not confined to analysis of silts and clays, but is commonly applied to the sand fraction of the sediments.

The earliest sedimentation tube widely used is the Emery tube (Emery, 1938). Particles released simultaneously at the top settle through a broad tube which narrows into a smaller diameter tube at the base. The heights of accumulation at known time intervals are measured by optical micrometer and the particle sizes calculated from these figures.

The introduction of a pressure transducer to determine the temporal variations of weight in the water column above a specific level permitted electrical recorders to be used. Initially, this modification involved simultaneous comparison of the weights of identical columns of water with and without sediment (Woods Hole Rapid Sediment Analyser: Zeigler, Whitney & Hayes, 1960), but improvements in transducer technology have since enabled the devices to be used in single tubes. The gradual decrease of weight of material in the column through time as sediment passes the transducer levels during settling enables estimation of the particle size ranges involved.

A further variation introduced a balance pan within and near the base of the tube. The varying weights of particles retained through time are recorded in the sedimentation balance method (Sengupta & Veenstra, 1968; Theide *et al.*, 1976).

In the sedimentation tube technique the particles are released simultaneously from the water surface, a process achieved by holding a 2–5 g sample on a platten by means of a wetting agent and lowering it into the water surface. The sample is kept small to minimize bunching of the cloud of settling particles. If the concentration exceeds 1% hindered settling occurs and this may decrease settling velocities by 5% below those of individual identical particles (Richardson & Zaki, 1954). The diameter of the tube is important here, broad tubes being required for analysis of coarse sediments (Channon, 1971).

Since most sedimentation tubes are at least 2 m in height problems arise of ensuring static water, for such columns develop internal convection currents unless maintained in thermally stable conditions.

The reliability of the sedimentation tube methods, as determined by good reproducibility of results from repeated analysis of the same samples and by analysis of multiple sub-samples from individual sediments, has been recorded by many workers. However, interpretation of the measurements obtained is less clear. Although fall velocities of individual particles may permit confident calculation of particle diameters, the behaviour of clusters of particles of a range of sizes is less easy to calculate. In most cases individual tubes require to be calibrated against well characterized particles such as single spheres (Zeigler & Gill, 1959) or clusters of spheres of single sizes and combinations of sizes (Schlee, 1966) before their performance may be known. Total calibration against the full range of particle size combinations and particle shapes is impracticable.

In practice the particle sizes in the sediment are computed from the weights settled at specific time intervals. The measured settling velocities are converted into 'equivalent sedimentation diameters', or the diameters of spheres settling at the same rate as the natural particles being tested (Gibbs, Matthews & Link, 1971).

One of the most relevant particle characteristics derived from sieve analysis is the size of the intermediate diameter. Investigation of settling of natural quartz sands enabled Baba & Komar (1981) to obtain a relationship:

$$V_m = 0.977\ V_s^{0.913}$$

in which V_s is the settling rate of the equivalent sphere of diameter, I, and V_m is the measured settling velocity. Combining this relationship with that of Gibbs *et al.* (1971) enabled Komar & Cui (1984) to compute intermediate grain diameters rather than equivalent sedimentation diameters using the expression:

$$I = \{0.111608\ V_s^2\rho + 2\sqrt{0.003114\ V_s^4\rho^2 + [g(\rho_s - \rho)(4.5\ \mu V_s + 0.0087\ V_s^2\rho)]}\}/g(\rho s - \rho)$$

in which ρ and ρ_s are the densities of water and the grains, and μ the dynamic viscosity of water.

In early comparisons of sizes determined by sieving and settling techniques Sengupta & Veenstra (1968) and Sanford & Swift (1971) concluded that in general the results were very similar but that settling over-estimated fine particle sizes and under-estimated coarse particles. Thus settling appeared to decrease the spread of sizes as estimated from sieving.

Following the application of correction factors to permit comparison of intermediate diameters from both sieving and settling techniques Komar & Cui (1984) demonstrated extremely close similarity between the results from the two forms of analysis applied to sediments from a wide range of depositional environments. However, they noted some deviations related to the presence of heavy minerals with densities considerably greater than quartz and also to mica flakes whose shapes differ from the spherical.

3.6 COULTER COUNTER

A second technique which provides the possibility of analysing sand, silt and clay sized particles in the size range 0.0005–0.85 mm depends on the detection of variation in electrical current as fluids containing the particles are passed through apertures of various diameters. The principle on which the Coulter counter operates is that the particles are suspended in an electrolyte which is drawn through the aperture. In the absence of particles an electrical current across the aperture remains constant, but when particles of low electrical conductivity, such as quartz, pass the aperture they cause current fluctuations, depending upon their volume. The stream of particles generates a series of pulses of current, up to 5,000 s^{-1}, which may be automatically recorded as numbers and particle volumes.

The system was originally introduced for size determination of fine particles (Sheldon & Parsons, 1967) but has been extended into the sand range (McCave & Jarvis, 1973). Using suitable techniques very fine increments of size may be detected (McCave, 1979b) by what is a very rapid and highly reproducible technique. Many samples may be processed in one day. The method is simple, but depends upon a relatively expensive apparatus which is not, but should be, available in all sedimentological laboratories, particularly where fine sediments are to be analysed. The Coulter counter is widely used in industrial laboratories and in medical situations but relatively rarely by academic sedimentologists.

3.7 GRAIN SIZE ANALYSIS OF LITHIFIED SEDIMENTS

Ancient sedimentary deposits which cannot be readily disaggregated present major difficulties to the potential size-analyst. Any grain size estimation must be undertaken from the rock itself or from polished or thin sections of the rock.

A thin section cut through a randomly packed set of spheres reveals circular outlines of a range of sizes. Knowing that all were of uniform original size permits calculation of that size from the density distribution observed. Without the knowledge that the grains were initially uniform it would have been impossible to reconstruct the original population (Krumbein, 1934). In natural sediments the grains are rarely uniform in size and few are spherical, so that the unknown elements in the analysis render the mathematical problem of interpreting the original grain size populations from thin section insoluble (Blatt, Middleton & Murray, 1980).

Nevertheless, sedimentologists dealing in the lithified materials frequently wish to make statements about the textural characteristics of their rocks. In the only systematic study of this topic, Friedman (1958) compared thin section analysis of artificially cemented and sectioned sands with sieving analyses of the same materials. The correction factors derived empirically were expressed in graphical form (Fig. 3.4).

Because there is a preferred grain orientation in most sediments, the particles lying with long and intermediate axes parallel to the depositional surface, thin sections for size analysis of such sediments should be cut parallel to bedding or cross-bedding surfaces. The size of the smaller axes encountered, i.e. intermediate axes of the grains encountered at grid intersections as in point-counting, should be determined. The use of gridded photomicrographs printed to a convenient size permits analysis at the desk rather than down the microscope.

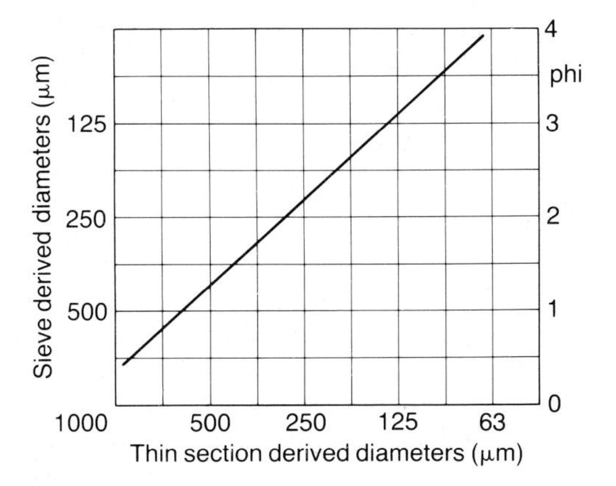

Fig. 3.4. Correlogram between thin section and sieve derived size data (Friedman, 1961).

3.8 GRAIN SIZE ANALYSIS

3.8.1 Scales of size

Since sediment particles range in size from several metres to less than one micrometer, a scale using uniform divisions by size places too much emphasis on coarse sediment and too little on fine particles. In consequence, a geometric scaling was introduced to place equal emphasis on small differences in fine particles and larger differences in coarse particles. Although there is fairly general agreement on the terms to be applied to sediment particles of various sizes the definitions used for the bounding sizes is not uniform. Most workers take values based on the Udden-Wentworth scale (Table 3.4), which is a ratio scale in which the grade boundaries differ by a factor of 2. One grade coarser is twice the size of its predecessor and one grade finer is half the size. The grade boundaries are established at 4, 2, 1, 0.5 and 0.25 mm etc. Even with agreement on the form of the scale and grade boundaries in the coarser ranges different authors still place the silt–clay boundary variously at 2 μm (Briggs, 1977; Friedman & Sanders, 1978), which is a size commonly used by soil scientists, or at 4 μm, as in the original Udden-Wentworth system (Tanner, 1969; Pettijohn, 1975) as is more normal amongst geological sedimentologists.

In order to plot the results of grain size analysis using the ratio scale it is necessary to use logarithmic scale graph paper for size so that visual equality is given to the scale divisions. This led Krumbein (1934) to introduce the phi transformation, which recognized the logarithmic equality of scale divisions. He expressed grain size such that

$$\phi = -\log_2 d$$

where d is the grain diameter in millimetres. This permitted the use of arithmetic graph papers for plotting. According to the phi transformation a coarse particle 4 mm in diameter has an equivalent diameter of -2ϕ, whereas a finer grain 0.125 mm in diameter equates with a $+3\phi$ diameter. The larger the particle phi number the finer the particle. The relationship between metric and phi scales presented in tables by Page (1955) is illustrated in Fig. 3.5.

The arithmetic form of the phi scaling also potentially simplifies statistical analysis of grain size data,

Table 3.4. Size scales of Udden & Wentworth

Udden-Wentworth (1922)	phi	mm	Friedman & Sanders (1978)	
			V. large	Boulder
	−11	2048		
			Large	
	−10	1024		
			Medium	
	−9	512		
			Small	
Cobbles	−8	256		
			Large	Cobbles
	−7	128		
			Small	
	−6	64		
			V. coarse	Pebbles
	−5	32		
			Coarse	
Pebbles	−4	16		
			Medium	
	−3	8		
			Fine	
	−2	4		
Granules			V. fine	
	−1	2		
V. coarse (sand)			V. coarse	Sand
	0	1		
Coarse (sand)		microns	Coarse	
	1	500		
Medium (sand)			Medium	
	2	250		
Fine (sand)			Fine	
	3	125		
V. fine (sand)			V. fine	
	4	62		
			V. coarse	Silt
	5	31		
			Coarse	
	6	16		
			Medium	
	7	8		
			Fine	
	8	4		
Clay			V. fine	
	9	2		
			Clay	

but confusion concerning the properties of the phi scale dimensions led to an accepted redefinition in which

$$\phi = -\log_2 \frac{d}{d_o}$$

where d_o is the diameter of a 1 mm grain. This stresses that the phi system is one of dimensionless numbers and so may be correctly used for statistical analysis to derive factors such as the standard deviation, skewness and kurtosis of grain size distributions (McManus, 1963; Krumbein, 1964).

3.8.2 Graphic presentation

The simplest method of presenting grain size data graphically is by means of the histogram. In this the

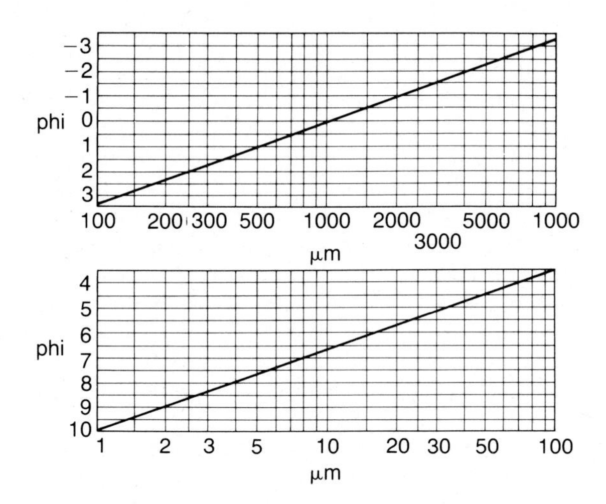

Fig. 3.5. Interrelationship between metric and phi scales of size.

grain size is the independent variable, which is therefore plotted on the horizontal axis and the vertical axis is used for the dependent variable, i.e. the determined weight percentage. The weight percentage of sediment retained on each sieve is plotted as a column rising to that value and extending from the relevant sieve mesh to the next coarser mesh used. The output from most Coulter counters is also given in this format. The resulting graphical construction is simplified if the size scale is represented by equal increments, usually represented by phi, half-phi or quarter-phi divisions. A line graph, which is the unique frequency curve of the sediment, may be formed by linking the mid-points of the histogram columns (Fig. 3.6). Ideally the frequency curve should be formed from the most detailed possible size analysis of the sediment. When sieving is used the sieves are arranged at $\frac{1}{4}$ phi scale intervals. The frequency curve approximates to a continuous functional relationship between grain size and weight percentage variation. The resultant curve closely resembles the Gaussian or 'normal' probability curve of statisticians, with high values near the centre of the size range and low value tails in the coarse and fine ends of the scale. Traditionally sedimentologists have regarded grain size populations as approximating to this form of distribution and have based their methods of data plotting and statistical analysis on this assumed relationship. In that the size scale is logarithmic in reality the distribution approximates to log normal. Because a limited amount of statistical

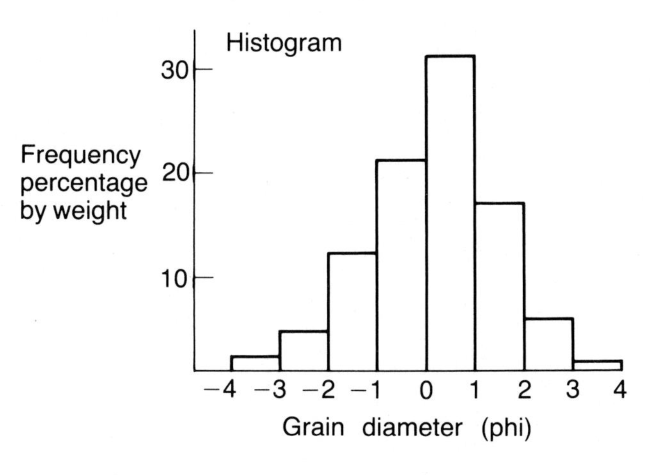

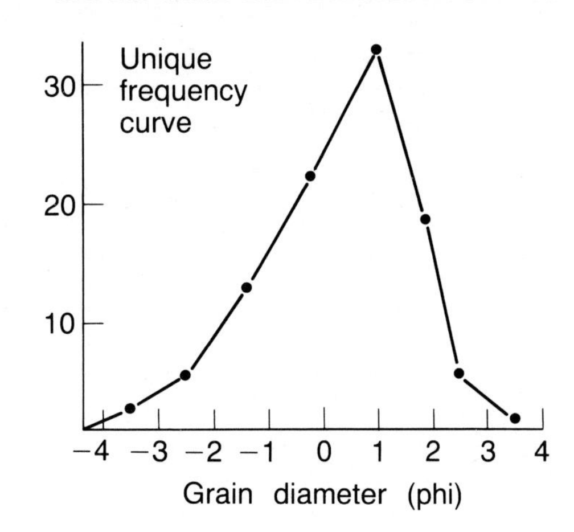

Fig. 3.6. Histogram and unique frequency plots.

information may be extracted from the frequency curve the same facts are most commonly used to construct the cumulative frequency curve.

In constructing the curve of *cumulative frequency by weight* the abscissa retains its size scaling. The weight percentage retained on the coarsest sieve is plotted at the appropriate grain diameter, the sum of weight percentages retained in the two coarsest sieves is plotted at the next finer mesh, the cumulative contents of the first three sieves at the next finer mesh and so on until the entire sediment has been accounted for at 100%. Although the ordinate scale of such curves may be arithmetic a second scaling is more commonly used which assumes the normal probability distribution. Plotting is therefore carried out on arithmetic probability graph paper. The ordinate scaling is derived by dividing the area beneath a normal distribution curve into columnar segments of equal area. Those near the centre of the distribution are long and relatively narrow whereas those towards the tails are low and proportionately broader (Fig. 3.7). Since the spacing patterns are the same for all normal distributions the divisions are carried forward to provide the probability ordinate scaling for the arithmetic probability graph which is used for plotting cumulative size distributions.

Whereas plotting on the arithmetic ordinate scale generally yields 'S' shaped grain size distribution patterns, using the probability ordinate scale straightens the curve (Fig. 3.8). Perfect log-normal grain size distributions plot as straight lines on the probability paper. However, natural sediments

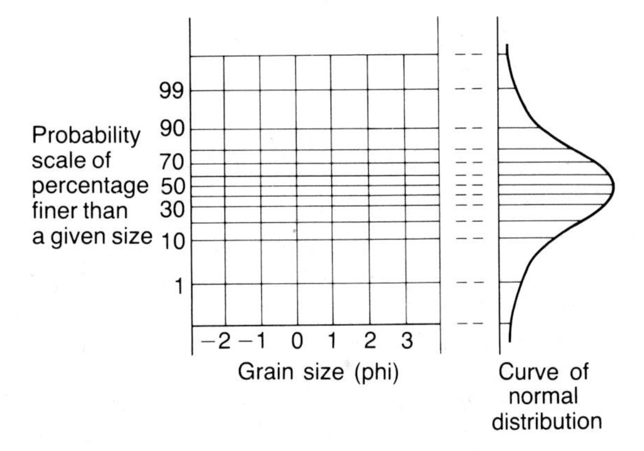

Fig. 3.7. Derivation of the graphical scale used in probability plots.

rarely conform to this pattern and the deviations from log-normality have provided much scope for discussion, interpretation and speculation.

By convention the geologist plots coarse grains to the left of the abscissa and fine to the right. In computing values for cumulative distributions it is normal to consider values in terms of the percentage coarser than a given grain size. Thus the higher values on the cumulative percentage by weight plot occur in the range characterized by the finer particles. This is the exact opposite of and neither more nor less correct than the practise adopted in civil engineering analysis of soils and aggregates.

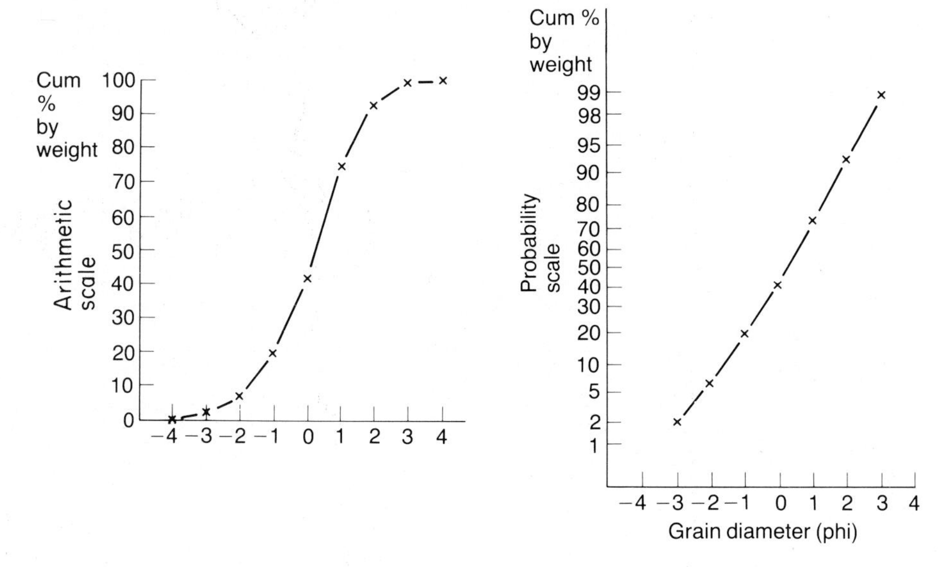

Fig. 3.8. Cumulative frequency distribution curves plotted or arithmetic and probability scales.

3.8.3 Curve characterization by statistical methods

Whilst data presented in graphical form have pictorial value, permitting crude general statements to be made about a sediment, they are of little value for detailed comparisons with other samples. In consequence, simple statistical techniques are used to characterize the grain size distribution data. An important difference between conventional statistics and statistics of grain size distributions is that in the former frequency is expressed as numbers whereas in the latter it is as weight percentage.

Two principal forms of analysis are normally used, *graphical* methods, in which values derived directly from plotted cumulative curves are entered into established formulae, and *moment* methods in which the characteristics of every grain in the sample analysed from the sedimentary deposit are used in the computation. In all cases the numbers obtained serve to define the position of the distribution plot, its slope, and the nature of any irregularities. The values of these parameters permit characterization of the curves, and enable numerical comparisons to be made between samples.

The parameters used fall into four principal groups: those measuring (a) average size, (b) spread of the sizes about the average, (c) symmetry or preferential spread to one size of the average, and (d) kurtosis or degree of concentration of the grains to the central size.

In discussion of particle sizes a standard form of notation is used. The size of particle for which 25 or 32% of the grains are coarser, P_{25} or P_{32} respectively in metric units, or ϕ_{25} and ϕ_{32} in the phi system may be read directly from the cumulative frequency curve. Grain sizes in ϕ corresponding to specified percentile values are entered into simple formulae to permit calculation of the various graphic parameters.

(a) MEASURES OF COMMON SIZE AND AVERAGE SIZE

Mode. On a size frequency histogram the size class in which the greatest percentage of grains is represented provides the modal class. On the size frequency distribution plot the highest point on the curve provides the modal value. The modal size is, therefore, the commonest grain size in a distribution.

The frequency curves of many sediments exhibit several peaks, the polymodal distributions indicating the presence of more than one population of grains. Some workers use modal values abstracted directly from the size frequency distribution by weight (Friedman & Johnson, 1982) but others advocate its derivation by recalculation from the gradient of the cumulative curve (Griffiths, 1967; Folk, 1974b).

Median (Md). Half of the grains are coarser and half finer than the median diameter, whose size is most readily determined from the 50% line of the cumulative distribution curve. Although useful for many unimodal sediments, in polymodal distributions the median may fall in the tails of two subpopulations of grains, in a size fraction which is scarce.

Mean (M). The best measure of average grain size, the mean is computed from sizes of particles spread through a range of percentile values. In its simplest form the Graphic Mean, M_z, is calculated from $\frac{1}{3}(\phi_{16} + \phi_{50} + \phi_{84})$ which assumes that three values alone are sufficient to give a useful mean. Where more confidence in the average value is required more percentage values may be used, e.g. M9 is computed by $\frac{1}{9}(\phi_{10} + \phi_{20} + \phi_{30} + \phi_{40} + \phi_{50} + \phi_{60} + \phi_{70} + \phi_{80} + \phi_{90})$ provided the percentile values used are evenly spread through the distribution. In practise this approximates to the mean of moment statistics.

(b) SPREAD ABOUT THE AVERAGE — SORTING (σ)

In many forms of analysis the full range of sizes present is of relevance. However, it is rarely possible to define the size of the largest or smallest particles precisely in a size distribution. Of more importance is an assessment of the spread of particles about the average, to define the dispersion or sorting of the sediment, as represented by the breadth of the frequency curve or the shape of the cumulative frequency distribution.

In all metric measures and their direct phi-based equivalents only the 25th, 50th and 75th percentiles are used, so that attention is only given to the central half of the grain population present. As a measure of the characteristics of the sediment they are of limited value, but as a method of characterizing the central segment of a curve the Trask Sorting coefficient $(P_{25}/P_{75})^{\frac{1}{2}}$ or its equivalent, the phi Quartile Deviation $\frac{1}{2}(\phi_{75} - \phi_{25})$ may have some value.

Ideally a measure of sorting should embrace a broader spectrum of the grains present, and the graphic standard deviation σ_ϕ of Inman (1952), computed from $\frac{1}{2}(\phi_{84} - \phi_{16})$, provides a measure of the spread of size of 68% of the population (one standard deviation on either side of the mean). Further extension to include 90% of the curve yields the inclusive graphic standard deviation σ_1 (Folk & Ward, 1957), which uses the 5th and 95th percentiles to define a spread corresponding to 1.65 standard deviations on either side of the mean. Taking the mean of the contributions from the 68 and 90% population fractions produced the expression:

$$\sigma_1 = \frac{1}{2}\left(\frac{\phi_{84} - \phi_{16}}{2} + \frac{\phi_{95} - \phi_5}{3.3}\right) \text{ or }$$

$$\frac{\phi_{84} - \phi_{16}}{4} + \frac{\phi_{95} - \phi_5}{6.6}.$$

The best sorted sediments approximate to a single size and have low σ_1 values (Table 3.5).

(c) PREFERENTIAL SPREAD — SKEWNESS (α)

In a normal distribution with a bell-shaped frequency curve the median and mean values coincide. Any tendency for a distribution to lean to one side, i.e. to deviate from normality, leads to differences between the median and mean values. These differences are used to characterize the asymmetry or skewness of the curve. The skewness has a positive or negative value when more fine or more coarse materials are present than in a normal distribution, seen as tails to the right or left respectively on frequency distribution plots. Again, although skewness may be computed for the central segment of the distribution, for most purposes broader spreads are used. Effectively skewness is determined from the value of the mean less the median, all divided by the range used in defining the mean. Many possible combinations of computation are available. Expressed simply:

$$\alpha\phi = \frac{1}{\sigma\phi}(M\phi - Md\phi).$$

When laid out fully the inclusive graphic skewness is obtained from:

$$SK_1 = \frac{\phi_{16} + \phi_{84} - 2\phi_{50}}{2(\phi_{84} - \phi_{16})} + \frac{\phi_5 + \phi_{95} - 2\phi_{50}}{2(\phi_{95} - \phi_5)}.$$

Table 3.5. Statistical formulae used in calculating grain size parameters

(a) Statistical measures in the metric and phi systems

	Median	Mean	Dispersion	Skewness	Kurtosis
Metric	$Md = P_{50}$	$M = \frac{P_{75} + P_{25}}{2}$	$QDa = \frac{P_{75} - P_{25}}{2}$	$Ska = \frac{P_{75} + P_{25} - 2Md}{2}$	$Kqa = \frac{P_{75} - P_{25}}{2(P_{90} - P_{10})}$
			$S_o = (P_{75}/P_{25})^{\frac{1}{2}}$	$Sk = \frac{P_{75} \cdot P_{25}}{Md^2}$	
Phi	$Md = \phi_{50}$	$M\phi = \frac{\phi_{16} + \phi_{84}}{2}$	$\sigma\phi = \frac{\phi_{84} - \phi_{16}}{2}$	$\alpha\phi = \frac{M\phi - Md\phi}{\sigma\phi}$	$\beta\phi = \frac{\frac{1}{2}(\phi_{95} - \phi_5) - \sigma\phi}{\sigma\phi}$
		$Mz = \frac{\phi_{16} + \phi_{50} + \phi_{84}}{3}$	$\sigma_1 = \frac{\phi_{84} - \phi_{16}}{4} + \frac{\phi_{95} - \phi_5}{6.6}$	$SK_1 = \frac{\phi_{16} + \phi_{84} - 2\phi_{50}}{2(\phi_{84} - \phi_{16})} + \frac{\phi_5 + \phi_{95} - 2\phi_{50}}{2(\phi_{95} - \phi_5)}$	$K_G = \frac{\phi_{95} - \phi_5}{2.44(\phi_{75} - \phi_{25})}$

(b) Descriptive terms applied to parameter values

Sorting (σ_1)		Skewness (SK_1)		Kurtosis (K_G)	
Very well sorted	<0.35	Very positively skewed	+0.3 to +1.0	Very platykurtic	<0.67
Well sorted	0.35–0.50	Positively skewed	+0.1 to +0.3	Platykurtic	0.67–0.90
Moderately well sorted	0.50–0.70	Symmetrical	+0.1 to −0.1	Mesokurtic	0.90–1.11
Moderately sorted	0.70–1.00	Negatively skewed	−0.1 to −0.3	Leptokurtic	1.11–1.50
Poorly sorted	1.00–2.00	Very negatively skewed	−0.3 to −1.0	Very leptokurtic	1.50–3.00
Very poorly sorted	2.00–4.00			Extremely leptokurtic	>3.00
Extremely poorly sorted	>4.00				

Skewness is a positively or negatively signed dimensionless number; it has neither metric nor phi value and lies within the range −1 to +1 (Table 3.5).

(d) CONCENTRATION OR PEAKEDNESS OF THE DISTRIBUTION — KURTOSIS

Not widely used, although frequently calculated, the kurtosis is related both to the dispersion and the normality of the distribution. Very flat curves of poorly sorted sediments or those with bimodal frequency curves are platykurtic, whereas very strongly peaked curves, in which there is exceptionally good sorting of the central part of the distribution, are leptokurtic. The measure of kurtosis, which is a ratio of the spreads of the tails and centre of the distribution (and is therefore also dimensionless) is given by the expression:

$$K_G = \frac{\phi_{95} - \phi_5}{2.44(\phi_{75} - \phi_{25})}.$$

A listing of commonly used formulae and verbal descriptions of the values obtained is presented in Table 3.5.

3.8.4 Moment methods

The second major approach to analysing grain size distributions, moment statistics, differs in concept but yields measures analogous to those of graphical methods. Neither technique is 'better' than the other. Graphic techniques are especially appropriate for analysis of open ended distributions whereas the moment methods should not be applied unless all grain sizes present lie within the defined grain size limits.

The principles of moment statistics are lucidly reviewed in Friedman & Johnson (1982). Effectively each grain in the population is taken into account in computing the characteristic parameters of the sediment. Whereas in mechanics the moment is calculated as the product of a force and the distance of its application from a fulcrum, in statistics the moment is computed from the product of the weight percentage in a given size class and the number of class grades from the origin of the curve. The first moment, the moment per unit frequency, is the mean, $\bar{x}$, of the distribution:

$$\bar{x} = \frac{\Sigma f m_\Phi}{100}$$

where $\bar{x}$ is the mean, f the frequency in weight per cent and m_Φ is the mid-point of each class interval.

Once the position of the mean is defined the spread of the distribution about it is calculated by summing the moments of each class interval about the mean. In each of the subsequent moment calculations the difference $(m_\Phi - \bar{x})$ is raised to higher powers.

The second moment has a value which is the square of the standard deviation of the distribution:

$$\sigma^2 = \frac{\Sigma f(m_\Phi - \bar{x})^2}{100}.$$

The third moment measures the symmetry of the distribution about the mean and yields a value which is the cube of the skewness:

$$\alpha^3 = \frac{\Sigma f(m_\Phi - \bar{x})^3}{100}.$$

The fourth moment yields a value for kurtosis raised to the fourth power:

$$\kappa^4 = \frac{\Sigma f(m_\Phi - \bar{x})^4}{100}.$$

Although the parameters obtained are analogous to those of graphical statistics their derivation employs the entire grain population and so they are more representative than the graphically derived values. Once again it should be stressed that, because all grains are used, this form of analysis should not be used unless the size distribution is fully known. Distribution of the 'pan residue' fraction from sieving among a specified number of phi classes 'to permit application of moment methods' is not good practise and may lead to misleading, if not meaningless, results being obtained. If a total of less than 1% of the population is undefined then the errors are unlikely to be great, but the reliability of the moments decreases sharply as the proportions of undefined materials increases, and the technique should not be used with a higher proportion of unknowns notwithstanding the convenience and availability of pocket calculators suitable to perform the arithmetic.

3.8.5 Alternative grain size distributions

The grain size distributions of some sediments yield linear log-normal cumulative plots, but others deviate from the linear patterns to a greater or lesser degree, especially in the tails of the distributions. In consequence some authors suggest that the conventionally applied log-normal probability law is not suitable for universal applications in grain size analysis.

Early work by Rosin & Rammler (1934) revealed that crushed coal exhibited regular grain size characteristics controlled by natural fracture patterns of the material. They established expressions from which Rosin paper was introduced for plotting grain sizes of crushed particulate matter (Kittleman, 1964). Some natural sediments such as pyroclastic debris, glacial tills and weathered crystalline rock products exhibit linear plots on Rosin paper, suggesting that they behave very much as crushed products, but the similarity vanishes swiftly with transport after which they rapidly come to resemble more commonly encountered materials transported and deposited from natural fluids.

Although Bagnold (1941) suggested that wind blown sands, which deviated from log-normality, might have some other probability function, it was not until recently that his suggested log-hyperbolic distribution has been closely examined as an alternative (Barndorff-Nielsen, 1977; Bagnold & Barndorff-Nielsen, 1980; Wyrwoll & Smyth, 1985). The log-hyperbolic distribution offers the possibility of a range of curve fittings, one limiting case of which is the log-normal distribution. In practise the natural logarithm of the frequency curve is plotted against grain size (Bagnold, 1968), and the four parameters of the best fitting log-hyperbolic curve are computed. Two of these relate to the curve position (grain size and peakedness) and the other two to inclination of the limbs of the curve. The distribution has been examined principally within the context of wind blown dune sands (Bagnold & Barndorff-Nielsen, 1980; Christiansen, 1984) for which it was claimed the hyperbolic function encompassed the extreme values of the size distribution tails better than the log-normal plot. However, comparative studies of Wyrwoll & Smyth (1985), who suggested that the 'improved' performance may be partly a function of shape-influenced deficiencies in the sieving technique, led to the conclusion that only a marginal improvement in fitting over the log-normal curve was achieved. They also pointed out that substantial computational skills were required to obtain the best-fit log-hyperbolic curves from which the parameters could be derived.

3.9 ENVIRONMENTAL INTERPRETATION FROM GRAIN SIZE DATA

The methods of analysis and presenting data are standard descriptive techniques, but it is always the aim of scientists to derive more than simple descriptions of their materials. The sedimentologist is no exception. The prospect of a universally applicable technique of abstracting from sediment grain size distributions measures diagnostic of depositional environment has encouraged generations of sedimentologists. The uncomfortable fact remains that though many have tried only modest success has been achieved and no overall method has yet been discovered. Indeed, some consider the quest, though honourable, most unlikely to succeed (Erlich, 1983) and potentially wasteful of effort through diverting attention away from attainable research targets. With this warning the reader is introduced to the final section of this chapter in which some of the approaches to the analysis of the distribution curves and the abstracted measures are outlined.

The simplest tool employed by the sedimentologist is the geographical map on which variations of individual parameters may be plotted, so that patterns of regional variation may be recognized. Systematic variation of mean grain size with environmental dynamics is normally detectable, and systematic changes in standard deviation and skewness may also be detected in suitable areas. Within a local or regional context recognizable patterns related to dynamics or supplied materials are not unexpected (Folk & Ward, 1957; Allen, 1971; Ryan & Goodell, 1972; Cook & Mayo, 1977).

Commonly, there are covariations of two or more measures, such as mean size and sorting, which Griffiths (1967) illustrated as being hydraulically controlled, so that in all environments the best sorted sediments had their mean sizes in the fine sand category. This energy related universal relationship has been confirmed by many subsequent studies.

One of the earliest attempts to characterize depositional environments by means of bivariate scattergrams was that of Stewart (1958) who plotted median against skewness and standard deviation for sediments from rivers, wave dominant zones and quiet water environments, and defined envelopes within which his analyses occurred (Fig. 3.9): Also, using graphic techniques, Schlee, Uchupi & Trumbull (1965) were unable to distinguish beach and dune sands from grain size data alone.

Following early studies by Friedman (1961, 1967) the use of scattergrams of sorting and skewness to provide discrimination between river, beach and dune sands has been repeatedly explored (Moiola & Weiser, 1968). The relatively fine grained unimodal sand deposits from many parts of the world yielded similar characteristics: beach sands were well sorted and negatively skewed, whereas river sands were less well sorted and usually positively skewed. Dune sands also had positive skewness but were finer than beach sands. However, the best separations reported were achieved using a combination of moment standard deviation and an unusual graphic skewness (Friedman & Sanders, 1978). Although general separation of points by environment was achieved there was always some overlap between adjacent fields. No clear cut distinction has been found in most forms of bivariate plot. Leaving the beach, dune and river sediments, which were mainly medium or fine sands, the greater spread of grain size enabled Landim & Frakes (1968) successfully to distinguish till from alluvial fan and outwash deposits using moment methods.

In few studies have the higher moments been used to advantage, the principal exception being that of Sly (1977, 1978) who used the skewness − kurtosis interrelationships to recognize differing environmental energy levels in deposits of the Great Lakes (Fig. 3.10). Deposits without pebbles, essentially of muddy materials from low energy environments, fall in fields A, B, C and D whereas those with pebbles, either lag deposits or high energy beaches, lie in fields E, F, G and H.

In some of the simplest forms of analysis ratios of the sand, silt and clay contents may have value. The

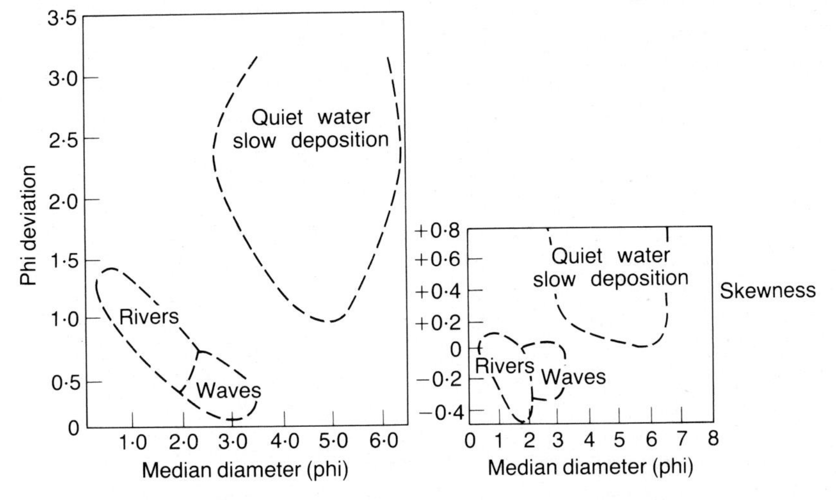

Fig. 3.9. Bivariate plots which Stewart (1958) believed were of environmental significance.

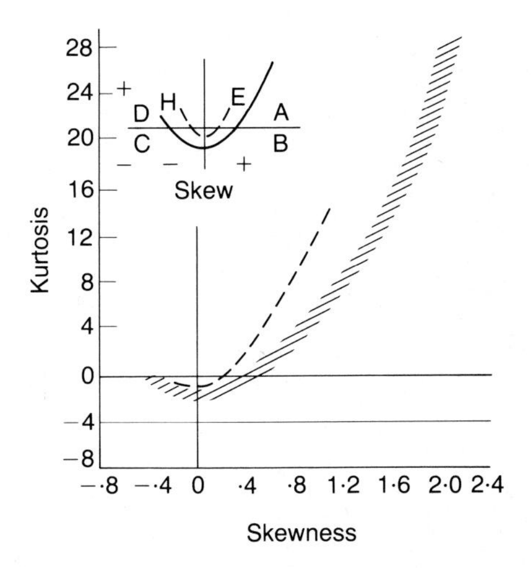

Fig. 3.10. Skewness – kurtosis plots of sediments from the Lake Ontario low energy ABCD and active environments EFGH.

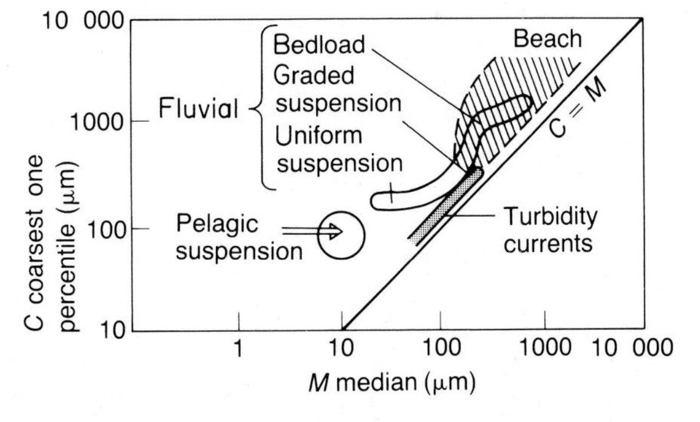

Fig. 3.11. The CM plot of Passega (1964) with envelopes with which he believed sediments should plot according to dynamics of the transport experienced.

silt-clay relationship has been explored in an environmental energy context for lake and marine systems. Pelletier (1973) demonstrated that the expected decrease of size with increasing depth of water was much stronger in lakes than in oceans.

Using information abstracted from the cumulative distribution curve, Passega (1957, 1964) suggested that the ratio of the coarsest one percentile, C, to the median diameter, M, could be used as an indicator of the dynamics of the depositional environment. The strongest currents should define the largest stable particle size. Counter-plotting gave the $C-M$ diagram within which he defined envelopes typifying deposits from traction currents, from beaches or from various forms of suspension (Fig. 3.11). In separate plots, $L-M$ diagrams, he also examined the role of L, the proportion of particles finer than 31 μm. With deposits of limited size variations the technique has some success but better discrimination is achieved where there are large size differences.

Many sediments are composed not of one single grain size population, but rather of a combination of sub-populations. Each sub-population may be defined by dynamic considerations or by supply characteristics. In the natural environment individual sub-populations may be present in varying proportions from site to site. Thus in allied localities it may be possible to trace material from one specific source using its modal size and spread, noting the increases and decreases in its contribution to local materials (Curray, 1960; McManus, Buller & Green, 1980). Elsewhere the effect of wave activity may be particularly intensely experienced by one size fraction which is either totally removed or greatly increased in proportion. The combination of the sub-populations in different proportions determines the grain size characteristics of the overall sediment.

Following the work of Moss (1962, 1963) who

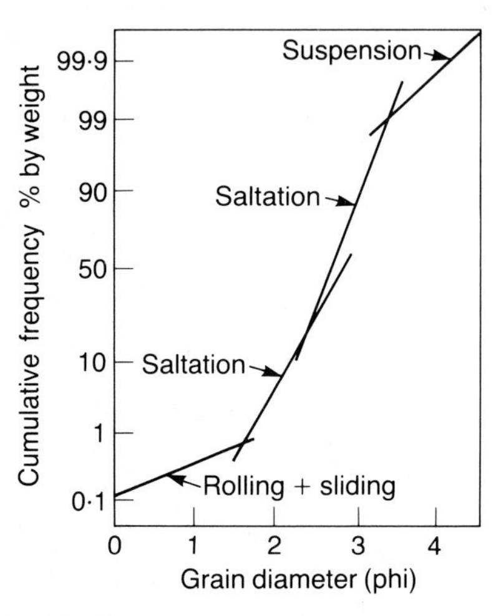

Fig. 3.12. The linear segments of cumulative size frequency curves as suggested by Visher (1969).

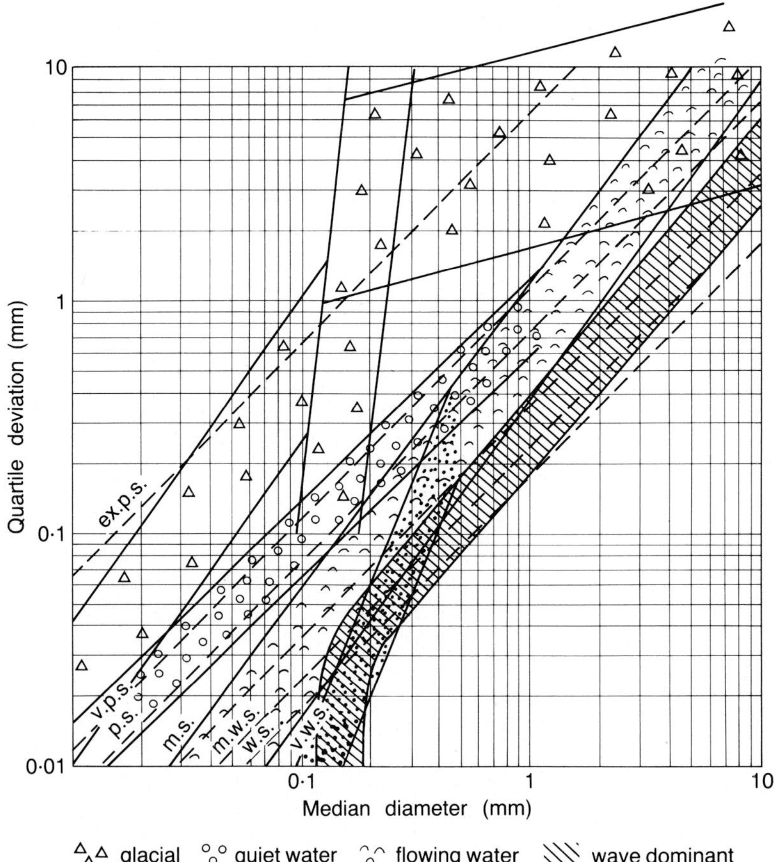

Fig. 3.13. Metric quartile-deviation – median diameter plots of Buller & McManus (1972) and their environmental envelopes together with the 'sorting' levels indicated by Folk & Ward (1957).

demonstrated three sub-populations of pebbles form most gravel deposits, Visher (1969) suggested that cumulative frequency curves could often be subdivided into two, three or four linear segments (Fig. 3.12). Because log-normal distributions follow linear trends on the cumulative probability plot he considered each segment to represent a separate subpopulation whose character was determined by the dynamics of transport: traction, saltation and suspension. Many plots required a two-fold saltation load to be considered. This apparently simplistic and attractive technique which has been readily espoused by many workers merits further attention. Replotting data from the cumulative curve to the frequency curve shows that linear segments do sometimes coincide with recognizable sub-populations but in other cases inflexions where linear segments join also coincide with population peaks. Experimentally combining two unimodal sediments produces not a two-fold linear pattern, but a three-fold one, the central segment forming in the size overlap zone. Nevertheless, this form of analysis may still merit further exploration.

In a further non-standard bivariate plot, Buller & McManus (1972) suggested using the central subpopulation, which frequently spans over half of the distribution, to provide a metric quartile deviation measure $\frac{1}{2}(P_{25} - P_{75})$ to plot against the median diameter. Using many hundreds of analyses they defined field envelopes each embracing 68% of all relevant data points from the river, beach, a eolian and quiet water deposits. Many of the envelopes partly overlapped, particularly in the medium to fine sand range, but good separations were achieved outside those sizes. Later envelopes for turbidites (Buller & McManus, 1973a), tills (Buller & McManus, 1973b), screes and pyroclastic (Buller & McManus, 1973c) deposits were added. Plotting the graphic sorting characteristics of the same sediment gradients enables recognition of the sorting characteristics which may be expected to occur with sediments from each depositional environment (Fig. 3.13). This method of analysis met with some success (Sedimentation Seminar, 1981) but has not been widely applied.

While the geological sedimentologist has developed one set of criteria by which sedimentary deposits are assessed, an entirely different approach to sands and gravels is taken by the construction industry which uses such materials for concrete, road metal aggregates, mortars and plastering.

The sedimentologist assesses materials with respect to their conformity to an ideal of sorting in which perfection is realized in a deposit in which all particles are identical. In an extremely well sorted sediment all particles approximate to the same size. For the civil engineer dealing with concrete design, sorting in the geological sense is undesirable. The *graded* aggregate, in which a defined spread of particle sizes is represented, is considered ideal. In a graded aggregate small particles occupy intergranular spaces, partly filling voids between larger particles. This permits reduction in quantities of cement and water required to give desirable properties such as workability, strength and durability potential.

The poorly sorted sample of the sedimentologist may be the well graded material of the engineer. In practice the engineer achieves design optimization by blending together 'coarse' and 'fine' aggregates of suitable strictly defined grading characteristics. As many of these are readily obtained by simple screening techniques applied to natural sedimentary deposits it is of interest to record the prescribed

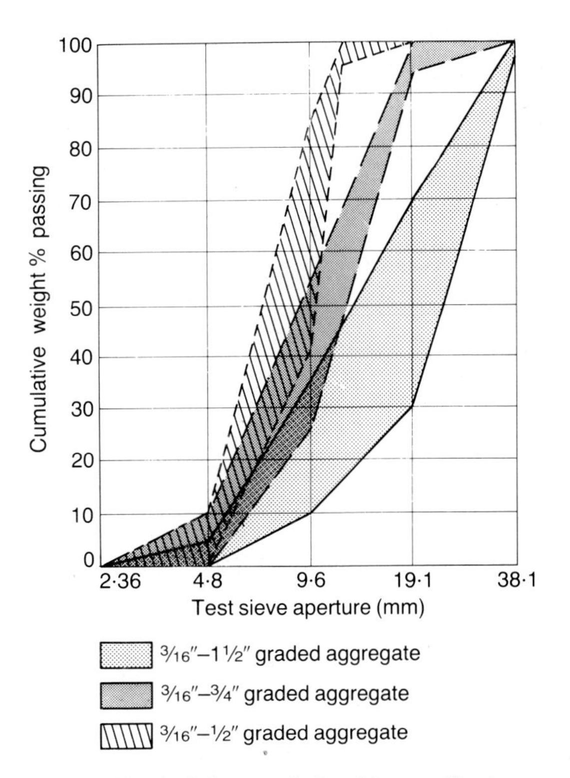

Fig. 3.14. Nominal diameter limits of size specifications for graded concrete aggregates, after BS 882 : 1973.

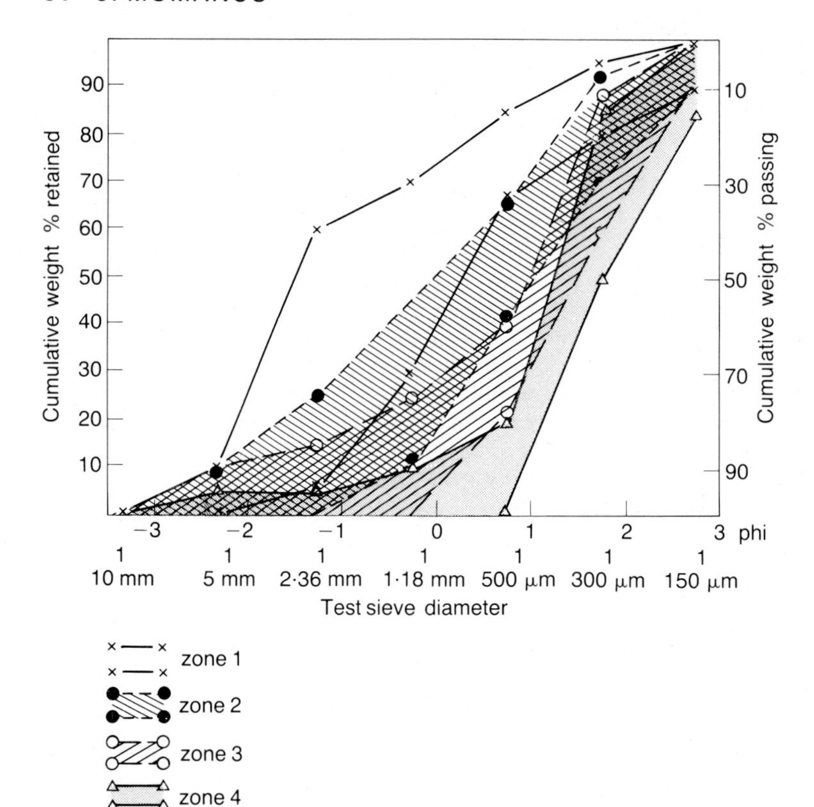

Fig. 3.15. Specification limits defined for the four zones of fine aggregate sizes, after BS 812: 1975.

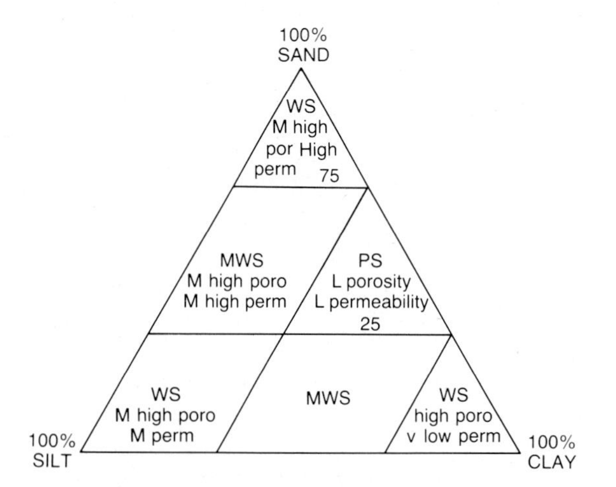

Fig. 3.16. Ternary diagrams interrelating sand, silt and clay components and their controls over porosity and permeability. WS, well sorted; P, S, poorly sorted; MSW, moderately well sorted.

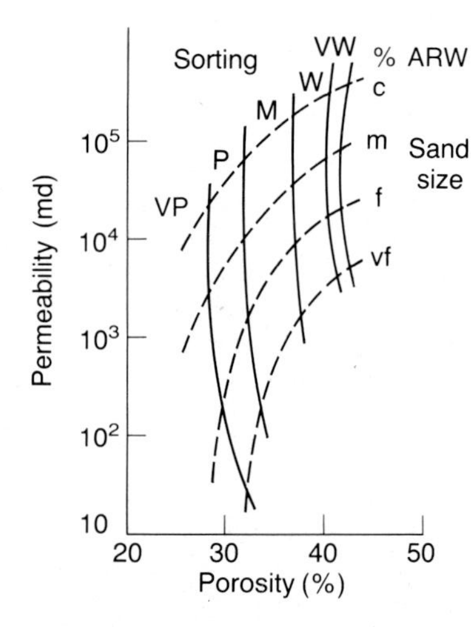

Fig. 3.17. Relationship between grain size, sorting, porosity and permeability in uncemented sands (after Selley, 1985).

limitations of the aggregate gradings. The gradings recommended for coarse aggregates arranged according to specified nominal diameters in BS 882: 1973 are indicated in Fig. 3.14 and those for the four zones of fine aggregates of BS 812: 1975 in Fig. 3.15. Materials whose grain size distributions fall within the limits specified for zone 3 fine aggregates are well suited for certain tasks.

According to Connor (1953), only restricted gradings of sand produce good mortars. Later, Swallow (1964) demonstrated that the best mortar sands had median diameters in the range 440–570 μm and graphic standard deviations of 200–275 μm. Nevertheless, a broader range of potentially useful mortar sands is indicated in BS 1199:1976.

A second property of sediments which is of economic importance is the porosity, the spaces between the solids, which may become filled with water or other fluids such as oil and gas. The porosity of a deposit is expressed as the percentage of voids in a given volume of rock. Two forms of pores are recognized, the *primary pores*, which develop during deposition of the sediment, and *secondary pores*, which develop after deposition, principally through solution of the rock particles themselves. The former are important in unconsolidated or cemented siliciclastic rocks whereas the latter is more commonly considered in the context of carbonate sands and limestones.

The intercrystalline or interparticle voids are present in all rocks. They are more open and better interconnected in most coarse clastic deposits than in fine sediments. The efficiency with which fluids pass through the rocks depends heavily on the interconnections, the spaces available in the pore throats and the sorting of the sediment. In very well sorted coarse uncemented sediments pore spaces are large and may have free interconnections whereas in poorly sorted sediments fine particles occupy void spaces, reducing the porosity and blocking many interconnections. Early cementation serves a similar purpose and may lead to closure of the throat passages. Compaction may cause fine particles to move into direct contact with their neighbours so a high initial porosity may become drastically reduced with burial.

Measurement of porosity is achieved by determining the volume of gas displaced from the pore spaces of a known bulk volume of rock. The latter is determined by displacement using total immersion in mercury, the former using a sealed pressure vessel and measuring the increased volume of gas present when pressure is decreased by a known amount.

The associated property of permeability is important for this is the factor which determines the ability of a fluid to pass through a porous medium. Measured as the rate at which fluids pass through cores of the rock or a bed of sand, the permeability is calculated according to Darcy's Law and the unit relevant in most natural sediments in the millidarcy. Again the permeability is largely dependent upon sediment sorting and the presence of cements,or authigenic growths in the pore throats. Particle shape may also be important. As grain size decreases so does the permeability in most cases, due to an increase in the associated capillary pressure resisting passage of the fluids.

The general interrelationships between grain size, permeability and porosity are summarized in a ternary diagram of sand, silt and clay contents in Fig. 3.16. The detailed impacts of sorting and grain size on porosity and permeability of unsorted sands is given in Fig. 3.17 (after Selley, 1985).

The analysis of sediment size distributions is thus not only of considerable academic interest but has highly practical significance to the construction industry, for extraction of water supplies and the production of hydrocarbons.

4 Microscopical techniques: I. Slices, slides, stains and peels

JOHN MILLER

4.1 INTRODUCTION

Since the pioneering work of Henry Clifton Sorby (1851), slices of rock ground thin enough to transmit light have been the staple material of sedimentary petrography. Sorby's work simply involved examination of thin sections in polarized light on a petrographic microscope. Today's sedimentologists are able to extract a great deal more information from thin sections by using special techniques to study fabrics, textures, mineralogy and geochemistry. In order to recover so much extra information, thin section preparations must be of consistent high quality.

To achieve the required precision in thin section making, a considerable degree of mechanization is required, and this is expensive. However, most laboratories which produce more than a modest number of thin sections are now equipped with appropriate automatic and semi-automatic systems. This does not, however, mean that high quality results can be achieved *only* by such methods: where a small volume of sections is involved, many of the techniques described in this chapter can easily be adapted for manual preparation of thin sections. A simple, manual slide-making process is outlined by Adams, MacKenzie & Guildford (1984, p. 97). Sections required for teaching purposes do not need to meet the stringent requirements of those used for research, but they should be of good standard; they are, after all, the primary material on which all geologists are trained.

Many workers proceed directly to examination of thin sections, but there is a great deal of information to be obtained from examining rock slices and cut faces, particularly in the case of limestones and sandstones. First, the three-dimensional arrangement of grains can be better studied in slices, whereas thin sections are essentially two-dimensional. Second, fabrics are studied more effectively at low magnifications and on faces larger than those possible on thin sections, where the microscope concentrates on detail by restricting the field of view. Perhaps more importantly, examination of cut faces enables appraisal of which areas should be made into thin sections. Careful selection of critical areas for sectioning saves time and expense, and allows subsequent petrographic study to be based on knowledge of the section's context. Suggestions are given below for preparation and study of rock faces, slices and slabs, and the way in which these preparations support examination of thin sections in a broad-based approach to sedimentary petrography.

4.2 PROCESSING SAMPLES

At the outset of any petrographic analysis, it is wise to begin by planning the kinds of sample preparation which will be required, and what information is to be sought from them. Samples can be collected which not only adequately represent the facies or lithosomes under study, but are appropriate in orientation, size, shape and freshness to the selected preparation techniques. Processing is then designed to minimize wastage and maximize data collection. A general scheme for processing samples for microscopical study is shown in Fig. 4.1. This flow chart is fairly complete; not every petrographic study will need to be so thorough, but particular paths are easily selected.

4.3 SLICES

In this section, I use the following terms: *faced sample* — a smooth cut across a rock sample, with the rest of the specimen left rough. *Slice* — a section through the sample with two parallel cut faces. Slices may be up to several centimetres thick. Those less than about 6 mm thick are sometimes called *plaquettes*. Usually, plaquettes are approximately the same area as thin sections: they may often be prepared as a first step in section-making.

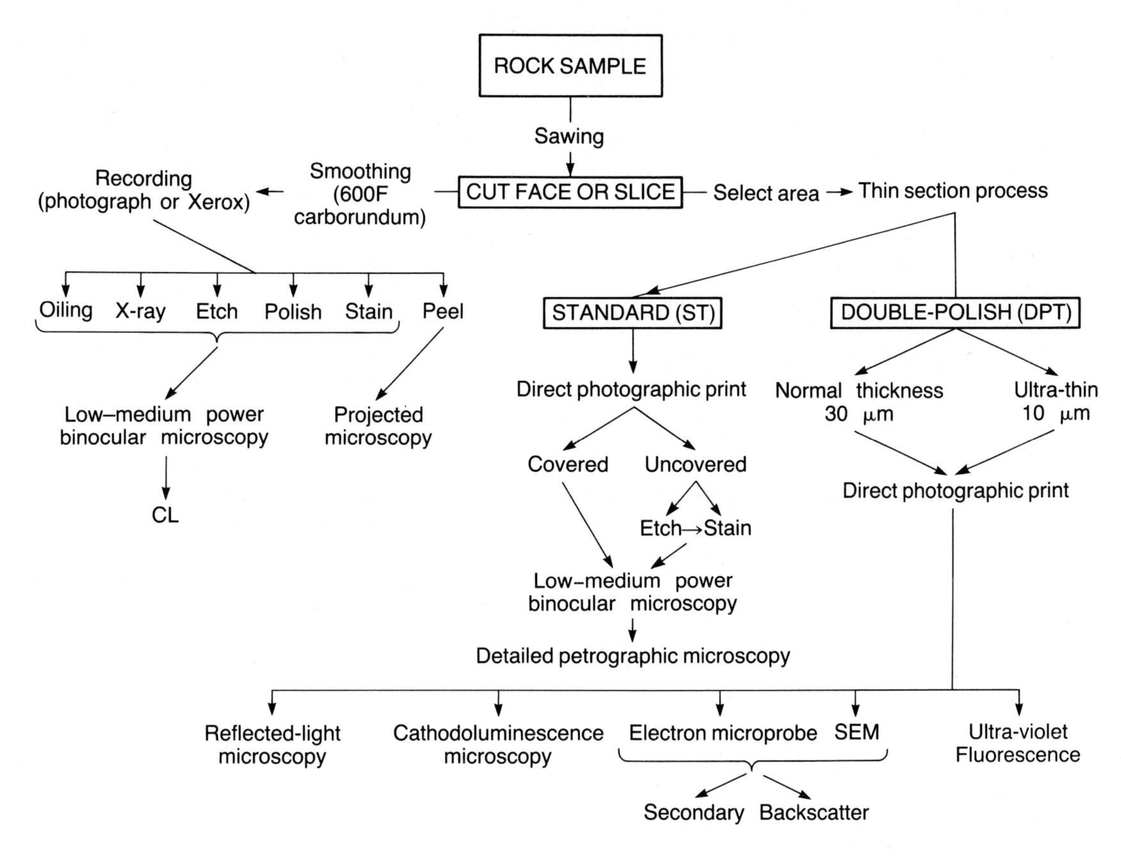

Fig. 4.1. Flow diagram showing various methods of sample preparation and possible techniques for microscopical study.

4.3.1 Slice preparation

Large samples (above the size of two clenched fists) must first be trimmed on a saw with a large radius blade (steel impregnated with diamond or a compressed carborundum disc), lubricated with water. Slabs or successive slabs with faces suitable for serial reconstruction of large-scale fabrics can be made at this stage. Further trimming can be carried out on smaller diamond saws, which are capable of producing thin slices. These saws normally use an oil/water emulsion as lubricant.

One or more faces are then lapped smooth either by hand or by an automatic lapping machine, using fine carborundum (silicon carbide) powder (e.g. 600F grade). Very coarse grits, e.g. 120F, should be avoided, because of their plucking and shattering effects, and their tendency to erode specimen edges preferentially, making it difficult to obtain true, flat faces.

At this stage, faces are suitable for general study by low power microscopy or for peel preparation (see Section 4.6), but better results will be obtained by washing and taking through 800F and even 1000F abrasives. For even better finishes, lapping is continued with silica gel suspensions, Aloxite or fine diamond pastes to a good surface polish. The secret of good, smooth surface finishes, particularly with polishing by hand, is careful washing and complete removal of abrasive from both specimen and lap before moving on to finer grades. It is also vital to keep all abrasives dust-free, and if possible to work in a dust-free environment.

It is best to avoid cutting soft or friable rocks such as shales, evaporites, coals and chalks on a lubricated saw. With evaporites and coals, cutting usually produces bad fracturing, and for all three rock-types the effect of aqueous lubricants may be disastrous. Slices are produced dry as in the early stages of thin section making (Section 4.4.2), with a glass backing. They are left a few millimetres thick and lapped flat or polished, using a lapping compound with paraffin oil lubricant.

Poorly consolidated sediments, or those for which porosity information is required, must first be vacuum-imbedded in resin (see Section 4.5). The resin may be stained to show the void space.

4.3.2 **Examination of cut faces**

Most of the techniques used to develop detail on cut faces of slices or plaquettes are non-destructive, so that a different technique can be used on the same face after washing or re-polishing.

SANDSTONES AND SILTSTONES

A light coating of clear mineral oil, even on polished surfaces, enhances contrast and increases resolution by filling surface irregularities. If the surface is to be stained or etched (see Section 4.5), glycerol may be used, as even traces of oil will inhibit both these processes, but glycerol can be washed away with water. With the aid of a zoom binocular microscope, small-scale sedimentary structures, such as graded bedding, can be viewed. The depth of field at low power (×1.5 to ×20 magnification) allows study of three-dimensional grain relationships so that packing and pore networks can be visualized, although fracture surfaces need to be examined for comparison.

Slices are also suitable for X-radiography. Slices from 3 to 10 mm thick give good results because 'soft' (5–50 kV) X-rays can be used for high contrast. This technique is particularly valuable for fine-grained clastic rocks such as siltstones, revealing otherwise unseen small-scale sedimentary structures. Very thin slices can be used to make radiomicrographs for examination on a grain-to-grain basis. A discussion of these X-radiographic techniques is given by Hamblin (1971).

CARBONATE ROCKS

Because of the inherent translucency of their cementing mineral crystals, many carbonates display their features well with low power biocular microscopy. Light coating of mineral oil (or glycerol if the faces are to be etched or stained later) can be used to enhance detail and contrast (Fig. 4.2a). However, highly polished surfaces are easier to photograph than oiled ones, as reflections are more easily controlled. Slices with two prepared faces give a three-dimensional context which aids identification of bioclasts and study of depositional and diagenetic fabrics. The partial transparency of the rock allows focussing up and down to assess grain packing. The determination of whether a fabric is grain- or mud-supported otherwise takes considerable experience to determine from thin sections, which are essentially two-dimensional.

Chalk fabrics are usually cryptic or apparently homogeneous when sliced normal to bedding. Bromley (1981) has shown how abundant biogenic and inorganic sedimentary textures can be revealed by lightly brushing thin mineral oil on a dry chalk surface and allowing it to 'develop' over about 30 minutes. Further enhancement of detail can be obtained from these surfaces by photographing them with Kodalith® or equivalent high-contrast orthochromatic line film.

Argillaceous, dark, fine-grained carbonates may also have cryptic fabrics. A technique used for revealing flaws in metal casings is often helpful in such cases. It relies on the preferential absorption on to clay minerals of an ultra-violet sensitive dye. There are three reagents, conveniently supplied in aerosol cans. A pre-smoothed face is gently etched with 5% hydrochloric acid. Zyglo® Penetrant is sprayed on the surface until completely wet, and allowed to soak in for at least 10 minutes. The surface is then flushed with Solvent spray, allowed to dry and then sprayed with Developer. All these operations should be carried out in a fume cupboard as the vapours are harmful, and surgical gloves should be used to protect the hands from the dye. Viewed under a low energy ultra-violet lamp, structures such as cross-lamination and bioturbation fluoresce bright blue and yellow.

Bockelie (1973), also working with cryptic argillaceous limestones, demonstrated how effectively a simpler technique combining light acid etching with photography using orthochromatic film could enhance fabric contrast and reveal burrow patterns.

Micritic carbonates such as those from mud build-ups and some calcretes may also have cryptic textures which can be revealed by gentle etching with a

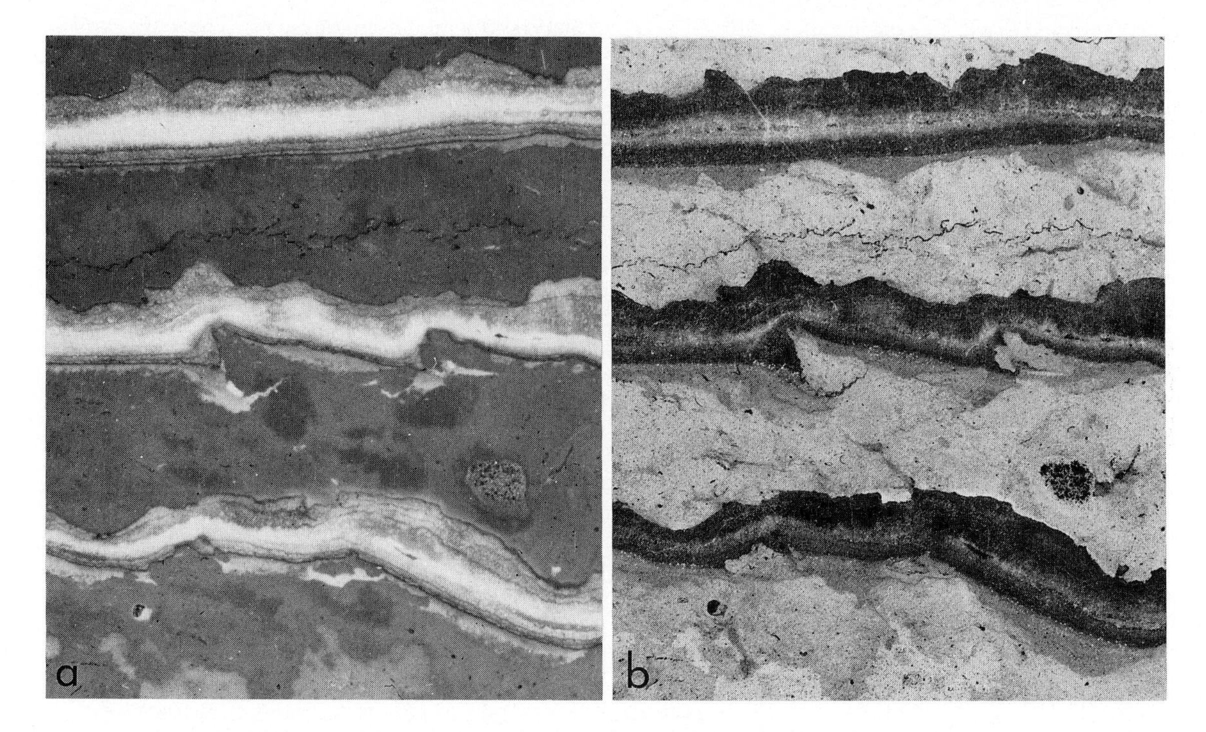

Fig. 4.2. Cut and smoothed face of Waulsortian facies limestone, Dinantian, Ireland, showing 'sheet spars' with fibrous calcite and interlayered carbonate mud. Approximately ×2. (a) Smoothed face photographed after coating with mineral oil. (b) Face after etching in dilute hydrochloric acid. Note the fine detail now revealed in the carbonate mud, and the differentiation of coarser geopetal floors in the spar-filled cavities. Photographs courtesy of A. Lees.

dilute acid (e.g. 5% hydrochloric acid). Geopetal sediments in stromatactoid cavities, for example, may be difficult or impossible to see on unetched, polished surfaces (Fig. 4.2a), but are clearly seen after etching (Fig. 4.2b), although other features are better revealed on the polished surface. The techniques of polishing and etching are thus complimentary.

Smoothed or polished slices can also be observed directly under cathodoluminescence (see Chapter 6); they may be etched or stained for this purpose.

OTHER ROCKS

With translucent or transparent rocks, the depth of field of low power zoom binocular microscopes can be used advantageously for three-dimensional study of included bioclasts and replacement fabrics in cherts, and orientation of fluid and crystal inclusions in evaporites. Various combinations of diffused transmitted light and bright incident light (delivered via fibre-optic tubes) can be tried. Contrast can be optimized and specific features emphasized by experimenting with coloured filters on the light sources.

4.4 THIN SECTION PREPARATION

4.4.1 Requirements for thin sections

An acceptable rock thin section is of 30 μm nominal thickness, bonded to a glass slide. Such standard thin sections (STs) can be produced by hand or with machine aid: they will usually be used for rapid bulk sample examination or in teaching collections where robustness outweighs their limited resolution. However, much more information can be obtained from sections made to a higher specification, with (1) both rock surfaces ground optically flat and polished, and (2) special care taken to produce an extremely thin and uniform bonding film between glass and slice. These double-polished thin sections (DPTs) can only be satisfactorily produced by machine; they are the best sections for research work and critical petrographic studies.

Recent advances in the technology of section-making show clearly that the common practice of using coarse abrasives results in unacceptable damage to samples. Diamond-impregnated buffing wheels, commonly used for rapid preparation of surfaces after cutting, cause just as much damage, shattering minerals and distorting grain/cement relationships. This stricture about severe abrasion applies both to ST and DPT preparation.

In the following account, details are given for preparation of high quality thin sections by automatic and semi-automatic means. The extra expense of making these sections is offset by their versatility: an uncovered DPT can be used for several different analytical purposes, and offers much more petrographic information. Machine-based systems give a considerable increase in throughput compared with hand preparation. A flow-chart for the process (Fig. 4.3) shows stages at which section-making can be simplified for less exacting requirements, or where special techniques can be followed for difficult rock types.

4.4.2 **High quality section-making process**

PREPARATION

Unconsolidated sediments, highly porous and friable rocks, as well as chips and cuttings, must be at least partially embedded in resin before slicing (also

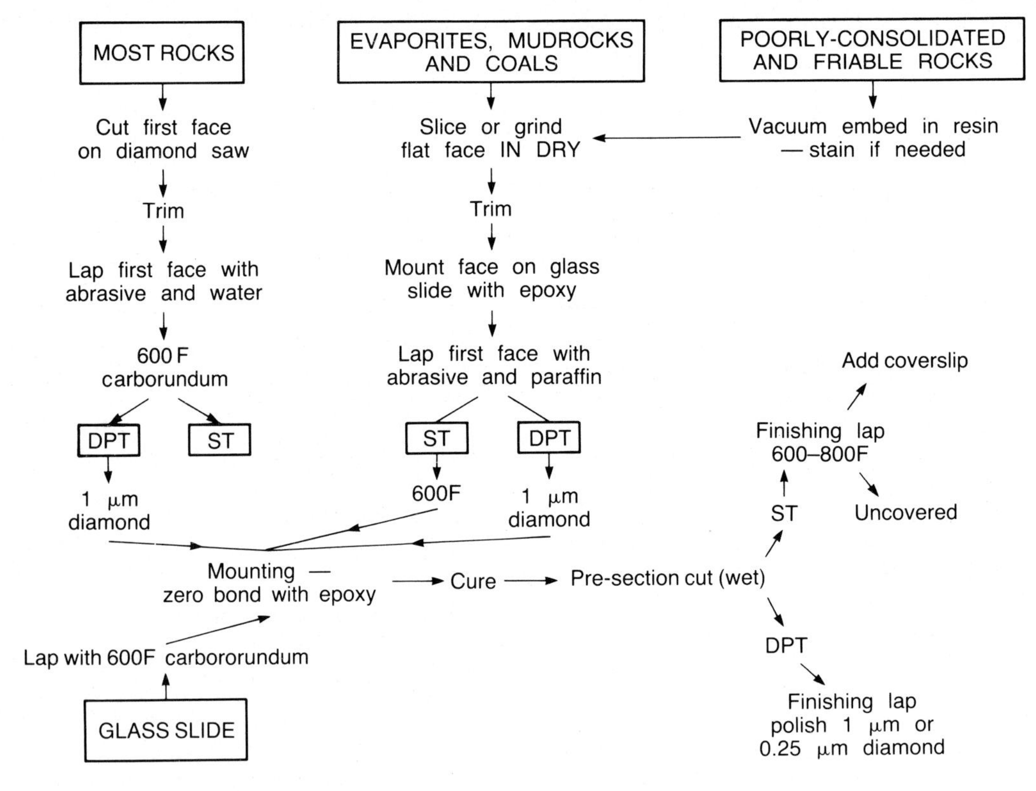

Fig. 4.3. Flow chart showing sequence of operations in production of standard thin sections (ST) and double- polished thin sections (DPT) for different types of rock samples.

see Section 8.5.5). If the sediments are wet, a water-miscible acrylic or polyester resin can be used. Otherwise, the specimen can be dehydrated by passage through successively more concentrated ethanol/water mixtures, through absolute alcohol into acetone, before placing in resin. If oil is present, remove as described in Section 8.5.1. Organic matter can be removed with chlorox or hydrogen peroxide. Poorly consolidated sediments and porous sandstones may require impregnation with resin under vacuum. Impregnation under pressure has been advocated, but comparative tests have shown that it may damage pores in sandstones, particularly those lined with delicate clay minerals. Vacuum (strictly, low pressure: 10–15 mm Hg) impregnation has been found consistent and satisfactory for most sediment types. Cold setting resins such as 'Epo-tek' are convenient for impregnation as they have a low initial viscosity and thus penetrate very well. It is also possible to dilute some Araldite resins with toluene for pre-casting soak. Depending on the nature of the sample, a degree of experimentation with resins of different types may be necessary before

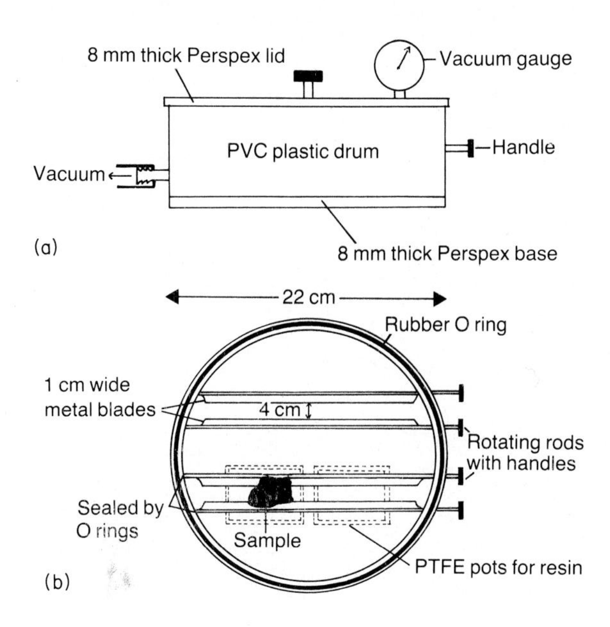

Fig. 4.4. A chamber for resin impregnation of porous or friable rock samples prior to slicing and thin section preparation. (a) Side elevation. (b) Plan view. Depending on the diameter of the chamber, one or more sets of rotating spindles can be used, according to the required throughout of samples. For procedure, see text.

satisfactory results are achieved. A simple method is described in Section 8.5.5, using a desiccator and mains water vacuum pump. The more elaborate chamber illustrated in Fig. 4.4, which can be constructed quite cheaply, is designed to overcome some problems in simpler chambers, where either the resin is drawn into the chamber after the sample is outgassed, or the sample is placed in the resin and both are evacuated together (as in Section 8.5.5). In the first case, the resin itself does not outgas thoroughly before coming into contact with the specimen, so gas and resin are both drawn into the pores; resin also solidifies in the inlet line. In the second case, frothing may be produced as both the sample and the resin outgas together. The apparatus shown in Fig. 4.4 is loaded with PTFE plastic pots, previously sprayed with a silicone release agent (e.g. silicone household polish) and containing the resin mixture. Samples, trimmed to suitable size, are suspended above the pots on the blades of the cross-rails. Evacuation is continued until the resin ceases to froth, when the samples will also have been outgassed. The cross-rails are then rotated so as to tip the samples into the resin pots. Samples can be left in the impregnation chamber overnight before curing under gentle heat. At the end of the impregnation period, care should be taken to release the vacuum slowly so as not to cause boiling.

Following impregnation, small samples may have to be cast into blocks which are more convenient for the sawing or for hand-holding while grinding a face on abrasive paper. Small moulds for casting blocks can be made from aluminium foil.

Estimation of porosity in rocks is aided by impregnation with stained resin. Special resin-miscible dyes are required, such as Epo-tek Blue Keystone or Waxoline Blue, mixed in a 10% dye:epoxy ratio. Blue resin areas in the final thin section depict the porosity clearly.

Generally speaking, the machine system of section-making is so gentle that only very difficult lithologies and loose sediments will need impregnation. The following account details the section-making process for standard rocks, with special procedures for friable or soluble lithologies described in Section 4.4.3.

CUTTING AND TRIMMING

Several sizes of thin sections can be made, depending on equipment available and the area required for examination. Useful and fairly standard sizes are 25 × 76 mm, 28 × 48 mm (best for electron microprobe and cathodoluminescence work) and 110 × 75 mm (good for fabric studies and coarse sediments). The selected area of a faced sample or hand specimen must be marked with water insoluble felt tip pen before cutting. Slices from 4 to 6 mm thickness are cut on a saw with a diamond impregnated wheel (Fig. 4.5a), usually with water or an oil/water mix as a lubricant.

Cutting should be done gently (slowly) to minimize shattering, especially with crystalline or well-jointed rocks. Several slices or plaquettes may be cut, which can be examined separately (see Section 4.2) or used as a back-up, should there be a failure at some subsequent stage in the section-making process. These relatively thick slices give good mechanical rigidity and wide zones of waste for first face lapping and final finishing, so that the completed section is made from that part of the slice unaffected

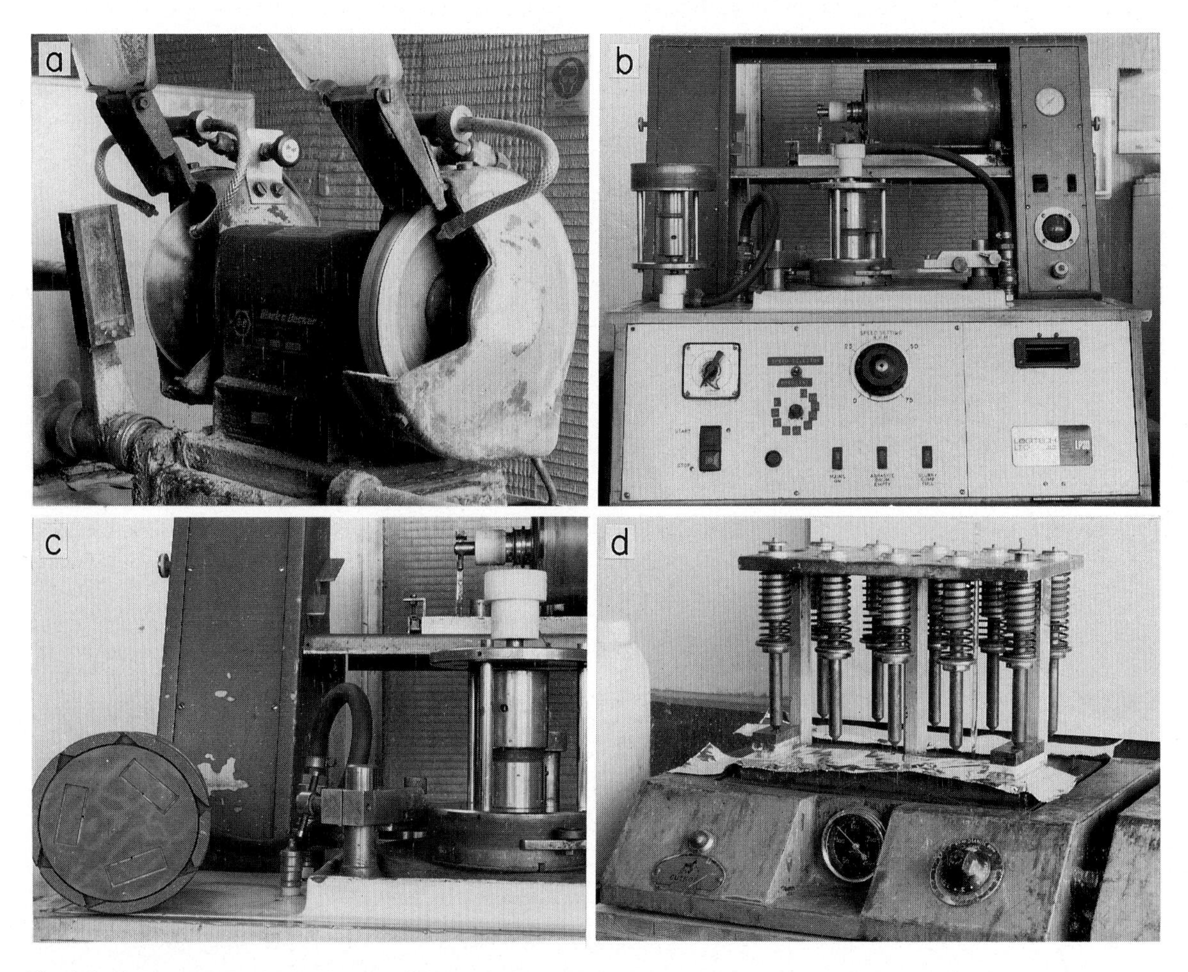

Fig. 4.5. Equipment used in semi-automatic preparation of thin sections. (a) Bench-mounted diamond-impregnated saw with diamond-impregnated buffing wheel (to right) for rapid sample trimming. Note safety guards. (b) Precision flat lapping machine (Logitech Ltd), showing vacuum chuck positioned on lap. (c) Close up view of sole face of vacuum chuck for precision flat lapping machine. Note the slots for slide mounting, and the diamond impregnated rim which helps to condition the lap. (d) Zero-bonding jig for applying controlled pressure to rock slice and section glass while resin cures under gentle heat from hot-plate below.

by saw damage. The slices are trimmed to avoid rock overhanging the slide. This discourages mounting adhesive from oozing on to the lower surface of the preparation, where it would affect attachment on the lapping machine's vacuum chucks.

Each slice is marked in waterproof felt-tip pen with the corresponding sample number and any 'way up' information.

FIRST FACE LAP

The key to success of the whole process lies in the production of a first face which is as smooth and flat as possible. This is accomplished on a precision free abrasive flat lapping machine (Fig. 4.5b). The slices are mounted on chucks with a sponge backing to accommodate different thicknesses and irregular specimens (Fig. 4.6). Completely omitting coarse abrasives, a face of the slice is ground with 600F Carborundum powder with water as a lubricant, for a minimum of 30 minutes (regardless of rock type), until the face approaches optical flatness.

At this stage, slices destined to become Standard Thin sections (STs) can be removed here for resin bonding to glass. Otherwise, they are lapped for a further period using 8 μm diamond paste if required for Double Polished Thin sections (DPTs) (Fig. 4.3). This extra lapping stage produces a polished lower surface, giving increased resolution in microscopy by reducing dispersion at the first face/resin junction. The face has enough 'key' to ensure bonding with resin. Higher degrees of polish produced by lapping with finer diamond pastes give no noticeable improvement in optical properties and can lead to adhesion problems, notably to lifting of the thin slice at later preparation stages. With a properly maintained precision lapping machine, it should easily be possible to maintain a flatness of at least 2 μm over

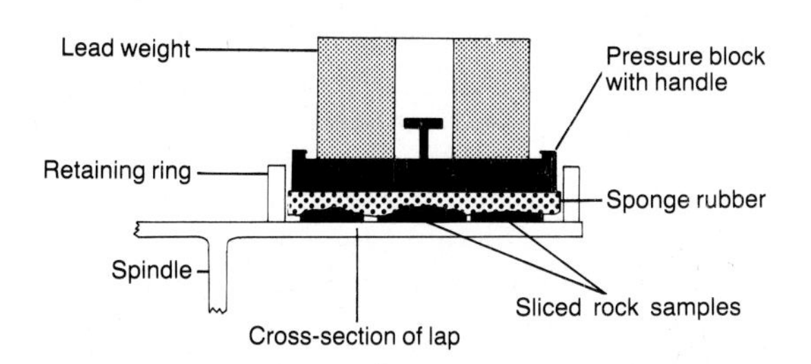

Fig. 4.6. Cross-section of chuck for holding sample on precision flat grinder when making the first face.

100 mm. Depending on slide size, up to about 24 slices can be flattened in one timed half hour run.

In all cases, the lap is run at a slow speed, 25–30 rpm, as opposed to the greater speeds recommended by the lapping machine manufacturers. This gives more control over the process, with the advantage of keeping the sections cool, further reducing damage and enabling the preparations to be used for fluid inclusion work.

GLASS SLIDE PREPARATION

For the precision machine finishing process, the glass slides must be of a precise and standard thickness. Commercial microscope slides rarely meet this requirement and so are first ground on the flat lapping machine as described above for rock slices. For STs, polishing with 1 μm diamond paste is usual. With DPT preparations, the 600F ground glass surface is used without final polish, as a slightly rough zone is required as a 'key' to ensure adhesion at the glass/resin interface.

BONDING SLICE TO GLASS SLIDE

Glass and rock surfaces must be cleaned of all abrasives, freed of grease by rinsing in acetone or petroleum ether, and dried. The prime requirement then is to produce an extremely thin but uniform glass/rock bond ('zero bond') with an adhesive whose refractive index is close to that of glass. It is vital to produce consistent and minimum cement thickness, otherwise there will be unpredictable variations in the thickness of the finished section.

Canada Balsam and Lakeside 70C have long been used as slide mountants, but they cannot produce a reliable bond for machine finishing, particularly with polished first faces. Moreover, they have poor stability, becoming brittle or discoloured with age. Being thermolabile, these resins are also unsuitable for sections to be studied in cathodoluminescence or electron microprobe equipment. Resins with all the required characteristics can be found in the Epoxy family. Allman & Lawrence (1972) give a useful discussion of the properties of Epoxy resins for geological use. Araldite MY750 mixed with Versamid polyamide hardener (1:1) is an excellent all-round bonding resin with good high temperature stability and the correct refractive index. If sections have been pre-impregnated in a cold-setting resin such as

Epo-tek, then they should be bonded with the same material.

Glass sides are first marked with the specimen number of the sample and any 'way up' indications which have been preserved, using a diamond scribe near one edge of the obverse glass face.

A small amount of mixed low-viscosity resin/hardener is applied to the first face, which is mounted on the lapped surface of the slide. The assembly is then placed in a spring-loaded jig (Fig. 4.5d) under a pressure of about 1 kG cm^{-2}. At first, the section 'floats', but as the adhesive rapidly squeezes out, the rock locks on to the glass and no longer slips. Bonds made in this way are extremely transparent, bubble-free and introduce no dimensional errors into final lapping.

Mounts must be left to cure under pressure. With the resin combination used above, the jig is placed on a hotplate and curing carried out at 80°C for about 15 minutes. For hand-prepared specimens, spring-loaded clothes pegs can be used to provide even pressure, and these are also very useful for large sections, where four or more pegs can produce a more even bond than the standard jig.

PRE-SECTION CUTTING

Bonded sections are next cut with a diamond blade to form a parallel second face, leaving a slice 300–400 μm thick on the slide. This initial thickness allows the final 30 μm section to be entirely in the undamaged layer. Cutting can be done automatically on a diamond-impregnated saw with a vacuum chuck, holding up to six slides simultaneously. Precision is not important here, as there is a machine finishing stage to follow.

FINISHING

At this stage, ST preparations are returned to the precision lapping machine for lapping to a final 30 μm thickness with 600F or 800F Carborundum. Thickness is automatically determined by pre-setting the vacuum slide-holding chuck (Fig. 4.5c); slides have reached the set thickness when they begin to 'float' on the lap, a process aided by the diamond-impregnated surround on the lower face of the chuck. Slide thickness can be checked with a micrometer or, less reliably, on a microscope, looking for the required interference colours, e.g. quartz greys. This optical method is of little use for limestones.

DPT polishing can be done on the same machine, using graduated diamond pastes, but in the Edinburgh workshop we have found that better results are obtained on a separate polishing machine. This also has the advantage of freeing the precision lapping machine for preparing successive batches of slides, as the polishing stage can be protracted. The polishing machine consists of a simple rotary polishing lap running at up to 30 rpm, on which is placed a precision lapping chuck (Fig. 4.7a) with several slides held by vacuum suction (Fig. 4.7b). Polishing is carried out on adhesive-backed paper discs (Engis papers) attached to the lap. It begins with a charge of 8 μm diamond slurry such as Hyprez compound. This does not need replenishing unless the surface is very hard (quartzite or chert).

Any faults in previous preparation stages will show up here: if the section is not flat, because of a poor first face or incorrect bonding, polishing will follow the irregularity and a wedged slice with eroded edges will result.

Lapping is continued until the section reaches 30–35 μm (about 10 minutes for soft rocks, up to 30 minutes for harder ones). As this is the stage at which material is actively removed, the sections should be watched carefully during this process. After washing and change of lap paper, polishing is continued with 1 μm diamond. Little or no material is removed during the final stages; the surface is being improved and so time is not critical. For critical work, final lapping is done with 0.25 μm diamond paste. This is needed for electron microprobe or back-scattered Scanning Electron Microscope work where a perfect surface is required. Experiments have shown that the most flawless surfaces are produced by carrying out the final lapping stages on a lead lap. This consists of a very flat sheet of metallic lead, about 8 mm thickness, bonded to a standard steel lap disc. The lead is prepared by engraving a spiral groove several millimetres deep, pitched so that the lead strip is narrower than the groove. Sections are lapped using paraffin oil as a lubricant. No material is removed, but the surface is perfected, so lapping time is not critical. This method can only be used for non-porous and crystalline rocks (particularly limestones), as the lead tends to be forced into any pores and, unlike powdered abrasives, it cannot be washed away. Because of its very soft surface, trueing and conditioning of the lead lap must be done frequently, and care taken to exploit the whole surface to avoid worn spots and dishing.

Fig. 4.7. (a) Polishing lap for final finishing of thin sections, showing a vacuum chuck and a weighted chuck in position. (b) Sole face of vacuum chucks showing how various sizes and types of thin sections can be polished simultaneously.

Ultra-thin sections are valuable for some purposes, particularly for very fine grained rocks, if it is necessary to measure grain size accurately (Halley, 1978) or to study fine structures such as algal filaments. Such sections are produced by extending the 1 μm diamond finishing stage, at which material is slowly removed and thus more easily controlled. Micrometer measurements must be made frequently, as it is very easy to polish the section away completely if the rock is soft. Optical methods of checking thickness by examining the interference colours will not work for these very thin slices, which produce very bright, high order colours.

COVERING

For research purposes it is best to leave thin sections uncovered, as they may then be subjected to a variety of treatments including staining, etching, cathodoluminescence and microprobe work. Coverslips are, of course, required for thin sections destined for teaching collections. DPT preparations need no oil for microscopic examination, but uncovered ST surfaces, being finished only to 600F or 800F Carborundum, require oil or glycerine before microscopic study.

All sections should be cleaned of abrasive residue before examination. This may be done by wiping DPTs with a soft, damp tissue, or by brief immersion in an ultrasonic bath. ST preparations may be left uncovered but sprayed with a polyurethane resin to give a protected surface for general microscopy (Moussa, 1976, 1978). The best medium for attaching standard glass coverslips is Canada Balsam. This has the required refractive index and the advantage that the slip can be removed if needed, after gentle heating on a hotplate. Covered slides should be cured in a warm place for at least 24 hours before use.

LABELLING

Standard thin sections required for teaching collections or archive purposes can now be given self-adhesive labels on which can be written all the details of the sample. However, paper labels should not be used for those sections destined for use in cathodoluminescence and scanning electron microscopes or microprobe, as they tend to char in the beam and may interfere with conductive coatings.

4.4.3 Variations in process for 'difficult' rock types

Certain rocks, such as shales, evaporites, coals, soft chalks and poorly-consolidated sediments require special treatment at various stages of the section-making procedure. Initial impregnation and embedding, if required, will have been completed as detailed above. With such difficult material, cutting on the usual diamond trimming saw is inadvisable, and water-based lubricants cannot be used.

Flat faces are produced by hand grinding on carborundum-impregnated papers ('wet and dry' paper) in the dry. With harder samples, a hacksaw will produce a face ready for lapping. The surface need not be perfectly flat or smooth. Evaporites can then be taken directly on to the precision grinder for first face lapping using paraffin oil as lubricant, but other rocks need a further stage. This involves mounting the ground face on to a glass slide of appropriate size, using epoxy resin: again this can be done quite crudely. Using a dry trimming saw (with the operator wearing a mask), excess rock is removed. This sawn surface will become the first face of the section, and it is lapped with paraffin oil and 600F Carborundum on the precision lapping machine. After the best flat surface is obtained, the paraffin oil is removed by placing the preparation on a warm plate for some hours. The sections are then zero bonded with epoxy as described in Section 4.4.2, to form a double glass mount with a few millimetres of rock sandwiched between.

If the first face is not satisfactory at this stage (wedged or with air bubbles), then one of the glass slides can be cut off and the section re-lapped and re-mounted. Otherwise, the sandwich is placed on the saw and the first glass slide cut off. The section is then returned to the Precision Lapping Machine with paraffin oil and 600F grit and taken down to 100 μm thickness, or 30 μm if it is to be covered. For best results it is transferred, after rinsing in paraffin or kerosene (inflammable!), to the polishing lap for final surfacing as for DPTs (see Section 4.4.2). Shale sections benefit from a long period with 0.25 μm powder, as this gives very slow removal of material without plucking of quartz grains. Sections of evaporites are either glass covered or stored in a desiccator.

4.5 ETCHING AND STAINING

4.5.1 Etching

Etching involves selective partial dissolution of a cut face. It may be done to emphasize certain textural characteristics or to prepare a surface for staining (see Section 4.52). Hydrofluoric acid is the etching agent for silicate-rich rocks; carbonates can be etched with various weak acids. In all cases surfaces to be etched must be grease-free (no fingerprints!).

Hydrofluoric acid treatment

HF is a dangerous chemical with a poisonous and corrosive vapour. It attacks glass, metal and some plastics. *Severe burns result after skin contact even with dilute solutions; the burns may not appear until hours after exposure. Splashes on skin should be neutralized immediately with a sodium bicarbonate solution, washed with copious amounts of water and treated with special HF Burn jelly, which should be available in all laboratories where HF is used. All operations involving HF should be carried out in a fume cupboard, and the operator protected with gloves and goggles. Surplus and discarded acid should be neutralized with alkali before disposal.*

Etching vessels should be shallow, tightly-closing and made of polythene or similar soft plastic. Small plastic food containers are suitable, with an improvised plastic sling-like support for plaquettes or thin sections suspended from the lid (Fig. 4.8). Exposed surfaces of glass slides, especially the undersides, should be painted with paraffin wax before placing in the container. This prevents HF from frosting the slide and reducing its transmissivity. The HF needs to be a strong solution (52–55%), with the vessel filled to a depth of about 5–6 mm. Etching times in HF vapour depend on the nature of the rock: feldspars, for example, require about 30 minutes etch before they are to be stained. Stubborn cases can be etched directly in the HF solution. Samples should be inserted and removed from the apparatus using plastic or rubber gloves, and care taken not to breath the vapour. Copious washing under clean tap water is required before the samples can be treated further, and wax scraped from slides. The latter process is

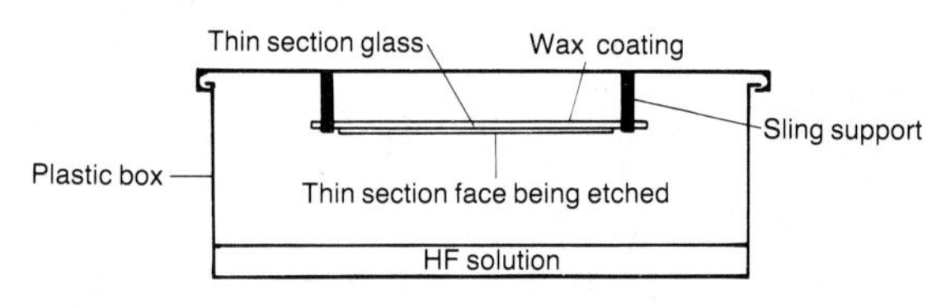

Fig. 4.8. Simple apparatus for etching thin sectons or small slices with HF vapour — *to be used in a fume cupboard only.*

made easier by placing the slide in a refrigerator for a few minutes.

ETCHING LIMESTONES

Several reagents are available for etching carbonate minerals. Hydrochloric acid (1–5% by volume) gives a vigorous reaction with a tendency to give a rather unselective etch. It affects minerals other than calcite and aragonite, especially if the solution is warm. Acetic acid (about 20% by volume) gives a more selective etch, bringing out details of grain textures and fabrics more subtly than HCl. Disodium ethyl diaminotetracetic acid (diNaEDTA) is an even more subtle etching agent: it has a chelating (complexing) action with comparatively little effervescence, thus reducing gas bubble damage of delicate structures. Miller & Clarkson (1980) used saturated EDTA solutions to reveal ultra-structure in calcitic trilobite cuticles.

Etching is simply carried out by placing the slice or thin section face up in the chosen etching solution. If the specimen is immersed face-down, bubbles cannot be released and remain attracted by surface tension to the surface, locally impeding further reaction and giving a very uneven etch.

The rate of etching varies according to the nature and strength of the solution, its temperature, and the grain size, crystallinity and mineral composition of the substrate. Etching times may vary from less than a minute to over half an hour. The optimum time must be determined by trial and error. Gentle agitation of the acid, and slight tilting of the face prevents the formation of bubble trains which trench the surface. For critical applications, the progress of etching can be observed with a binocular microscope (with a clear plastic bag tied over the optics to prevent corrosion from acid spray). This works best with the slower and more easily-controlled action of diNaEDTA, and is ideal for studying the location of delicate organic matter, such as shell membranes or algal and fungal filaments. Etching a thin section in a 10% solution of NH_4Cl for 2 minutes is also good for revealing filaments, which are then best seen in a combination of reflected and transmitted light.

Thin sections for etching should be left a little thicker than usual, after being taken down to at least the 600F Carborundum stage. In some instances, the complete removal of carbonate from a thin section can provide additional information (Lees, 1958). This leaves the insoluble components available for observation in a dry state by reflected light or transmitted light with crossed nicols. For this technique, the lower sample face is etched with 30% acetic acid for 5 min before rinsing, drying and bonding to the glass slide. This preliminary etch ensures that a replica is formed in resin of the carbonate phases, and that the non-carbonate components adhere to the preparation. The 30 μm section is then immersed in 30% acetic until all the carbonate is dissolved. The replica shows details to at least 5 μm resolution.

4.5.2 **Staining**

While the mineralogical composition of rocks can be determined by optical study of thin sections, this can be a tedious, time consuming and error prone process. For example, distinction between calcite and dolomite is difficult because they have similar optical properties. For similar reasons, small untwinned feldspar grains and quartz grains cannot easily be distinguished. Chemical staining involves a reaction which produces a coloured precipitate on a specific mineral surface and therefore makes that mineral more easily recognized. Usually the surface requires some preparation, often etching, to receive and retain the precipitate.

The success of staining slices and thin sections depends, as does etching, on various factors: strength and 'age' of reagents, temperature, pre-treatment, grain size, orientation, fabric, grain interactions, cleavage and surface finish. To achieve consistently good results with any staining procedure requires a considerable degree of skill. Much patient experimentation may be required. Staining times in particular are highly variable, and those given herein should only be used as a starting point.

Detailed guides for staining are given by Allman & Lawrence (1972) and Friedman (1971). In this section only the most successful techniques for common sedimentary minerals are given, with an emphasis on those which have been recently refined.

FELDSPARS

Houghton (1980) gave a method for staining plagioclase and alkali feldspars which is much more reliable than previous recipes.

Reagents: (a) Potassium rhodizonate (0.01 g K-rhodizonate in 30 ml distilled water, filtered before use). Maximum useful life one hour.

(b) Sodium cobaltinitrate (saturated solution; about 50 g per 100 ml distilled water). Six months shelf life in dark bottle.
(c) Barium chloride (5% solution in distilled water). Stable.
Procedure: (a) Etch over HF vapour (55% HF solution, see Section 4.5.1) for 25–35 s: increase time if HF is loosing potency.
(b) Remove slide from etching box with polythene forceps and drop into beaker with Na-cobaltinitrate solution. Leave for 45 s.
(c) Rinse slide twice in distilled or de-ionized water. Gently shake off excess water and blot end of slide by touching to a paper towel.
(d) Dip slide quickly into beaker with $BaCl_2$ solution for no more than 2 s.
(e) Dip slide immediately in distilled water and agitate for about 10 s.
(f) Place several drops of the rhodizonate solution on the wet surface of the section, tilting back and forth to spread the stain; leave until plagioclase grains become pink, then rinse in water.
(g) Dry with compressed air and examine with a microscope. If plagioclase is light grey or pale pink, dip in distilled water and repeat rhodizonate stage. The intensity of the pink plagioclase stain is proportional to the amount of calcium in the molecule: albite/oligoclase will stain lighter than a more calcic plagioclase. Pure Na-albite will not take up any of the rhodizonate stain. Alkali feldspars are stained yellow. The accuracy of the method decreases according to grain size: with finer grained specimens, the pink stain tends to pervade the surface and obscure the quartz grains.

Cathodoluminescence is a much better way of detecting feldspars in fine sands and siltstones (see Chapter 6).

GYPSUM AND ANHYDRITE

Hounslow (1979) gave a method for staining these minerals, useful in the field as well as on drill cores, soil samples, slabs and sections.
Reagents: (a) 1:1 nitric acid solution (10 ml concentrated nitric acid slowly added to 10 ml water in a 50 ml beaker).
(b) 10 g mercuric nitrate in 100 ml de-ionized water. Stir until dissolved. A fine, milky precipitate will form. *Caution!* Mercuric nitrate is extremely poisonous and may be absorbed through the skin.
(c) Slowly add the 1:1 nitric acid to the mercuric nitrate with constant stirring until the milky precipitate just dissolves or is just about to — approximately 8 drops of acid is usually sufficient.
(d) Test the reagent by adding a few drops to small crystals of Na_2SO_4 or K_2SO_4 in a watch glass. A yellow precipitate should form immediately. If not, the reagent is too acid and a further gram of mercuric nitrate added to produce a slight cloudiness should produce the yellow precipitate.
(e) Filter the solution when the test is successful and store in a dark bottle.
Procedure: (a) Immerse the section, slab or plaquette in the reagent contained in a suitable dish or tray for a few seconds or until the yellow precipitate forms.
(b) Gently rinse the stained face in de-ionized or distilled water and allow to dry before microscopic examination.
Both minerals are stained yellow; other minerals are unstained.

ARAGONITE

The most sensitive stain for aragonite was developed by Fiegl (1937). Despite its sensitivity, it is not always reliable.
Reagents: (a) 1 g commercial grade Ag_2SO_4 is added to a boiling solution of 11.8 g $MnSO_4.7H_20$ in 100 ml distilled water.
(b) Cool and filter the suspension; add 2 drops 10% sodium hydroxide solution.
(c) Stand for 2 hours and then filter into a dark storage bottle.
Procedure: Immerse the plaquette or thin section in the Fiegl's solution for about 10 min at 20°C. Sometimes a hot solution works better.
Aragonite is stained black while calcite remains unstained.

DOLOMITE

Dolomite is often distinguished from calcite by its failure to stain with solutions which react with calcite (see below). It can also be recognized after etching, especially on a universal stage: dolomite remains at 30 µm, giving fourth- and fifth-order pinks and greens, while calcite is reduced to 10 µm or so, giving second- and third-order reds, blues and yellows (Dickson, 1966, p. 491). Titan yellow in alcohol and alkaline Alizarin red S are two reliable specific stains for dolomite, but have the disadvantage for thin sections of needing to boil.

Titan yellow — reagent: (a) 0.2 g Titan yellow is dissolved in 25 ml methanol in a small beaker over a hot water bath. Replace any alcohol lost by evaporation.
(b) 30 g sodium hydroxide pellets dissolved cautiously in 70 ml distilled water.

Titan yellow — procedure: (a) Add 15 ml of the 30% NaOH solution to the Titan yellow solution and bring to the boil.
(b) Immerse the thin section or plaquette in boiling Titan yellow reagent for at least 5 min until a deep orange-red colouration is formed. Add a few drops of methanol to compensate for evaporation.

Alkaline Alizarin red S — reagent: (a) Dissolve 0.2 g Alizarin red S in 25 ml methanol in a small beaker over a hot water bath.
(b) Carefully dissolve 30 g sodium hydroxide pellets in 70 ml water.

Alkaline Alizarin red — procedure: (a) Add 15 ml of the 30% NaOH solution to the Alizarin red solution and bring to the boil.
(b) Immerse the section or plaquette in boiling reagent for at least five minutes. Add a few drops of methanol to compensate for evaporation.
(c) Rinse the section in distilled water.
Dolomite is stained purple, calcite is unstained in the alkaline solution.

MG-CALCITE

Choquette & Trusell (1978) devised a method of making an alkaline Titan yellow stain permanent, enhancing the value of this technique. The stain can detect the presence of Mg-calcite containing more than about 3% $MgCO_3$.
Reagents: (a) Stain: 1.0 g Titan yellow powder, 8.0 g NaOH and 4.0 g diNaEDTA dissolved in 1 litre distilled water at room temperature. Store in dark bottle; shelf life at least two years.
(b) Fixer: 200 g NaOH pellets dissolved slowly in 1.0 litre distilled water. *Caution!* heat and fumes and evolved, so best carried out in fume cupboard. Solution is corrosive to flesh. Store in polythene bottles as 5 M caustic soda etches glass. Shelf life indefinite.
Procedure (*Caution!* use surgical gloves as solutions are corrosive): (a) Thin sections must be epoxy mounted: Lakeside and Canada Balsam are soluble in the stain solution.
(b) Etch uncovered, grease-free thin section or faced sample for about 30 s in 5% acetic acid solution.
(c) Dry surface in stream of warm air.
(d) Immerse specimen in stain solution for about 20 min.
(e) Dry stained surface in stream of warm air; do not touch surface in any way.
(f) Immerse stained surface in fixer solution for about 30 s.
(g) Air dry again and cover. As the residual caustic would attack Lakeside or Canada Balsam, epoxy cement is required for cover slips.
Choquette & Trusell (1978) found that the intensity of the stain was proportional to the Mg content of calcite. Calcite with 5–8% $MgCO_3$ takes on a pink to pale red colour, and 'high-Mg' calcite takes on a deep red colour. *c*-axis normal sections of crystals stain more vividly than parallel sections, as do very fine-grained substrates. This stain is very valuable for its minute selectivity on a scale discernible with a petrographic microscope.

CALCITE

Friedman (1959) gave a number of stains for calcite. One of the simplest and most reliable is Alizarin red S. Many carbonate workers routinely acid etch one-half of a thin section and stain it in acidified Alizarin red S solution before covering. While this helps to distinguish between red-stained calcite and unstained dolomite and quartz, it gives very little other information about diagenesis and mineralogy. It also makes microphotography of that slide difficult. A far more valuable scheme was given by Dickson (1965, 1966) which in one operation differentiates between ferroan phases in calcites and dolomites, and is capable of revealing subtle growth zones in cement crystals.

Dickson's method — reagents: (a) Etching solution — 15 ml of 36% HCl dissolved in 500 ml distilled water, then topped up to 1000 ml with distilled water.
(b) Staining solution part 1: 0.2 g Alizarin red S dissolved in 100 ml 1.5% HCl solution.
(c) Staining solution part 2: 2 g potassium ferricyanide crystals dissolved in 100 ml 1.5% HCl solution. This solution must be freshly made for each staining session.
(d) Mix staining solutions parts 1 and 2 in the ratio 3 of dye to 2 ferricyanide. The combined solution lasts for only one staining session.

Dickson's method — procedure (*Caution!* use rubber gloves): (a) Immerse thin section (grease free surface) or faced sample in etching solution, face uppermost, for 10–15 s at 20°C (times are only approximate, and may vary widely with grain size, etc.). Cold solutions give very poor results. Experimentation is necessary to achieve the optimum etch. Weak etching gives a thin, patchy stain. Over-etching, particularly with Alizarin red S, produces a very dense stain which tends to spread a precipitate even over dolomite and quartz if they are present.
(b) Immerse specimen in the combined solution for 30–45 s (again experimentation is required). For thin sections, best results are obtained if the stain is warmed. As HCN fumes can be evolved from the mixed stain, the safest way of doing this is to immerse the slide in a dish of hot water before placing in the bath of stain which is at room temperature. *Avoid skin contact with the stain: ferricyanide is poisonous*. Dickson suggested a further stage, briefly dipping the stained slide or surface into Alizarin red solution alone, but this is not often needed and indeed may detract from the first stain.
(c) Wash the specimens gently in two changes of distilled water (not tap water, which contains Fe and Ca) for only a few seconds at a time — the stains are relatively soluble.
(d) Dry the specimen surface quickly in a stream of warm air (from a hair drier or equivalent). Handle carefully; the stain is a thin surface film and is easily damaged.
This method produces the following colour differentiation in carbonate minerals:

Calcite	Varying through very pale pink to red.
Ferroan calcite	Varying through mauve, purple to royal blue with increasing Fe content.
Dolomite	No colour.
Ferroan dolomite	Pale to deep turquoise with increasing Fe content.

Dickson (1966) pointed out some useful features of this stain coupling. Alizarin red S can differentiate slightly different types of calcite: for example, different kinds of bioclasts stain with differing intensities depending on their crystallite size and structure. The optic orientation of sparry cement crystals can be discerned, as sections normal to the *c*-axis are stained very pale pink, whereas sections parallel to that axis stain deep pink. This is due to different rates of etching. The potassium ferricyanide component is very sensitive, and will detect iron in calcite with 1% ferrous carbonate in solid solution.

Although it is convenient to stain sections by Dickson's method before microprobe analysis so that the mineral phases can easily be distinguished, Sommer (1975) showed that microprobe determinations of iron in carbonates stained with potassium ferricyanide showed depletion by an order of magnitude compared with unstained crystals.

CLAY MINERALS

Stains for particular clay minerals usually require powdered and acid-extracted samples (Allman & Lawrence, 1972, pp. 109–111) and are therefore beyond the scope of this chapter. It is often useful, however, to be able to visualize the distribution of clays in a sandstone or argillaceous carbonate, since clays are often related to otherwise obscure fabric in these rocks. The Zyglo® ultra-violet sensitive dye described in Section 4.3.2 is useful here. A simpler, but not always predictable, technique is to soak grease-free surfaces in aqueous solutions of 0.5 g malachite green, congo red or methylene blue in 250 ml water. The surface is lifted out of the solution occasionally and gently washed until the stain is found to be satisfactorily developed. The dye colour chosen should be complementary to the rock colour so as to provide the greatest degree of contrast. An alcoholic solution of dye (equal parts ethanol or methanol and water) of methylene blue works well on limestones, including chalks, with the alcohol acting as a wetting agent and aiding penetration.

POLYSACCHARIDE STAIN FOR BIOTURBATION

Burrowing in sandstones and siltstones may often be cryptic, and sometimes difficult to distinguish from water escape structures. Risk & Szczuczko (1977) developed a reliable method for enhancing the morphology of burrowing in siliciclastics. It is based on the tendency of many burrowing organisms to secrete polysaccharide mucus as a burrow lining. The presence of this carbohydrate can be detected by a periodic acid — Schiff (PAS) reaction. The method is not suitable for use with carbonates because of the HCl in the Schiff reagent.
Procedure: (a) Cut a face or slice using water only as a lubricant; avoid all grease and oil; do not touch the cut face with uncovered hands. Massive, unjointed

material is best in order to avoid carbohydrate contamination.

(b) Polish with 600F or 800F Carborundum, again using water only.

(c) Wash the slab thoroughly with tap water.

(d) Place the face downwards in 1% periodic acid (by weight) in distilled water. Support the slab in a dish with glass rods at the edges. Gently swirl the liquid. Leave for 30 min. For very large blocks, the solution can be painted on.

(e) Rinse thoroughly for 30 s in running water.

(f) Place face down in Schiff's reagent (commercially available) for 30 min, in the dark.

(g) Wash for several minutes in running water.

(h) Examine with binocular microscope. Store specimens in the dark, where they will remain unfaded for at least several years.

The PAS reaction is not porosity-controlled. Specimens which are bioturbated will have a light purple background with dark stains marking the burrows. Arthropod burrows do not stain as they do not normally have mucoid linings.

4.6 PEELS

Making a replica of an etched surface on transparent plastic film is a very rapid and cheap way of preparing a sample for microscopic examination. A solvent is flooded on to the prepared face. The lower side of the film is softened and, as the solvent evaporates, the film settles down into the irregularities of the etched face and produces the replica. There is very little sample wastage. Serial peels can easily be produced by re-grinding surfaces after each peel, to a fixed distance if required. This is much quicker than making serial thin sections for three-dimensional visualization of fabrics like void spaces or for reconstructing fossils in full relief.

Peels are mostly used for limestones and calcite-cemented clastic rocks, although cherts and siliceous clastics can be successfully treated. Best results are obtained with non-porous samples; surface cavities cause the film to bulge into the void, causing blisters and a risk of tearing the completed peel. Porosity can be reduced by vacuum impregnation with resin before facing the samples (the resin used must not be soluble in acetone when cured). Larger holes should be filled with decorator's filler paste (such as Polyfilla), which is allowed to harden before the face is rubbed down.

Replicas can be made of quite large surfaces, even up to hundreds of square centimetres in area, but the difficulty of producing good surfaces (and satisfactory solvent distribution cross them) increases rapidly with samples larger than about 0.5 m × 0.5 m.

Although simple in principle, manufacture of good peels is a craft learned only through considerable experience. Beginners should not be satisfied with their earliest efforts, but should persevere, using a single rock sample and experimenting with etching times, pre-polishing, staining etc., until high quality peels can be consistently obtained.

4.6.1 Peel material and solvents

Davies & Till (1968) advocated making peel sheets by pouring solutions of ethyl cellulose in trichloroethylene on to glass plates. While very thin sheets can be prepared in this way, they are fragile, non-uniform in thickness, and tedious and time-consuming to prepare. Commercially produced cellulose acetate is now available in sheets and rolls in a wide range of thicknesses. It is uniform in thickness and free from blemishes and density variation. 0.1 mm thick film (polished on both sides) is suitable for making peels from a wide range of rocks and produces excellent resolution. Thicker films, while less liable to crinkle and curl on porous or high relief specimens, are much more difficult to mount flat, and give lower contrast and microscopic resolution.

The film is vulnerable to the collection of static charges and therefore to contamination. It should be stored in the roll or interleaved with tissue paper in a dust-free (and preferably slightly damp) drawer. Film left lying around the laboratory becomes dusty and scratched, and makes poor quality replicas.

Acetate sheets are soluble in methyl acetate, ethyl acetate, ethyl lactate, diacetone alcohol and tetrachloroethane. However, the cheapest and most useful solvent for laboratory use is commercial grade acetone. *Acetone has a damaging effect on cell membrane lipids, and inhalation of the vapour should be minimized. It also dissolves fat from the skin with extreme rapidity, leading to dermatitis, so hands should always be protected by surgical gloves.*

4.6.2 Stained peels

Katz & Friedman (1965) demonstrated how effective stained peels were in carbonate petrography. No special preparation is required: dry, stained surfaces are taken immediately to the acetone flooding stage (see below). Different stains can be used on the same specimen since earlier stains are removed by re-polishing. This provides a very quick way to obtain a set of peels representing many of the mineralogical and fabric properties of a specimen.

4.6.3 **Procedure**

PREPARING THE CUT FACE

Sawn faces should be lapped to at least 600F Carborundum stage. For greater detail, further smoothing by polishing with 1 μm Aloxite or diamond abrasives is recommended. Etching such a polished surface gives a relief which is more subtly fabric-selective than etches controlled by an imperfect surface.

ETCHING THE SMOOTHED FACE

Limestones can be etched with hydrochloric acid, acetic acid or diNaEDTA depending on the degree of detail required (see Section 4.5.1). Large slabs are conveniently etched (and stained, if required) in plastic photographic developing trays; the cut face preferably being uppermost in the solution unless the sample is too large, in which case the face is held downwards and gently swirled to remove bubble trains.

Price (1975) showed that good peels could be made from cherts and other siliceous rocks after etching with HF. The rock slice is placed face down in a polythene tray into 30% HF solution for 3 or 4 min. The vapour alone does not give a deep enough etch for replicas, but is sufficient for stain preparation. All the usual safety precautions for using HF must be observed (Section 4.5.1)

WASHING THE PREPARED SAMPLE

Hold samples in a bath with slowly running water for several minutes. In hard water areas, use distilled or de-ionized water only, to avoid residues after drying. If HF was used, the specimen must be totally immersed. Avoid touching the etched face, or agitating violently, as the delicate relief may be damaged. Drain the sample and dry in a stream of warm air. Drying may be accelerated by flooding the surface with acetone several times; this carries off some water when it evaporates. Dried and etched surfaces should be peeled immediately, as they collect dust rapidly.

ORIENTATION OF THE SPECIMEN

The specimen should be supported so that the etched face is almost horizontal, tilting by only a few degrees. Some workers use a sand tray to hold specimens, but this should be avoided because of the likelihood of contaminating the surface. Plasticene is a good medium for supporting all sizes of specimen as it is easily moulded and re-used.

PREPARATION OF FILM

Cut a fresh piece of acetate film, free of scratches, to a shape which allows about 1 cm overlap beyond the specimen edge. If the sheet is heavily charged with static, this can be removed by discharging with a piezo-electric antistatic pistol available from audio equipment shops. The discharged sheet is much easier to manage and does not attract dust and lint.

APPLYING THE ACETATE FILM

The specimen should be cold and in a cool room, otherwise the acetone will evaporate too quickly from its surface. Gently flood the etched face with acetone from a washbottle. Too much acetone will cause wrinkling, too little will cause air bells. Place the acetate film gently across the lower margin of the specimen where a pool of acetone will have collected. Holding the sheet in a curve (Fig. 4.9), unroll it

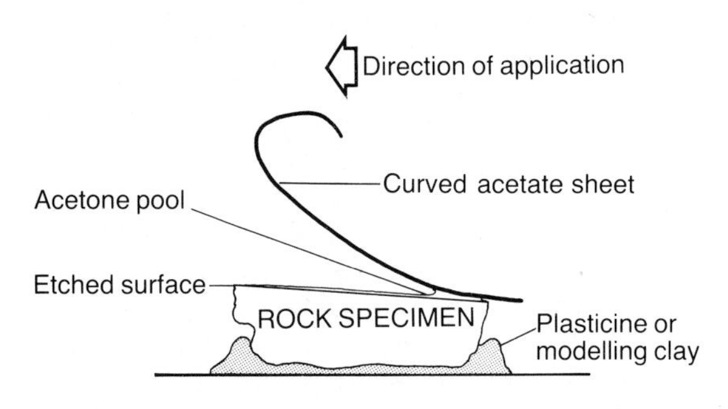

Fig. 4.9. Side view of sample prepared for acetate replica process, showing how the curved acetate film is applied at the edge of the slightly tilted sample before being unrolled across its face, pushing the acetone before it.

across the specimen, pushing the solvent in front of it. Extra liquid exuding from the far edge can be evaporated away quickly by blowing across the specimen; this avoids wrinkling at the edge of the film, which would make it difficult to flatten the peel for subsequent study.

For a few moments, the film should just float on the liquid, then it will draw down as evaporation starts. If there are wrinkles or silvery air bubbles at this stage, do not try to remove the film, as a very messy, sticky surface will be left. Discard the bad peel after it has dried, then repolish and etch the face.

DRYING THE PREPARATION

The specimen should be left for at least half an hour for the acetone to evaporate. This time can be shortened by blowing warm (not hot) air evenly across the surface: uneven drying causes premature lifting. A 250 W infra-red reflector bulb placed about 0.3 m above the specimen dries the film in about 10 min.

TAKING THE PEEL

Gently lift back the film from one edge of the specimen and steadily pull it off the surface in one continuous motion. If a tear develops, peel from the opposite corner and work towards the damage. Films which have been left to dry too long (e.g. a day or more) may be very difficult to remove without tearing. It is a matter of experience to judge the best time to remove a peel, but flicking the edge upwards immediately releases the peel slightly when ready, producing a diagnostic 'dry' sound.

TRIMMING AND MOUNTING OF PEELS

With a sharp pair of scissors, trim the excess film from the peel (but leave a small area as a finger-hold), and immediately either mount it between two sheets of glass, or mark it with specimen number and store it between the leaves of a thick book. The slightly damp, hygroscopic pages of a book provide excellent storage for peels which are either awaiting permanent mounting or are only needed for temporary examination. With some materials, coal balls for example, part of the rock surface is actually incorporated in the peel. Dipping the peel briefly in 2% HCl 'clears' the film before mounting.

Do not leave peels lying about in the laboratory before mounting or filing them: they attract dust and lint and are easily scratched.

While it may be difficult to keep them flat, examining peels unmounted has the advantage of being able to use oblique incident and transmitted light combinations to increase contrast, and for viewing the relief more effectively to gain a three-dimensional effect. Unmounted peels can also be examined with high power objectives which have smaller working distances than glass mounting plates. However, small peels, or cut-out sections of large ones, can be mounted between two standard microscope slides for examination with a low power binocular or stereomicroscope. Larger glass mounting plates can be obtained from photographic dealers. Very large peels may have to be mounted between specially cut pieces of thin window glass. In all cases, the glass should be thoroughly clean and grease free before sandwiching the peel and binding the edges with adhesive plastic tape. The best tape to use is matt 'invisible' mending tape, as standard Sellotape rapidly becomes brittle and splits. Handle the peels only by their edges to avoid grease marks, and lay them on clean, lint-free paper or cloth while they wait for mounting, to avoid dust and scratch collection. Mounted peels should be stored in a dust-free box or drawer, and stained peels are best kept in the dark.

Peels and thin sections should be regarded as complementary. Some peels show no signs of the 'phantom textures', e.g. in recrystallized limestones, which are otherwise visible in thin sections. On the other hand, the etched surfaces of peels frequently reveal details not seen in thin sections. Stained peels, when made with great care, have very high resolution and can be as good as a DPT for point counting and studies of diagenesis.

4.7 EXAMINATION OF MICROSCOPICAL PREPARATIONS

Having gone to some trouble to obtain high quality sample preparations, it is important to realize their potential by using appropriate techniques when studying them. Many workers begin immediately with detailed study under medium to high power on a petrographic microscope which has a very small field of view. There is much to be gained from preliminary survey of the whole preparation at low power, to provide a context for later detailed work.

4.7.1 Photographic map

A suggested first step in the use of any preparation, be it peel or thin section, is to make one or more photographs of the entire preparation (Fig. 4.10). This 'map' has many advantages: it preserves a record of the sample in its initial state before treatment; it gives an overall view of the section or peel which can be used as a guide for subsequent detailed work, and it offers a convenient and rapid way of comparing samples.

The simplest and cheapest way of photographing a slide or peel is to use it as a photographic negative. Place the slide or peel mount into the negative carrier of a photographic enlarger and project the image on to an international A4 size sheet of printing paper of appropriate grade, determine the exposure and print two copies. One of these can form part of a reference archive, the other is a working

Fig. 4.10. Negative print of thin section, showing good contrast and detail obtained by projecting a thin section directly on to bromide paper. Waulsortian facies, Dinantian, Crow Hill, near Clitheroe, Lancashire, England. Lime muds are pale grey, fibrous calcite spars mid-grey and clear blocky calcite late cavity fill cements are black. Cavities are partially or completely filled with peloidal geopetal sediment. Bioclasts include abundant sponge spicules, fenestellid bryozoan fragments, gastropods and echinoderms. Scale is 5 mm.

copy, upon which notes and details can be written, or over which can be placed transparent overlays to mark grains, microprobe analysis sites etc.

Small thin sections will fit into the carrier of a 35 mm enlarger: larger preparations and peels require a plate enlarger. The negative print obtained has the advantage of increased contrast, enhancing many details in the preparation. If a positive image is required, a whole plate negative film sheet can be substituted for photographic paper at the first stage: this intermediate negative is developed and then contact printed. Alternatively, the thin section or peel can be placed on the stage of a 35 mm transparency copying machine, and a black and white or colour photograph taken by flash for later printing or projection.

With very large slices or bulky faced samples, a very simple method of producing a 'map' image is to place the face carefully on to the glass of a xerographic machine (Ireland, 1973). The image can be produced on a transparent sheet or on paper as required. Contrast can be adjusted by varying the toner applied.

4.7.2 Low power examination and drawing

Lees (1962) drew attention to a noticeable gap which existed in the instrumentation for examination of transparent thin sections or surface replicas such as peels, particularly with sizes larger than a standard thin section. He advocated the use of an industrial measuring projector for this purpose. The Watson Manasty 'Shadomaster' Model VMP (no longer in production) was chosen as being the best (Fig. 4.11). Viewers for microfilm are a good substitute, but lack the dimensional stability for accurate measurement which is a feature of the 'Shadomaster' projectors.

In the 'Shadomaster', samples are placed on a viewing stage and illuminated from above with a quartz-halogen projection lamp and a condenser. Below the stage, one of several interchangeable magnifying lens systems focusses the image on to an inclined plane mirror at the base of the instrument. The final image is formed on a large frosted glass plate held at a convenient viewing angle. The resolution is good, and grains less than 10 μm can be seen at ×100. For work at these magnifications, the frosted glass plate can be replaced by clear glass with a matt drafting plastic film taped over. Accurate measurements of grains can be made if the stage is fitted with a vernier mechanical stage.

This instrument has many advantages. Several people can examine a preparation at the same time. Very large peels or sections can be scanned at low magnifications. Tracings of fabrics and grains can be quickly and accurately made at a range of magnifications. Point counting can be done by preparing transparent overlay coordinate grids at various sizes; eye strain is found to be less than with a petrographic microscope. Grids can also be prepared for determination of roundness and sphericity. Polarizing filters can easily be interposed. Photographs may be prepared by placing negative sheets or printing papers face down on the glass viewing plate. One disadvantage is that the viewing room must be darkened while the instrument is in use.

Fig. 4.11. 'Shadomaster' industrial measuring projector used for examining thin sections and acetate peels. The quartz-iodide lamp at the top projects a beam through the specimen and sub-stage lens system, which is reflected by a mirror in the base of the instrument on to a large ground-glass screen. A grid overlay is shown, used for point-counting.

An alternative for viewing and drawing small peels or segments of larger peels is to mount them in 35 mm transparency mounts (preferably with glass both sides) and to project them on to white paper or card, on which tracings can be made. Larger peels can be sandwiched between two glass plates and placed in an episcope for projection.

4.7.3 **Petrological microscope**

In order to optimize viewing conditions, the microscope should be correctly adjusted and all optical surfaces clean. In particular, care should be taken to set the condenser correctly — this is probably the single greatest cause of poor results in microscopy. Curry, Grayson & Hosey (1982) gave details of how to adjust substage condensers for standard and Kohler illumination.

The next important factor is adjusting the contrast to suit each specimen. Thin sections of carbonate rocks in particular require a high contrast, and this may be obtained by stopping down the substage iris somewhat. Complete stopping-down, however, produces spurious interference effects at grain edges and reduces resolution. Neutral density filters should then be interposed to reduce light intensity. Reducing the illuminator lamp intensity may not be desirable as this alters the colour temperature of the incident light.

Highly recrystallized and dolomitized carbonates present particular difficulties. Delgado (1977) noted that details such such as bioclasts and voids visible in hand specimens or peels of such rocks vanished under the petrological microscope. He discovered that diffusing the transmitted light resulted in a dramatic increase in the resolution of such cryptic features. The best diffuser was produced by coating the undersurface of a thin section by burning magnesium ribbon some 0.2–0.25 m below the slide. At this distance, a 0.07 m length of Mg ribbon produces a coating of about 20 μm thickness. The coated glass surface is protected by another glass slide before mounting on the microscope stage. (*Caution! use goggles — the bright flame of burning magnesium can cause retinal damage!*) Experiments need to be done with and without the condenser in place, depending on the specimen. Best results are obtained with magnifications of ×10 or less.

In all cases, it is advantageous to take time to experiment with lighting, filters, iris and condenser setting etc., so as to optimize the results from each rock type and method of preparation.

4.7.4 **Photomicrography**

Most research petrological microscopes have photomicrographic attachments, while simpler microscopes can be used with 35 mm SLR cameras fitted with appropriate adaptor rings. Curry *et al.* (1982) give useful hints about using these systems. In general, the most appropriate films are fine grained, high contrast types. This usually means they have slow speed and concomitantly long exposure times, so it is necessary to ensure that the microscope is free from vibration, best done by working on a stone or concrete optical bench. The photomicroscope should be rigidly supported.

Since the optics of the microscope are arranged for white light, where each wavelength is brought to a slightly different focus, optimum resolution in monochrome photographs is obtained by employing a medium density green filter (selecting a wavelength in the middle of the range). This also has the effect of producing some differential contrast within the specimen. Stained peels or thin sections may be enhanced by using a complementary filter on the illuminator.

Correct colour photographs are difficult to achieve. One of the main problems is to match the colour temperature of the film with that of the incident light, assuming correct exposure. The problem is particularly acute with pictures taken with crossed polars, where often the identification of the minerals depends on the subtle rendition of characteristic interference colours. Daylight colour film (around 5500 K colour temperature) can only be used directly with daylight and a substage mirror. If a substage illuminator is fitted, then a colour compensation filter is required (Kodak Wratten no. 80A). With tungsten light film (3200 K), no filter is needed if the substage illuminator has a tungsten lamp. That does not, however, guarantee that exact colour rendition will be obtained, as the colour temperature of tungsten lamps depends on the voltage at which they are run and on the age of the lamp. For critical applications, a colour temperature meter should be used, or the manufacturer's graphs consulted.

4.8 MANUFACTURERS AND SUPPLIERS

Acetate film for peels: Charles Tennant & Company Limited, 214 Bath Street, Glasgow G2 4HR, Scotland.

Automated and semi-automated thin-sectioning equipment: Logitech Limited, Erskine Ferry Road, Old Kilpatrick, Dunbartonshire, Scotland.

Araldite, Epo-tek and Versamid: Ciba-Geigy Plastics and Additives Company, Duxford, Cambridge CB2 4QA, England.

Blue keystone resin stain: Epoxy Technology Inc., PO Box 567, 14 Fortune Drive, Billerica, Mass 01821, USA.

Carborundum abrasives: Sohio Electro Minerals Company (UK) Limited, Mosley Road, Trafford Park, Manchester M17 1NR, England.

Diamond powders, engis papers, silicon carbide discs, lapping machines, lead lap faces: Engis Limited, Park Wood Trading Estate, Sutton Road, Maidstone, Kent ME15 9NJ, England.

Stains and reagents: BDH Chemicals Limited, Fourways, Carlyon Industrial Estate, Atherstone, Warwickshire CV9 1JQ, England.

Zyglo® penetrant 2L.22A, Solvent ZC 7 and developer aerosols: Magnaflux Limited, South Dorcan Industrial Estate, Swindon, Wiltshire SN3 5HE, England. (Used for detecting cracks in industrial castings.)

5 Microscopic techniques: II. Principles of sedimentary petrography

GILL HARWOOD

5.1 INTRODUCTION

Sedimentary petrography is the analysis of both depositional and diagenetic fabrics from thin sections, and includes mineralogic composition, grain and sediment provenance, fabric studies and determination of the sequence of diagenetic events. Petrographical studies of thin sections form the basis of much modern research on sedimentary rocks, whether siliciclastic, volcanic, carbonate or evaporitic, and the information gained can greatly supplement data from field- or core-based studies.

Perhaps the first question to be answered is why do so many sedimentologists concentrate on sedimentary petrography? A brief literature review of the popular sedimentological journals indicates an increasing emphasis on petrography within Europe, whereas within the United States and Canada sedimentary petrology has for long been a necessary component of integrated sedimentological research projects. Many such studies are related to hydrocarbon evaluation — both for source rocks and for reservoir potential. A summary of the interrelationships together with some of the applications of aspects of sedimentary petrography is given in Fig. 5.1. It should be emphasized that much information gained from petrofabric research can be used to back up field studies and forms an aid to provenance studies, including facies analysis and construction of depositional models; these applications are commonly lost amongst an ever-growing number of papers solely on aspects of diagenesis.

5.1.1 Techniques and tools

Although some techniques used in sedimentary petrography are quantitative, many are solely qualitative and are principally descriptive. Many research projects concentrate on *either* siliciclastic *or* carbonate sediments; few investigate combined carbonate-siliciclastic systems and fewer yet reference volcanogenic sediments, mudrocks or evaporites. This chapter attempts to portray fabrics which may develop within any sediment, independent of mineralogy. Although some specific fabrics are included, the aim is to produce a general 'guide' for all sedimentary petrographers, with references leading to more detailed texts. Examples used in photomicrographs are predominantly from '*grain-supported*' rocks (sandstones, grainstones/packstones) as many petrofabrics are more readily apparent within these sediments. This is not to say that these features are not present within more clay-rich or carbonate mud-rich sediments. They are, however, commonly more difficult to discern, and to photograph, in clay-rich and mud-rich sediments. Their examination profits from the use of ultra-thin sections (Chapter 4), employed initially in studies of carbonate diagenesis (Lindholm & Dean, 1973), so that initial research into mudrock diagenesis can now be carried out with a good petrological microscope.

A good petrological microscope is the essential tool for sedimentary petrography, with refinement by the addition of subsidiary components where necessary. One tool commonly ignored by many sedimentary petrographers is a Shadowmaster®, or similar equipment, where a thin section is projected on to a ground glass screen to produce an image (a shadowgraph). This can be extremely useful for overviews of a thin section, grain shape and size determinations and for fabric analysis.

At this point a word of warning should be included: it is little use spending days/weeks describing and analysing one thin section if you are not sure whether this section is representative of the sedimentary sequence you are evaluating. It is always tempting to sample the more 'interesting' sections of a sedimentary sequence and to find, on return to the laboratory, that you have few or no samples of the more mundane lithologies which comprise the greater portion of the sequence. In practice it is very difficult to decide what *is* representative; perhaps the best method is to make a general study of several sections from the same sequence before proceeding to detailed description and analysis. In most cases it is more important to note the regional trends; thus

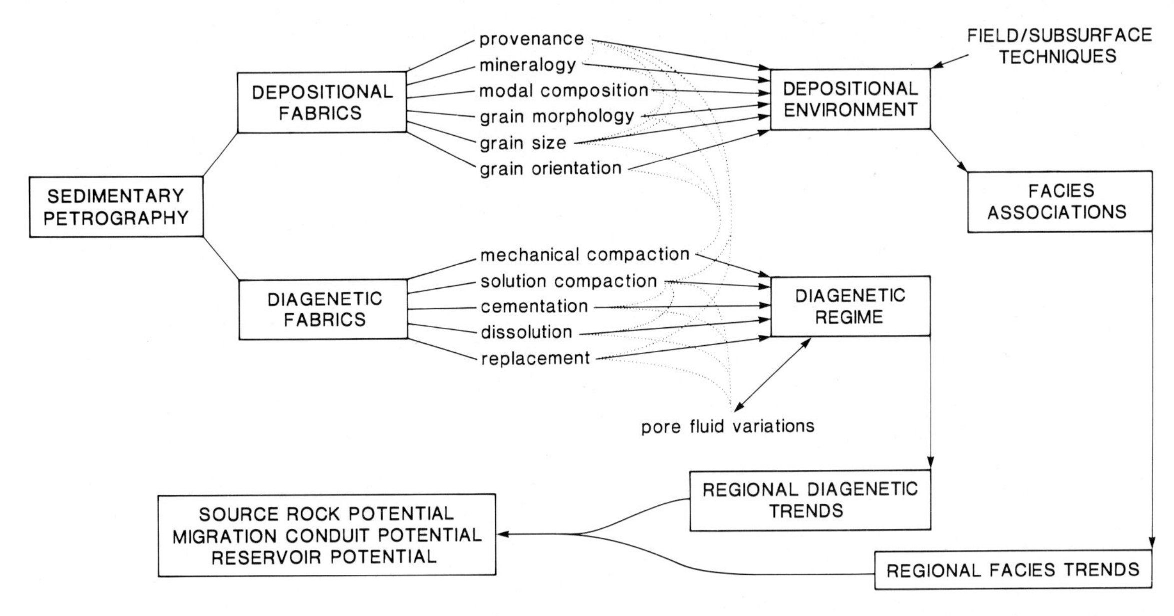

Fig. 5.1. Relationship of sedimentary petrography and petrographic fabrics to other branches of sedimentology.

care must be taken to ensure that these are not obscured by local variations.

5.1.2 Components and petrofabrics

The general fabric of a sedimentary rock will depend on three major components: (1) the original detrital grains,

(2) authigenic minerals, both cements and replacive minerals, which have formed since deposition, and

(3) pore spaces.

Detrital grains may be of any size (from clays to pebbles or larger) and of varying shape. Sand or larger sized grains commonly form a framework within the sediment (Fig. 5.2a, b) between which the finer matrix (silt and clay sized particles) may have accumulated (Fig. 5.2a). In well-winnowed sediments there is generally little or no matrix, leaving resultant pore spaces which may be partially or totally occluded (or filled) by cement during diagenesis (Fig. 5.2b). Where there are higher proportions of matrix the grains do not form a framework, but float within the finer grained matrix, forming a matrix-supported sediment (Fig. 5.2c). Whereas these may appear to be easy to recognize in theory, in practice it can be difficult to determine whether a sediment is grain supported or matrix supported in a hand specimen, but particularly when seen in two dimensions within a thin section (e.g. Fig. 5.2d).

Authigenic minerals grow after sediment deposition, during diagenesis; they include both cements and replacive minerals. *Cements* are the crystals which grow into existing pore spaces. They may, or may not, totally occlude the available pore space (e.g. Fig. 5.2b). The form of many cement crystals can be indicative of the environment in which they grew (see Section 5.3.3). *Replacive* minerals grow, as their name suggests, in the place of pre-existing minerals and not into pore spaces. They are commonly alteration products of the primary detrital grains, but may also form from the introduction of additional ions by circulating pore fluids, as in many instances of dolomitization of a precursor carbonate.

Pore spaces are the voids not filled by grains or matrix within a sediment. Impregnation of the sample with a coloured epoxy resin (commonly blue) makes the pore spaces more easily visible in thin section (Chapter 4) and allows distinction between a true pore space and a void where a grain or crystal has been plucked out during the process of making a thin section.

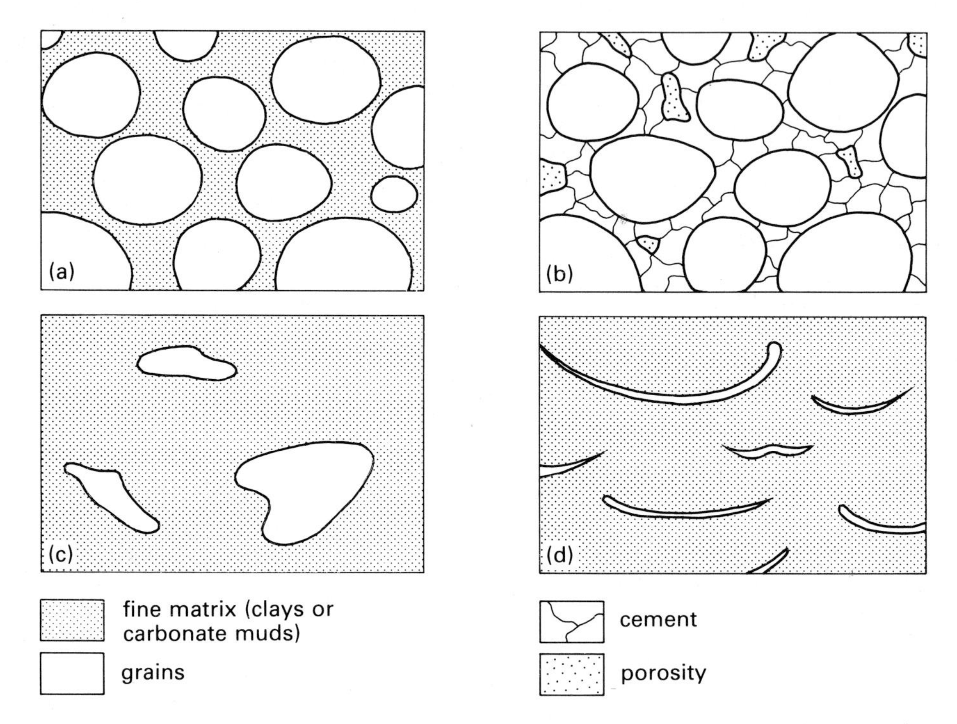

Fig. 5.2. Fabrics of sedimentary rocks: (a) Grain-supported sediment; finer matrix has accumulated between a framework of grains.
(b) Grain-supported sediment; pore spaces between grains are partially occluded by cement, although there is some remaining porosity.
(c) Matrix-supported sediment; grains appear to 'float' in finer-grained matrix.
(d) In three-dimensional samples, this specimen can be seen to be grain-supported, with matrix between skeletal fragments. In the two dimensions of a thin section, however, it appears to be matrix-supported.

The combination of these three components forms the fabric of any sedimentary rock. Using a more genetic approach, however, petrofabrics are divided into two major types, each the result of different processes:

(1) Depositional fabrics — fabrics which result from processes which were active during deposition of the sediment. These include grain mineralogy, grain morphology, grain orientation and provenance studies.

(2) Diagenetic fabrics — fabrics which result from processes which occurred *after* deposition of the sediment. These include compaction, cementation, mineral replacement and dissolution of pre-existing phases, and their study can lead to the construction of a diagenetic history of the sediment.

5.2 DEPOSITIONAL FABRICS

5.2.1 Grain identification

Mineral identification in thin sections can be backed up by XRD analysis of powdered rock samples (Chapter 7). There are many excellent texts on mineral identification, all with much more detail than can be included here. The properties of some commoner minerals in sedimentary rocks are given in Table 5.1; examples of these minerals in thin sections are shown in modern coloured guides to sedimentary constituents (Scholle, 1978, 1979; Adams *et al.*, 1984) plus the older standard descriptive texts (Carozzi, 1960; Krumbein & Pettijohn, 1961; Milner, 1962a, b; Pettijohn, 1975).

A minor, but important constituent of many sandstones are the heavy minerals, with a specific gravity in excess of 2.9. These are usually studied by separating them from a crushed rock with heavy liquids, notably tetrabromoethane, **which is extremely toxic.** The procedure is well described in Krumbein & Pettijohn (1961), Hubert (1971) and Friedman & Johnson (1982). It is also possible to use a magnetic separator. Identification of the heavy minerals is by microscopic examination of the mounted grains (see references cited above for properties of the minerals). The value of the heavy minerals is in the information they give on a sediment's provenance, although their dissolution during diagenesis may result in modification to the original assemblage (Morton, 1985b).

5.2.2 Modal composition

The modal composition of a sediment is a measure of the proportions of the different major depositional components. These have three end members; (i) detrital, dominantly siliciclastic (or terrigenous) components derived from outside the depositional area, (ii) allochemical components produced within, or adjacent to, the depositional area, and (iii) orthochemical components resulting from chemical precipitation within the area. Five major classes of sedimentary rock are defined using a triangular plot of these components (Fig. 5.3; Table 5.2) (Folk, 1974b). This classification is based *solely* on composition immediately after deposition; cements and authigenic minerals are not included. Difficulties are encountered where considerable dissolution has taken place during diagenesis of the sediment (Section 5.3.5).

To obtain the modal composition of any sediment one needs to know (i) the nature of the contained grains and (ii) the proportions of these grains present within the sediment. The nature of the contained grains can be obtained from grain mineralogy (Section 5.2.1) plus the origin of lithic clasts (Table 5.3). The proportion of the different grains present is more difficult to ascertain and is done using two main techniques, point counting and visual estimates. Several systems of computer enhancement and image analysis which can be used with a petrographic microscope are beginning to come on the market and promise relative ease of modal composition evaluation.

POINT COUNTING TECHNIQUES

Point counting is the most accurate method of establishing the modal composition of a sediment, albeit a time-consuming method (Chayes, 1956; Galehouse, 1971a). Point counting of many thin sections is tedious; however, its uses generally outweigh the time spent in data accumulation. There are various methods of point counting, the most obvious being the spot identification of each grain as it appears under the cross wires of the microscope (e.g. Van der Plas & Tobi, 1965; Soloman & Green, 1966). However, in a sediment with a large proportion of lithic clasts this may involve changing many times to a different magnification (generally larger, to identify the whole clast). One method (Ingersoll *et al.*, 1984) obviates this by identifying minerals within the clast, rather than the complete clast, although this produces a modal composition based on total mineralogy rather than based on detrital grain mineralogy.

During point counting, the thin section is placed on a mechanical stage screwed to the rotating stage of the microscope, which is then clamped. The mechnical stage is connected to a counting unit, which, when a master key is pressed, moves the section a given distance along a traverse. This distance can be varied, dependent on the grain size of the sediment. Six (or more) additional keys can be marked for different components within the section. One of these keys is pressed each time that component is visible under the cross hairs of the microscope. The master key both advances the stage and records the total number of points counted. Grains present, but not encountered under the cross hairs, are noted as additional minerals (Table 5.4). Several traverses are made for each thin section, and a total of some 250–300 points per section need to be counted to obtain sufficiently accurate percentages of the components present. All data are rigorously tabulated (Table 5.4). The position of the traverses should be noted as this enables the counting of additional points, should this be necessary; it also enables checking of the original data. Although point counting is time consuming, any petrographer should practice this method on several thin sections before proceeding to much more rapid methods, such as the use of visual estimates.

Table 5.1. Summary of common minerals in sedimentary rocks and their optical properties. Compiled from Kerr (1959) and Tucker (1981)

Group/ mineral	Crystal system	Colour	Cleavage	Relief
Quartz	Trigonal	Colourless	None	Low +ve
Cherts		Colourless	None	Low +ve
Feldspars				
Microcline	Triclinic	Colourless	Present	Low −ve
Orthoclase	Monoclinic	Colourless	Present	Low −ve
Plagioclase	Triclinic	Colourless	Present	Low −ve
Micas				
Muscovite	Monoclinic	Colourless	Prominent Planar	Moderate +ve
Biotite	Monoclinic	Brown-green	Prominent Planar	Moderate +ve
Clay minerals				
Chlorite	Monoclinic	Green/blue green	Planar	Moderate +ve
Kaolinite	Triclinic	Colourless	Planar	Low +ve
Illite	Monoclinic	Colourless	Planar	Low +ve
Smectite	Monoclinic	Colourless	Planar	Low −ve
Glauconite	Monoclinic	Green	Planar	Moderate, masked by colour
Zeolites	—	Most colourless	—	Low, most −ve
Carbonates				
Aragonite	Orthorhombic	Colourless	Rectilinear	Moderate−high
Calcite	Trigonal	Colourless	Rhombic	Low−high
Dolomite	Trigonal	Colourless	Rhombic	Low−high
Siderite/ Ankerite	Trigonal	Colourless/ pale brown	Rhombic	Low−high
Evaporites				
Gypsum	Monoclinic	Colourless	Planar	Low
Anhydrite	Orthorhombic	Colourless	Rectilinear	Moderate
Celestite	Orthorhombic	Colourless	Planar	Low−moderate
Halite	Cubic	Colourless	Rectilinear	Low
Baryte	Orthorhombic	Colourless	Planar	Low−moderate
Iron minerals				
Pyrite	Cubic	Opaque	—	—
Magnetite	Cubic	Opaque	—	—
Haematite	Cubic	Opaque, brown tinge	—	—
Chamosite/ Berthierine	Monoclinic	Green	—	Moderate
Collophane	Non-crystalline	Browns	—	Moderate
Bitumens	Non-crystalline	Opaque	—	—

Table 5.1. (*Continued*)

Birefringence	Other features	Cement	Form and occurrence
Grey		Overgrowth cement common	As detrital grains, cements and as diagenetic (replacive) quartz
Grey		Chalcedony and megaquartz occlude voids	Acicular chert (chalcedony), megaquartz and microquartz; all forms diagenetic unless as detrital grain
Grey	Crosshatch twins	May form overgrowth cements	Present as detrital minerals but commonly have cloudy appearance as alter to clay minerals
Grey	Simple twins		
Grey	Multiple twins		
Bright colours	Parallel extinction	—	Common detrital mineral
Bright colours, masked by colour	Parallel extinction, pleochroic	—	Common detrital mineral
Grey }	Fine grained }	Common as burial cements	Present as detrital minerals, as alteration products of silicates and as cements
Grey }			
Grey }			
Grey }			
Grey, masked by colour	Commonly replaces pellets		Characteristic of low sedimentation rates, may infill foram tests, etc.
Commonly grey		Common as cements where volcanics	Associated with volcanogenic sediments
High colours		Commonly acicular.	
Very high colours }	Distinguished by staining	Many cement morphologies	Present as detrital grains, cements and replacement fabrics in carbonates.
Very high colours }			
Very high colours }			Common replacement mineral in ironstones
Grey	—	—	—
Bright colours		Common burial cement	Commonly crystalline due to replacement of evaporite sequences
Grey	—	Burial cement	Commonly partially replacive
Isotropic	Only present if section prepared in oil	Common burial cement	
Grey	—	—	May be associated with sulphide mineralization; difficult to distinguish from celestite under microscope
— }	Distinguished in reflected light	{ Yellow	{ Common authigenic minerals
— }		{ Grey-black	
— }		{ Red-grey	
Grey, masked by colour	—	—	Occurs in ooids in ironstones and as partial pore fills
Isotropic	—	—	Replacive textures, commonly in carbonates
—			Occurs in pore spaces and in fluid inclusions
	Distinguished in fluorescence	—	

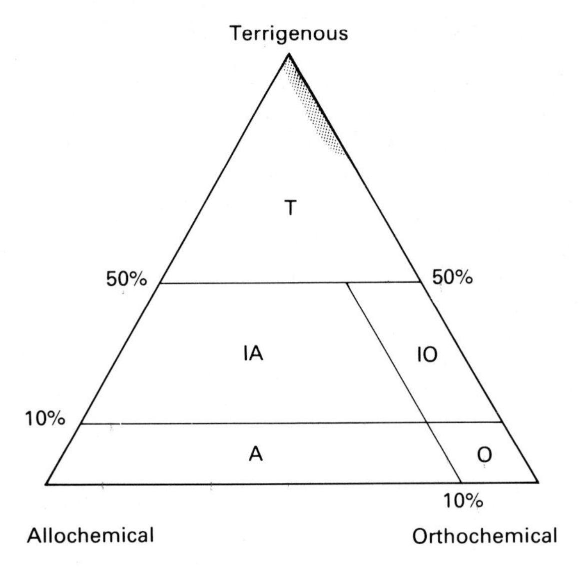

Fig. 5.3. The five basic classes of sedimentary rock types (after Folk, 1974b). Descriptions of the classes are in Table 5.2.

VISUAL ESTIMATION TECHNIQUES

Many petrographic texts include visual comparators to be used in estimations of modal composition/ mineral percentages, etc. These may include grains of different shapes and sizes (Terry & Chilingar, 1955), or be computer-generated random percentages (Folk, Andrews & Lewis, 1970) (Fig. 5.4). Care must be taken in making visual estimates, and results are usually much less accurate than with point counting techniques. The method is both simple and quick, but should only be used for estimations and not where accurate results are required. Computer-aided techniques will help by providing higher accuracy results, but will not bypass the need for previous detailed petrographic study.

ABUSES AND USES

With both visual estimation and point counting techniques it must be remembered that values are being computed from the two dimensions of the thin section into the three dimensions of a sediment for modal composition. For inhomogeneous sediments, particularly those with a distinct fabric, total measurement of modal composition should be a combination of the measurements from three perpendicular thin sections. It is also important that the thin sections counted are representative of the sequence studied, thus necessitating preliminary overviews of the thin sections.

Each of the major components of a sedimentary rock (Fig. 5.3) can be further subdivided during detailed modal analysis (Table 5.3). Expression of modal composition is commonly shown on a triangular diagram, using three appropriate major

Table 5.2. Explanation of symbols used in the classification of sedimentary rocks of Fig. 5.3 (from Folk, 1974b)

Symbol from Fig. 5.3	Examples and comments	Approx, % of strat. record
T Terrigenous rocks	Most mudrocks, sandstones and conglomerates Most terrigenous rocks fall into the shaded area of Fig. 5.3	65–75
IA Impure allochemical rocks	Very skeletal shales, sandy skeletal or ooid-rich carbonates	10–15
IO Impure orthochemical rocks	Clayey carbonate mudstones	2–5
A Allochemical rocks	Skeletal, ooid-rich, pellet or intraclastic carbonates	8–15
O Orthochemical rocks	Carbonate mudstones, anhydrite, chert Collectively IA and IO are classed as impure chemical rocks and A and O as pure chemical rocks	2–8

Table 5.3. Summary of the components of sedimentary rocks and their symbols, for use with triangular compositional diagrams similar to those of Fig. 5.5 (modified from Folk, 1974b)

Symbol	Components	Origin		
Q_i	Igneous quartz	Quartz (Q)	Terrigenous components	Detrital
Q_m	Metamorphic quartz			
P	Plagioclase	Feldspar (F)		
K	Orthoclase/microcline			
R_S	Sedimentary rock fragment	Rock fragments (or lithic clasts) (L or RF)		
R_v	Volcanic rock fragment			
R_{met}	Metamorphic rock fragment			
R_{plut}	Plutonic rock fragment			
S_k	Skeletal fragments	Carbonates	Allochemical	
P_{el}	Pellets/peloids			
O	Ooids			
G_{rns}	Unidentifiable grains			
M_{dep}	Some carbonate muds			
M_{ic}	Some carbonate muds	Chemical/diagenetic	Orthochemical	*in situ*
E_{vap}	Evaporites			
C_{hert}	Cherts			

components (Fig. 5.5a, b). Care should be taken that the same classification method is used for all samples plotted, particularly if data are obtained from other workers, as different classifications can result in variation of the position of plotted points (and hence composition) (Zuffa, 1985). Modal compositions have commonly been used by sandstone or carbonate petrographers but Zuffa (1986) has demonstrated the importance of consideration of siliciclastic *plus detailed* carbonate components. It is also important to note the presence of later dissolution (McBride, 1985) as this can greatly affect the resultant modal composition (Section 5.3.4). The use of multivariate analysis (common in igneous geochemistry) has been little applied to modal compositions of sediments, although it offers considerable potential. This lapse may be due to the lack of point counting in recent years, concomitant with lack of communication between various branches of the geological sciences. In addition to modal composition, point counting can be used to assess the relative proportions of grains and matrix, grains and cement and cement or grains and porosity.

5.2.3 Grain morphology, size, sorting measurements and orientation

Several authors have considered the difficulties of evaluating these parameters from thin sections of lithified sediments as opposed to those of unconsolidated or disaggregated sediments (Rosenfeld, Jacobsen & Ferm, 1953; Friedman, 1962; Klovan, 1966; Folk, 1966, 1974b). Although these parameters are commonly measured for unconsolidated or disaggregated sediments, the geological significance of the difference in grain size within sediments is not fully understood (Folk, 1974b). In addition, disaggregation of sediments may lead to inaccurate results if overgrowth cements are present (Section 5.3.3). The size of detrital grains within a sediment is proportional to the available materials and to the amount of energy imparted to the sediment. The degree of sorting of a sediment is dependent on the size range of the supplied grains, the mode of deposition (for example, whether wave reworking or current deposition), the current characteristics (predominantly strength) and the time

Table 5.4. Data sheet for use when point counting thin sections (or grain mounts). Modified from Galehouse (1971a)

Thin section/sample number ____________ Date ____________
Rock name ____________ Operator initials ____________
Stage type ____________
Counter type ____________
Initial slide position ____________
Horizontal interval ____________
Vertical interval ____________

Mineral		Percentage	Counter reading		Number of Points	Comments
Number	Name		Start	Finish		
1						
2						
3						
4						
5						
6						
7						
8						
9						
10						
11	Opaque					
Total						

Accessory minerals:
(present but not at point count sites)

involved in deposition. Field or core studies can give information on some of these features and thin section petrography can aid further evaluation, but care must be taken to avoid diagenetic fabrics in any such analysis. Shadowgraphs are particularly useful in grain morphology and orientation studies, as they enable overviews of a large area of the thin section.

GRAIN MORPHOLOGY

Recognition of detrital grain morphology (form, sphericity, roundness and, in some cases, surface textures) in thin section is only possible where grain surfaces have not been modified during diagenesis. Parameters such as roundness and sphericity are

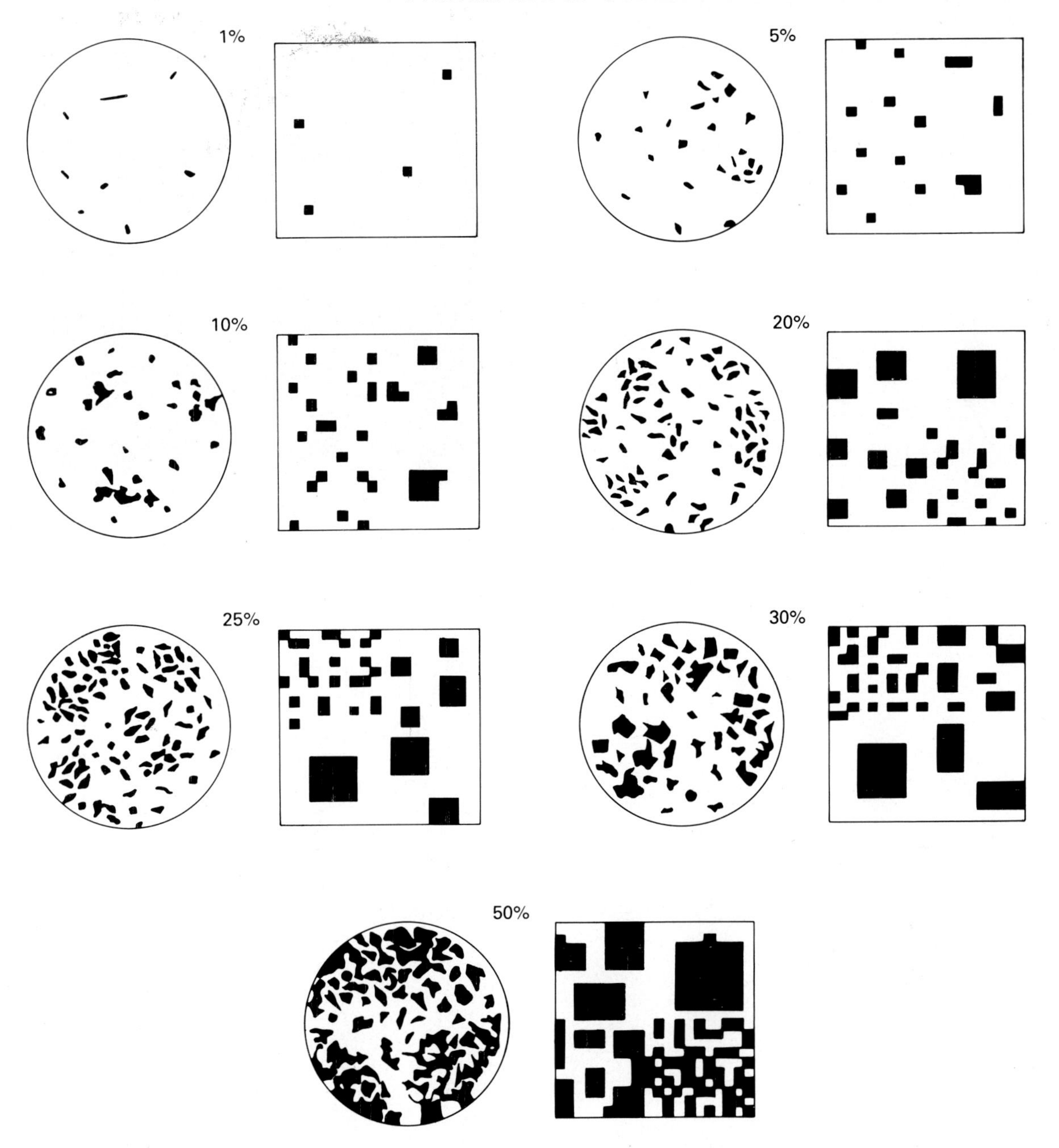

Fig. 5.4. Percentage estimation comparison charts (from Folk *et al.*, 1970). This chart combines the visual comparator of Terry & Chilingar (1955) with computer-generated visual comparators.

readily measurable from thin section. Form (or shape), however, is dependent on the orientation of the thin section; determination of grain form therefore involves measurement of shape of many grains in different orientations within one thin section, and, preferably, the use of three perpendicular thin sections.

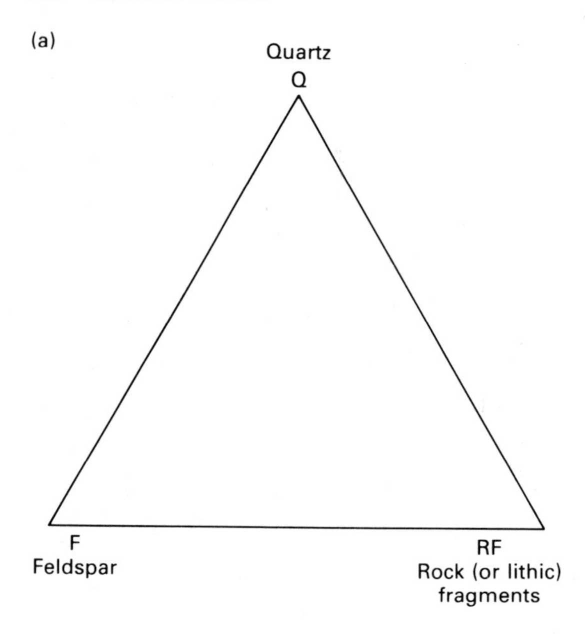

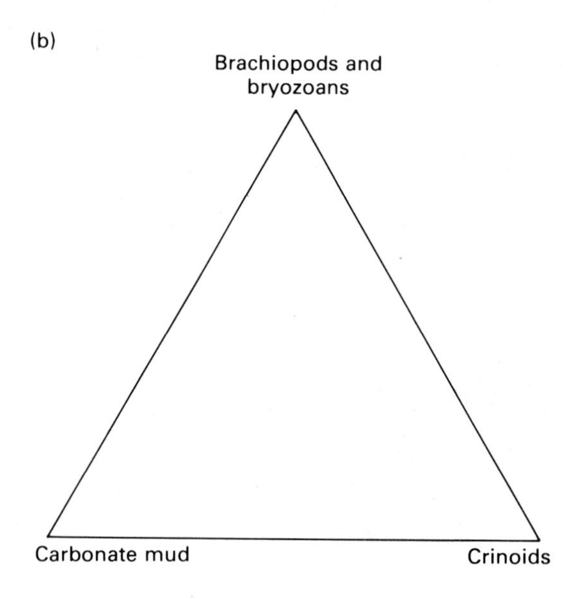

Fig. 5.5. Component diagrams for demonstrating modal composition of (a) siliciclastic/terrigenous sediments and (b) carbonate sediments. Components can be modified to suit the individual study.

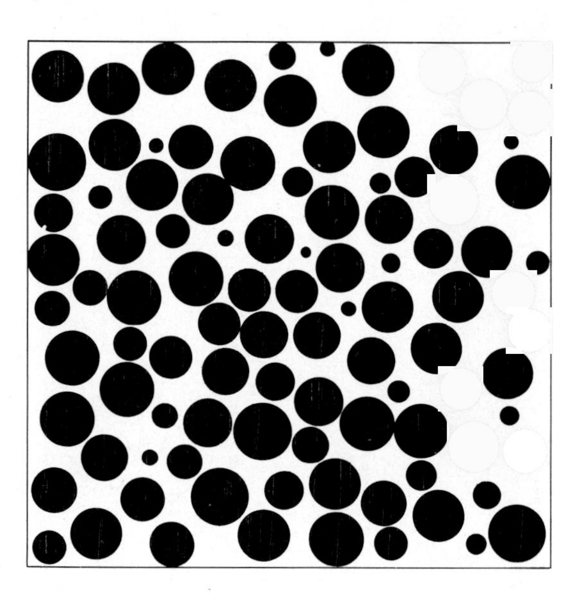

Fig. 5.6. A random section through a solid comprising a framework of spherical grains of equal size (from Harrell, 1984).

GRAIN SIZE AND SORTING MEASUREMENTS

A thin section is a random two-dimensional slice through a three-dimensional solid. Any slice through a solid comprising spherical grains of equal size will show spheres of many, apparently, different sizes (Fig. 5.6). Both grain size and sorting cannot, therefore, be measured directly from a thin section and can only be estimated. This can be done by direct measurements of the maximum dimension of a hundred grains through a calibrated eye piece graticule, or on a screen, followed by calculation of the standard deviation of these dimensions and application of a conversion equation which adjusts this value for the effects of random sectioning (e.g. Harrell & Eriksson, 1979). Alternatively, visual estimation using a comparator can be used, a procedure which is considerably less time consuming and much easier, although such estimates may lack accuracy. Harrell (1984) compared actual and apparent sorting images for sets of spheres with log normal distribution, and demonstrated that the apparent sorting is less accurate for well sorted sediments, but, that using his visual comparators (Fig. 5.7), the apparent sorting is very nearly the same as for actual sorting within the sand and gravel size ranges. It

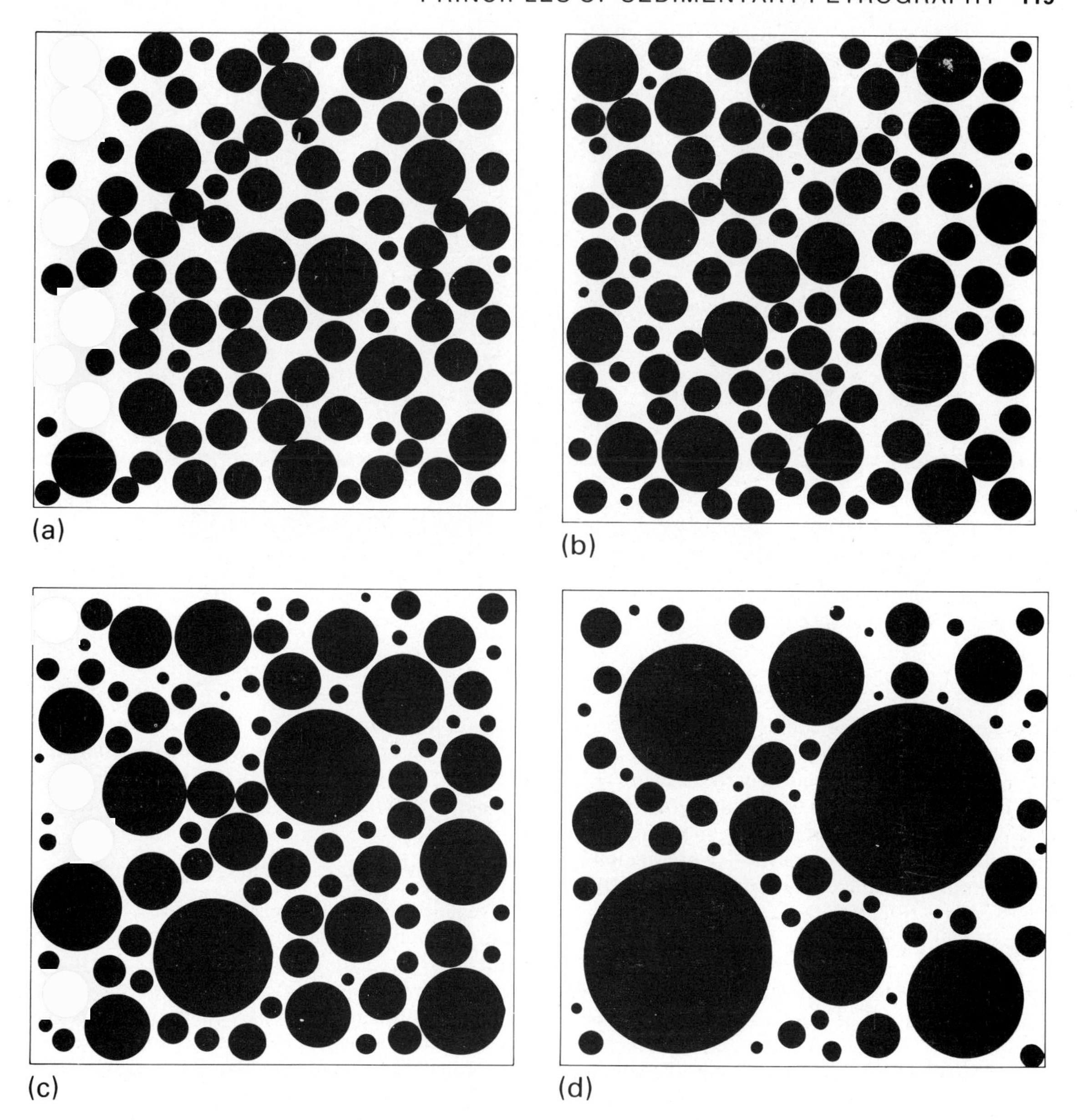

Fig. 5.7. Visual comparators for random sections through log normally distributed sets of spherical grains. Actual sorting in (a) is 0.35 ϕ, in (b), 0.50 ϕ, in (c), 1.00 ϕ and in (d) is 2.00 ϕ. Apparent sorting is 0.69, 0.77, 1.16 and 2.08 ϕ respectively (from Harrell, 1984).

should be emphasized that use of sorting and grain size comparators is only valid for sediments which have not undergone compaction (see Section 5.3.2). Abnormal sorting or textural inversion (e.g. Fig. 5.8) are common in sediments and indicate unusual environmental conditions, possibly mixing of sediment from two different environments, storm mixing of material in a high energy environment or multiple sources of sediment supply. These conditions are not always apparent from field/core studies and

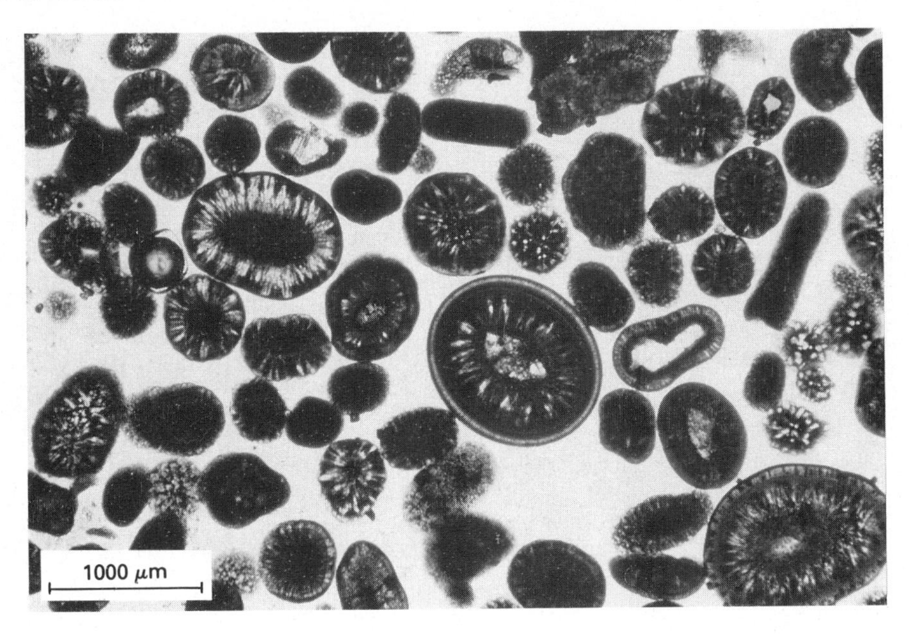

Fig. 5.8. Bimodal distribution in modern ooid sands. Radial ooids from Great Salt Lake are mostly 600 μm in diameter, but larger grains average over 1,000 μm. Recent, Utah, USA.

illustrate the uses of the additional information obtainable from thin section petrography. In interpretation, it should be remembered that there is very little size reduction of quartz sand grains during transport, but that selective sorting may mean that the small grains are carried farther, or are winnowed and deposited by weaker currents (Folk, 1974b).

GRAIN ORIENTATION

Measurement of grain orientation (e.g. Fig. 5.9) necessitates the use of oriented field samples plus measurements from three perpendicular thin sections. The orientation of the long axis of the grains can be measured directly from the microscope or shadowgraph, or it can be measured by a photometric method which integrates the extinction behaviour of all the grains in the field of view (Sippel, 1971). Orientations determined can be plotted on rose diagrams, balloon diagrams or, better, stereonets as for field palaeocurrent analysis (Chapter 2). Compaction after sedimentation may have considerably reduced any angle of cross lamination measured from grain orientation; the significance of this can be estimated from compaction features visible in the same thin sections (Section 5.3.2). More detailed accounts of grain orientation methods from thin sections are given in Bonham & Spotts (1971) and Gibbons (1972).

5.2.4 Provenance studies

Provenance studies of sandstone sequences are possible using the modal composition plus maturity indicators (Fig. 5.10; Table 5.5) (Valloni & Maynard, 1981; Dickinson, 1985; Valloni, 1985). Sandstone sequences are the result of both provenance and tectonic environment, modified by climate, depositional environment and later diagenetic events. Such broad groupings are only possible when many separate thin section analyses are grouped together; they are not applicable where the study is of one sandstone sequence from one particular area of a sedimentary basin.

In addition to studies using the modal composition of a sediment, other provenance determinations can be made from individual quartz grains. This technique was initiated by Krynine (1940, 1946) and can detail sources not recognized in more regional composition-based provenance studies.

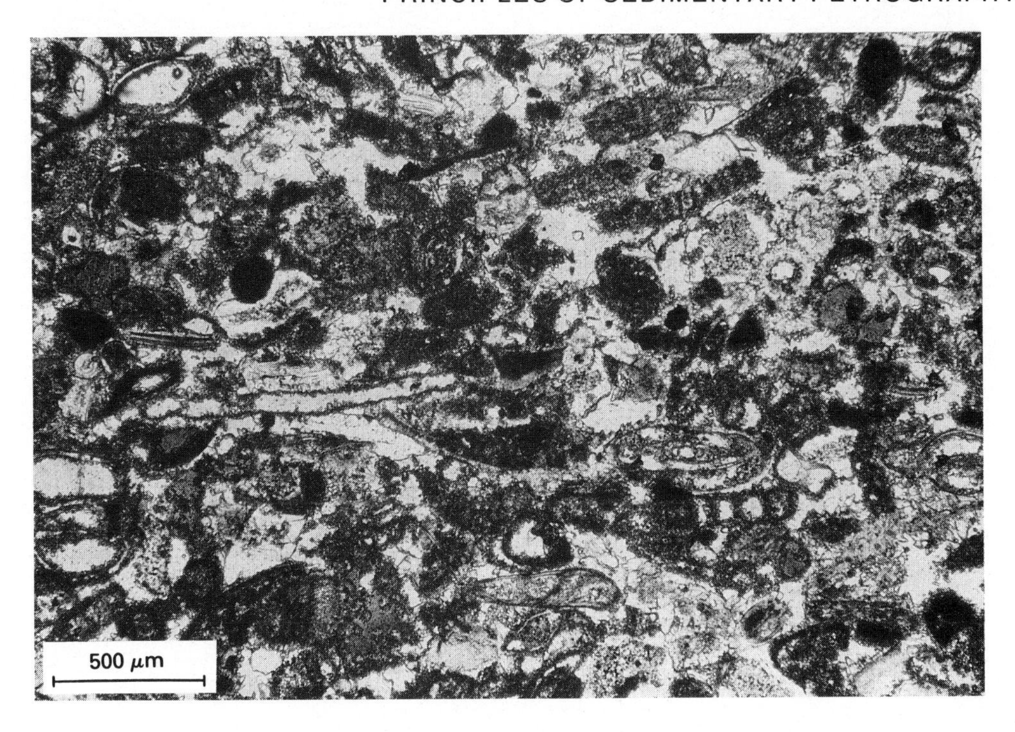

Fig. 5.9. Grain orientation in a carbonate turbidite. Elongate grains have preferential near-horizontal orientation with some imbrication. Miocene, Site 626, ODP Leg 101, Straits of Florida. Sample courtesy of ODP.

QUARTZ PROVENANCE STUDIES

Quartz grains may be single crystals (monocrystalline) and show straight extinction or undulose extinction (best measured on a universal stage). They may also be composite (or polycrystalline) with straight or undulose extinction within individual subcrystals and may, or may not, show elongation within the subcrystals. Quartz grains may also contain trails of inclusions, commonly fluid or gas filled. Subparallel trails of very small inclusions, or Boehm lamellae, are the products of intense strain, and indicate strain to have occurred within the quartz lattice, either *in situ* or before incorporation in the present sediment. The two may be distinguished by measuring the strain directions in several quartz grains; a random strain direction indicates strain prior to deposition, whereas a constant strain direction indicates stress was applied to the sediment. Quartz grains may also contain solid inclusions, the mineralogy of which may give evidence of provenance (e.g. sillimanite inclusions are excellent evidence for a metamorphic source area). A summary of quartz grain features is given in Fig. 5.11 and photomicrographs in Fig. 5.12. A more detailed discussion of quartz grains as indicators of provenance is given by Folk (1974b) who includes further references.

5.2.5 Depositional fabrics — a conclusion

In conclusion to this section on depositional fabrics it can be seen that a considerable amount of information can be gained from petrographic studies which can be used to supplement field studies and gain insight into the origins of the sediment. In particular provenance studies, both from modal compositions and from quartz grain variations, are not possible without petrographic analysis; these form an important aid in the interpretation of ancient sediments and their depositional environments. Grain size, form, roundness and sphericity

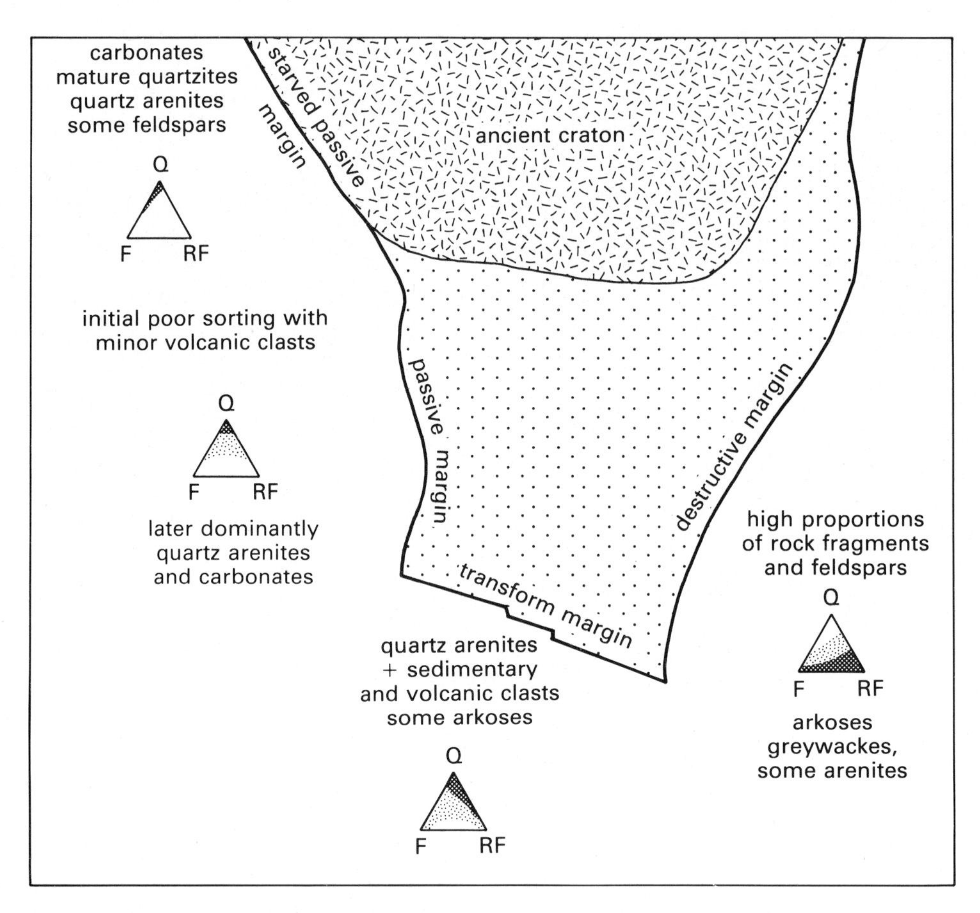

Fig. 5.10. Average modal compositions of groups of sandstones from different tectonic environments. This technique is only valid where many different sandstone modal compositions are available and cannot be used for single sandstone samples (modified from Folk, 1974b). Component details are documented in Table 5.5.

Table 5.5. Explanation of original quartz, feldspar and rock fragment end members used for the triangular plots of Fig. 5.10. Compiled from Folk (1974b)

Symbol and mineral	Explanation	Association
Q Quartz	All detrital quartz grains plus metaquartzite clasts (but not chert)	Mature sediments with little tectonic or igneous activity
F Feldspar	All detrital feldspars (whether orthoclase, microcline or plagioclase) plus granite and gneiss clasts	Post-tectonic granitic intrusions plus uplift of basement metamorphics
RF Rock fragments	All other rock fragments (chert, slate, schist, volcanics, carbonates, sandstone, mudrock, etc.)	Rapid uplift and erosion rate; more detailed associations possible on prevalence of rock fragment types

data are supplemented from grain orientations measured and mineralogical composition established.

5.3 DIAGENETIC FABRICS

Sedimentary rocks exhibit a vast range of diagenetic fabrics from compaction to cementation, dissolution to complete replacement. These diagenetic fabrics can be important indicators both of the depositional environment of the sediment and of the chemistry of a variety of fluids which have been flushed through the sediment during burial. Mineralogical changes can also be used to indicate temperature and depth of burial; this is of particular importance when correlated with organic temperature/depth indicators in studies of petroleum generation and migration. The sequence of diagenetic events within a sediment governs its potential as a hydrocarbon reservoir, as a host for mineralization, as a conduit for mineralizing fluids/hydrocarbons, or as a source for hydrocarbons. Research on diagenetic fabrics supplements information from field/core based work and from studies of depositional fabrics.

Diagenetic fabrics can be subdivided into two major types — those due to compaction and those due to chemical alteration, whether cementation, dissolution or replacement (Fig. 5.13). A sediment may exhibit one or more phases of any, or all, of these processes; the task of the sedimentary petrographer is to unravel the different events, place these in a paragenetic sequence and then build on these in interpreting the burial history of the sediment. Sound petrographic interpretations are essential before geochemical or isotopic analysis of the different diagenetic components is attempted.

Increased research into diagenesis has resulted in the introduction of many new terms; to aid the reader a brief review of general nomenclature relating to diagenetic environment and to porosity fabrics is included here before description of the various diagenetic fabrics.

5.3.1 General nomenclature used in diagenetic studies

DIAGENETIC ENVIRONMENTS

Three major diagenetic regimes were defined by Choquette & Pray (1970) (Fig. 5.14) and are summarized, with slight refinements, in Table 5.6. In most texts, simplification has led to the terms 'eogenetic', 'mesogenetic' and 'telogenetic' being replaced by 'near-surface', 'burial' and 'uplift' (or 'unconformity-related'). Near-surface diagenetic environments are further subdivided depending on whether the pore spaces are saturated with fluids, within the *phreatic* zone, or whether they lie above the water table so are partially fluid filled and partially gas filled, within the *vadose* zone (Fig. 5.15). The origin of the pore fluids, whether marine or meteoric (or fresh), also controls the diagenetic reactions which take place within these zones (Sections 5.3.3, 5.3.4 and 5.3.5).

POROSITY NOMENCLATURE

Porosity nomenclature is based on two classic papers: Choquette & Pray (1970) and Schmidt, McDonald & Platt (1977). Porosity within a sediment may be primary, a result of depositional voids between or within grains, or secondary, the result of dissolution, shrinkage or fracturing within the subsurface (Choquette & Pray, 1970). Primary porosity is commonly intergranular (or interparticle) (Fig. 5.16a), although rare intragranular porosity may be present within rock fragments, skeletal material and other detrital grains (Fig. 5.16b); intracrystalline primary porosity can be significant within clays and clay coatings and some dolostones. Secondary porosities can be due to the dissolution of detrital grains (Fig. 5.16c, d), of cements and authigenic minerals (Fig. 5.16e) plus shrinkage of certain sediments (particularly glauconitic sediments), and fracturing after cementation. Both primary and secondary porosity types may be mutually interconnected (Fig. 5.16a), therefore resulting in high permeabilities, or isolated (Fig. 5.16b), with consequent low permeabilities. The *effective* porosity of a sediment is dependent on the degree of interconnection of the pore spaces, which are thus utilized during fluid migration through the sediment.

Submicroscopic porosity (in places mistakenly termed 'chalky' porosity) is present in chalks, mudrocks and some altered detrital grains. Efficient impregnation with coloured resin within these sediments will demonstrate the presence of submicroscopic porosity under a good petrographic microscope; further

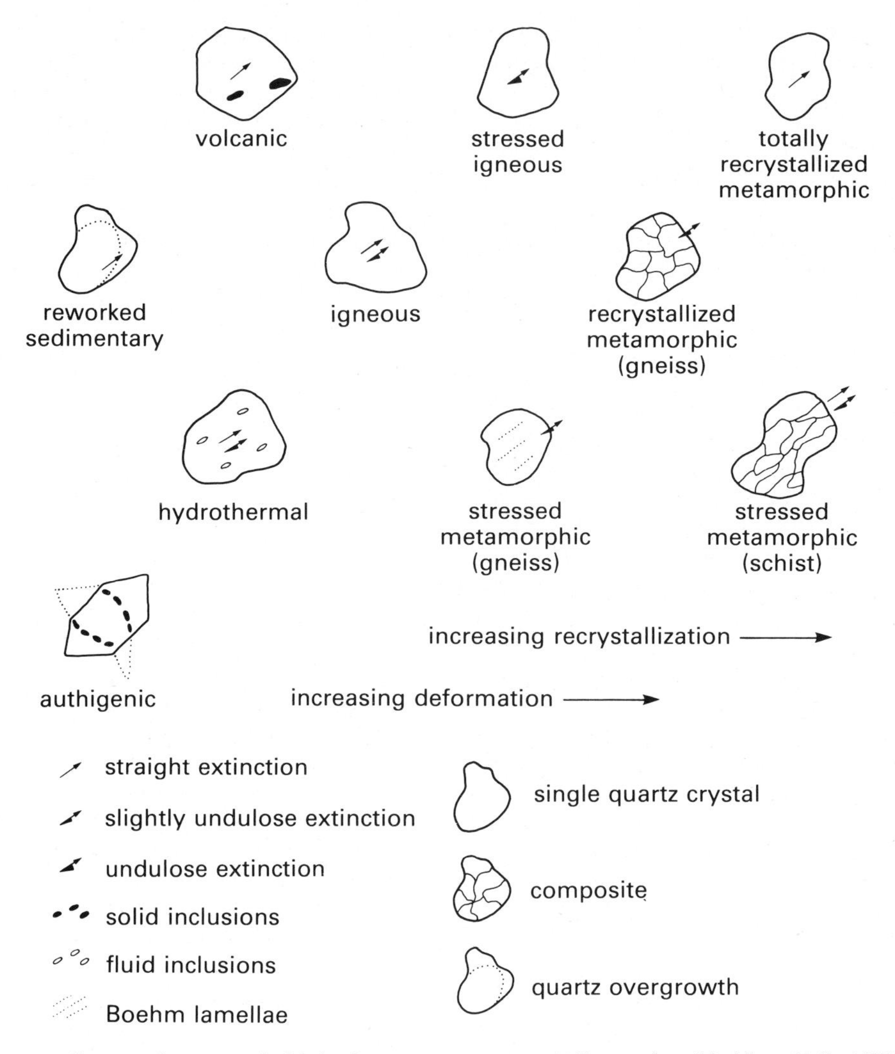

Fig. 5.11. Summary diagram of quartz grain fabrics for use as provenance indicators (modified from Folk, 1974b).

research on its fabrics and genesis requires electron microscopy.

5.3.2 Compaction fabrics

In thin section grainy sediments commonly show evidence of compaction, the result of progressive increase in overburden during burial. Before cementation, compaction takes place dominantly by mechanical processes, with slippage between individual grains, grain reorientation and subsequent fracture of some grains. With increase in overburden pressure, more stress is placed on the grain to grain contacts and compaction starts to proceed by chemical processes.

COMPACTION BY MECHANICAL PROCESSES

Compaction by mechanical processes involves movement by slippage between grains and the breakage, or fracture, of individual grains; as such it is sometimes termed *brittle compaction*. Any increase in overburden pressure in an unconsolidated sediment tends to result in a denser configuration of that sediment by movement between the individual

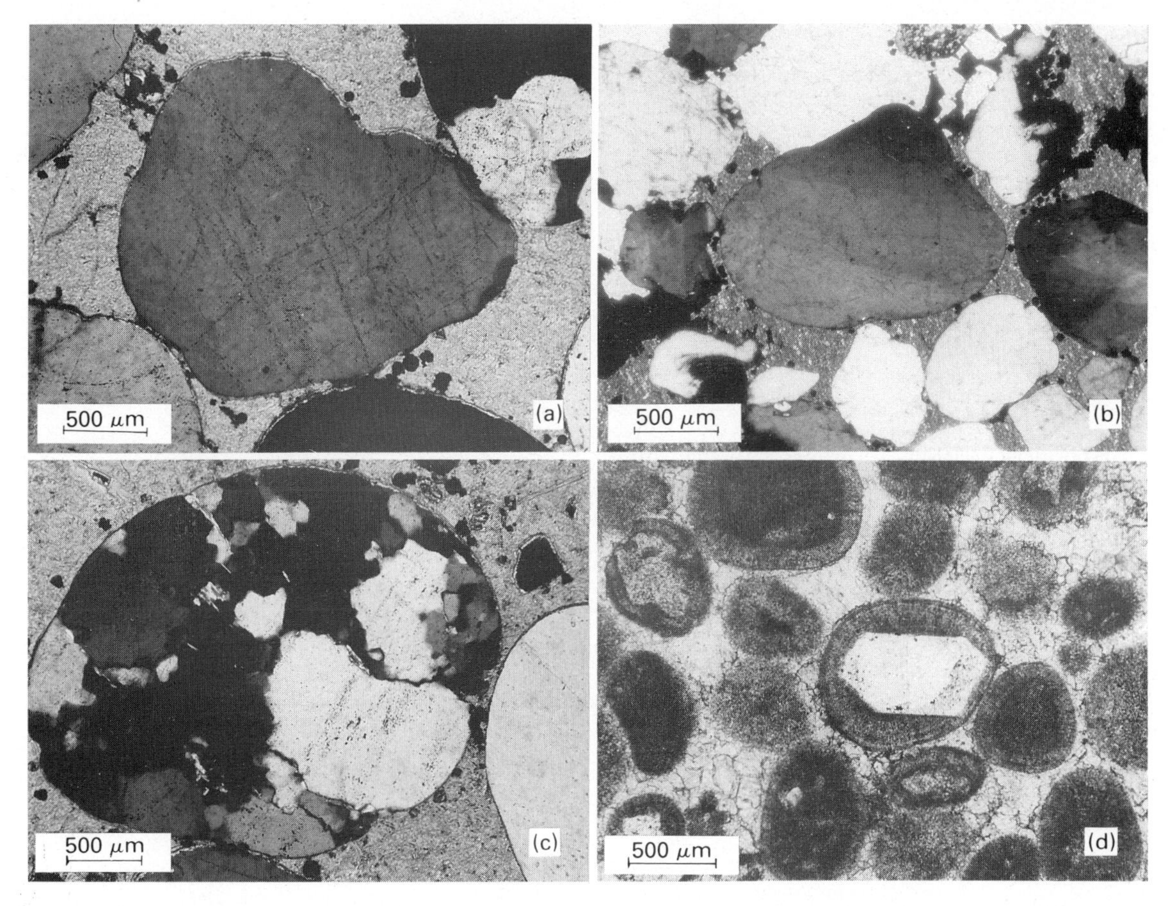

Fig. 5.12. Provenance of quartz grains: (a) Single, subrounded, monocrystalline quartz grain, showing straight extinction, from an igneous or vein quartz source. Crossed polars. Permian, Yellow Sands, Co. Durham, UK.
(b) Single, rounded, monocrystalline quartz grain, with undulose extinction, from an igneous, or stressed igneous, source. Crossed polars. Permian, Yellow Sands, Co. Durham, UK.
(c) Single, rounded, composite quartz grain, showing numerous subcrystals from a metamorphic source. Crossed polars. Permian, Yellow Sands, Co. Durham, UK.
(d) Authigenic, euhedral quartz grain in carbonate grainstone. Inclusions of carbonate are present in the outermost portions of the quartz crystal. Jurassic, Gilmer Formation, east Texas, USA.

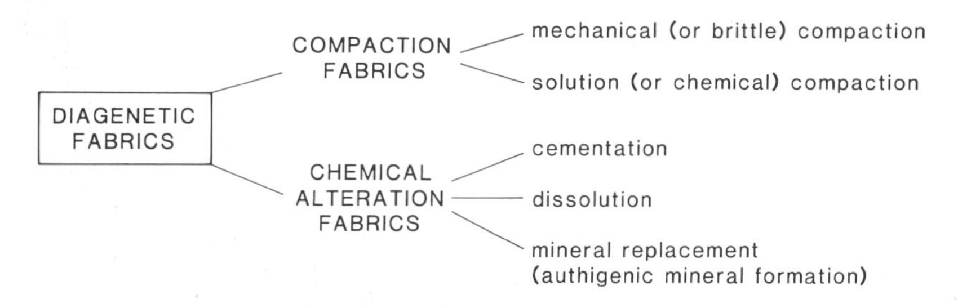

Fig. 5.13. Major classes of diagenetic fabrics.

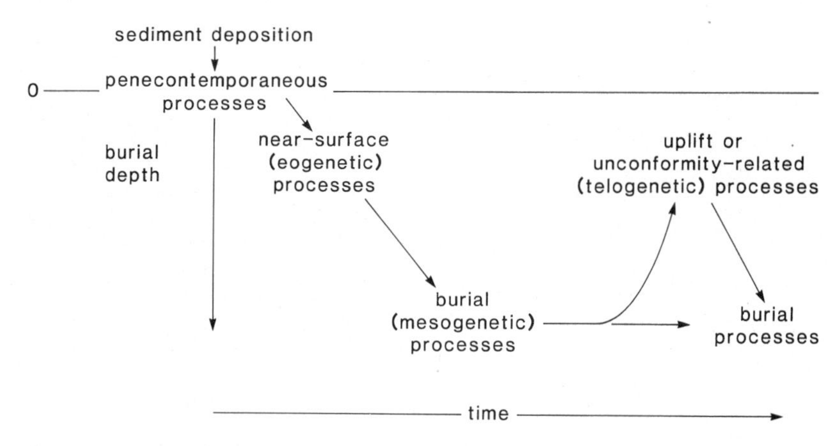

Fig. 5.14. Major diagenetic regimes (see also Table 5.6).

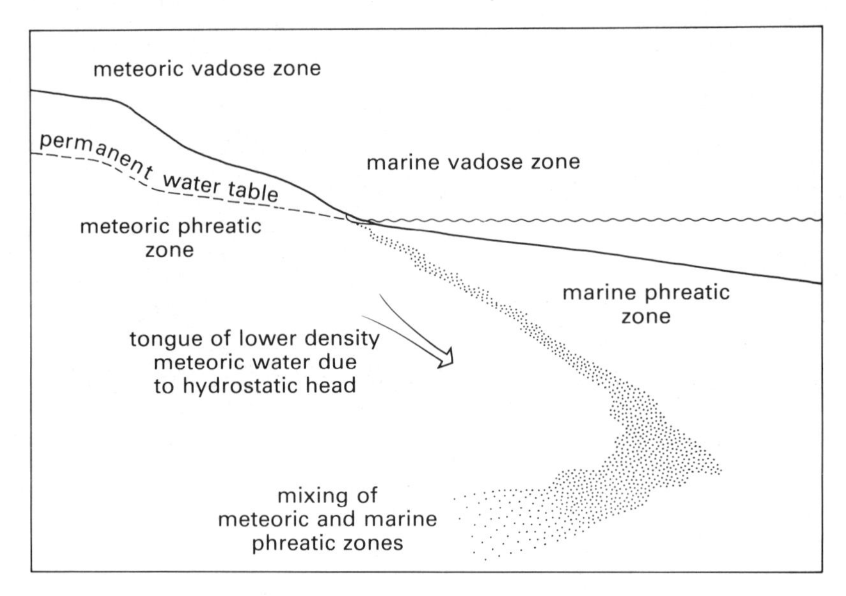

Fig. 5.15. Theoretical cross-section showing near-surface diagenetic environments.

grains and the consequent grain reorientation. Most spherical, cylindrical, disc-shaped or ellipsoidal grains (the regular grains of Vinopal & Coogan, 1978) simply slip against each other. However, any larger grains within the sediment commonly form bridges between the smaller grains creating a shelter porosity (Fig. 5.17a). If these larger grains are platey or concavo-convex (the radical grains of Vinopal & Coogan, 1978), brittle fracture may take place with continued increase of overburden pressure, resulting in destruction of the shelter porosity and tighter packing of the grains (Fig. 5.17b). The combination of grain size plus grain morphology therefore dictates which grains are susceptible to brittle fracture; these grains are commonly thin walled skeletal fragments (Fig. 5.18a), including bryozoans (Meyers, 1980), mica flakes (Fig. 5.18b) and plant material (Fig. 5.18c) (Ting, 1977). Such mechanical compaction is common during early diagenesis, prior to cementation.

Mechanical compaction and brittle failure may also take place after cementation, both as a result of shrinkage, expansion, and subsequent fracturing beneath a soil zone (Fig. 5.16e), and, commonly, as a

Table 5.6. Summary of the different diagenetic environments of sedimentary rocks. Modified from Choquette & Pray (1970)

Diagenetic environment	Description
Penecontemporaneous (syndepositional)	Diagenetic processes which occur within the depositional environment
Eogenetic (near-surface)	Diagenetic processes which occur within the zone of action of surface-related processes and surface-promoted fluid migration
Mesogenetic (burial)	Diagenetic processes that take place during burial, away from the zone of major influences of surface-related processes
Telogenetic (uplift or unconformity related)	Diagenetic processes which are related to uplift and commonly result from surface-related fluid migration

result of secondary porosity creation during burial diagenesis in both carbonate and siliciclastic host sediments (Fig. 5.18d, e). Mechanical compaction is relatively easy to envisage in near-surface regimes and can be readily demonstrated experimentally (Vinopal & Coogan, 1978; Shinn & Robbin, 1983). It is, however, more difficult to envisage the formation of a substantial amount of secondary porosity followed by fracturing during deep burial; this may necessitate a reduction in hydrostatic pressure (perhaps a relaxation of overpressuring), a change in pore fluid chemistry (cessation of secondary porosity formation) allowing subsequent fracture, or slight tectonic movement. Little research has yet been carried out on controls of brittle fracture within the subsurface, although Bjŏrlykke (1983) attributed cement dissolution (and consequent secondary porosity generation) to overpressuring, with subsequent cementation in the overlying strata a consequence of pressure release.

COMPACTION BY CHEMICAL PROCESSES I: SOLUTION COMPACTION BETWEEN INDIVIDUAL GRAINS

Compaction by chemical processes, or pressure solution, takes place in two ways: (i) in an uncemented sediment, by dissolution at isolated stressed grain to grain contacts and (ii) within a cemented sediment, by dissolution being concentrated along a particular surface, usually irregular, termed a *stylolite*. Pressure solution has been defined as the preferential dissolution of mineral material at points of stress (Wanless, 1979).

Once the grains within an uncemented sediment have assumed their densest configuration by slippage on grain surfaces, grain reorientation and fracture of radical grains, overburden pressure is transferred through grain to grain contacts, commonly point contacts (Fig. 5.19a). Progressive increase in overburden pressure increases the stress at grain to grain contacts, with resultant deformation of the crystal lattice at these contact points plus changes in the chemical potential within the immediate area of the contact (Fig. 5.19b). Continuation of stress causes dissolution of the contact area. Pressure solution is controlled not only by the degree of crystal strain, but also by the solubility of the specific crystal, its orientation, the saturation state of the surrounding fluid, the thermodynamics and kinetics of dissolution and ion concentration gradients in addition to the mechanism of solute transport (whether by diffusion or fluid flow) (Wanless, 1983). Progressive solution compaction leads to alteration of the grain to grain contacts, from the original point contacts, through planar (or tangential) contacts to interpenetrating (concavo-convex) and sutural grain to grain contacts (Fig. 5.19c–e). In sediments with monomineralic, or monocompositional, grains, the grain to grain contacts in, or near to, the principle stress direction commonly show similar degrees of solution compaction, with, for example, planar or interpenetrating contacts (Fig. 5.20a). However, sections transverse to the principle stress direction may exhibit a much lesser degree of solution compaction (Fig. 5.20b). The amount of solution is partially dependent on the orientation of the grains, as many anisotropic minerals dissolve more readily

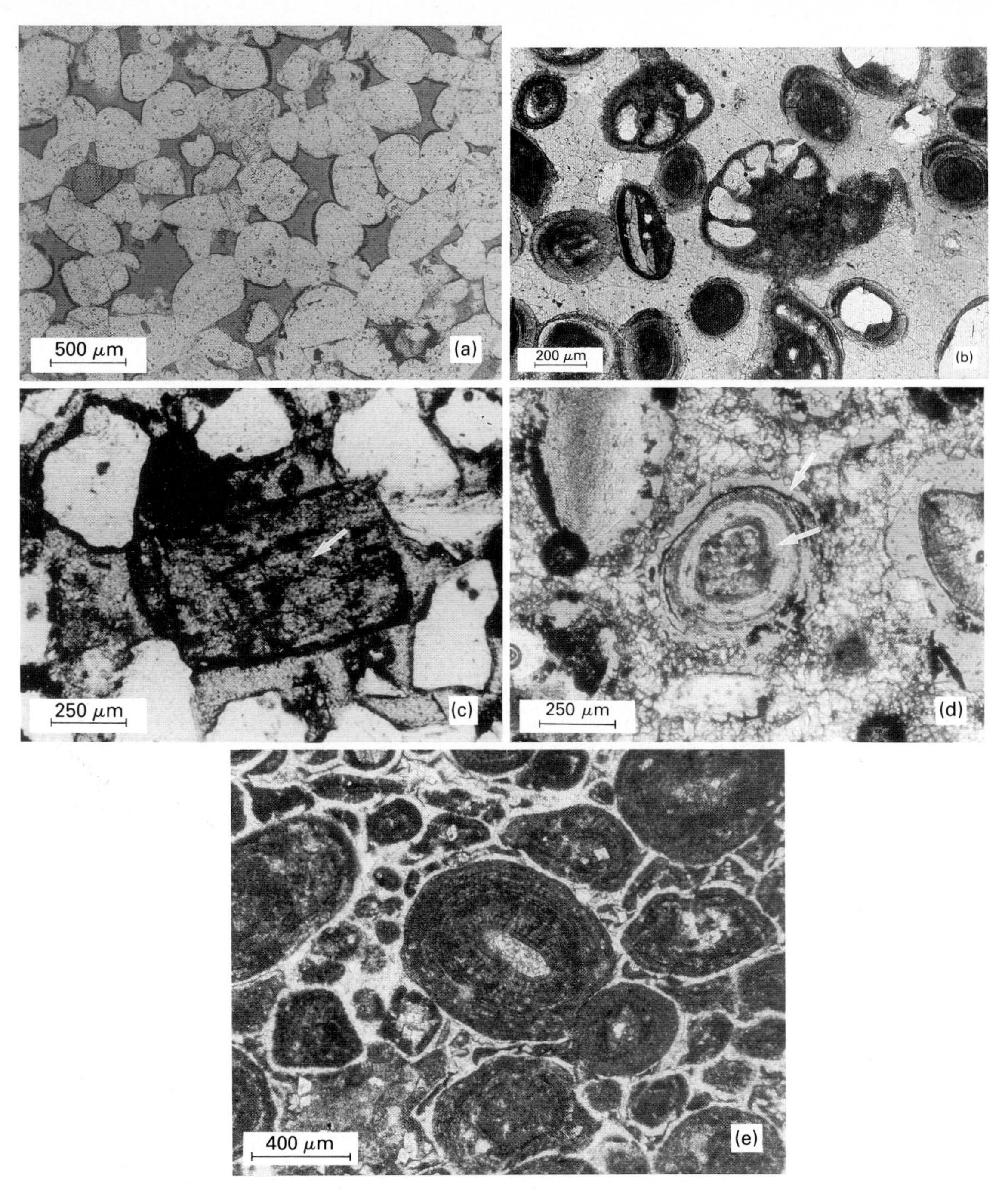

Fig. 5.16. Porosity types: (a) Intergranular porosity in aeolian sandstone; sandstone has been impregnated with coloured epoxy to demonstrate porosity. Cretaceous, Loch Aline Sandstone, Argyll, UK.
(b) Intragranular porosity within foraminifer test in poorly cemented carbonate sands; sample has been impregnated with coloured epoxy to demonstrate porosity. Holocene, Shark Bay, western Australia.
(c) Secondary porosity (arrowed) resulting from feldspar dissolution; clays indicate original shape of grain. If the clays were removed by flushing, an oversized pore would result. Sandstone was cemented prior to feldspar dissolution or pore space would not have been preserved. Cretaceous, Woodbine Formation, subsurface Arkansas, USA. Photomicrograph courtesy of Dr W. Ward, UNO.
(d) Secondary, intragranular, porosity (arrowed), resulting from partial dissolution of aragonitic ooid coatings. Surrounding cement, which maintains sediment fabric, is low Mg calcite. Pleistocene, Shark Bay, western Australia.
(e) 'Expansion' porosity, later occluded by calcite, in fitted-texture grainstone. Fractures surrounded individual grains after fitted-texture had been developed. Jurassic, Smackover Formation, east Texas, USA.

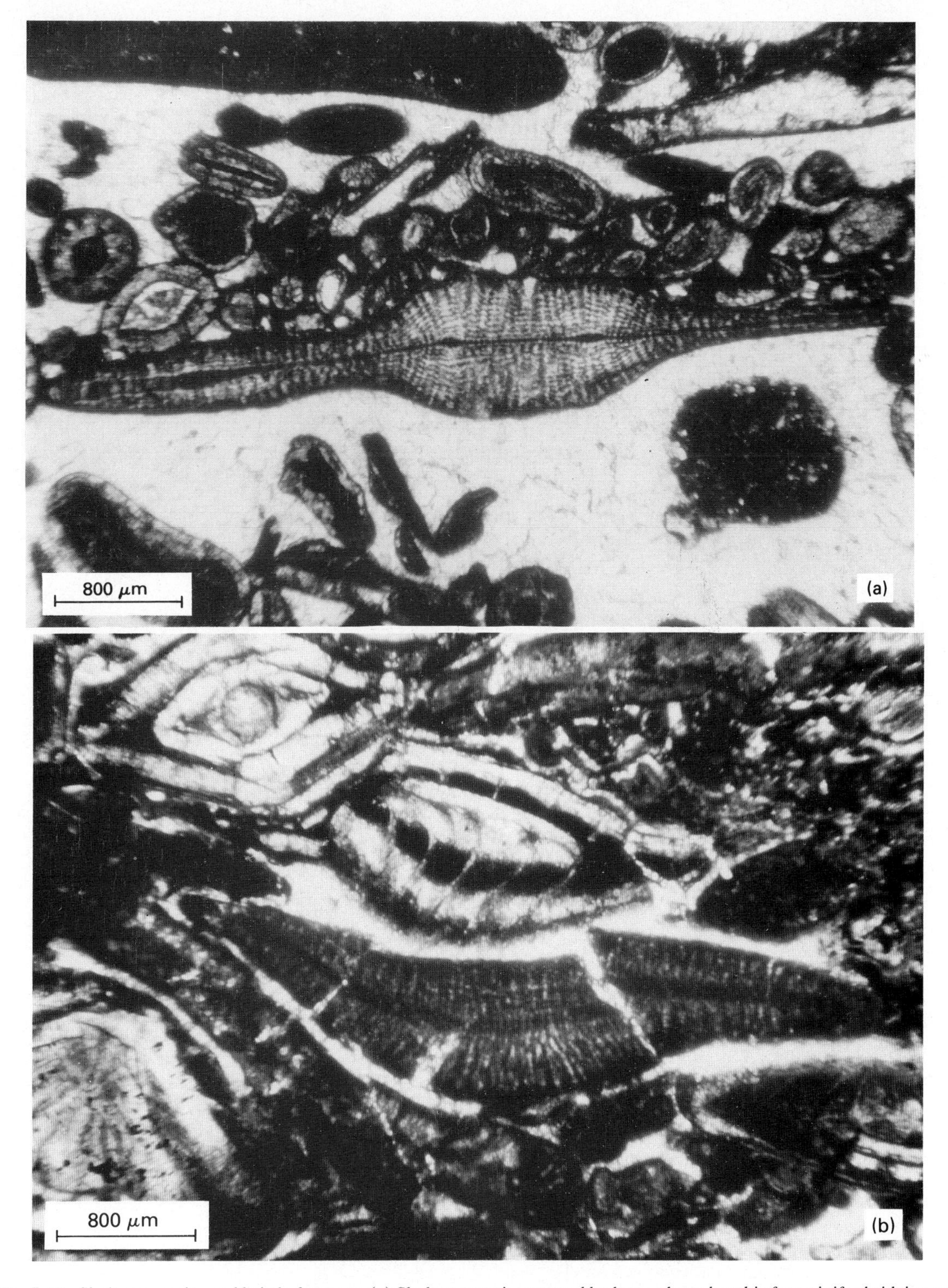

Fig. 5.17. Shelter porosity and brittle fracture: (a) Shelter porosity created by large platey benthic foraminifer bridging pore space. Grains (ooids, superficial ooids and peloids) are packed on top of foraminifer, whereas shelter porosity (here later occluded by calcite cement) is maintained below. Eocene carbonates, east India.
(b) Brittle fracture of foraminifer, a result of increase in overburden pressure before cementation. Brittle fracture has destroyed shelter porosity. Fractures have since been healed by calcite cement. Eocene carbonates, east India.

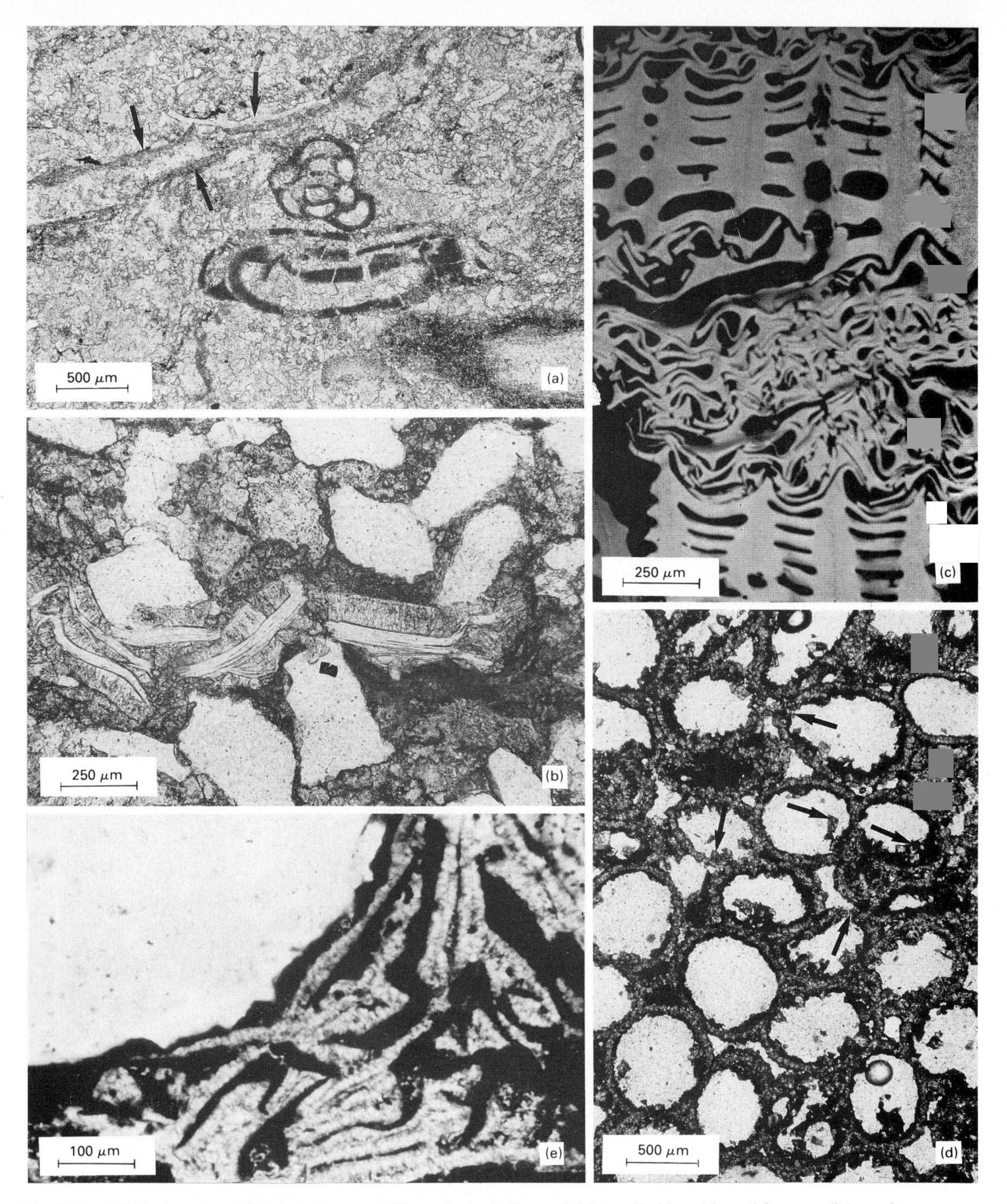

Fig. 5.18. Brittle fracture: (a) Brittle fracture of large foraminifer and thin-walled brachiopod (arrowed) in carbonate wackestone. Permian, Cadeby Formation, Yorkshire, UK.
(b) Brittle fracture and deformation of detrital mica flake in sandstone; acicular carbonate cement has later grown around mica. Lower Carboniferous, Lower Crag Point Sandstone, Northumberland, UK.
(c) Brittle fracture in plant material; section through conifer wood from peat showing collapse of earlier, central cells. Reflected light, oil immersion. Palaeocene, North Dakota, USA (from Ting, 1977).
(d) Brittle fracture after formation of secondary porosity. Flushing by fluids during diagenesis has caused ooids to dissolve, leaving framework of dolomitized circumgranular cements which have since fractured (arrowed). Jurassic, subsurface Smackover Formation, east Texas, USA.
(e) Brittle fracture of clay coatings. Clay coatings originally formed around altering feldspar grain, which has since completely dissolved (cf. Fig. 5.16c) leading to collapse of clay coatings. Surrounding pore space maintained by earlier cementation of quartz sand grains. Cretaceous, Tuscaloosa Formation, subsurface Arkansas, USA. Photomicrograph courtesy of Dr W. Ward, UNO.

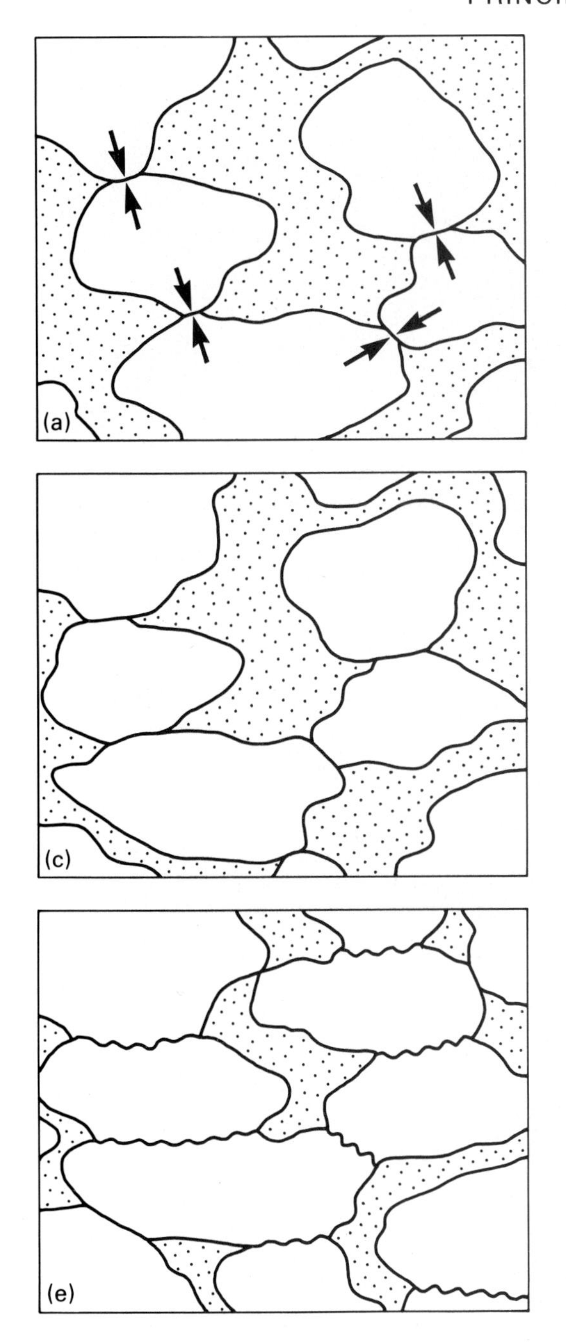

Fig. 5.19. Solution compaction between individual grains (porosity is stippled throughout): (a) Point grain to grain contacts (arrowed).
(b) Stressed grain to grain contacts (large arrows), leading to formation of dislocations in crystal lattice and subsequent dissolution, with lateral fluid transport of solutes (small arrows).
(c) Planar grain to grain contacts.
(d) Interpenetrating grain to grain contacts.
(e) Sutural grain to grain contacts.

along a certain crystallographic axis (Fig. 5.20c). Solution amounts are also dependent on the relative stability of different grains; stabilities may differ between mineral grains (Fig. 5.20d), or between grains and mudrock clasts where the clasts undergo ductile deformation so that, in places, they simulate a dispersed clay matrix (Scholle, 1979, p. 162).

Solution compaction between grains reduces the effective intergranular porosity within a grainy sediment (Fig. 5.19b) and is therefore important in

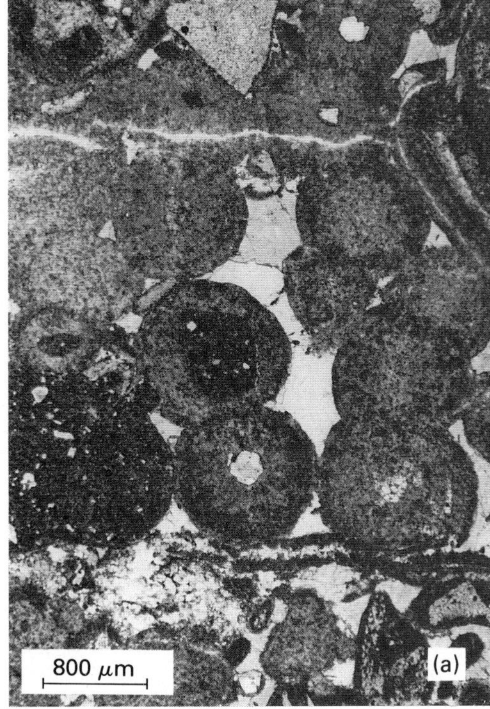

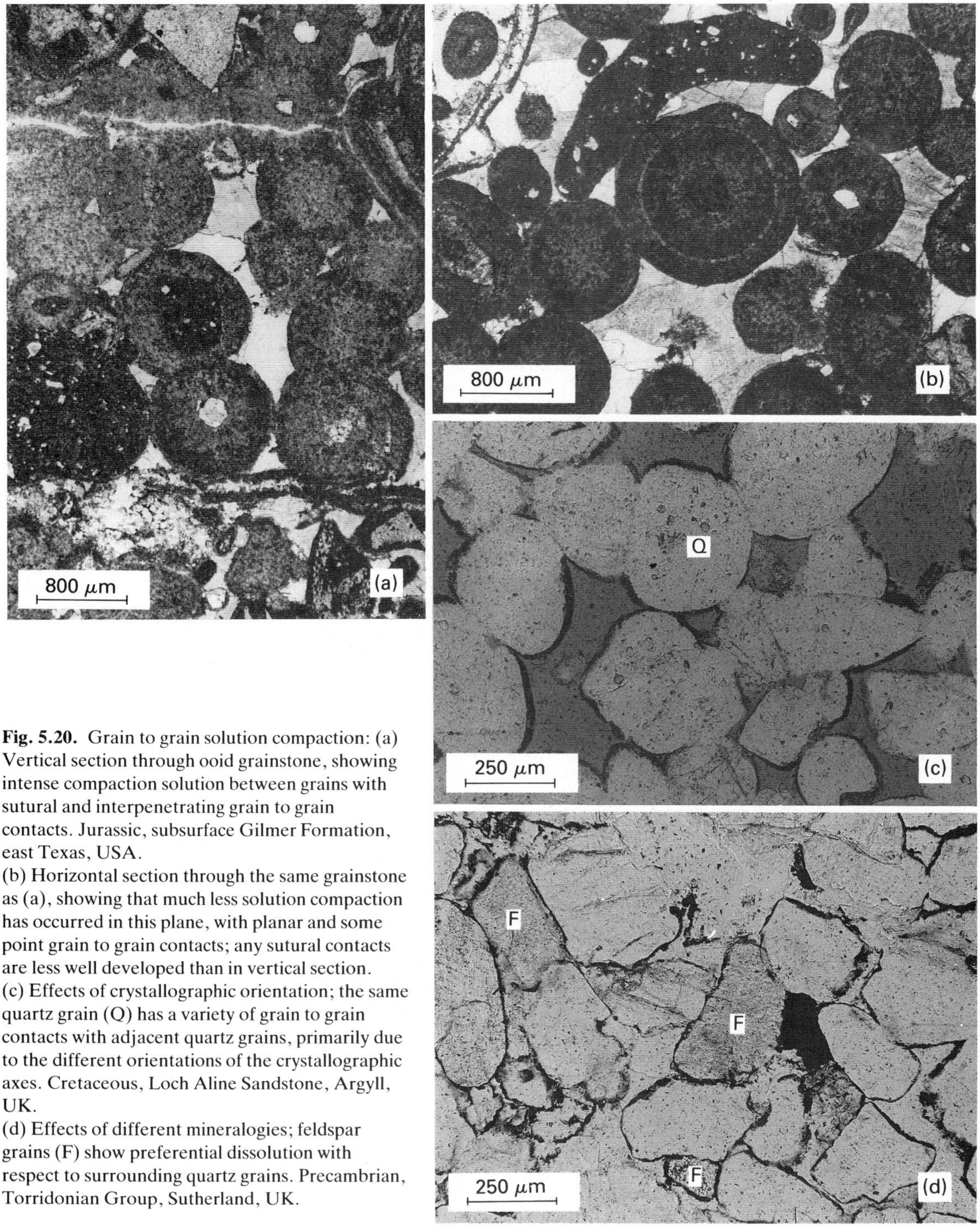

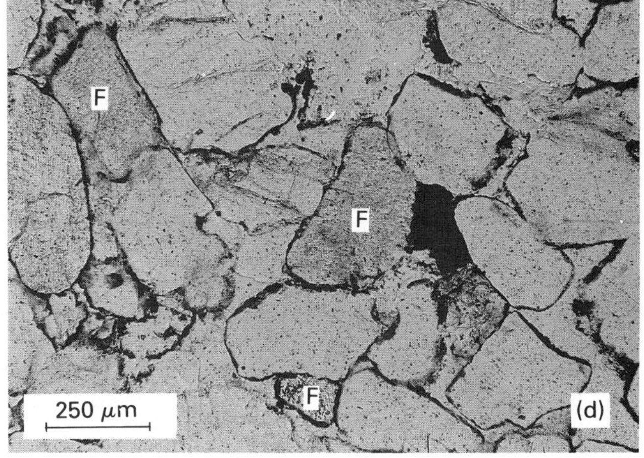

Fig. 5.20. Grain to grain solution compaction: (a) Vertical section through ooid grainstone, showing intense compaction solution between grains with sutural and interpenetrating grain to grain contacts. Jurassic, subsurface Gilmer Formation, east Texas, USA.
(b) Horizontal section through the same grainstone as (a), showing that much less solution compaction has occurred in this plane, with planar and some point grain to grain contacts; any sutural contacts are less well developed than in vertical section.
(c) Effects of crystallographic orientation; the same quartz grain (Q) has a variety of grain to grain contacts with adjacent quartz grains, primarily due to the different orientations of the crystallographic axes. Cretaceous, Loch Aline Sandstone, Argyll, UK.
(d) Effects of different mineralogies; feldspar grains (F) show preferential dissolution with respect to surrounding quartz grains. Precambrian, Torridonian Group, Sutherland, UK.

reservoir evaluation; in some cases solution compaction may account for most of the overall porosity reduction. Considerable solution compaction between grains indicates that cementation did not take place until relatively late in the diagenetic history of the sediment. This may be a factor of environment of deposition, or a consequence of rapid sedimentation rates, particularly in carbonate environments. The relative timing of solution compaction and cementation is important in reservoir evaluation (Fig. 5.21).

During solution compaction, the area immediately adjacent to the stressed grain to grain contact is taken into solution and may be flushed away from the immediate area by pore fluid migration. These solutes are thought to be the origin of cements within the subsurface, either as quartz overgrowth cements (e.g. Sibley & Blatt, 1976) or as calcite cements. They may be precipitated locally (Thompson, 1959) or carried for some distance in solution (Bathurst, 1975; Wanless, 1979). For example, Moore (1985) attributed the formation of zoned burial cements to be the result of former grain to grain solution compaction within the surrounding Jurassic grainstones, whereas James, Wilmar & Davidson (1986) concluded that the Nugget Sandstone is an overall exporter of silica, despite the development of some quartz overgrowth cements.

As cementation proceeds in clean sandstones or grainstones, or the matrix in clayey-sandstones and packstones becomes lithified, pressure is transmitted through the whole rock and there is no excess of stress at the grain to grain contacts. Solution compaction between individual grains therefore ceases, but continues over a much wider area along stylolites, seams and sutures within the sediment. Stylolites are commonly formed within carbonate sediments where cementation takes place early in the sediment's diagenetic history.

COMPACTION BY CHEMICAL PROCESSES II: SOLUTION COMPACTION IN CEMENTED SEDIMENTS

Stylolites and solution seams are present within many sediment types (Heald, 1955, 1959; Thompson, 1959; Weller, 1959; Trurnit, 1968; Brown, 1969; Bathurst, 1975; Logan & Semeniuk, 1976; Mimran, 1977; Wanless, 1979; Gillett, 1983 and others). Both stylolites and solution seams transect the cemented sediment and develop perpendicular to the axis of maximum stress, which may be either overburden pressure or tectonic stress. Because stylolites and solution seams are zones of preferential solution within a sediment, small fragments of less soluble minerals tend to accumulate along the solution seam. Thus, in immature sandstones and carbonates, clay minerals may become concentrated along a seam. Both Weyl (1959) and Oldershaw & Scoffin (1967)

Table 5.7. Common diagenetic processes which affect detrital grain types, after McBride (1985). For each mineral, processes are listed with the most common first

Grain type	Processes	Grain type	Processes
K-feldspar	calcitized zeolitized dissolved kaolinized albitized	Plagioclase	calcitized dissolved albitized zeolitized
Carbonate rock fragments	dissolved dolomitized recrystallized	Volcanic rock fragments	calcitized dissolved chloritized zeolitized
Micaceous metamorphic rock fragments	mashed dissolved	Chert rock fragments	dissolved calcitized
Claystone/shale/siltstone rock fragments	mashed	Muscovite	kaolinized dissolved
Biotite	dissolved argillized	Unstable heavy minerals	dissolved calcitized

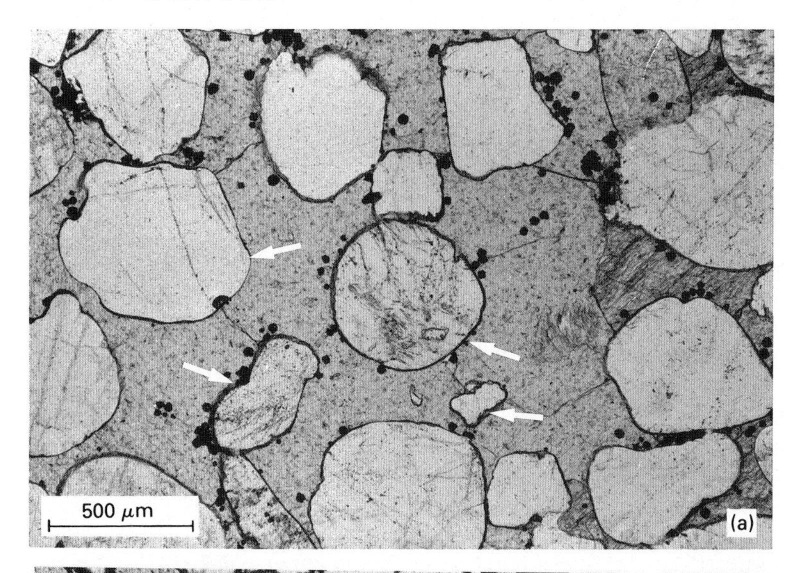

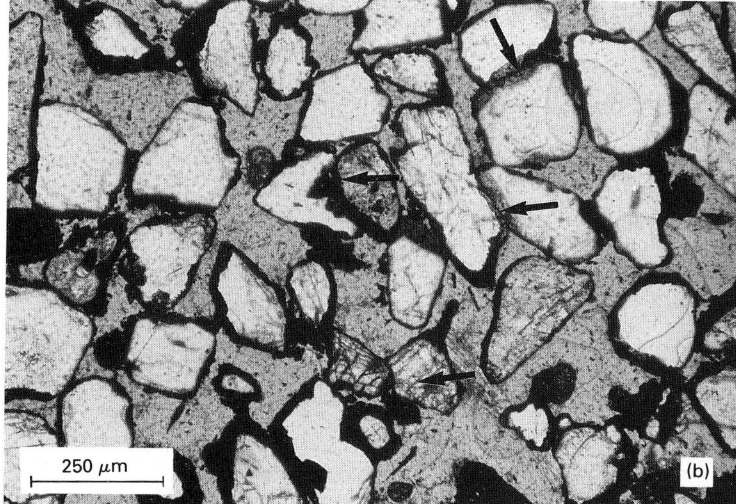

Fig. 5.21. Relative timing of compaction and cementation:
(a) Floating grains (arrowed) and point grain to grain contacts in calcite-cemented sandstone. Cement contains oxidized pyrite crystals (opaque). Cementation occluded porosity before little compaction had taken place. Permian, Yellow Sands, Co. Durham, UK.
(b) Point and planar (arrowed) grain to grain contacts in poorly cemented sandstone. Grains have been slightly compacted before cementation with a thin coating of iron-rich calcite; considerable intergranular porosity remains. Recent, Lafourche Delta, Louisiana, USA.
(c) A combination of early, echinoderm overgrowth (E) and later, intergranular (I) cements. Echinoderm overgrowth cement grew before much compaction as echinoderm fragment has few contacts with other grains. Away from areas of overgrowth cement, considerable grain to grain compaction has taken place, resulting in sutural grain contacts (arrowed). Intergranular cement occluded remaining porosity. Jurassic, subsurface Gilmer Formation, east Texas, USA.
(d) Interpenetrating grain to grain contacts in ooid grainstone. The presence of only thin early cements (arrowed) resulted in interpenetration between adjacent grains with loss of intergranular porosity. Remaining porosity was occluded by calcite cement. Jurassic, subsurface Smackover Formation, Louisiana, USA.

documented how stylolites and solution seams preferentially develop in clay-rich layers in carbonate sediments, with the solution residue accumulating in stylocumulates (Logan & Semeniuk, 1976) (Fig. 5.22a).

Wanless (1979) proposed that there were three basic styles of solution compaction, dependent on the maturity of the sediments and the responsiveness of the various units (Fig. 5.23). He demonstrated that pervasive (or non-seam) solution compaction may result in considerable reductions in thickness (commonly up to 50% and more, rarely as much as 80%) of finely crystalline carbonate units. The more visible solution seams (whether high-amplitude stylolites, microstylolite swarms, or fine clay seams) may also represent considerable amounts of dissolution in the subsurface. With high-amplitude stylolites, the minimum dissolution is equivalent to the maximum amplitude of the stylolite (Fig. 5.22b, c).

Until the last ten years most authors thought that

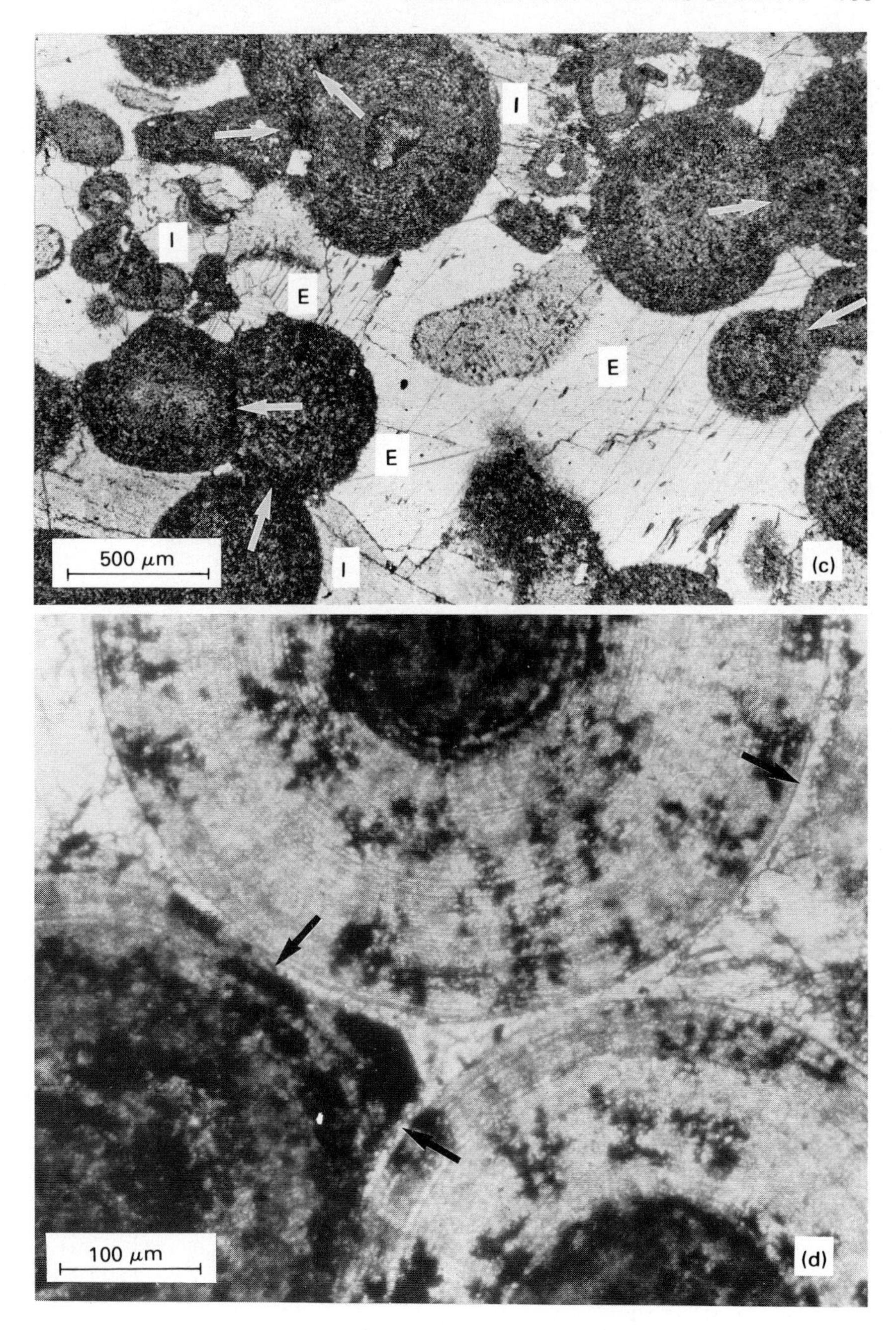

solution seam and stylolite formation took place during deep burial diagenesis. Recently, however, Shinn & Robbin (1983) have demonstrated the formation of stringers, superficially very similar to solution seams, forming in Recent Bahamian muds under relatively low pressures (6,784 psi, equivalent to burial depths of approximately 2,000 m) over short periods of time. These stringers were composed of organic matter, principally sea grasses. In Mississippian carbonates of the United States,

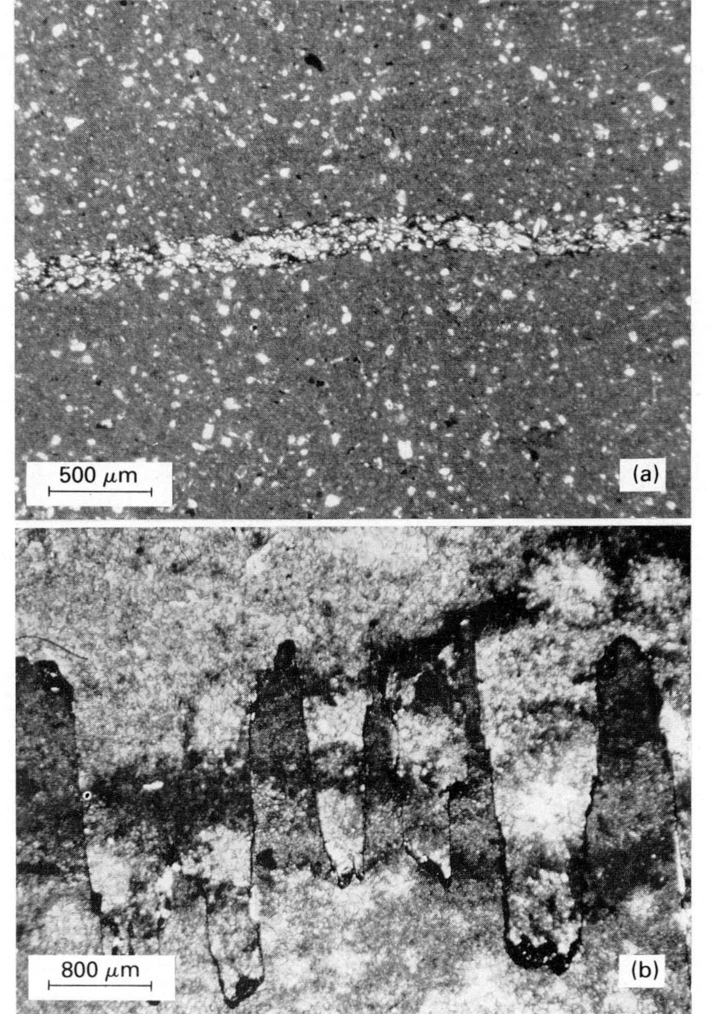

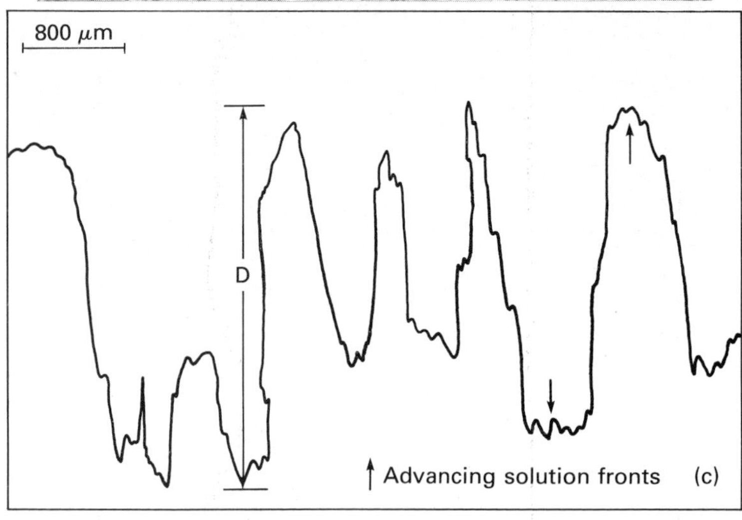

Fig. 5.22. Solution compaction in cemented sediments:
(a) Thin, low-amplitude, stylocumulate in silty carbonate mudstone. Quartz silt and clay minerals are concentrated along the stylocumulate. Jurassic, subsurface Gilmer Formation, east Texas, USA.
(b) High-amplitude stylolite in carbonate mudstone. Fine residue (probably clay minerals) emphasizes the line of the stylolite. Permian, Yates Formation, New Mexico, USA.
(c) Tracing of stylolite from (b). Minimum dissolution that has taken place is distance D, (here 2,400 μm [2.4 mm]), the greatest amplitude between adjacent peaks of the stylolite.
(d) Low magnification photomicrograph of the ooid grainstone of Fig. 5.21(c), showing high-amplitude stylolite (arrowed) cutting both echinoderm overgrowth cements and post-compaction, intergranular cements. Jurassic, subsurface Gilmer Formation, east Texas, USA.
(e) Medium- to low-amplitude microstylolites in silty carbonate mudstone. Jurassic, subsurface Gilmer Formation, east Texas, USA.

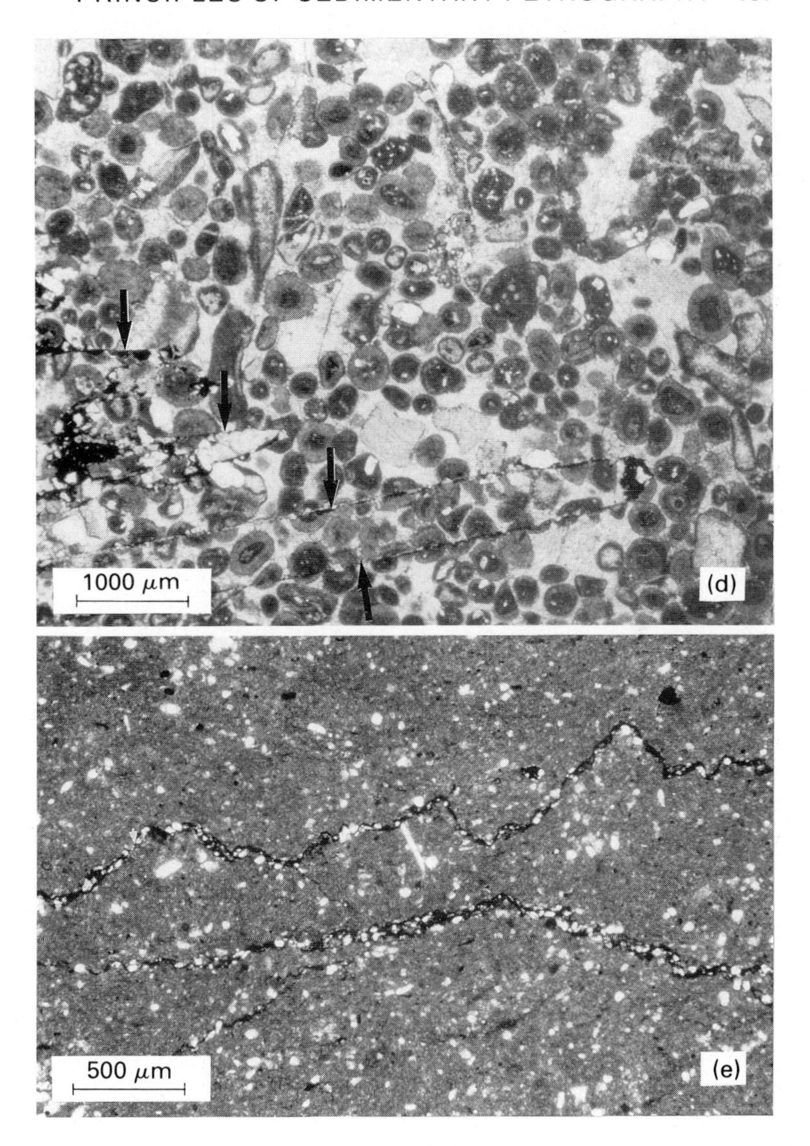

Meyers (1980) and Meyers & Hill (1983) have demonstrated mechanical and chemical compaction at less than 2,000 m burial depth with initiation of chemical compaction at depths of 'tens to hundreds of metres'. Earlier, Beall & Fischer (1969) had observed that the transition from mechanical compaction to solution compaction occurs at around 250 m sediment depth in some North Atlantic DSDP sites. Harris (Meyers, 1984 personal communication) has found that high amplitude stylolites formed at burial depths of less than 1,000 m in the Burlington Formation (Mississippian). This combined evidence indicates that *some* stylolite and solution seam formation can take place during shallow burial; these solution features should not, therefore, be used as evidence for deep burial.

The timing and mode of formation of stylolites remains an area of contention in sedimentary petrography, although many stylolites show evidence of at least some component of their formation relative-

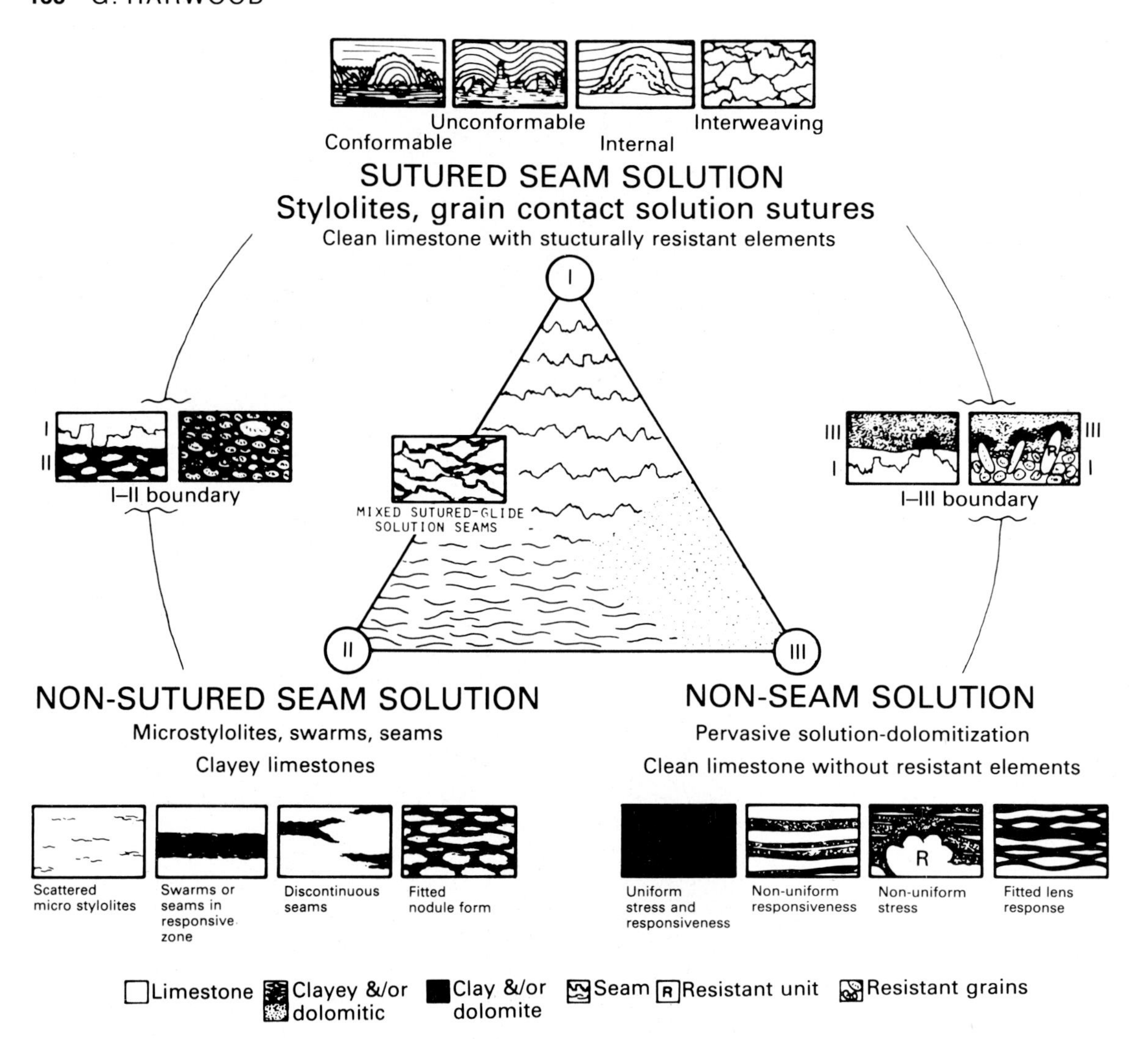

Fig. 5.23. Basic styles of solution compaction in carbonates (from Wanless, 1979). These styles can also be used to describe solution compaction in siliciclastic sediments.

ly late, if not necessarily deep, in the diagenetic history of their containing sediment (Fig. 5.22d).

However and whenever they form, stylolites and solution seams are important in reservoir evaluation. Koepnick (1985) has demonstrated how stylolites act as migration barriers; dense stylocumulates in particular are barriers to fluid migration. Stylolites with little insoluble residue and high amplitudes may, however, be pathways of preferential fluid migration, and may remain open, or be later filled by cements. Solution compaction along seams may involve dissolution of considerable volumes of minerals, commonly quartz and calcite; these solutes may locally source cementation, lessening the permeability of the zone surrounding the solution seams (Wong & Oldershaw, 1981). However, some authors (notably Scholle, 1971 and Wachs & Hein, 1974) are of the opinion that pressure solution is too late to account for the origin of most burial cements. More recently, Scholle & Halley (1985) concluded that, although solution compaction features are widespread in many rocks, as is evidence of late cementation, connecting the two processes in most instances requires a 'leap of faith'.

ESTIMATION OF AMOUNTS OF COMPACTION

Although there are many qualitative studies demonstrating the wide distribution of mechanical and chemical compaction fabrics, published quantitative results are considerably rarer. Measurement of stylolite amplitudes (Fig. 5.22b, c) or of solution seam thickness provides only a minimum estimate of the amount of sediment removed (Stockdale, 1926; Mossop, 1972) and becomes more difficult with low amplitude stylolites and solution seams (Fig. 5.22f). Volume reductions in carbonates of approximately 40% were detailed by Dunnington (1967) and Park & Schot (1968, p. 72); similar calculations are lacking for other sediments. Perhaps a future valid contribution to this field could be made by a combination of experimental simulation of mechanical and chemical compaction in conjunction with the use of computer imagery in volume reconstruction; at present we only know compaction is extremely effective at reducing both absolute volume and porosity in many sediments.

5.3.3 **Cementation**

Many different minerals form cements. Quartz and calcite are perhaps the most common but chlorite, clay minerals, hematite, dolomite, siderite, aragonite, phosphates and evaporite minerals (particularly halite) also occur as cements, as do zeolites, particularly in volcanogenic sediments (Table 5.1). More rarely, hydrocarbons may also form a cement. Although the mineralogy may differ, the relationships between the various generations of cements and their adjacent grains enable the timing of cementation relative to other diagenetic events to be evaluated (Fig. 5.21). Many cements can be clearly recognized using a combined transmitted-reflected light petrological microscope. However, cathodoluminescence (Chapter 6) can give additional information on crystal growth directions and can aid elucidation of many cement relationships; it is an important adjunct for the study of many carbonate, quartz and feldspar cements. Staining (Chapter 4) and blue light emission spectroscopy (Dravis & Yurewicz, 1985) may also aid evaluation of some diagenetic sequences, particularly in carbonates. In addition, scanning electron microscopy, both of solid specimens and of etched thin sections (e.g. Sandberg, 1985), can be used to confirm further deductions based on thin section petrography (see Chapter 8).

Conformable cements may form as an overgrowth on existing grains within a sediment, in many places growing in optical continuity with a substrate of the same mineralogy (e.g. Fig. 5.21c). Alternatively, cements may nucleate on grains as separate, or *disconformable*, crystals (Fig. 5.24a), or they may grow from a few nucleation points so that they enclose the grains as poikilotopic cements (Fig. 5.24b); they may be the same mineralogy as the grains they enclose. The morphology of these separate, disconformable crystal cements is, in many cases, dependent on the diagenetic environment and contained fluids in which they form (e.g. Folk, 1974a; Longman, 1980), although growth kinetics (Given & Wilkinson, 1985) are also an important factor. A summary of the many descriptive terms used in petrologic identification of cement morphology is given in Fig. 5.25.

CEMENT TYPES I: GRAIN OVERGROWTH CEMENTS

Grain overgrowth cements (or conformable cements) commonly develop on quartz and feldspar detrital grains and on some skeletal carbonate fragments (Fig. 5.26). Quartz overgrowth cements can be clearly recognized using a petrological microscope if clay- or hematite-rich coatings (commonly inaccurately termed 'dust' rims) are trapped between overgrowth cement and grain (Fig. 5.26a, b). More rarely, two coatings may become trapped, indicating two stages of development of the overgrowth cement. Where no such coatings are present, quartz overgrowth cements are much more difficult to recognize. Such cements are, however, typically inclusion-free (or, at least, inclusion-poor) and thus, with care, may be distinguished from the parent grain. Further, if sufficient intergranular pore space is present within the sediment, the overgrowth cements will develop planar euhedral crystallographic faces (Fig. 5.26c, d), although it should be remembered that continued enlargement of the overgrowth cements will lead to compromise boundaries between the cements. Such compromise boundaries may be difficult to distinguish from compacted fabrics: cathodoluminescence can, in some cases,

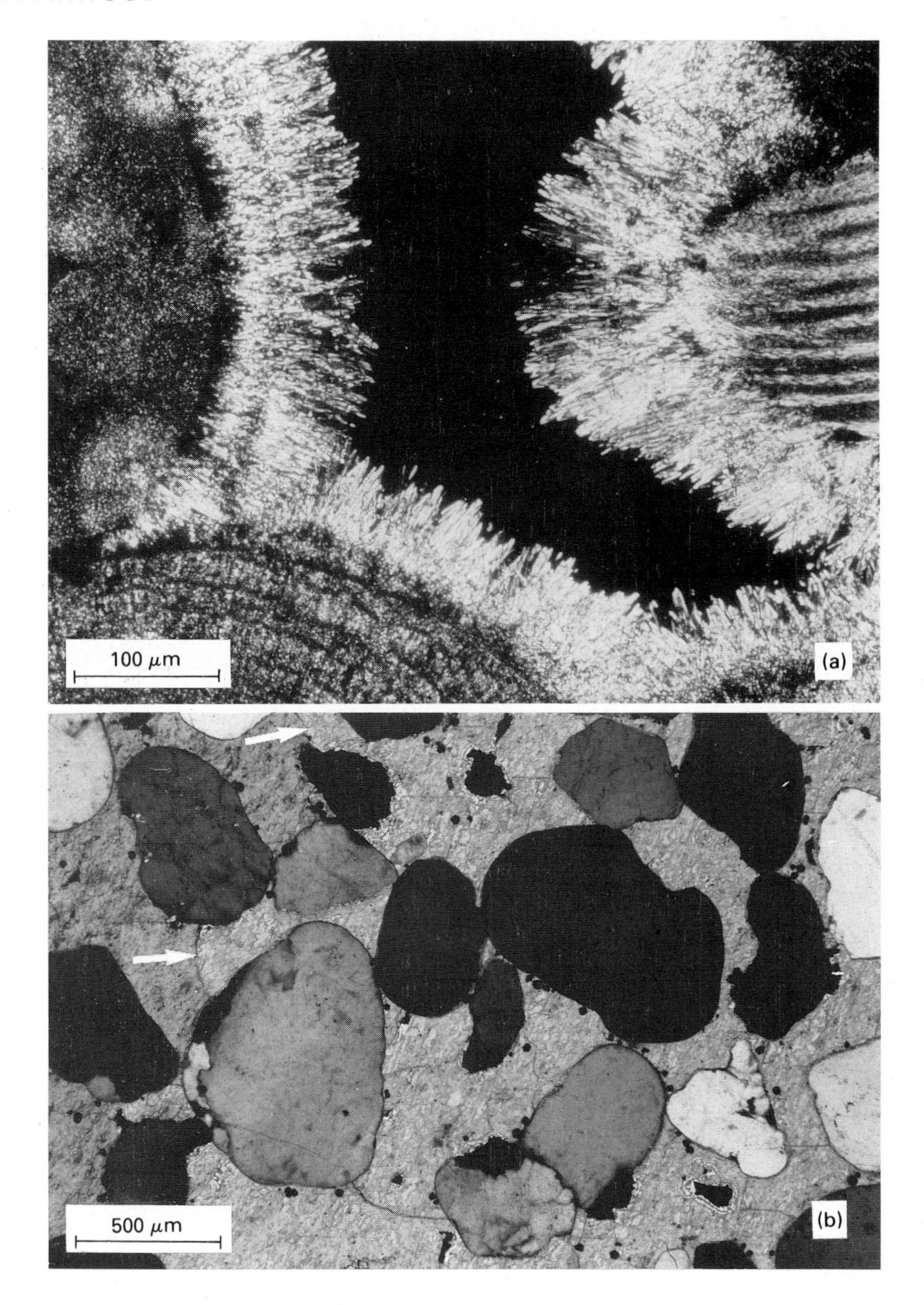

Fig. 5.24. Disconformable crystal cements: (a) Multiple nucleation sites, resulting in circumgranular, acicular, aragonite cements growing on skeletal fragments in modern, shallow-water, carbonate sands. Crossed polars. Recent, US Virgin Islands.
(b) Few nucleation sites, resulting in poikilotopic calcite cement enclosing quartz grains; ιwo calcite cement crystals only are present, their boundary is arrowed. Crossed polars. Permian, Yellow Sands, Co. Durham, UK.

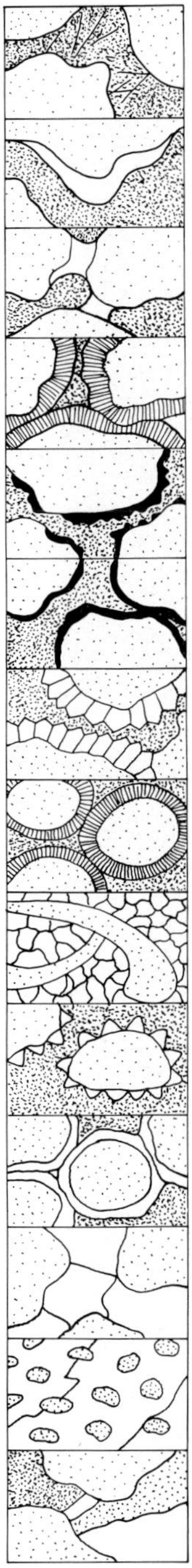

Cement terminology	Less correct terminology	Description
needle (or fine acicular)	whisker needle fibre	thin (<10 μm) cements of single or en echelon crystals; form in or near soil zone
pendant or microstalactitic		cement forms on 'droplets' beneath grains within vadose zone
meniscus		cement forms at or near grain-to-grain contacts; meniscus characterizes vadose zone
acicular	fibrous	thin straight long form (~ 10 μm × 300 μm +), typical of marine phreatic zone
pelodial		dark, microcrystalline irregular coating to grains and pores; marine phreatic organ
'micritic' or microcrystalline		microcrystalline cement coats grains and may form 'bridges' between grains; marine phreatic environment
columnar		broad cements (~ 20 μm +), commonly longer than broad. Common in mixed meteoric/marine phreatic environments
circumgranular isopachous acicular		equal thickness acicular cements surround grains; characterize marine phreatic environments
equant	blocky	equidimensional cement crystals, (commonly ~ 100 μm +); form in meteoric phreatic or burial phreatic environments
circumgranular equant		equidimensional cements surround grains; typify meteoric phreatic regime
overgrowth		cement is in optical continuity with grain substrate; form during burial diagenesis
sparry		coarse cement crystals (~ 300 μm), commonly equidimensional; typify burial environments
poikilotopic		coarse cement crystals enclose grains; form in phreatic, commonly burial, regime
baroque (or saddle)		coarse cement crystals with undulose extinction; characterize deep burial environments

cement   grain   porosity

Fig. 5.25. Summary diagram of descriptive terms used for cement morphologies. For scale, grains shown have an average diameter of 500 μm.

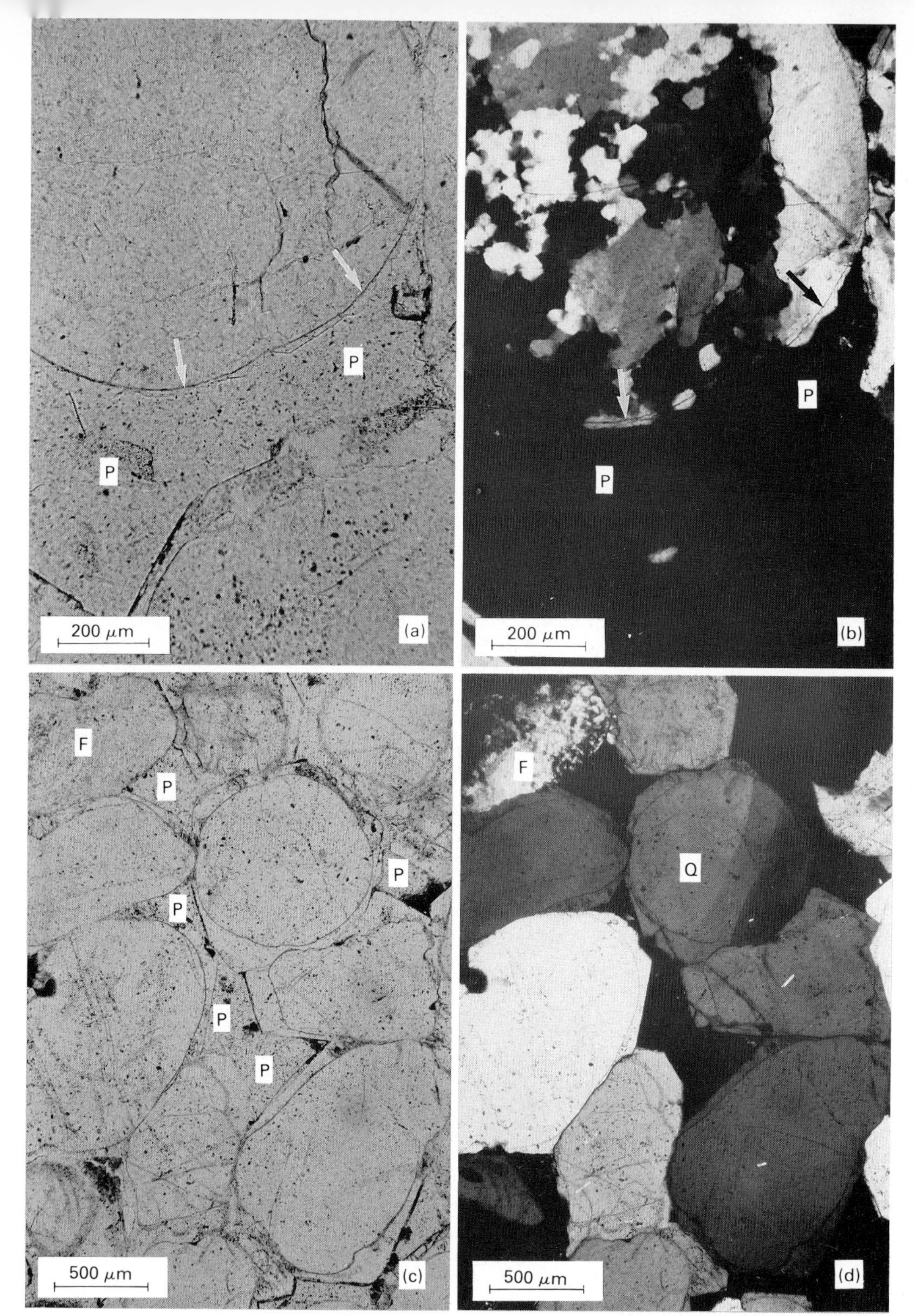

Fig. 5.26. Grain overgrowth cements I: (a) Thin quartz overgrowth cements (arrowed) adjacent to open pore (P). Overgrowth cements grew on rounded and subrounded grains, Permian, Penrith Sandstone, Cumbria, UK.
(b) Crossed polars photomicrograph of (a), showing composite nature of detrital quartz grain plus optical continuity of overgrowth cements with their substrate. Haematite-rich coatings (arrowed) define detrital grain surfaces.
(c) Planar, euhedral faces of quartz overgrowth cements, which have grown into pore spaces (P). Cement thickness varies around pore space and is partially a function of crystal orientation and subsequent growth direction. Haematite-rich coatings on detrital grain surfaces are clearly visible. Permian, Penrith Sandstone, Cumbria, UK.
(d) Crossed polars photomicrograph of (c). Note optical continuity of overgrowth cement on strained, detrital quartz grain (Q). Altered feldspar grain (F) has no overgrowth cement.

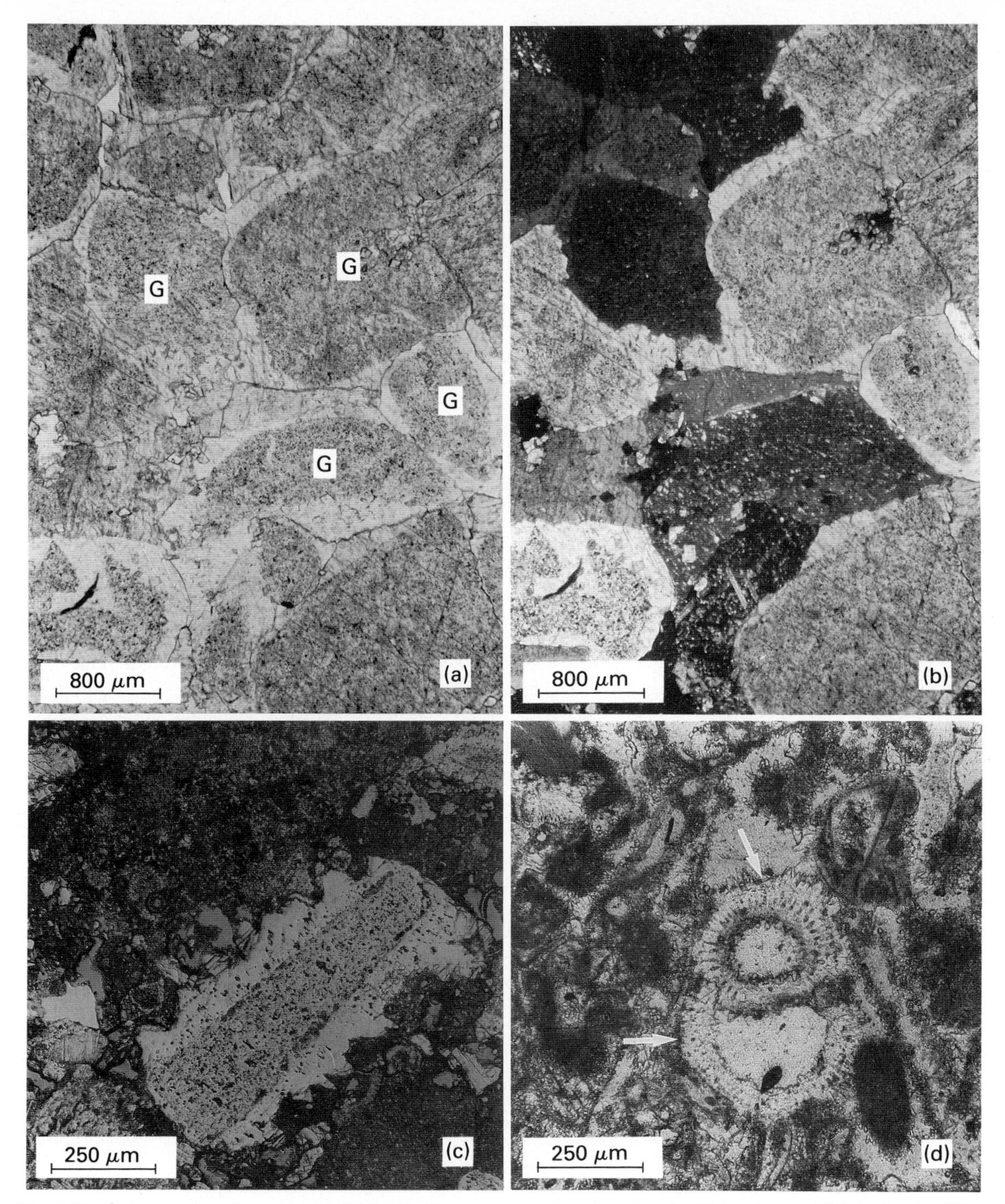

Fig. 5.27. Grain overgrowth cements II: (a) Echinoderm overgrowth cements in crinoidal grainstone. Original grains (G) appear greyer than overgrowth cements as they include traces of the original internal pores plus some microdolomite crystals. Overgrowth cements occlude intergranular porosity and form compromise boundaries. Mississippian, Burlington Formation, Illinois, USA. Sample courtesy of W.B. Meyers, SUNY.
(b) Crossed polars photomicrograph of (a), demonstrating optical continuity of overgrowth cements and occlusion of intergranular porosity.
(c) Echinoid overgrowth cement on echinoderm spine from winnowed carbonate tubidite. Overgrowth cement grew in deep water (>1000 m) sediment. Miocene, Site 626, ODP Leg 101, Straits of Florida. Sample courtesy of ODP.
(d) Radiating overgrowth cements (arrowed) on planktonic foraminifer. Cements grow in optical continuity with radial calcite of foraminifer test in deep water (>1000 m) sediment. Miocene, Site 626, ODP Leg 101, Straits of Florida. Sample courtesy of ODP.

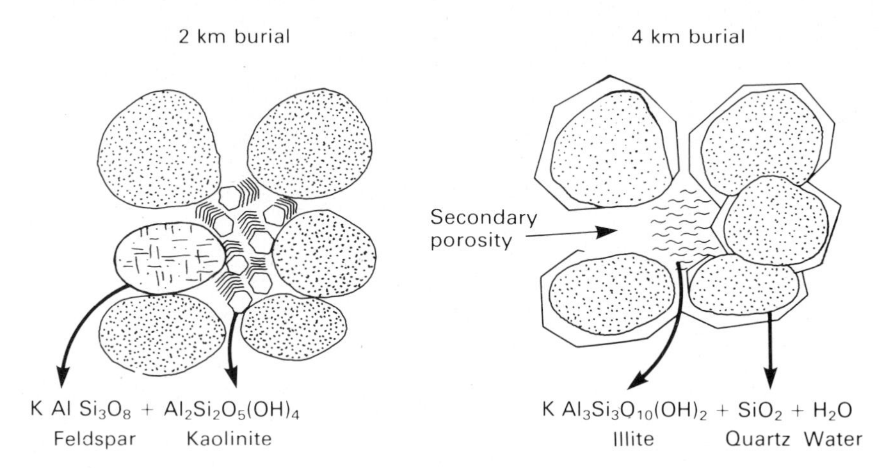

Fig. 5.28. Chemical reactions between feldspar and kaolinite are triggered by continued burial to produce illite, quartz overgrowth cements plus secondary porosity (from Bjørlykke, 1983).

be used for this purpose (Sippel, 1968). Scholle (1979, pp. 112–3) used photomicrographs and SEM photographs to demonstrate the progressive development of quartz overgrowth cements with concomittant reduction of porosity. Quartz grains with overgrowth cements may be reworked into younger sediments (see Fig. 5.11) where their importance is commonly underestimated (Sanderson, 1984). Reworked grains with overgrowth cements are recognized by the lack of interlocking overgrowths, the presence of quartz overgrowth cements on isolated grains and, less commonly, the presence of rounded or broken terminations (Scholle, 1979). Caution also has to be expressed where thin sections contain possible volcanic quartz grains; these can exhibit zonation and rounding which closely mimic quartz overgrowth cements. However, the presence of microlites and vacuoles in the outer rim distinguishes these volcanic fabrics from those of overgrowth cements.

Overgrowth cements on feldspars and carbonate echinoderm fragments are similar to quartz overgrowth cements in that they contain few or no inclusions or are commonly in optical continuity with the parent grain (Fig. 5.27a, b). Chemically, feldspar overgrowth cements are commonly the pure sodium or potassium feldspar end members and are rarely calcic; clear overgrowth cements may surround partially altered feldspar grains (e.g. Heald, 1956). Echinoderm grains contain a regular pattern of internal pores which contrast with the surrounding, inclusion-free overgrowth cements (e.g. Figs 5.21 and 5.27a, b). The echinoderm grains may also contain small solid inclusions of dolomite, euhedral or subhedral, the microdolomites of Meyers & Lohmann (1978), which result from the alteration of the original high Mg calcite skeleton to low Mg calcite and dolomite.

Overgrowth cements on echinoderm fragments are commonly a relatively early phase in the diagenetic history of a sediment, both in shallow water carbonates (Fig. 5.27a, b) (Meyers, 1980) and in deep water carbonate turbidites (Fig. 5.27c) (Schlager & James, 1978; Reid & Mazzullo, 1985, personal communication). In both shallow and deep water examples the origin of the overgrowth cement calcite is generally thought to be from aragonite dissolution during early diagenesis, although modifications of sea water chemistry with ocean depth and with shallow burial may be in part responsible. Overgrowth cements also form on some more complex carbonate grains, including planktonic foraminifers (Fig. 5.27d). It is noticeable that overgrowth cements only develop on biogenic carbonate grains, and do not form where the skeletal wall has been heavily bored by fungi or algae (thus forming a 'micrite' envelope).

The timing of the development of quartz overgrowth cements is more variable. Many sandstones with overgrowth cements exhibit little solution compaction at grain-to-grain contacts (Fig. 5.26c, d); the source of silica in such sediments can only partially be derived from the solution compaction and may largely result from the dissolution of opaline silica.

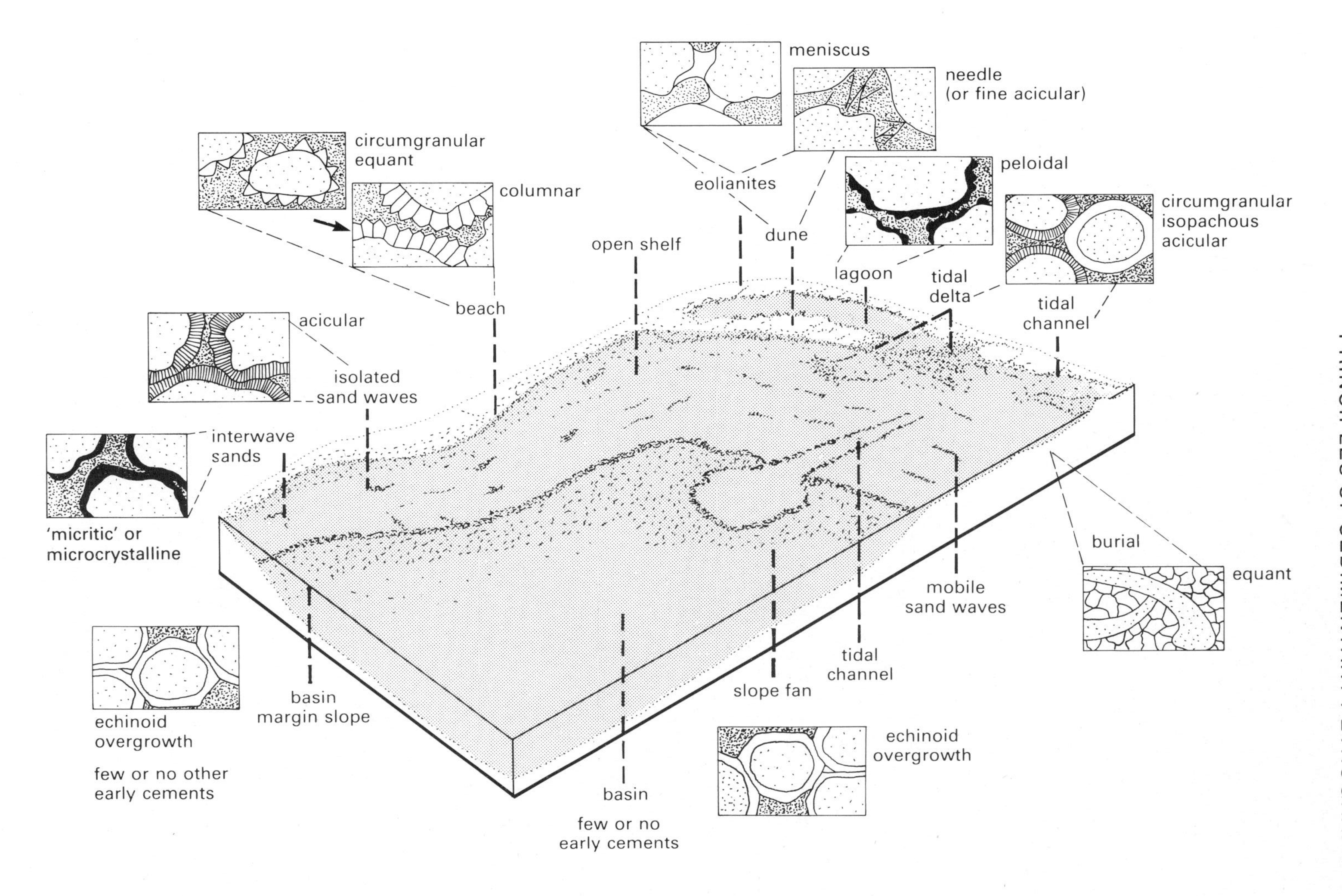

Fig. 5.29. Summary diagram of carbonate cement types plus their common location of formation.

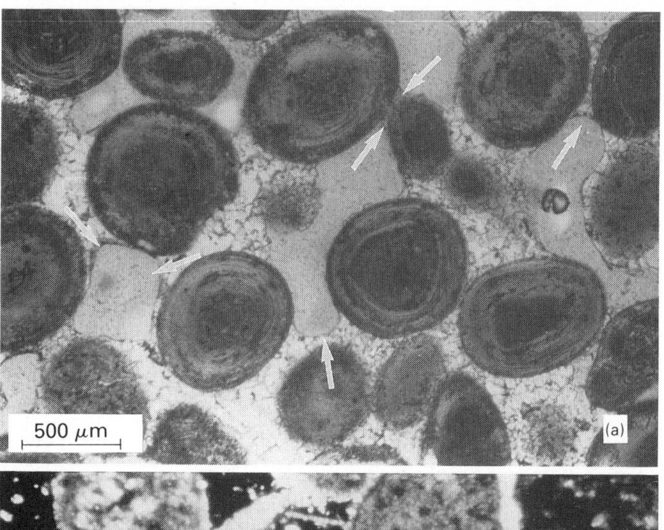

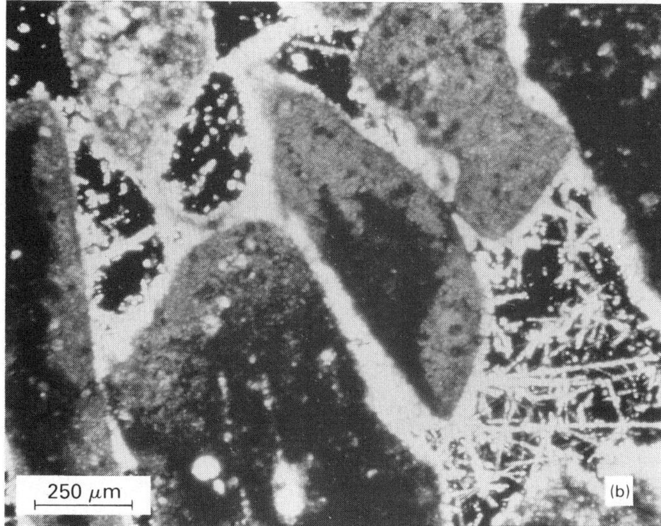

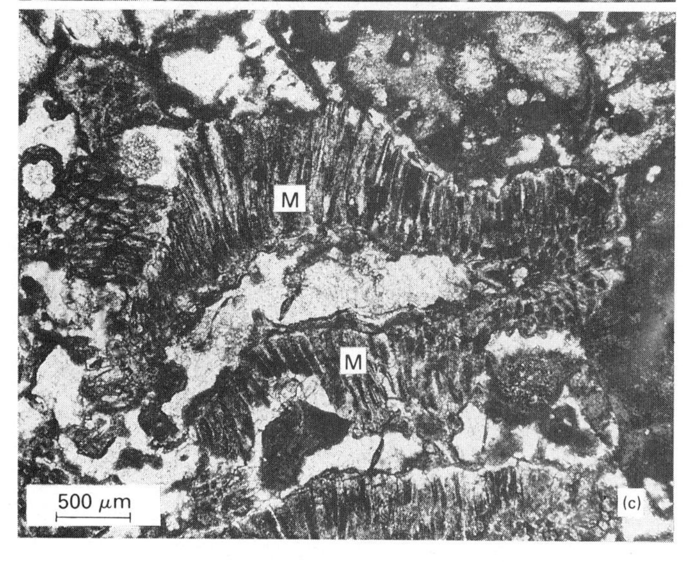

Fig. 5.30. Carbonate cements I: (a) Low Mg calcite meniscus cements (arrowed) forming in ooid grainstone subaerial dune. Holocene, Joulters Cay, Bahamas. Sample courtesy of C.H. Moore, LSU.
(b) Low Mg calcite needle cements in aeolian dune. Pleistocene, Yucatan, Mexico. Sample courtesy of W.B. Ward.
(c) *Microcodium* (M), a low Mg calcite growth form associated with root zones. Pleistocene, Yucatan, Mexico.
(d) Acicular, isopachous, circumgranular cements in shallow marine grainstone; acicular cements may have been aragonite, since converted to low Mg calcite with a high degree of textural preservation, or originally low or high Mg calcite. Mississippian, Moyvoughly Beds, Co. Meath, Ireland. Modern acicular, circumgranular, aragonite cements are shown in Fig. 5.24(a).
(e) High Mg calcite, peloidal, circumgranular cements (opaque) coating grains, with later acicular aragonite cements growing into intergranular porosity. Holocene, San Salvador, Bahamas.
(f) Microcrystalline, aragonite, circumgranular cements on aragonite grains. Cements are not isopachous and form distinct bridges between grains; textures are very different from 'micrite envelopes'. Crossed polars. Recent, US Virgin Islands.

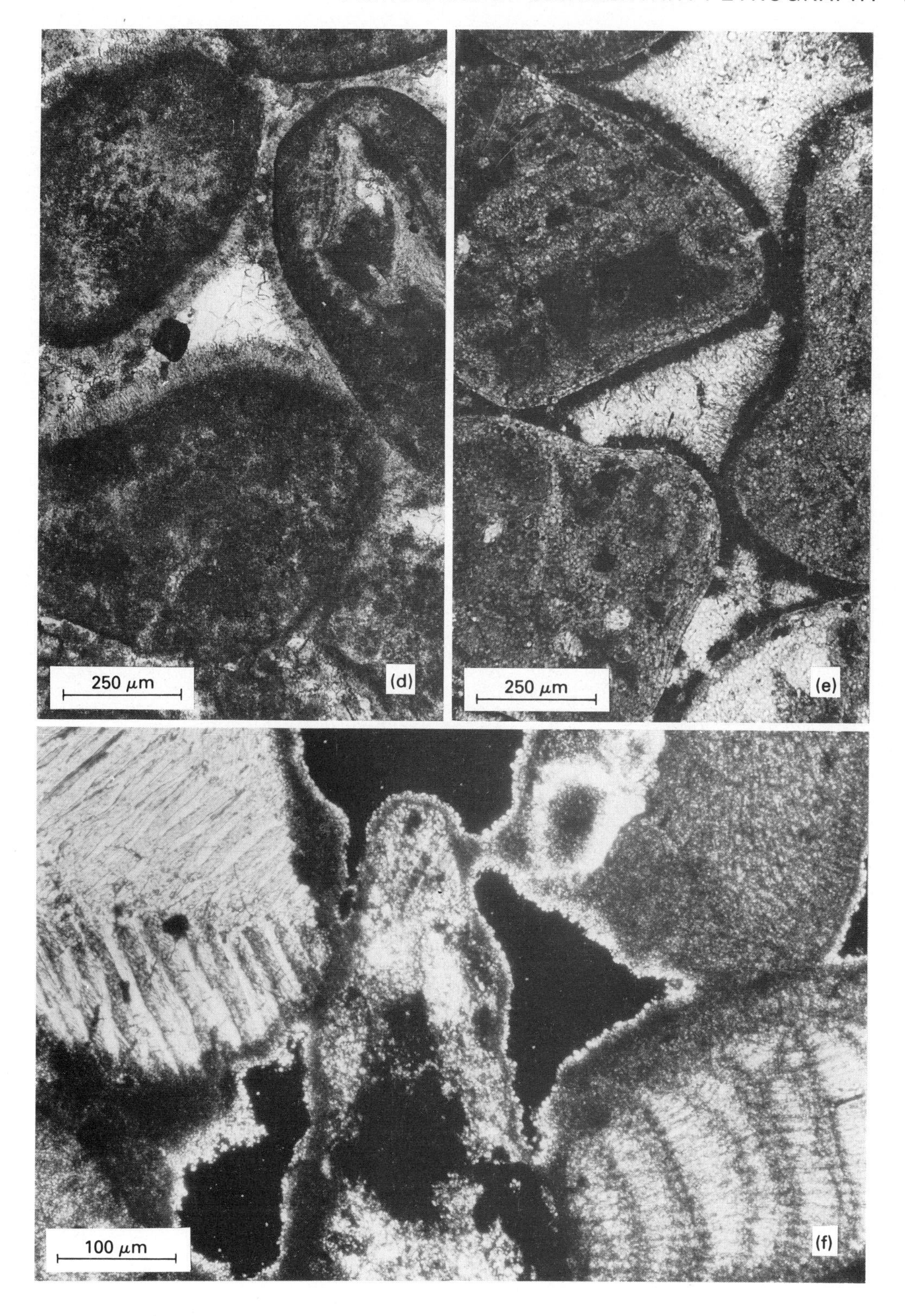
250 μm
(d)
250 μm
(e)
100 μm
(f)

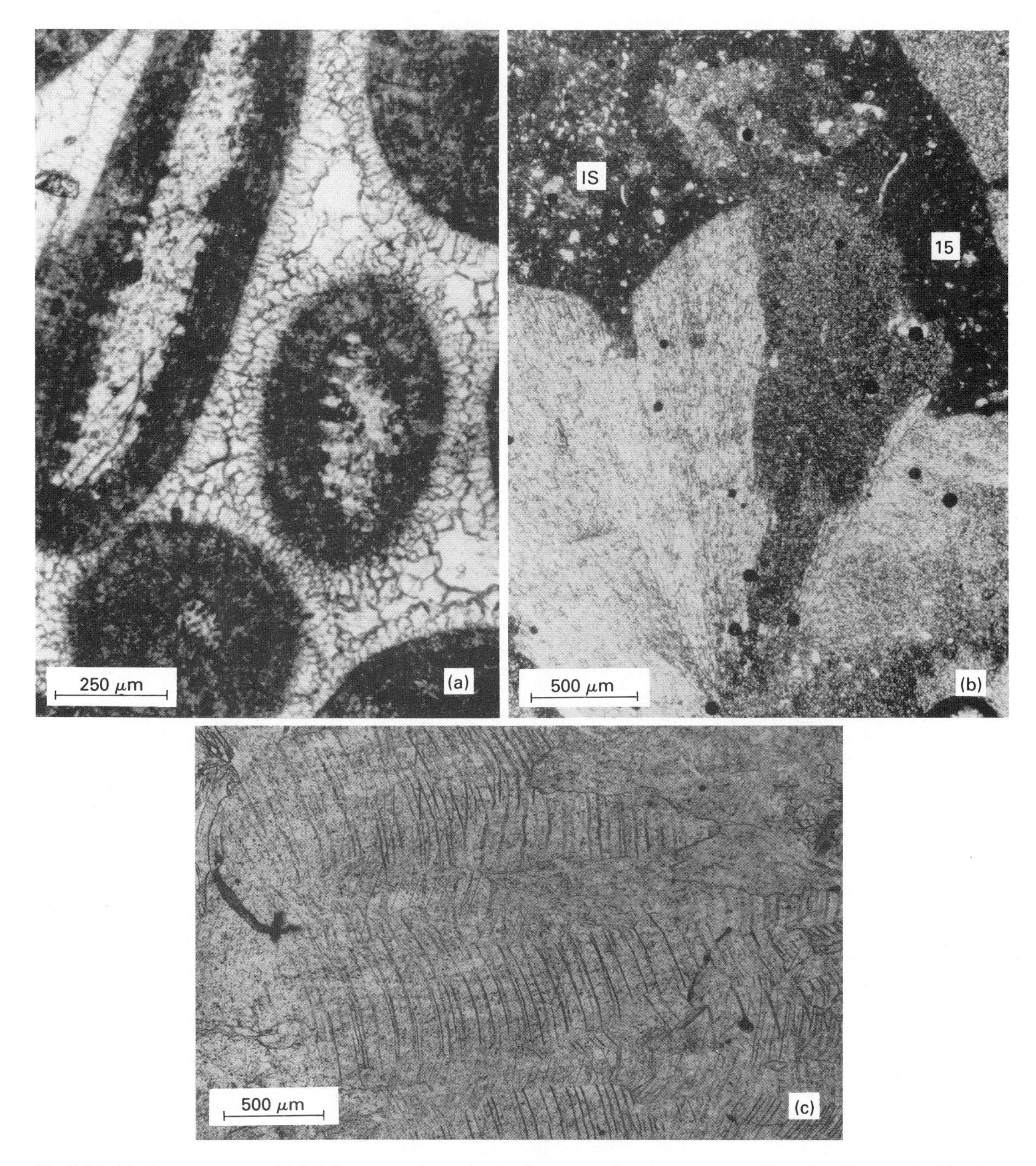

Fig. 5.31. Carbonate cements II: (a) Columnar, isopachous, circumgranular cements on ooids and ooid-coated echinoderm fragments. Remaining porosity is occluded by equant low Mg calcite cements. Lower Carboniferous, Brofiscin Oolite Formation, Glamorgan, UK. Photomicrograph courtesy of K. Hird and M.E. Tucker, University of Durham, UK.
(b) Acicular, aragonite, 'ray' cements which grew within a reef cavity, later filled by internal sediment (IS). 'Ray' fans exhibit sweeping extinction. Crossed polars. Recent, Jamaica. Photomicrograph taken courtesy of C.H. Moore, LSU.
(c) Radiaxial acicular calcite cements, showing characteristic curved cleavages of crystals; crystals are now low Mg calcite but

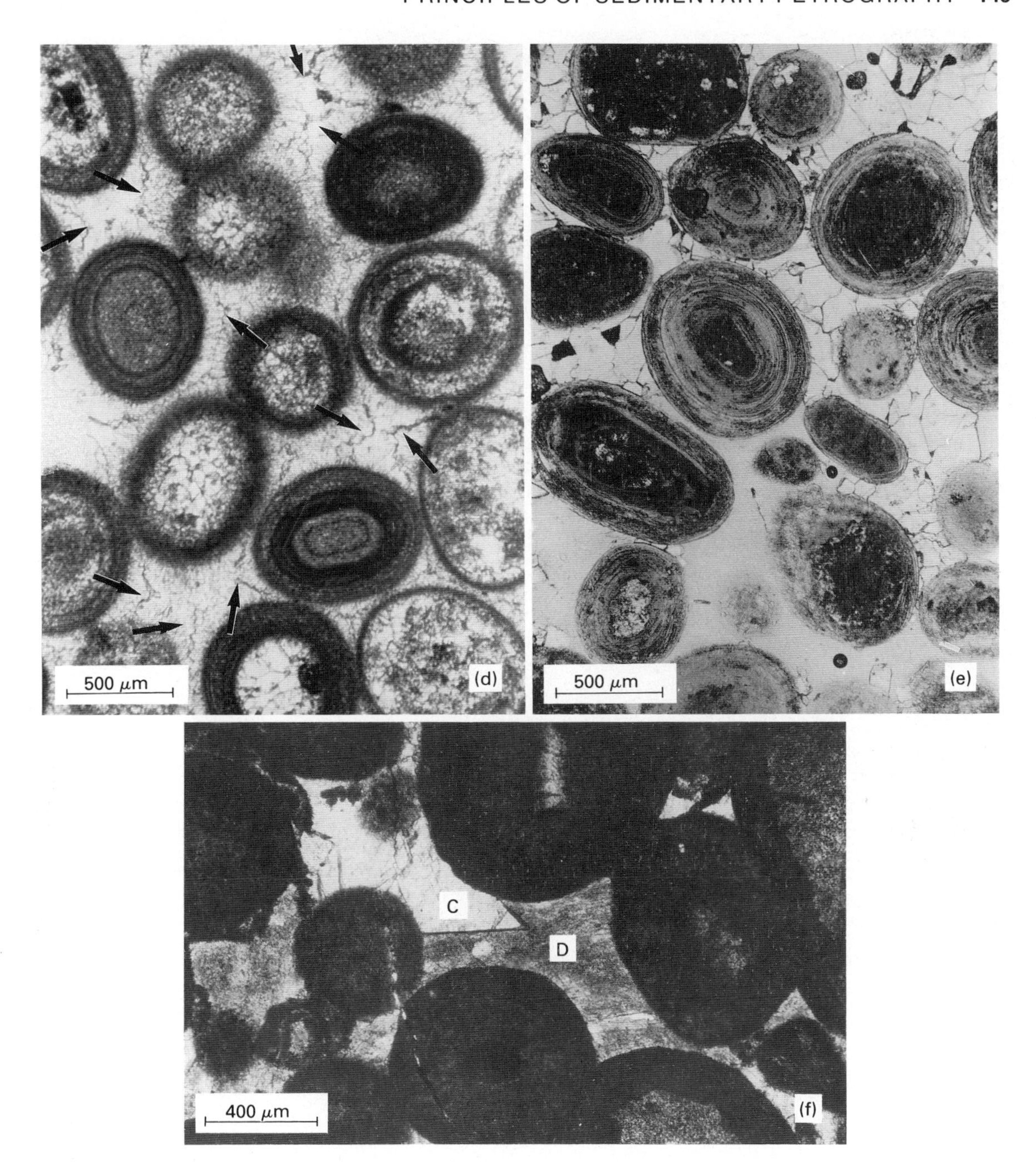

analysis (M.C. Akhurst, personal communication) indicates they were originally high Mg calcite. Growth direction was from right to left. Mississippian, Carbonate mound core, Co. Galway, Ireland.
(d) Compromise boundaries (arrowed) formed by columnar cements competing for growth in intergranular pore spaces in ooid grainstone (see also Harwood & Moore, 1984). Jurassic, subsurface Smackover Formation, east Texas, USA.
(e) Equant low Mg calcite cements in exposed ooid grainstone. Holocene, Joulters Cay, Bahamas.
(f) Baroque (or saddle) dolomite cement (D) in ooid grainstone. Dolomite is inclusion-rich and shows undulose extinction. Calcite spar cement (C) occludes remaining pore space. Crossed polars. Jurassic, subsurface Gilmer Formation, east Texas, USA.

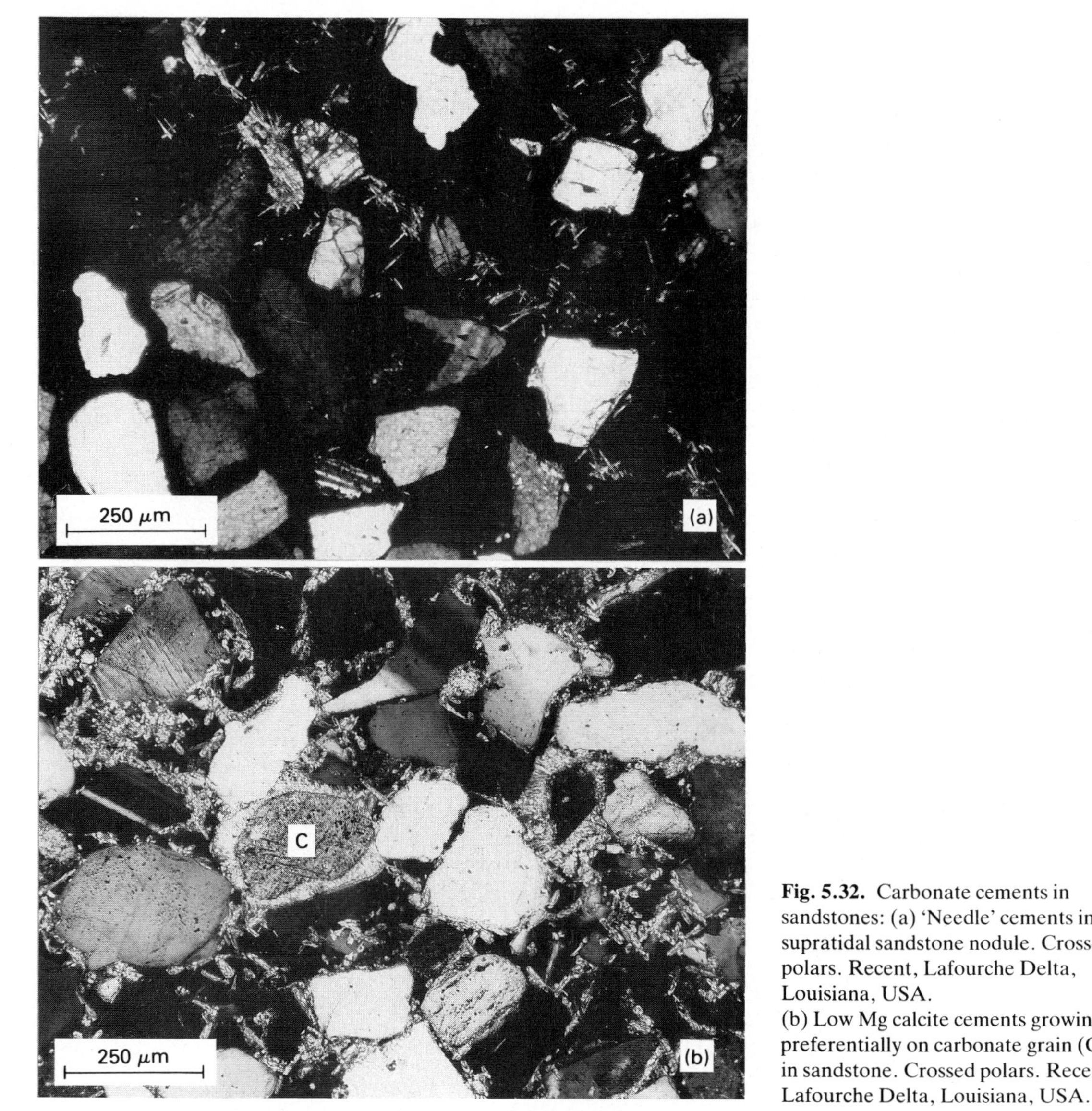

Fig. 5.32. Carbonate cements in sandstones: (a) 'Needle' cements in supratidal sandstone nodule. Crossed polars. Recent, Lafourche Delta, Louisiana, USA.
(b) Low Mg calcite cements growing preferentially on carbonate grain (C) in sandstone. Crossed polars. Recent, Lafourche Delta, Louisiana, USA.

In other sandstones, quartz overgrowth cements do not form until after considerable solution compaction; here the solution compaction can be the major source of silica, although reactions between clay minerals in immature sandstones may also contribute silica (Fig. 5.28) (Bjørlykke, 1983).

CEMENT TYPES II: DISCONFORMABLE CRYSTAL CEMENTS

Disconformable crystal cements nucleate as separate crystals on sediment grains and, as such, are distinct from overgrowth cements. The size and number of these crystals are largely dependent on

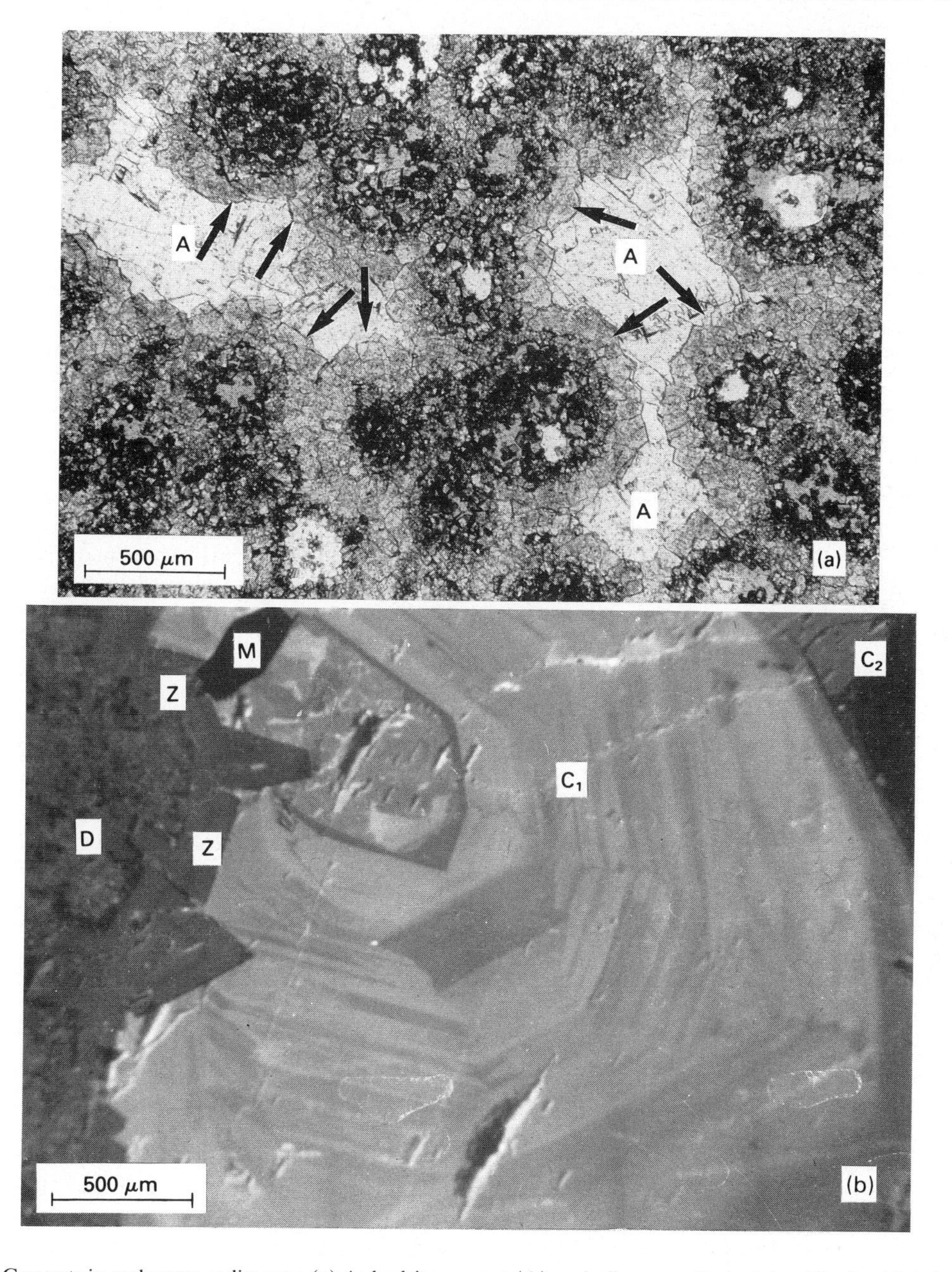

Fig. 5.33. Cements in carbonate sediments: (a) Anhydrite cement (A) occluding porosity in dolomitized ooid grainstone. Dolomitized circumgranular cements (arrowed) coat ooids which have been partially dolomitized before dissolution of remaining calcite. Opaque coating in ooid centres is bitumen, from hydrocarbon migration after anhydrite cementation (see also Harwood & Moore, 1984). Jurassic, subsurface Smackover Formation, east Texas, USA.
(b) Multiphase carbonate cements. Dolomitized algal laminae (D) adjacent to void. Void fill was in several stages; first, zoned dolomite cement (Z), second, rare marcasite (M), third, complexly zoned large calcite spar (C_1) and last, void-occluding calcite spar (C_2). Photomicrograph taken under cathodoluminescence. Permian, subsurface Cadeby Formation, Selby Coalfield, Yorkshire, UK.

the composition of the enclosing pore fluids and on the availability of the substrate to provide nucleation sites. Thus circumgranular (or rim) cements (Fig. 5.24a) form where there are many nucleation sites on a sediment grain, whereas poikilotopic cements (Fig. 5.24b) are the result of few nucleation sites and, in general, slower crystal growth. Characterization of the diverse morphologies of cements formed in modern diagenetic environments of rapid cementation has aided recognition of ancient analogues, particularly for near-surface carbonate diagenetic environments (for summaries see Longman, 1980; James & Choquette, 1983, 1984; Harris, Kendall & Lerche, 1985). These, plus the carbonate cements formed during burial diagenesis are summarized in Fig. 5.29 with examples in Figs 5.30, 5.31 and 5.32; relevant additional comments on these diagenetic environments are included below.

VADOSE AND SHALLOW MARINE CARBONATE CEMENTS

Although the cement morphologies summarized in Fig. 5.29 occur principally in carbonate sediments, quartz meniscus cements have been described from sandstones (e.g. Scholle, 1979, p. 114) and may comprise partial overgrowth cements. Near-surface carbonate cements also occur in quartzose sandstones but do not, however, form so readily on quartz grains (Fig. 5.30b, c). This may be partly a factor of reduced possible nucleation sites and partly a result of the decreased supply of calcium carbonate. Siliciclastic sediments in general remain almost uncemented until buried to depths of 1000 m or more, whereas carbonate sediments are commonly cemented at or near the sediment–water interface, or in subaerial environments. Indirect evidence of vadose diagenesis in carbonates comes from 'fitted-textures' (Fig. 5.16e), where expansion and or dissolution around grain margins has occurred beneath a soil zone.

A note of caution needs to be added concerning cementation in the meteoric vadose zone. Although there is an ever-increasing literature on cement morphologies and diagenesis within this zone, especially within soils, it is important to remember that many such cements are the exception rather than the rule. In many modern examples, diagenetic reactions in the meteoric zones are minimal. Both modern and ancient zones show intense reaction within a few centimetres of the soil surface, perhaps the area with least preservation potential, but very little alteration of the original sediment below that depth. Thus the presence of many ancient meteoric vadose zones may be difficult to determine from petrographic evidence and from field/core or geochemical techniques.

A further complication arises as many ancient shallow marine carbonate cements were of different mineralogies to those occurring in modern carbonates (e.g. Figs 5.24a, 5.30d, 5.31b, c). However, although the mineralogies may differ, the cement morphologies are commonly similar, thus enabling interpretation of ancient early diagenetic environments.

BURIAL CEMENTS

Carbonate burial cements are commonly coarsely crystalline, and include baroque dolomite and calcite spar (Fig. 5.31f), in places zoned. Burial cements in other sediments are composed of a complex range of mineralogies, not all of which are readily distinguishable with a petrological microscope (Table 5.1). Some are finely crystalline (clay minerals, hematite, limonite), others are coarsely crystalline (similar to carbonate burial cements) and commonly poikilotopic, whereas yet others are both coarsely crystalline and, to some extent, appear to replace the host sediment (anhydrite, halite, phosphates, glauconite) (Fig. 5.33a). Quartz overgrowth cements are a further common burial cement, although their presence alone is not an indication of deep burial. In siliciclastic sediments careful attention to petrographic detail, particularly the compaction history, is necessary to determine whether a cement formed during deep burial (>2,000 m) or at an earlier stage in the diagenetic history.

At whatever stage in the diagenetic history of the sediment they form, cements are a record of the fluctuating pore water history of a sediment. Numerous, clearly identifiable stages of cementation in a sediment are evidence of changes in pore water chemistry and saturation states, even if all the precipitated phases are of a single mineralogy (e.g. Fig. 5.33b). More important, perhaps, is the evidence given by the termination of precipitation of a cement phase, indicating a decrease in saturation, or supersaturation, of that phase within the pore fluid and, in places, dissolution of pre-existing cement phases when undersaturated pore fluids are introduced into a sediment. Analysis of the chemistry and isotopic compositions of the various cement phases can give

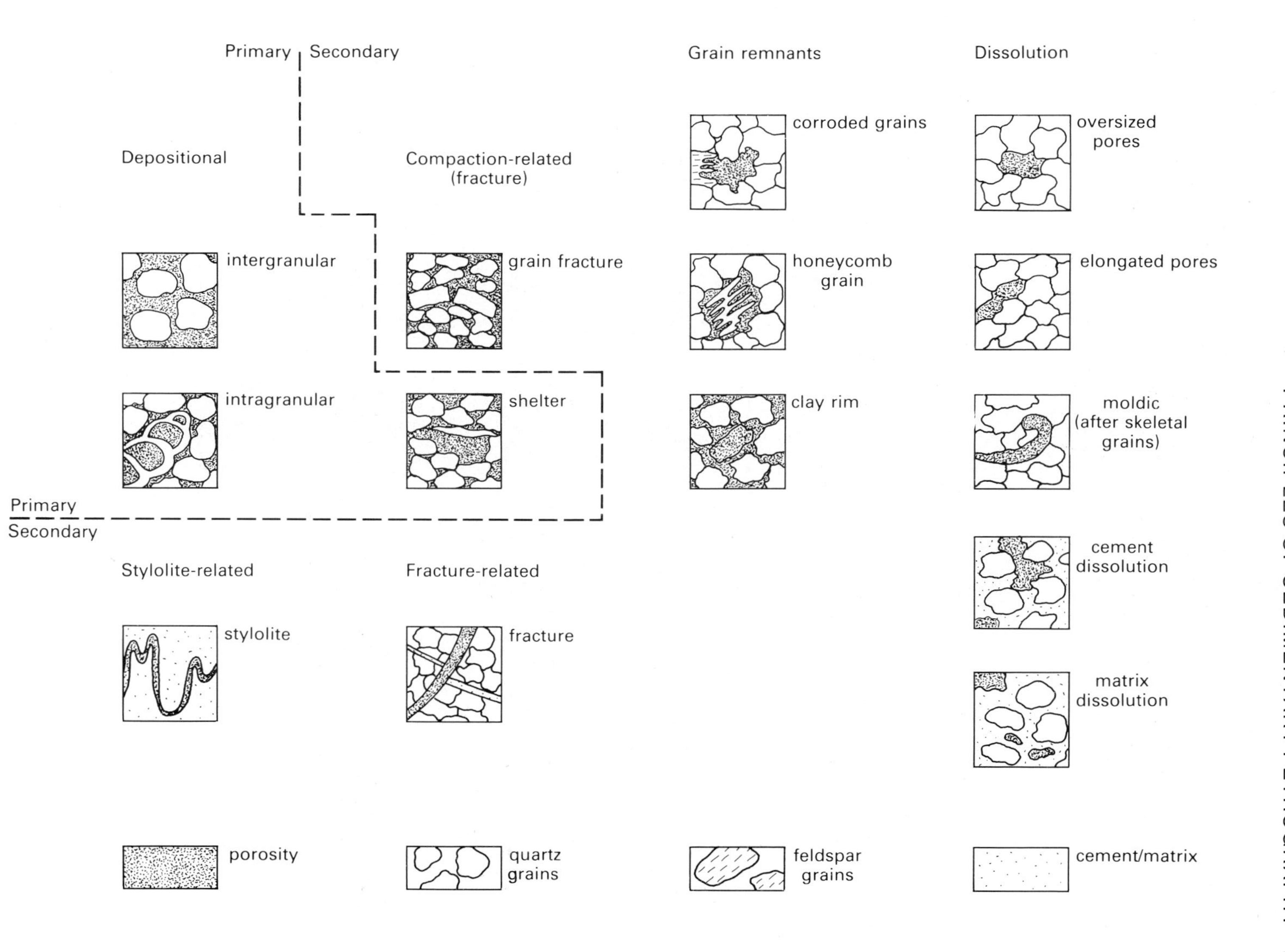

Fig. 5.34. Summary diagram of porosity fabrics in sedimentary rocks (based on Schmidt *et al.*, 1977; Hayes, 1979; McBride, 1980; Shanmugan, 1985).

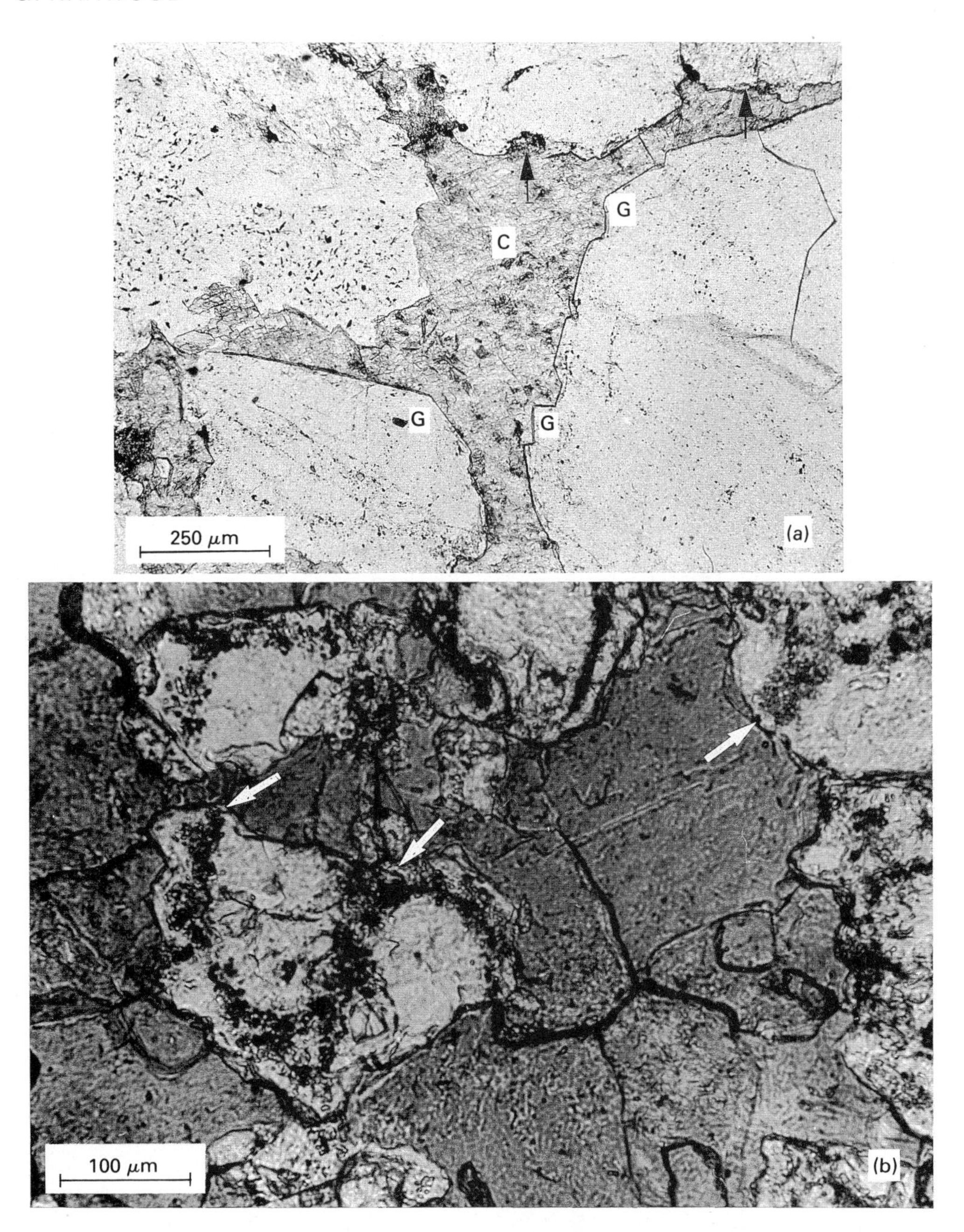

Fig. 5.35. Corrosion and dissolution of grains: (a) Corrosion of detrital quartz grain margins (arrowed) by migrating pore fluids prior to precipitation of poikilotopic calcite cement (C). Adjacent grain margins (G) show no corrosion but planar margins from euhedral overgrowth cements. Permian, Penrith Sandstone, Cumbria, UK.
(b) Corrosion of precursor dolomite. Dolomitization of former carbonate grains with micrite envelopes has resulted in coarsely crystalline dolostone with remnants of micrite envelopes preserved. Subsequent corrosion of dolomite crystals has led to truncation of micrite relics (arrowed) with later porosity occlusion by calcite. This apparent replacement of dolomite by calcite is termed calcitization or, less correctly, dedolomitization. Section stained with Alizarin red S. Permian, Cadeby Formation, Yorkshire, UK.

(c) Corrosion along cleavages in dolomite, probably an effect of recent weathering. Dolomite within 5 mm of pore space has been corroded along cleavage planes with subsequent precipitation of iron oxides, defining amount of dissolution. Further away from pore space (bottom right corner of photomicrograph) dolomite remains uncorroded. Permian, Cadeby Formation, Yorkshire, UK.
(d) Corrosion and secondary porosity formation along cleavage planes in feldspar. Impregnation has shown intragranular porosity (arrowed) within feldspar grains. Carboniferous, Upper Crag Point Sandstone, Northumberland, UK.

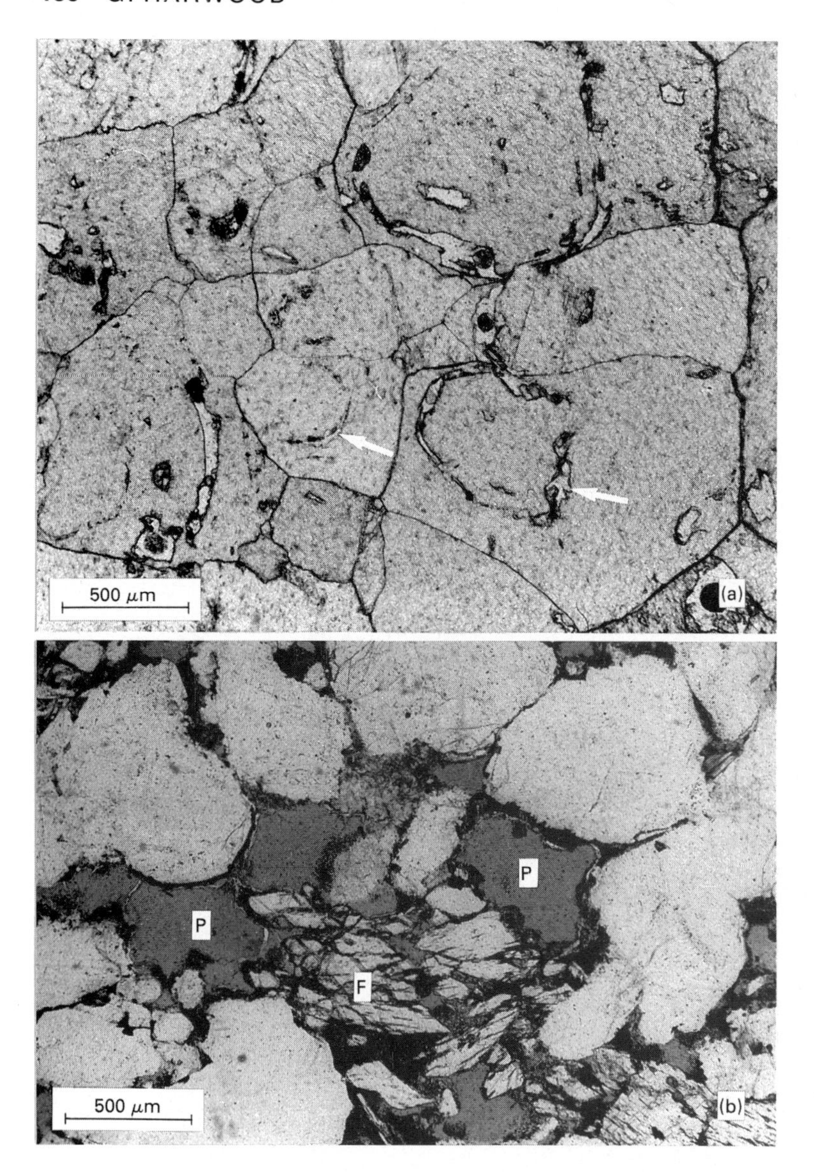

clear indications of concomittant variations in the nature of the pore fluids (Chapter 9) (e.g. in a carbonate system, Dickson & Coleman, 1980).

5.3.4 **Dissolution fabrics**

Dissolution in the subsurface generates secondary porosity (Choquette & Pray, 1970) (Section 5.3.1). This may not necessarily be large-scale dissolution (e.g. Fig. 5.16d), but may result only in slight corrosion of the detrital grains or pre-existing cements in the sediment. Dissolution may be very subtle, with dissolution of calcic plagioclase whereas albite and potassium feldspar are not affected (Boles, 1984). In carbonates preferential solution of some grain types (e.g. Harris & Kendall, 1986) occurs in what would otherwise appear to be a chemically homogeneous sediment. The delicate textural changes commonly encountered in studies of dissolution fabrics emphasize the importance of effective impregnation of the sediment prior to thin section making; this is not only essential to identify plucked grains, or grains

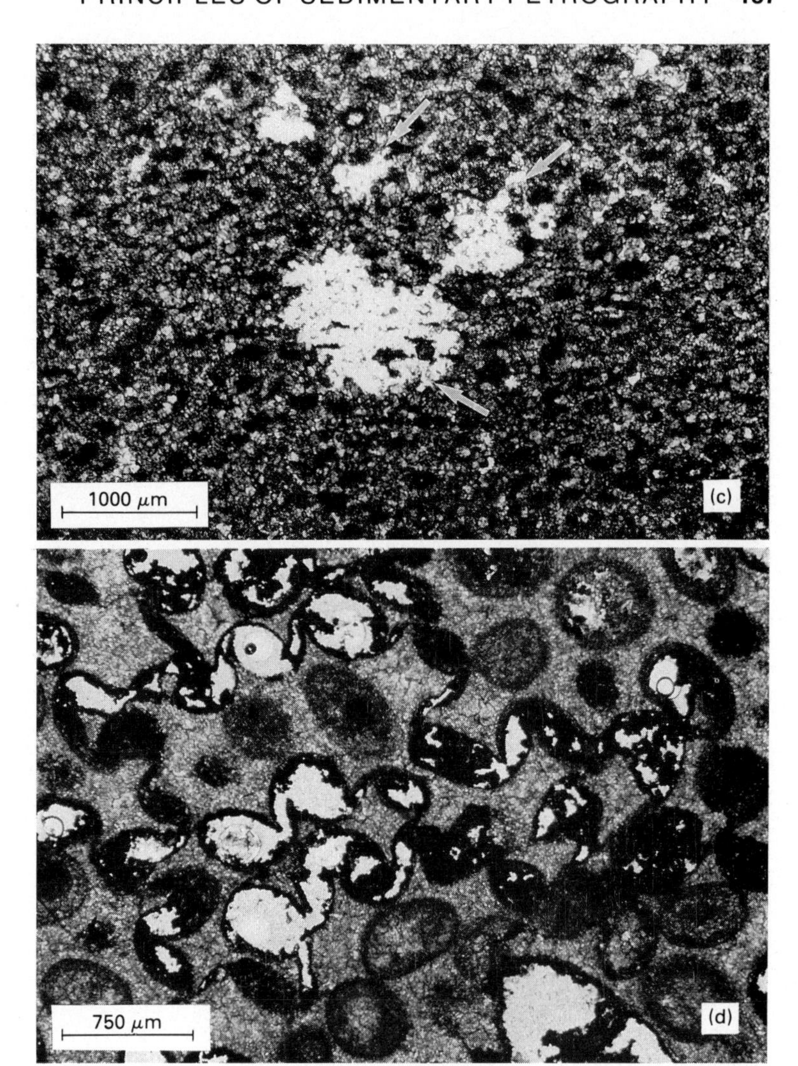

Fig. 5.36. Dissolution of grains and cements: (a) Dissolution of zones (arrowed) within carbonate cement crystals in former dolomite. Permian, Cadeby Formation, Yorkshire, UK. (b) Dissolution along cleavages in detrital feldspar grain (F) has resulted in isolated grain remnants which have collapsed within pore spaces. Surrounding quartz grains were cemented, largely by overgrowth cements, before feldspar dissolution. Complete dissolution of other detrital grains has led to formation of pore spaces (P), slightly oversized. Carboniferous, Upper Crag Point Sandstone, Northumberland, UK. (c) Isolated pore spaces (arrowed) formed from dissolution of replacive anhydrite in peloid grainstone. Lack of interconnection between pore spaces means they are ineffective as migration conduits for pore fluids. Porosity has subsequently been occluded by calcite. Permian, Cadeby Formation, Yorkshire, UK. (d) Interconnected secondary porosity formed by collapse after selective dissolution of some ooids. Fracture between oomolds produced linked 'chains' which were later used for hydrocarbon migration; black bitumen relics remain in pore spaces. Jurassic, subsurface Smackover Formation, east Texas, USA.

partially plucked from their matrix as opposed to secondary porosity, but is also necessary for reflected light petrography of grain and cement margins when investigating corrosion fabrics. Common dissolution fabrics (based on the work of several authors, e.g. Schmidt *et al.*, 1977; Hayes, 1979; McBride, 1980; Shanmugan, 1985) are shown diagramatically in Fig. 5.34.

CORROSION FABRICS

Initial dissolution leads to the corrosion of grain margins and cement terminations. This is common in many sediments and takes place adjacent to pore spaces as the result of flushing by undersaturated pore fluids and not, as is sometimes mistakenly stated, as the result of 'corrosive' growth of a later mineral phase (Section 5.3.5). Such corrosion may be *selective*, only effecting certain mineralogies, or *non-selective*, so that all existing minerals adjacent to pore spaces are corroded. Corrosion fabrics show grain margins and cement terminations to have a pitted or corrugated character (Fig. 5.35a) which in ore mineralogy is termed a 'caries' texture. If there is no later cementation, corrosion produces a more effective connection between pore spaces and thus enhances existing porosity, whether this be primary or secondary. More commonly, however, the adjacent pore space is later occluded by cement (Fig. 5.35b) which may render the corrosion, if minor, more difficult to distinguish. Corrosion may also take place during the precipitation of what, superficially, appears to be a single cement phase; in carbonates this can be detailed using cathodoluminescence petrography (Chapter 6). Corrosion fabrics are very common in evaporite sediments, or sediments where there is a phase of evaporite mineral cementation, as these minerals are very soluble and hence susceptible to slight changes in fluid salinity (e.g. Schreiber, 1978).

PENETRATIVE DISSOLUTION FABRICS

More progressive selective dissolution of cements and grains within a sediment leads to attack of the whole mineral, commonly commencing by corrosion along cleavage planes (Fig. 5.35c, d). As dissolution proceeds, a honeycomb texture may result, particularly where a mineral has near rectilinear cleavages. Slight changes in mineral chemistry are also prone to selective dissolution (Fig. 5.36a), as are zones of less stable mineralogies (e.g. Ward & Halley, 1985). Continued dissolution will result in collapse of remaining grain fragments (Fig. 5.36b) and, in time, may remove the entire detrital grain (Fig. 5.16c), forming oversized pores. In addition, subsequent collapse of grain coatings, whether original or produced by earlier diagenetic alteration, may occur (Fig. 5.18d). Selective dissolution is restricted to less stable mineralogies and, in places, to strain boundaries within grains (McBride, 1985); it results in intragranular, or partial intragranular, secondary porosity (Schmidt *et al.*, 1977). This enhanced porosity will only be effective if there is either a considerable proportion of the mineral, or minerals, susceptible to dissolution, so forming a connection between isolated intragranular pore spaces, or if there is an existing intergranular porosity. For example, the common dissolution of isolated evaporite minerals forms isolated, secondary pore spaces (Fig. 5.36c) which rarely enhance the effective porosity of a sediment. This contrasts with Fig. 5.36(d), where selective dissolution of some ooids, together with subsequent brittle collapse of the supporting framework, has formed an interconnecting fracture porosity.

It is very important to detail the amount of mineral dissolution that has taken place in the subsurface, especially where measurements of modal composition (Section 5.2.2) or provenance estimates (Section 5.2.4) are being made. McBride (1985, 1986) detailed dissolution and grain modification from the zone of weathering to the deep subsurface of heavy mineral grains, rock fragments and feldspars, resulting in an apparent composition very different from the original (Fig. 5.37). Dissolution of diagnostic minerals, together with alteration of others, can significantly modify original grain composition, particularly as the original grain can rarely be confidently identified (e.g. Fig. 5.36b). Modal composition and provenance studies can therefore only be partially accurate in many siliciclastic formations. Furthermore, the amount of dissolution/alteration must be assessed separately for each formation studied.

5.3.5 Alteration and replacement fabrics

Dissolution in the subsurface is commonly associated with mineral alteration and replacement. Within the subsurface, many minerals are altered to form either new minerals, or a new suite of minerals (commonly clay minerals) which grow on, or adjacent to, the site of the precursor grain or cement

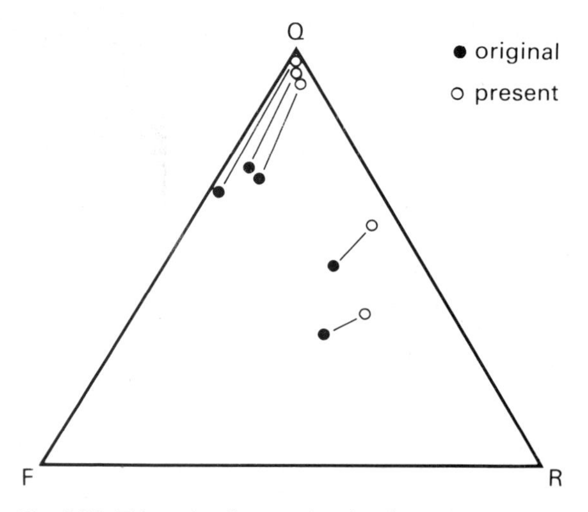

Fig. 5.37. Triangular diagram showing the present composition of five sandstones, after dissolution and alteration, and their reconstructed composition, assuming 15% of the grains which occupied oversized pores were rock fragments and 85% were feldspar (from McBride, 1985).

(e.g. Fig. 5.28). In contrast, replacive fabrics result both from the growth, during diagenesis, of individual, isolated crystals, or clusters of these minerals, and from large-scale or complete replacement of a pre-existing fabric. Replacement fabrics also result from the recrystallization of the alteration products of metastable minerals. These fabrics result from temperature and pressure increases and from interaction with migrating pore fluids, although alteration does not necessarily require the introduction of additional ions within these fluids.

ALTERATION FABRICS

During the temperature and pressure increases through burial and during migration of pore waters, many previously-stable minerals, or minerals which were deposited and buried too rapidly to become in equilibrium with surface temperatures and pressures, are brought into a regime where they are unstable in the prevailing conditions. These minerals break down to form more stable products, generally in the site of the original mineral (e.g. Fig. 5.16c); common examples are heavy minerals where, in US Gulf Coast sandstones, the degree of alteration has been related to burial depths (Fig. 5.38) (data of Milliken in McBride, 1985). Alteration of feldspars to clay minerals (Fig. 5.16c) and of aragonite to low Mg calcite (Fig. 5.39a) also occurs. Unless later modified (e.g. Fig. 5.18d), alteration fabrics are recognizable as they mimic the form of the precursor mineral, even although the original texture is commonly destroyed to some extent.

INDIVIDUAL MINERAL REPLACEMENT FABRICS

Replacive minerals grow in the place of earlier mineral phases. They commonly cut across pre-existing grain and cement boundaries (Fig. 5.39b) and thus can be seen to post-date all phases they traverse. Most individual replacive minerals have euhedral, planar crystallographic faces although they may contain inclusions, remnants of their precursor phases (e.g. Figs 5.12d and 5.39c, d). Many have crystal forms much larger than their precursors; they therefore *destroy* the original fabric (Fig. 5.39b) (e.g. Assereto & Folk, 1980). However, some fabric retention may be achieved through abundant inclusions, although this is not always immediately apparent (Fig. 5.40a, b). True fabric retention is rare with individual replacement minerals, although it

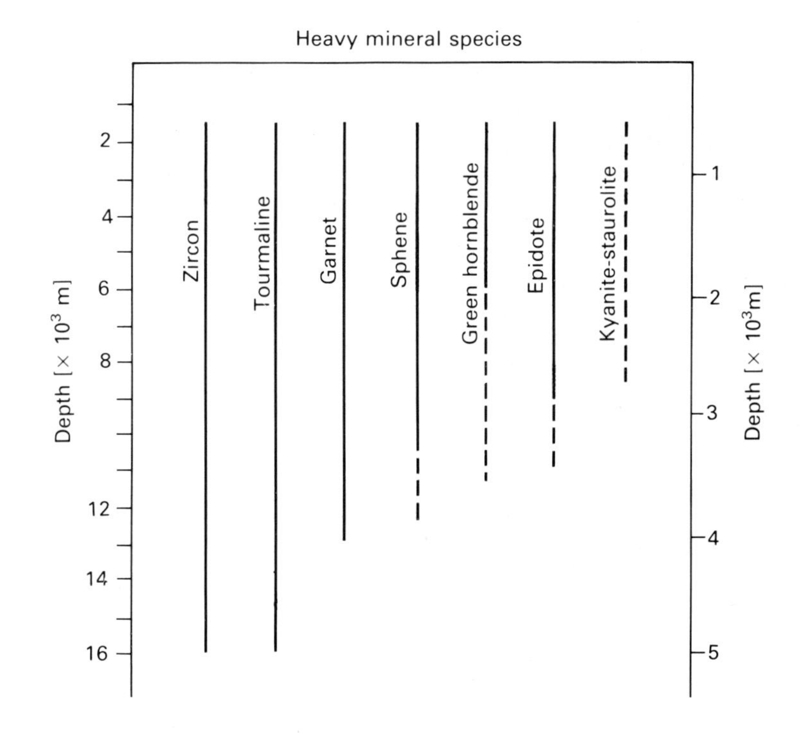

Fig. 5.38. Depth plot of the distribution of seven heavy mineral species in Plio–Pleistocene sandstones from the northern Gulf of Mexico basin, USA. Solid line = abundant; dashed line = rare (data from Milliken in McBride, 1985).

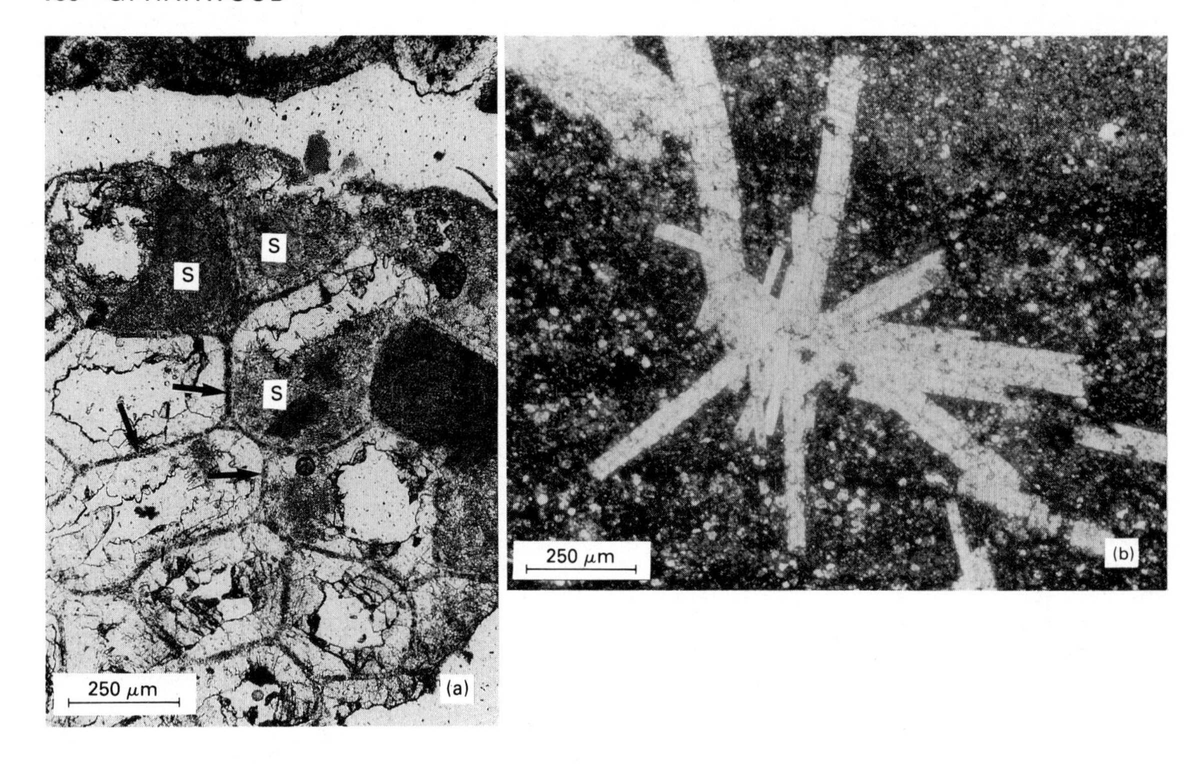

does occur where the replacive phase mimics the precursor mineral.

LARGE-SCALE REPLACEMENT FABRICS

Large-scale replacement can be either fabric-retentive (Fig. 5.40c) or fabric-destructive (Fig. 5.40d). Large-scale replacement occurs where sediments are reactive; it is common in evaporites (where fabrics are similar to many metamorphic textures), carbonates, volcaniclastics and, to a lesser degree, greywackes. Volcaniclastic sediments are extremely susceptible to alteration and dissolution. Mathisen (1984) documented secondary porosity enhancement of 40% in Plio–Pleistocene volcaniclastic sandstones from Indonesia in addition to alteration of many of the original grains and cements (e.g. feldspars, volcanic fragments, zeolites); this has occurred in burial depths of only 400–900 m. In greywackes there is commonly little secondary porosity generation, but considerable mineral alteration and recrystallization takes place within both rock fragments and matrix (Brenchley, 1969).

Where there is a monomineralic, or near-monomineralic, precursor fabric, as in carbonates and evaporites, replacement is commonly by a single mineral. Replacement within evaporite sequences is common, especially where potassium-rich minerals are present. Evaporites also replace large amounts of adjacent carbonate sediment (Fig. 5.40e), although the volume of carbonate replaced is difficult to ascertain. Harwood (1986) estimated that some 2–3 m of the uppermost Permian Cadeby Formation has been replaced by (former) anhydrite within the subsurface of Yorkshire, but in other carbonate formations within the Zechstein Basin more considerable amounts of replacement have probably taken place. One common form of large-scale replacement in carbonate sediments is dolomitization (Fig. 5.40c, d).

TIMING OF DISSOLUTION, ALTERATION AND REPLACEMENT

The timing of dissolution, alteration and replacement can be established relative to other diagenetic events. For example, Loucks, Dodge & Galloway

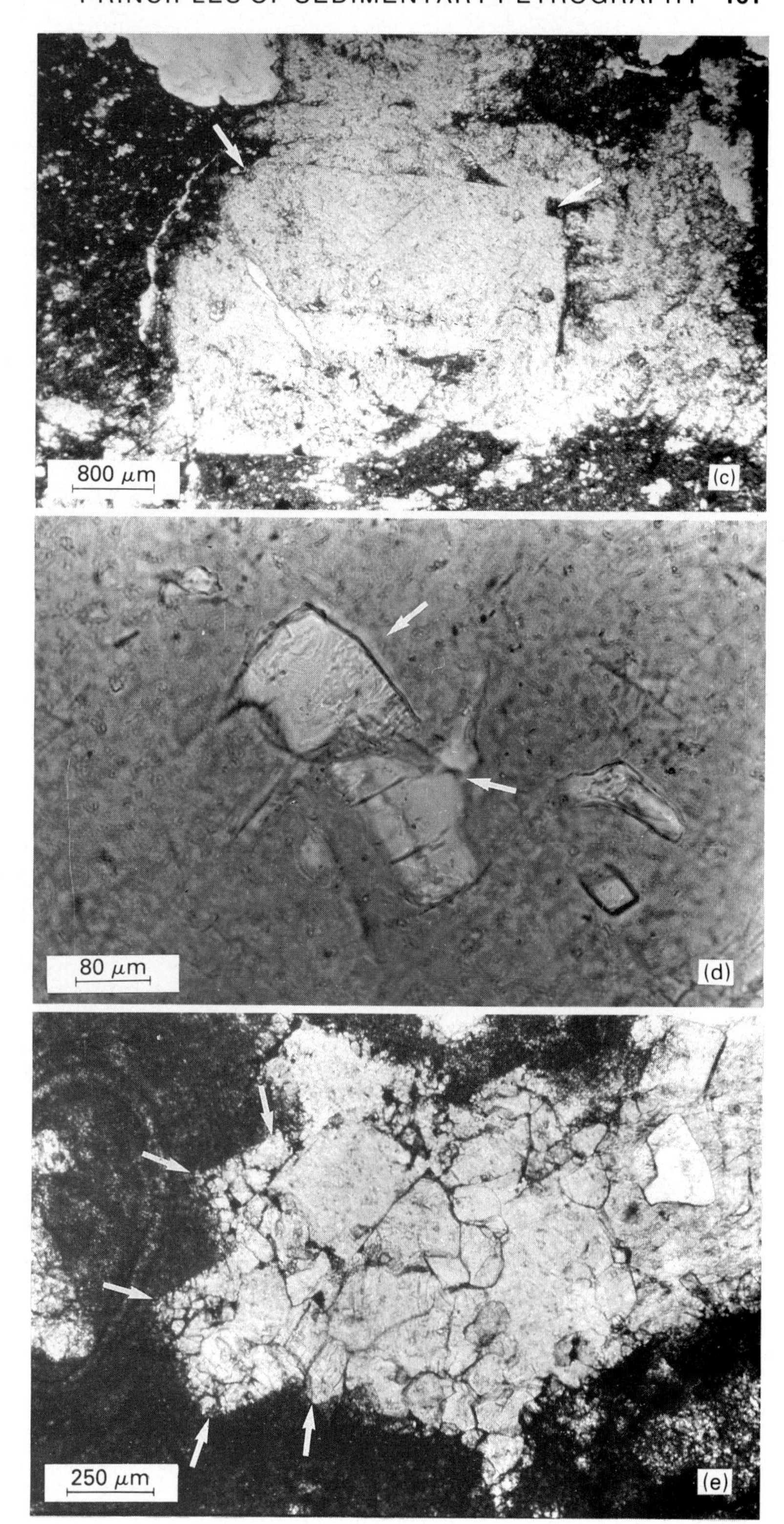

Fig. 5.39. Mineral replacement fabrics: (a) Coarsely crystalline calcite which has replaced aragonite of scleractinian coral structure in clast from debris flow deposit. Relics of aragonite skeleton walls are visible (arrowed) within calcite crystals. Calcite partially fills intraskeletal pore spaces apart from where internal sediment (S) is present. Miocene, Site 626, ODP Leg 101. Sample courtesy of ODP.
(b) Laths of anhydrite replacing dolomitized silty mudstone. Anhydrite laths cross-cut sedimentary laminae, indicating replacive nature. The morphology of the cluster of laths is a common form of anhydrite replacement of carbonates in sedimentary basins. Permian, Z2 carbonate, UK sector of North Sea.
(c) Calcite replacing anhydrite. Rectilinear outline of crystal with castellated margins (arrowed) is typical of anhydrite, but crystal has since been replaced by single calcite crystal (see also Harwood, 1980). Permian, Cadeby Formation, Yorkshire, UK.
(d) Corroded relic of anhydrite (arrowed) within calcite of (c); rectilinear cleavages of anhydrite visible. Section stained with alizarin red S. Calcite stains pink but anhydrite remains clear.
(e) Contrasting fabric where replacive anhydrite, identifiable by rectilinear outlines (arrowed), has been subsequently dissolved with the pore space later filled by calcite spar. Fabric of calcite, from small crystals near the void margins to larger crystals in the void centre (competitive growth, Bathurst, 1975) indicates dissolution preceded calcite cement precipitation, and that this is a 'dissolution-precipitation' texture, rather than a *replacement* texture. Jurassic, subsurface Smackover Formation, east Texas, USA.

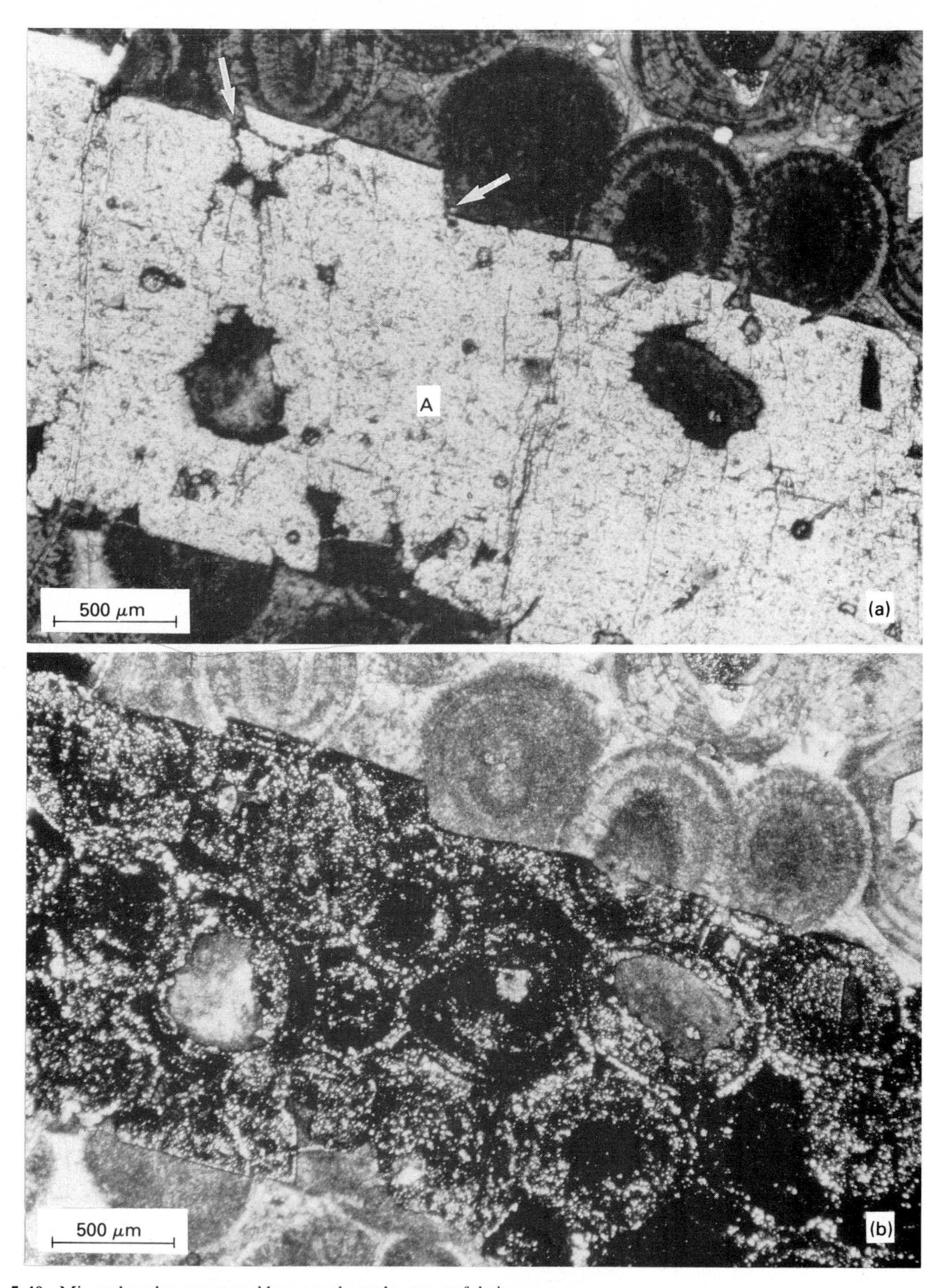

Fig. 5.40. Mineral replacement and large-scale replacement fabrics:
(a) Large lath of anhydrite (A) replacing ooid grainstone. Ooids had little or no early circumgranular cements and grain to grain compaction occurred with formation of interpenetrating grain to grain contacts before precipitation of intergranular calcite cements. Anhydrite replacement took place after compaction and cementation. Anhydrite lath has minor rectilinear outlines (arrowed). Jurassic, subsurface Smackover Formation, Arkansas, USA.
(b) Crossed polars view of (a), with anhydrite lath in extinction position. Trails of small calcite inclusions, not apparent in normal transmitted light, define original fabric of ooid grainstone. The timing of anhydrite replacement, after compaction, is confirmed as ooid outlines are continuous from grainstone into anhydrite lath.

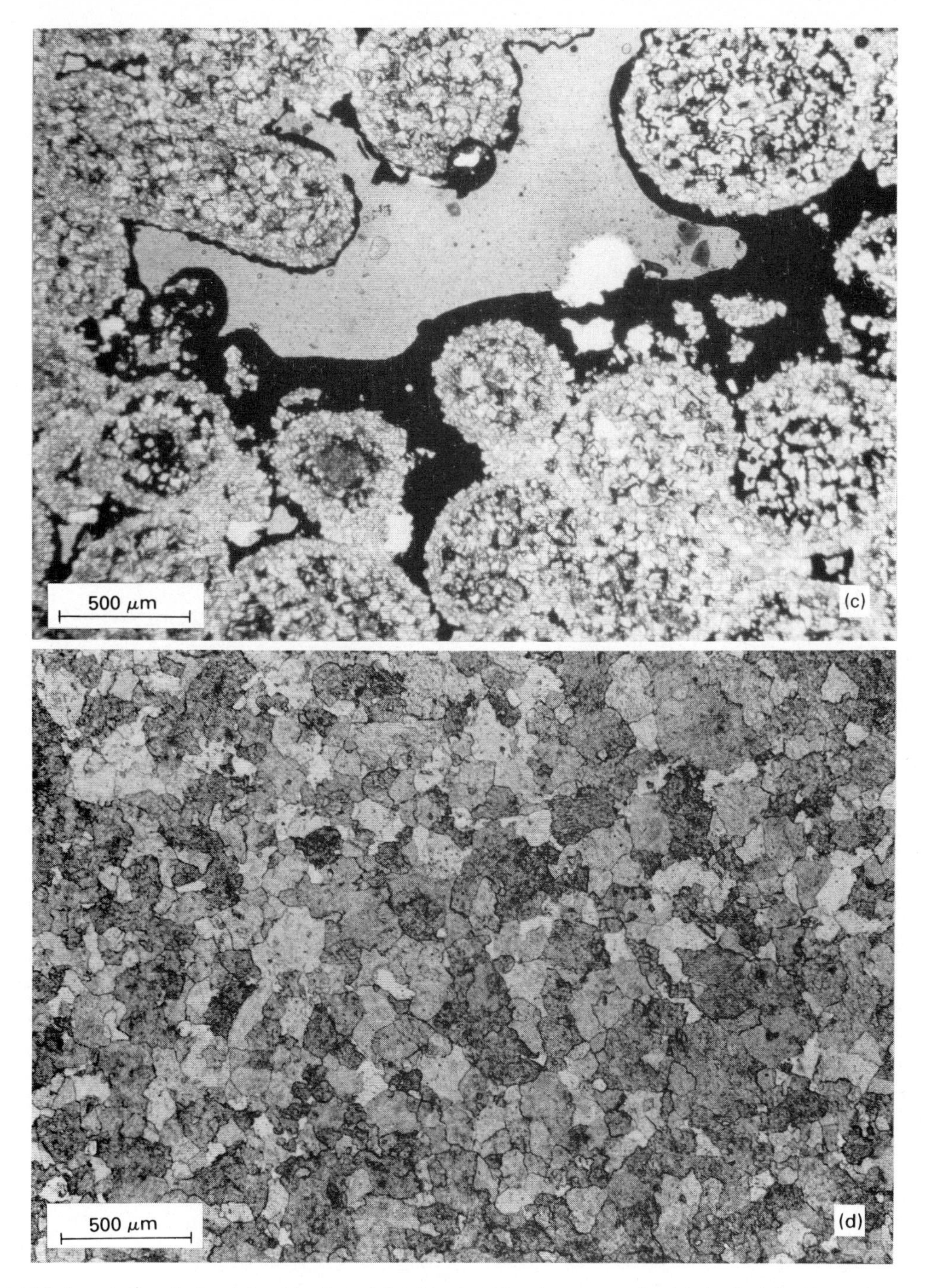

(c) Partial fabric retention during dolomitization; dolomitized ooid grainstone retains fabric of original isopachous circumgranular cements and form of ooids, although textures within the ooids are obliterated with formation of partial oomoldic porosity. Intergranular porosity has been partially occluded by bitumen. Jurassic, subsurface Smackover Formation, east Texas, USA.
(d) Fabric destruction during dolomitization; dolomitization has produced coarse sub- to euhedral dolomite crystals with no vestige of original fabric and considerable variation in dolomite crystal size. Crystal size also varies according to position within sediment (Harwood, 1986). Permian, Cadeby Formation, Nottinghamshire, UK.

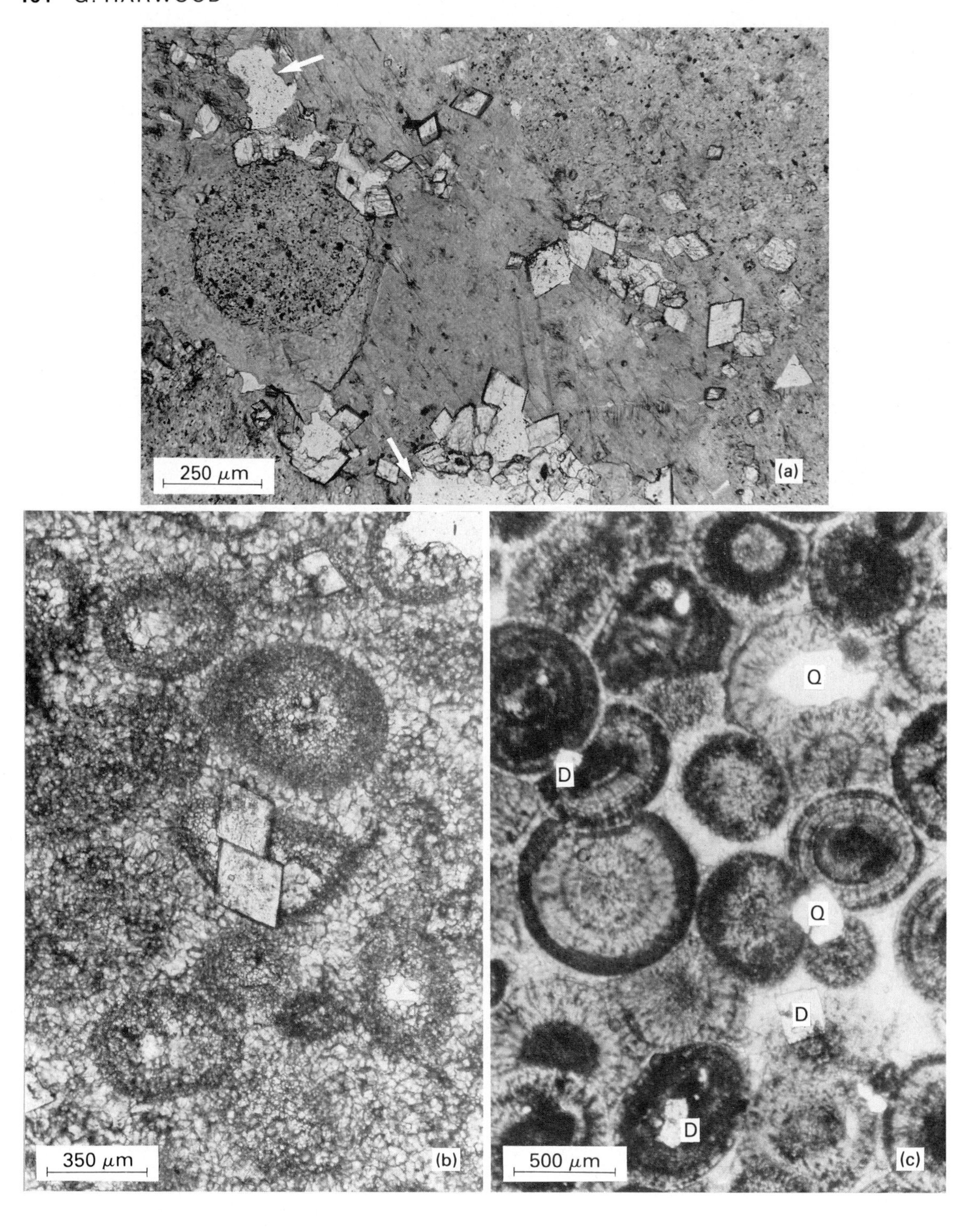
250 μm
(a)
350 μm
(b)
Q
D
Q
D
D
500 μm
(c)

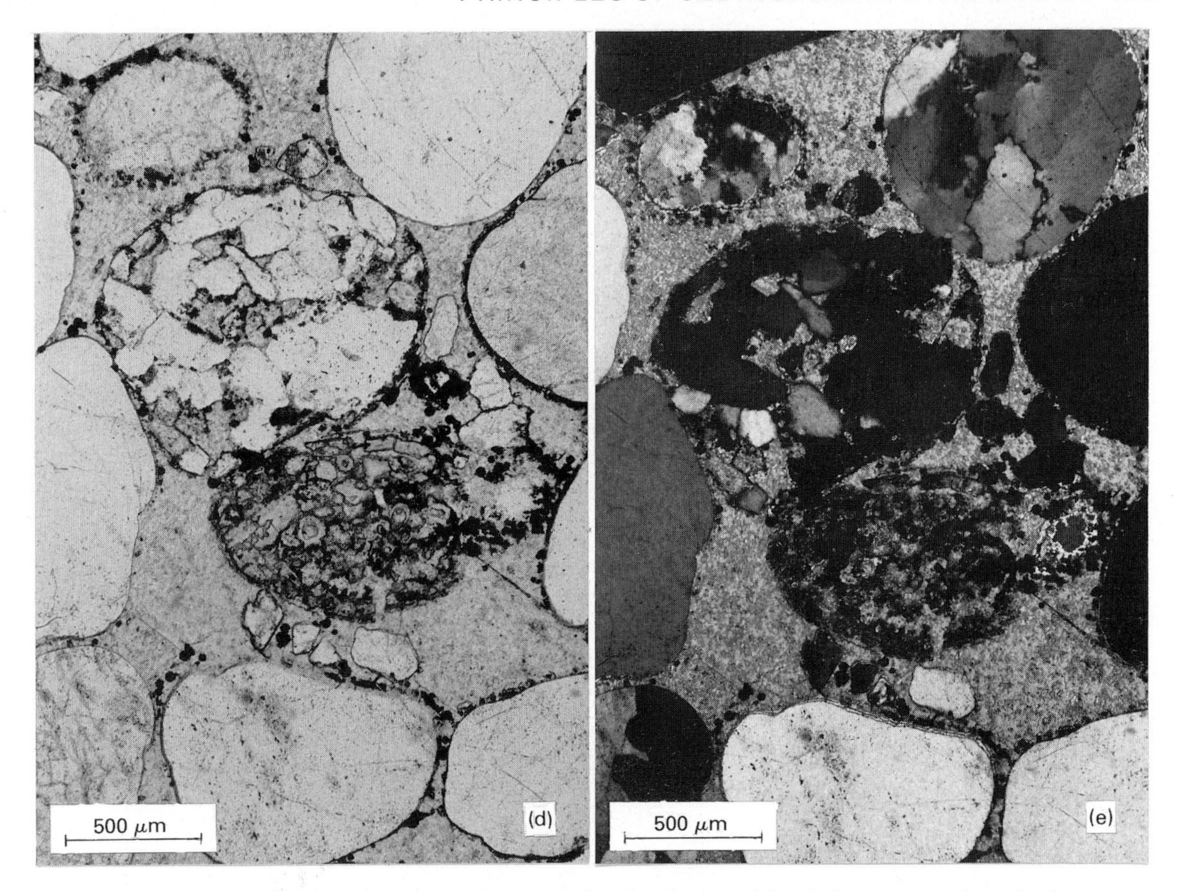

Fig. 5.41. Timing of replacement: (a) Dolomite rhombs in crinoid grainstone; dolomite has grown replacively after formation of crinoid overgrowth cements in crinoid grainstone of low porosity and permeability. Voids (arrowed) are the result of plucking during section-making and thus show false porosity. Section stained with alizarin red S. Mississippian, Burlington Formation, Illinois, USA. Sample courtesy of W.B. Meyers, SUNY.
(b) Dolomite rhombs in ooid grainstone; dolomite has formed after cementation of ooid grainstone and after partial dissolution and compaction of ooid (arrowed) (from Harwood & Moore, 1984). Jurassic, subsurface Smackover Formation, east Texas, USA.
(c) Dolomite and euhedral authigenic quartz growth after compaction and cementation. Ooid grainstone with interpenetrating grain to grain contacts and later intergranular calcite cements (see also Fig. 5.40a, b) with later replacive dolomite rhombs (D) and authigenic quartz crystals (Q). The relative timing of dolomite and quartz growth is not determinable as the two are not found intergrown. Jurassic, subsurface Smackover Formation, Arkansas, USA.
(c) Replacement of certain parts of detrital sedimentary grains; calcite of intergranular cement has replaced components of the derived sediment grains, leaving quartzose portions unreplaced. Permian, Yellow Sands, Co. Durham, UK.
(e) Crossed polars view of Fig. 5.40(d), showing that calcite within detrital grains is in optical continuity with poikilotopic intergranular calcite. Former mineralogy of replaced components cannot be determined.

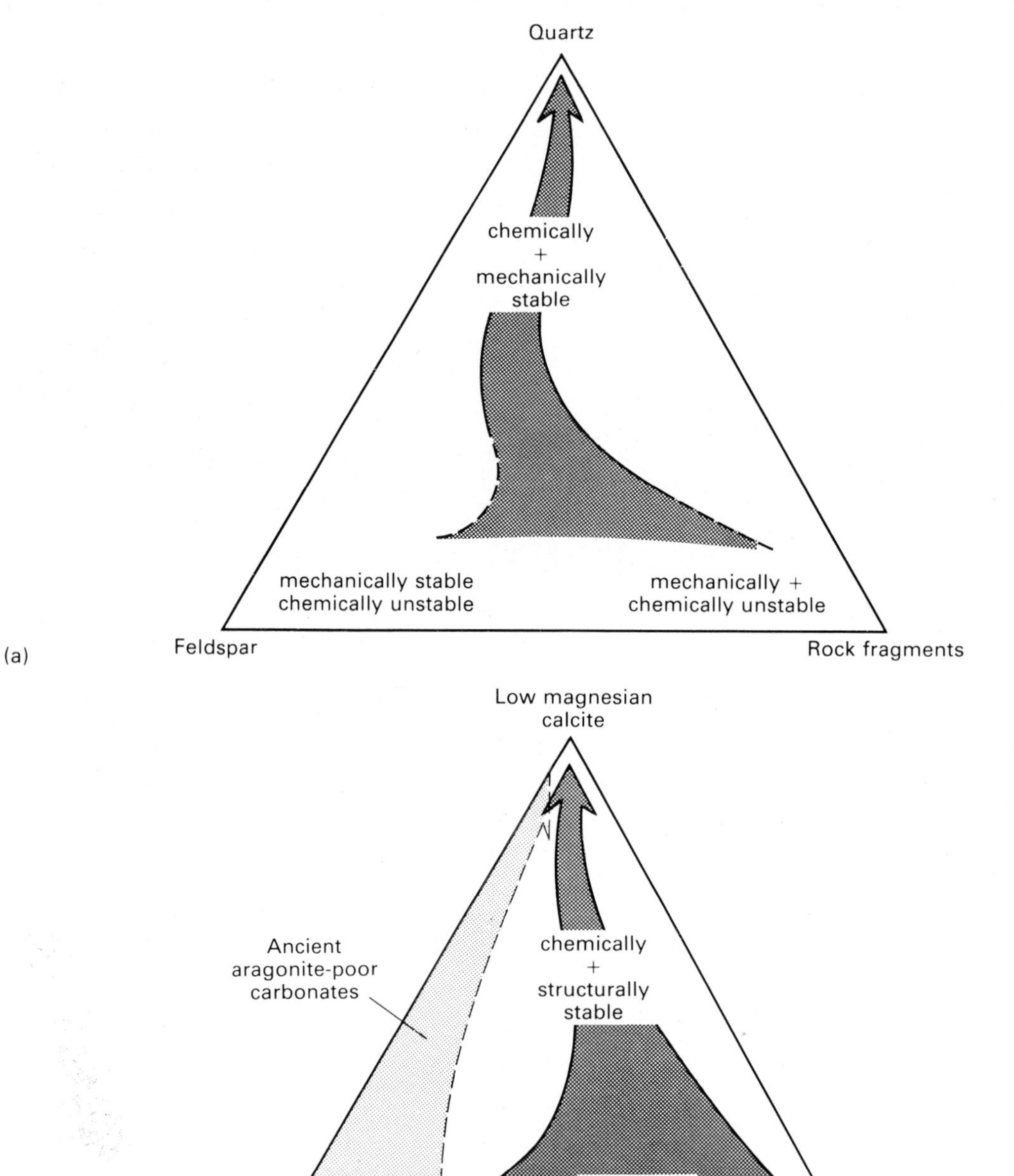

Fig. 5.42. Diagenetic potential of sedimentary rocks: (a) Triangular diagram of siliciclastic end members, demonstrating mechanical and chemical resistance to diagenetic alteration (modified form Bjørlykke, 1983). Width of arrow increases with increased diagenetic potential.
(b) Similar diagram for carbonate mineralogical end members, showing high diagenetic potential of aragonite and high Mg calcite dominated sediments. Carbonate minerals have similar rigidities, but the aragonite-low Mg calcite transformation is by dissolution-reprecipitation (albeit on a fine scale in some cases) whereas that of high Mg calcite to low Mg calcite appears to involve no structural recrystallization, but simply magnesium 'stripping'. It should be remembered that many Palaeozoic carbonates were dominated by high Mg calcite or low Mg calcite, with little or no aragonite. Width of arrow increases with increased diagenetic potential.

(1984) demonstrated feldspar dissolution to have taken place after the formation of quartz overgrowth cements in some US Gulf Coast Lower Tertiary sandstones, whereas, in similar sandstones, Siebert, Moucure & Lahann (1984) showed feldspar dissolution occurred after calcite cementation but before hydrocarbon maturation and migration. Moore & Druckman (1981) detailed late, non-selective dissolution of carbonate sediments, in advance of hydrocarbon migration. Although existing carbonate phases may be dissolved by non-organic acids within the pore fluids, feldspar dissolution necessitates aluminium complexing, only possible with some organic constituent of the pore fluids (Surdam, Boese & Crossey, 1984; Ednam & Surdam, 1986; Meshri, 1986).

Such dissolution related to hydrocarbon migration has been documented to take place within the deep subsurface, but other selective and non-selective dissolution can take place at various depths, depending more on the efficiency of pore fluid migration and pore fluid chemistry. Considerable dissolution is unconformity-related, where active meteoric circulation can remove less stable mineralogies, particularly evaporites, some carbonate phases and weathered silicates.

If a grain or cement retains its form during alteration it is sometimes very difficult to determine when that alteration took place, unless it can be linked to other dissolution/replacement phases. Only if the alteration product is otherwise modified is it possible to fit the alteration precisely into the diagenetic history of the sediment. Similar problems occur when a sediment has been completely replaced, particularly when replacement has destroyed the original texture. Dravis & Yurewicz (1985) found that blue light emission spectroscopy aided identification of predolomitization fabrics; Goodall & Hughes (1985, personal communication) successfully used this technique in evaporite petrography. Cathodoluminescence petrography (Chapter 6) may also help to 'see through' complete replacement fabrics. Individual mineral replacement fabrics are, on the whole, easier to place in a diagenetic sequence (e.g. Fig. 5.41a, b, c) as relationships with pre-existing diagenetic events are visible.

5.3.6 Diagenetic potential

Many sediments are more susceptible than others to change during diagenesis. The diagenetic changes which take place within a supermature quartz sandstone are limited; quartz overgrowth cements may form, but, unless pore fluids are extremely aggressive, neither dissolution, alteration nor replacement of detrital quartz grains are likely to occur. The presence of feldspar grains within a sandstone renders it more susceptible to change during diagenesis as alteration, replacement or dissolution of these grains may take place. Feldspar grains therefore have a higher diagenetic potential than quartz grains, particularly to chemical change; mechanically, they are almost as rigid as quartz grains when fresh. Other minerals (e.g. pyroxenes, amphiboles, micas and many rock fragments) have a considerably higher potential for alteration and dissolution and are also less rigid. On a quartz–feldspar–rock fragments triangular diagram these more susceptible minerals plot together in one corner. Sandstones with compositions which plot in this area are therefore much more susceptible to diagenetic change than those with compositions nearer a quartzose sandstone (Fig. 5.42); they thus have a much higher *diagenetic potential*.

Carbonate and evaporite sediments also have a high diagenetic potential. Many modern carbonate sediments are a mixture of aragonite and high Mg calcite, minerals which are theoretically unstable at surface temperatures and pressures. During early diagenesis these minerals rapidly change to low Mg calcite, either by dissolution/reprecipitation or by alteration. However, many ancient carbonates, particularly those in the Palaeozoic, have a somewhat lower diagenetic potential, being initially dominantly high and/or low Mg calcite. Evaporite sediments have a yet higher diagenetic potential than modern carbonate sediments, producing a near-metamorphic fabric. The diagenetic fabric of a sedimentary rock is therefore largely dependent on the original sediment composition, although fluctuating pore fluid compositions may govern the rate, the extent and the relative timing of diagenetic reactions.

5.4 CONCLUSIONS

Petrographic examination of thin sections can therefore reveal much information about the depositional and diagenetic history of the sediment. Recording of these data in a systematic manner is vital to its interpretation. This is best done by systematically working through a given series of points on a data

Sedimentary Petrography

Description Form

Petrographer ....
Date ....
Specimen ....
....

Terrigenous sediments

Hand specimen Colour: Fresh— .... Weathered ....

Structures:

Grain size: Max. .... mm
Min. ....
Mode(s) ....

Roundness ....
Sphericity ....

Sorting: ....
Other features:

Thin section

μm

Apparent grain size: Max. .... Min. .... Mode(s) ....

Roundness ....
Sphericity ....
Sorting ....

Detrital composition %

Quartz ....
K feldspar ....
Plagioclase ....
Rock fragments ....
Muscovite ....
Biotite ....
Heavy minerals ....
Opaque minerals ....
Carbonates ....
Matrix ....
Others ....

Diagenetic components

Cement ....
Other authigenic minerals ....
....

Compaction (type and degree)

Quartz varieties: %

monocrystalline < ....

polycrystalline < ....

Matrix components:

....
....
....
....

Porosity: Type %

Primary ....
....

Secondary ....
....
....

(a)

Components on triangular diagrams

1. Origin of components

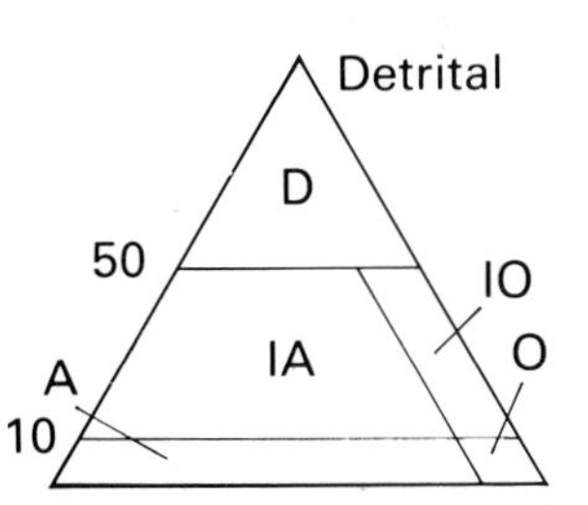

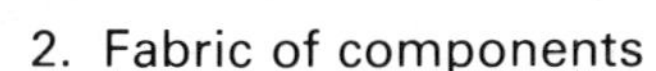

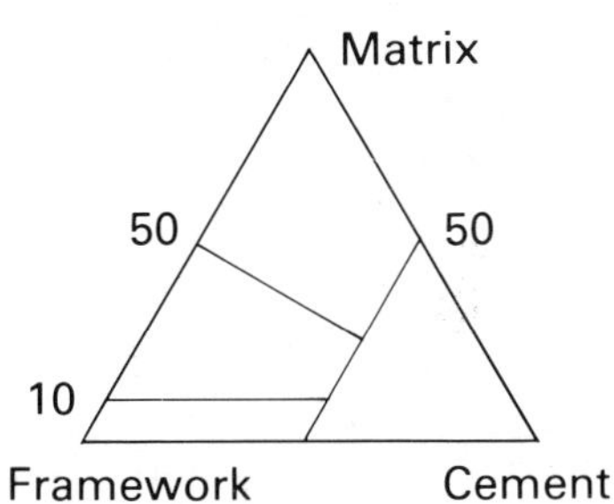

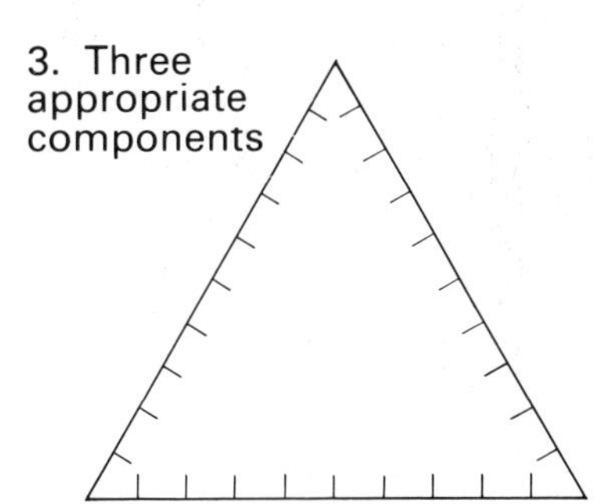

Sketch of key fabric(s)

Diagenetic events

Comments and interpretations

Fig. 5.43. Data sheets for systematically recording petrographic observations; sheet A is for terrigenous sediments and sheet B for carbonates. Sheets can be combined if all observational criteria are not itemized.

Sedimentary Petrography
Description Form

Petrographer ....
Date ....
Specimen ....

Carbonate/evaporite sediments

Hand specimen Colour: Fresh .... Weathered ....

Induration:

Structures:

Grain size: Max. ....
Min. ....
Mode(s) ....

Rock type (Durham class.) ....
Skeletal clasts ....

Sorting: ....
Other features

Thin section

Detrital grains

Skeletal types	Size (μm)	Roundness	Fragmentation	%
....	....	....	....	....
....	....	....	....	....
....	....	....	....	....
....	....	....	....	....
....	....	....	....	....
....	....	....	....	....
....	....	....	....	....
Intraclasts	....	....	....	....
Ooids	....	....	....	....
Pellets	....	....	....	....
Peloids	....	....	....	....
Quartz grains	....	....	....	....
Feldspar grains	....	....	....	....
Other detrital minerals	....	....	....	....

Matrix

Evaporites	%	Size	Comments
Gypsum	....	....	....
Anhydrite	....	....	....
Halite	....	....	....
Others	....	....	....
	....	....	....
	....	....	....

(b)

Diagenetic components

Cement type	%
....	....
....	....
....	....
....	....

Other authigenic minerals

....	....
....	....
....	....
....	....

Porosity

	Type	%
Primary	....	....
	....	....
Secondary	....	....
	....	....

Compaction (type and degree)

Components on triangular diagrams

1. Origin of components

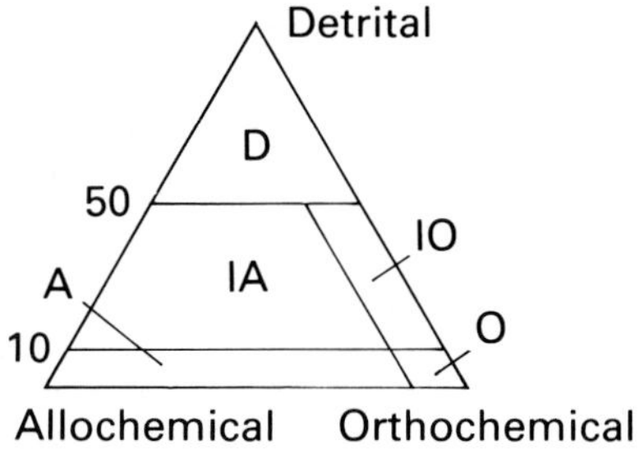

(I = impure)

2. Fabric of components.

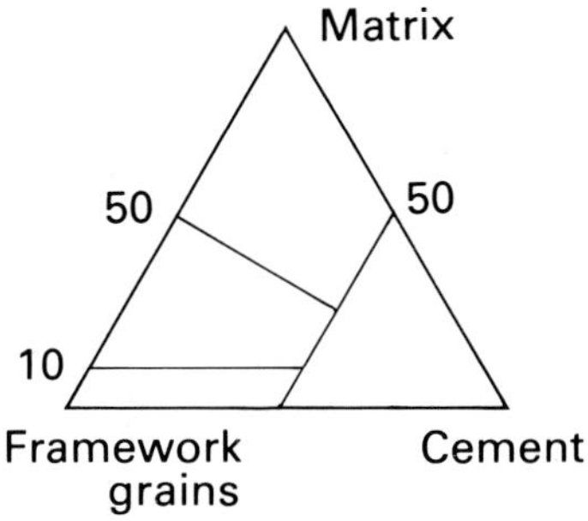

3. Three appropriate components

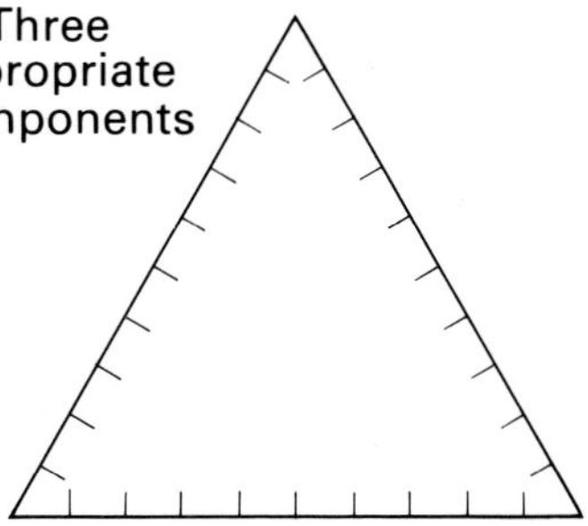

Sketch of key fabric(s)

Diagenetic events

Comments and interpretation

Location/well ....

Elevation/depth ....

Grain supported ☐
Matrix supported ☐

Diagram/photomicrograph

Carbonate minerals %
- Calcite ....
- Fe-calcite ....
- Dolomite ....
- Fe-dolomite ....

Terrigenous minerals
- Quartz ....
- K feldspar ....
- Plagioclase ....
- Mica ....
- Others (biotite, amph. etc.) ....

Other minerals
- Anhydrite ....
- Gypsum ....
- Halite ....
- Chert ....
- Others ....
- ....
- ....

Cement ....
....
....
....
....
....

Comments

Rock name ....

Sedimentary structures

Skeletal components

Matrix

Porosity

(c)

Fig. 5.44. Abridged data sheet of a type used by many petroleum company research laboratories.

sheet (Figs 5.43a, b and 44), particularly at the start of a project. Data compiled in this way can then be used by co-workers to combine results into a more regional synthesis than otherwise possible. Familiarity with a certain sediment type, however, can lead to abbreviated data sheets, containing only those aspects present in those sediments, which, in turn, may lead to unusual or less common fabrics being overlooked. Unfortunately, experience is the best guide to fabric interpretation, and this can only be obtained by study of many different sediment types.

The combination of depositional and diagenetic

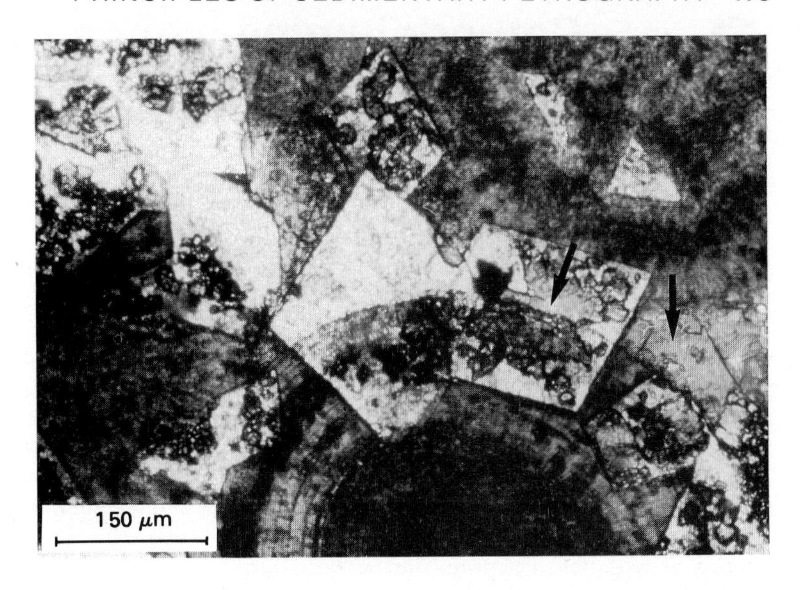

Fig. 5.45. Things are not always as they seem! Dolomite rhombs replacing ooid grainstone have hollow centres with irregular outlines. These centres were originally interpreted as being the result of calcitization of the dolomite, thus accounting for the corrosive nature of the secondary calcite/dolomite margins. Closer investigation, however, demonstrates that the textures of the ooids and intergranular cements are continuous through the centres of the dolomite crystals and, when viewed in colour, the differences in stain colour persist from outside of to within the dolomite rhombs (arrowed). The dolomite therefore has only partially replaced the original carbonate and the hollow centres are a primary feature of the dolomite rhombs, representing unreplaced grains and cements. Interpenetrating and sutural grain to grain contacts between the ooids demonstrate that the diagenetic history of this sediment can be itemized as: (i) mechanical followed by dissolution compaction, (ii) precipitation of late intergranular ferroan calcite spar cements, and (iii) partial dolomitization. Jurassic, Smackover Formation, subsurface Arkansas, USA.

fabrics therefore creates a complex texture within a sediment, representing both sedimentation events and events which have taken place during diagenesis. These events are closely linked to the setting and evolution of the containing sedimentary basin with burial diagenetic events representing different stages of basin history. Petrofabric evaluation should therefore form an integral part of any study of basin evolution. As modern weathering masks many of these textures, petrographic studies are best carried out on subsurface samples wherever possible, particularly if samples are to be chemically analysed later. The study of petrographic fabrics forms the basis for later analytical research, both chemically and isotopically. Diagenetic fabrics are also linked to subsurface porosity evolution and hydrocarbon migration and closer future links with organic geochemists will evaluate potential interrelationships between these events. Perhaps the one point that should be emphasized is that extreme care should be taken in documentation and interpretation of petrofabrics, as one can be easily misled by what appears to be a straightforward thin section (Fig. 5.45)!

6 Cathodoluminescence microscopy

JOHN MILLER

6.1 INTRODUCTION

Luminescence is the emission of light from a solid which is 'excited' by some form of energy. The term broadly includes the commonly-used categories of fluorescence and phosphorescence. Fluorescence is said to occur where emission ceases almost immediately after withdrawal of the exciting source and where there is no thermal cause, whereas in phosphorescence the emission decays for some time after removal of excitation. The distinction between these so-called types of luminescence is somewhat arbitrary and confusing; for example, many minerals have very long post-excitation decay times. Confusion is avoided by using the term luminescence, and specifying the activating energy as a descriptive prefix. Thus roentgenoluminescence is produced by X-rays, photoluminescence by light (e.g. ultraviolet) and cathodoluminescence (CL) results from excitation by electrons. Thermoluminescence results from heating.

Luminescence has been known in geological materials since 1604, when the alchemist Cascierolo described light emission from barite. Nowadays the phenomenon is very familiar: CL provides our television pictures from excitation of chemical phosphors by the cathode-ray tube's electron beam. CL of minerals was first investigated systematically by Crooks (1880), who sealed diamonds, rubies and other gems into a discharge tube and observed their brilliant emission colours. This work was not followed up thoroughly by geologists until the advent of the analytical electron microprobe (Smith & Stenstrom, 1965). The small beam size and poor optics of the microprobe were restrictive, however, and it was quickly realized that a simple device specifically designed for CL petrography was required. Long & Agrell (1965) and Sippel & Glover (1965) both published descriptions of such a device, which could be attached to a petrographic microscope for examination of thin sections. There are now several commercial luminescence machines available, mostly developments of these early designs.

Ultra-violet fluorescence microscopy is often used in examination of hydrocarbon residues in sediments (e.g. Burruss, Cercone & Harris, 1985). Polished thin section surfaces are required, and a special microscope with UV source and quartz lenses is needed, such as used for immunological work in many biological laboratories. Various wavelengths of UV can be selected by means of filters, and filters can be interposed when viewing the emission. Hydrocarbon inclusions show strong luminescence, the colour varying with the gravity of the oil. Recrystallized organic-rich fossils, such as renalcid micro-organisms in reefs, may show up very well under UV, whereas they may be invisible in transmitted light and CL. Dravis & Yurewicz (1985) have shown that in some limestones, cement generations and fine crystal growth zoning can be revealed by UV. Certainly UV microscopy is attractive because it does not require elaborate vacuum arrangements, but inorganic materials such as calcite often show only very weak UV luminescence, so UV microscopy is not a general substitute for CL work.

Sedimentologists have now found a wide variety of applications for CL, and it has become a very important tool for petrographic analysis. Nevertheless, despite the growing volume of publications reporting CL results, there is still much fundamental work to be done on understanding the precise causes and significance of luminescence phenomena in geological situations. It is therefore essential to have at least a general knowledge of the physical background of CL to avoid serious misinterpretation.

6.2 PHYSICAL EXPLANATION OF LUMINESCENCE

6.2.1 Excitation factors

An explanation of luminescence was not forthcoming until the development of quantum theory. Quantum approaches to luminescence are outlined by Nickel (1978), Marfunin (1979) and Walker

(1985). The energy of a beta-ray (electron) is sufficient to excite an atom or molecule and cause a quantum jump, with the input energy being totally absorbed. After a short delay time (10^{-8} s) the excited atom or molecule returns to its former energy state and may emit radiation in the form of light, alpha-, beta- or gamma-rays. The wavelength of the emitted radiation is always longer than that of the exciting radiation. For geological purposes, we are usually only interested in emission in the visible spectrum, although it should be noted that a proportion of emission in some minerals may be in the infra-red or ultra-violet (e.g. in feldspars).

The intensity of CL is a function of current density at the specimen and the voltage (accelerating potential) of the applied electron beam. The relationship of luminescence intensity to electron beam intensity is not a simple linear one, however, and moreover varies within the mineral families (Coy-yll, 1970).

Coy-yll (1970) also showed that, for a given crystalline solid, there is a point at which increasing the electron beam current ceased to produce greater luminescence intensity; he termed this the *saturation level*. Increasing beam energy beyond this level actually produces a decrease in luminescence intensity; this is the *inhibition phase*. For a given current, saturation occurs with beams of about 8 kV potential in feldspars but at more than 16 kV with quartz. In practical terms this means that attempting to elicit more luminescence from a mineral by simply increasing beam power can have the reverse effect if the saturation level is exceeded. Overly powerful beams are also likely to have a local heating effect, causing thermoluminescence. Thermoluminescence and CL spectra may then combine, giving a false impression of the luminescence colour. Intense, narrowly-focussed beams may also cause local heating to the point of incandescence, particularly with micas, which flare and collapse. Further damage to preparations caused by local heating includes blistering of thin section bonding resins and even cracking of the glass slide.

Within that part of a sample bombarded by electrons (the 'reaction volume'), there are many effects other than cathodoluminescence. Secondary, backscattered and auger electrons are produced, as well as X-rays. With suitable detectors, in scanning electron microscopes or microprobes, some of these effects can be used for analytical or imaging purposes. However, in cold cathode luminescence machines (see below), many ions are produced in addition to the electrons. These also enter the reaction volume and can induce damage in the uppermost layers of a thin section by remobilizing elements and homogenizing luminescence activators (see Section 6.2.2). Therefore, a thin section remaining in the beam for a long time may show a decrease in luminescence intensity as ionic bombardment and thermal diffusion take place. According to Remond, Le Gressus & Okuzumi (1979), this change in CL is irreversible and the affected section must be repolished, removing the damaged layer. Certain minerals are very susceptible to irradiation damage, and may become permanently coloured after removal from the electron beam. For example, halite is 'electron-stained' blue in a few minutes, and Dickson (1980) reported a stable purple colouration in electron-ombarded fluorospar.

6.2.2 Luminescence centres

An excellent summary of the present knowledge of CL generation in minerals is given by Walker (1985), and only a simplified outline follows here.

At any temperature, a real crystalline solid is in a state of dynamic unrest, with the various electrons and atoms in its lattice vibrating about their mean positions. This state of energy is called 'ground state'. Few crystals are perfect. During normal growth, they may acquire defects from distorted internal structure between mosaic crystals, suffer omission defects where grounds of atoms or molecules are missing from lattice sites, have charge displacements (for example where there are abnormally ionized atoms), or undergo mechanical damage such as formation of distorted surfaces and cracks. Upon excitation by an electron beam, such local sites of crystal imperfection are more liable to absorb energy from the beam than are neighbouring lattice sites. The domains of imperfection become luminescence centres; they preferentially trap energy from the cathode beam which induces 'jumps' in their lattice electron orbitals. On subsequent decay to the resting state, photons are emitted and luminescence occurs.

The excitation–emission process is often temperature dependent. Cooling samples with liquid nitrogen, for example, increases CL efficiency in some minerals such as quartz, producing much greater luminescence intensity for a given beam energy. Other minerals, such as calcite, tend to emit less on cooling. Also, cooling can produce spectral shifts in emission wavelength, so although it may have potential uses for geological applications where CL emission at room temperature is meagre, the full effects of cooling are at present poorly known and the technique cannot be recommended as a regular practice for petrographic purposes.

Luminescence centres may occur in two general forms: intrinsic and extrinsic. Intrinsic centres are due to lattice imperfections such as growth distortions and other electronic lattice defects, acquired independently of the composition of the precipitating medium. Extrinsic centres are those acquired from the parent medium during crystallization, such as impurities in surface, regular lattice and interstitial sites, or compositional inhomogeneities between different parts of a crystal. In practice, it may be very difficult to determine the principal cause of CL in a mineral, as often there are complex interactions between extrinsic and intrinsic centres.

Extrinsic luminescence centres may behave differently depending on their response to electron beam excitation, and the following types can be distinguished:

(1) Activator centres: those where radiactive transitions (luminescence) are highly probable as the energized centre returns to its ground state.
(2) Trap centres: those where additional energy is required to raise the energy state sufficiently to produce luminescence on transition to the ground state.
(3) Quencher centres: where even the excited state of the centre is close to a radiationless transition level, so little or no luminescence is emitted.

Extrinsic luminescence centres are the best known, and perhaps the easiest to detect by geochemical analysis, whereas the causes of intrinsic CL require detailed crystallographic and solid state physical investigation. Transition metal ions are the commonest activator impurities causing extrinsic CL, substituting in crystal lattices for normal ions with appropriate ionic radii. Rare earth elements, such as Eu^{3+}, Sm^{3+} and Dy^{3+}, are also implicated as inducing activator centres (Mariano & Ring, 1975).

In geological materials, luminescence is commonly controlled by the balance of activator and quencher centres. For example, Mn^{2+} is the main activator causing luminescence in calcite, whereas Fe^{2+} is a quencher in the same mineral. However, there is no guarantee that activator or quencher elements will always have the same effect in different minerals; for example, Fe^{3+} is an activator in some feldspars, despite the quenching activity of Fe in calcite.

The colour (wavelength) and intensity of CL may also vary according to the sites in the crystal lattice where activator ion substitutions occur. Because of interactions between activators and quenchers, and between intrinsic and extrinsic luminescence centres, it is often difficult to be certain of the cause of luminescence in a given mineral. Nickel (1978) and Amieux (1982) summarized known emission colours and activators for many minerals but, in our present state of knowledge, these data should only be used as a guide.

Depending on the activator, a single excited mineral may radiate at several different wavelengths; for example, apatite CL may be green, yellow, pink, violet or white, with a variety of rare earths acting as activators. Activator and quencher ions may produce their effects at extremely low concentrations, below the detection limits of the electron microprobe. Analytical techniques such as atomic absorption spectrophotometry and neutron activation analysis may be required for quantitative determination at such low levels. The surest method of identification of CL-activating ions is by analysing luminescence excitation spectra (Walker, 1983, 1985) and combining this with studies of emission spectra (Mariano & Ring, 1975).

6.3 EQUIPMENT

There several ways of generating electron beams under vacuum: by stripping electrons from a hot filament; utilizing field effect or corona discharge from metal points; or by generating a plasma across a low pressure gas using a cold metal cathode disc. Zinkernagel (1978) and Ramseyer (1983) described a hot filament luminescence apparatus especially suitable for observation of very dull emission from quartz. While this equipment gives good results with such difficult cases, it is bulky and expensive, difficult to construct and maintain, and requires a high vacuum from diffusion pumps. Cold cathode machines are much less complex, easily mounted on a standard petrographic microscope and perfectly suited to most sedimentological requirements. They offer the advantages of cheapness, ease of operation and simple maintenance, and are ideal for routine petrographic purposes. Commercial CL machines (Figs 6.1 and 6.2) use the cold cathode method at present.

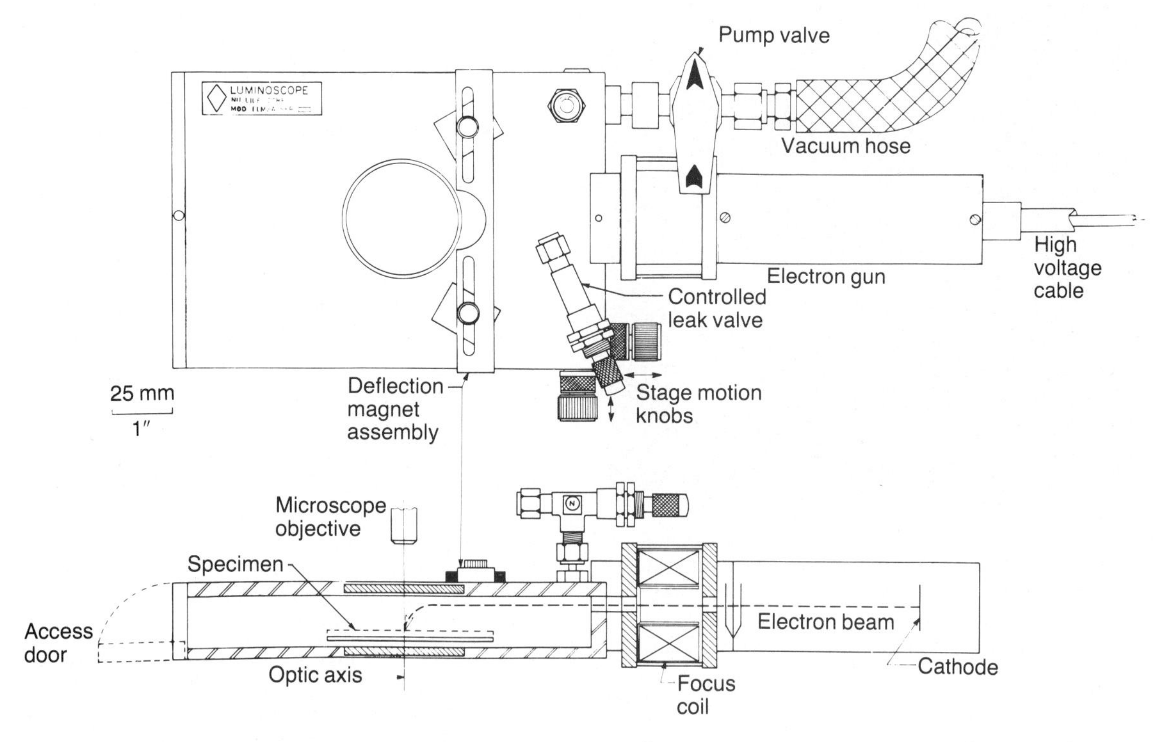

Fig. 6.1. Schematic drawings of the Nuclide ELM-2A Luminoscope® specimen chamber in top view and cross-section. Courtesy of D. Marshall, Nuclide Corporation.

With the cold cathode system (Fig. 6.1), there are always sufficient positive ions in the vicinity of the specimen to prevent it acquiring a space charge, eliminating the need for pre-treatment of samples by evaporating a conductive coating on to them. Since a steady leak of gas must be maintained for plasma generation, a small rotary vacuum pump is adequate. In contrast to hot filament cathode systems, a sudden loss of vacuum causes no damage to the equipment. Beams can also be established at low accelerating voltages, thus reducing the risk of specimen damage.

There are, however, a number of disadvantages to cold cathode systems. Delicate arrangements are required to maintain the controlled gas leak and these are prone to wear and maintenance problems. The discharge plasma produces a blue glow which may interfere with luminescence observation, particularly with meagre emitters such as quartz: this problem may be overcome by using helium rather than air as the leak gas. Because of the presence of gas (usually air) in the system, a stream of positive and negative ions is also generated and this may induce sample damage. All of these effects mean that cold cathode equipment, while eminently convenient for qualitative work, is unsuitable for making reliable and precise observations on CL intensity and emission wavelengths (see Section 6.5). For this purpose a new generation of hot cathode machines must be developed.

6.3.1 Operation

Figure 6.1 shows a schematic cross-section of a typical CL apparatus. The electron beam is generated by applying a potential up to 30 kV between a small tungsten steel cathode disc and an annular anode which lie within an evacuated glass tube. At low pressures, a plasma is established with a discharge of positive ions travelling towards the cathode. An electron beam (together with some negative ions) is emitted in the opposite direction, attracted towards the collimating anode and passing through its central hole. In the Nuclide Luminoscope® (Fig.

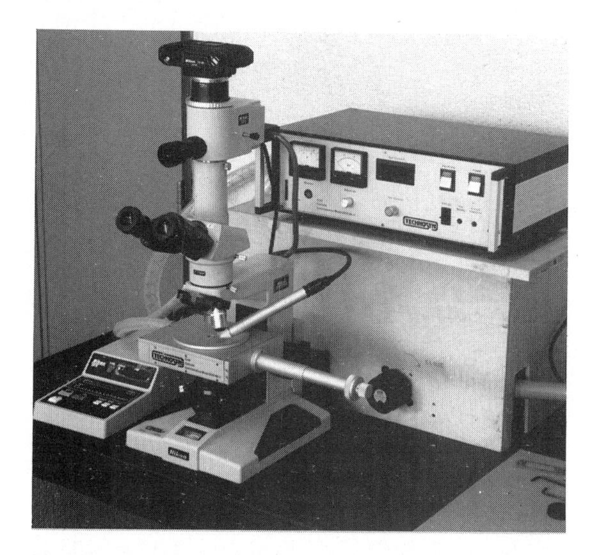

Fig. 6.2. Technosyn luminescence chamber on a Nikon binocular microscope with photomicrographic attachment. Note the oblique electron gun and the projecting X–Y specimen traverse controls to the right of the chamber. The HT and vacuum control box is behind the microscope and the photographic exposure meter lies to the left.

6.1), the beam then passes through a focussing coil, which carries a variable DC voltage to give control over the beam size impinging on the specimen. Narrow beams with high current density can be used for observation with high magnification lenses or on samples with poor CL; the beam can be also expanded to cover an area of several square centimetres for macro-photography of small slabs or large thin sections. The electrons are deflected down on to the specimen by a pair of adjustable magnets attached to a movable carriage on the top of the specimen chamber. In the Technosyn equipment, the gun is arranged so that the beam fires obliquely down on to the specimen (Fig. 6.2) and these magnets are not needed. A focussing coil is not required, and as the beam cross-section is relatively constant, the current density at the specimen surface is also relatively constant. There are some advantages to a lateral gun position and focussing device; larger viewing windows can be accommodated so that greater sample areas may be excited by de-focussing the beam, and the occurrence of sample damage by direct bombardment is lessened.

Samples are placed on a glass tray supported on an X–Y bearing carriage. They can thus be moved in any direction across the microscope stage. Viewing may be by transmitted light, CL, or a combination of the two, observed through a lead glass window which absorbs X-radiation. The equipment should be used in a well-darkened room, and a red photographic safelight usefully provides illumination for operating and note-taking without disturbing the worker's 'night-vision', which is essential for observation of low intensity CL.

During excitation, samples are heated and begin to outgas, producing more ions and concomitantly increasingly the electron beam current, which if unchecked will reach a preset level within the power unit and cause the high tension to be switched off to protect the gun. The art of achieving a steady-state beam depends on balancing the controlled gas leak and the sample outgassing with the continuous pump evacuation at a given beam voltage.

Sample changing is rapid. The beam is turned off, the vacuum valve closed and the chamber vented to air. Several samples can often be placed in the chamber at once, depending on their size. For viewing with higher magnification lenses, which are necessarily of shorter working distance, a recessed window is usually required to accommodate the smaller working distance of the lens.

6.3.2 Microscopes for cathodoluminescence

It is of capital importance that the microscope upon which CL equipment is mounted should be of the highest possible transmissivity. Because most petrological microscopes are designed for use with transmitted light which can be provided at any required intensity, manufacturers mostly pay little attention to this parameter. Luminescence is rarely more than 1% efficient and thus of comparatively low intensity. Marshall (1980) investigated the behaviour of several microscopes with CL samples. He found that the controlling factors were the magnifications and numerical apertures of eyepieces and objectives, transmission of individual optical elements, and other details of the optical path, including prisms, splitters and polarizers. The optimum system for CL should have as few optical elements and lens/air surfaces as possible, with good coatings on lens surfaces. Therefore, monocular microscopes with 'straight through' ray paths have the greatest transmissivity. It is recommended that several microscopes meeting the criteria outlined above should be tried, using a fairly dull CL subject such as a sandstone sample, before deciding on one which will be dedicated to the CL equipment.

For those setting up a CL unit for the first time, and considering purchasing a new microscope, the inexpensive Olympus POS Student Microscope has one of the best transmissivities. However, it is a simple monocular design and suffers slightly from pin-cushion distortion, but makes a good general-purpose CL microscope where low cost is important. If long-term viewing comfort and high quality flat-field optics are required, a stripped-down version of the Nikon binocular petrological microscope is recommended (Fig. 6.2), where there is a 'straight-through' path for photomicrography.

The objective lenses required for cathodoluminescence must have long working distances, so choice is often limited and prices high. Measuring or Universal Stage objectives are usually suitable. However, these are complex multi-element systems, and their transmissivity decreases rapidly with increased magnification. Some of these lenses appear to have selective absorption at certain wavelengths, and there can be distinct changes in perceived CL colour when such lenses are interchanged during viewing of a single sample. It is important to carry out some experimentation with various objectives and objective-eyepiece combinations so as to be familiar with the optical properties of the microscope system in use.

To maintain optimum transmissivity, it is also vital to keep viewing windows clean. Lower window surfaces rapidly become coated with a brown layer of pyrolized hydrocarbons, derived partly from heated resin on thin sections and from vacuum pump oil mist. The window coating problem is considerably reduced by using a foreline trap on the vacuum system, but the trap filter material must be regularly replaced, otherwise pump-down times are increased. Window coatings not only impair transmissivity but, as they build up, selectively absorb certain wavelengths. Acting as a coloured filter, such coats give a misleading impression of CL colours.

Viewing windows are best cleaned with a pinch of non-abrasive laboratory glass detergent and a damp wad of soft tissue or cloth, taking care to avoid the sealing gasket around the window. Stubborn deposits may require application of organic solvents such as acetone or petroleum ether, but as these have harmful effects on skin and may also dissolve or harden the window sealing gaskets, they should only be used as a last resort.

6.3.3 Radiation precautions

At the voltages used for CL, electron beams are capable of generating appreciable amounts of X-rays. Luminescence chambers are designed to cope with this and are tested up to specified voltages with metal targets as samples. However, it is important to check radiation levels regularly with a geiger counter at worst-case operating levels, paying particular attention to joints and gaskets, particularly around the viewing window and anode. In European countries the mains supply voltage may fluctuate considerably, often being above any rated working voltage of the CL power pack, so that actual beam voltages may be appreciably higher than those set on the instrument panel. A potential X-ray hazard could then exist. If there is any suspicion of fluctuating voltages, instruments should be fitted with stabilized power supplies.

With the Nuclide Luminoscope®, a standard glass coverslip may be used in the recessed viewing window when the thicker lead glass window does not allow the shorter working distance objectives to focus. This unprotected window provides an X-ray hazard. It may be covered with a lead foil annulus around the lens barrel, but a much better arrangement is to manufacture brass collars for the high magnification lenses. These should fit into the well of the recessed window and have a flange which overlaps the edge of the well. Focussing movements of the objective can then be accommodated without risking radiation scatter.

6.4 SAMPLE PREPARATION

CL observations are usually made on uncovered thin sections. They must be bonded with thermo-stable epoxy resin (see Chapter 4). Bonding agents such as Lakeside 70 and Canada Balsam are unsuitable mountants for CL work as they are volatile, causing blistering and rapid charring under the electron beam. Some epoxy resins also show luminescence (usually yellow) which can interfere with observations.

Polished surfaces give the best optical resolution, and the double polished thin sections described in Chapter 4 are ideal. If great detail is not required, surfaces can be quite crudely prepared. Etching of limestones with dilute acid gives an apparent increase in luminescence intensity, but this is due to points of high relief catching the oblique beam, with shadowing giving an increase in contrast but at the expense of resolution. Nevertheless, both etched and stained surfaces (see Chapter 4) can be used. Stained surfaces, however, rapidly fade and acquire a brown discolouration in the beam, so transmitted-light examination and photography of stained samples should be carried out *before* irradiation.

Mechanically strained or damaged crystals tend to become strong CL emitters, and their luminescent patterns can be mistaken for geologically significant features. Such artefacts can easily be generated by rapid or crude preparation of thin sections, using coarse abrasives and forcing rocks through diamond-impregnated saw blades. The gentle and conservative method of high quality thin section preparation described in Chapter 4 is particularly recommended for samples to be investigated with CL. Scratches, cleavage shattering and other features of mechanical damage in crystals luminesce brightly and can be recognized by experienced observers.

All sample surfaces to be placed in the CL chamber should be dry, clean and free of contamination, including fingerprints as well as dust and lint. Fibres tend to cast shadows in the beam, wave about or become incandescent, blemishing time exposure photographs. A blast from a compressed air canister or a jet of petroleum ether across the surface helps to give a contamination-free surface before samples are irradiated. It is also important to remove all traces of abrasive powders used in sample preparation, because they usually have a bright luminescence themselves and can be mistaken for heavy mineral grains in the sample. Even the smallest particles of diamond are revealed by their brilliant green emission.

CL is essentially a surface phenomenon. Beam penetration is proportional to accelerating voltage, and is only a few tens of microns at 18 kV, so 30 μm thin sections are adequate. Thicker sections may give better heat dissipation for prolonged CL examination. Ultra-thin sections are not recommended, as the beam may be able to penetrate to the resin or glass and produce anomalous luminescence. Evaporated conductive coatings are needed only for use with high voltage hot cathode cathodoluminescence instruments. Slides which have been carbon-coated for probe work, however, can be examined directly in cold cathode equipment; the thin coating offers no impediment to the electron beam.

Depending on the size of the specimen chamber, rock slices can also be examined under CL. The standard Nuclide Luminoscope® chamber in particular is large enough to take slices over 5 mm in thickness, and there is a special deep version which can take sawn core samples for rapid examination, possibly on site. Best results are obtained from smoothed surfaces, but even broken faces can give useful information. Pump-down times are correspondingly increased for these larger and more porous samples, which should be well dried before evacuation.

Single crystals and loose sediment can also be viewed with little preparation. A rapid and simple technique for checking the mineralogy of placer sands is simply to sprinkle some sand on to the specimen tray glass or on a standard slide and evacuate it in the chamber (Ryan & Szabo, 1981). If there is a tendency for grains to become charged and leap about disconcertingly in the beam, a dried aqueous grain suspension usually has sufficient adhesion.

6.5 INTERPRETATION AND DESCRIPTION OF CL RESULTS

CL colour and brightness are dependent on several factors, including beam voltage, current, surface current density (a function of beam focus), nature of sample surface, time for which sample has previously been irradiated, degree of heating (thermal quenching or emission of thermoluminescence) and the geochemical composition of the sample itself. Marshall (1978) has suggested a standard for reporting CL results so that comparisons between the results of different workers may be more valid. With the cold cathode equipment, so many variable factors are involved that there is some doubt whether a true standard can be established (Fairchild, 1983). However, it certainly facilitates communication and comparison if the observation conditions are reported as Marshall suggested. Experience has also shown that the same samples can have different appearances under CL in machines from different manufacturers. Details such as fine zonation in calcite cements may be visible in one case but not in another. The reasons for this are not fully understood; the metal used for the cathode disc may be a factor, and the angle at which the beam impinges on the specimen surface may be another. Cold cathode systems may generate some CL as a result of ion bombardment, and the nature of the ion content of the plasma is likely to be machine-dependent. Therefore, the type of equipment should be specified as well as the observation conditions.

Problems in communication of results arise from the difficulty of describing CL colours and intensities. Authors refer to 'bright' or 'dull' luminescence, but such terms are relative to *their* sample suites and are also highly subjective. The descriptions depend on the sensitivity of the observer's eye to specific wavelengths, instrumental variation (such as chamber pressure, voltage, beam focus and beam current) and the wavelength-dependent transmission properties of the optical system, including the gradual build-up of brown pyrlysis deposits on the viewing window. Furthermore, human eyes are uneven in their perception of the visible spectrum, with peaks and troughs in sensitivity across the frequency range.

A significant proportion of the population (mainly males) is also affected with 'colour-blindness' of one kind or another, often to a subtle degree which may not be apparent in normal life. Reporting of colour perception is also subject to variations in linguistic understanding of words for describing colours. These are learned at an early age and reflect the bias and background of our parents. Upon asking a number of people to describe the CL colour of a given sample, I have received responses including 'bright red', 'dull orange', 'purple-brown' and 'crimson'; a red-green colour-blind person perceived 'grey'. To help overcome this potentially serious problem, observations should be made under internally consistent operating conditions and descriptions of CL characteristics chosen to be as unambiguous as possible. Careful photographic recording is most important to support subjective descriptions, but this has problems in itself (see below).

What steps can be taken to minimize subjectivity in recording and reporting CL results? Several petrographic uses of CL involve comparisons of CL intensities and colours between samples or within different areas of a single sample. Pierson (1981) suggested that a sample with the brightest observed luminescence be kept in the chamber and used as a reference against which the intensity of other sample areas could be less arbitrarily judged. He also advocated the use of a colour chart (fig. 21-1 in Pauling, 1970) for determining CL colour. This colour chart is quite inadequate for registering the subtle range of CL colours, and standard rock colour charts do not contain an appropriate range or scale for comparison either. Until special colour charts or reference sample scales become available, CL operators can do little more than ensure internal consistency, or attempt to construct a system for quantification of CL intensity and wavelength.

6.5.1 Quantification of CL results

There are many difficulties in designing ways of obtaining meaningful and consistent measurements of CL intensity from extant cold cathode CL equipment. The current density at the specimen must remain fixed during the observation period, and there is no direct way of measuring this in present CL devices. Variables such as window and microscope transmissivity must be calibrated. Since one usually needs to measure emission from only a small area of the viewing field rather than the whole field, a fibre-optic probe inserted in the light path would probably be the best method, with the light-pipe output directed on to a highly sensitive photoelectric cell or photomultiplier, whose output would also have to be calibrated against some standard phosphor. No commercial equipment for this application exists at present.

Production of a CL emission spectrum is the only way to record and compare CL 'colours' objectively. Mariano & Ring (1975) took spectra from feldspars on a Nuclide Luminoscope®. They used a fibre optic probe of 1250, 700 or 450 μm diameter measurement area, transmitting the CL through an external light pipe to the entrance of a grating monochromator. The output of the monochromator was detected by a photomultiplier, amplified and then plotted on a wavelength-synchronized $x-y$ recorder. Spectra were produced by a scanning spectroradiometer which was calibrated against a US Bureau of Standards incandescent standard. Plots of luminescence intensity against wavelength thus obtained must be corrected for any variations in spectral response of optical fibre, monochromator and photomultiplier. Although beam conditions should be kept steady while spectral measurements are made, this is less critical than with CL intensity measurements, as CL frequency is not dependent on beam current or voltage.

So far, quantitative measurements have mainly been done in order to discover the nature of the emitting centres, but there is considerable scope for using emission spectra from different growth zones in crystals as 'fingerprints' in establishing carbonate cement stratigraphies (see below) on a more certain and detailed basis.

Clearly, at the present state of the art, both communication and interpretation of CL results in petrography are dominantly subjective. There is an urgent need to develop quantitative and thus more objective methods for assessing CL intensity and wavelength. This poses a formidable problem for instrumentation, because of the very low light levels for most CL emission, and the difficulty of keeping beam parameters constant while making the optical measurements. Hot cathode devices appear to provide much better capabilities for quantification of CL results, and the next generation of instruments may well be built on this principle. Electron microprobes and scanning electron microscopes are now appearing with custom-designed CL attachments, allowing spectra to be taken and computer-assisted image enhancement processes to be used. Element analysis and CL can thus be more closely and conveniently linked.

Because of the subjectivity referred to above, interpretations of CL results should therefore be framed critically and cautiously. Many assumptions are involved: apart from variables inherent in instrumentation, our understanding of activating factors in many minerals is incomplete. It would be most unwise at present to rely solely on luminescence interpretations of geological phenomena. The technique should only be used *in conjunction with* other standard petrographic approaches. Its greatest strength lies in revealing fabrics and 'mapping' compositional variation, rather than in providing direct geochemical information, wherein the greatest uncertainties exist.

6.6 APPLICATIONS

Cathodoluminescence has several general applications in sedimentology:

(1) Rapid visualization of mineral distribution, where minerals have closely similar optical properties or are very fine-grained. For example, yellow-orange CL of calcite generally distinguishes it from the darker red-crimson of dolomite, and feldspars are very bright blues, reds or greens compared with subdued violets and browns for quartz grains. Complex intergrowths of minerals such as halite and sylvite (blue-grey and silver-grey CL respectively) can be easily visualized. It is essential to be very cautious and check mineral identifications with other techniques, as CL has complex origins and emission colours are rarely diagnostic.

(2) Fabric and textural characteristics are often more easily visible, especially in recrystallized carbonate rocks (Fig. 6.3). Fossils in neomorphosed limestones or carbonate-cemented sandstones reappear and can give otherwise unavailable stratigraphic and environmental information (Figs 6.4c, d, 6.5a, b). Point counts carried out on such rocks under CL may differ significantly from those performed under transmitted light, particularly in the greater proportion of bioclasts detected. CL makes it much easier to determine if sparry fabrics are of neomorphic or cement origin by displaying growth zones in crystal aggregates (Dickson, 1983). Sparry crystal aggregates might display the outline of primary void systems under CL (Fig. 6.3a, b), providing more information on porosity evolution than available with transmitted light alone.

(3) Small-scale features, which are difficult or impossible to see in transmitted light microscopy, may be well-displayed with CL. Fine veins, grain fractures (caused by extension or compaction), and authigenic mineral overgrowths readily become apparent (Fig. 6.5c, d).

(4) Provenance studies are more exact, since many grain suites carry characteristic luminescence 'fingerprints' which can be related to their source rocks (Fig. 6.5f). The origin of quartz grains can be determined to some degree, and mixing of quartz grains, feldspars and heavy minerals from different sources more easily detected, e.g. Stow & Miller (1984), Richter & Zinkernagel (1975). This aspect of CL petrography has received little attention, but it could be extremely valuable in petroleum exploration, giving an extra feature for use in correlation of borehole core sequences and interpretation of their depositional environments.

(5) Diagenetic and geochemical studies are enhanced by the very fine CL resolution of growth histories in crystals (Fig. 6.4). Taken together with other evidence, these geochemical variations may indicate variations in groundwater chemistry and burial depths, greatly increasing the resolution of diagenetic histories (e.g. Grover & Read, 1983).

(6) Mechanically induced post-depositional changes in sediments such as compaction, stylolitization and structural deformation can more easily be detected and evaluated with the aid of CL.

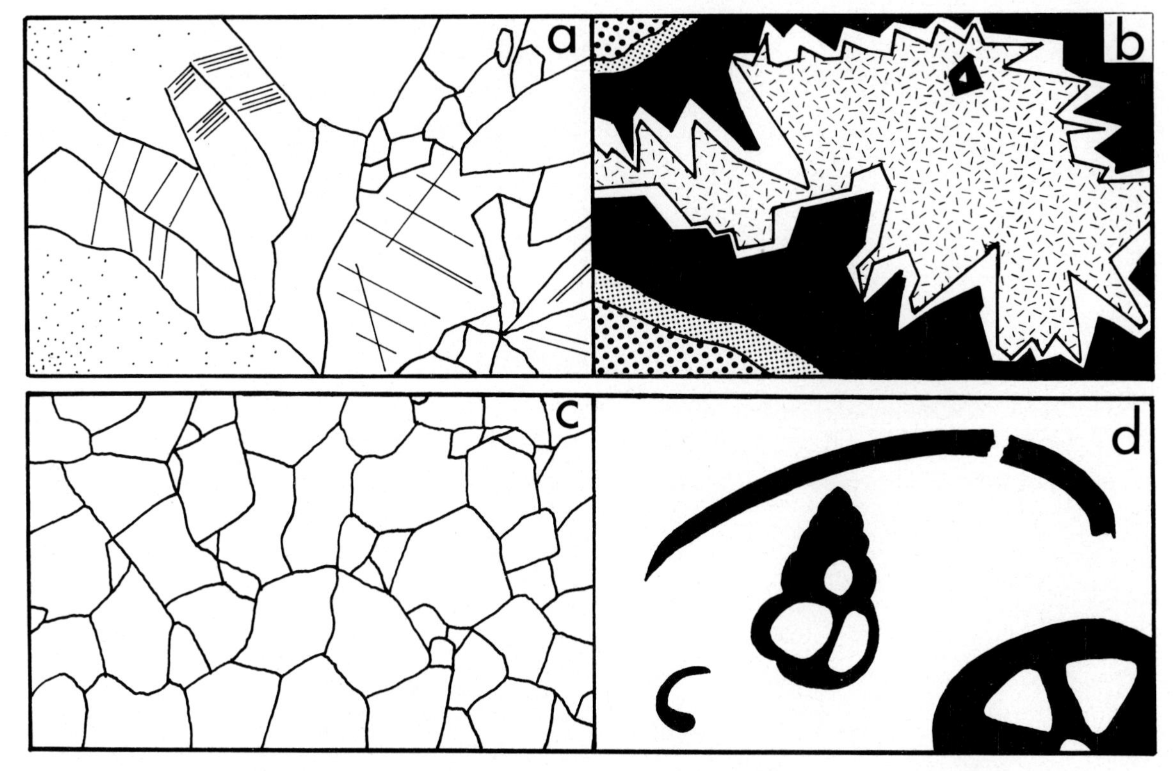

Fig. 6.3. Drawings made from photographs of CL in limestones from the Dinantian (Lower Carboniferous) of Ireland. (a) Coarse sparry calcite mass seen in transmitted light. (b) Same view but with cathodoluminescence, showing a void developed in micrite (coarse stipple), with a cement sequence of radial fibrous spar (light stipple), non-luminescent ferroan calcite (black) and brightly luminescent outer zone (white). The void fill is completed by dolomite (hashures). (c) Medium-grained blocky spar mosaic seen under transmitted light is revealed under cathodoluminescence (d) as a neomorphosed, brightly luminescent biomicrite with gastropods, bivalves and foraminifera which are weakly luminescent.

6.7 EXAMPLES OF CL USE IN SEDIMENTOLOGY

Exhaustive discussions of the CL properties of minerals are beyond the scope of this chapter, and the reader is referred to the reviews of Nickel (1978), Amieux (1982) and Walker (1985). Some of the most important petrographic uses of CL are outlined below for the more common sedimentary rock types.

6.7.1 Carbonate rocks

PRINCIPLES

Carbonate minerals give bright and stable luminescence at low accelerating voltages, so limestones and dolomites have attracted much attention from sedimentologists working with CL (Amieux, 1982).

Chemically pure calcite may show a blue CL which is probably due to an intrinsic lattice defect. Mn^{2+} is the primary activator and produces yellow to red emission, dominating any low intensity blue peak. Fe^{2+} is the commonest quencher ion, but Ni^{2+} has a similar effect. Complete quenching by Fe produces a black luminescence distinct from non-luminescence (Amieux, 1982); Pierson (1981) found this to occur in dolomites at 1.5 wt.% Fe. Mostly, however, Mn and Fe tend to co-precipitate in the lattice of carbonate minerals and varying degrees of quenching occur, reducing the intensity of Mn^{2+} emission and inducing a brownish colouration according to the quenching. Amieux (1982, fig. 5) has

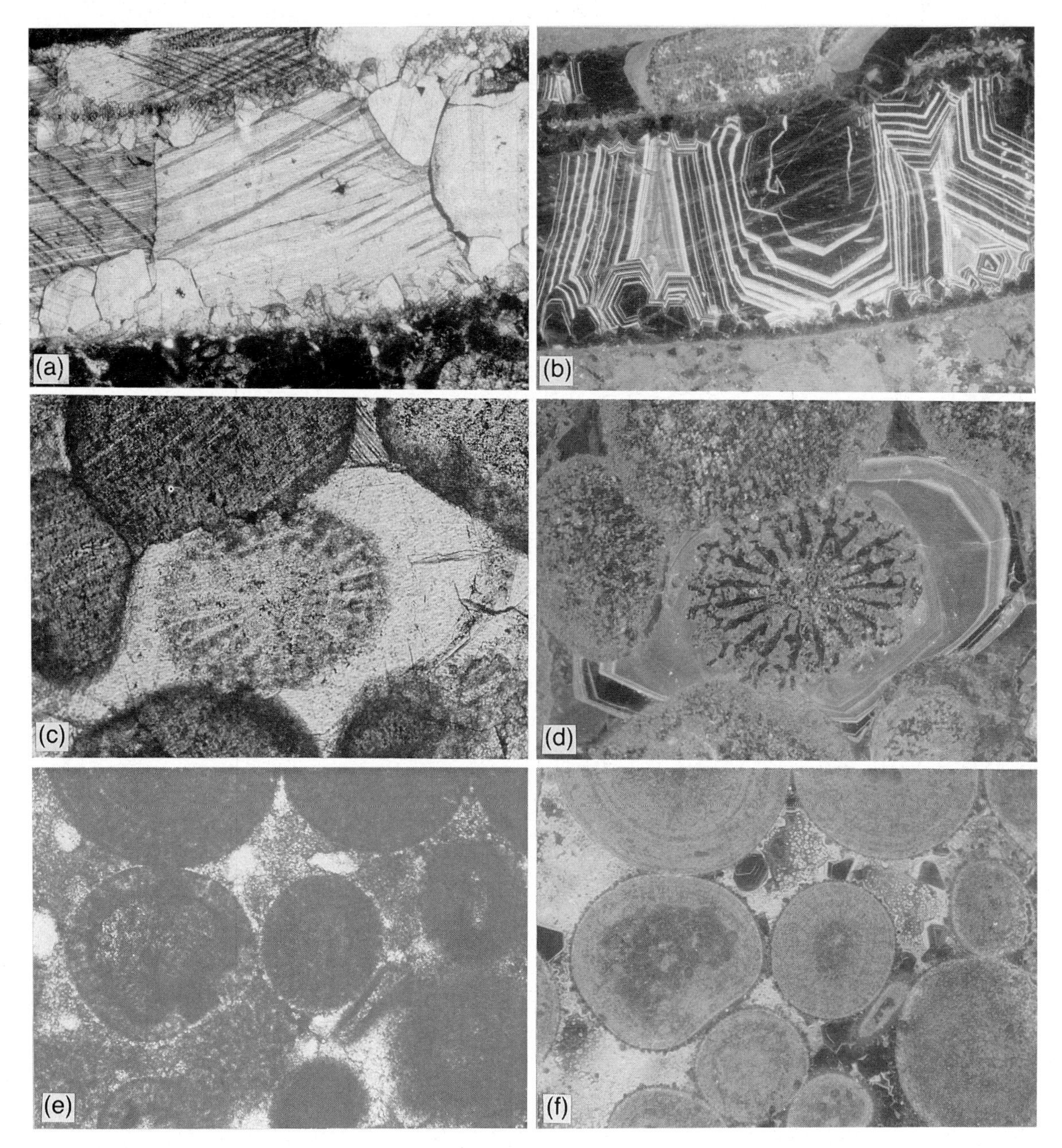

Fig. 6.4. Paired photomicrographs of limestone thin sections, with transmitted light view to the left and cathodoluminescence view to the right. All from the Dinantian (Lower Carboniferous) of South Wales. (a), (b) Pwll-y-Cwm Oolite. Calcite cement fill of bivalve mould, showing details of crystal growth by fine luminescent and non-luminescent growth bands. (c), (d) Blaen Onneu Oolite. Syntaxial calcite overgrowth on an echinoid spine, showing preferential nucleation on the crystallographically suitable substrate. CL reveals detail of the internal structure of the recrystallized spine. Changes in CL intensity in the overgrowth cement are due to varying concentrations of Fe^{2+} quencher. (e), (f) Gilwern Oolite. Details of the internal structure of ooids is better revealed by CL. Thin, non-luminescent calcite cement fringes occur on the ooids, followed by brightly-luminescent microspar associated with calcrete formed during subaerial exposure. Photographs by courtesy of Dr M. Raven.

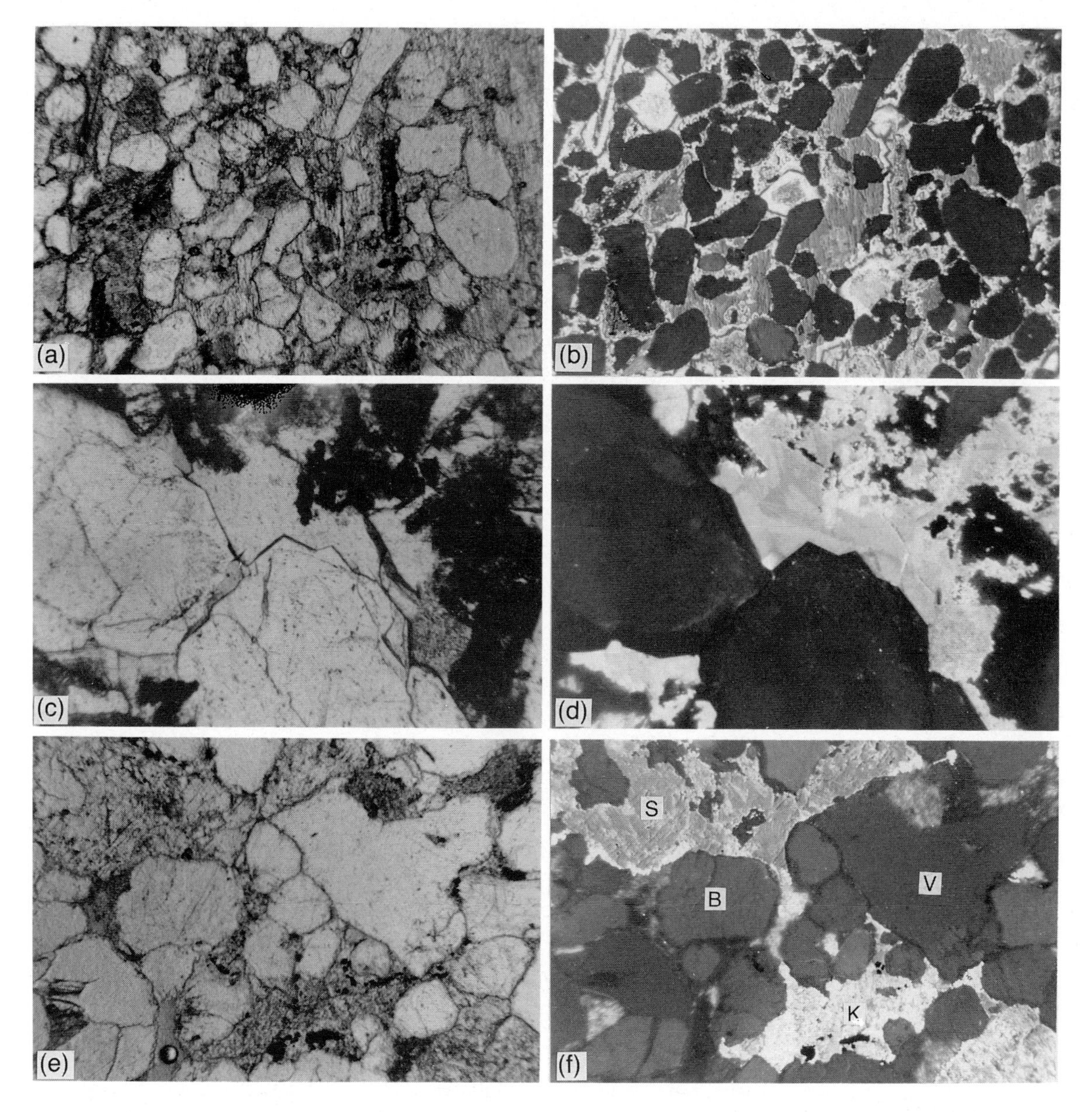

Fig. 6.5. Paired photomicrographs of sandstones from North Sea cores, transmitted light view to the left and CL view to the right. (a), (b) Medium-grained sandstone with a carbonate cement is seen under CL to have a fairly high fossil content, mainly echinoderms: crinoid fragments and an echinoid spine (top left) provide substrates for large, zoned overgrowths which have occluded the primary porosity. (c), (d) Coarse sandstone with non-luminescent authigenic quartz overgrowths on violet-luminescing quartz grains. A subsequent zoned calcite cement is being dissolved by kaolinite (white on picture, royal blue CL). (e), (f) Violet luminescing (V) and brown (B) quartz grains showing a mixture of metamorphic and igneous sources. A sparry calcite cement (S) is suffering dissolution by brightly-luminescing kaolinite (K).

related Mn/Fe ratios and their emission colours to geochemical environments, particularly to redox potential changes during crystallization.

Trace elements such as Sm, Dy, Eu, Er, Ce and Pb act as sensitizers or 'co-activators', facilitating Mn^{2+} activation (Mukherjee, 1948; Schulman *et al.*, 1947; Machel, 1985). Empirical studies, comparing Mn and Fe contents of calcite and dolomite with their luminescence, have indicated that the Mn/Fe ratio is the main controlling factor influencing CL intensity in these minerals, at Fe^{2+} concentrations below 1% (Fairchild, 1983). Cross-plots of Fe/Mn are given by Fairchild (1983) and Grover & Read (1983). These observations are supported by the correspondence of dull CL with blue ferricyanide stained areas in calcites, where the Prussian Blue precipitate is specific for the Fe^{2+} ion.

The CL colours produced in calcite and dolomite by Mn^{2+} range from yellow to dark reds and pinks. Sommer (1972a) determined that with increasing substitution of Mn^{2+} into Mg^{2+} rather than Ca^{2+} sites, the colour tends towards red (see also Amieux, 1982). Broadly, therefore, low Mg calcites give yellow CL and high Mg calcites are orange to red. Dolomite is characteristically a 'brick-red' colour. Heavy quenching by Fe was considered by Amieux to produce a dull brown-maroon CL quite distinct from black non-luminescence.

Ten Have & Heynen (1985) investigated the incorporation of Mn^{2+} into calcite crystals in over 50 crystal synthesis experiments from gels and solutions. Normal temperatures and pressures were used. They reported that 15–30 ppm Mn were sufficient to induce luminescence in the synthetic calcites (Fe less than 200 ppm). Dolomites from the Jurassic of the Middle East and Miocene of the Bahamas were also analysed. These were found to luminesce when Mn was higher than 30–35 ppm and Fe less than 300 ppm. A higher level of activator was thus required in dolomite compared to calcite, due to the more efficient luminescence of Mn^{2+} when present in the Ca^{2+} lattice sites than when substituting Mg^{2+} sites. In dolomite, Mn^{2+} largely occupies the Mg^{2+} sites (Sommer, 1972b). The minimum amount of Mn^{2+} activator for CL is thus lower in calcite than in dolomite.

Many of the synthetic calcites produced in the experimental work displayed zonation in CL comparable to that found in many natural carbonate crystals. Variations in CL intensity in the zones were related to variations in their Mn content. Ten Have & Heynen suggested that differential Mn uptake during crystal growth may be caused by two processes: (1) Changes in the bulk chemical composition of the precipitating solution, such as an increase in its Mn content during crystal growth. (2) Changes in the rate of crystal growth related to variations in the supersaturation level of the solution. While the crystals were growing slowly, the lattice was able to 'sweep out' impurities such as the Mn activator ions and thus exhibit less luminescence. More Mn was incorporated during rapid crystal growth.

Both processes resulted in similar CL zonation patterns. This implies that variations in CL intensity need not necessarily reflect changes in bulk pore fluid composition. However, the experimental growth rates are probably much higher than those encountered under geological conditions, but this effect should be born in mind.

CEMENT STRATIGRAPHY

One of the commonest applications of CL in carbonate rocks is in revealing successive stages or zones of void-filling cements with far greater precision than that possible with optical microscopy. Cement stratigraphy (Fig. 6.4) involves the application of stratigraphic principles at an inter-granular and void level, correlating stages of cementation within a given basin of sedimentation.

Meyers (1974, 1978) pioneered cement stratigraphy, demonstrating that 'zones' of carbonate cements, as revealed by their CL and other petrographic characteristics, could be correlated both vertically and regionally in the Mississippian of the Sacramento Mountains, New Mexico. Grover & Read (1983) performed a similar study in the Ordovician of Virginia, and related the sequence to diagenetic changes resulting from changes in formation water composition during shallow and deep burial. Walkden & Berry (1984) correlated CL zones in overgrowth cements from the Upper Dinantian of northern Britain, and ascribed them to cyclic replenishment of vadose and shallow meteoric water tables in calcretized marine limestones.

Miller & Gillies (in press) discussed the principles and practise of cement stratigraphy, especially the confused and varied terminology existing in the literature. In order to improve communication and comparison of CL results, they suggested standard terms for cement sequences, using numbered *stages*, *sub-stages* and *zones*, divided on the basis of unconformities, crystal habit and composition. Further, they noted a tendency to concentrate on cement stratigraphies to the exclusion of many other diagenetic events, such as dissolution, compaction, stylolitization, neomorphism and mineralization. Instead, they emphasized that the detailed cement sequences revealed by CL could be used as a time framework upon which to locate other diagenetic events, providing a complete 'diagenetic stratigraphy'. This provides a powerful tool for analysis of post-depositional changes in sedimentary rocks.

Uncertainties exist in our present interpretation of CL and its relation to the geochemistry of carbonate precipitation from pore fluids. Carbonate minerals are also inherently unstable and liable to both obvious and subtle alteration during diagenesis. Therefore, additional criteria, such as evidence of dissolution or fracturing, changes in crystal habit, inclusions, carbon/oxygen isotope ratos and accurate geochemical analyses *of individual luminescent zones* using microprobe, AAS and neutron activation techniques must be integrated with CL results. This involves very precise, tedious work, often using very small samples of cements for analysis. At the very least, the precaution of staining thin sections with Alizarin Red S and potassium ferricyanide (Chapter 4) is required as a rapid and sensitive method of determining whether variation in CL intensity is likely to be due to fluctuations in Mn activator or to the presence of Fe as a quencher. Generally, CL work is done in conjunction with microprobe analysis performed on the same thin sections. CL cannot be used *alone* for other than general interpretations of cement histories.

Studies of cement sequences must pay particular attention to veining phases. Many post-burial cement generations can be linked to specific sets of veins, some of which are extremely fine and are only visible with CL. Commonly, later stages of cements precipitated during veining episodes may induce neomorphism of earlier cements, replacing them partially or completely, or may merely precipitate films of new material along crystal interfaces between existing cements. In all cases this leads to a pronounced change in the CL of the affected cements which is an important element in the diagenetic event stratigraphy. Large thin sections (see Chapter 4) are important for this work, as veins and their conjunction with voids are often missed in small samples. It is also necessary specifically to select veined material for study, hitherto not a regular practise in traditional carbonate petrography.

Studies of stable isotopes in carbonate minerals and skeletal materials can provide valuable environmental and diagenetic information. Prior examination of material to be sampled for stable isotope analysis with CL enables selection of areas uncontaminated by neomorphism or inter-crystal infiltration as described above.

6.7.2 **Sandstones**

QUARTZ

The causes of CL in quartz are not yet fully determined (Walker, 1985). Alpha quartz shows visible CL with two broad emission bands in the blue and red (Zinkernagel, 1978). Various authors have attributed these emissions to Ti and Mn respectively but, as Walker (1985) pointed out, the spectral bands are present in highly pure synthetic silica, and it seems certain that emission is intrinsic rather than due to impurities. CL from natural quartz is characteristically thermally unstable. Heating collapses the blue spectral peak, causing the violet luminescence to shift towards red. This presents a major problem for observation of quartz grains under an electron beam: its heating effect can quickly cause the red colouration to be assumed. High accelerating voltages and current densities speed the change, which is permanent. It may be very difficult to obtain accurate colour photographs of quartz luminescence because the shift to red may occur during the time of exposure.

Zinkernagel showed that natural quartz grains have a luminescence which apparently reflects their source (or, more correctly, their thermal history). Broadly, violet CL is typical of igneous sources, but brown grains originate in certain types of metamorphic rocks (Fig. 6.5f). This scheme of provenance using CL was further elaborated by Matter & Ramseyer (1985). Quartz overgrowths and other forms of authigenic quartz are commonly non-luminescent under normal CL conditions (Fig. 6.5d), while zoned CL is typical of grains derived from hydrothermal vein-quartz. CL is thus a very sensitive method of determining presence and timing of pore occlusion by quartz authigenesis, particularly where there are no obvious dust-rims on grain cores.

Owing to the thermal instability described above, with violet grains easily becoming brown while in the electron beam, much care and some skill is needed to distinguish quartz CL colours successfully. Given this, CL affords a rapid means of detecting different quartz grain populations in sandstones and in ascertaining their provenance. The different shades of violet and brown in a grain population probably represent various grain sources, since there is no evidence that crystal orientation exerts significant control over CL wavelength and intensity in quartz. Of course, it is wise to check the other properties of the grains, such as inclusions and type of extinction, to gain confirmation of CL determinations.

Sippel (1968) and Sibley & Blatt (1976) have applied CL in investigations of silica authigenesis in sandstones. Burley, Kantorowicz & Waugh (1985) suggested how CL may profitably be combined with other techniques in elucidating diagenesis of clastic rocks.

FELDSPARS

The causes of CL in this mineral group are relatively well-known from investigations on lunar rocks (Geake *et al.*, 1977) and carbonatites (Mariano, Ito & Ring, 1973). Feldspars have low CL thresholds and are thus obvious even at low accelerating voltages. Typical CL colours are brilliant blue (possibly Ti^{2+} activation), red to infra-red (Fe^{3+} substitution for Al^{3+}) and green (Mn^{2+} substitution probably for Ca^{2+}). Feldspar CL is polarized and some intensity variation is seen upon rotation of a polaroid sheet interposed between the viewing window and the microscope eyepiece. This property often helps to distinguish feldspars from other minerals of similar CL colour. In general, authigenic feldspars are non-luminescent or very dully luminescent (Kastner, 1971), serving to differentiate detrital cores from their overgrowths.

In terms of provenance, there is no evidence to suggest that feldspar CL is related to specific igneous or metamorphic origins. However, the presence of feldspar 'suites' containing consistent proportions of the rarer red and/or green emitter grains amongst the common blue emitters can be used to trace particular sources of clastic feldspar supply (e.g. Stow & Miller, 1984).

Because even the smallest feldspar grains luminesce brightly, a far more accurate estimate of the percentage of feldspar in a clastic sediment is obtained by point-counting under CL. This is particularly true for fine sandstones and siltstones. Even grains which are badly weathered or altered still register the typical emission.

HEAVY MINERALS

Many accessory minerals of clastic rocks contain at least traces of rare earth or transition metal impurities and are therefore liable to luminesce. Nickel (1978) listed the known CL characteristics of such minerals. Apatites give particularly bright CL, with europium a common activator (Mariano & Ring, 1975). Apatite emissions range from bluish-violet, lilac, pink and orange to yellow. Portnov & Gorobets (1969) have proposed that certain of these colours are related to particular igneous or metamorphic origins. If this can be confirmed, it would be an invaluable adjunct to provenance studies. Again, the small size of many heavy mineral grains means that they are often overlooked and easily underestimated in transmitted light petrography of clastic rocks. Invariably, use of CL produces an often surprising re-evaluation of the importance of accessory grains.

CEMENTS AND DIAGENESIS

Carbonate cements are commonly involved in occlusion of sandstone porosity (Fig. 6.5b, f). Their mineralogy and sequence of development are more accurately followed by using CL than by transmitted light alone (see above). Other diagenetic features such as corrosion of quartz grains by carbonate, relative timing of quartz overgrowth formation and cementation and the onset of dolomitization can also be rapidly documented with CL. Authigenic kaolinite or dickite commonly form late stage cements in sandstones (Fig. 6.5f, Burley *et al.*, 1985, fig. 7f), and they usually have a brilliant royal blue CL, whose origin is unknown. The extent and timing of these cements and their relationship to compaction and tectonic events, such as extension veining, is much more easily appreciated under CL than with transmitted light microscopy.

A good example of the combined use of CL, SEM and transmitted light petrography in documenting the provenance and diagenetic history of a Jurassic reservoir sandstone is given by Olaussen *et al.* (1984).

6.8 PHOTOGRAPHIC RECORDING OF CL

Prior to the realization that high transmissivity microscopes were mandatory for successful CL work, photography of CL was extremely difficult and the results highly variable. Minerals giving very low intensity emission, such as quartz grains, required many minutes or even hours of exposure with fast films. Generally, these films were working at luminance ranges beyond their zones of reciprocity failure, resulting in poor resolution, low contrast and indifferent colour rendition. Good microscope transmissivity, aided by straight-through light paths between objective and film, means that slower (and thus finer-grained) films can now be used, working well within their optimum exposure ranges.

Most petrological microscopes can be fitted with photomicrography accessories which allow automatic exposure determination (Fig. 6.3). The large viewing window of the Nuclide Luminoscope® also allows the use of a 55 mm *f*/3.5 macro lens directly mounted on a camera body to take pictures of whole thin sections or slabs under CL. Such pictures are invaluable for mapping CL zones in carbonate cements whch are to be sampled for AAS and stable isotope analysis by drilling or scraping from individual cement stages. They also show up fabrics and small-scale sedimentary structures in sandstones and siltstones.

Recent advances in film technology have increased the range of films suitable for CL recording. For routine work, particularly with limestones, where the colour range is restricted and can be adequately represented in grey tones, black and white films are satisfactory. Invariably, good quality prints will show up more detail than was apparent to the observer viewing the CL, because of fatigue and insensitivity of the eye at low light levels. Suitable films are medium-speed types (ISO 125–400) which possess fine grain, good contrast range and some exposure latitude, such as Ilford FP4 or XP1. Normally, paired exposures are made, one showing the transmitted light view and one the CL view. CL negatives often have a somewhat restricted contrast range, and these 'flat' negatives may be enhanced by printing on harder paper than the transmitted light frames.

Sandstones are difficult subjects; quartz CL is very subdued but feldspars are very bright, often with a high UV component. It is almost impossible to get brilliant and dark grains properly exposed on the same frame. Correct exposure for dull quartz usually results in serious over-exposure of feldspars, with development of halation areas around the grains, while correct exposure of feldspars produces marked under-exposure of quartz grains. A series of half-stop under- and over-exposure gradations will usually produce one frame which is satisfactory for each. Correction can also be done by changing the film speed between exposures on an automatic centre-weighted averaging meter — setting a slower film speed will trick the meter into underexposing a bright grain. A light meter with a movable spot, as in the Nikon system (Fig. 6.3) can provide direct exposure readings for particular grains.

It is also important, with porous rocks, to remove carefully all traces of the diamond polishing abrasives from the slide: industrial diamond grains have a brilliant green CL which produces halation spots, giving the impression of a larger grain, which is apt to be mistaken for an apatite or other heavy mineral grain.

Colour pictures are somewhat more problematic, in that the precise colours of CL are very difficult to reproduce exactly. Daylight films give the best colour balance, with a blue compensating filter for the transmitted light (tungsten lamp) frames. The choice lies between print films with a narrower contrast range and variable colour balance depending on printing filtration, and reversal films with a higher contrast and resolution range but where subsequent correction is not possible. Agfa, Kodak and Fuji produce ranges of colour slide and print films ranging from ISO 25 to 1000, with an HR suffix indicating the films' high resolution capability by virtue of the very thin emulsions used in the colour separation layers. The slower the speed, the finer the grain and the greater the resolving power of these films. ISO 1000 films may be needed to collect CL images of thermally unstable quartz grains which rapidly redden under excitation. Each worker is advised to experiment with a range of films, both colour and monochrome, and then to become thoroughly familiar with those which are best suited.

CL depends so often on rendition of a range of colours, some soft and subtle, some brilliant, that optimum reporting of CL studies in publications may demand expensive colour plates. While monochrome CL images are adequate for limestones, as noted above, sandstone CL recorded in monochrome is unsatisfactory (Fig. 6.5). Several publications have colour plates, e.g. Zinkernagel (1978), Richter & Zinkernagel (1981), Amieux (1982), Olaussen *et al.* (1984), Dickson (1985), Burley *et al.* (1985) and Matter & Ramseyer (1985). They are very few compared with the many thousands of monochrome illustrations which appear each year. The cost of providing adequate colour reproduction in publication could dissuade some petrographers from using CL, particularly with sandstones.

6.8 CONCLUSIONS

There is little doubt that the availability of relatively cheap and convenient machines for viewing CL is one of the most important developments in sedimentary petrography in the last twenty years. The technique significantly enhances both the quality and quantity of information to be derived from thin sections and rock surfaces, and provides a powerful tool for the construction of extremely detailed diagenetic event stratigraphies.

However, a great deal of development work remains to be done in determining the causes of CL in minerals and in understanding the geochemical significance of changes in activator concentration in minerals precipitated during diagenesis. It is clear even at this early stage that CL is not a substitute for existing petrological techniques but, when combined with these approaches, it provides a rapid and powerful way of eliciting more data from geological samples. There are now overwhelming arguments for considering CL to be a standard technique and not an esoteric diversion.

6.9 MANUFACTURERS OF CL EQUIPMENT

Nuclide Corporation, AGU Division, 916 Main Street, PO Box 315, Acton, MA 01720, USA.

TECHNOSYN Ltd, Coldhams Road, Cambridge CB1 3EW, England.

7 X-ray powder diffraction of sediments

RON HARDY and MAURICE TUCKER

7.1 INTRODUCTION

X-ray diffraction (XRD) is a basic tool in the mineralogical analysis of sediments, and in the case of fine-grained sediments an essential one. It has the advantage, with modern instrumentation, that almost complete automation can be achieved to give fast, precise results. This chapter discusses the basic theory of the technique, in so far as it is needed to set up systems of analysis, and goes on to describe several specific applications to sediments with appropriate preparative techniques and guides to interpretation.

7.2 THEORY OF X-RAY DIFFRACTION

For a comprehensive treatment of theory, the reader is referred to the classic text of Klug & Alexander (1974). Only a simplified treatment for the understanding and evaluation of basic analytical procedures is given here. A very useful introduction can be obtained from the Philips organization in the UK called 'An introduction to X-ray powder diffractometry' by R. Jenkins and J.L. de Vries. The address is Pye Unicam Ltd, York Street, Cambridge CB1 2PX, tel. Cambridge (0223) 358866.

7.2.1 Production of X-rays

In a normal commercially available laboratory diffractometer (assumed throughout this chapter), X-rays are produced by bombardment of a metal anode (the target), by high energy electrons from a heated filament in a Röntgen X-ray tube (Fig. 7.1). The resulting radiation emerges from a thin, usually beryllium, window and consists of (Fig. 7.2):
(i) A broad band of continuous radiation (white radiation) produced by the electrons from the filament converting their kinetic energy to X-rays on collision with the atoms of the anode target.
(ii) A number of discrete lines of varying intensity called the characteristic radiation which represents the energy released by rearrangements of the orbital electrons of atoms of the anode target following ejection of one or more electrons during the excitation process. These lines are known as K, L, M lines etc.; the type of line produced is determined by the orbital electrons taking part in the rearrangement (Fig. 7.3).

Various anode materials are commonly used as

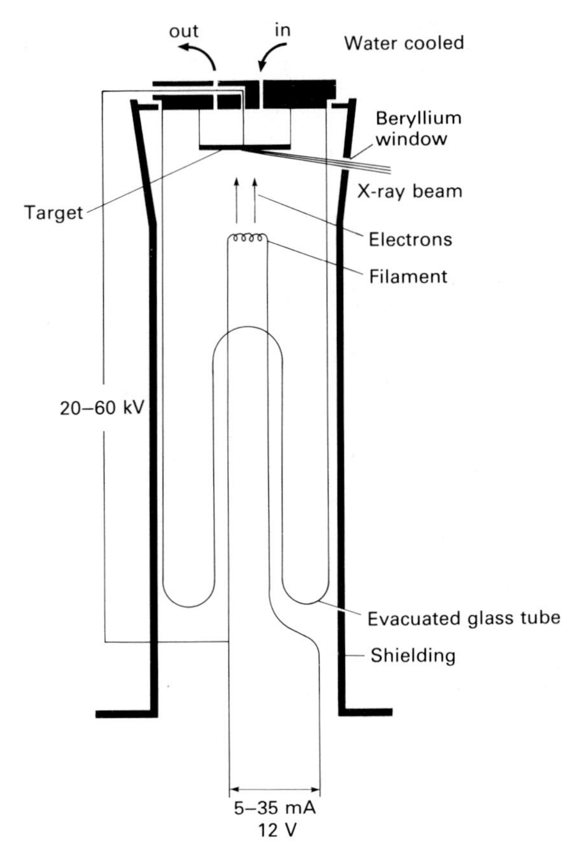

Fig. 7.1. A section of a Röntgen X-ray tube (based on Phillips & Phillips, 1980).

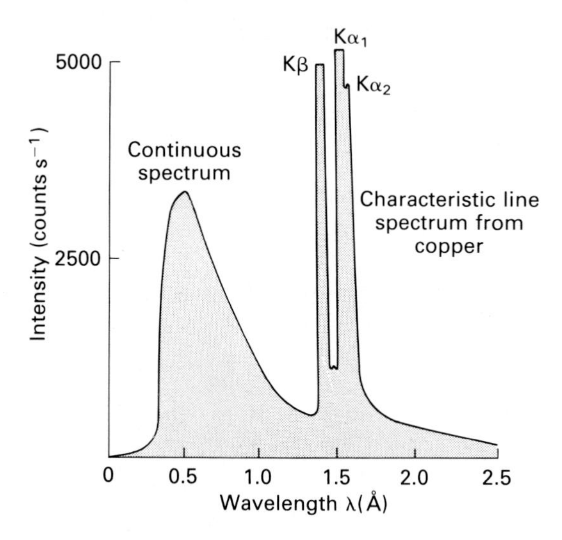

Fig. 7.2. The spectrum from a copper anode X-ray tube.

targets in X-ray tubes, they include: Cu, Cr, Fe, Co, Mo and Ag. Each has advantages and disadvantages and these will be discussed later in connection with the analysis of sediments.

Considerable advantages are to be gained by the use of purely monochromatic radiation, i.e. radiation of a single wavelength and this is achieved by the use of a β filter, or alternatively by using a crystal monochromator.

The β filter consists of a thin metal foil which is positioned in the X-ray beam generally close to the beryllium window of the X-ray tube. The type of metal foil selected is such that the intensity of the *K*β radiation of a particular X-ray tube is effectively removed from the spectrum (Fig. 7.4). This is achieved by using the variation of mass absorption of metals with wavelength (Fig. 7.5) in such a way that the high absorption of the *K* Absorption Edge of the metal falls between the *K*β and *K*α radiations of the X-ray tube.

Table 7.1 lists the usual combinations of target element and filters. It should be remembered that as well as greatly reducing the *K*β radiation intensity, the β filter also reduces the overall intensity of the continuous radiation and the intensity of the *K*α radiation.

Further improvements can be achieved in the production of monochromatic radiation with the use

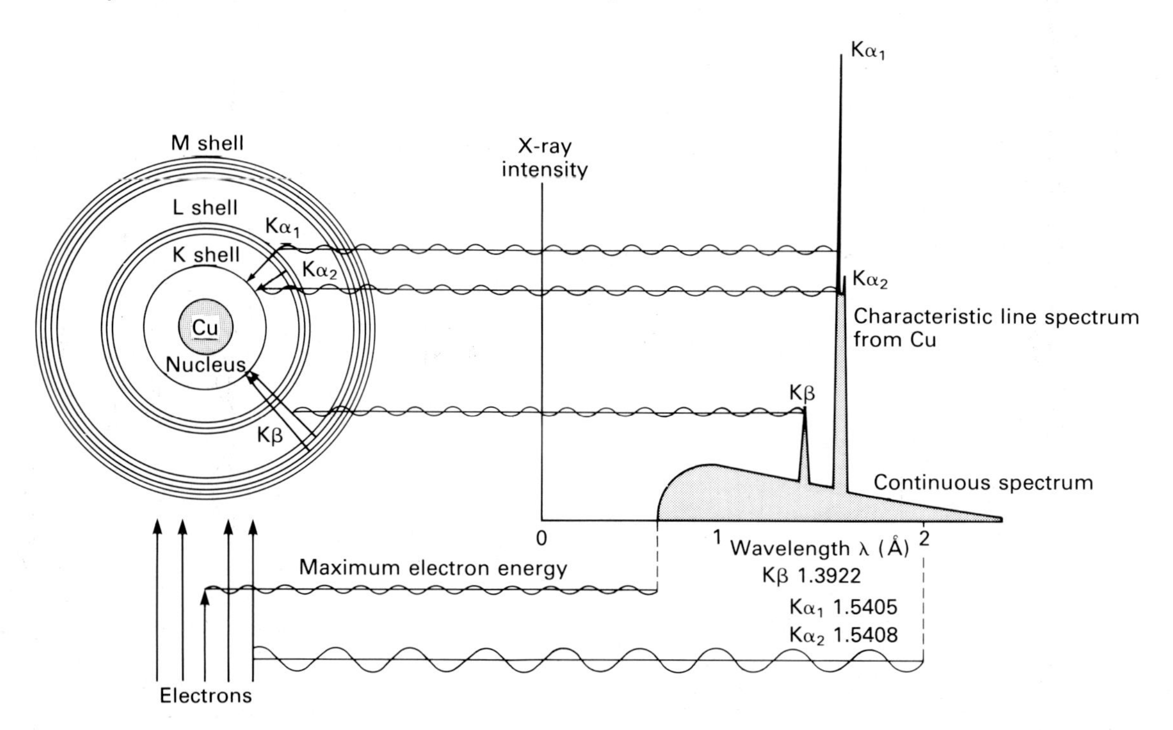

Fig. 7.3. The generation of X-rays of the line spectrum of copper due to the transfer of electrons into the *K* shell and the generation of the continuous spectrum due to complete or partial electron collisions (from Phillips & Phillips, 1980).

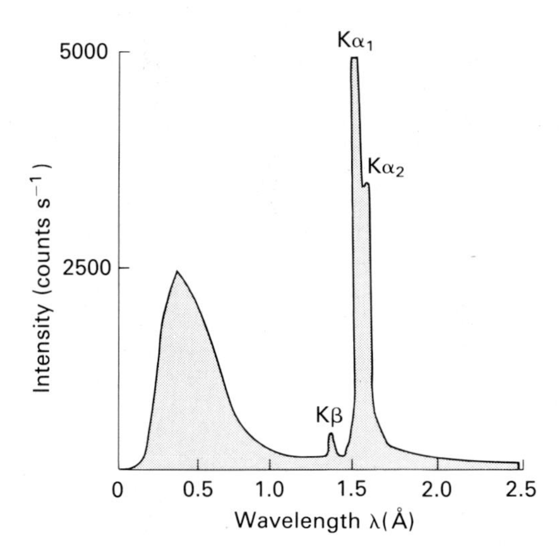

Fig. 7.4. The copper X-ray spectrum after passing through a nickel filter.

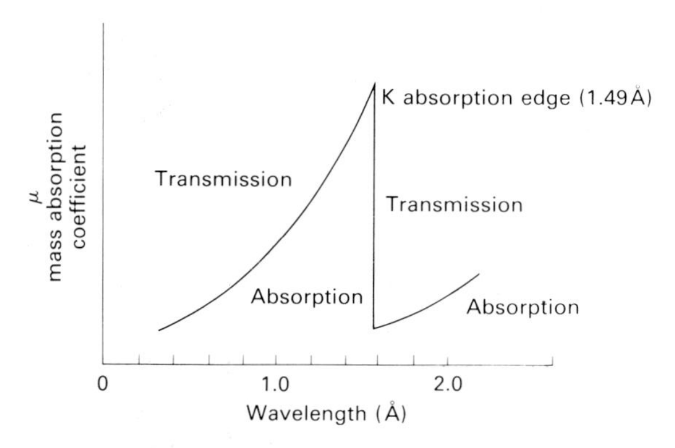

Fig. 7.5. The variation of mass absorption coefficient of nickel at different X-ray wavelengths.

of electronic filters (pulse height analysis) associated with the detection and recording devices of the diffractometer (Fig. 7.6).

In the case of a graphite curved crystal monochromator (Fig. 7.7), adjustments of the graphite crystal can be made such that only the desired wavelengths, i.e. $K\alpha$, pass to the detection device, thus filtering out the unwanted $K\beta$ and continuous radiation without the use of β filters and pulse height analysis.

Table 7.1. X-ray tube targets and suitable filters

Target element	$K\alpha$(Å)	$K\beta$(Å)	Filter	K Absorption Edge (Å)	Thickness (μm)*
Cr	2.291	2.085	V	2.269	16
Fe	1.937	1.757	Mn	1.897	16
Co	1.791	1.621	Fe	1.744	18
Cu	1.542	1.392	Ni	1.488	21
Mo	0.710	0.632	Zr	0.688	108

* This thickness reduces $K\beta/K\alpha$ to 1/600.

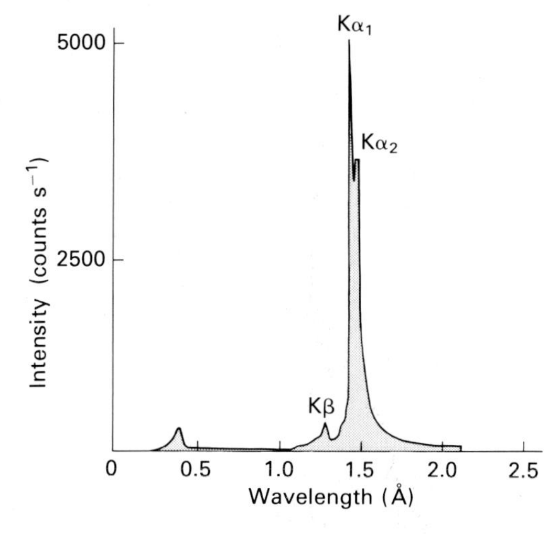

Fig. 7.6. The copper X-ray spectrum after passing through a nickel filter and subjected to pulse height analysis.

7.2.2 Diffraction of X-rays by a sample

Figure 7.8 shows a schematic representation of a typical diffractometer. The X-rays are first collimated to produce a subparallel beam, the amount of divergence being controlled by the size of the divergent slit, i.e. a 4° large divergence slit for high angle work to a 1/12° small divergence slit for low angle work.

The divergent beam is then directed at the sample, which is motor driven to rotate at a regular speed in degrees per minute. When mineral planes in the sample attain an appropriate angle, they will diffract the X-rays according to Bragg's Law,

i.e. $n\lambda = 2d \sin \theta$

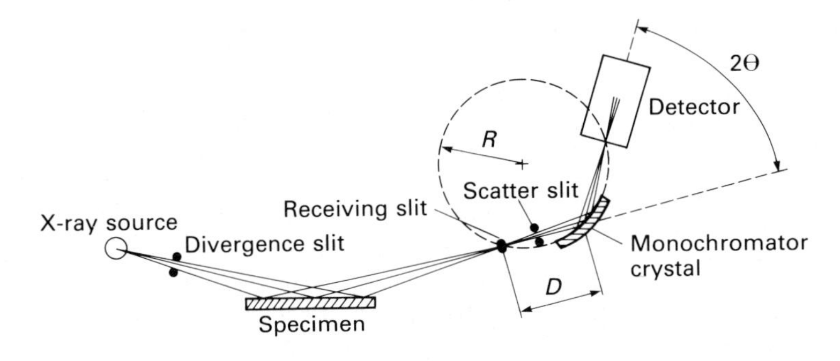

Fig. 7.7. Schematic representation of the geometry of a curved crystal monochromator (from Philips UK information sheet).

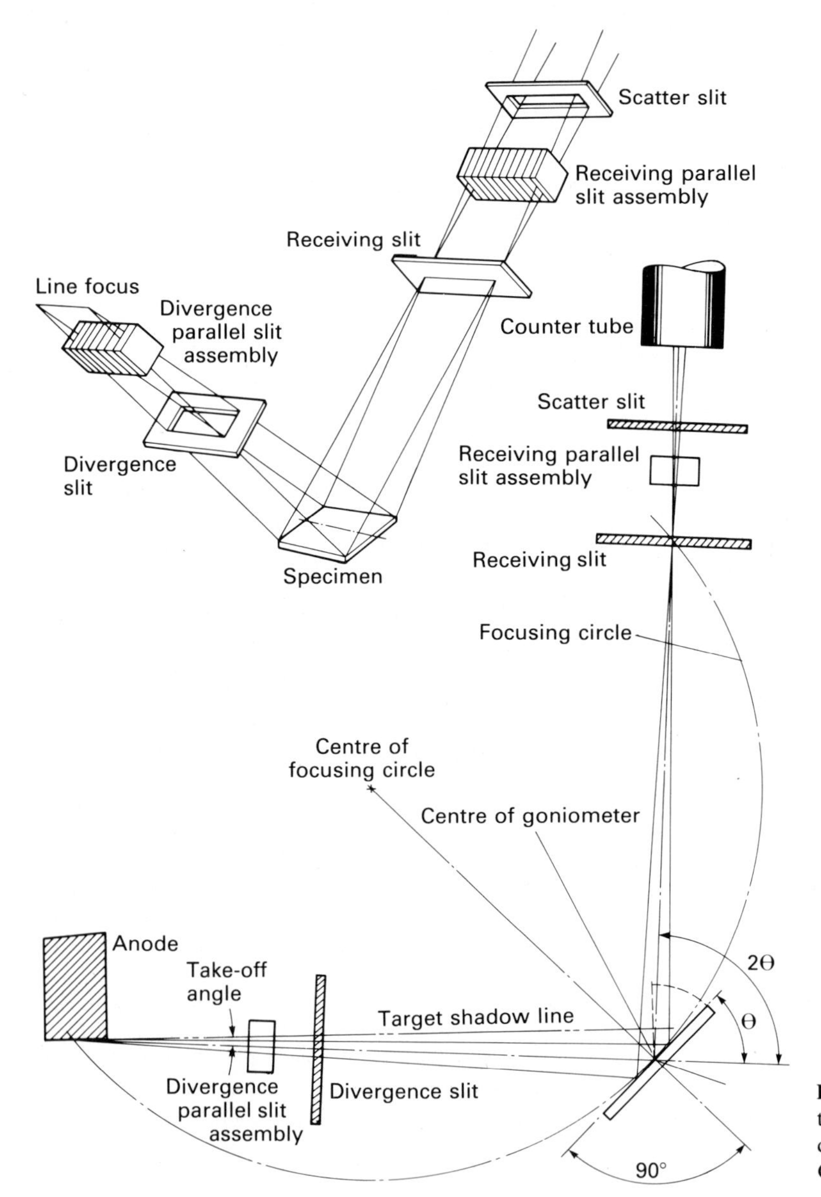

Fig. 7.8. Schematic representation of the geometry of a typical diffractometer (from Philips UK Goniometer Manual).

where n is an interger, λ is the wavelength of the X-rays, d is the lattice spacing in ångstroms and θ is the angle of diffraction (Fig. 7.9). The diffracted beam is passed through a receiving slit and collimator and, then a scatter slit is introduced to reduce any scattered X-rays other than the diffracted beams from finally entering the detector. In the case of a monochromator being used (Fig. 7.7), the beam passes straight from the receiving slit to the crystal monochromator and then to the detector. The signal produced by the X-ray photons on the detector is first amplified, and then passed on to the electronic recording equipment.

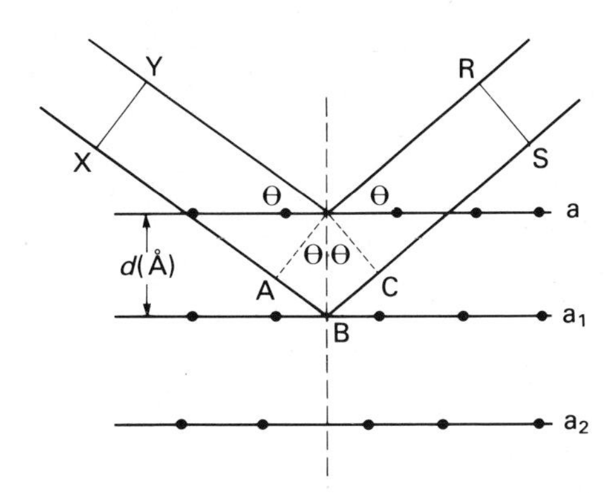

Fig. 7.9. An illustration of Bragg's Law: a, a_1 and a_2 are lattice arrays of atoms that can be regarded as an infinite stack of parallel, equally spaced planes. If a wavefront X−Y is incident on $a - a_1$ the reflection path from the lower plane (a_1) is longer, i.e. $AB+BC = \Delta =$ difference in paths of wavefronts

$$d \sin \theta + d \sin \theta = 2d \sin \theta = \Delta.$$

For diffraction to occur Δ must equal a whole number of wavelengths: $2d \sin \theta = n\lambda$ (Bragg's Law).

7.2.3 Output devices

Output from a diffractometer can be in either analogue or digital form. The conventional analogue recording is a strip chart whose speed in millimetres per minute is synchronized with that of the detector in degrees of 2θ per minute so that the x-axis is calibrated in °2θ (Fig. 7.10). Deflections recorded can be easily converted into lattice (d) spacings of the minerals present by applying Bragg's Law. Most diffractometers possess a digital counter which can record actual X-ray intensities in counts per second.

Early XRD was carried out using powder cameras which used a film recording technique. This is now largely restricted to specialized applications such as structural determinations and to instances where the amount of sample available is very small, the order of a few micrograms. It is a very time-consuming method and not easily susceptible to automation.

Use of the diffractometer has been transformed by the advent of microprocessor control, microcomputers and automatic sample loaders. These have led to completely automated instrument operation and data processing.

7.3 XRD ANALYSIS OF SEDIMENTS

XRD is particularly useful in the analysis of fine-grained material which is difficult to study by other means, but there are a whole range of applications to the various components of sediments, as defined by either size and/or mineralogy.

7.3.1 Whole rock analysis

The most basic application of XRD to sediments is in the analysis of whole rock samples. To obtain satisfactory results the original grain size must be reduced to a mean particle diameter of 5−10 μm with, preferably, a limited size range. Care must be taken in particle size reduction not to damage the crystallite by strain as this can lead to diffraction line broadening, which is in any case related to particle size (see Klug & Alexander, 1974, chapter 9). Careful grinding by hand can give satisfactory results but mechanical devices such as the McCrone Micronizing Mill (Walter McCrone Associates, Chicago) have been shown to produce very reproducible particle size distributions without damaging the crystal lattice, both very important criteria in quantitative work. Care must also be taken to obtain a representative sample — this can be achieved by applying appropriate sampling techniques, e.g. riffle boxes, quartering, etc.

Having ground a representative sample to the appropriate grain size, the next stage is to present the sample to the X-ray beam. There are basically two methods to do this; one method is to produce a cavity mount, and the other is to produce a smear mount.

The cavity mount holder (Fig. 7.11) is usually made of aiuminium and the sample is packed in from the rear of the cavity in the manner described by Klug & Alexander (1974, p. 373). This type of

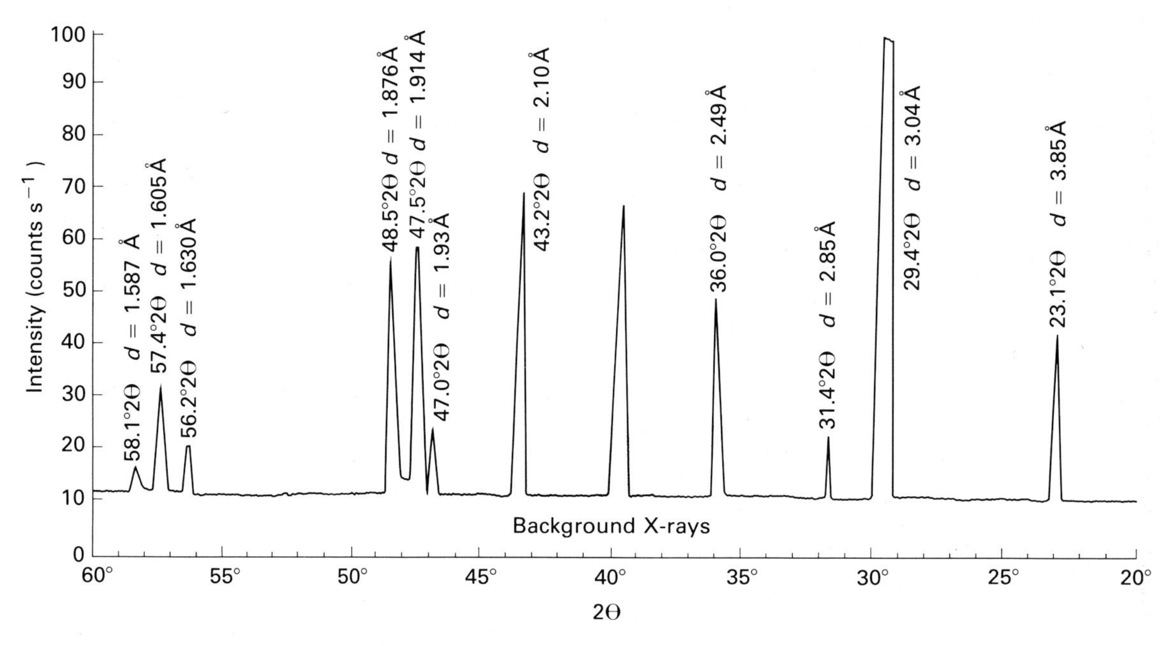

Fig. 7.10. A strip chart of diffractometer trace of the X-ray diffraction pattern of the mineral calcite. The 2θ angle increases from right to left on the horizontal scale and the intensity of the diffracted peak above background is given by the vertical scale. The X-ray tube target was copper.

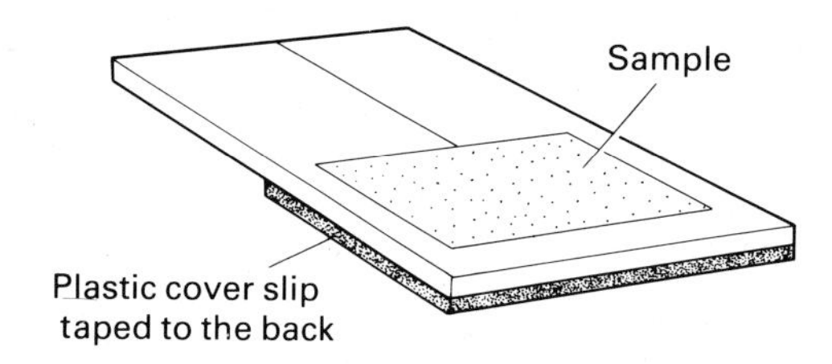

Fig. 7.11. Aluminium cavity mount.

mount produces an almost random sample especially if care is taken not to put any form of pressure on the crystallites when filling in the cavity.

The smear mount is prepared by smearing the sample with a volatile organic solvent, such as acetone, on to a glass slide. This type of mount produces a partially oriented sample which can produce problems during interpretation, e.g. estimating correct relative intensities.

The cavity mount or smear mount is then inserted into the diffractometer and the appropriate conditions set and recorded: (1) X-ray radiation used; (2) generator voltage and current; (3) the slit combination used; (4) scanning speed; (5) recording device parameters (if a chart recorder used); time constant, rate meter and chart speed should be recorded; (6) the start and the finish positions.

If a graphite crystal monochromator is not available, the choice and target of the X-ray tube used depends to a large extent on the predicted nature of the sample. For clay mineral analysis and for most sediment samples with low amounts of iron minerals present, a copper anode is preferred. If the sediments contain high levels of iron minerals, the use of a copper anode will cause iron fluorescence radiation from the sample as well as the diffraction peaks from the lattice planes. This results in very high backgrounds and hence a poor peak to background ratio (Fig. 7.12). The iron fluorescent radiation (Fe $K\alpha$) is caused by the incoming copper $K\alpha$ radiation from the X-ray tube being the correct energy to displace K orbital electrons from the iron atoms in the sample. In this case, an iron or a cobalt anode X-ray tube is used. Similarly, in the analysis of manganese sediments, an iron or a chromium anode is used as both copper and cobalt anodes cause manganese fluorescence radiation to occur.

If a graphite curved crystal monochromator is available, then as only the desired Cu $K\alpha$ wavelengths are adjusted to enter the detector, the Fe $K\alpha$

wavelengths are effectively filtered out, thus allowing the use of a copper anode in the determination of iron-rich sediments (Fig. 7.12).

7.3.2 Qualitative analysis

Assuming that a simple qualitative analysis is required the only task is the interpretation of the chart record (see Fig. 7.10). First the peaks must be identified, then measured in terms of 2θ and converted to lattice spacings. A peak (Fig. 7.13a) is commonly defined as being any reflection reaching a height 2σ above the adjacent average background (N). The laws of X-ray production mean that $\sigma = \sqrt{N}$ to a close approximation.

For most experimental purposes the position of the diffraction peak is taken as the position of the point of greatest intensity; problems arise in the case

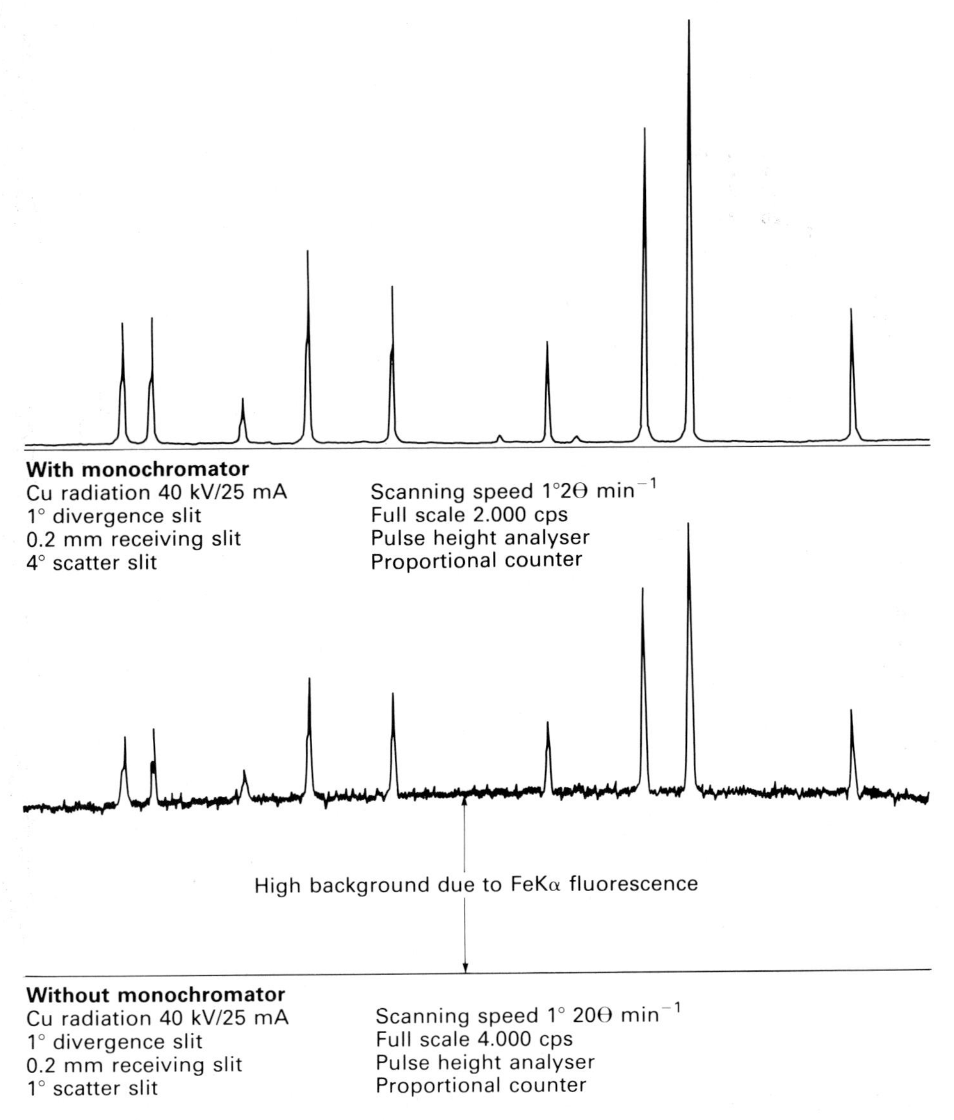

Fig. 7.12. A comparison of diffractometers with and without a curved crystal monochromator on the mineral haematite using copper radiation (from Philips UK Information Sheet).

Table 7.2. (a) Conversion charts of 2θ to ångstroms (D-spacing) for copper *K*α radiation (Computed by M.J. Smith, University of Durham)

CuKalpha , lambda = 1.5418A 2.0 of 2THETA to 51.9

	.0	.1	.2	.3	.4	.5	.6	.7	.8	.9
2	44.172	42.068	40.156	38.411	36.810	35.338	33.979	32.721	31.553	30.465
3	29.450	28.500	27.609	26.773	25.986	25.244	24.543	23.879	23.251	22.655
4	22.089	21.551	21.038	20.549	20.082	19.636	19.209	18.801	18.409	18.034
5	17.673	17.327	16.994	16.674	16.365	16.068	15.781	15.504	15.237	14.979
6	14.730	14.489	14.255	14.029	13.810	13.598	13.392	13.192	12.999	12.810
7	12.628	12.450	12.277	12.109	11.946	11.787	11.632	11.481	11.334	11.191
8	11.051	10.915	10.782	10.653	10.526	10.402	10.282	10.164	10.048	9.936
9	9.826	9.718	9.612	9.509	9.408	9.309	9.213	9.118	9.025	8.934
10	8.845	8.758	8.672	8.588	8.506	8.425	8.346	8.268	8.192	8.117
11	8.043	7.971	7.900	7.830	7.762	7.695	7.628	7.563	7.500	7.437
12	7.375	7.314	7.255	7.196	7.138	7.081	7.025	6.970	6.916	6.862
13	6.810	6.758	6.707	6.657	6.607	6.559	6.511	6.463	6.417	6.371
14	6.326	6.281	6.237	6.194	6.151	6.109	6.067	6.026	5.985	5.946
15	5.906	5.867	5.829	5.791	5.754	5.717	5.680	5.644	5.609	5.574
16	5.539	5.505	5.471	5.438	5.405	5.372	5.340	5.309	5.277	5.246
17	5.216	5.185	5.155	5.126	5.096	5.068	5.039	5.011	4.983	4.955
18	4.928	4.901	4.874	4.848	4.822	4.796	4.770	4.745	4.720	4.695
19	4.671	4.647	4.623	4.599	4.575	4.552	4.529	4.506	4.484	4.462
20	4.439	4.418	4.396	4.375	4.353	4.332	4.311	4.291	4.270	4.250
21	4.230	4.210	4.191	4.171	4.152	4.133	4.114	4.095	4.077	4.058
22	4.040	4.022	4.004	3.986	3.969	3.952	3.934	3.917	3.900	3.883
23	3.867	3.850	3.834	3.818	3.802	3.786	3.770	3.754	3.739	3.723
24	3.708	3.693	3.678	3.663	3.648	3.633	3.619	3.604	3.590	3.576
25	3.562	3.548	3.534	3.520	3.507	3.493	3.480	3.466	3.453	3.440
26	3.427	3.414	3.401	3.389	3.376	3.363	3.351	3.339	3.326	3.314
27	3.302	3.290	3.278	3.267	3.255	3.243	3.232	3.220	3.209	3.198
28	3.187	3.175	3.164	3.153	3.143	3.132	3.121	3.110	3.100	3.089
29	3.079	3.069	3.058	3.048	3.038	3.028	3.018	3.008	2.998	2.988
30	2.979	2.969	2.959	2.950	2.940	2.931	2.921	2.912	2.903	2.894
31	2.885	2.876	2.867	2.858	2.849	2.840	2.831	2.823	2.814	2.805
32	2.797	2.788	2.780	2.771	2.763	2.755	2.747	2.739	2.730	2.722
33	2.714	2.706	2.698	2.691	2.683	2.675	2.667	2.659	2.652	2.644
34	2.637	2.629	2.622	2.614	2.607	2.600	2.592	2.585	2.578	2.571
35	2.564	2.557	2.550	2.543	2.536	2.529	2.522	2.515	2.508	2.501
36	2.495	2.488	2.481	2.475	2.468	2.462	2.455	2.449	2.442	2.436
37	2.430	2.423	2.417	2.411	2.404	2.398	2.392	2.386	2.380	2.374
38	2.368	2.362	2.356	2.350	2.344	2.338	2.332	2.327	2.321	2.315
39	2.309	2.304	2.298	2.292	2.287	2.281	2.276	2.270	2.265	2.259
40	2.254	2.249	2.243	2.238	2.233	2.227	2.222	2.217	2.212	2.206
41	2.201	2.196	2.191	2.186	2.181	2.176	2.171	2.166	2.161	2.156
42	2.151	2.146	2.141	2.137	2.132	2.127	2.122	2.117	2.113	2.108
43	2.103	2.099	2.094	2.090	2.085	2.080	2.076	2.071	2.067	2.062
44	2.058	2.053	2.049	2.045	2.040	2.036	2.032	2.027	2.023	2.019
45	2.014	2.010	2.006	2.002	1.998	1.993	1.989	1.985	1.981	1.977
46	1.973	1.969	1.965	1.961	1.957	1.953	1.949	1.945	1.941	1.937
47	1.933	1.929	1.926	1.922	1.918	1.914	1.910	1.907	1.903	1.899
48	1.895	1.892	1.888	1.884	1.881	1.877	1.873	1.870	1.866	1.863
49	1.859	1.855	1.852	1.848	1.845	1.841	1.838	1.834	1.831	1.828
50	1.824	1.821	1.817	1.814	1.811	1.807	1.804	1.801	1.797	1.794
51	1.791	1.787	1.784	1.781	1.778	1.774	1.771	1.768	1.765	1.762

CuKalpha , lambda = 1.5418A 52.0 of 2THETA to 100.0

	.0	.1	.2	.3	.4	.5	.6	.7	.8	.9
52	1.759	1.755	1.752	1.749	1.746	1.743	1.740	1.737	1.734	1.731
53	1.728	1.725	1.722	1.719	1.716	1.713	1.710	1.707	1.704	1.701
54	1.698	1.695	1.692	1.689	1.687	1.684	1.681	1.678	1.675	1.672
55	1.670	1.667	1.664	1.661	1.658	1.656	1.653	1.650	1.647	1.645
56	1.642	1.639	1.637	1.634	1.631	1.629	1.626	1.623	1.621	1.618
57	1.616	1.613	1.610	1.608	1.605	1.603	1.600	1.598	1.595	1.593
58	1.590	1.588	1.585	1.583	1.580	1.578	1.575	1.573	1.570	1.568
59	1.566	1.563	1.561	1.558	1.556	1.554	1.551	1.549	1.546	1.544
60	1.542	1.539	1.537	1.535	1.533	1.530	1.528	1.526	1.523	1.521
61	1.519	1.517	1.514	1.512	1.510	1.508	1.506	1.503	1.501	1.499
62	1.497	1.495	1.492	1.490	1.488	1.486	1.484	1.482	1.480	1.478
63	1.475	1.473	1.471	1.469	1.467	1.465	1.463	1.461	1.459	1.457
64	1.455	1.453	1.451	1.449	1.447	1.445	1.443	1.441	1.439	1.437
65	1.435	1.433	1.431	1.429	1.427	1.425	1.423	1.421	1.419	1.417
66	1.415	1.414	1.412	1.410	1.408	1.406	1.404	1.402	1.400	1.399
67	1.397	1.395	1.393	1.391	1.389	1.388	1.386	1.384	1.382	1.380
68	1.379	1.377	1.375	1.373	1.372	1.370	1.368	1.366	1.365	1.363
69	1.361	1.359	1.358	1.356	1.354	1.352	1.351	1.349	1.347	1.346
70	1.344	1.342	1.341	1.339	1.337	1.336	1.334	1.332	1.331	1.329
71	1.328	1.326	1.324	1.323	1.321	1.319	1.318	1.316	1.315	1.313
72	1.312	1.310	1.308	1.307	1.305	1.304	1.302	1.301	1.299	1.298
73	1.296	1.294	1.293	1.291	1.290	1.288	1.287	1.285	1.284	1.282
74	1.281	1.279	1.278	1.277	1.275	1.274	1.272	1.271	1.269	1.268
75	1.266	1.265	1.263	1.262	1.261	1.259	1.258	1.256	1.255	1.254
76	1.252	1.251	1.249	1.248	1.247	1.245	1.244	1.242	1.241	1.240
77	1.238	1.237	1.236	1.234	1.233	1.232	1.230	1.229	1.228	1.226
78	1.225	1.224	1.222	1.221	1.220	1.218	1.217	1.216	1.215	1.213
79	1.212	1.211	1.209	1.208	1.207	1.206	1.204	1.203	1.202	1.201
80	1.199	1.198	1.197	1.196	1.194	1.193	1.192	1.191	1.189	1.188
81	1.187	1.186	1.185	1.183	1.182	1.181	1.180	1.179	1.177	1.176
82	1.175	1.174	1.173	1.172	1.170	1.169	1.168	1.167	1.166	1.165
83	1.163	1.162	1.161	1.160	1.159	1.158	1.157	1.155	1.154	1.153
84	1.152	1.151	1.150	1.149	1.148	1.147	1.145	1.144	1.143	1.142
85	1.141	1.140	1.139	1.138	1.137	1.136	1.135	1.134	1.132	1.131
86	1.130	1.129	1.128	1.127	1.126	1.125	1.124	1.123	1.122	1.121
87	1.120	1.119	1.118	1.117	1.116	1.115	1.114	1.113	1.112	1.111
88	1.110	1.109	1.108	1.107	1.106	1.105	1.104	1.103	1.102	1.101
89	1.100	1.099	1.098	1.097	1.096	1.095	1.094	1.093	1.092	1.091
90	1.090	1.089	1.088	1.087	1.086	1.085	1.085	1.084	1.083	1.082
91	1.081	1.080	1.079	1.078	1.077	1.076	1.075	1.074	1.073	1.073
92	1.072	1.071	1.070	1.069	1.068	1.067	1.066	1.065	1.065	1.064
93	1.063	1.062	1.061	1.060	1.059	1.058	1.058	1.057	1.056	1.055
94	1.054	1.053	1.052	1.052	1.051	1.050	1.049	1.048	1.047	1.046
95	1.046	1.045	1.044	1.043	1.042	1.041	1.041	1.040	1.039	1.038
96	1.037	1.037	1.036	1.035	1.034	1.033	1.032	1.032	1.031	1.030
97	1.029	1.029	1.028	1.027	1.026	1.025	1.025	1.024	1.023	1.022
98	1.021	1.021	1.020	1.019	1.018	1.018	1.017	1.016	1.015	1.015
99	1.014	1.013	1.012	1.012	1.011	1.010	1.009	1.009	1.008	1.007
100	1.006									

(b) Conversion charts of 2θ to ångstroms (D-spacing) for cobalt $K\alpha$ radiation

CoKalpha , lambda = 1.7902A 2.0 of 2THETA to 51.9

	.0	.1	.2	.3	.4	.5	.6	.7	.8	.9
2	51.288	48.846	46.626	44.599	42.741	41.032	39.454	37.993	36.636	35.373
3	34.194	33.091	32.058	31.086	30.172	29.311	28.497	27.727	26.997	26.305
4	25.648	25.023	24.427	23.859	23.317	22.799	22.304	21.830	21.375	20.939
5	20.521	20.119	19.732	19.360	19.002	18.656	18.324	18.002	17.692	17.393
6	17.103	16.823	16.552	16.289	16.035	15.789	15.550	15.318	15.093	14.874
7	14.662	14.456	14.255	14.060	13.871	13.686	13.506	13.331	13.160	12.994
8	12.832	12.674	12.519	12.369	12.222	12.078	11.938	11.801	11.667	11.536
9	11.409	11.283	11.161	11.041	10.924	10.809	10.697	10.587	10.479	10.374
10	10.270	10.169	10.069	9.972	9.876	9.782	9.690	9.600	9.511	9.424
11	9.339	9.255	9.173	9.092	9.012	8.934	8.857	8.782	8.708	8.635
12	8.563	8.493	8.423	8.355	8.288	8.222	8.157	8.093	8.030	7.968
13	7.907	7.847	7.788	7.729	7.672	7.615	7.560	7.505	7.451	7.397
14	7.345	7.293	7.242	7.191	7.142	7.093	7.044	6.997	6.950	6.903
15	6.858	6.812	6.768	6.724	6.681	6.638	6.595	6.554	6.512	6.472
16	6.432	6.392	6.353	6.314	6.276	6.238	6.201	6.164	6.127	6.091
17	6.056	6.021	5.986	5.952	5.918	5.884	5.851	5.818	5.786	5.754
18	5.722	5.691	5.660	5.629	5.599	5.569	5.539	5.509	5.480	5.452
19	5.423	5.395	5.367	5.340	5.313	5.286	5.259	5.232	5.206	5.180
20	5.155	5.129	5.104	5.079	5.055	5.030	5.006	4.982	4.958	4.935
21	4.912	4.889	4.866	4.843	4.821	4.799	4.777	4.755	4.734	4.712
22	4.691	4.670	4.649	4.629	4.608	4.588	4.568	4.548	4.529	4.509
23	4.490	4.471	4.452	4.433	4.414	4.395	4.377	4.359	4.341	4.323
24	4.305	4.288	4.270	4.253	4.236	4.219	4.202	4.185	4.168	4.152
25	4.136	4.119	4.103	4.087	4.071	4.056	4.040	4.025	4.009	3.994
26	3.979	3.964	3.949	3.934	3.920	3.905	3.891	3.877	3.862	3.848
27	3.834	3.820	3.807	3.793	3.779	3.766	3.753	3.739	3.726	3.713
28	3.700	3.687	3.674	3.662	3.649	3.636	3.624	3.612	3.599	3.587
29	3.575	3.563	3.551	3.539	3.527	3.516	3.504	3.493	3.481	3.470
30	3.458	3.447	3.436	3.425	3.414	3.403	3.392	3.381	3.371	3.360
31	3.349	3.339	3.329	3.318	3.308	3.298	3.287	3.277	3.267	3.257
32	3.247	3.238	3.228	3.218	3.208	3.199	3.189	3.180	3.170	3.161
33	3.152	3.142	3.133	3.124	3.115	3.106	3.097	3.088	3.079	3.070
34	3.062	3.053	3.044	3.036	3.027	3.018	3.010	3.002	2.993	2.985
35	2.977	2.968	2.960	2.952	2.944	2.936	2.928	2.920	2.912	2.904
36	2.897	2.889	2.881	2.873	2.866	2.858	2.851	2.843	2.836	2.828
37	2.821	2.814	2.806	2.799	2.792	2.785	2.778	2.770	2.763	2.756
38	2.749	2.742	2.735	2.729	2.722	2.715	2.708	2.701	2.695	2.688
39	2.681	2.675	2.668	2.662	2.655	2.649	2.642	2.636	2.630	2.623
40	2.617	2.611	2.605	2.598	2.592	2.586	2.580	2.574	2.568	2.562
41	2.556	2.550	2.544	2.538	2.532	2.526	2.521	2.515	2.509	2.503
42	2.498	2.492	2.486	2.481	2.475	2.470	2.464	2.459	2.453	2.448
43	2.442	2.437	2.432	2.426	2.421	2.416	2.410	2.405	2.400	2.395
44	2.389	2.384	2.379	2.374	2.369	2.364	2.359	2.354	2.349	2.344
45	2.339	2.334	2.329	2.324	2.319	2.315	2.310	2.305	2.300	2.296
46	2.291	2.286	2.281	2.277	2.272	2.268	2.263	2.258	2.254	2.249
47	2.245	2.240	2.236	2.231	2.227	2.222	2.218	2.214	2.209	2.205
48	2.201	2.196	2.192	2.188	2.184	2.179	2.175	2.171	2.167	2.163
49	2.158	2.154	2.150	2.146	2.142	2.138	2.134	2.130	2.126	2.122
50	2.118	2.114	2.110	2.106	2.102	2.098	2.094	2.091	2.087	2.083
51	2.079	2.075	2.072	2.068	2.064	2.060	2.057	2.053	2.049	2.046

CoKalpha , lambda = 1.7902A 52.0 of 2THETA to 100.0

	.0	.1	.2	.3	.4	.5	.6	.7	.8	.9
52	2.042	2.038	2.035	2.031	2.027	2.024	2.020	2.017	2.013	2.010
53	2.006	2.003	1.999	1.996	1.992	1.989	1.985	1.982	1.978	1.975
54	1.972	1.968	1.965	1.962	1.958	1.955	1.952	1.948	1.945	1.942
55	1.939	1.935	1.932	1.929	1.926	1.922	1.919	1.916	1.913	1.910
56	1.907	1.903	1.900	1.897	1.894	1.891	1.888	1.885	1.882	1.879
57	1.876	1.873	1.870	1.867	1.864	1.861	1.858	1.855	1.852	1.849
58	1.846	1.843	1.840	1.838	1.835	1.832	1.829	1.826	1.823	1.821
59	1.818	1.815	1.812	1.809	1.807	1.804	1.801	1.798	1.796	1.793
60	1.790	1.788	1.785	1.782	1.779	1.777	1.774	1.771	1.769	1.766
61	1.764	1.761	1.758	1.756	1.753	1.751	1.748	1.746	1.743	1.740
62	1.738	1.735	1.733	1.730	1.728	1.725	1.723	1.720	1.718	1.716
63	1.713	1.711	1.708	1.706	1.703	1.701	1.699	1.696	1.694	1.691
64	1.689	1.687	1.684	1.682	1.680	1.677	1.675	1.673	1.671	1.668
65	1.666	1.664	1.661	1.659	1.657	1.655	1.652	1.650	1.648	1.646
66	1.643	1.641	1.639	1.637	1.635	1.633	1.630	1.628	1.626	1.624
67	1.622	1.620	1.617	1.615	1.613	1.611	1.609	1.607	1.605	1.603
68	1.601	1.599	1.597	1.595	1.592	1.590	1.588	1.586	1.584	1.582
69	1.580	1.578	1.576	1.574	1.572	1.570	1.568	1.566	1.564	1.563
70	1.561	1.559	1.557	1.555	1.553	1.551	1.549	1.547	1.545	1.543
71	1.541	1.540	1.538	1.536	1.534	1.532	1.530	1.528	1.527	1.525
72	1.523	1.521	1.519	1.517	1.516	1.514	1.512	1.510	1.508	1.507
73	1.505	1.503	1.501	1.500	1.498	1.496	1.494	1.493	1.491	1.489
74	1.487	1.486	1.484	1.482	1.480	1.479	1.477	1.475	1.474	1.472
75	1.470	1.469	1.467	1.465	1.464	1.462	1.460	1.459	1.457	1.456
76	1.454	1.452	1.451	1.449	1.447	1.446	1.444	1.443	1.441	1.439
77	1.438	1.436	1.435	1.433	1.432	1.430	1.428	1.427	1.425	1.424
78	1.422	1.421	1.419	1.418	1.416	1.415	1.413	1.412	1.410	1.409
79	1.407	1.406	1.404	1.403	1.401	1.400	1.398	1.397	1.395	1.394
80	1.393	1.391	1.390	1.388	1.387	1.385	1.384	1.382	1.381	1.380
81	1.378	1.377	1.375	1.374	1.373	1.371	1.370	1.368	1.367	1.366
82	1.364	1.363	1.362	1.360	1.359	1.358	1.356	1.355	1.354	1.352
83	1.351	1.350	1.348	1.347	1.346	1.344	1.343	1.342	1.340	1.339
84	1.338	1.336	1.335	1.334	1.333	1.331	1.330	1.329	1.327	1.326
85	1.325	1.324	1.322	1.321	1.320	1.319	1.317	1.316	1.315	1.314
86	1.312	1.311	1.310	1.309	1.308	1.306	1.305	1.304	1.303	1.302
87	1.300	1.299	1.298	1.297	1.296	1.294	1.293	1.292	1.291	1.290
88	1.289	1.287	1.286	1.285	1.284	1.283	1.282	1.280	1.279	1.278
89	1.277	1.276	1.275	1.274	1.273	1.271	1.270	1.269	1.268	1.267
90	1.266	1.265	1.264	1.263	1.261	1.260	1.259	1.258	1.257	1.256
91	1.255	1.254	1.253	1.252	1.251	1.250	1.249	1.247	1.246	1.245
92	1.244	1.243	1.242	1.241	1.240	1.239	1.238	1.237	1.236	1.235
93	1.234	1.233	1.232	1.231	1.230	1.229	1.228	1.227	1.226	1.225
94	1.224	1.223	1.222	1.221	1.220	1.219	1.218	1.217	1.216	1.215
95	1.214	1.213	1.212	1.211	1.210	1.209	1.208	1.207	1.206	1.205
96	1.204	1.204	1.203	1.202	1.201	1.200	1.199	1.198	1.197	1.196
97	1.195	1.194	1.193	1.192	1.191	1.191	1.190	1.189	1.188	1.187
98	1.186	1.185	1.184	1.183	1.182	1.182	1.181	1.180	1.179	1.178
99	1.177	1.176	1.175	1.175	1.174	1.173	1.172	1.171	1.170	1.169
100	1.168									

(c) Conversion charts of 2θ to ångstroms (D-spacing) for chromium $K\alpha$ radiation

CrKalpha , lambda = 2.2909A 2.0 of 2THETA to 51.9

	.0	.1	.2	.3	.4	.5	.6	.7	.8	.9
2	65.633	62.508	59.667	57.073	54.695	52.508	50.489	48.619	46.883	45.267
3	43.758	42.347	41.024	39.781	38.611	37.508	36.467	35.482	34.548	33.663
4	32.821	32.021	31.259	30.533	29.839	29.176	28.542	27.935	27.354	26.796
5	26.260	25.746	25.251	24.775	24.316	23.874	23.448	23.037	22.641	22.257
6	21.887	21.528	21.181	20.845	20.520	20.205	19.899	19.602	19.314	19.035
7	18.763	18.499	18.242	17.993	17.750	17.514	17.284	17.059	16.841	16.628
8	16.421	16.218	16.021	15.828	15.640	15.456	15.277	15.102	14.930	14.763
9	14.599	14.439	14.283	14.129	13.979	13.833	13.689	13.548	13.410	13.275
10	13.143	13.013	12.886	12.761	12.638	12.518	12.401	12.285	12.172	12.060
11	11.951	11.844	11.738	11.635	11.533	11.433	11.335	11.238	11.143	11.050
12	10.958	10.868	10.779	10.692	10.606	10.522	10.438	10.357	10.276	10.197
13	10.119	10.042	9.966	9.891	9.818	9.745	9.674	9.604	9.535	9.466
14	9.399	9.333	9.267	9.203	9.139	9.077	9.015	8.954	8.894	8.834
15	8.776	8.718	8.661	8.605	8.549	8.494	8.440	8.387	8.334	8.282
16	8.230	8.180	8.129	8.080	8.031	7.983	7.935	7.888	7.841	7.795
17	7.750	7.705	7.660	7.616	7.573	7.530	7.487	7.445	7.404	7.363
18	7.322	7.282	7.242	7.203	7.164	7.126	7.088	7.050	7.013	6.977
19	6.940	6.904	6.868	6.833	6.798	6.764	6.730	6.696	6.662	6.629
20	6.596	6.564	6.532	6.500	6.468	6.437	6.406	6.376	6.345	6.315
21	6.286	6.256	6.227	6.198	6.169	6.141	6.113	6.085	6.058	6.030
22	6.003	5.976	5.950	5.923	5.897	5.871	5.846	5.820	5.795	5.770
23	5.745	5.721	5.697	5.672	5.649	5.625	5.601	5.578	5.555	5.532
24	5.509	5.487	5.464	5.442	5.420	5.399	5.377	5.356	5.334	5.313
25	5.292	5.271	5.251	5.230	5.210	5.190	5.170	5.150	5.131	5.111
26	5.092	5.073	5.054	5.035	5.016	4.998	4.979	4.961	4.943	4.925
27	4.907	4.889	4.871	4.854	4.836	4.819	4.802	4.785	4.768	4.751
28	4.735	4.718	4.702	4.686	4.669	4.653	4.637	4.622	4.606	4.590
29	4.575	4.559	4.544	4.529	4.514	4.499	4.484	4.469	4.455	4.440
30	4.426	4.411	4.397	4.383	4.369	4.355	4.341	4.327	4.313	4.300
31	4.286	4.273	4.259	4.246	4.233	4.220	4.207	4.194	4.181	4.168
32	4.156	4.143	4.131	4.118	4.106	4.093	4.081	4.069	4.057	4.045
33	4.033	4.021	4.009	3.998	3.986	3.975	3.963	3.952	3.940	3.929
34	3.918	3.907	3.896	3.885	3.874	3.863	3.852	3.841	3.830	3.820
35	3.809	3.799	3.788	3.778	3.768	3.757	3.747	3.737	3.727	3.717
36	3.707	3.697	3.687	3.677	3.667	3.658	3.648	3.638	3.629	3.619
37	3.610	3.601	3.591	3.582	3.573	3.564	3.554	3.545	3.536	3.527
38	3.518	3.509	3.501	3.492	3.483	3.474	3.466	3.457	3.448	3.440
39	3.431	3.423	3.415	3.406	3.398	3.390	3.382	3.373	3.365	3.357
40	3.349	3.341	3.333	3.325	3.317	3.309	3.302	3.294	3.286	3.278
41	3.271	3.263	3.256	3.248	3.241	3.233	3.226	3.218	3.211	3.204
42	3.196	3.189	3.182	3.175	3.168	3.160	3.153	3.146	3.139	3.132
43	3.125	3.118	3.112	3.105	3.098	3.091	3.084	3.078	3.071	3.064
44	3.058	3.051	3.045	3.038	3.032	3.025	3.019	3.012	3.006	3.000
45	2.993	2.987	2.981	2.974	2.968	2.962	2.956	2.950	2.944	2.938
46	2.932	2.926	2.920	2.914	2.908	2.902	2.896	2.890	2.884	2.878
47	2.873	2.867	2.861	2.855	2.850	2.844	2.838	2.833	2.827	2.822
48	2.816	2.811	2.805	2.800	2.794	2.789	2.784	2.778	2.773	2.767
49	2.762	2.757	2.752	2.746	2.741	2.736	2.731	2.726	2.721	2.715
50	2.710	2.705	2.700	2.695	2.690	2.685	2.680	2.675	2.670	2.666
51	2.661	2.656	2.651	2.646	2.641	2.637	2.632	2.627	2.622	2.618

CrKalpha , lambda = 2.2909A 52.0 of 2THETA to 100.0

	.0	.1	.2	.3	.4	.5	.6	.7	.8	.9
52	2.613	2.608	2.604	2.599	2.594	2.590	2.585	2.581	2.576	2.572
53	2.567	2.563	2.558	2.554	2.549	2.545	2.540	2.536	2.532	2.527
54	2.523	2.519	2.514	2.510	2.506	2.502	2.497	2.493	2.489	2.485
55	2.481	2.477	2.472	2.468	2.464	2.460	2.456	2.452	2.448	2.444
56	2.440	2.436	2.432	2.428	2.424	2.420	2.416	2.412	2.408	2.404
57	2.401	2.397	2.393	2.389	2.385	2.381	2.378	2.374	2.370	2.366
58	2.363	2.359	2.355	2.352	2.348	2.344	2.341	2.337	2.333	2.330
59	2.326	2.323	2.319	2.315	2.312	2.308	2.305	2.301	2.298	2.294
60	2.291	2.287	2.284	2.281	2.277	2.274	2.270	2.267	2.264	2.260
61	2.257	2.254	2.250	2.247	2.244	2.240	2.237	2.234	2.230	2.227
62	2.224	2.221	2.218	2.214	2.211	2.208	2.205	2.202	2.199	2.195
63	2.192	2.189	2.186	2.183	2.180	2.177	2.174	2.171	2.168	2.165
64	2.162	2.159	2.156	2.153	2.150	2.147	2.144	2.141	2.138	2.135
65	2.132	2.129	2.126	2.123	2.120	2.117	2.115	2.112	2.109	2.106
66	2.103	2.100	2.098	2.095	2.092	2.089	2.086	2.084	2.081	2.078
67	2.075	2.073	2.070	2.067	2.064	2.062	2.059	2.056	2.054	2.051
68	2.048	2.046	2.043	2.040	2.038	2.035	2.033	2.030	2.027	2.025
69	2.022	2.020	2.017	2.015	2.012	2.010	2.007	2.005	2.002	2.000
70	1.997	1.995	1.992	1.990	1.987	1.985	1.982	1.980	1.977	1.975
71	1.973	1.970	1.968	1.965	1.963	1.961	1.958	1.956	1.953	1.951
72	1.949	1.946	1.944	1.942	1.939	1.937	1.935	1.933	1.930	1.928
73	1.926	1.923	1.921	1.919	1.917	1.914	1.912	1.910	1.908	1.906
74	1.903	1.901	1.899	1.897	1.895	1.892	1.890	1.888	1.886	1.884
75	1.882	1.879	1.877	1.875	1.873	1.871	1.869	1.867	1.865	1.863
76	1.861	1.858	1.856	1.854	1.852	1.850	1.848	1.846	1.844	1.842
77	1.840	1.838	1.836	1.834	1.832	1.830	1.828	1.826	1.824	1.822
78	1.820	1.818	1.816	1.814	1.812	1.810	1.808	1.807	1.805	1.803
79	1.801	1.799	1.797	1.795	1.793	1.791	1.789	1.788	1.786	1.784
80	1.782	1.780	1.778	1.776	1.775	1.773	1.771	1.769	1.767	1.766
81	1.764	1.762	1.760	1.758	1.757	1.755	1.753	1.751	1.749	1.748
82	1.746	1.744	1.742	1.741	1.739	1.737	1.736	1.734	1.732	1.730
83	1.729	1.727	1.725	1.724	1.722	1.720	1.719	1.717	1.715	1.714
84	1.712	1.710	1.709	1.707	1.705	1.704	1.702	1.700	1.699	1.697
85	1.695	1.694	1.692	1.691	1.689	1.687	1.686	1.684	1.683	1.681
86	1.680	1.678	1.676	1.675	1.673	1.672	1.670	1.669	1.667	1.666
87	1.664	1.663	1.661	1.659	1.658	1.656	1.655	1.653	1.652	1.650
88	1.649	1.647	1.646	1.644	1.643	1.642	1.640	1.639	1.637	1.636
89	1.634	1.633	1.631	1.630	1.628	1.627	1.626	1.624	1.623	1.621
90	1.620	1.618	1.617	1.616	1.614	1.613	1.611	1.610	1.609	1.607
91	1.606	1.605	1.603	1.602	1.600	1.599	1.598	1.596	1.595	1.594
92	1.592	1.591	1.590	1.588	1.587	1.586	1.584	1.583	1.582	1.580
93	1.579	1.578	1.577	1.575	1.574	1.573	1.571	1.570	1.569	1.567
94	1.566	1.565	1.564	1.562	1.561	1.560	1.559	1.557	1.556	1.555
95	1.554	1.552	1.551	1.550	1.549	1.547	1.546	1.545	1.544	1.543
96	1.541	1.540	1.539	1.538	1.537	1.535	1.534	1.533	1.532	1.531
97	1.529	1.528	1.527	1.526	1.525	1.524	1.522	1.521	1.520	1.519
98	1.518	1.517	1.515	1.514	1.513	1.512	1.511	1.510	1.509	1.507
99	1.506	1.505	1.504	1.503	1.502	1.501	1.500	1.499	1.497	1.496
100	1.494									

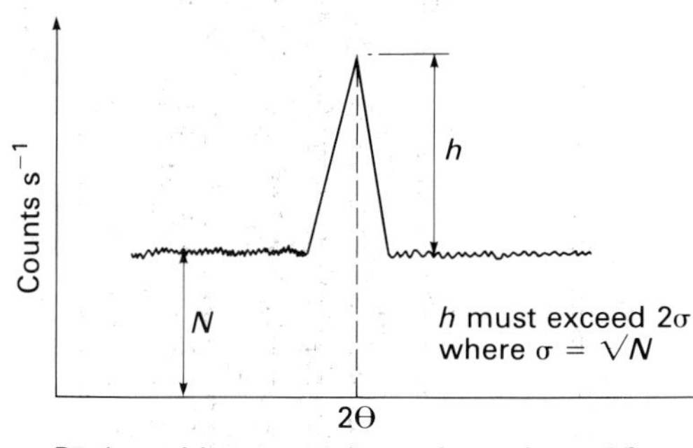

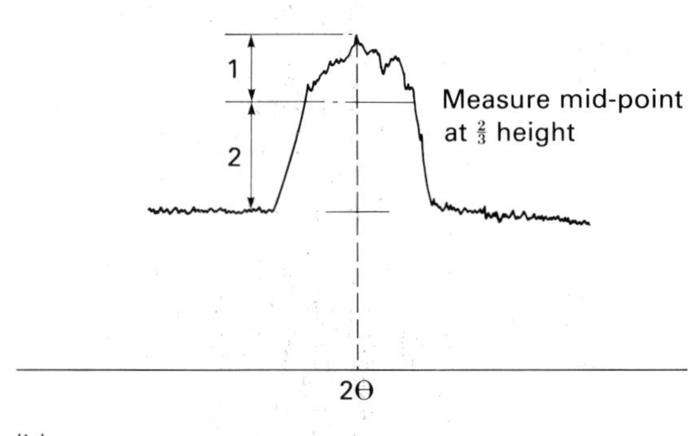

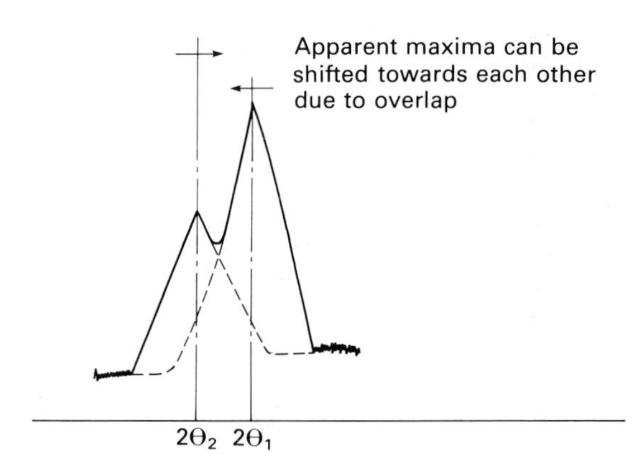

Fig. 7.13. (a) Definition of a peak on an X-ray chart record. (b) Ragged X-ray diffraction peak. (c) Overlapping X-ray diffraction peaks.

of ragged peaks (Fig. 7.13b) and overlapping peaks (Fig. 7.13c). A reasonably successful method of estimating the position of a ragged peak is to take the mid-point position at 2/3 of the height above background. One should bear in mind, however, that the reason that the peak is ragged in the first place is likely to be related to a poorly defined crystal structure. Overlapping peaks can be deconvoluted by graphical or computational means, but in practice a visual estimate of the position of the maxima is adequate for qualitative analysis if one bears in mind the tendency of the peaks to move towards a common 'centre of gravity'.

Instrumental conditions also play an important role in the correct measurement of peak position. As the time constant is a pulse averaging circuit, if a large value of the time constant is used together with a fast scanning speed, errors will result due to the formation of asymmetric peaks, in which the point of greatest intensity is shifted to a higher value of 2θ. Generally the product of the time constant and the scanning speed is kept as low as possible, but is never greater than 4.

The diffraction angles (2θ) of the peaks are converted to lattice spacings (d) by means of conversion tables (Fang & Bloss, 1966), by conversion charts (Parrish & Mack, 1963) or by computing into a calculator/computer the equation $d = \frac{\lambda}{2 \sin \theta}$. Table 7.2 shows 2θ to lattice spacing conversion tables for copper $K\alpha$, cobalt $K\alpha$ and chromium $K\alpha$ radiations. The lattice spacings (d) are in ångstroms (Å), still commonly used by X-ray mineralogists, though the nanometer (nm) is the technically correct SI unit. Once all peaks have been measured together with an estimate of their relative intensities, it only remains to assign them to mineral phases.

As all minerals and indeed all crystalline materials possess a unique X-ray diffraction pattern, a comparison of diffraction patterns of unknown mineral phases with a set of standard patterns will lead to their identification. This method is very similar to human fingerprint identification. These standard patterns have been compiled by an international organization called the Joint Committee on Powder Diffraction Standards (JCPDS), which collects and updates powder diffraction data. In principle, by a systematic searching of the JCPDS Powder Diffraction Index for Minerals, it is possible to identify almost any mineral that may be present, provided that sufficient peaks are present for that mineral (usually a minimum of three). This operation can of course, be carried out by computers and the JCPDS provide just such a commercially available computer system. For further information contact JCPDS — International Centre for Diffraction Data, 1601 Park Lane, Swarthmore, Pennsylvania, PA 19081, USA.

For sediments, however, the total number of possibilities is fairly limited, and can be limited further if one has some idea of its provenance. With this knowledge a reverse search of the Index, beginning with the most likely minerals and eliminating the peaks as they are identified, is a useful short-cut. Alternatively one can make up a template on chart recorder paper, of the diffraction patterns of commonly-encountered minerals, and then use this as an overlay for the unknown chart records. Brown & Brindley (1980, table 5.18, pp. 348–355) computed a comprehensive table of consecutive lattice spacings for the common clay and sedimentary minerals. Table 7.3 lists the main diffraction spacings of some selected minerals likely to occur in sediments together with their relative intensities (I/I_i), *hkl* values and 2θ conversion angles for Cu and Co $K\alpha$ radiations.

7.3.3 Quantitative analysis

As the intensity of the diffraction pattern (generally estimated either as peak height or preferably peak area) of a mineral in a mixture is proportional to its concentration, it is possible to make rough estimates of the relative proportions of the minerals in a sample by measuring their relative peak heights or areas. This method, however, is unreliable due to the different 'diffracting abilities' of minerals from different crystal systems. An example of this would be to compare the intensities of halite (cubic system) with illite (monoclinic system) which would be seen to be completely unrelated.

Most 'quantitative' systems of sediment analysis depend upon the preparation of calibration curves. This entails the measurement of the intensity of selected diffraction lines plotted against the quantities of the respective minerals in known mixtures. Generally an internal standard is added in constant amount to the known mixtures and unknown samples to compensate for any errors introduced during the sample preparation or caused by machine drift, etc. The internal standard method has been used widely since being described by Alexander & Klug (1948). Various substances have been proposed as internal standards. Generally it is a completely new substance to the system, such as potassium permanganate $KMnO_4$, halite NaCl, or cerium oxide CeO_2, but it is an advantage to have a substance which has a similar response to that of the analysed minerals. Griffin (1954) used the mineral boehmite (γ-AlOOH) in his study of clay minerals because they both have a similar response to X-rays.

The principal alternative to the addition of an internal standard is to use a mineral already present in the sample (e.g. quartz) as the internal standard. A relatively simple version of this latter method (after Hooton & Giorgetta, 1977) is recommended here.

The equation used is derived from the Klug & Alexander basic equation:

$$W_i = \frac{\eta_i I_i \mu_m}{K_i} \tag{1}$$

where I_i = intensity of diffraction pattern of component i,

K_i = a constant depending on the nature of the component and the geometry of the apparatus,

W_i = weight fraction of component i in the sample,

η_i = density of component,

μ_m = mass absorption of the total sample.

As η_i and K_i are both constant for any mineral (in a given instrument) they can be represented in the equation by the symbol H_i giving

$$W_i = H_i I_i \mu_m. \tag{2}$$

If all components are taken into account, equation (2) becomes ${}_1^n\Sigma W_i = \mu_m {}_1^n\Sigma H_i I_i = 1$ the sum of all the weight fractions.

Rearranging: $\mu_m = {}_1^n\Sigma \frac{1}{H_i I_i}$ and substituting into equation (2)

$$W_i = \frac{H_i I_i}{{}_1^n\Sigma H_i I_i}. \tag{3}$$

The method can be simplified by determining relative H values instead of absolute values, i.e. to select one mineral as a reference standard, in this case, quartz. If equation (2) is now rewritten as

$$W_{MIN} = H_{MIN} I_{MIN} \mu_m \quad \text{(unknown component)}$$

$$W_{QUARTZ} = H_{QUARTZ} I_{QUARTZ} \mu_m \quad \text{(reference component).}$$

Table 7.3. Dominant X-ray diffraction peaks of selected minerals likely to occur in sediments in order of decreasing intensities (I/I_1) including 2θ angles for Cu and Co $K\alpha$ radiations. Minerals are listed in order of decreasing d (Å) value of the most intense line (from JCPDS, 1974).

Mineral							
RECTORITE	d (Å)	24.7	12.4	3.10	4.94	1.902	3.54
Regular interstratified	I/I_1	100	50	18	8	4	2
mica–montmorillonite	hkl	001	002	008	005	0013	007
	$2\theta_{Cu}$	3.58	7.13	28.79	17.95	47.82	25.16
	$2\theta_{Co}$	4.15	8.28	33.56	20.88	56.15	29.30
SMECTITE	d (Å)	15.7	4.58	2.56	1.53	3.63	2.95
Saponite	I/I_1	100	100	100	100	60	60
	hkl	001	020	200	060	004	005
	$2\theta_{Cu}$	5.63	19.38	35.05	60.51	24.52	30.30
	$2\theta_{Co}$	6.54	22.54	40.93	71.61	28.55	35.33
SMECTITE	d (Å)	15.0	4.50	5.01	3.02	1.50	1.493
Montmorillonite	I/I_1	100	80	60	60	50	50
	hkl	001	020	003	005	060	—
	$2\theta_{Cu}$	5.89	19.73	17.70	29.58	61.85	62.17
	$2\theta_{Co}$	6.84	22.95	20.58	34.48	73.27	73.67
VERMICULITE	d (Å)	14.2	1.528	4.57	2.615	2.570	2.525
	I/I_1	100	70	60	50	50	45
	hkl	002	060	020	200	132	202
	$2\theta_{Cu}$	6.22	60.60	19.42	34.29	34.91	35.55
	$2\theta_{Cu}$	7.23	71.72	22.59	40.03	40.76	41.53
CORRENSITE	d (Å)	14.0	7.08	3.53	29.0	4.72	4.62
Regular interstratified	I/I_1	100	60	60	30	30	30
chlorite–montmorillonite	hkl	002	004	008	001	006	020
	$2\theta_{Cu}$	6.31	12.50	25.23	3.05	18.80	19.21
	$2\theta_{Co}$	7.33	14.53	29.38	3.54	21.86	22.34
ILLITE	d (Å)	10.0	4.48	3.33	2.61	1.53	2.42
	I/I_1	100	90	90	60	60	40
	hkl	002	020	006	200	060	133
	$2\theta_{Cu}$	8.84	19.82	26.77	34.36	60.51	37.15
	$2\theta_{Co}$	10.27	23.05	31.18	40.11	71.61	43.42
MUSCOVITE	d (Å)	9.97	3.331	4.99	1.999	2.564	4.49
	I/I_1	100	100	55	45	25	20
	hkl	003	009	006	0015	112	100
	$2\theta_{Cu}$	8.87	26.76	17.77	45.37	35.00	19.77
	$2\theta_{Co}$	10.30	31.18	20.67	53.20	40.86	23.00
GYPSUM	d (Å)	7.56	3.059	4.27	2.679	2.867	3.79
	I/I_1	100	55	50	28	25	20
	hkl	020	$14\bar{1}$	$12\bar{1}$	022	002	031
	$2\theta_{Cu}$	11.70	29.19	20.80	33.44	31.20	23.47
	$2\theta_{Co}$	13.60	34.03	24.20	39.04	36.38	27.32

Table 7.3. (*Continued*)

CHLORITE	d (Å)	7.19	4.80	3.60	14.3	2.88	2.56
Penninite	I/I_1	100	100	100	60	60	40
	hkl	002	003	004	001	005	$13\bar{2}$
	$2\theta_{Cu}$	12.31	18.48	24.73	6.18	31.05	35.05
	$2\theta_{Co}$	14.30	21.49	28.79	7.18	36.21	40.93
KAOLINITE	d (Å)	7.17	1.489	3.579	1.620	4.366	1.589
	I/I_1	100	90	80	70	60	60
	hkl	001	060	002	133	$1\bar{1}0$	$13\bar{4}$
	$2\theta_{Cu}$	12.34	62.36	24.88	56.83	20.34	58.04
	$2\theta_{Co}$	14.34	73.90	28.97	67.08	23.66	68.57
CHLORITE	d (Å)	7.07	14.1	3.541	4.726	2.845	2.576
Thuringite	I/I_1	100	90	60	30	30	30
	hkl	002	001	004	003	005	131
	$2\theta_{Cu}$	12.52	6.27	25.15	18.78	31.44	34.83
	$2\theta_{Co}$	14.55	7.28	29.28	21.84	36.68	40.67
LEPIDOCROCITE	d (Å)	6.26	3.29	2.47	1.937	1.732	1.524
	I/I_1	100	90	80	70	40	40
	hkl	020	120	031	200	151	231
	$2\theta_{Cu}$	14.15	27.10	36.37	46.90	52.86	60.77
	$2\theta_{Co}$	16.44	31.57	42.49	55.05	62.24	71.94
Regular interstratified	d (Å)	4.53	15.0	4.97	30.0	2.54	1.509
montmorillonite–chlorite	I/I_1	100	90	75	60	50	50
	hkl	020	002	006	001	—	060
	$2\theta_{Cu}$	19.60	5.89	17.85	2.94	34.34	61.44
	$2\theta_{Co}$	22.79	6.84	20.75	3.42	41.27	72.77
GOETHITE	d (Å)	4.18	2.69	2.452	2.192	1.721	2.490
	I/I_1	100	30	25	20	20	16
	hkl	110	130	111	140	221	040
	$2\theta_{Cu}$	21.25	33.31	36.65	41.18	53.22	36.07
	$2\theta_{Co}$	24.73	38.87	42.82	48.20	62.68	42.14
ARAGONITE	d (Å)	3.396	1.977	3.273	2.700	2.372	2.481
	I/I_1	100	65	52	46	38	33
	hkl	111	221	021	012	112	200
	$2\theta_{Cu}$	26.24	45.90	27.25	33.18	37.93	36.21
	$2\theta_{Co}$	30.56	53.84	31.74	38.72	44.34	42.30
QUARTZ	d (Å)	3.343	4.26	1.817	1.541	2.458	2.282
	I/I_1	100	35	17	15	12	12
	hkl	101	100	112	211	110	102
	$2\theta_{Cu}$	26.67	20.85	50.21	60.03	36.56	39.49
	$2\theta_{Co}$	31.06	24.26	59.03	71.02	42.71	46.19

Table 7.3. (*Continued*)

Mineral							
ALKALI FELDSPAR	d (Å)	3.31	3.77	4.22	3.24	3.29	2.992
Orthoclase	I/I_1	100	80	70	65	60	50
	hkl	220	130	$20\bar{1}$	002	$20\bar{2}$	131
	$2\theta_{Cu}$	26.94	23.60	21.05	27.53	27.10	29.86
	$2\theta_{Co}$	31.38	27.47	24.49	32.07	31.57	34.82
PLAGIOCLASE FELDSPAR	d (Å)	3.196	3.780	6.39	3.684	4.03	3.663
Albite	I/I_1	100	25	20	20	16	16
	hkl	002	111	001	130	$20\bar{1}$	$1\bar{3}0$
	$2\theta_{Cu}$	27.92	23.54	13.86	24.16	22.06	24.30
	$2\theta_{Co}$	32.53	27.40	16.10	28.12	25.67	28.29
CALCITE	d (Å)	3.035	2.285	2.095	1.913	1.875	2.495
	I/I_1	100	18	18	17	17	14
	hkl	104	113	202	108	116	110
	$2\theta_{Cu}$	29.43	39.43	43.18	47.53	48.55	36.00
	$2\theta_{Co}$	34.31	46.12	50.59	55.80	57.03	42.05
DOLOMITE	d (Å)	2.886	2.192	1.786	1.804	2.015	1.389
	I/I_1	100	30	30	20	15	15
	hkl	104	113	009	018	202	030
	$2\theta_{Cu}$	30.99	41.18	51.14	50.60	44.99	67.42
	$2\theta_{Co}$	36.14	48.20	60.16	59.50	52.75	80.24
SIDERITE	d (Å)	2.79	1.734	3.59	0.931	2.13	1.963
	I/I_1	100	80	60	70	60	60
	hkl	104	018	012	$30\underline{12}$	113	202
	$2\theta_{Cu}$	32.08	52.79	24.80	111.80	42.44	46.25
	$2\theta_{Co}$	37.43	62.16	28.88	148.07	49.70	54.26
HEMATITE	d (Å)	2.69	1.690	2.51	1.838	1.484	1.452
	I/I_1	100	60	50	40	35	35
	hkl	104	116	110	024	214	300
	$2\theta_{Cu}$	33.31	54.28	35.77	49.60	62.59	64.14
	$2\theta_{Co}$	38.87	63.96	41.78	58.29	74.19	76.12
PYRITE	d (Å)	1.633	2.709	2.423	2.212	1.916	3.128
	I/I_1	100	85	65	50	40	35
	hkl	311	200	210	211	220	111
	$2\theta_{Cu}$	56.34	33.07	37.10	40.79	47.45	28.54
	$2\theta_{Co}$	66.48	38.59	43.36	47.74	55.70	33.26

Dividing: $$\frac{W_{MIN}}{W_{QUARTZ}} = \frac{H_{MIN}}{H_{QUARTZ}} \times \frac{I_{MIN}}{I_{QUARTZ}} \quad (4)$$

A series of binary calibration mixtures of quartz and other major sedimentary minerals can then be prepared by accurate weighing and the intensities of the chosen analytical peaks measured. Curves can then be plotted, the slope of which gives H_{MIN}/H_{QUARTZ} if we assume $H_{QUARTZ} = 1$; a series of H

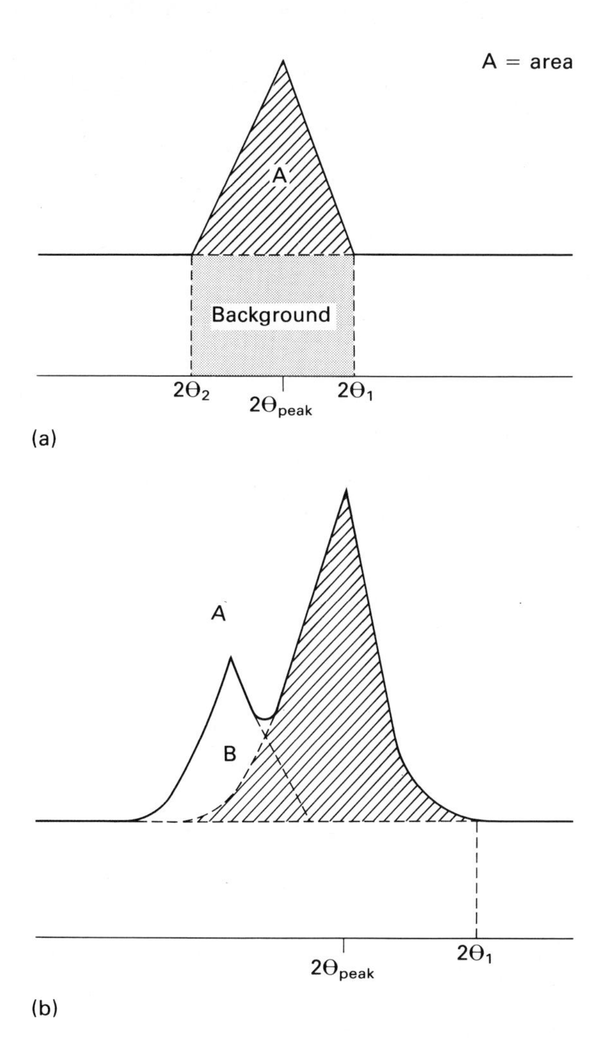

Fig. 7.14. Measurement of intensity of a diffraction peak by integration of area. (a) Single peak. (b) Overlapping peaks.

values can then be obtained for all the major sedimentary minerals which can then be substituted together with the intensities of the minerals into equation (3).

The measurement of intensity in all quantitative methods is best done using a digital counter and integrating a whole area of peak (Fig. 7.14a). This is achieved by starting the counter and the X-ray scan from position $2\theta_1$ simultaneously, and allowing both to run until position $2\theta_2$ is attained and then stopping them simultaneously; the total integrated counts (N_I) and the time are then recorded. N_I includes counts not only due to the peak intensity, but also the background. The background counts are estimated by fixed counting at each of the positions $2\theta_1$ and $2\theta_2$, for exactly half the total time of the integrated scan. The sum of these two readings equals the background N_B which is substracted from N_I to give the number of counts of the integrated peak, i.e. $N_I - N_B$ = AREA.

However, if a digital counter is not available either a simple polar planimeter can be used or a simple geometrical calculation can be made, i.e. peak height above background multiplied by the width at half peak height — both produce reasonable estimates of area. In certain cases the use of peak heights is unavoidable, especially where partially overlapping peaks make area measurements difficult (Fig. 7.14b). There are mathematical techniques available using computer programs which can separate overlapping peaks; an example of this method using Lorenzian profile calculations is given in Martinez & Plana (1987).

There are many more published methods of quantitative analysis but all suffer from similar problems which affect the measurement of intensities, in particular: (i) the crystallinity of some species can vary widely, e.g. micas, kaolinites (Fig. 7.15); (ii) chemical variations, e.g. solid solution series, which can cause peak shifts as well as varying intensities; (iii) the ordering of the component chemical elements within the structure, e.g. dolomite; (iv) large mass absorption differences between standards and unknown samples.

7.4 CLAY MINERAL ANALYSIS

XRD provides the most efficient method for the determination of clay minerals in mudrocks, sandstones and limestones. A knowledge of the clay mineralogy can be most useful for provenance studies and it can also give information on the burial history for the formation. Although some clay minerals are evident in whole rock diffractograms, the most satisfactory method is to extract and separately analyse the clay fraction (usually defined as less than 2 microns, <2 μm). It is particularly important to do this in the case of very fine grained and poorly crystalline clays, which are unlikely to give recognizable diffraction patterns in a whole rock scan. Also chemical pre-treatments are sometimes

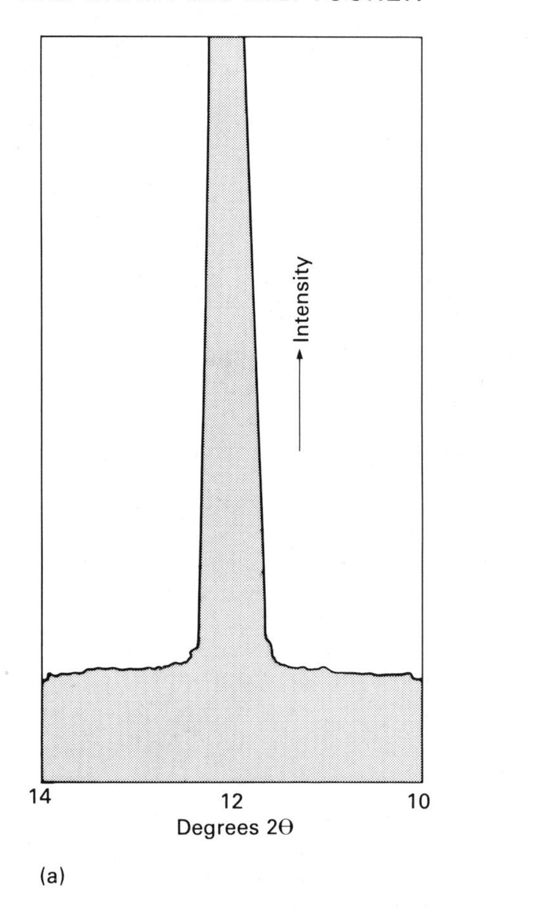

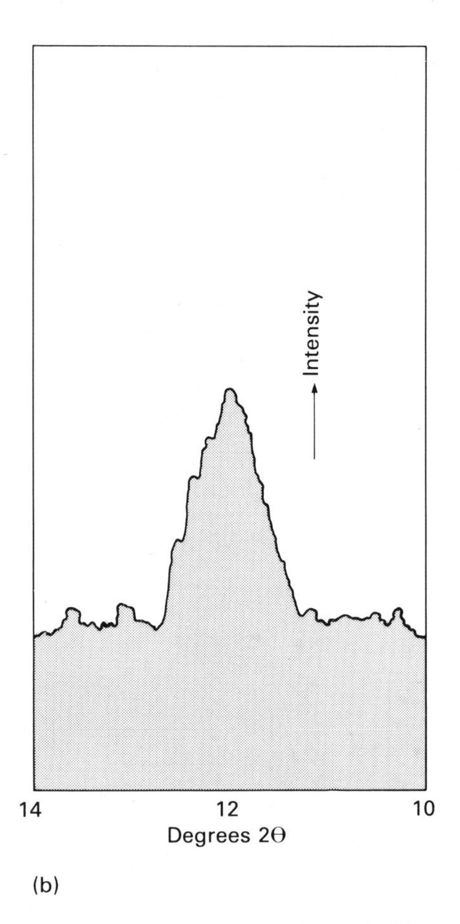

Fig. 7.15. X-ray diffraction scans of the (001) lattice spacing of: (a) a well crystalline kaolinite from Cornwall; (b) a poorly crystalline kaolinite from a fireclay from N England, using Cu *Kα* radiation and similar analytical conditions.

necessary to remove large amounts of certain non-clay and cementing materials and in so doing concentrate the clay mineral fraction in the samples.

If high levels of carbonate material are expected this can be removed by acid attack. However, certain clay minerals, in particular chlorites and some smectites, are susceptible to attack from dilute mineral acids, e.g. dilute hydrochloric acid. Instead, weak organic acids are recommended, in particular the method described by Hein, Scholl & Gutmacher (1976) in which the samples are heated in a buffered solution of sodium acetate and glacial acetic acid (Morgan's solution).

Similarly, if large quantities of organic matter are expected, this may be removed by pre-treatment with 35% hydrogen peroxide (H_2O_2) after Jackson (1958, 1979). Large amounts of iron compounds may be removed using sodium dithionite ($Na_2S_2O_4$) together with a citrate-chelating agent after the method of Mehra & Jackson (1960).

7.4.1 Separation

Complete disaggregation is desirable before size separation is attempted, and this is best effected by ultrasonic means (Gipson, 1963), although, for poorly lithified samples, vigorous shaking with distilled water will sometimes suffice. The sediments should be broken (e.g. by percussion) to particles of approximately 10 mm diameter or less, placed in a beaker, covered with distilled water and immersed in an ultrasonic bath (Fig. 7.16a). Alternatively, an ultrasonic probe can be inserted into the beaker (Fig. 7.16b).

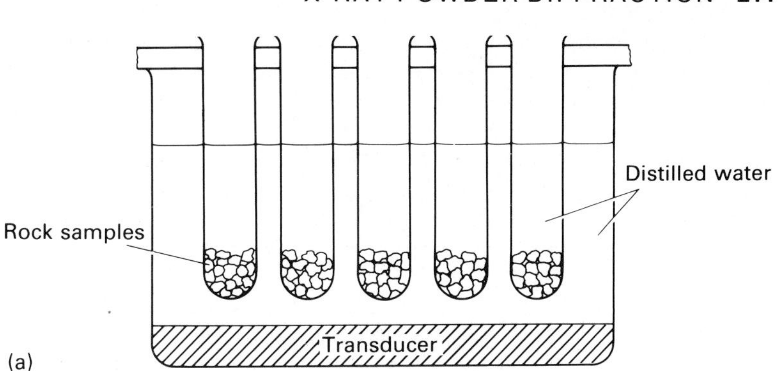

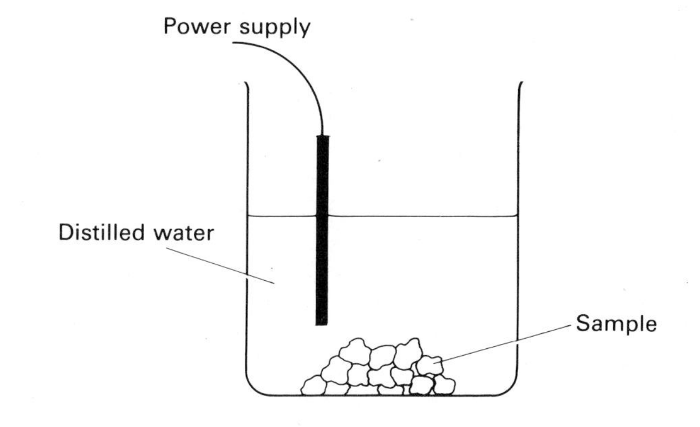

Fig. 7.16. Ultrasonic disaggregation. (a) Ultrasonic bath. (b) Ultrasonic probe.

After the sample is disaggregated, the resulting suspension should be transferred to a settling column (e.g. a measuring cylinder) and allowed to stand for an appropriate time until the required size fraction can be removed. Stokes' Law is used to calculate the settling velocity (v) of the clay particles:

$$v = \frac{2ga^2(d_1 - d_2)}{9\mu}$$

where g is the acceleration due to gravity, a is the sphere radius, d_1 is the density of the particles, d_2 that of the settling medium (usually water) and μ its viscosity. This law is empirical only, and applies to spherical particles in a non-turbulent medium at a constant temperature, where the suspension is sufficiently dilute to avoid particle–particle interaction. Because of the platey nature of clay particles, they must be regarded as having an equivalent spherical diameter (esd) when applying Stokes' Law. A centrifuge may be used as an alternative to a settling column, the effect being to increase the value of g in the equation. The usual size fraction separated for routine clay analysis is $<2\ \mu m$, and Table 7.4 gives settling times for this and other fractions using both a settling column and a centrifuge. The method as described by Galehouse (1971) is recommended here.

A major problem in the separation of clays is flocculation; this may be avoided in a number of ways: (i) by repeated washing in distilled water to remove any electrolyte present; (ii) by the addition of a deflocculant, commonly sodium hexametaphosphate solution (calgon); (iii) by the addition of ammonium hydroxide NH_4OH.

In principle, any chemical additive, either as a pre-treatment or a deflocculant, is to be avoided, however, since this may have an undesirable effect on the clay minerals themselves.

7.4.2 Preparation

After extraction by pipette, at the required depth from the measuring cylinder, the less than 2 μm fraction is prepared for XRD analysis as an oriented and also as an unoriented mount. There are various schools of thought on making oriented samples;

Table 7.4. (a) Pippette withdrawal times calculated from Stokes' Law for spherical particles (SG = 2.65) in a settling column of water at different temperatures (after Galehouse, 1971)

Diameter in micrometres finer than	Withdrawal depth (cm)	Temperature — Elapsed time for withdrawal of sample in hours (h), minutes (m), and seconds (s)					
		20°	21°	22°	23°	24°	25°
62.5	20	20 s	20 s	20 s	20 s	20 s	20 s
44.2	20	1 m 54 s	1 m 51 s	1 m 49 s	1 m 46 s	1 m 44 s	1 m 41 s
		Restir	Restir	Restir	Restir	Restir	Restir
31.2	10	1 m 54 s	1 m 51 s	1 m 49 s	1 m 46 s	1 m 44 s	1 m 41 s
22.1	10	3 m 48 s	3 m 42 s	3 m 37 s	3 m 32 s	3 m 27 s	3 m 22 s
15.6	10	7 m 36s	7 m 25 s	7 m 15 s	7 m 5 s	6 m 55 s	6 m 45 s
7.8	10	30 m 26 s	29 m 41 s	28 m 59 s	28 m 18 s	27 m 39 s	27 m 1 s
3.9	5	60 m 51 s	59 m 23 s	57 m 58 s	56 m 36 s	55 m 18 s	54 m 2 s
1.95	5	4 h 3 m	3 h 58 m	3 h 52 m	3 h 46 m	3 h 41 m	3 h 36 m
0.98	5	16 h 14 m	15 h 50 m	15 h 28 m	15 h 6 m	14 h 45 m	14 h 25 m
0.49	5	64h 54 m	63 h 20 m	61 h 50 m	60 h 23 m	48 h 59 m	57 h 38 m

(b) Time required at various centrifuge speeds and temperatures for sedimentation of particles using a standard 100 ml tube, 100 mm suspension depth, 90 mm of fall and 1000 mm^3 (1 cm^3) of sediment (from Jackson, Whitting & Pennington, 1950)

Limiting particle diameter (μm)	Density of particles g cm^{-3}	Centrifuge speed (rpm)	Centrifuge time in minutes at two temperatures	
			20°C	25°C
5	2.65	300	3.3	2.9
2	2.65	750	3.3	2.9
0.2	2.50	2400	35.4	31.4

three principal methods are: (i) smearing the clay slurry on to a solid substrate, e.g. a glass slide; (ii) pipetting the clay suspension into a glass beaker containing a glass slide, then placing the beaker in an oven at 60°C and allowing the distilled water to evaporate and the clays to settle on to the slide; (iii) sedimenting a suspension under a vacuum on to a porous substrate, e.g. a ceramic tile (Fig. 7.17).

The objectives in the various methods are to achieve an orientation of the clay plates parallel to the surface of the substrate, and to avoid differential settling of the clay mineral species (average diameters of kaolinite particles, for example, tend to be larger than those of smectite).

This problem of differential settling of the less than 2 μm fraction in the initial extraction stage can cause serious problems, especially where quantitative results are being used to interpret or substantiate geological theories, and the problems are discussed by Towe (1974).

Unoriented samples are more difficult to achieve. In most cases, cavity mounts are employed for bulk analysis, and unfortunately pressure is usually applied at some point in order to fill the cavity. Because of the predominantly platey nature of clay particles, they tend to reorient under pressure, and some way must be found to avoid this. The methods recommended are: (i) adding a material whose shape will prevent

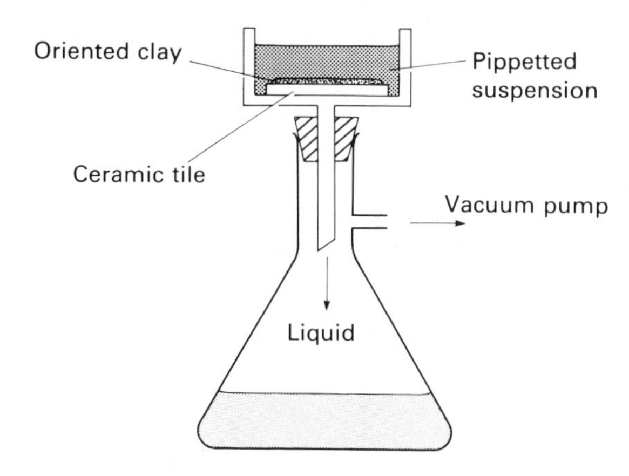

Fig. 7.17. Vacuum settling of pippetted suspension on to a ceramic tile.

orientation but will not complicate the diffraction pattern, e.g. non-crystalline; (ii) embedding the sample in a resin (e.g. Araldite) and regrinding to give approximately equal particles which will not orient.

7.4.3 Qualitative analysis

For a comprehensive description of clay mineral analysis by XRD the reader is referred to Carroll (1970), Thorez (1974) and Brindley (1980). A large proportion of sediments, however, can be treated in a fairly simple fashion by a combination of examining the basal spacings on the oriented samples, and the other lattice spacings on the unoriented samples. These are summarized comprehensively in flowchart form by Starkey, Blackmon & Hauff (1984), see Table 7.5, but the principal features are described here.

After the sample has been air-dried, a pattern is obtained from approximately 2°–34°2θ with copper radiation (Fig. 7.18). The four principal clay mineral groups give the basal spacings: kaolinite: 7 Å, illite: 10 Å, smectite: 12–15 Å, chlorite: 14 Å, and mixed-layer minerals at intermediate or higher values.

In order to distinguish smectite minerals from chlorite, where a possible overlap occurs, the sample is treated with an organic compound which systematically intercalates itself into the lattice. Ethanediol (ethylene glycol) is generally used for this purpose, although glycerol may also be employed. Ethylene glycol has the effect of expanding smectite to a basal spacing of about 17 Å. The method described by Brunton (1955) in which the oriented slides are introduced into a desiccator containing half a pint (0.25 litre) of ethylene glycol, instead of the usual drying agent, and then placing the desiccator in an oven at 60°C for about 4 hours is recommended here.

Further treatment involving heating the sample in a furnace also helps to distinguish the individual clays. Heating to 375°C collapses smectite (and illite-smectites) to 10 Å while leaving the other clays unaffected. Heating to 550°C destroys kaolinite and certain chlorites. The effects of these tests on the various clay minerals are summarized in Table 7.6.

The distinction between the basal spacing of kaolinite at (7 Å) and the second-order chlorite reflection also at 7 Å is not satisfactorily determined by heating to 550°C. The problem can be resolved by a convenient method suggested by Schultz (1964) who dissolved the chlorite by HCl treatment (6N at 60°C for 16 h) before obtaining another diffractogram. Any remaining 7 Å peak is therefore attributable to kaolinite although, as certain chlorites are known to offer variable resistance to acid attack (Kodama & Oinuma, 1963), care must be taken.

To determine if a clay is a dioctahedral or trioctahedral-type, the *d* spacing of the (060) reflection is recorded. If a dioctahedral clay is present the spacing falls between the values 1.48–1.50 Å, if the clay is trioctahedral the spacing is 1.53–1.55 Å. However, if several clays are present in a sample this becomes a very difficult exercise.

Mixed-layer clays have spacings generally intermediate between those of their components (Fig. 7.19) and an excellent treatment of this topic is presented by Reynolds & Hower (1970) and Reynolds (1980). Of particular importance is the position of the basal spacing after glycolation of the oriented mount which, in the case of illite-montomorillonite mixed layer clays, occurs between 10 and 17 Å, the exact position depending on the composition of the mixed layering (Fig. 7.20). It should be noted that regular interstratified mixed layer clays give a rational sequence of higher orders (see Table 7.3) and randomly interstratified clays give an irrational sequence (Fig. 7.19c, d).

For identification of more complex phases, Table 7.5 gives most of the necessary information. To obtain data on non-basal spacings, the unoriented samples are used.

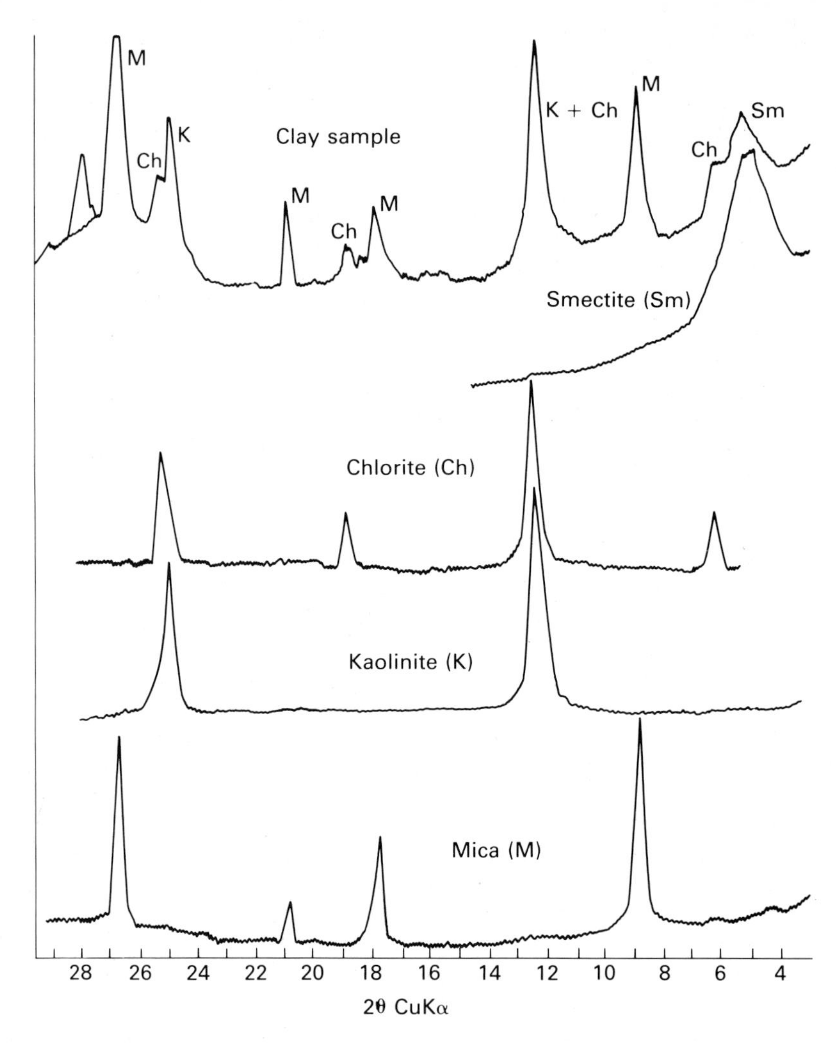

Fig. 7.18. Diffractometer traces of a typical clay sample and component clay minerals (after Gibbs, 1967).

7.4.4 Quantitative analysis

The quantitative analysis of clays is subject to similar, but more severe, problems as for whole-rock analysis. The X-ray response for a particular clay mineral is strongly dependent on, among other things, grain size, crystallinity, structure and chemical composition. The problems have been discussed at length in the literature: Johns, Grim & Bradley (1954), Schultz (1960), Bradley & Grim (1961), Hinckley (1963), Gibbs (1967) and Carroll (1970), etc., and calculations or assumption of 'diffracting ability' factors and crystallinity indexes to correct the measured intensities have been proposed with varying degrees of success.

For rapid, reproducible semi-quantitative results, expecially where large numbers of samples are involved, methods proposed by Schultz (1964), Biscaye (1965) and Weir, Ormerod & El-Mansey (1975) are particularly useful for comparative purposes. In these methods the intensitites of the individual clay components are measured on an oriented mount, after various treatments (air dried, glycol solvated, and heating to 375 and 550°C in order to isolate the clay components) and the sum normalized to 100%.

Table 7.6. X-ray identification of the principal clay minerals (<2 μm) in an oriented mount of a separated clay fraction from sedimentary material (from Carroll, 1970)

Mineral	Basal *d* spacings (00*l*)	Glycolation effect; 1 hr, 60°C	Heating effect, 1 hr
Kaolinite	7.15 Å (001); 3.75 Å (002)	No change	Becomes amorphous at 550–600°C
Kaolinite, disordered	7.15 Å (001) broad; 3.75 Å broad	No change	Becomes amorphous at lower temperatures than kaolinite
Halloysite, $4H_2O$	10 Å (001) broad	No change	Dehydrates to $2H_2O$ at 110°C
Halloysite, $2H_2O$	7.2 Å (001) broad	No change	Dehydrates at 125–150°C; becomes amorphous at 560–590°C
Mica, 2M	10 Å (002); 5 Å (004) generally referred to as (001) and (002)	No change	(001) becomes more intense on heating but structure is maintained to 700°C
Illite, 1Md	10 Å (002), broad, other basal spacings present but small	No change	(001) noticeably more intense on heating as water layers are removed; at higher temperatures like mica
Montmorillonite Group	15 Å (001) and integral series of basal spacings	(001) expands to 17 Å with rational sequence of higher orders	At 300°C (001) becomes 9 Å
Vermiculite	14 Å (001) and integral series of basal spacings	No change	Dehydrates in steps
Chlorite, Mg-form	14 Å (001) and integral series of basal spacings	No change	(001) increases in intensity; <800°C shows weight loss but no structural change
Chlorite, Fe-form	14 Å (001) less intense than in Mg-form; integral series of basal spacings	No change	(001) scarcely increases; structure collapses below 800°C
Mixed-layer minerals	*Regular*, one (001) and integral series of basal spacings	No change unless an expandable component is present	Various, see descriptions of individual minerals
	Random, (001) is addition of individual minerals and depends on amount of those present	Expands if montmorillonite is a constituent	Depends on minerals present in inter-layered mineral
Attapulgite (palygorskite)	High intensity *d* reflections at 10.5, 4.5, 3.28, 2.62 Å	No change	Dehydrates stepwise
Sepiolite	High intensity reflections at 12.6, 4.31, 2.61 Å	No change	
Amorphous clay, allophane	No *d* reflections	No change	Dehydrates and loses weight

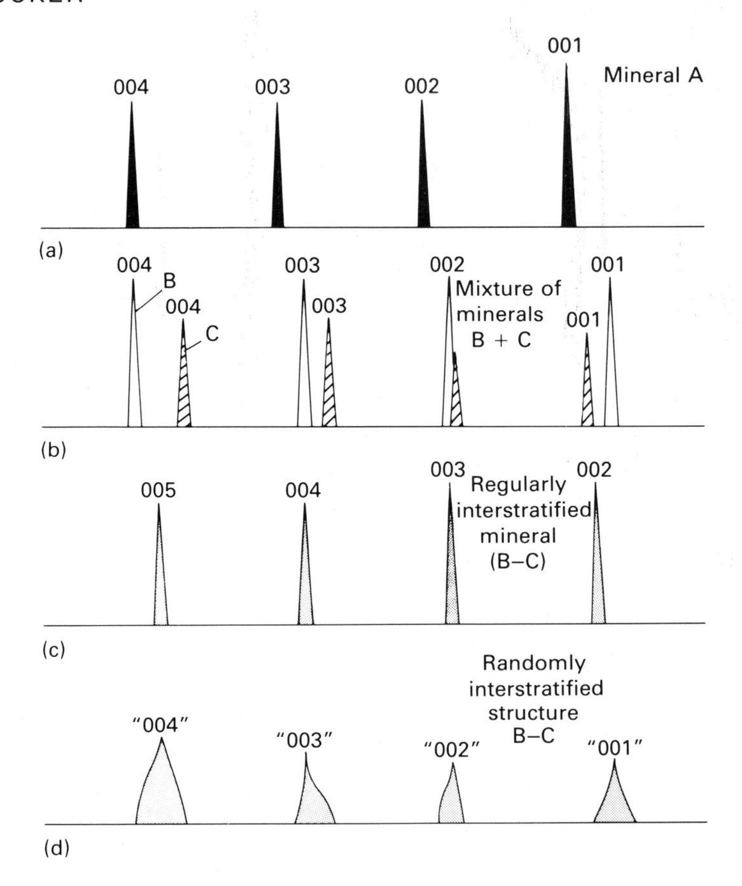

Fig. 7.19. Diagram showing the XRD peaks of: (a) single clay mineral; (b) mixture of clay minerals; (c) regular interstratified clay minerals (rational sequence); and (d) randomly interstratified clay minerals (irrational sequences) (from Thorez, 1976).

The method proposed by Weir *et al.* (1975) uses the equation:

$$\frac{I_{\mathrm{KAOLINITE}}}{2.5} + I_{\mathrm{ILLITE}} + I_{\mathrm{SMECTITE}} + \frac{I_{\mathrm{CHLORITE}}}{2.0} = 100\%.$$

The divisors have the function of correcting for the relatively greater X-ray responses of kaolinite and chlorite.

This method has the serious disadvantage common to all normalizing techniques in that an error in one component affects all the others, and also no account is taken of the presence of X-ray amorphous material.

If one wants a truly quantitative technique a more rigorous approach must be adopted. The introduction of an internal standard is certainly necessary and a material such as boehmite (γ-AIOOH) which has a similar mass absorption coefficient and X-ray response to clays, but a separate diffraction pattern, can be used. Just such a method is described by Griffin (1954) in which a known amount of boehmite (generally 10% by weight) is added to mixtures of standard clay minerals to produce calibration charts of percentage mineral against the relative intensity of the measured diffraction line to that of the 10% boehmite diffraction line.

Two difficulties are: (i) the mixing of platey particles satisfactorily, and (ii) variability of composition and degrees of crystallinity within clay groups.

The first problem can be alleviated by using a commercial mixing/grinding device (e.g. McCrone mill); the second problem can only really be overcome by preparing calibration curves from the actual clays under investigation. This technique was used by Gibbs (1967) in which he extracted, using a variety of techniques, the individual clay mineral

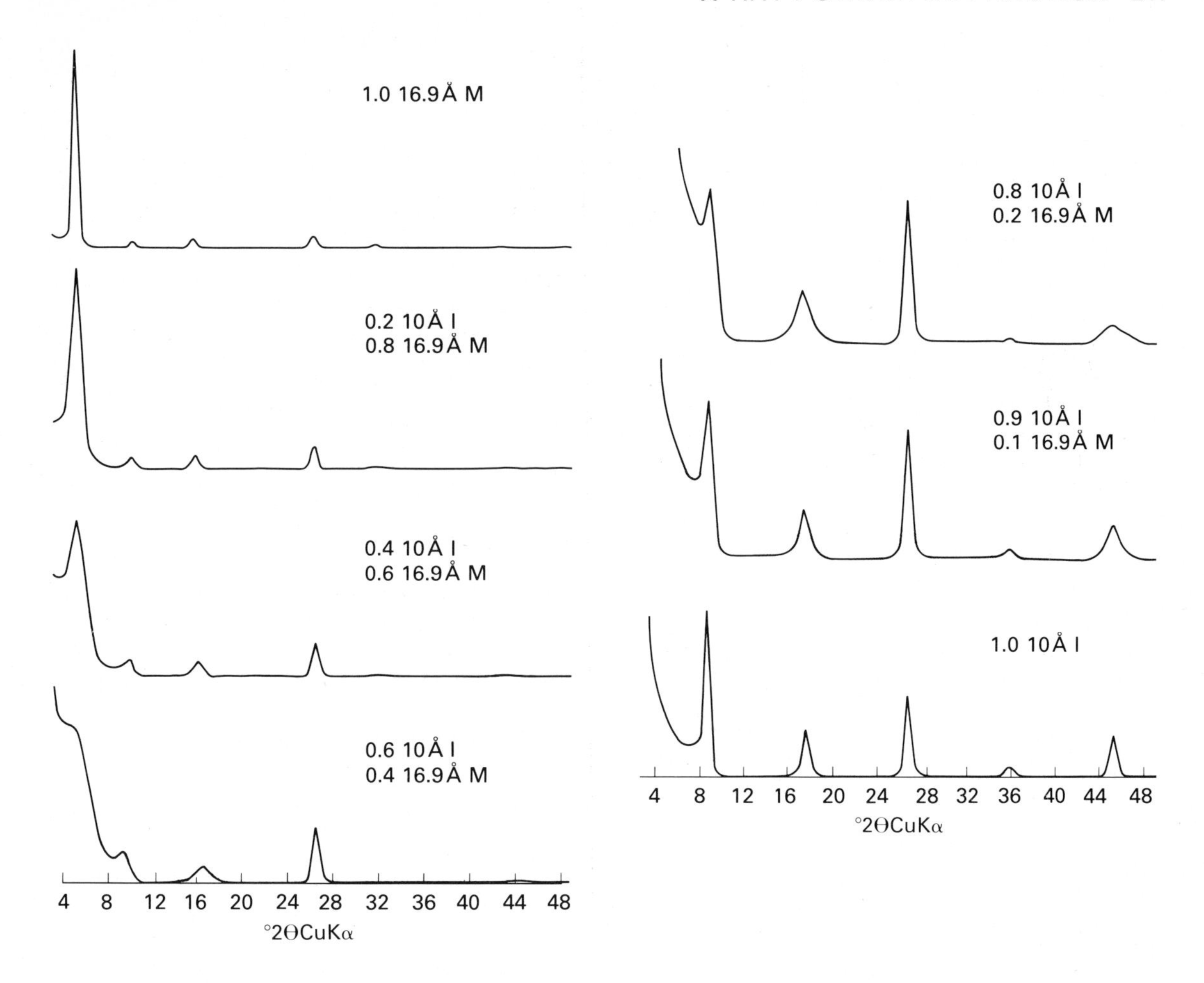

Fig. 7.20. Calculated diffraction profiles assuming random interstratification of 10 and 16.9 Å layers (illite and montmorillonite, glycolated). Fraction of montmorillonite layers are 1.0, 0.8, 0.6, 0.4, 0.2, 0.1 and 0 (from Reynolds & Hower, 1970).

components from selected samples, which were then used in the preparation of calibration curves. It is particularly useful in the analysis of large numbers of similar samples.

For more general purposes, 'typical' pure clays are used and these can be obtained from various commercial sources, an example being the Source Clays Repository. This organization was established by the Clay Mineral Society of the USA to provide workers with reference clay materials ('Source Clays'). For further information contact the curator of the sample collection, Professor William D. Johns, Source Clays, Department of Geology, University of Missouri, Columbia, Missouri 65211, USA. Another example is OECD, the Organization for Economic Co-operation and Development which also provides reference clays. Further details from Mlle S. Caillere, Laboratoire de Mineralogie du Museum National d'Histoire Naturelle, 61 rue de Buffon, Paris 5e, France. Data for the clay samples of the two organizations, together with useful information, are summarized in Van Olphen & Fripiat (1979).

7.4.5 Illite crystallinity

The X-ray diffraction response of illite is often used to give an indication of the diagenetic and low-grade metamorphic history of a sedimentary rock (e.g. Weaver, 1960; Kubler, 1968; Dunoyer de Segonzac, 1970; Frey, 1970; Gill, Khalaf & Massoud, 1977; Stalder, 1979). In essence, there is an increase in the degree of crystallinity and a change in the chemical composition of illite in the late diagenetic–early metamorphic realm. The degree of illite crystallinity, the sharpness ratio of Weaver (1960), is measured from the ratio of the height of the illite 001 peak at 10 Å to the height above the base line at 10.5 Å. In a study of clay minerals as an indicator of degree of metamorphism in Carboniferous sediments of the South Wales Coalfield, Gill *et al.* (1977) obtained a sharpness ratio of 2.0 in areas of no metamorphism, rising to 6.0 in a region of low grade metamorphism, where coals were up to anthracite rank.

In some studies (e.g. Dunoyer de Segonzac, 1970), it appears that there is also a change in the chemistry of the illite, with an increase in the Al/Fe+Mg ratio with increasing metamorphic grade. This has been quantified by calculating an intensity ratio (IR) of the illite 5 Å and 10 Å peaks, i.e. $I_{(002)}/I_{(001)}$.

7.4.6 Use of XRD data on mudrocks

For the study of mudrocks, X-ray diffraction analysis is the basic tool to determine the nature and proportions of the clay minerals present and from there to make deductions on, for example, the likely provenance of the sediment, the conditions of deposition and palaeoclimate, and the diagenesis and burial history. Clays in sandstone and limestones are also best identified by XRD. A brief note of these aspects of clay mineralogy is presented here, but a full review is not intended or warranted; there are many textbooks and papers reviewing these topics (e.g. Millot, 1970; Velde, 1977; Eberl, 1984) and the interested student should peruse the current issues of *Clay Minerals*, *Clays and Clay Minerals*, *Journal of Sedimentary Petrology*, *Sedimentology* and the Proceedings of the International Clay Conference, held every 3 years or so, and generally published as a book (e.g. Van Olphen & Veniale, 1982).

Clay minerals in a sediment or sedimentary rock have three origins: (1) inheritance, (2) neoformation and (3) transformation. In the first, the clays are detrital and have been formed in another area, perhaps at a much earlier time, and they are stable in their present location. In the second, the clays have formed *in situ*, and they have either been precipitated from solution or formed from amorphous silicate material. With transformation, inherited clays are modified by ion exchange or cation rearrangement. In the study of mudrocks, it is clearly important to be certain of the origin of the clays if meaningful interpretations are to be made. Inherited clays will give information on the provenance of the deposit and probably the climate there, whereas neoformed clays reflects the pore fluid chemistry, degree of leaching and temperature that existed within the sample at some stage. Transformed clays will carry a memory of inherited characteristics from the source area, together with information on the chemical environment to which the sample was later subjected.

There are three major locations where these three processes of clay mineral formation take place: (1) in the weathering and soil environment, (2) in the depositional environment and (3) during diagenesis and into low grade metamorphism. In the weathering environment, the types of clays produced, mainly by neoformation if fresh rock is being weathered, depend on the climate, drainage, rock type, vegetation and time involved. In very broad terms, illite is typical of soils where the degree of leaching is minimal, as in temperate and higher latitudes. Chlorite also forms under these conditions but, since it is more easily oxidized, it occurs preferentially in acid soils. Illite and chlorite are also in soils of arid regions where chemical processes are limited. Smectites are produced where the degree of leaching is intermediate or where soils are poorly drained. They also form in alkaline arid-zone soils. Mixed-layer clays mostly form through the leaching of pre-existing illite and mica. Kaolinite and halloysite are characteristic of acid tropical soils where leaching is intensive. Further leaching leads to vermiculite and then gibbsite through removal of silica. Some less common clay minerals are developed in particular soil environments, such as palygorskite and sepiolite in calcretes, Mg-rich soils and silcretes (e.g. Meyer & Pena dos Rois, 1985).

Clay minerals are generally little altered during transportation, by wind or water. In the marine environment, where most clays are eventually deposited, pelagic muds of the ocean basins have a clay mineralogy which is largely a reflection of the climate and weathering pattern of the source areas on adjacent land masses (e.g. Griffin, Windom & Goldberg, 1968). Thus kaolinite is most common in low latitudes, particularly off major rivers draining regions of tropical weathering, and illite and chlorite are more common in higher latitude marine muds. Smectites are a common alteration product of volcanic ash so that their distribution on the seafloor does reflect oceanic volcanicity as well as the composition of river-derived and wind-borne mud. On continental shelves, there are local variations in the clay mineralogy of muds, reflecting the proximity to rivers and deltas. Thus clays can be used to demonstrate sediment dispersal patterns in estuaries, bays and on shelves (e.g. Knebel *et al.*, 1977, studying the muds of San Francisco Bay, and Baker, 1973, following the path of suspended sediment from the River Columbia, off the NW United States coast).

There are differences in the grain sizes of the clay minerals which affect the distribution. Kaolinite is the coarsest (up to 5 μm), illite intermediater (0.1–0.3 μm) and smectite finer still, although the last commonly occurs as floccules several micrometres in diameter. As a result of these size differences a kaolinite-rich zone may occur inshore of smectite–illite mud. There is the potential here for using the regional distribution of clays within a sedimentary basin to infer the direction of sediment transport and distance from source area. Studies of the clays on the Atlantic Ocean floor off the Amazon River revealed decreases in kaolinite and 10 Å mica (illite + muscovite), and increases in montmorillonite with increasing distance from the river mouth (Gibbs, 1977). These trends are explained as the result of physical sorting of the clays by size.

Transformation of terrestrial clays and neoformation appears to be minor on the seafloor, although chamosite and glauconite do form in sediment-starved locations. Glauconite may form from alteration of degraded micaceous clays by absorption of K^+ and Fe^{2+} or from neoformation within pre-existing particles such as carbonate grains, clay minerals or faecal pellets (Odin & Matter, 1981). There is, however, much evidence for chemical alteration of clays; Na^+, Mg^{2+} and K^+ may all be adsorbed, commonly exchanged for Ca^{2+} (e.g. Sayles & Mangelsdorf, 1979).

Non-marine mudrocks of lacustrine, fluvial and glacial environments will also be largely of the inherited, detrital type, with little transformation and neoformation taking place. In some lakes, however, with quite extreme salinities and chemistries, clay minerals may be transformed or neoformed. Sepiolite, palygorskite, attapulgite and corrensite (mixed layer chlorite-montmorillonite) occur in Mg-rich, alkaline lake sediments for example. Where there is volcanic ash in a lake sediment or soil, then clay minerals are commonly formed by alteration of the ash, along with other minerals such as zeolites.

In the burial diagenetic environment, there is a progressive alteration of the clay minerals with rising temperature. One of the first transformations is smectite to mixed-layer smectite–illite and then this goes to illite. Chlorite also develops at depth, and into the zone of incipient metamorphism kaolinite is converted to illite and chlorite. These changes are discussed by Perry & Hower (1970), Hower *et al.* (1976), Iman & Shaw (1985), Jennings & Thompson (1986) and others and frequently they are related to vitrinite reflectance and timing of hydrocarbon generation. In addition, there is an increase in the crystallinity of illite, as explained in Section 7.4.5. Clearly, in interpreting the clay mineralogy of the mudrock the burial history of the formation has to be taken into account. Burial diagenetic changes in clays may largely account for the uneven distribution of clays through time, with smectites and to a lesser extent kaolinite being less abundant in older, especially Palaeozoic and Precambrian mudrocks, which are composed largely of illite and chlorite (Dunoyer de Segonzac, 1970).

From this brief discussion, it can be seen that the clay mineralogy of a sediment or sedimentary rock may reveal information on the palaeoclimate in the source area and on the nature of the source area itself. Samples taken vertically through a sequence of mudrocks may show variations in the clay mineral assemblage which reflect changing climate or changing source area. Jacobs (1974), for example, discussed variations in the clay minerals of Cainozoic seafloor muds of the Southern Ocean in terms of the impending Pleistocene glaciation of the Antarctic continent and its effect on weathering and erosion rates. In a detailed study of just 2 m of rock, Spears & Sezgin (1985) recorded a decrease in kaolinite and increases in illite, chlorite, vermiculite and illite to illite/smectite ratio across a Coal Measures marine band. They attributed this to a decrease in the contribution of mature clay from within the basin and to an increase in the proportion of less weathered clay from outside the basin, through time and during the marine transgression. With the Triassic

Keuper Marl (Mercia Mudstones), Jeans (1978) was able to distinguish a detrital (inherited) illite and chlorite assemblage occurring throughout the sequence, from an assemblage of neoformed Mg-rich clays (sepiolite, palygorskite, chlorite, smectite and corrensite) which occurs at particular horizons. The neoformed clays are the result of changes in water chemistry within the Keuper Basin, probably arising from the influx of marine waters. Retallack (1986) demonstrated that there was a change in the clay mineralogy of Cretaceous to Tertiary paleosols of the NW United States, with decreasing kaolinite and increasing illite and chlorite up through the sequence resulting from a change in climate from more humid to more arid.

Clays can be neoformed within sandstones during diagenesis, and there are many case studies documenting this (see, e.g. Hancock & Taylor, 1978, discussing the Jurassic Brent Sand of the North Sea, which shows a change from kaolinite to illite with increasing depth, and an improvement in the illite crystallinity — Fig. 7.21). Authigenic clays in sandstones may show much better-defined diffraction peaks than those in adjacent shales (e.g. Wilson & Pittman, 1977).

7.5 XRD of carbonates

X-ray diffraction is commonly used in the study of modern carbonate sediments, limestones and dolomites. XRD data can give information on the chemical composition of carbonate minerals, namely the Mg content of calcite and the Ca or Mg excess of dolomite. They can also be used to determine the ordering of dolomite crystals. The percentages of the various $CaCO_3$ minerals in a mixture can be calculated from X-ray peaks, but estimates of the dolomite content of dolomitic limestones are less precise. The main peaks of the common carbonate minerals, calcite, aragonite and dolomite are given in Table 7.3.

7.5.1 Magnesium in calcite

The magnesium ion can substitute for calcium in the calcite lattice, and, since the Mg^{2+} ion is smaller

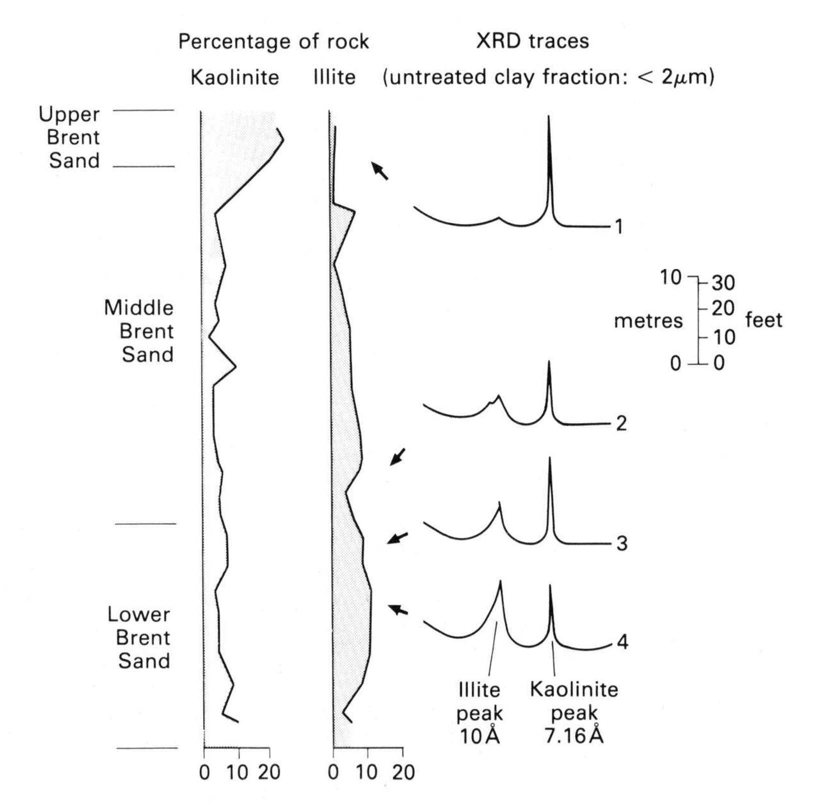

Fig. 7.21. Clay mineralogy of the Middle Jurassic Brent Sand from the North Sea. Petrographic data show that kaolinite is replaced downwards by increasingly abundant illite. Representative X-ray diffraction traces (right) show changing emphasis from kaolinite to illite downhole, together with downward improvement in illite crystallinity (sharper peak). Note some detrital mixed-layer illite/montmorillonite in sample 2 (second peak on flank of 10 Å peak) (from Hancock & Taylor, 1978).

than the Ca^{2+} ion, this results in a decrease in the d_{104} lattice spacing (Goldsmith & Graf, 1958a). Several authors have produced a graph showing the mole % $MgCO_3$ against d_{104} in ångstrom units and 2θ for Cu $K\alpha$ radiation and Fig. 7.22 from Goldsmith, Graf & Heard (1961) is the one most commonly used. The precise position of the d_{104} peak is obtained by using an internal standard such as halite, quartz or fluorite which has a major peak close to the main calcite peak (Table 7.3).

On the basis of magnesium content, calcite occurs in two forms: low Mg calcite (LMC) with 0 to 4 mole % $MgCO_3$ and high Mg calcite (HMC) with more than 4 mole %, but with a range of 11–19 mole % being most common. In modern tropical carbonate sediments, HMC is mainly of biogenic origin, coming from calcareous red algae, echinoderms, bryozoans and some benthic foraminifera. HMC micritic and bladed cements are common in reefs. LMC is also mostly of biogenic origin, coming from planktonic foraminifera, coccoliths and some molluscs. In addition to the 'vital' effect, the Mg^{2+} content of biogenic calcite is determined by water temperature: lower Mg^{2+} occurs in skeletons precipitated in cooler waters. With marine calcite cements, too, those precipitated in colder, commonly deeper waters, have lower Mg^{2+}. Although HMC grains and cements generally lose their Mg^{2+} during diagenesis, so that most ancient calcite is low in magnesium, a memory of the original high Mg is commonly retained. In limestones, original LMC will typically have 0–1 mole % $MgCO_3$, whereas original HMC may have a little more (2–3 mole %).

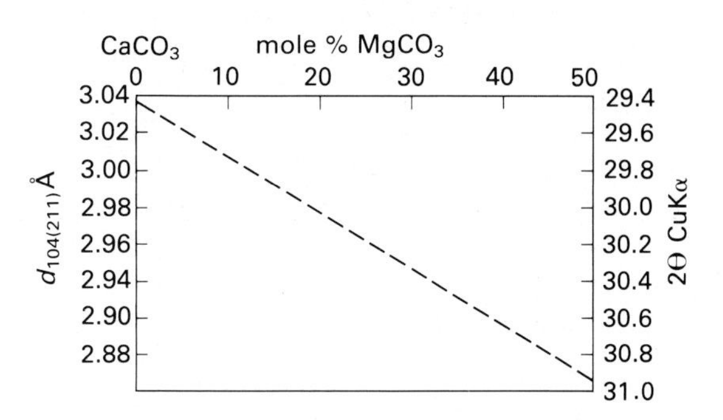

Fig. 7.22. Displacement of the d_{104} peak of calcite with increasing $MgCO_3$ to dolomite (based on Goldsmith *et al.*, 1961).

7.5.2 Mixtures of $CaCO_3$ minerals

Many modern carbonate sediments consists of a mixture of three $CaCO_3$ minerals, aragonite, low Mg calcite and high Mg calcite. Aragonite is a major constituent, coming from the shells and skeletons of green algae, corals and molluscs, and forming ooids and many marine cements. X-ray diffraction is the quickest method for determining the mineralogical composition of modern carbonates, although there has been discussion over whether peak height or peak area analysis should be used.

Careful grinding of the sample is important since this process determines the particle size of the powder and the structural damage to the minerals, which respond differently to the grinding. These differences can affect the XRD peak intensities (Gavish & Friedman, 1973). An original very fine crystal size or overgrinding producing too fine a powder results in a decrease of peak intensity. This effect is shown more by calcite than aragonite (e.g. Fig. 7.23 and Milliman, 1974), but even then it varies with the particular calcite skeleton being analysed. Excessive and hard grinding is to be avoided as this may lead to mineralogical changes in the sample through the heat generated and pressure applied. In general, marine biogenic calcites have similar peak intensities, largely because the crystallite size is similar. The exception is echinoid material, which shows a greater peak intensity because of its larger crystal size. Reagent grade calcite also shows a greater peak intensity, and so should be avoided as a standard. A particle size of less than 63 μm is generally acceptable, obtained by grinding the sample for several minutes until all the powder will pass through a 200-mesh sieve. If aragonite is present this can be used as an internal standard to determine precisely the displacement of the high Mg calcite peak. If both LMC and HMC are present, then the two peaks will overlap; one will be a shoulder to the other. To determine the amounts of each mineral present, the ratio of the aragonite peak intensity to all peak intensities is calculated: IA/IA+ILMC+IHMC (Fig. 7.24). Peak intensity can be measured by either peak height or peak area and in both cases a base-line is taken just above the background level. Gavish & Friedman (1973) considered that peak height analysis was more reliable for separating LMC and HMC. They showed that the particle size and amount of structural damage through excessive grinding affected the peak areas more than the peak heights. Milliman & Bornhold (1973), on the other

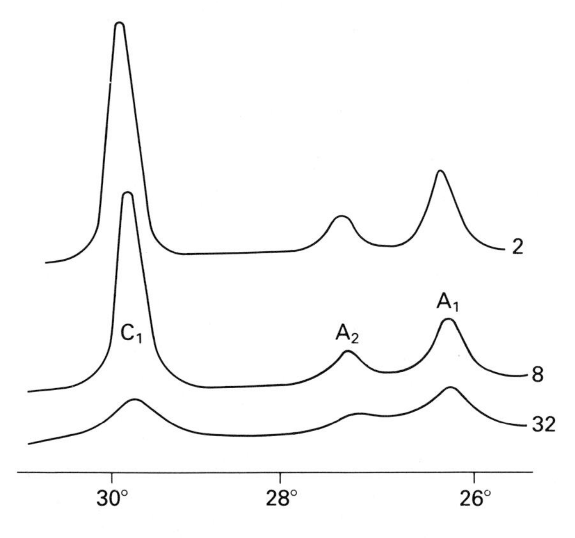

Fig. 7.23. X-ray diffractograms (A_1 and A_2 aragonite; C_1 calcite) showing the effect of extended grinding on the peak intensity of a 50–50 mixture of mollusc aragonite and oyster calcite. Grinding times (circled numbers) are in cumulative minutes (from Milliman, 1974).

hand, were able to demonstrate that the peak heights and asymmetries of HMC grains varied between different algal genera but that total peak intensity was relatively constant and that, for aragonite and calcite mixtures, peak area was more reliable. In peak area analysis, the areas can be determined by (a) tracing off the peaks, cutting out their shapes and weighing these, (b) using a planimeter or (c) accurate measuring of the height and width of each peak. The accuracy of this method is probably around 5%.

From the ratio for aragonite peak intensity, a calibration curve is needed to convert this to a percentage, since the XRD response of similar amounts of calcite and aragonite is not the same. Aragonite gives lower intensity peaks for a given percentage compared to calcite. Milliman (1974) presented a standard curve for aragonite determination from peak area analysis (Fig. 7.25). From this, the percentage of aragonite which corresponds to the ratio can be read off, and the amount of calcite present can be found by subtraction. If both calcite minerals are in the sample, then a small calculation is needed to find the percentage of each (Fig. 7.24).

A method for determining weight percentages of carbonate minerals in a modern sediment by spiking was devised by Gunatilaka & Till (1971). They prepared a spike mixture by hand picking grains from the sediment, identifying their mineralogy by XRD, and then mixing them together in a determined proportion. The spike mixture was added to the unknown in 1:1 ratio and from a comparison of the diffraction traces of the spike and spike + unknown, calculation of peak areas and use of an equation, the percentages of the various minerals present can be deduced quite accurately. The advantage of this procedure is that the standard (the spike) is made from components which are present in the sediment, so that the diffraction behaviour will be similar.

7.5.3 XRD and dolomites

The mineral dolomite, $CaMg(CO_3)_2$, is commonly not stoichiometric, but has an excess of Ca^{2+}, up to Ca:Mg 58:42, or less commonly an excess of Mg, up to Ca_{48} Mg_{52}. The effect of Ca^{2+} substitution for Mg^{2+} is to increase the cation lattice spacing (Fig. 7.22) and XRD is often used to determine this and give the Ca/Mg ratio, by measurement of the position of the d_{104} peak relative to a standard. Apart from reference to Fig. 7.22, the Ca excess can be calculated from the equation of Lumsden (1979) relating mole % $CaCO_3$ (N_{CaCO_3}) to the d_{104} spacing measured in ångstrom units (d): $N_{CaCO_3} = Md + B$, where M is 333.33 and B is −911.99. The d_{104} spacing for 50.0% $CaCO_3$ is taken as 2.886 Å and for 55.0% $CaCO_3$ as 2.901 Å, based on Goldsmith & Graf (1958a).

Iron can substitute for the cations in dolomite to give ferroan dolomite (>2 mole % $FeCO_3$) and ankerite with much higher values reaching 25 mole % $FeCO_3$. In view of the slightly larger size of the Fe^{2+} ion relative to Mg^{2+}, with more than a few mole % $FeCO_3$ there is a noticeable increase in the lattice spacing of d_{104} (Goldsmith & Graf, 1958b; Runnells, 1970; Al-Hashimi & Hemingway, 1974). In addition, the intensities of the XRD reflections are commonly weaker in ferroan dolomites. Al-Hashimi & Hemingway (1974) presented a calibration curve for ferroan dolomites (Fig. 7.26), by first analysing the iron content with atomic absorption (Chapter 6). Clearly, care has to be exercised in interpreting dolomite XRD data, and note taken of the iron content.

XRD of dolomites also gives information on the

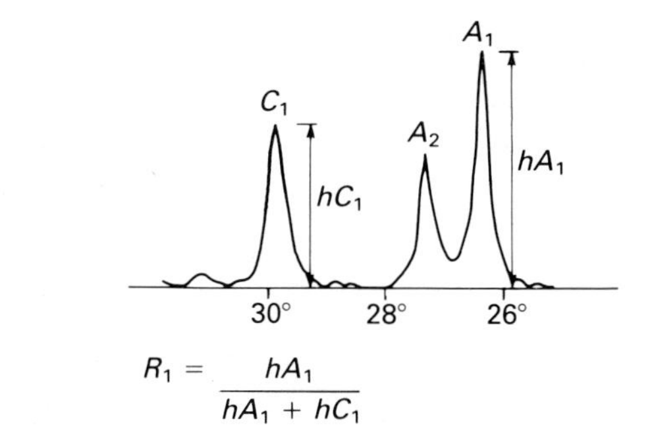

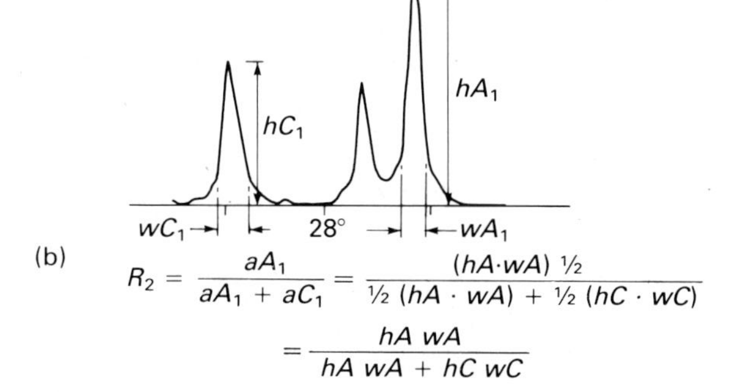

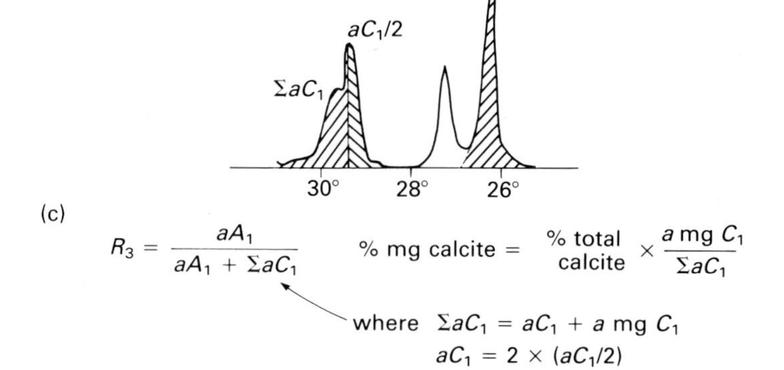

Fig. 7.24. Calculation of carbonate mineralogy by X-ray diffraction. In peak height analysis (a), the simple ratio $h_A/(h_A + h_C)$ is calculated. A geometric calculation can be used (b) with simple calcite–aragonite mixtures, in which both peaks are assumed to be triangles, and the area of each is calculated. Where the calcite curve is composed of two or more types of calcite, a more complex analysis is required (c). The intensity of the free half of the major peak (the right side of the low Mg calcite curve) is calculated and multiplied by 2. This intensity is then compared to the total intensity in order to differentiate between it and the other calcites present. The integrated peak intensity is calculated either by planimeter analysis or by weighting (from Milliman, 1974).

ordering of the crystals. As a result of the segregation of the cations into separate sheets in the dolomite crystals, a set of superstructure reflections corresponding to d_{021}, d_{015} and d_{101} is revealed with XRD (Fig. 7.27), which is not present in the structurally similar calcite. The sharpness and relative intensities of these ordering peaks can be used to give a measure of the degree of ordering of the dolomite crystal. The greater the ratio of the heights of the ordering peak 015 to diffraction peak 110, the higher the degree of order.

Dolomites which are non-stoichiometric are generally less well ordered than 'ideal' dolomite, through the occurrence of some Ca ions in the Mg sheet (or vice versa). It is theoretically possible for a 50:50 Ca:Mg carbonate to have no ordering reflec-

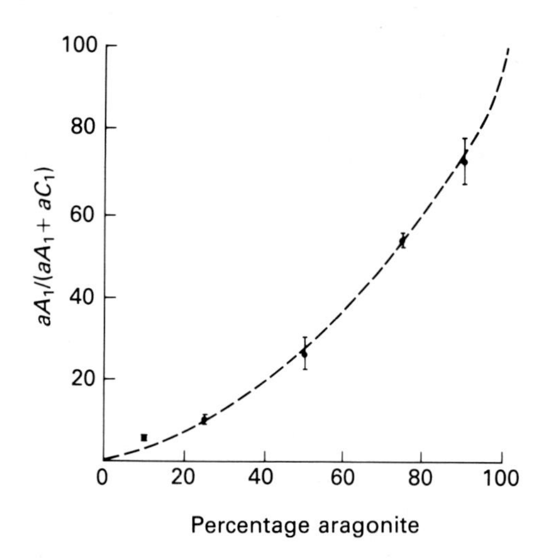

Fig. 7.25. Standard curve for aragonite determination using peak area analysis. Each data point represents eight to ten analyses of various calcite and aragonite standards. In most instances the standard deviation (represented by vertical bars) is considerably less than 5% (from Milliman, 1974).

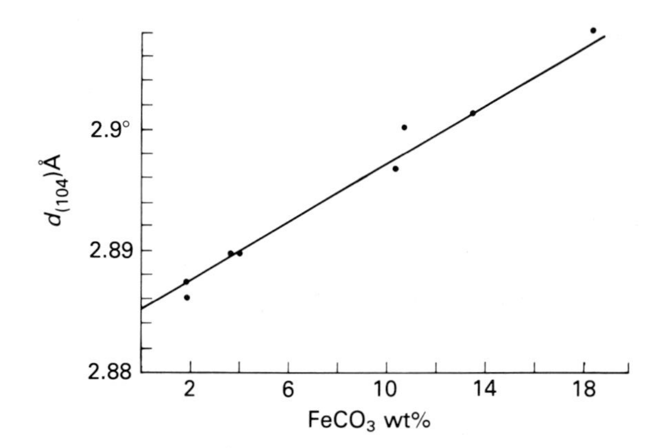

Fig. 7.26. Calibration curve relating d_{104} spacing of dolomite to the $FeCO_3$ content (from Al-Hashimi & Hemingway, 1974).

tions if the cation sheets in the lattice are equal mixtures of Ca and Mg. In practice, all naturally occurring dolomites are ordered to an extent (otherwise, strictly, the mineral is not dolomite) with most modern dolomites showing poor ordering reflections, compared with many ancient dolomites. The term protodolomite was introduced by Goldsmith & Graf (1958a) for dolomite manufactured in the lab with no or only poor ordering reflections. However, it has occasionally been used for modern naturally-occurring dolomites with weak ordering peaks, although the concensus now (Land, 1980) is that the word should be restricted to synthetic dolomites, and if a naturally occurring Ca–Mg carbonate has the ordering reflections, no matter how weak, then it is a dolomite. Dolomite with an excess of Ca can simply be referred to as calcian dolomite.

XRD data from dolomites can be useful in providing a more detailed knowledge of the crystal structure and chemistry, and they can be used to distinguish between different types of dolomite within one carbonate formation. For example, Fig. 7.28 presents data from the Lower Carboniferous of South Wales where peritidal dolomicrites have different degrees of order and Ca/Mg ratios from mixing-zone dolomites and from burial 'vein' baroque dolomites. Lumsden & Chimahusky (1980) and

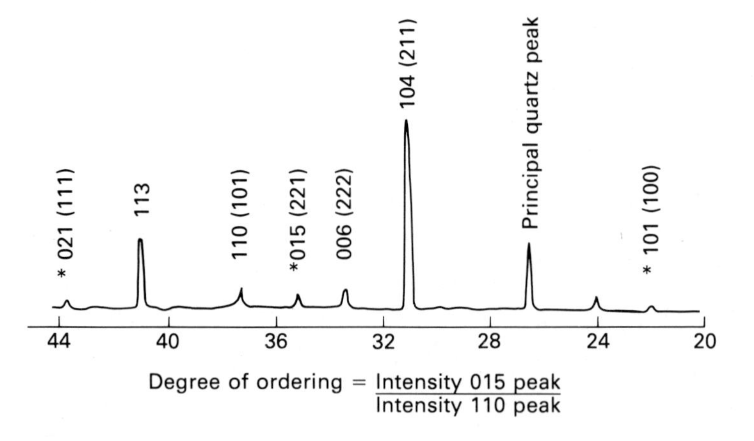

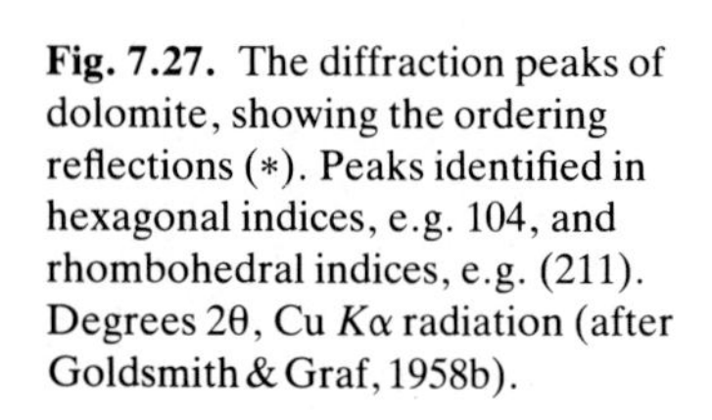

Fig. 7.27. The diffraction peaks of dolomite, showing the ordering reflections (*). Peaks identified in hexagonal indices, e.g. 104, and rhombohedral indices, e.g. (211). Degrees 2θ, Cu $K\alpha$ radiation (after Goldsmith & Graf, 1958b).

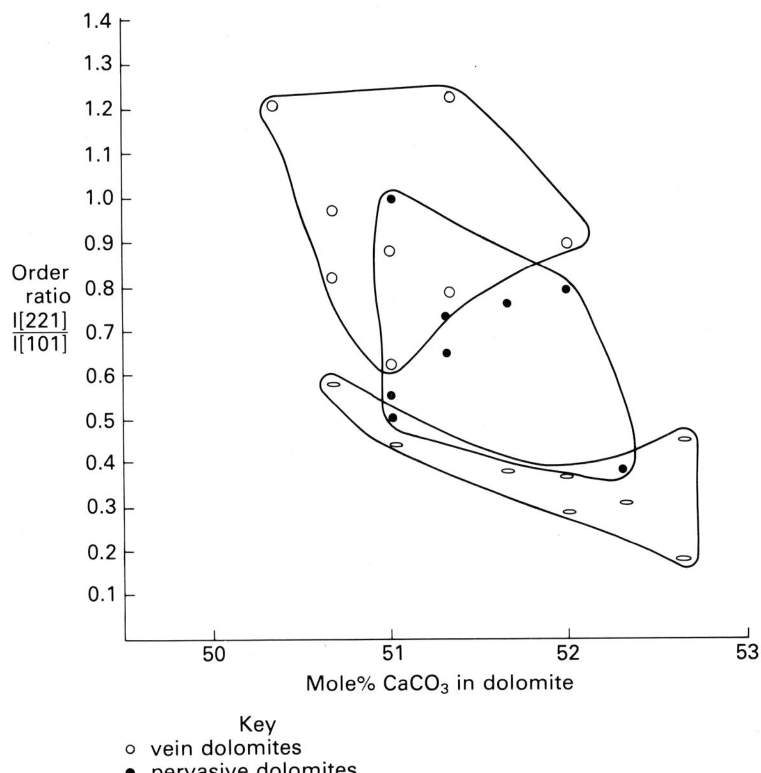

Fig. 7.28. XRD data from some Lower Carboniferous dolomites from South Wales showing stoichiometry (as mole % $CaCO_3$) plotted against ordering. The three types of dolomite, peritidal dolomicrites, pervasive mixing-zone dolomites and burial dolomites developed along joints, plot in different areas, with some overlap (from Hird, 1986).

Morrow (1978, 1982) identified three broad groups of dolomite, based on stoichiometry, texture and whether associated with evaporites or not (Fig. 7.29): I — coarsely-crystalline, sucrosic dolomites whch are generally nearly stoichiometric (mode 50.0–51.0% $CaCO_3$), II — fine-grained dolomites associated with evaporites which are also nearly stoichiometric (mode 51.0–52.0% $CaCO_3$) and III — finely crystalline dolomites not associated with evaporites which are generally Ca-rich (54–56% $CaCO_3$). Groups II and III are usually early diagenetic, near-surface in origin. The underlying cause of these associations is thought to be the salinity and Mg/Ca ratio of dolomitizing solutions, with a climatic control important for groups II and III. Where there is an evaporite association (group II), indicating an arid climate, then pore fluids are likely to have had a high Mg/Ca ratio from precipitation of gypsum–anhydrite and aragonite. It is contended that the abundance of Mg^{2+} ions in the fluids would result in near stoichiometric dolomite. The calcian dolomites of group III are thought to have formed from solutions with lower Mg/Ca ratios, such as occur in mixing-zones, which are more active during humid climatic times. Group I dolomites are generally of late diagenetic burial origin and the near stoichiometry could reflect slow growth from dilute solutions, possibly aided by elevated temperatures.

Using new and published XRD data, Sperber, Wilkinson & Peacor (1984) obtained two pronounced modes at 51 and 55 mole % $CaCO_3$ in Phanerozoic dolomites which ranged from 48 to 57 mole % $CaCO_3$. They also found a bimodal distribution in the percentage of dolomite in carbonate rocks: a mode at 97% dolomite (dolostones) and at 20% (dolomitic limestones), indicating that carbonate rocks are either partially or completely dolomitized. Sperber *et al.* (1984) suggested that the dolomitic limestones, which generally contain rhombs of calcian dolomite, originated in diagenetically closed systems during high Mg calcite dissolution – low Mg calcite and dolomite precipitation, so that in these

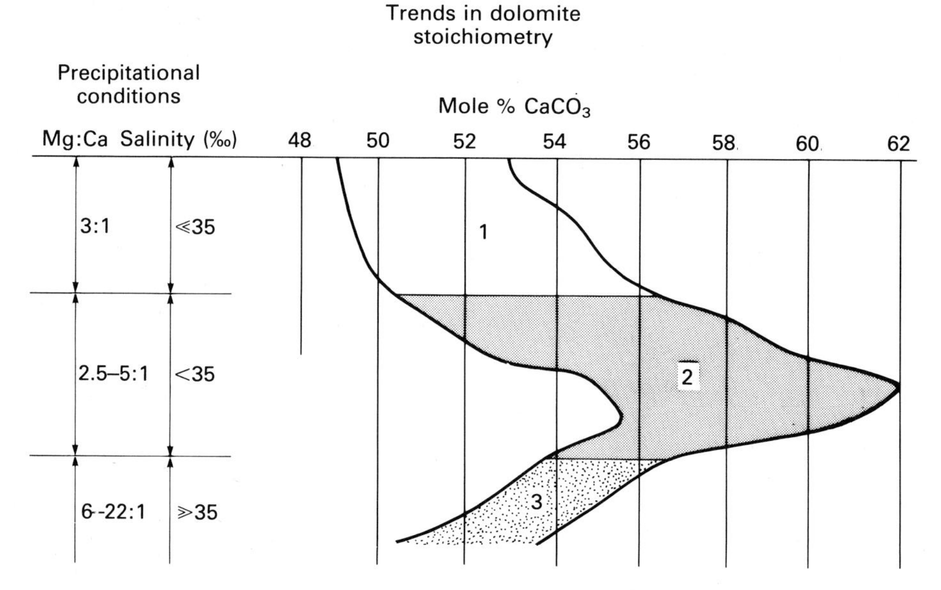

Fig. 7.29. Diagram of dolomite stoichiometry data and inferred precipitational conditions. Group 1 is composed of ancient sucrosic and sparry dolomites. Group 2 is composed of finely crystalline modern and ancient dolomites not associated with evaporites and Group 3 are finely crystalline modern and ancient dolomites associated with evaporites (from Morrow, 1982).

rocks there was an internal supply of Mg^{2+}. The dolostones, on the other hand, consist of more stoichiometric dolomite, and are thus considered to have originated in more open diagenetic systems. A trend towards more stoichiometric dolomite in older dolostones is evident from the data of Lumsden & Chimahusky (1980) and Sperber *et al.* (1984) and there is also a broad correlation of increasing stoichiometry with increasing crystal size. Both these features are consistent with dolomites undergoing solution-reprecipitation through diagenesis and the formation of more stoichiometric, better ordered dolomite from a less stoichiometric, poorly ordered original precipitate.

7.5.4 Calcite-dolomite mixtures

XRD charts have long been used to calculate the proportions of calcite and dolomite in a sample. As with $CaCO_3$ minerals, the ratio of the peak area of dolomite to calcite plus dolomite is converted to a percentage using a calibration curve. Several of these have been published (e.g. Tennant & Berger, 1957; Weber & Smith, 1961; Bromberger & Hayes, 1965; Royse, Waddell & Petersen, 1971; Lumsden, 1979; Fig. 7.30) and they have generally been constructed from a set of standard mixtures. However, there are several potential errors in the method which need to be considered and have been identified by comparing XRD data with dolomite–calcite ratios measured from stained thin sections (Lumsden, 1979; Gensmer & Weiss, 1980). One major problem is the stoichiometry of the dolomite. The position of the d_{104} peak depends on the Ca excess and at higher $CaCO_3$ contents (>55%); this peak overestimates the dolomite content by some 10–12% (Lumsden, 1979). For this reason, it is best to use the d_{113} peak of dolomite which is not affected by non-stoichiometry. The d_{104} peak position is also affected by iron content (Fig. 7.26) and Al-Hashimi & Hemingway (1974) determined calibration curves for calcite–dolomite, calcite–ferroan dolomite and calcite–ankerite mixtures (Fig. 7.31). A difference in particle size of powdered samples relative to the crystallite size of the standard mixtures can introduce another error, since particle size influences peak area ratios. In preparing a calibration curve one should clearly try and have crystallites of similar size to those in the unknown samples. In addition, calcite and dolomite crystals may break up at a different rate during

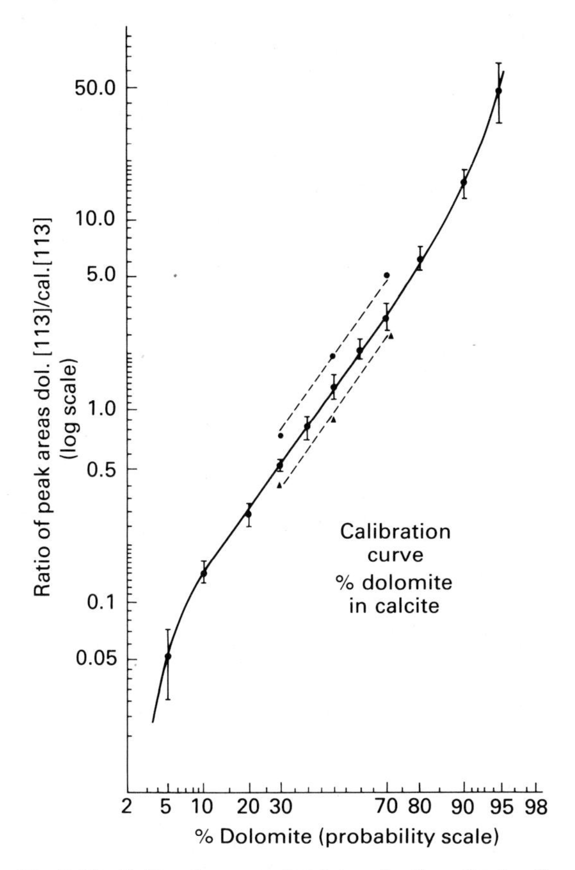

Fig. 7.30. Calibration curve for determination of dolomite in calcite, using dolomite/calcite ratio of area of [113] peaks. Main line connects points obtained with standards prepared using calcite ground for 30 minutes. Vertical bars give range of one standard deviation. Dashed line connecting squares is for standards prepared with calcite ground for 15 minutes, i.e. relatively coarse crystalites. Dashed line connecting triangles is for standards prepared with chemical calcite, i.e. relatively fine crystallites (from Lumsden, 1979).

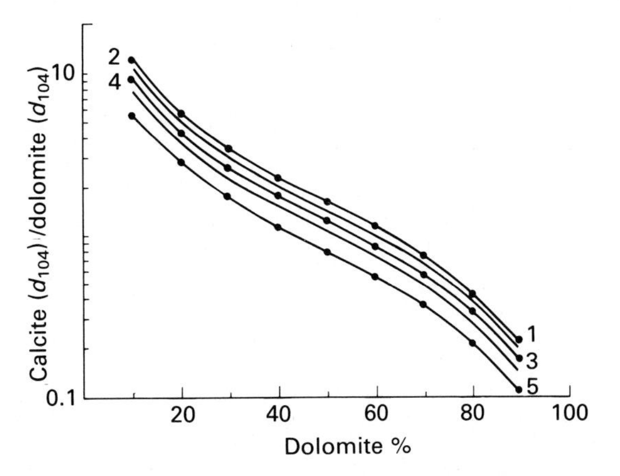

Fig. 7.31. Calibration curves: curve 1 for calcite-ankerite ($FeCO_3$ = 22%); curve 2 for calcite + highly ferroan dolomite ($FeCO_3$ = 15%); curve 3 for calcite + medium ferroan dolomite ($FeCO_3$ = 10%); curve 4 for calcite + low ferroan dolomite ($FeCO_3$ = 5%); curve 5 for calcite + dolomite ($FeCO_3$ = 0) (from Al-Hashimi & Hemingway, 1974).

grinding or be of different grain size. Gensmer & Weiss (1980) found that grinding samples for between 10 and 20 minutes to give a particle size of around 5 μm overcame this problem. A further error arises if there is much quartz in the sample. The quartz d_{102} peak (d = 2.285 Å) interferes with the calcite d_{113} peak (d = 2.285 Å) and if there is more than around 20% quartz then the intensity of the calcite peak is increased, so that there appears to be more present than there actually is.

Recently, several methods have been described for quantitative X-ray analysis of mixtures of minerals. Fang & Zevin (1985) presented a standardless method whereby sample absorption is measured, as well as peak intensity. Martinez & Plana (1987) have devised a method for a mixture of carbonate minerals using the technique of fitting Lorentzian profiles.

7.6 XRD OF SILICA MINERALS

The metastable silica minerals which occur in young cherts, that is opaline silica and opal-CT, are best identified by X-ray diffraction. Opaline silica, sometimes referred to as Opal-A (Jones & Segnit, 1971) is a highly disordered, nearly amorphous hydrous silica, with a very fine crystallite size of 11–15 Å. The XRD pattern of opal is a diffuse band between 6 and 2.8 Å, with a maximum at about 4.0 Å (see Fig. 7.32). The tests of radiolarians, diatoms and sponge spicules, which are the main constituents of deep sea siliceous oozes, are composed of opaline silica. Precious opal is also opal-A. Disordered cristobalite, sometimes called lussatite and termed opal-CT by Jones & Segnit (1971), is more obviously crystalline, although still hydrous, and consists of irregularly stacked layers of low-(α-) cristobalite and

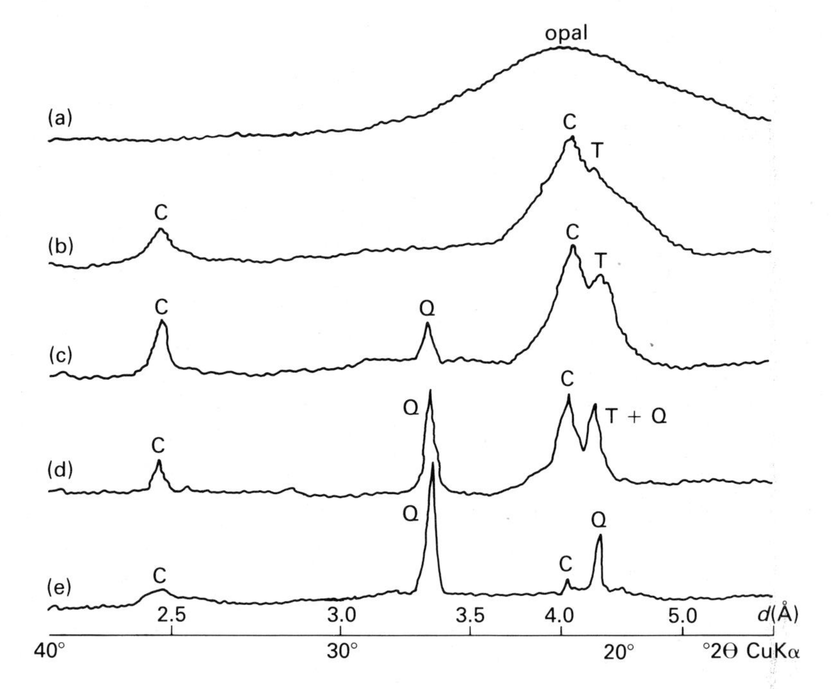

Fig. 7.32. Typical X-ray diffraction patterns of silica minerals arranged in order of increasing maturity: (a) opaline silica, such as precious opal, (b) immature chert or porcelanite of Middle Eocene age from the Equatorial Pacific, (c) Monterey chert from California, (d) granular chert and (e) vitreous chert from the Lower Tertiary of Cyprus. C = cristobalite, T = tridymite, Q = quartz.

low-tridymite. The XRD pattern consists of broadened, but well-defined peaks of low cristobalite, especially d_{101} at 4.1 Å and d_{200} at 2.5 Å, and tridymite peaks, notably d_{100} at 4.25–4.3 Å. Disordered cristobalite is an intermediate mineral in the diagenesis of opaline silica to quartz chert. It occurs in porcelanites, most common opals, bentonitic clays and silica glass. In fact, with increasing depth of burial, there is a decrease in the d_{101} spacing of opal-CT, from a maximum of 4.11 Å to a minimum of 4.04 Å. The cristobalite also becomes more ordered (sharper peak) relative to tridymite (Mizutani, 1977).

Chalcedony is a fibrous variety of quartz, disordered, with crystals much larger than a micrometre. It has the same peaks as the other varieties of quartz, cryptocrystalline (<1 μm), microquartz (1–10 μm) and megaquartz (>10 μm), i.e. a major peak at 3.343 Å and a secondary peak at 4.26 Å, which coincides with the main tridymite peak. These varieties of quartz comprise the older, diagenetically 'mature' cherts, such as occur in Mesozoic and older successions. In fact, there is also a progressive increase in the crystallinity of the quartz, until it has peaks as strong and sharp as quartz of metamorphic and igneous origin. Papers noting the X-ray diffraction patterns of siliceous sediments and rocks can be found in Hsü & Jenkyns (1974), notably von Rad & Rosch (1974) and in Garrison *et al.* (1981). Williams *et al.* (1985) have recently reviewed the changes in silica minerals during diagenesis.

8 Use of the Scanning Electron Microscope in sedimentology

NIGEL TREWIN

8.1 INTRODUCTION

This chapter is intended to introduce the reader to the use of the SEM in sedimentology. In the space available it is not possible to go into details and theory of machine operation, and neither is it intended to evaluate the great range of SEM machines and ancilliary instrumentation now on the market. When purchasing an SEM it is essential to consider carefully the features required and to ensure that the machine is capable of taking any additions such as analytical equipment and a backscattered electron detector at a later date. In general it is advisable to buy the best SEM one can afford as small models have limited versatility.

The techniques and examples described here are those most commonly in use in general sedimentological studies involving the SEM and will be found sufficient for most studies undertaken in undergraduate and postgraduate courses in sedimentology. If more detail is required on either theory or practice the reader is referred to Smart & Tovey (1982), McHardy & Birnie (1987) and to the volumes edited by Hayat (1974–78). Types of backscattered electron detectors are described by Pye & Krinsley (1984). These publications illustrate the great advances that have been made in SEM techniques from the early days of SEM development in Cambridge. With the advent of the first commercially available SEM in 1965 (Cambridge Scientific Instruments) there has been a rapid expansion of techniques and applications to suit a great variety of scientific disciplines, and within the broad area of sedimentology the SEM is used for a wide variety of investigations as described by many authors in Whalley (1978).

The SEM is of great value in any situation requiring examination of rough surfaces at a magnification range from ×20 up to ×100,000. It is the great depth of focus which enables excellent photographs to be taken of features too small and rough to be within the focus range of a binocular microscope, and features so small and delicate that they would be destroyed in the making of thin sections. The major contribution of the SEM to sedimentology has been in the general field of diagenesis where rock texture, pores and delicate pore fillings can be examined on fresh broken surfaces (Wilson & Pittman, 1977; Blanche & Whitaker, 1978; Waugh, 1978a, b; Hurst & Irwin, 1982; Welton, 1984). With an analytical facility (essential for most diagenesis work) individual grains can be analysed and identified *in situ* and information gained on diagenetic sequences.

Examination of broken surfaces is frequently appropriate for porous rocks; non-porous rocks can be examined on polished and etched surfaces, the method of etching being determined by the desired feature to be observed. Backscattered electron (BSE) images of polished surfaces provide atomic number contrast so enabling recognition of different mineral phases and permitting the use of automatic image analysis systems (Dilks & Graham, 1985). Applications of BSE imagery are illustrated by Pye & Krinsley (1984, 1986b) and White, Shaw & Huggett (1984). Individual grains can be examined for details of surface texture with a view to elucidating their transport history (Krinsley & Doornkamp, 1973; Bull, 1981); grains from heavy mineral separations can be rapidly examined, analysed and photographed. Individual grains or loose sediments can be embedded in resin, polished and etched to reveal internal structures and origins; this is particularly useful with carbonates (Hay, Wise & Stieglitz, 1970).

In sediments and rocks with an organic component, detailed information can be gained on the physical and chemical breakdown of bioclasts (Alexandersson, 1979). Biological destruction by boring algae (Lukas, 1979) and fungi and also by organisms grazing grain surfaces can be deduced (Farrow & Clokie, 1979). The extensive use of the SEM in the study of microfossils is outside the scope of this book. The SEM can also be extremely useful as an auxiliary tool for checking sizes of objects, such as the crystal size of clay separations intended for XRD work.

8.2 BASIC FUNCTION AND MODES OF OPERATION OF THE SEM

The SEM (Fig. 8.1) comprises an electron gun which produces a stream of electrons to which an accelerating voltage of 2–30 kV is applied. The beam passes through a series of two or more electromagnetic lenses to produce a small (10 nm or less) demagnified image of the electron source on the specimen. For most geological work a tungsten hairpin filament is used as the electron source with a vacuum of about 10^{-5} torr. Brighter images and better definition can be obtained with a LaB_6 gun at 10^{-6} torr and for ultrafine definition a field emission source operating at 10^{-9} torr can be used, but this requires special seals and pumping arrangements to achieve the required vacuum.

Before passing through the final electromagnetic lens a scanning raster deflects the electron beam so that it scans the surface of the specimen. The scan is synchronized with that of the cathode ray tube and a picture built up of the scanned area of the specimen. Contrast in the cathode ray picture is due to variation in reflectivity across the surface of specimen.

When the electron beam strikes the surface of the specimen (Fig. 8.2) some electrons are reflected as backscattered electrons (BSE) and some liberate low energy secondary electrons (SE). Emission of electromagnetic radiation from the specimen occurs at various wavelengths but those of principle interest are visible light (cathodoluminescence) and X-rays.

The backscattered (BSE) and secondary (SE) electrons reflected and emitted from the specimen are collected by a scintillator which emits a pulse of light at the arrival of an electron. The light emitted is then converted to an electrical signal and amplified by the photomultiplier. After further amplification the signal passes to the cathode ray tube grid.

The scintillator is usually held at a positive potential of 5–10 kV to accelerate low energy emitted electrons sufficiently for them to emit light when they hit the scintillator. The scintillator has to be shielded to prevent the high potential of the scintillator deflecting the primary electron beam. The metal shield includes an open metal gauze which faces the specimen and allows passage of most electrons to the scintillator surface.

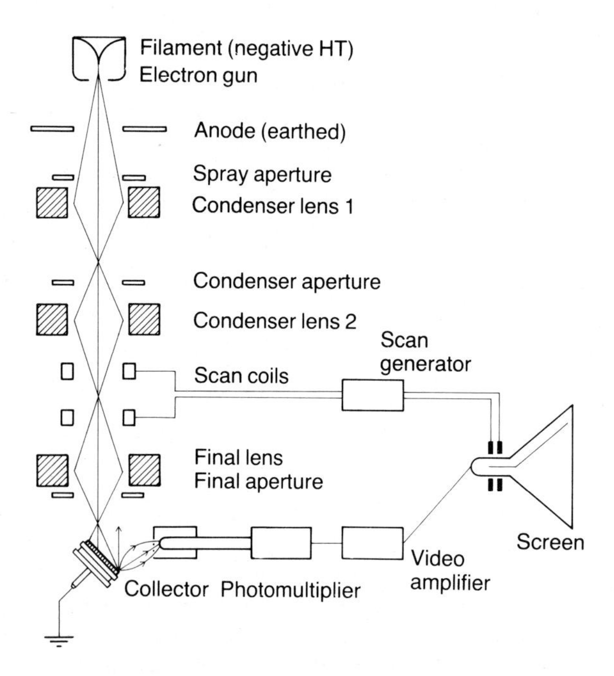

Fig. 8.1. Schematic diagram to illustrate basic functional features of a Scanning Electron Microscope.

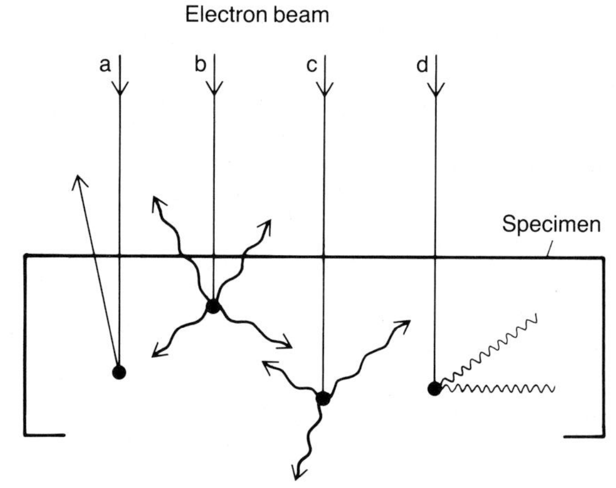

Fig. 8.2. Types of emission generated by an electron beam striking the surface of a specimen in the SEM.
(a) High energy electron produced by single reflection.
(b) Low energy secondary electrons generated close to surface.
(c) Secondary electrons generated at such a depth within specimen that all are absorbed.
(d) Generation of X-rays or cathodoluminescence.

8.2.1 Emission (SE) and reflection (BSE) modes of operations

During normal operation both reflected and emitted electrons are collected (Fig. 8.3), and the scintillator shield and gauze are kept at a positive potential. When operated in the reflective (BSE) mode the low energy secondary electrons (SE) are prevented from reaching the collector by application of a negative potential (−100 to −300 V) to the shield and gauze and only the high energy reflected electrons are collected. The standard type of collector used is a highly directional scintillator-photomultiplier which can be placed well away from the specimen since the low energy secondary electrons are attracted to the scintillator by the applied voltage. This type of detector is not very efficient at collecting high energy reflected electrons and a purpose-built BSE detector is required. Three types of BSE detectors are in common use.

These are: (1) Solid State Conductor Devices; (2) Wide-angle Scintillator-Photomultiplier Detectors such as the Robinson Detector (Fig. 8.4); (3) Multiple Scintillator-Photomultiplier Detectors.

The basic function of these BSE detectors is described by V.N.E. Robinson (1980) and Pye & Krinsley (1984); all designs of BSE detectors subtend a large solid angle with respect to the specimen to collect as many high-energy electrons as possible (see also Section 8.3.4).

Contrast in the image produced is due to a number of factors:

(a) *General topography and orientation of the specimen surface.* Reflected electrons (BSEs) generally have high enough energy to travel in straight lines and thus a suitable angular relation must exist between beam, surface and collector, hence the advantage of wide angle detectors. Low energy secondary electrons (SEs) have paths which are easily bent by the field created by the potential on the collector shield. Thus information can be collected from areas which do not have a direct line of sight to the collector.

Tilting the specimen towards the collector increases the efficiency of collection, and hence brightness, particularly in the reflective mode. Thus brightness of image depends on orientation of surfaces on the specimen.

(b) *Chemistry of the specimen surface.* The efficiency of reflection of electrons from a flat surface inclined towards the detector is dependent on the chemistry of the surface. The reflection coefficient increases as the atomic number increases (Thornton, 1968) and hence particles with higher atomic numbers appear brighter. Operated in the normal reflection-emission mode (SEs + BSEs) this effect is usually masked by topographic effects, but can be most effectively utilized with a BSE detector to produce images of polished surfaces which, under favourable conditions, can reveal differences in atomic number down to 0.12 (Hall & Lloyd, 1981).

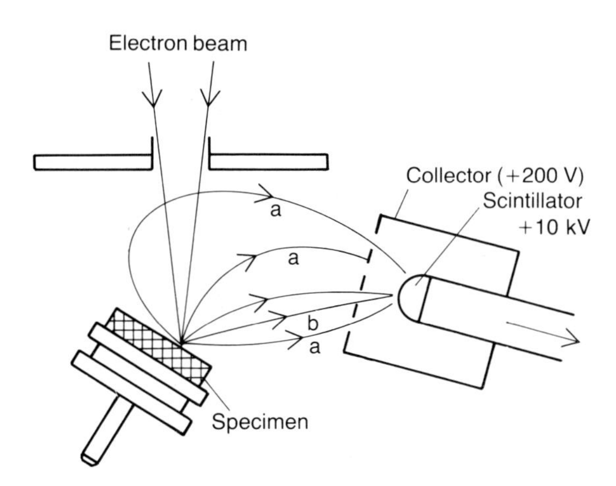

Fig. 8.3. Collection of low energy emitted electrons (a paths) by scintillator in positively charged collector shield. Reflected electrons (b path) are also collected.

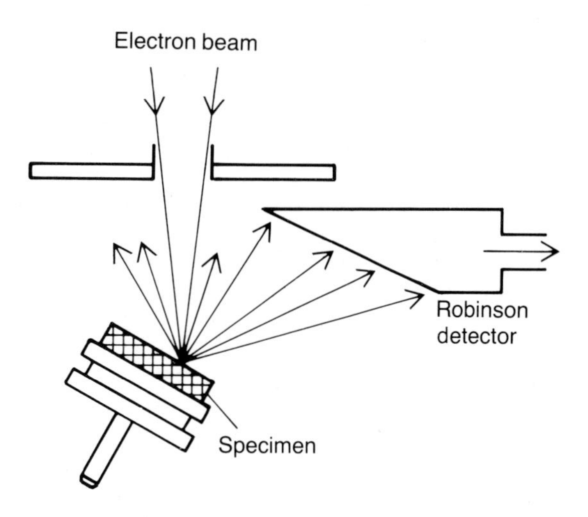

Fig. 8.4. Collection of high energy reflected electrons by a wide-angle scintillator-photomultiplier type of backscattered electron collector such as the Robinson detector.

(c) *Differences in electrical potential on the specimen surface*. The trajectories of low energy secondary electrons are affected by small variations in surface charge on the specimen. Areas of the specimen which are relatively negative appear bright on the image and this results in the effect known as charging which occurs if the point of beam impact is not effectively earthed by either the natural conductivity of the specimen or its applied coating. Charge accumulations of only a few volts give rise to charging features with low energy electrons, but with high energy reflected electrons a charge build-up of several kilovolts is required; thus BSE images are charge free.

For the examination of most geological materials the SEM is operated in the combined SE-BSE mode. Increasing use is being made of BSE imaging of polished surfaces with the advent of efficient back-scattered detectors. Although on many instruments it is possible to obtain a back-scattered image, with the normal detector the quality is usually poor since the detector is not designed for the collection of reflected electrons.

8.2.2 X-ray mode of operation

The X-rays given off when the electron beam strikes the specimen can be used for X-ray micro-analysis of the target area, by comparing either the wavelengths or energies of the X-rays with emissions from standards. When incident electrons collide with atoms in the specimen, electrons may be displaced from inner to outer electron shells. X-rays are emitted when an electron falls back to the inner shell. X-rays are emitted relevant to K, L and M electron shells within the constraints of electron beam energy.

In analysis by wavelength discrimination the X-rays are passed to a spectrometer to be diffracted by a crystal from which those that satisfy the Bragg equation of $\lambda = 2d \sin \theta$ pass on to the detector to be counted. The lighiest element detected by wavelength discrimination is boron, $Z = 5$.

Energy discriminating systems (EDS) are in more general use and have a solid state lithium drifted silicon single crystal detector, which must be kept cool with liquid nitrogen. The entire X-ray beam is passed to the detector and the resulting signal is divided by a multichannel pulse height analyser. With a conventional beryllium detector window, the lightest element readily detected is sodium ($Z = 11$), but with ultra-thin windows and windowless detectors lighter elements such as C, O and N can be detected.

Most samples to be analysed require a conductive coating (Section 8.6.4). Carbon is most suitable, but the coating must not contain any elements for which an analysis is required. Ultra-thin gold coating is frequently used but has the disadvantage that gold produces many spectral lines. Further details of X-ray analysis of sediment and relevant references are given in Chapter 9 of this volume.

8.2.3 Cathodoluminescent mode of operation

The CL operation mode utilizes the visible light emitted when electrons strike the specimen. Materials vary greatly in their luminescent properties and minor traces of impurities can affect the luminescent properties. Emission is also altered by lattice strain resulting from crystal defects and residual strain.

Collection of the emitted light is achieved by focussing it with a lens into a light guide and on to a photomultiplier. Picture contrast is due to a variety of factors. Topography of the specimen is important but the resulting features of greater interest are due to chemical variations, crystal defects and crystal orientation. Further details on this technique using the SEM can be found in Muir & Grant (1974), Grant (1978), Krinsley & Tovey (1978) and Smart & Tovey (1982). The multiple scintillator-photomultiplier type of BSE detector can be adapted for cathodoluminescence work. In the Philips Multi-Function Detector, cathodoluminescence detector elements can be substituted for BSE detector elements (Pye & Windsor-Martin, 1983). Ruppert *et al.* (1985) illustrated contrasting BSE and CL images of a sandstone, and further examples are shown in Fig. 8.8. The reader is also referred to Chapter 6 on cathodoluminescence in this volume.

8.3 SEM ANCILLARY EQUIPMENT AND TECHNIQUES

8.3.1 Photographic equipment and techniques

EQUIPMENT

The choice of a camera for use with an SEM is largely a matter of user preference. Many machines are equipped with a good quality 35-mm single lens reflex camera for routine work. This is the cheapest mode of operation, using a high speed film such as Ilford HP5 or Kodak Tri-X. For many purposes a slower film such as Ilford FP4 is satisfactory and may be preferable if long scan times of 5 minutes or more are used. It may be preferred to have a larger negative size to reduce the photographic enlargement phase and to use a '120' or a 70-mm camera.

Some machines are designed to take a Polaroid camera and this system has the great advantage that an instant picture is obtained and the operator can ensure that a picture of the desired quality has been obtained before moving on to examine other areas of the specimen. The great disadvantages are that picture quality is lost if the print has to be rephotographed to produce copies and it is an expensive process.

An ideal system is a Polaroid positive/negative film which produces both a print and a large size negative. The necessary photographic solutions have to be kept fresh and at hand, but where both rapid results and a quality negative are required the system is excellent.

TECHNIQUES

The excellent resolution and great depth of field of the SEM have made it very popular for providing illustrations for a wide variety of geological studies. The quality of SEM photographs in geological journals is highly variable, ranging from excellent to examples displaying many of the faults illustrated and discussed in Section 8.6.2. In some cases the paper quality of the journal is responsible for poor reproduction.

For most sedimentological work simple views of surfaces are all that is required and with most machines a zoom facility enables the operator to compose the photograph satisfactorily. If it is proposed to make a multiple plate of photographs it is helpful to take the shots at the same magnification or at multiples of a basic magnification. The incorporation of a micron marker in the photograph is essential and easier to 'read' than a photograph captioned as × 4350! Ensure that the micron marker contrasts with its background, or enhance and label the marker for publication purposes.

MONTAGES

Photographic montages may have to be constructed to cover large areas at high magnification. The construction of both controlled and uncontrolled montages is discussed by Smart & Tovey (1982). Uncontrolled montages are made merely by matching features in the overlap area of the photographs (Fig. 8.5), but because the area scanned is generally trapezoidal severe distortions in orientation develop if large numbers of micrographs are joined. It is preferable that photographs are taken at magnifications >2000 and that less than *c.* 10 photographs are joined together.

Controlled montages (Fig. 8.5) are prepared by photographing on a regular grid pattern and noting x and y shift co-ordinates of the centre of each photograph and then mounting the photographs on a regular grid. Matching will not be perfect but distances and orientations will be better preserved. All photographs must be taken with ample overlap (*c.* 25%), without altering the magnification, and must be enlarged by the same factor in processing.

Distortion effects in montage construction are greatest at low magnifications and high tilt angles of the specimen. The mounting and printing of photographs for montages requires skill. Contrast of adjacent photographs must be reasonably close to

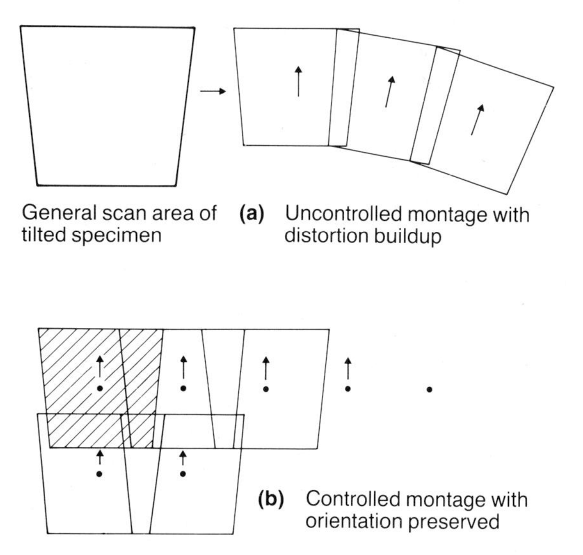

Fig. 8.5. Production of (a) uncontrolled and (b) controlled photographic montages.

provide a pleasing match. When mounting the photographs, joins can be made less obvious by matching along prominent features such as sharp grain margins to avoid contrast differences.

It may be possible to take montage photographs in the emission mode without tilting the specimen, thus avoiding the worst features of distortion, and some machines have a tilt compensation device which helps reduce distortion. Generally a balance has to be struck between distortion and picture quality. The value of montages is that fine detail and its distribution can be evaluated over large areas: the ultimate example must be the 3 × 3 m montage by Wellendorf & Krinsley (1980) consisting of over 1100 SEM photographs taken at × 4300 to cover a single quartz grain! Unfortunately the montage loses a little detail in the required reduction to 135 mm diameter for publication!

STEREOSCOPIC MICROGRAPHS

Any specimen stage which enables the specimen to be tilted permits the taking of stereoscopic pairs. Methods that can be employed are discussed by Tovey (1978) and Smart & Tovey (1982). The most common method used is the 'tilted convergent' method whereby the specimen is tilted by a small angle (±2–5°) about a mean tilt angle. This method ensures that the specimen is facing the detector in both positions. The illumination is effectively from the side in this method and the photographs must be rotated by 90° for stereoscopic viewing.

Techniques vary depending on the type of specimen stage, but in general one must ensure that the same area is photographed on each occasion. This can be achieved by sketching the first view on a transparent overlay on the screen. The magnification is then reduced to less than × 200 while keeping track of a convenient reference feature near the centre of the screen. The stage can then be tilted by the desired amount and the reference feature brought back to its previous position using the X,Y controls. The magnification can then be increased to the correct figure and the desired feature checked with its outline previously drawn on the screen overlay. Any refocussing should be done with the Z control to avoid image rotation and magnification changes due to alteration of working distance. The second photograph can then be taken.

Detailed discussion of stereographic techniques can be found in Clarke (1975) including methods for accurate calculation of dimensions of surface relief features. Surprisingly little use has been made of stereographic techniques in sedimentology, but examples can be seen in Smart & Tovey (1982), and Whitaker (1978) used stereo pairs to illustrate diagenetic textures in the Brent Sandstone (Middle Jurassic, North Sea). Much more use could be made of this technique for illustrating diagenetic textures on rough surfaces.

8.3.2 Alpha-numeric displays

Some machines have built in alpha-numeric displays on which a variety of information can be recorded directly on to the photographic negative, for other machines such a display can be added as an optional extra. Most simple displays permit the operator to identify the photograph with one or more serial numbers, so saving lengthy cross-checking with notes on long negative sequences, and also provide information on operating kV, scale bar length and laboratory or user's name or number.

The information display may only appear on the recording screen, thus the operator must ensure that vital areas of the object to be photographed will not be covered by the display.

8.3.3 Analytical equipment

For most sedimentological work it is essential to have some form of analytical facility on the SEM for rapid checking of grain compositions. An Energy Dispersive X-ray analysis system (EDS) is commonly used which generally permits analysis at a specific spot of about 1 μm diameter (Fig. 8.7) or to collect emissions from a specific line on the screen or the whole screen area. Many systems also have the facility whereby maps of the distribution of particular elements in the screen area can be built up and these can be compared with both SE and BSE images as illustrated by Loughman (1984), White *et al.* (1984), Gawthorpe (1987) and Fig. 8.9.

Depending on a suitable system being available, quantitative analysis can be performed on the SEM but for most sedimentological work a good qualitative system is satisfactory. The majority of sedimentological SEM work involves the study of rough surfaces, and surface topography affects factors such as the intensity and depth of X-ray production, as well as the absorption effect. Rough surfaces can also obstruct the path of X-rays to the collector and backscattered electrons and X-rays can produce radiation from areas outside the primary target area of the beam (Fig. 8.6).

To eliminate this situation, the specimen should be reasonably flat and tilted to face the detector. Even in this situation there will be areas (e.g. within deep pores) from which low count rates result in small peaks with much background noise. Background noise may mask relatively low intensity peaks.

Another problem is that of extraneous X-rays arriving at the detector. These unwanted X-rays are produced by backscattered electrons and X-rays from the target area of the beam which strike other surfaces which have a line of sight to the detector. The result can be that in an attempted analysis of a grain in a quartzose sandstone, Si may be detected from extraneous X-rays from quartz rather than from the grain being analysed. Some of the effects of analysis of rough surfaces are illustrated by Fig. 8.7. Whereas some of these factors can be compensated for as described by Bomback (1973), it is essential to do quantitative analysis using polished surfaces which can be compared with similarly prepared standards (Chapter 9).

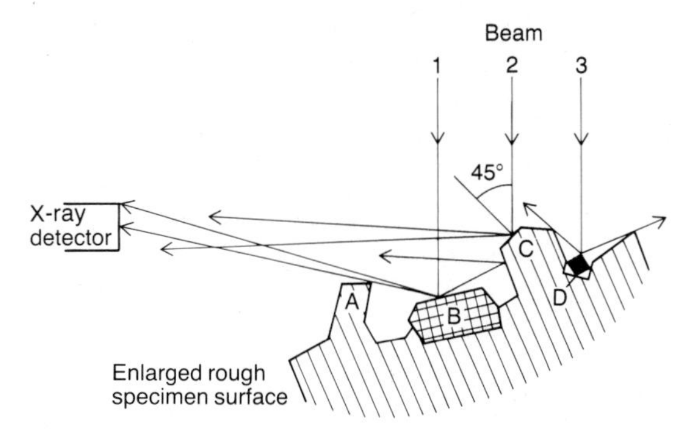

Fig. 8.6. Analysis of rough surfaces in the SEM; examples of angles to the detector. Beam position 1 — grain B to be analysed. Count rate reduced by partial obstruction A, and extraneous X-rays reach detector from C.
Beam position 2 — surface of C at virtually ideal 45° position with respect to detector.
Beam position 3 — grain D has no line of sight to detector. No counts from grain.

8.3.4 Back scattered electron detectors

The purpose of these detectors has been explained (Section 8.2.1) and such a detector is an excellent addition to an SEM for the production of charge-free images of rough surface material and for images of polished surfaces to show atomic number contrast (Fig. 8.9).

The solid state conductor type of BSE detector consists of an annular silicon diode mounted between the specimen and the final lens. Unlike early versions, the modern ones can operate at TV scan rates. This form of detector appears to be the cheapest, but produces good results as shown by Pye & Krinsley (1984) and Dilks & Graham (1985).

The wide-angle scintillator-photomultiplier type of detector is typified by the Robinson detector (Robinson, 1975) which performs well but has a rather bulky detector head. There can be space problems with relatively small SEM chambers if an EDS detector is also required to be in position.

Multiple scintillator-photomultiplier detectors have up to four detection elements mounted between the specimen and final lens. This is a highly efficient (but more expensive) system and in the case of the Philips Multi-Function Detector can be modified to perform cathodoluminescence work.

For further details of the operation and uses of these types of detector refer to V.N.E. Robinson (1980) and the well-illustrated account by Pye & Krinsley (1984).

8.3.5 Charge-Free Anticontamination System (CFAS)

The CFAS system can be used in conjunction with a back-scattered electron detector such as the Robinson detector and operates by maintaining a small residual gas pressure of between 0.01 and 0.5 torr in

Fig. 8.7. General view of sandstone to illustrate the use of EDS analysis at five points. The sample was gold coated, hence the gold peaks on the EDS traces. Analysis identifies at point (1) quartz overgrowths; (2) K-feldspar with overgrowth; (3) ferroan calcite of diagenetic origin growing into pore space and overgrowing quartz; (4) kaolinite in pore space either growing around the ferroan calcite or being forced apart by growth of the carbonate; (5) siderite rhomb on quartz grain surface. Minor Si peaks associated with siderite and calcite analyses are probably due to secondary reflection.

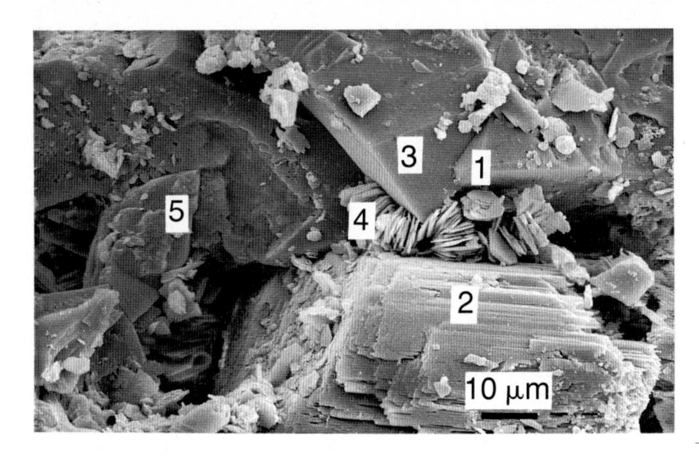
3
1
5
4
2
10 μm

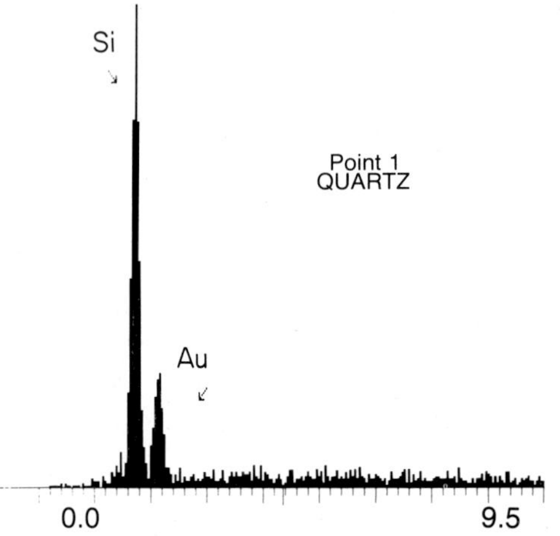
Si
Au
Point 1
QUARTZ
0.0
9.5

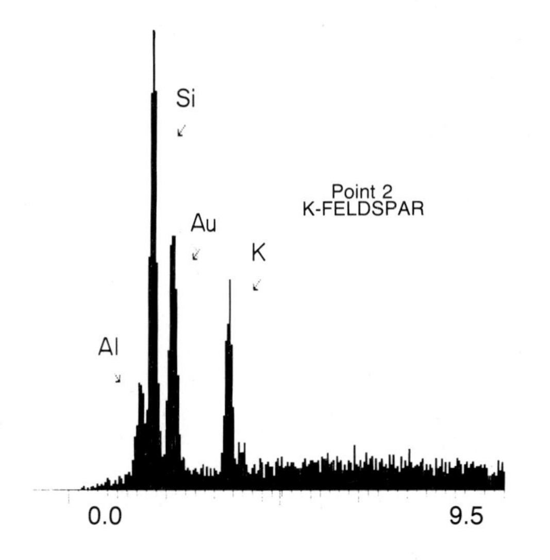
Si
Au
K
Al
Point 2
K-FELDSPAR
0.0
9.5

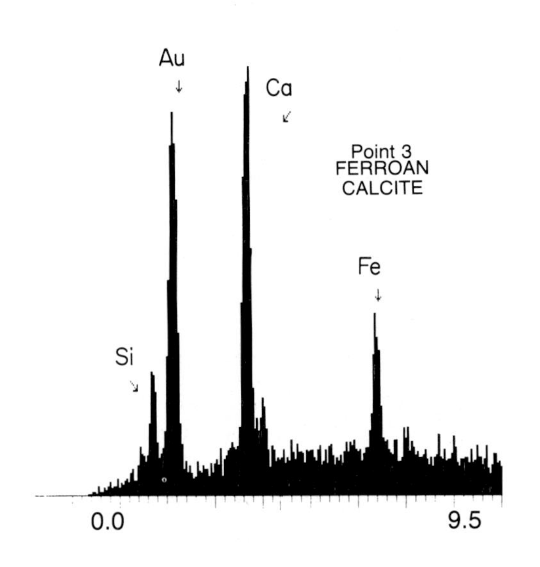
Au
Ca
Point 3
FERROAN
CALCITE
Fe
Si
0.0
9.5

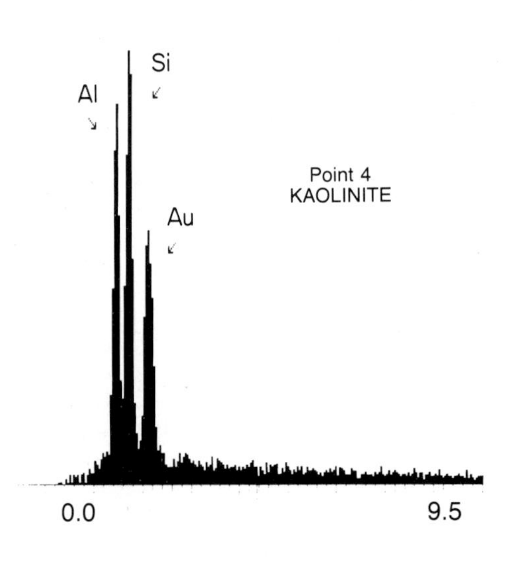
Si
Al
Au
Point 4
KAOLINITE
0.0
9.5

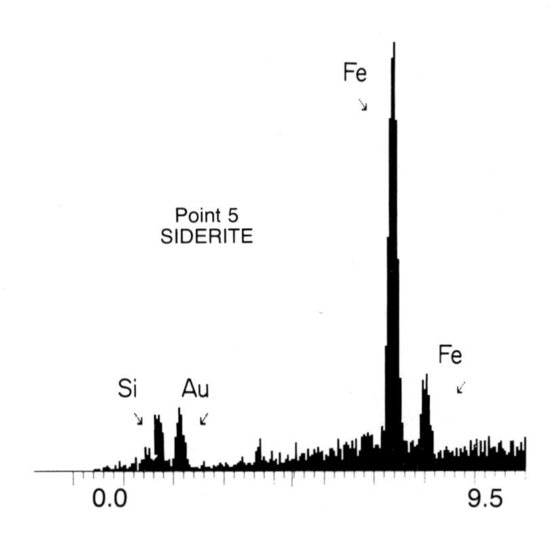
Fe
Point 5
SIDERITE
Fe
Si
Au
0.0
9.5

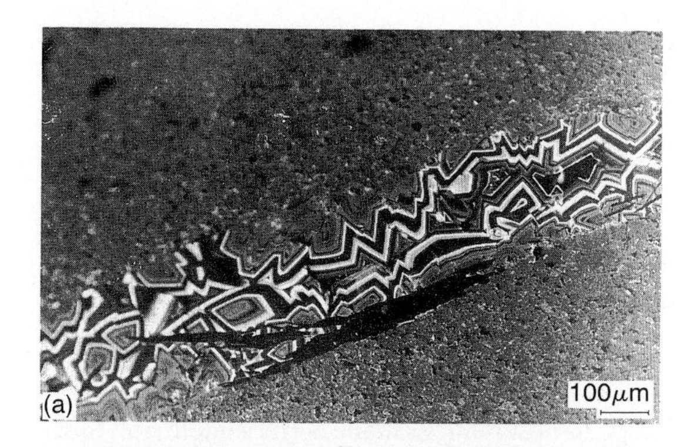
100µm
(a)

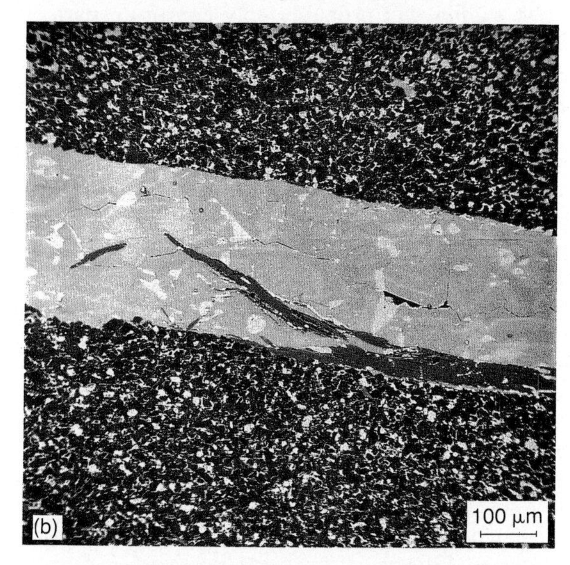
100 µm
(b)

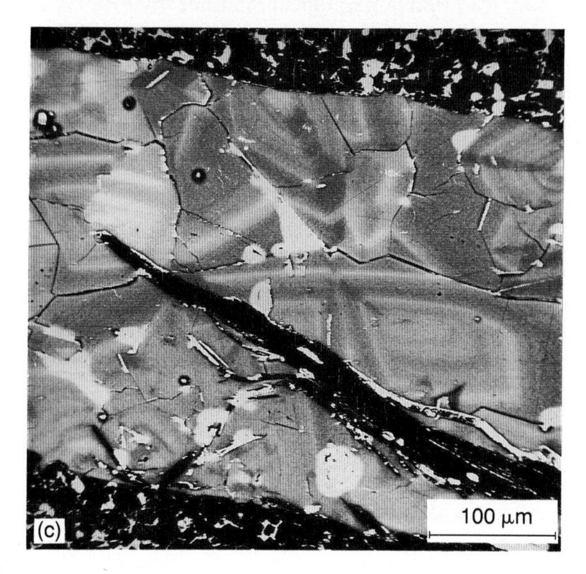
100 µm
(c)

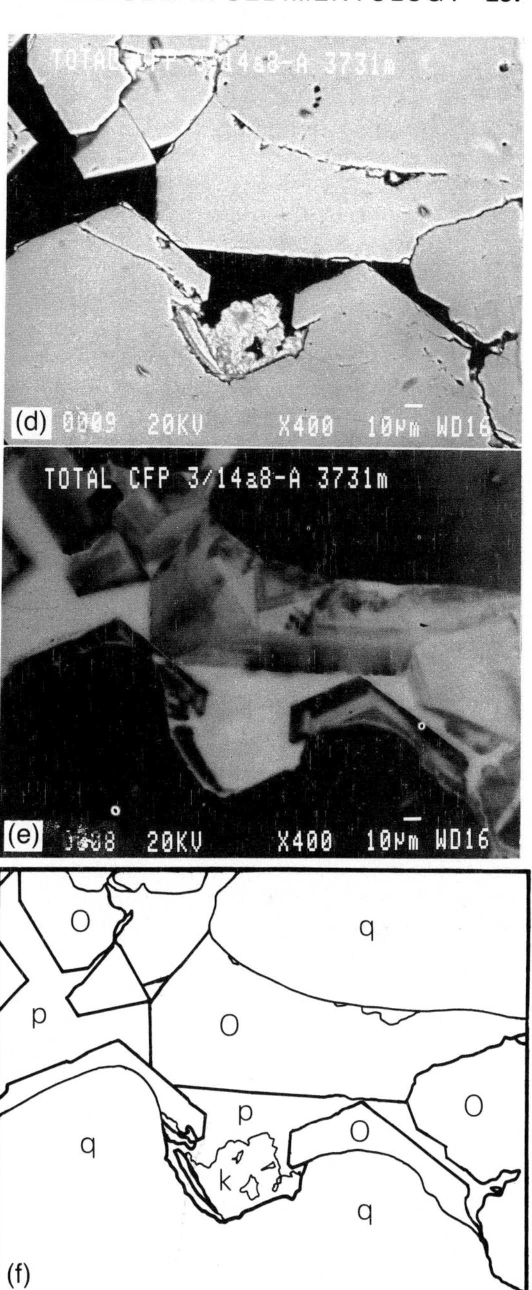
(d) 0009 20KV X400 10µm WD16
TOTAL CFP 3/14a8-A 3731m
(e) 20KV X400 10µm WD16
O
q
p
O
p
O
O
q
k
q
(f)

the specimen chamber. It is operated by having separate diffusion pumps on the chamber and column, gas leakage through the 200 μm final aperture is not significant. In fitting the CFAS system the secondary electron detector HV supply is disconnected, thus it is not possible to switch directly from normal operation to CFAS operation.

The advantage of the system is that specimens can be put in as received without the need for any special preparations and coatings. Gas ionization in the chamber is always equal to charge accumulation and thus charging is automatically eliminated. Specimens which are damp or oily and which would normally contaminate the column can be examined.

This system is excellent for geological use on any samples requiring rapid examination for mineralogical composition, especially rapid search for small heavy-element grains (B.W. Robinson, 1980; Robinson & Nickel, 1979). An added advantage is that filament life is reported to be increased as compared with conventional operation, but more frequent cleaning of the column may be found necessary.

8.3.6 Image analysis

The use of image analysis of BSE images produced from flat surfaces promises to be an extremely useful technique. Dilks & Graham (1985) have described the application of a Kontron SEM-IPS image analyser coupled to a SEM and demonstrate how powerful and fast the technique can be to machine-discriminate minerals and porosity on the basis of levels of grey of a calibrated BSE image. Using the system, rocks can be screened for particular minerals and point counted for a range of defined minerals. Grain size distributions of components of the rock can be produced, and pore size and two-dimensional pore throat size analysed. The limitation of the technique is the inability to discriminate between two minerals which have the same image intensity.

8.4 SAMPLE COLLECTION

8.4.1 Surface material and outcrops

Field sampling of material for SEM studies requires that some elementary precautions are taken and sufficient observations made at the sample site. Since SEM studies will probably only constitute part of any laboratory study, sufficient material must be gathered for all possible aspects, such as grain-size analysis and thin section preparation.

Unconsolidated material in which the fabric is not required to be preserved can be collected as a bag sample (dry) or bottle sample (wet). Samples containing live or dead organic material may need to be preserved in alcohol if the samples cannot be immediately treated. Aluminium foil is useful for wrapping small samples to prevent contamination and the wrapped samples can be safely transported in plastic bottles. Cloth or paper is to be avoided since fibres are likely to be shed into the sample.

Impregnation of unconsolidated sediments in the field provides a later opportunity to examine fabric details using thin sections or the SEM. For suggested sample preparation see Section 8.5.2.

Consolidated rock material is generally collected in the traditional geological manner by use of a hammer. However, such violence can have the effect of producing cracks, fracturing grains and crushing porous material. Samples must be sufficiently large to avoid such effects and (unless it is the object of

Fig. 8.8. Comparison of cathodoluminescence and backscattered electron imagery.
(a, b, c) The rock is a Precambrian dolostone from the Dalradian of Scotland in which phlogopite mica crystals grew during metamorphism. Subsequently veins developed, tending to propagate parallel to the phlogopite crystals, and were filled by chemically-zoned calcite crystals. (a) (cathodoluminescence image) shows the fine-grained dolomite host-rock luminescing dully, the phlogopite crystals not luminescing, whilst the calcite shows an intricate zonal pattern invisible in plane polarized light. (b) This backscattered electron image is of a similar area to (a). Faint zonation is visible in the calcite which is overall brighter than the phlogopite and the dolomite because of its higher average atomic number. Bright flecks in the dolomite and calcite are artefacts. (c) shows a magnified backscattered image in which the atomic number contrast has been electronically amplified to show the zonation more distinctly. Microprobe analysis shows Fe, Mn and Mg zonation in the calcite.
(d,e,f) Rock is a porous sandstone from the Jurassic Brent Group, N North Sea. (d) BSE image with quartz (grey) and porosity (black). Cathodoluminescence image (e) negative and interpretation (f) showing detrital quartz grain shapes (q) and zoned quartz overgrowths (o) representing different stages of quartz cementation. Minor kaolinite (k) is also present within pore space (p).

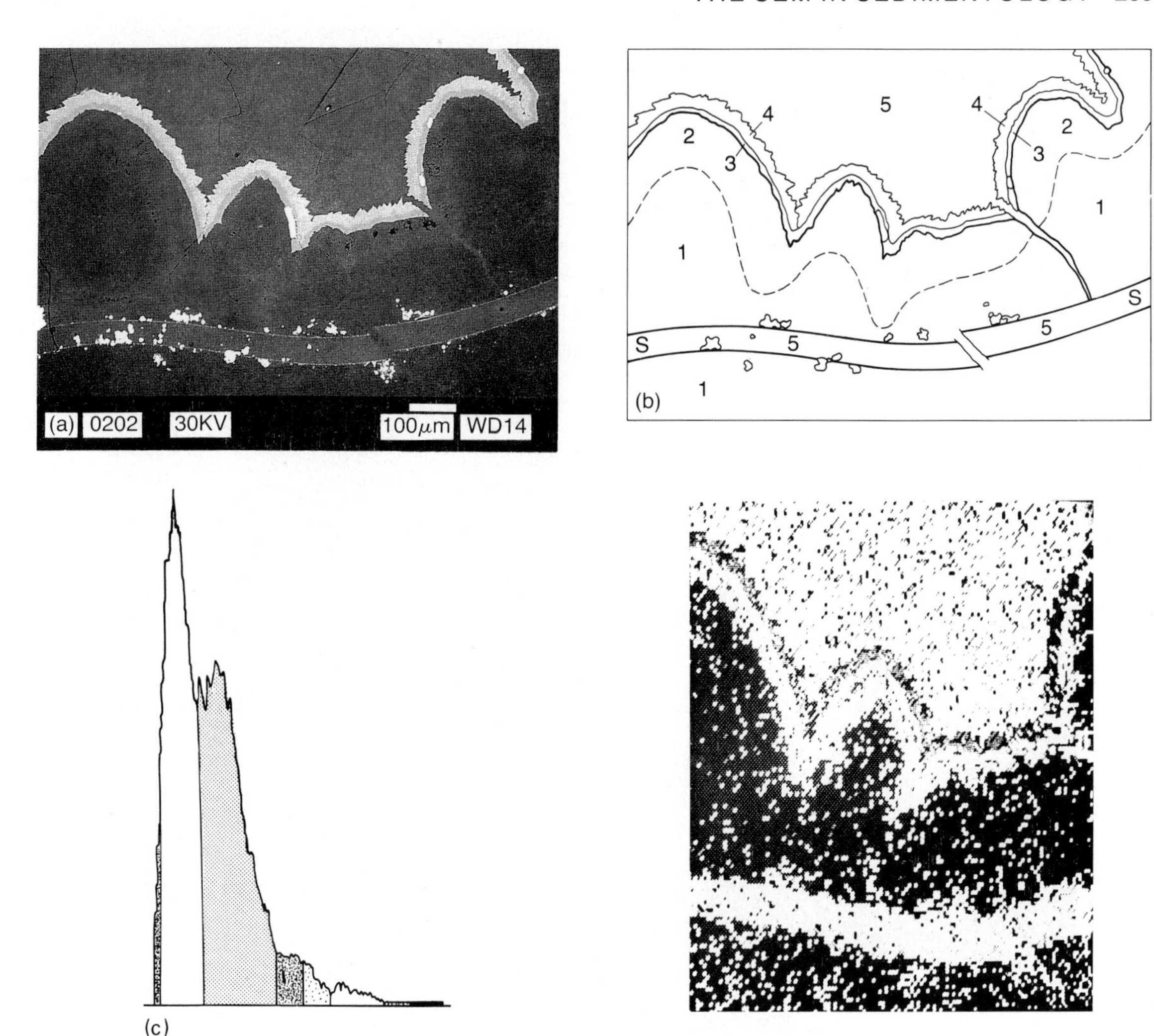

Fig. 8.9. Backscattered electron images and element distributions. (a–c) Growth of carbonate cements within a specimen of *Dactylioceras* cf. *commune* from the Toarcian of NE Yorkshire. (a) Backscattered electron image of polished, carbon-coated thin section. Cement stages are interpreted in (b) in which S is a section of an ammonite septum defined by apatite rims and with associated pyrite framboids. Cement growth stages: (1) high Mg calcite (filling fracture in septum); (2) low Mg calcite; FeS_2; (3) high Mg, Fe calcite; (4) high Fe calcite; (5) high Mn, Fe calcite (replaces septum) fills void. (c) Histogram and X-ray intensity (α element abundance) map showing the different concentration of Mg across part of the area of (a). Cement growth stages are clearly picked out in the elemental map (black and white print derived from a colour image, hence both high and low Mg concentrations appear pale in this image).
(d–h) Chemical variations in a dolomite crystal from the Pendleside Limestone, Lower Carboniferous, Bowland Basin, northern England. (d) Subhedral dolomite crystal in silicified limestone. The light zones rimming the crystal suggest the presence of elements with higher atomic numbers compared with the dull core. Note the irregular contact between the dull and light zones. Scale bar = 50 μm. Back-scattered electron photomicrograph (BSEM). (e) DIGIMAP of view in (d) for Ca. Note relatively uniform distribution. Scale bar = 25 μm (same for f–h). (f) DIGIMAP of view in (d) for Mg. Depletion in Mg in the crystal rim relative to the core is indicated by the darker nature of the rim. (g) DIGIMAP of view in (d) for Fe. Note the very low Fe content in the black core compared with the grey rim. (h) DIGIMAP for Mn showing similar distribution to Fe (g).

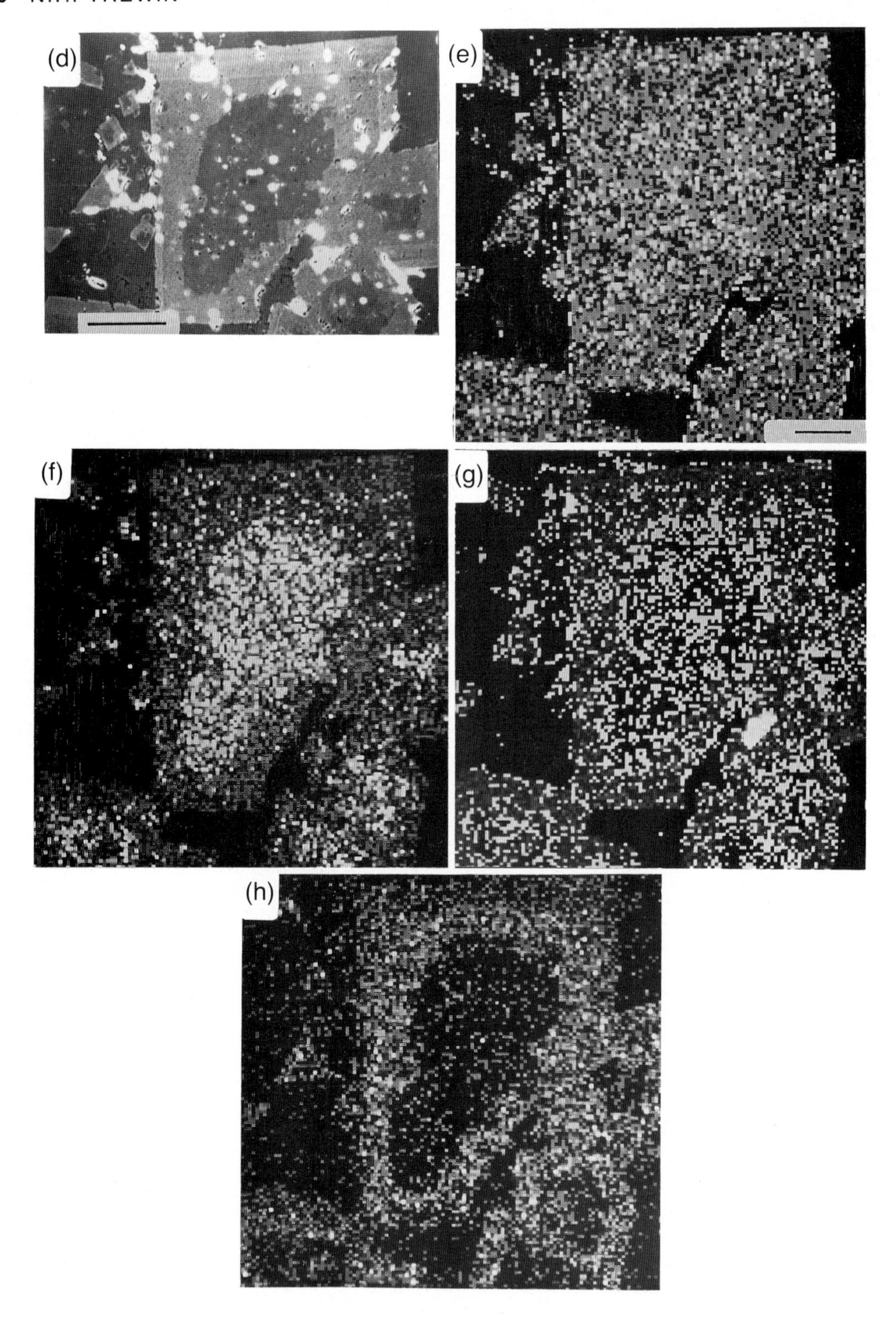
(d)
(e)
(f)
(g)
(h)

study) to be unaffected by weathering. Since the final sample to be used in any SEM study must be clean it is not practicable merely to collect small rock chips in the field. Many pitfalls exist in preparation and ample material should be collected. It is advisable to collect oriented specimens so that both way-up and geographical orientation are known.

Field collection of porous rocks should be done with regard to possible fluids in the pore space. For example, on a shore section the pores may be filled with fresh or salt water and may also be regularly air-filled as tides fall or cliffs dry. Such possibilities of long continued interchange of gas and fluid and of flow through the pore system should be considered in any study of surface collected material.

Collection of material by use of a portable drill producing a short, small diameter core is also likely to be unsatisfactory if it is hoped to observe delicate growths within pore space. The forces of fluid pressure and vibration are likely to affect the fabric within pore space. Joslin (1977) described a tool for collecting samples of poorly consolidated soft sediments (e.g. chalk). An 80-mm core barrel with toothed end is gently drilled into the rock with a hand operated drill (brace). The core obtained can be mounted direct on to a stub in the field after trimming if no further treatment is required prior to mounting.

8.4.2 Subsurface material

Material from subsurface cores must be examined and sampled with as much knowledge as possible of the conditions under which the core was taken and any subsequent laboratory treatment that may have taken place.

Preserved samples (pore fluids sealed immediately the core is brought to the surface) are obviously the best to work with and will preserve most in the way of original texture. Such samples are suitable for advanced techniques such as critical-point drying (Section 8.5). If the core is fresh and still wet when sampled it can be advantageous to prevent further drying by sealing samples or keeping them in simulated formation fluid.

If the core has already been air dried there is little point in using advanced techniques of preparation; the damage to delicate structures such as illite has probably already taken place. Should the core have been slabbed it may already have been dried and rewetted.

Whatever the state of a core any sample for SEM examination should be taken from as near the centre of the core as possible when the core is still in 'the round'. This will hopefully avoid drilling mud which may have invaded pore space at the margin of the core. A zone of invasion or else of flushing can frequently be seen in freshly cut cores. If the core to be sampled has been slabbed the sawcut and adjacent areas should also be avoided for similar reasons. Penetration of organic material such as coccolith plates into pores during injection of sea water into reservoir rocks has been described by Hancock (1978a); it is thus advisable to consider the possibility that any rock with large pores and high permeability may have had material introduced to the pore space by the passage of fluids through the sample.

Sidewall cores are less satisfactory as they are usually taken by shooting a short tube into the wall of the borehole with an explosive charge. This process frequently results in visible crushing of fabric and fracturing of grains. Care should be taken to avoid any mudcake that may be present on the walls of the borehole.

Cuttings from boreholes are the least satisfactory, but can be cleaned and may provide useful information. However they are normally only obtained from relatively well lithified, low porosity material and not from the most porous and permeable rocks in the borehole. In all cases where borehole material is used electric logs should be consulted for information on other sedimentological and petrological aspects.

8.5 SAMPLE TREATMENT

8.5.1 Porous rocks to be viewed on fractured surfaces

The following preparation techniques are generally used for viewing broken surfaces of porous rocks which fracture around grains rather than through grains. The method of preparation of material can have a marked effect on the fabric of delicate clays such as filamentous illite present within pore space. McHardy, Wilson & Tait (1982) have shown how critical-point drying can preserve features which would be destroyed by air-drying and freeze-drying techniques. The value of using critical-point drying is that the technique prevents the passage through the pores of an air–liquid interface or an interface between immiscible liquids. It is the passage of such an interface which can cause collapse and reorientation of clays within pore space. If it is known that the specimen has already been air dried, advanced techniques of preparation cannot restore the situation. Further useful details of preparation techniques for clays are given by McHardy & Birnie (1987).

AIR-DRIED SAMPLES

Samples previously air dried and which contained water as the pore fluid can be examined without further treatment, but it is frequently advisable to wash the specimen with distilled water to remove any chlorides or other soluble salts which may have crystallized in pore space due to the evaporation of saline pore water. This is best achieved by soaking a sample, not more than 1 cm thick, in distilled water until it is chloride free. (The time taken will depend on permeability and salinity of the evaporated pore water.) This technique is also useful for modern marine sediments and rock material collected from shore sections.

Air drying, combined with removal of water soluble material, is also a technique which can be used to study the soluble salts and the effects of their removal. This has applications in the study of weathering of rocks and the formation of evaporites.

If delicate material such as fibrous illite is suspected to exist in pore space and either time or lack of apparatus precludes use of critical-point drying, better results may be obtained by air drying samples from amyl acetate. The advantage of this method is that amyl acetate has a much lower surface tension than water and so the forces exerted at the liquid–gas interface as the specimen dries are reduced.

OIL-SATURATED SAMPLES

Rocks that are oil-saturated, or which contain oil residues after evaporation of lighter hydrocarbons, must have the hydrocarbon removed prior to SEM examination. If oil is not removed it is vaporized by the electron beam and will make the SEM column dirty and reduce picture quality. Oil also prevents the efficient coating of the sample with gold, resulting in charging of the specimen and poor resolution pictures.

Oil can be removed by use of solvents such as acetone, xylene, toluene, chloroform and trichloroethane. Chloroform and trichloroethane are the most efficient at removing oil. It is possibly safer to use trichloroethane than chloroform, but full safety precautions (gloves and fume cupboard) should be taken in handling all these solvents.

Method. A sample measuring about $10 \times 10 \times 5$ mm is immersed in the solvent and sealed in an air-tight jar (to prevent fumes). The solvent is changed every 12 hours. The process is repeated until all the oil has been removed; this can be judged by lack of a residual oil film on evaporation of a few drops of the used solvent on a glass slide. With reasonably permeable rocks two changes of trichloroethane are normally sufficient to remove the oil. The sample is then allowed to air dry at room temperature for 24 hours. If residual oil appears on the specimen surface after drying, a final washing in solvent will usually remove the last traces. A fresh fracture surface can then be produced for observation in the SEM. Comparisons of oil saturated and oil-free samples treated by the above method show no apparent difference in clay morphology or orientation.

Removal of oil from rocks with poor permeability can be exceedingly difficult and the samples may have to be flushed of oil by application of a pressure difference across the sample. Oil can be trapped in pores which are no longer connected, and in cases such as this only the surface of the sample can be effectively cleaned and only small specimens should be used to minimize the amount of oil that may be released from the rock when under vacuum in the SEM chamber.

In commercial core analysis laboratories oil is frequently removed using a Soxhlet extractor. The sample is first flushed with methanol to remove water, a process which may take from 8 hours to 3 days depending on permeability. The sample is then dried in air until the bulk of the methanol has evaporated and then put back in another extractor with xylene or toluene and flushed until clean. Cleanliness is generally judged by a lack of fluorescence under UV light. The sample can then be air dried as required. There is the possibility that the more active flushing in this method and the intermediate drying stage could affect clay fabrics.

Drying of wet specimens can be accomplished at room temperature, by use of an oven or by application of a vacuum. All these methods involve a volume decrease of clay or mud specimens with the result that depositional textures of clays are probably not retained. If specimens are oven dried, a low temperature (50°C) should be used to avoid loss of structural water in clays. If washed cuttings from oil wells are being examined it is essential to ensure that the cuttings were not dried at a high temperature after the washing process.

FREEZE DRYING

Freeze-drying techniques are used extensively on biological material and have been applied successfully to clay–water systems in soils. This method is favoured by Erol, Lohnes & Demirel (1976) for preparation of swelling clays such as Na-montmorillonites. It does not involve replacement of water with an organic liquid which could cause swelling and ion exchange in the clays.

Method. The method described by McHardy *et al.* (1982) used 10-mm pieces of rock (Jurassic, Magnus Field reservoir from the North Sea) which were either directly frozen with the contained pore water or frozen after the samples were washed free of chloride. Samples were plunged into liquid nitrogen or freon for at least 30 minutes, they were then vacuum-dried from the frozen state in an Edwards Modulyo freeze-drier. The drying phase takes several days.

For the method to be a success the specimen must be cooled very rapidly so that the water is frozen in place and ice crystals do not form. The rate of cooling required is several hundred degrees per second and this will only be accomplished on the outer surface of the rock, and possibly to a depth of only 0.1 mm (Greene-Kelly, 1973). Another problem is encountered in the drying phase when it is necessary to warm the specimen to a temperature greater than −40°C to achieve sublimation of the ice in a reasonable time, but at this temperature rapid growth of ice crystals will occur which will displace matrix clays.

The main applications of freeze-drying of samples are in the study of soils and swelling clays. McHardy *et al.* (1982) found this method inferior to critical-point drying for preservation of delicate illite in pore space.

CRITICAL-POINT DRYING

The advantage of the critical-point drying technique is that it eliminates the surface-tension forces applied to delicate crystals during the drying stage by avoiding the passage of an air–water interface through the pore space. This results in better preservation of delicate structures, such as fibrous illite (McHardy *et al.*, 1982). Tovey & Wong (1978) favoured this method for the preparation of wet sediment samples. The disadvantage of the method is that it is time-consuming, requires special apparatus, and has disadvantages if swelling clays are present (see above).

The method of preparation used by McHardy *et al.* (1982) is as follows. Pieces of rock of 10-mm size which contained simulated formation water had the water replaced by methanol by the successive passage through the specimens of 1:3, 1:1 and 3:1 methanol–water mixtures and finally 100% methanol. The methanol was then replaced daily until free from chloride which was taken to indicate that all pore fluid had been replaced by methanol. The samples were then transferred to the critical-point drying apparatus (Polaron model E3000 with a liquid transfer boat with integral drain valve). The boat with six to eight samples in methanol was loaded into the pressure chamber which was pre-cooled by cold mains water. The chamber was flushed with liquid CO_2 for about 10 min and allowed to stand for several hours with the CO_2 level above the samples at all times. To ensure replacement of methanol by liquid CO_2 the chamber was flushed with CO_2 for 5–10 min twice a day for a week. Finally the pressure apparatus was heated to 37°C by water circulation from a 40° thermostat to bring it above the critical point of CO_2 (32°C) and the temperature held for 30 min to ensure equilibrium was reached in the samples. The CO_2 was then slowly vented from the chamber at 2 ml s^{-1} which was slow enough to prevent condensation of CO_2 in the chamber. The dried samples were then ready for production of a clean fracture face and mounting as described below in Section 8.6.

Critical-point drying has been found useful for the treatment of illite-bearing sandstones (McHardy *et al.*, 1982; McHardy & Birnie, 1987) and has also been used in studies on pore-size distribution in soils (Lawrence, 1977; Murray & Quirk, 1980) and should be considered as a preparation method well worth trying with porous samples that have their original pore fluids preserved. Farrow & Clokie (1979) used critical-point drying in a study of boring algae to reveal algal filaments within the borings. The samples were first fixed with glutaraldehyde, dehydrated through a series of alcohols and brought to acetone and then critically point dried.

8.5.2 Non-porous rocks

Rocks which have low porosity or in which grains and cement give similar resistance to fracturing, as is the case in many quartz-cemented sandstones and well-cemented limestones or dolomites, may not reveal much information when rough broken surfaces are examined. In such cases it is frequently necessary to view a polished and etched surface to obtain useful information.

This technique is particularly useful in the examination of carbonate cemented sandstones, limestones and dolomites, since by suitable choice of etching agent different minerals can be attacked and their role as grains or cement elucidated. The technique can also be used on unconsolidated sediments which have been resin-impregnated in the field or laboratory.

GENERAL METHODS

A slice of the rock to be examined is cut to a suitable thickness (say 3–5 mm). The specimen can thus be oriented in any required manner and part of the rock slice used to make a thin section or polished section for examination by other techniques.

The rock slice or chip should be polished to a reasonable standard with respect to the features required to the viewed (polishing details in Lister, 1978). The specimen slice should be cut or broken to the required size for insertion in the SEM chamber and mounted on the stub prior to etching to minimize the possibility of damage to the delicate etched surface. Some SEM chambers will accommodate a polished thin-section; this enables the same area to be examined using a variety of techniques and is an obvious advantage.

The etching process can be varied to suit the particular situation. For limestones, dilute hydrochloric, acetic or formic acid can be used for low-relief etching. It is best to use as dilute a solution as possible to avoid vigorous gas generation which could damage delicate features. Sandberg & Hudson (1983) used 0.25% formic acid and an etch time of 60–90 s to reveal aragonite relics in calcite-replaced shells. Hay *et al.* (1970) used 1% HCl and found 30 s of etching sufficient to reveal the ultrastructure of fine carbonate grains of biogenic origin. Despite the above examples of gentle etching, Wilkinson, Janecke & Brett (1982) used 50% glacial acetic acid and an etch time of 60 s with success on Ordovician limestones.

Deep etching can be employed profitably in rocks with a soluble cement. A calcite-cemented sandstone can be etched down from a flat broken surface to reveal the pre-cement morphology and to observe relics of carbonate replaced grains. Thus it is possible by use of suitable acids to manufacture examples of secondary porosity in carbonate cemented rocks. Particularly interesting results can be achieved with dolomite/calcite rocks and partially silicified limestones. It is useful to prepare a number of stubs of each rock to be examined so that ample material is available for experimentation with different acids and etch times. In this way photographs can be taken of examples 'before' and 'after' treatment. Polished surfaces of quartzitic sandstones can be etched with hydrofluoric acid to reveal rock texture and internal structures of grains such as heavy minerals.

With all etching techniques the specimen must be well washed with distilled water following etching to remove any traces of the acid or its reaction products, which otherwise may crystallize on the specimen surface during drying.

8.5.3 Mudrocks and fine grained sediment

Freshly fractured surfaces of mudrocks can be examined for features such as mineral orientation, shape and size, and organic content. Care must be taken to ensure that specimens are properly dried, and only small samples should be used since the high porosity of mudrocks combined with their low permeability results in large specimens degassing for long periods when placed in a vacuum for coating or observation.

Well-lithified mudrocks can be prepared as uncovered polished thin sections or as polished chips mounted in plastic; these can be examined with advantage in BSE images as shown by Krinsley, Pye & Kearsley (1983), White *et al.* (1984) and Huggett (1986). Preparation of the polished surface for high resolution SEM work has been discussed by Pye & Krinsley (1984). Conventional polishing methods can cause smearing of the surface which is apparent at magnifications greater than × 2500. Very gentle etching with HF may improve the surface but etching must not proceed to the point where unacceptable topographic differences are produced between grains. Ion-beam etching is a method which can be employed to remove a surface layer of even thickness to leave a 'cleaner' and flat surface (Smart & Tovey, 1982).

O'Brien, Nakazawa & Tokuhashi (1980) recorded a marked difference in texture in turbidite and hemipelagic muds when viewed on broken surfaces parallel to bedding. Hemipelagic muds show a preferred grain orentation due to dispersed settling of clays and turbidites exhibit a random texture caused by deposition of flocculated clays.

Recent sediments, both natural and artificial, have been prepared by freeze-drying techniques (Osipov & Sokolov, 1978) to examine the microtexture of high porosity (to 99%) samples of various clays.

Loose sediments of fine grain size can be examined by mounting grains directly on a stub using double-sided sticky tape, or a thin layer of glue on the stub. Sediment can be embedded in resin as described below (Section 8.5.5) and observed on polished and etched surfaces; this technique is particularly useful for carbonates (Hay *et al.*, 1970).

8.5.4 **Sedimentary grains**

The SEM is extensively used for examination of individual sedimentary particles. The treatment given to the particles prior to mounting on a stub is dependent on the type of material being examined and the object of the study. It is essential to ensure that the preparation method does not destroy or modify the features to be observed and does not create features which are artefacts of the preparation technique. If the grains to be examined have to be released from a partly-consolidated rock the minimum of force should be used and grinding of the sample must be avoided. The method employed will depend on the study to be undertaken, but ultrasonic and freeze-thaw techniques can be used as well as gentle mechanical and chemical methods.

SURFACE TEXTURES OF GRAINS

It is generally considered that only small numbers of detrital quartz grains need be examined from a sample to obtain representative surface textures (Krinsley & McCoy, 1977). The method described by Krinsley & Doornkamp (1973) and used by Wang, Piper & Vilks (1982) is as follows.

Five grams of the sample are boiled for 10 min in concentrated HCl and then washed with distilled water. If iron oxide coatings are present the grains are then boiled in stannous choride solution for 20 min and rewashed in distilled water. Organic debris can be removed by a strong oxidizing solution. Krinsley & Doornkamp (1973) used 1.5 g each of potassium dichromate and potassium permanganate dissolved in 15 ml of concentrated H_2SO_4. Wang *et al.* (1982) boiled their samples in 30% hydrogen peroxide for 10 min. The grains are given a final wash in distilled water and dried before mounting on a stub.

This preparation technique might be considered rather violent, but the fact that populations of quartz grains do exhibit features characteristic of their environments (Section 8.9) after such treatment suggests that relevant features are not destroyed in the process. It is usually stressed that ultrasonic cleaning should not be used as it may damage the grain surfaces, but Le Ribault (1978) processed samples through HCl, washing with distilled water then drying, sieving and selecting grains before subjecting them to ultrasonic cleaning and an alcohol wash prior to drying and mounting.

GENERAL PURPOSE GRAIN MOUNTS

Grain assemblages to be mounted for general purpose examination or analysis will include minerals that would be damaged or destroyed by the above technique. Careful washing of the grains in distilled water with gentle agitation can be sufficient to remove loose surface particles, but still leave adherent particles and coatings (which may themselves be of interest) on the grain surfaces.

The treatment employed will largely depend on the object of the study, but treatments can readily be devised to remove organic matter (hydrogen peroxide) or calcite (formic or acetic acid). It is frequently instructive to examine samples to which various treatments have been applied. Which treatment is employed should be clearly stated in any publication, a factor often ignored in many SEM studies of sedimentary grains.

8.5.5 **Impregnation of pores and borings (see also Section 4.4.2)**

It is frequently useful to impregnate porous media to reveal details of pore geometry or details of borings within grains. The impregnation allows the framework of the grains or rock to be dissolved to leave the pore network or boring preserved as a cast which can be examined with the SEM. Various materials have been used for impregnation of rock pores including coloured lakeside cement, Wood's metal and wax, but plastics are now used which are hard, strong and stable. Various epoxy resins are available which will dissolve a suitable dye (required for thin section work), have low viscosity and high wetting characteristics together with high final hardness and negligible shrinkage or expansion on setting.

Some techniques involve alternate use of vacuum and pressure (Pittman & Duschatko, 1970) to achieve impregnation. Impregnation can be achieved with no more vacuum than that provided by a water vacuum pump connected to a mains water supply and has been found satisfactory for impregnation of porous rocks with permeabilities above 30 md and for impregnation of fungal borings less than 1 μm in diameter.

PROCEDURES PRIOR TO IMPREGNATION

Prior to impregnation the sample may have to be treated in the manner described in Section 8.5.1 to remove oil, since the presence of oil makes impregnation difficult as it hinders the ability of the resin to wet the grain surfaces and may restrict impregnation depth to 1 or 2 mm. If organic matter is present, as may be the case in recent borings, it should be removed by use of chlorox or hydrogen peroxide. If it is desired to retain organic matter the methods of Golubic, Brent & Lecampion (1974) (summarized by Lukas, 1979) can be used.

If sediment grains are to be embedded and/or impregnated it is advantageous to clean them by washing in distilled water and by brief use of an ultrasonic cleaner to remove loose surface material which may be partly blocking the cavities to be impregnated. Samples should be dry before impregnation.

In the following simple method, developed by I.S.C. Spark at Aberdeen, the epoxy resin 'Epofix' is used. The temperature of the resin is critical for good impregnation to take place. Below 40°C impregnation is difficult to achieve as the resin has too high a viscosity (550 cP at 25°C as against 150 cP at 60°C). Above 70°C the resin sets very rapidly when the hardener is added since the reaction is exothermic. A temperature of between 55 and 65°C is usually satisfactory. At 60°C the dye ('Waxoline blue') dissolves more easily in the resin than at lower temperatures but it still has a pot life of about 45 min. This limits the impregnation time but it is generally found to be adequate. The degree of vacuum applied is important to achieve good impregnation. Too high a vacuum (less than 2 mm Hg) results in the resin boiling, but too low a vacuum results in lack of impregnation. A vacuum pump connected to the mains water supply provides a sufficient vacuum. To avoid waste of resin it is usually convenient to impregnate material in batches of 10 or more samples at a time.

EQUIPMENT AND MATERIALS

Glass vacuum desiccator connection to mains water vacuum pump. Cylindrical polythene moulds 40-mm diameter (suspplied by Struers).
Oven (60°C).
Vacuum grease.
'Prepo' release powder.
'Waxoline Blue' dye powder.
'Epofix' resin and hardener.

First ensure that the samples to be impregnated will fit into the moulds and are clean and dry and have had any oil removed. Seal the contact between the mould body and base with vacuum grease to prevent leakage, and smear the whole internal mould surface with 'Prepo' release powder to prevent adhesion of the hardened resin to the mould. Clean off any loose powder.

Two hundred ml of resin (for 10 samples) are mixed with a level teaspoon of 'Waxoline blue' dye and placed in an oven at 60°C and gently stirred at 5 min intervals until the dye has dissolved (30 min approx.). Twenty-five ml of hardener is added and well mixed in for 1 min. The resin is then immediately poured into the 10 labelled moulds and the samples dropped in. (Do not put the samples in first as a flat based sample may not be wetted or may adhere to grease.) Put the moulds plus samples in the desiccator and apply the vacuum. Every 4 min release air into the chamber for 1 min and repeat this process for 30 min. Then remove the moulds and allow to set and harden for 24 hours at room temperature; do not allow the resin to set under vacuum as small remaining air bubbles will not dissolve in the resin. Finally release the impregnated samples from the moulds and cut suitable sections for treatment and examination by SEM. (It is not essential to use a dye for the SEM work but it is useful if a thin section is also required.)

A similar method of impregnation is described by

Walker (1978) which he utilized for preparation of casts of chalk porosity (Fig. 8.10). Walker's method uses the epoxy resin Araldite AY-18 which has the advantage of low viscosity which is retained for up to a week after mixing. Curing is started by heating the resin to 80°C. Using the method, spectacular detailed impregnation of pores only 0.1–0.2 μm across was obtained and even pores in foraminifera walls were impregnated. Patsoules & Cripps (1983) have described an impregnation method also used on chalk in which Trylon CL 223 PA resin is used. A more elaborate impregnation chamber is described in Section 4.42 and Fig. 4.4.

The 'Aberdeen' method described previously has the great advantage of cheapness, simplicity and speed of operation, but Walker's (1978) method probably achieves a greater penetration of pores. Commercially available impregnation units are now on the market which provide better vacuum conditions and permit a greater number of samples to be impregnated at one time under more easily controlled conditions. However, there is still considerable scope for experimentation with impregnation techniques. Examples of impregnated rocks and grains are illustrated in Fig. 8.21 (pore casts) and Fig. 8.24 (microborings).

8.6 SAMPLE MOUNTING, COATING AND STORAGE

8.6.1 General considerations

When the required preparation techniques have been performed on the sample it may be of suitable size and shape for mounting directly on to a stub for examination or it may require some final shaping. Prior to mounting the sample the stubs to be used must be labelled; this is simply done by scratching or engraving a number on the side or underside of the stub.

Specimens are usually individually mounted on 10-mm diameter stubs designed to fit the particular make of machine being operated but larger 25-mm stubs can be used to take several specimens. It is generally possible to design an adaptor to make different stubs transferable between machines if this becomes desirable.

It is useful to have a known reference point on the stub surface so that the sample can be placed in a known orientation in the sample holder within the chamber. This can greatly facilitate the finding of a particular field of view when a sample is re-examined at a later date, provided X and Y co-ordinates of the area are also noted.

Using 10-mm stubs one can examine one stub at a time, but if changing the sample in the SEM chamber involves bringing the column to air it is advisable to have a mount which allows three stubs to be inserted and examined without the need for a specimen change. Apart from the obvious advantage of time saving, the reduction in number of specimen changes gives longer filament life and aids maintenance of cleanliness. However, it is essential to ensure that any modifications made to sample holders do not interfere with take-off angles to detectors and that the modified sample holder cannot touch the detector when the stage is tilted.

8.6.2 Mounting rock chips and slices

It is essential to ensure that there is a good contact between the specimen and the stub. For this reason samples should be as flat-based as possible. Samples should not seriously overlap the edges of the stub since the underside of the sample will not be coated in the sputter-coating process. The sample should also have as flat a surface as possible, particularly if analytical work is to be carried out. It is also an advantage to try and make the samples of approximately the same thickness. Figure 8.11 illustrates correct and incorrect sample configuration.

Porous samples can be trimmed using small pliers or whittled to size using a variety of probes; dental picks are very useful for this operation. Once the freshly-fractured surface to be observed has been exposed it is essential to prevent transference of dust to this surface and to protect it from damage. Some samples can be mounted prior to producing the fresh fracture; this is ideal, but frequently difficult to achieve. If the chosen rock chip has an irregular base this can be gently filed down or rubbed on glass until flat, the method employed depending on the friability of the specimen. Cut slices of relatively non-porous material are best cut to size and mounted on the stub prior to any etching techniques which produce a delicate surface.

The glue used to stick the sample on the stub should be stable under high temperature and vacuum conditions and have good adhesion to gold and carbon coatings. The silicone rubber glue 'Loctite' has these properties and has the added advantage of

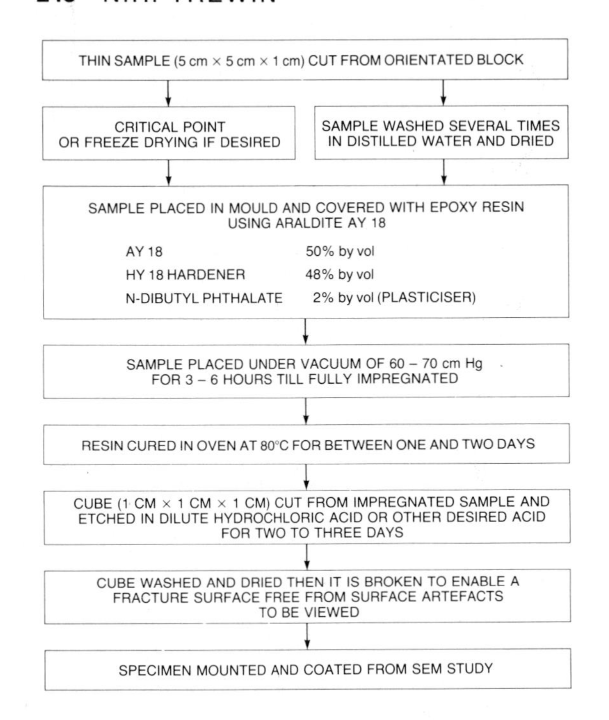

Fig. 8.10. Impregnation technique for the production of epoxy resin pore casts. Modified from Walker's (1978) technique for the production of casts of chalk porosity.

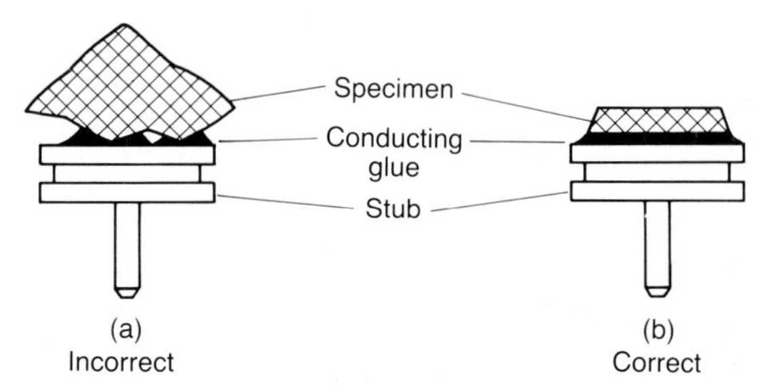

Fig. 8.11. Correct and incorrect configuration for specimen mounting on a stub. Note lack of overhangs, flat surface and complete bond in (b).

being initially very viscous and fast setting so that fracture surfaces can be easily mounted in a suitable orientation even when the sample base is unavoidably uneven. An advantage can sometimes be obtained by using a conducting glue to help carry away any charge from the specimen surface. If for unavoidable reasons an irregular specimen has to be mounted which is poorly coated at the sides it can be partly coated with conducting glue or carbon-dag to help prevent charging of the specimen. Glues of the Polyvinylacetate type must be allowed to harden fully for a day before coating is attempted as they continue to de-gas for some time after initial drying. Epoxy resin glues such as 'Araldite' can be used but must also be allowed to cure properly.

8.6.3 Mounting of grains and loose sediments

Individual grains can be mounted using double-sided sticky tape or alternatively the stub can be coated with a thin layer of glue or conducting silver paint and the grains dropped or pressed on to the surface. The glue layer must be thin enough to prevent the grains sinking into the glue. Loose sediments from which a random sample of grains is to be mounted can be placed on a flat surface and the glued stub gently pressed on to the grains. Glue types are discussed in the section above.

Grains mounted on double-sided tape have a smaller area in contact with the stub than do those mounted with a spot or thin layer of glue and thus have a greater tendency to charge up in the electron beam due to poor conductivity to the stub. This is enhanced by the shadow area beneath the grain which is poorly covered in the coating process.

When individual grains are being mounted it is advisable to have some sort of reference point or grid on the stub surface so that a simple map of the stub surface can be made to aid the location of individual grains (Fig. 8.12a).

Roughly spherical grains, of which as much of the surface as possible is required to be viewed, can be mounted on modified stubs of rod shape or stubs with a vertical semicircular rim as in Fig. 8.12(b).

Many studies require surface textures and/or grain size to be determined with the SEM and frequently very small numbers (10–20) of grains have been used. Most workers dealing with sand grain surface textures use 15–20 grains. Culver *et al.* (1983) concluded that 30 is an adequate number and it appears that operator bias in recognition of features would outweigh anything gained by examination of more samples. Tovey & Wong (1978) gave evidence that a sample of 50 grains chosen at random is reasonably representative of the grain size distribution of a sample determined by sieve analysis. Thus a random method of selection of grains for mounting must be

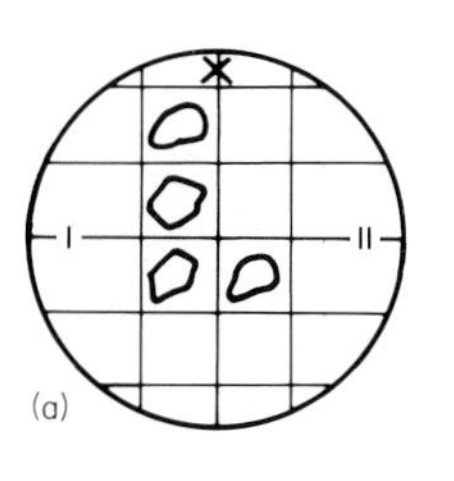

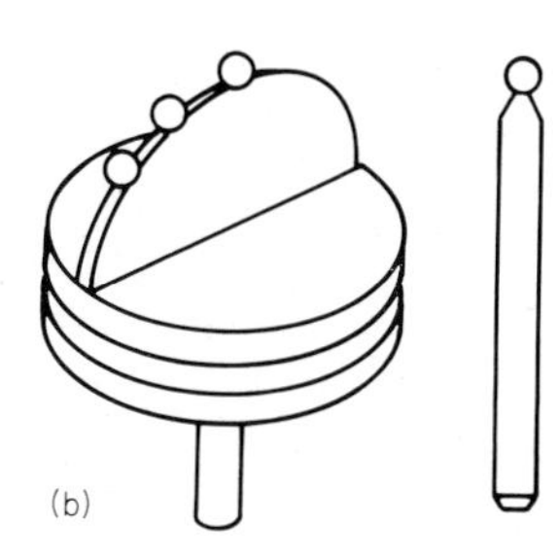

Fig. 8.12. (a) Stub with simple orientation marks and grid engraved on surface to facilitate location of specific grains. (b) Modifications of specimen stubs to permit viewing of greater proportions of the surfaces of spherical grains (modified from fig. 6.2.1 of Smart & Tovey, 1982).

used to ensure an adequate sample of grains for the intended project.

8.6.4 **Coating the specimen**

When the specimen has been mounted and after the glue is fully dried the sample must be coated with a conductive layer to take away the electrical charge which builds up on the specimen surface due to bombardment by the electron beam. The two most commonly used types of coating are gold and carbon, but various metal alloys such as platinum-palladium and gold-palladium can be used.

Samples are normally gold coated using one of the many models of sputter coater now on the market. Coating units with a carbon evaporation power supply are also readily available. The quality of the coating achieved depends on the quality of vacuum obtained; thus it is worth spending time obtaining a good vacuum rather than one which is only marginally satisfactory before commencing coating.

The thickness of gold coating to be applied should be considered in relation to the type of work to be undertaken and the type of specimen. Porous specimens with grains which are only in poor contact with each other generally require a longer coating time in order to eliminate charging effects on the specimen at any particular operational voltage (and consequently may give better results if examined in backscattered mode). The merits of different thicknesses and types of coating are summarized below: (1) Gold coating. 100 Å or more. (2) Gold coating. 25 Å or more. (3) Carbon coating. (4) Uncoated samples.

(1) The thicker gold coating is advantageous in eliminating any charge effects and is useful for obtaining good resolution pictures. The disadvantage of a thick gold coating is found if energy dispersive analysis is to be performed on the sample, since large gold peaks are obtained which can obscure peaks due to other elements (e.g. sulphur). Reduction of peak heights for elements such as Na, K and Mg also occur. In addition has been found that if the coating is increased to 200 Å the Na or K in feldspar may not give peaks so leaving only Si, Al and the gold peaks. Thus thick gold coatings should be used only for obtaining better pictures when analysis is not required. Figure 8.7 illustrates analytical results from a gold coated specimen.

(2) The thin gold coating of 25 Å has the disadvantage that charging effects may be a problem, particularly at high magnifications, and it is usually necessary to apply a conductive medium (carbon-dag or conducting glue) to the sides of the specimen mounted on the stub. Thin coating has the advantage that gold peaks on EDS analysis traces are very small or absent and elements such as sulphur can be easily detected and the absorption effects on other elements are greatly reduced.

(3) Carbon coating is of great advantage in that carbon is outside the detection range of the analyser and does not affect peak intensities on EDS traces. Carbon coating is achieved by vaporizing carbon rods or carbon fibre; it is more difficult to achieve an even coating with carbon than with gold and charging effects may be experienced. However, if good carbon coating apparatus is available it is a preferable method if analysis is envisaged.

(4) Uncoated samples. Sometimes it is desired not to coat a sample, possibly for a rapid evaluation or if the specimen is a curated mineral or fossil specimen. In this case the sample can be viewed and analysed but charging will build up rapidly on the specimen. If, however, a back scattered electron detector (e.g. Robinson detector) together with a CFAS (Charge-Free Anti-Contamination System) working at a low vacuum is available the sample can be observed, analysed and photographed without specimen charging, and with the added advantage that atomic number contrast images are produced (Robinson & Nickel, 1979).

8.6.5 **Storage and handling of stubs**

Storage of SEM stubs can be the responsibility of individual workers (most academic establishments) or can be a technical responsibility of the SEM laboratory. In either case the requirement is for dust tight boxes, lidded trays or a cabinet in which stubs are firmly held and can be labelled in such a way that file numbers can be read without having to pick up the stub. It is advantageous to have small storage units taking tens of samples rather than large units with hundreds of specimens in a single container — this results in minimal disturbance for each stub. Stubs can also be stored individually in glass or plastic tubes by having a hole in a cork to take the stem of the stub. It is commonly found that a coated specimen stored for a period of maybe only a few days, but more normally months, will charge badly when re-examined. This applies particularly to rock samples containing expandable clays. Bohor & Hughes (1971) recommended keeping clay samples individually in glass tubes as above with the addition of a small amount of desiccant in the tube to keep the sample in stable, dry conditions, and to prevent swelling of clays causing rupture of the conductive coating. Samples with hygroscopic salt content are also prone to charging following storage.

Storage units can be made sturdy enough for transmittal by post if the stubs are firmly secured in their cavities. When handling stubs it is advisable to use a pair of suitably shaped tweezers which will hold the stub firmly. It is easy to damage a coated stub with careless handling.

8.7 PROBLEMS OF SEM OPERATION

8.7.1 General

It is not possible or desirable to attempt to provide a trouble-shooting manual for the SEM operator in a short chapter such as this. Details differ between machines of different makes, and technology is advancing so rapidly that any technical specifications will soon be out of date. It is therefore assumed that all machine functions are operating correctly, and only those features most commonly found to affect image quality are listed here. In all cases the operation manual supplied by the makers of the SEM should be followed in the setting up and running of the machine. It much be stressed at the outset that many of the problems encountered in day to day running of an SEM are a direct result of lack of cleanliness and poor specimen quality.

8.7.2 Possible reasons for poor image on screen

(a) *Dirty column, apertures or filament housing.* SEM models where specimens can be changed without bringing the column to air have a distinct advantage in this respect. In other models dust can more easily enter the chamber during specimen changes and so contaminate the column. A spare clean column should be kept available. If the final aperture in the column becomes dirty and starts charging then distortion of the image and sudden shifts of image can occur.

(b) *Unstable filament.* The tungsten filament is heated close to its melting point to provide the electron beam, and can behave in an unstable manner, resulting in rapid changes in filament current and hence rapid fluctuations in picture brightness. Unless the filament current is stable there is no point in attempting to take photographs. A new filament frequently takes an hour to settle down and may become unstable towards the end of its life.

(c) *Filament not correctly centred.* If the filament is not correctly centred a low intensity picture with poor focus will result. Occasionally a filament will go off centre during operation and it is advisable to check the centering of the filament following specimen changes.

(d) *Specimen faults.* A specimen that is charging badly will give a poor picture with bright spots and photographs will have bright horizontal lines emanating from charged areas (Fig. 8.14). Charging of the specimen could be the result of poor specimen coating, poor connection of the sample to the stub, or poor earthing of stub or stage. Samples which were wet or contained volatiles such as oil when coated, give poor pictures due to poor coating. Samples mounted with volatile glue may also give noisy pictures, as do specimens with a large quantity of organic matter. Microporous samples can take a significant time to de-gas in the vacuum and performance may improve if the specimen is left in the vacuum for an hour before observation.

(e) *Image not sharp.* If the image cannot be sharpened by use of the focus control, and charging is not a problem, it is possible that the spot size is too large, but if spot size if reduced too much the image will become excessively noisy. Reduction in beam current or photo-multiplier gain may also help sharpen the image.

(f) *Astigmatism effects*. If at high magnifications (× 10,000) there is distortion giving oblique elongation of features when the focus is adjusted the astigmatism controls are probably in need of adjustment. Serious astigmatism may be due to poor alignment of the column or dirt in one of the column apertures.

(g) *Extraneous 'noise' problems*. Many buildings are not particularly stable and vibration due to traffic, other machinery, high wind and the SEM pump itself can be a problem. The SEM should be set up in the most stable and quiet environment possible. If vibration is suspected compare operation at different times of day and place the pump as far from the machine as possible. Vibration results in straight edges having a saw-tooth appearance at high magnifications.

8.7.3 Possible reasons for poor photographic results

Photography of images on the SEM should result in a slightly improved image to that seen in the slow scan mode on the screen, and a greatly improved image over that seen on TV mode. Photographic instructions should be followed for the individual machine and minor adjustments to practice are generally made to suit the individual machine and provide acceptable negatives for the type of investigation undertaken.

Brightness and contrast are usually judged from the amplitude and position of the waveform of the image on the viewing screen. The optimum position and amplitude will depend on film speed and aperture used on the camera; thus it is worthwhile experimenting with both brightness and contrast to achieve the best negatives or Polaroid prints. It is a simple matter to fix a reference scale at the side of the viewing screen on which can be marked the optimum positions of the waveform for different film speeds and apertures as shown in Fig. 8.13.

The following factors can be responsible for poor photographic results:

(a) *Incorrect exposure* resulting in either a pale or dark negative or one with too much or too little contrast. This can be easily corrected by experiment as described above.

(b) *Camera not focussed* on photo-screen. If all photographs appear out of focus, despite the viewed image having been good, this is a probable reason.

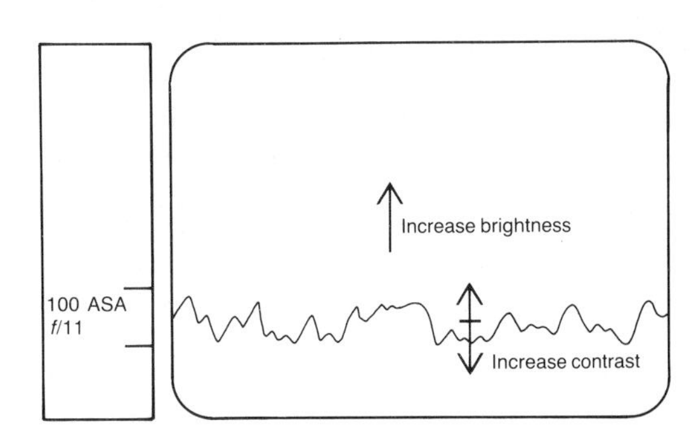

Fig. 8.13. Appearance of waveform on viewing screen adjusted to lie within marks determined by experiment to be suitable for 100 ASA film speed and aperture *f*/11. To increase contrast the waveform should be expanded, and to increase overall brightness the waveform should be moved up the screen.

(c) *Unsuitable views*. It is not possible to take excellent photographs of all views! The depth of focus required can be too great and the contrast may not be adjustable to a suitable level; however, the provision of a gamma control on some instruments

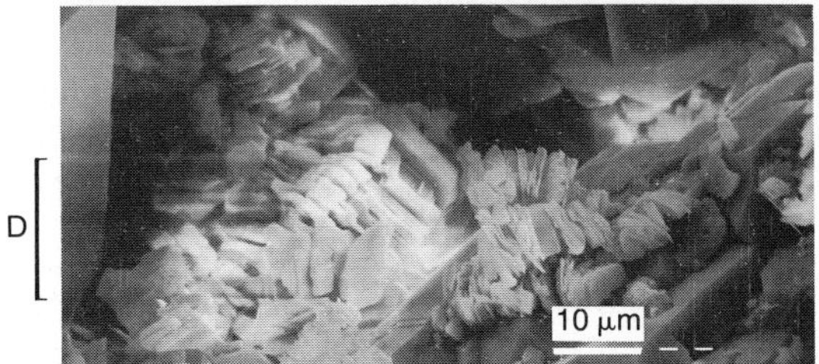

Fig. 8.14. Faults commonly observed on photographs.
A. Bands due to fluctuation of filament current.
B. Picture dislocation due to vibration.
C. Bright, washed-out area, due to over-exposure, or poor specimen coating.
D. Specimen charging.

aids in producing acceptable contrast for photography in otherwise poor areas.

(d) *Charging effects*. If the specimen is affected by charging, light or dark lines will be produced emanating from the charged areas (Fig. 8.14).

(e) *Unstable filament current*. Fluctuations in filament current will result in abrupt changes in brightness, forming bands on the photograph (Fig. 8.14).

(f) *Vibration*. Vibration at high magnifications results in a wavy image on the screen; sharp knocks to the instrument while a photograph is being taken can result in fault-like dislocations in pictures as in Fig. 8.14.

(g) *Dirty photo-screen*. In some instruments there is a strong tendency for the photo-screen to become electrically charged and attract dust particles which adhere to the surface. The patches where dust adheres appear as diffuse darker areas on photographic prints since the dust has cut out some light from the camera. The photo-screen should be inspected and cleaned regularly to prevent the accumulation of dust.

8.8 EXAMPLES AND REVIEWS

8.8.1 Introduction

The SEM is now regarded as a standard instrument for use in sedimentological studies to provide information and illustrations of small-scale three-dimensional surfaces. In recent years there has been a marked increase in the number of sedimentological papers utilizing the SEM and it is not the intention to review all the current applications, but to concentrate on some aspects where SEM use is essential to the study. It should be stressed that the SEM is seldom used in isolation from other techniques; thus in the examples discussed it must be appreciated that the contribution from XRD, thin sections, cathodoluminescence, isotopes and biological studies may also be essential.

The relative ease with which publishable quality photographs can be obtained using the SEM both in SE and BSE modes gives great scope for illustration of reports and papers. There is a strong tendency when viewing material to photograph the unusual and the beautiful at the expense of the general features. Sedimentologists are attracted by well-rounded aeolian quartz grains, vermicular kaolinite and aragonite needles, and these are frequently illustrated, whilst less attractive and less easily interpreted features tend to be ignored. SEM use is now entering a quantitative phase and every effort should be made to quantify observations. The recent developments in quantitative analysis of BSE images of polished surfaces (Dilks & Graham, 1985) are most important and should contribute to many branches of geology. Studies combining several of the available SEM techniques are most valuable, and the current trend is away from repetitive illustration and towards integrated studies using SE, BSE, cathodoluminescence and element mapping.

The four topics below have been selected to provide a wide range of examples and also because they account for much of the sedimentological use of the SEM in the current literature. They are topics in which SEM work is essential to achieve the best understanding.

Quartz grain surface textures — Section 8.9.
Diagenesis of sandstones — Section 8.10.
Limestones and dolomites — Section 8.11.
Endolithic microborings — Section 8.12.

Other topics for which the SEM is a useful tool but which are not covered by specific sections are the examination of fine grained non-carbonate muds and mudrocks as illustrated by O'Brien (1987) in a study on the effects of bioturbation on the fabric of shales. The examination of heavy mineral assemblages is greatly aided by use of the SEM with EDS system for analysis. Surface textures of grains other than quartz are also providing useful information as in the study by Hansley (1987) comparing natural etch features on garnets with those produced experimentally using organic acids. The SEM has also proved valuable in the investigation of detailed features of framboidal pyrite as illustrated by Love *et al.* (1984). Examples of other applications can be found in Whalley (1978) and in the volumes of the proceedings of the Annual Scanning Electron Microscopy Symposium.

8.9 QUARTZ GRAIN SURFACE TEXTURES

8.9.1 Introduction

The main impetus in the study of surface textures of quartz grains is the belief that environmental interpretations can be made on the basis of characteristic surface textures. The main problem facing a new recruit to this field of study is the proliferation of names of surface features and the subjective manner in which many have been described. The atlas of Krinsley & Doornkamp (1973) remains a useful source of illustration of textures but many have been described since and illustrations are scattered in the literature. Good illustrations are provided by Cater (1984) and by Higgs (1979) who also gave definitions and sources of original descriptions and summarized the relationships of surface textures to depositional environment. Bull (1981) provides a valuable review of the use of surface textures in environmental interpretation in more detail than can be given here and provides an excellent reference list. It is essential to stress from the outset that surface textures are produced by *processes* of erosion, transport and deposition, many of which are duplicated in different environments.

8.9.2 **Processes, textures and procedures**

Margolis & Krinsley (1974) discussed the physical and chemical conditions responsible for the production of surface textures, and related textures to the crystallography of quartz. Some 22 surface textures were recognized and their relative abundance in different environments discussed. Higgs (1979) recognized 30 main textures and Culver *et al.* (1983) utilized 32 features in a useful comparative study where five different operators examined the same coded samples. This study revealed considerable variance between operators in recognition and scoring of the textures, but nevertheless the operators correctly classified the samples in terms of environment in 49 out of 50 cases. Cater (1984) used 22 surface textures and usually estimated the percentage of the grain surface covered by each feature.

As can be seen from the list of features (Fig. 8.15,

	% GRAINS SHOWING FEATURE
● ABUNDANT	>75
• COMMON	25 - 75
○ SPARSE	25 - 75
□ RARE	<5

SURFACE FEATURE

ENVIRONMENT	1 SMALL IRREGULAR PITS (<10μ)	2 MEDIUM IRREGULAR PITS	3 LARGE IRREGULAR PITS (>100μ)	4 SMALL CONCHOIDAL FRACTURE (<10μ)	5 MEDIUM CONCHOIDAL FRACTURE	6 LARGE CONCHOIDAL FRACTURE (>100μ)	7 STRAIGHT STEPS	8 ARCUATE STEPS	9 FRACTURE PLATES/PLANES	10 PARALLEL STRIATIONS	11 IMBRICATED GRINDING FEATURES	12 ADHERING PARTICLES	13 MEANDERING RIDGES	14 STRAIGHT SCRATCHES	15 CURVED SCRATCHES	16 V's	17 ANGULAR OUTLINE	18 ROUNDED OUTLINE	19 LOW RELIEF (<0.5μ)	20 MEDIUM RELIEF	21 HIGH RELIEF (>10μ)	22 ORIENTED ETCH PITS	23 ANASTOMOSING ETCH PATTERN	24 SOLUTION PITS	25 SOLUTION CREVASSES	26 SCALING	27 SILICA GLOBULES	28 SILICA FLOWERS	29 SILICA PELLICLE	30 CRYSTALLINE OVERGROWTH
CRYSTALINE SOURCE ROCK				●	●	●	●	●	•			●					●		●											
SUBAQUEOUS – FLUVIATILE – LOW ENERGY	●	●	●											○	○	○	•	•		●	•						•		•	○
SUBAQUEOUS – FLUVIATILE – MEDIUM ENERGY	●	●	●											○	○	○	•	•		●	•						•		•	
SUBAQUEOUS – FLUVIATILE – HIGH ENERGY	●	●	●											•	•	•	•	•		●	•									
SUBAQUEOUS – FLUVIATILE – TORRENTIAL	●	●	●	•	•	•	•	•						•	•	•	●	○	•	●	•									
SUBAQUEOUS – DELTAIC – SUBAERIAL	●	●	●											○	○	○	•	•		●	●							•		•
SUBAQUEOUS – DELTAIC – MARSH	●	●	●											○	○	○	•	•		●	●								•	○
SUBAQUEOUS – DELTAIC – CHANNEL – LANDWARD	●	●	●											○	○	○	•	•		●	●						•		○	•
SUBAQUEOUS – DELTAIC – CHANNEL – MEDIAL	●	●	●											○	○	○	•	•		●	●						○			○
SUBAQUEOUS – DELTAIC – CHANNEL – SEAWARD	●	●	●											•	•	•	•	•		●	●	•	○							
SUBAQUEOUS – MARINE – INTERTIDAL	●	●	•	○	○		○	○						●	●	●	○	●	○	●	●	○					○			
SUBAQUEOUS – MARINE – SUBTIDAL	●	●	•											•	•	•	○	●		●	●	•	○							
AEOLIAN – COASTAL	●	●	•	○	○	○	○	○				•	•	•	•	•	○	●	○	●	●									
AEOLIAN – HOT DESERT	●	●	•		○	○						•	○				○	●		●	●		○	○	○	○				
GLACIAL				●	●	●	●	●	●	•	•	●	○	●	•		●		●											
PEDOLOGIC – SILICA DISSOLVED – TEMPERATE	●	●	●														•	•		●	●	•	○			○				
PEDOLOGIC – SILICA DISSOLVED – TROPICAL	●	●	●														•	•		●	●	●	•	●	●	●				
PEDOLOGIC – SILICA PRECIPITATED – TEMPERATE	●	●	●														•	•		●	●						●	○	●	•
PEDOLOGIC – SILICA PRECIPITATED – TROPICAL	●	●	●														•	•		●	●						●	•	●	•
SUBSURFACE DIAGENETIC	●	●	●									•					•	•		●	●		•	•	•		●		●	•

Fig. 8.15. Quartz grain surface textures characteristic of various sedimentary environments. Redrawn from Higgs (1979).

adapted from Higgs, 1979) many of the names are subjective and rely on visual interpretation by the operator. Contrasting surface textures of quartz grains from beach environments are illustrated in Fig. 8.16.

The statistical analysis employed by Culver *et al.* (1983) showed that a combination of features should be used to distinguish between samples and that the use of a single microtexture to distinguish the environment is invalid. This feature was stressed by Krinsley & Donahue (1968) but much work in the 1970s failed to live up to their standards.

The same surface features can develop in very different environments. Manker & Ponder (1978) showed that grains from fluvial environments develop some features also found in aeolian dune and beach environments, thus confirming the need to observe an assemblage of features. Features can also be inherited from a previous environment as in the case of 'glacial' features such as conchoidal fractures and angular edges being still recognizable in grains transported by turbidites for 120 km from the shelf edge over the Laurentian Fan (Wang *et al.*, 1982).

Before proceeding with any study of surface textures, be they on quartz or any other mineral, it is pertinent to consider the likely history of the grains in question. The major features are summarized in Table 8.1. Many studies either omit to consider these basic geological principles or make the assumption that the transport regime produced all the observed features and that the depositional regime reflected the transport regime.

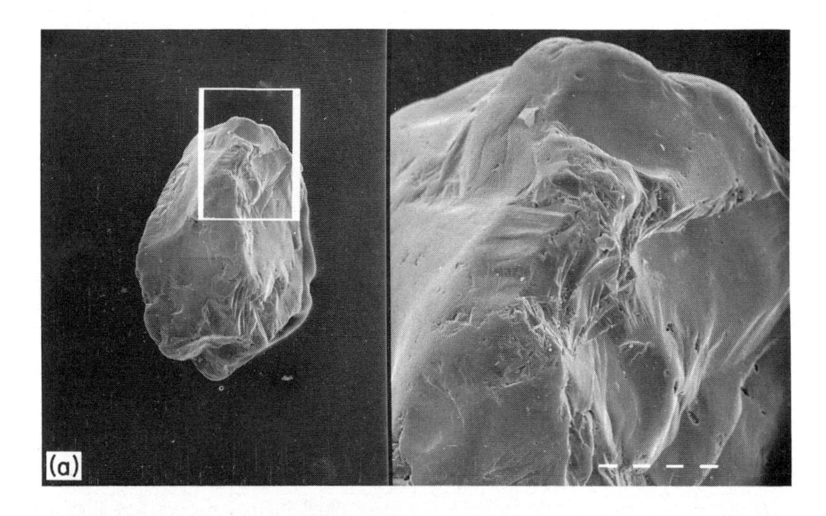

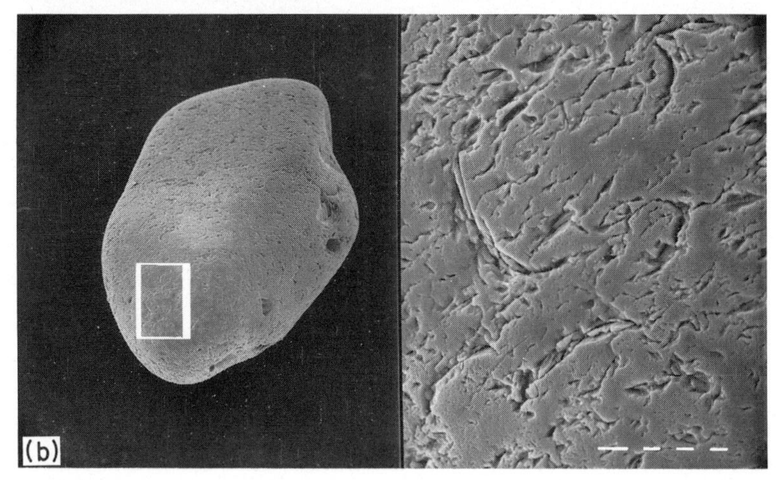

Fig. 8.16. Examples of contrasting quartz grain surface textures from recent beach environments.
(a) Aberdeen beach, North Scotland. Grain with conchoidal fractures partly smoothed by abrasion. Probably a grain of glacial origin reworked into beach sediment. Split screen enlargement area × 5 marked scale (first scale bar 100 μm).
(b) Rottnest Island, West Australia. Rounded grain, probably reworked from older sediments but typical of this sub-tropical beach environment. Surface etched along percussion cracks and crystal imperfections. Split screen enlargement area × 10 marked scale (first scale bar 100 μm).

Table 8.1. Origins of quartz grain surface textures in the geological cycle.

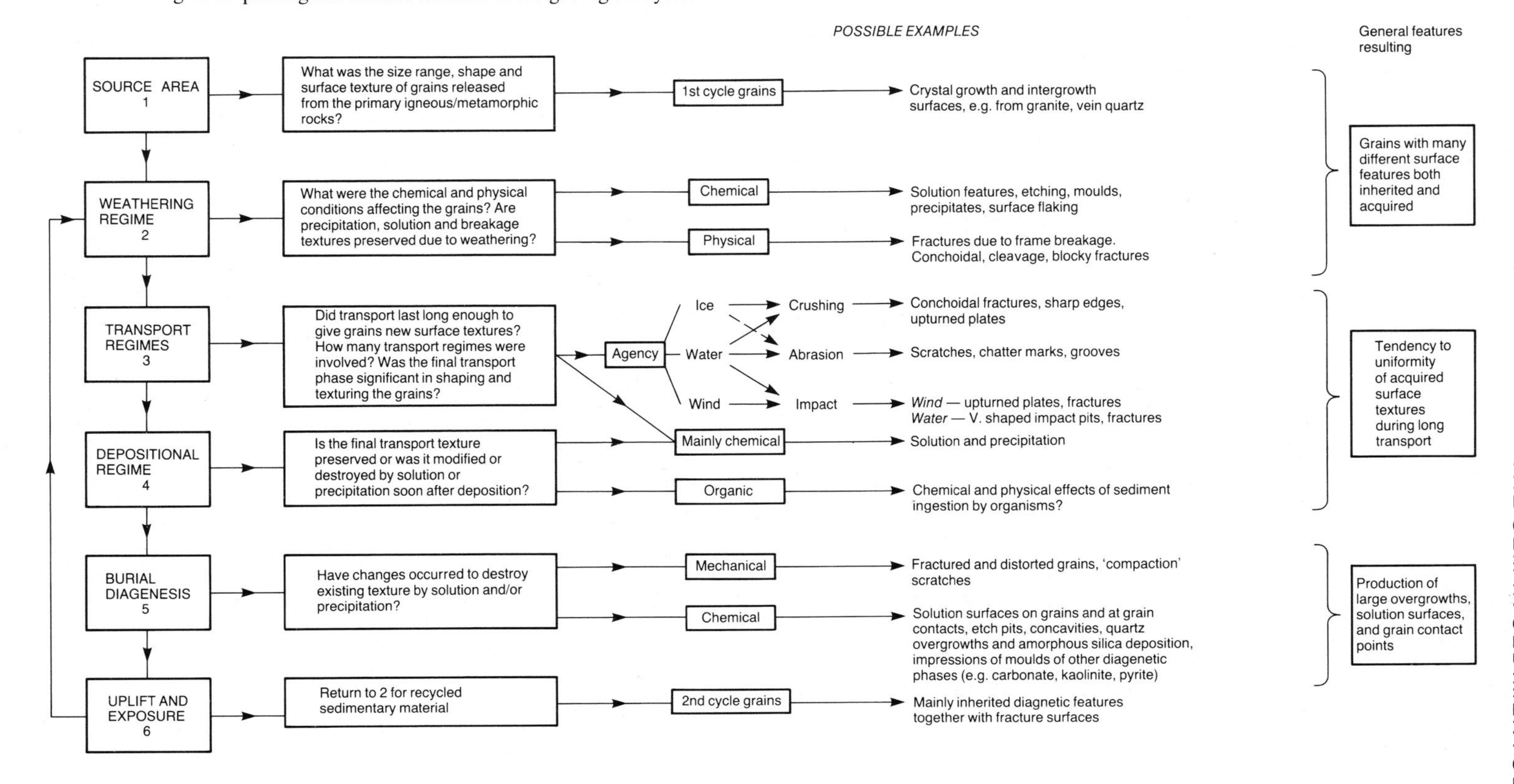

It is clear that in many cases it will be impossible to answer all the questions posed in Table 8.1 and the fact that 'environments' can be recognized from surface textures implies that in many cases the final transport regime does impart a new surface texture to the grain and that the texture remains for a significant time after deposition. The cases which do not 'work' are seldom reported so a bias towards successful interpretations builds up in published work.

The only way to arrive at a satisfactory conclusion in studies of surface textures is to examine sufficient numbers of grains (30–40) to record all features seen and employ a suitable statistical technique in processing the data. Bull (1978) used cluster analysis in cave sediments with effect and Culver *et al.* (1983) used canonical variate analysis. By the very nature of the study, statistical treatments must be employed to test for significant combinations of features. Clearly, as remarked by Bull (1978), one conchoidal fracture does not indicate glacial modification.

In the description of surface features it is now essential to refer to an identifiable surface texture without prejudice as to its origin, thus, for example, conchoidal fractures should not be referred to as being of glacial origin prior to consideration of all the evidence. Conchoidal fractures are common on glacial grains due to the frequency of crushing, but crushing also occurs in other environments, for example during bed load transport by high energy streams, and conchoidal fractures are also recorded as a mechanical weathering feature of a Carboniferous sandstone by Wilson (1978) and of granite and gneisses by Krinsley & Doornkamp (1973). Pye & Sperling (1983), in weathering experiments on the production of silt using a climatic cabinet, discovered that salt weathering is effective in producing angular quartz silt with conchoidal and blocky textures.

One factor which needs standardization in studies on surface textures is that of grain size. Some workers pick grains 'at random' while others specify 'sand size'. Wang *et al.* (1982) used 15–18 grains of 2.0–0.4 mm diameter, Mazzullo & Ehrlich (1983) fine sand of 0.180–0.125 mm and Manker & Ponder (1978) used grains of 'approximately the same size (1.0 mm) and shape'.

Tovey & Wong (1978) discussed selection of grains advocating either a random selection or one based on size fractions of the sample. It would appear that more studies need to be done on grain size effects on surface texture, particularly with regard to water transport. Transport by rolling, saltation or by suspension should result in different surface textures in the same way that rounding is affected. Hence grains of the different transport populations (Visher, 1969) should be examined. Krinsley & Doornkamp (1973) noted that features changed with grain size and considered 200 μm a suitable divide between small and large grains; this may in many cases reflect a generalized break between suspension and traction populations of grains. Larger grains >400 μm tend to show records of abrasion and grains <200 μm are biased towards showing chemical effects (Margolis & Krinsley, 1974). Middleton & Kassera (1987) have shown that there is a considerable variation in the density of V-shaped impact pits with grain size in intertidal sands, and stress the need to adopt standardized techniques for such studies. The pit-density recorded varied with the magnification used for the photographs from which the pits were counted; thus the scale at which observations are made is most important. Manickam & Barbaroux (1987) have described seasonal variations in surface textures of suspended sand grains from the River Loire; mechanical features are dominant on samples collected in winter floods and chemically produced features during low summer flows.

The correlation of environment with surface textures as summarized by Higgs (1979) is a useful approach but, as stressed above, the mechanism of transport is most important in surface texture production and the mechanism need not be environmentally confined.

8.9.3 Experimental work

Some experimental work has been done to reproduce surface textures in the laboratory. Krinsley & Doornkamp (1973) report the reproduction of textures similar to those on glacially transported grains by freeze and thaw experiments, and the production of V-shaped pits and grooves by water transport. Most experimental work has been performed on the production of aeolian surface textures. Kaldi, Krinsley & Lawson (1978) mounted individual grains from different environments on a stub together with crushed quartz and produced the characteristic 'upturned plate' textures on the grains in only 24 hours of abrasion by quartz in a 'wind bottle' with a 'wind speed' of 20 km hr^{-1}. Features such as small V-shaped pits produced in a beach environment were virtually destroyed and replaced by the new 'aeolian' texture. In the experiments, 'before and after' photographs of the same area of each grain could be studied. Wellendorf & Krinsley (1980) related artificially produced upturned cleavage plates to quartz crystallography.

Krinsley & Wellendorf (1980) took this experimentation further in recognizing that both the size and spacing of platelets produced by aeolian bombardment are influenced by the impact velocity. Such studies lead the way to possibilities of interpreting energy levels of the environment.

8.9.4 Ancient deposits

The vast majority of papers on surface textures utilize material from contemporary environments. Extension of the technique to the interpretation of ancient sedimentary environments requires that the surface texture survives diagenesis, probably including some degree of lithification. In most cases quartz solution, overgrowths and cementation will have destroyed surface textures produced during transport but many examples do exist, such as aeolian textures preserved on Triassic sand grains (Krinsley, Friend & Klimentidis, 1976), and Rehmer & Hepburn (1974) recovered grains with typical glacial textures from the Palaeozoic Squantum 'Tillite' of Massachusetts.

Mazzullo & Ehrlich (1983) identified two grain populations in the Ordovician St Peter Sandstone in Minnesota which retained surface textures of aeolian and fluvial origin. The sandstone was deposited in a marine environment, thus little reworking took place, and they postulated that the sand grains bypassed the active beach environment from their fluvial and aeolian source areas.

Higgs (1979) examined textures on Lower Cretaceous to Palaeocene grains from the western North Atlantic continental margin. He concluded that many inherited features were present due to derivation of the sands from crystalline rocks undergoing acid weathering and that deposition was in both marine and non-marine environments. Such general conclusions could probably be reached more easily than by the study of surface textures, but on occasion surface texture studies may provide valuable information.

The study by Cater (1984) on quartz grains in 10 samples from a 145 m thick Neogene carbonate sequence in the Finestrat Basin of Spain provides an example of surface texture analysis applied to a sequence lacking indigenous fauna, and in which characteristic first cycle grains could be recognized. Hill & Nadeau (1984) used surface features of Wisconsin sands from the Canadian Beaufort Shelf to reconstruct depositional environments in a situation where it was not possible to produce sedimentological logs of boreholes. In these cases surface texture analysis provides useful additional environmental evidence to that available from field sedimentology, but diagenetic features and grain reworking are recognized as problems in interpretation. One limitation of such studies is that the time-consuming nature of the practical work does not enable great numbers of samples to be analysed, and grain surface texture studies inevitably have to be regarded as an addition to basic field sedimentology rather than an independent topic.

It appears difficult to distinguish many surface textures of diagenetic origin from those gained during transport and deposition. The reader might compare illustrations of diagenetic surface textures, as in Burley & Kantorowicz (1986), with the fluvially transported grains figured by Manickam & Barbaroux (1987).

8.9.5 Conclusion

The study of surface textures has certainly been shown to be of practical value, but a great deal more controlled experimentation with source rock disintegration, weathering and transport media still needs to be done. More comparative studies of experimental and natural systems using the same starting material would also be most useful. Perhaps the greatest need is for a new 'atlas of surface textures' to provide a fresh impetus to the studies. It is apparent that many workers have evolved their own terms which are not adequately defined. Attempts have been made to redefine features, such as the proposed terminology for cracks and hollows of Baynes & Dearman (1978), but unless a generally accepted and manageable list is adopted the technique will possibly die or remain in the hands of a few faithful adherents.

For the practical sedimentologist the grain surface texture technique is a useful adjunct to other studies leading to environmental discrimination. It should be used in conjunction with detailed grain size and shape analysis, field sedimentology and palaeontology. Surface textures are seldom employed as a primary discriminator between fossil environments due to the complexity of textures, lack of generally accepted standards, and the numerous possibilities for textural destruction and modification outlined in Table 8.1.

8.10 SANDSTONE DIAGENESIS

8.10.1 General

The SEM with EDS is an essential tool for the examination of porous sandstones. Impregnated and stained thin sections, and XRD analysis of clays, feldspars and carbonates can provide much information, but to investigate the morphologies and detailed textures of grain overgrowths and diagenetic minerals the SEM is essential. Details of fine-grained clays and other grain coatings and pore-fillings cannot generally be obtained from thin sections. With the SEM, mineralogy, textures and diagenetic sequences can be better elucidated and porosity and permeability can be related to diagenetic and depositional textural features. It is the study of oil and gas reservoir rocks which has provided the greatest boost for SEM studies in the past ten years and the SEM studies have now sparked off considerable experimental diagenesis work.

Burley *et al.* (1985) have provided a useful review of clastic diagenesis which clearly shows the important role of SEM studies and their relation to other essential experimental methods. The contribution of the SEM in studies of sandstone diagenesis and the essential use of other techniques to provide a balanced study is well illustrated by Huggett (1984b, 1986) on the controls and diagenetic sequence of Coal Measures sandstones, Kantorowicz (1985) on the Middle Jurassic of Yorkshire and Burley (1986) on Jurassic sandstones of the Piper and Tartan fields of the North Sea. Increasing use is being made of combined SE and BSE studies of sandstones as illustrated by Pye & Krinsley (1986a) on Rotliegend sandstones, and of combined BSE and cathodoluminescence (Ruppert *et al.*, 1985 and Fig. 8.8). Future progress will involve more experimental work and comparison of natural and laboratory-produced features in attempts to simulate diagenetic conditions. Huang *et al.* (1986) documented experiments on the conversion of feldspar to illite, and Hansley (1987) on experimental etching of garnets by organic acids. Chemical approaches to diagenesis are also discussed in the volume edited by McDonald & Surdam (1984).

8.10.2 Practical considerations

IDENTIFICATION OF MINERALS

The identification of minerals, particularly the clay minerals, is frequently a problem; even with the help of EDS analysis positive identification is not always possible with the SEM. Reliance on morphology alone is most dangerous, particularly with illites, chlorites and smectites which commonly exist as mixed-layer structures. The *SEM Petrology Atlas* (Welton, 1984) provides a useful combination of pictures and spectra for most common minerals. Without analytical backup the SEM is of limited use, frequently leaving the operator with severe problems in the identification of fine-grained phases.

With the use of an analytical facility it is possible to build up a file of spectra of known common minerals to compare against unknowns. Some phases may require separation for quantitative or XRD analysis (if not available with the SEM), but for many studies qualitative analysis is sufficient to confirm identification.

Reference to the literature provides many excellent examples of typical morphologies. Clays are illustrated by McHardy & Birnie (1987), and Wilson & Pittman (1977) illustrated the common morphologies of clays in porous sandstones. Clays and many other minerals are well covered by Scholle (1979) and Welton (1984). Authigenic feldspar is illustrated by Stablein & Dapples (1977), Waugh (1978a, b) and Ali & Turner (1982). Authigenic quartz overgrowths are commonly illustrated (Waugh, 1970; Pittman, 1972) and some forms of iron and titanium oxides are shown by Ixer, Turner & Waugh (1979) and by Walker, Waugh & Crone (1978).

Illustrations here (Fig. 8.17) show some of the most frequently observed morphologies of a few common minerals.

DETRITAL AND AUTHIGENIC PHASES

The distinction between authigenic and detrital material is normally based on the general rule that authigenic minerals display characteristic crystal forms with evidence of *in situ* growth within pore

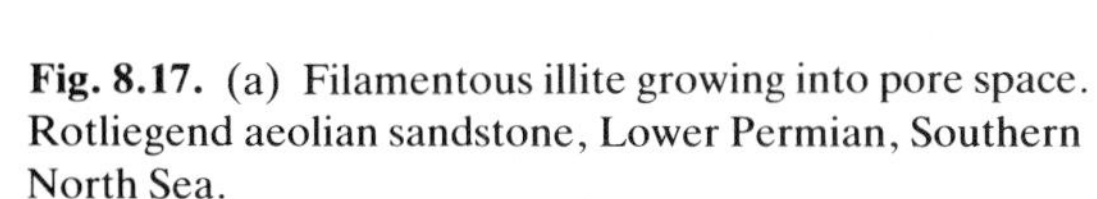

Fig. 8.17. (a) Filamentous illite growing into pore space. Rotliegend aeolian sandstone, Lower Permian, Southern North Sea.
(b) Quartz overgrowth which partly post-dates growth of siderite rhombs and fine-grained illitic clay. Biggada Sandstone Member, Hermite No. 1, NW Australian Shelf.
(c) Books of coarse kaolinite in pore space. Well developed quartz overgrowths partly post-date kaolinite formation as shown by 'impressions' of kaolinite in quartz. Mungaroo Formation, Flinders Shoal No. 1, NW Australian Shelf.
(d) Diagenetic kaolinite with vermicular habit, typical of 'freshwater' diagenesis, probably resulting from alteration of a feldspar grain. Upper Jurassic paralic facies, Piper Formation, Claymore Field area, North Sea (enlarged area × 5 indicated scale). (Photograph by I.S.C. Spark.)
(e) K-feldspar grain with diagenetic overgrowth. Needle-like projections in a grain contact area may represent incomplete development of overgrowth or dissolution at the grain contact. Lower Cretaceous, Claymore Field area, North Sea. (Photograph by I.S.C. Spark.)
(f) Microquartz crystals covering part of a single grain surface, such coatings can inhibit the development of quartz overgrowths and help preserve porosity. Lower Cretaceous, Claymore Field area, North Sea.

space or on grain surfaces. Some authigenic material such as iron hydroxide grain coatings and amorphous silica do not have a characteristic shape but tend to plaster detrital grains (Fig. 8.18).

Detrital grains may be clearly recognizable or may be so enclosed in clays of diagenetic origin that their surfaces are obscured (Fig. 8.18). Detrital clays pose the greatest problem for identification. The distinction between detrital and authigenic matrix clays is usually based on the assumption that detrital clays have poor crystal form, and may show distortion due to compaction and a tendency to wrap around larger grains. However, many 'muddy' or 'dirty' sandstones have a matrix comprising a mixture of detrital and authigenic components which may include detrital clays which have become overgrown or recrystallized during burial (Wilson & Pittman, 1977). Thus, as recognized by Cummins (1962) it has not been possible to distinguish between 'detrital matrix' and authigenic clays in greywackes, but Morod (1984) uses the SEM to demonstrate a diagenetic origin for the matrix of some Upper Proterozoic

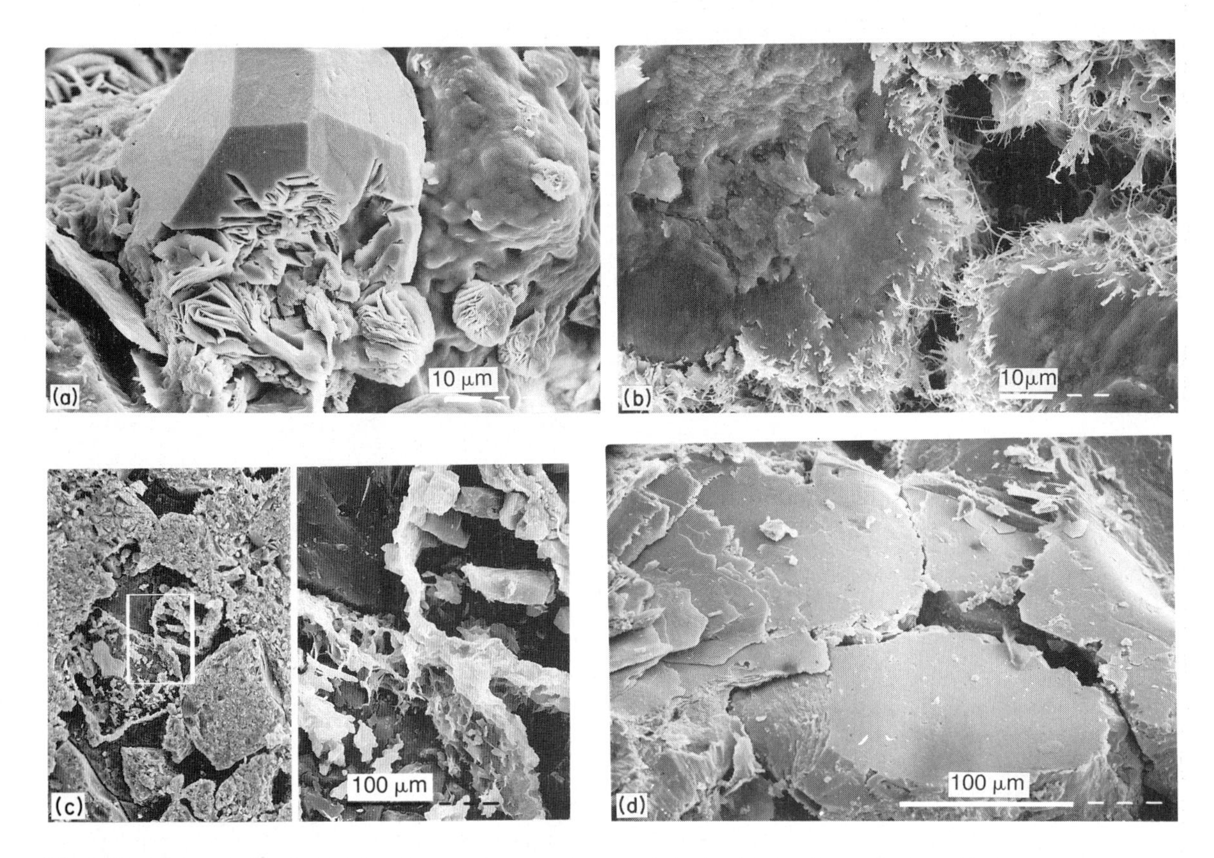

Fig. 8.18. (a) Detrital grain (right) with amorphous coating containing Si, Al, K, Mg and Fe. Mg-chlorite rosettes grow on the grain coating and pre-date development of isolated euhedral authigenic quartz. Permian aeolian sandstone, Corrie Shore, Arran, Scotland.
(b) Sandstone with amorphous Fe-rich coatings on detrital grains from which illite has grown into pore space. Quartz grain contact solution surface at top left. Rotliegend aeolian sandstone. Lower Permian, southern North Sea.
(c) Cut and acid-etched surface of calcite cemented sandstone. Cut grains stand out from the (darker) etched calcite cement. Detail shows two grains extensively replaced by calcite but with preserved illitic clay rims. Carboniferous sandstone, Canning Basin, West Australia.
(d) Solution contact between detrital mica and quartz grains; the platy mica is broken away to reveal the smooth flat contact-solution surface of the quartz grains beneath. Upper Jurassic sandstone, Claymore Field area, North Sea. (Photograph by I.S.C. Spark.)

greywackes from Sweden. Detrital clays can also infiltrate pore space as shown by experiment and in nature by Walker *et al.* (1978).

Apart from completely new-formed authigenic minerals in sandstones there are several minerals which are commonly overgrown during diagenesis. Quartz most commonly exhibits overgrowths ranging from micron-sized features scattered on grain surfaces, to cases where the sand grain is converted to a perfect bi-pyramidal quartz crystal (Waugh, 1970). Feldspars also commonly have overgrowths (Waugh, 1978a, b; Ali & Turner, 1982). These overgrowths may be in the form of smooth faced terminations or may grow in a 'skeletal' open form when they bear a strong resemblence to etch features (Fig. 8.17e). Stablein & Dapples (1977) illustrated well the different forms overgrowths may take on different crystal faces of the same grain.

DISSOLUTION AND REPLACEMENT OF GRAINS

Dissolution and replacement features are common in sandstones and most frequently involve feldspar, carbonates and clays.

Dissolution of feldspar grains resulting in creation of secondary porosity (Schmidt & McDonald, 1979a, b) may leave small etched relics of feldspar or a skeletal relic of clays. Alteration of feldspar to aggregates of kaolinite is frequently responsible for the patchy distribution of kaolinite seen in thin section and with the SEM. Dissolution of ferromagnesian minerals such as amphibole (hornblende) and pyroxene (augite) is well documented by Walker *et al.* (1978) and Waugh (1978a) who illustrated spectacularly etched crystals.

Carbonates frequently cement sandstones and also replace detrital grains. The replacement is frequently highly selective resulting in replacement of particular minerals. Feldspar and quartz are most frequently affected, but many examples exist where one or the other is attacked preferentially. Replacement features of carbonate-cemented sandstones can be revealed by etching the carbonate to reveal surfaces of replaced grains and relics of part replaced grains (Fig. 8.18c). Burley & Kantorowicz (1986) illustrated features of quartz grain surfaces resulting from replacement textures caused by carbonate cements, so allowing recognition of some sandstone from which carbonate cement had been removed. However, it must be stressed that carbonate cements of a passive nature may leave no clear evidence of their previous presence.

Clays also replace detrital grains, particulalry volcanigenic grains, feldspars and other chemically unstable grains. Clay transformations during burial diagenesis such as the smectite-illite transformation result in modification of clay morphologies and chemistry. Kaolinite is frequently replaced by illite at depth resulting in distinctive modification of kaolinite morphology (Hancock & Taylor, 1978; Jourdan *et al.*, 1987).

MECHANICAL DEFORMATION

During early burial mechanical compaction results in reorientation of grains to produce closer grain packing, mechanically weak grains (e.g. shale fragments, glauconite) may be deformed and squeezed into pore space and micas are frequently bent around framework grains. Evidence of a mechanical deformation phase is not always obvious in SEM studies of 'clean' sandstones, but it is frequently evident in fine grained muddy and micaceous sandstones, and is well shown in studies of mudrocks using backscattered electron images.

DISSOLUTION AT GRAIN CONTACT

Chemical compaction, resulting in dissolution at grain contacts, is seen in most sandstones, usually only being absent where early introduction of cement produced a rigid frame. The surfaces of grain to grain contacts of quartz display pitted and grooved solution surfaces (Fig. 8.18b) which contrast with surfaces of overgrowths on detrital grains.

Solution surfaces are found at many different mineralogical contacts, but are often impressively developed at quartz/mica contacts where quartz is preferentially dissolved (Fig. 8.18d) and similarly at quartz/organic carbon contacts.

As pointed out by Sellwood & Parker (1978), and others in the same discussion, this phenomenon of 'pressure solution' does not bear a simple relation to depth of burial for any particular facies, there being a dependence on evolution of pore fluids and the presence of grain coatings to catalyse reactions. Organic acids are extremely important in diagenetic reactions by increasing aluminium mobility and aiding feldspar dissolution (e.g. Surdam *et al.*, 1984).

The early cementation of a sandstone will also protect grain contact points by reducing stress at grain contacts. Pore fluid pressure may largely support the overburden weight and inhibit chemical compaction in over-pressured formations.

EVIDENCE OF ORDER OF DIAGENETIC EVENTS

The order of development of authigenic minerals is usually stratigraphically based on the assumption that the younger phases grow on the older. Complications arise in the case of grain replacements when it is not always possible to tell when replacement took place relative to other diagenetic events. Despite the simplicity of the basic premise there are several factors which can combine to produce ambiguous evidence of the order of diagenetic events. Two mineral phases may grow at the same time and a later mineral phase may grow to enclose previously formed material. Some minerals have favoured substrates for growth and diagenetic minerals will not be evenly distributed on grain surfaces; thus grains both with and without a particular overgrowth or coating may occur in close proximity.

In rocks with low permeability diagenetic phases are not evenly distributed, possibly due to differing local flow rates and chemistry of pore water, ion availability from altered grains, or substrate availability. Such variations can occur on a large scale as with the obviously different diagenetic histories between carbonate cemented concretions in sandstones and the adjacent uncemented rock. Thus a wide study of many specimens is required to establish a full diagenetic sequence for a particular formation.

8.10.3 Applications of SEM studies of sandstones

Apart from the identification of fine grained minerals, matrix, cements, grain coatings and pore fillings the most important SEM applications are the elucidation of diagenetic sequences with progressive burial, the relation of diagenesis to depositional facies and the explanation of factors relating to porosity and permeability of reservoir sandstones. Work on reservoir sandstones has now advanced to the stage where the SEM is used in studies of 'artificial diagenesis' during reservoir treatment, such as the effects of steam injection on Cretaceous tar sands of Alberta described by Hutcheon (1984).

DIAGENETIC SEQUENCE

The SEM is an essential tool in the elucidation of diagenetic sequences involving the development of authigenic minerals. Hurst & Irwin (1982) have summarized some of the sequences recorded in recent years and their summary illustrates well the variation to be found in the order in which authigenic minerals develop under different circumstances. Hurst & Irwin listed important factors influencing diagenesis; their list can be modified as follows:

1 Temperature
2 Pressure
3 Detrital mineralogy, roles of stable and unstable grains
4 Organic geochemistry
5 Availability of carbonate, both biogenic and non-biogenic
6 Pore water composition, migration of pore fluids and gases, and evolution
7 Sediment texture, porosity and permeability
8 Sedimentary facies, including enclosing formations
9 Tectonics
10 Time

In view of the numerous variable influences, projects involving interpretation of diagenetic sequence require a great deal of background information. Evidence should certainly be sought on the following points:

Depositional environment. Marine or fresh water and if fresh water, vadose or phreatic? Likely initial pore water type, and later pore water modifications?

Mineralogy. Is the original detrital mineralogy preserved or recognizable; what major changes have taken place?

Texture. What was the original texture (grain size, sorting) and how has it been modified? How has porosity and permeability been altered?

Burial history. What is the burial history in terms of maximum burial depth, when did this occur, and has more than one cycle of burial taken place?

Pore fluid and gas. Has the rock been flushed by different pore fluids at any time, what was the final pore filling prior to collection (e.g. air, gas, oil, fresh water, saline water?

From these considerations it may be possible in well-explored areas to estimate likely pressures and temperatures and reconstruct with some accuracy the geological history of the formation.

Diagenetic studies can be organized to examine

depth and/or facies-related phenomena but in all cases a wide range of comparable material (grain size, sorting, composition etc.) is needed, and the clay mineralogy of any enclosing shales should also be examined to determine likely detrital and diagenetic clays present. The following notes on specific facies types can provide only a basic idea of the diagenetic variation in sandstones.

RED BEDS

Walker (1967) and Walker *et al.* (1978) have demonstrated how the red pigmentation in first-cycle desert sandstones forms in response to the alteration of Fe-bearing minerals such as hornblende and augite by solution. The iron released is probably deposited as ferric hydrate which gradually converts to haematite with burial and ageing. The iron oxide or hydroxide rims to grains are frequently associated with clays which may initially be mixed-layer illite/smectites but are converted to illite during burial as in the extensively studied Rotliegend (Lower Permian) sandstones of the southern North Sea.

Numerous studies on the Rotliegend sandstone (Rossel, 1982; Glennie, Mudd & Nagtegaal, 1978; Hancock, 1978b; Nagtegaal, 1979; Pye & Krinsley, 1986) illustrate the general diagenetic features and some of the variable factors. General features are iron oxide grain coatings of early diagenetic origin which include illite probably representing original illite/smectite (Rossel, 1982). Second stage diagenesis results in feldspar overgrowths, dolomite cement and replacement, and also quartz overgrowths. Feldspar is frequently converted to kaolinite prior to the final stage when the characteristic 'hairy' or ribbon illite (Fig. 8.17a) is produced along with some chlorite. Many variations on this theme are recorded such as early calcite, halite and gypsum facies controlled cements (Glennie *et al.*, 1978). Late anhydrite cements are found near faults (Glennie *et al.*, 1978) or close to overlying Zechstein evaporites (Hancock, 1978b).

Burley (1984) produced a detailed diagenetic history of the Triassic Sherwood Sandstone Group and recognized distinct stages of diagenesis related to depositional environment, burial and subsequent uplift. The value of SEM work is greatly enhanced by the use of other techniques (isotopes, XRD, microprobe analysis and petrography) both in this paper and in Burley (1986) on reservoir sandstones of the Piper and Tartan fields in the North Sea.

'MARINE' SANDSTONES

Sandstones with original marine pore waters have varied diagenetic histories (Hurst & Irwin, 1982) which may commence with illite/smectite (Hawkins, 1978) or chlorite (Tillman & Almon, 1979) or concretionary carbonate, but frequently are cemented by quartz overgrowths. The earliest phase of quartz cement, particularly in sandstones adjacent to shales, may be of randomly oriented microquartz crystals on grain surfaces (Fig. 8.17f), which subsequently inhibit the formation of quartz overgrowths (Spark & Trewin, 1986).

Following establishment of a rigid frame, dissolution of unstable grains may result in secondary porosity or production of kaolinite by feldspar alteration. The kaolinite in marine sandstones frequently forms pore filling cements of well formed euhedral crystals of 4–10 μm size (Hurst & Irwin, 1982). Carbonate is frequently introduced after the initial quartz overgrowth phase both as cement and as grain replacement, particularly of feldspars, and this phase of diagenesis may eliminate most porosity. Deeper burial may, however, result in carbonate dissolution producing secondary porosity (Schmidt & McDonald, 1979a, b) which can be available for occupation by migrating hydrocarbons. Production of organic acids during the maturation of organic matter is frequently thought responsible for secondary porosity development by silicate (frequently feldspar) dissolution.

Entry of hydrocarbons effectively stops diagenetic reactions (Hawkins, 1978) but reactions can continue in adjacent water-saturated sandstones to produce further quartz, kaolinite or illite cements, resulting in diagenetic contrasts at oil–water contacts.

'FRESHWATER' SANDSTONES

There is a full range in pore water compositions between highly saline and fresh water giving a great variety of environmental controls to diagenesis, but sandstones such as fluvial sandstones with fresh pore waters display some apparently distinctive features (Hurst & Irwin, 1982). The major feature is a tendency for kaolinite development (at the expense of feldspar) to precede quartz overgrowth formation and for the kaolinite to form coarse vermicular crystals with ragged or skeletal edges (Fig. 8.17d) which may have grown more rapidly than the characteristic 'marine' form of kaolinite. However, it is important to recognize that a great many sandstones deposited under shallow marine conditions are subsequently flushed by fresh water from adjacent land areas during early burial or later uplift.

Carbonates may be developed at more than one stage in the diagenetic history, ranging from pedogenic carbonate of caliche soils, or early pre-compaction carbonate nodules through to late diagenetic calcite, dolomite or siderite precipitation post-dating quartz cementation and occuring at depths of 2–3 km. Flushing of originally marine sandstones by fresh water will frequently superimpose new diagenetic features, thus formations involved in uplift and reburial can possess complex diagenetic histories.

POROSITY AND PERMEABILITY

The permeability of a sandstone is related to the size and shape of pore throats which connect larger pore volumes in a sandstone, the tortuosity of pores, and the specific surface area within pore space. The SEM is ideal for examination of pore geometry by means of pore casts and also ideal for viewing the factors which serve to reduce porosity and permeability in sandstone. One of the most obvious examples of permeability reduction by the formation of authigenic minerals within pore space is that provided by the illite diagenesis of the Rotliegend sandstone of the southern North Sea where delicate illite crystals bridge and block pore throats (Fig. 8.18b), so reducing permeability. The Brent Sandstone (Middle Jurassic, North Sea) displays similar features with illite also responsible for permeability reduction (Hancock & Taylor, 1978; Blanche & Whitaker, 1978), and the Magnus reservoir (Upper Jurassic, North Sea) is similarly affected (McHardy *et al.*, 1982). Kaolinite is not usually so detrimental to permeability as illite since it has a larger grain size, and smaller surface area (Fig. 8.17c) so that pore tortuosity and water absorption on the clay surface are not so great. Kaolinite also tends to be more patchy in its distribution in the rock. SEM studies can be valuable in the assessment of microporosity in reservoirs as in the case of porcelaneous cement of opal, microquartz and montmorillonite described from Miocene turbidite sandstones of the Los Angeles Basin by Sears (1984).

As oil migrates into a reservoir diagenesis is arrested as in the case of the Brent Sandstone (Hancock & Taylor, 1978; Sommer, 1978) where it can be shown that oil migrated into place synchronously with illite formation, there being a downward increase in illite within the reservoir. Thus the relative timing of generation and migration of hydrocarbons during diagenesis can be determined and predictions can be made on reservoir quality with respect to facies, geographical area and depth of burial.

Porosity can be observed with the SEM and, by use of image analysis equipment on polished sections, particularly in the backscattered mode, can be determined quantitatively. Pore surfaces are best examined on rough broken rock surfaces. Pore throat size and pore connection is most easily studied using pore casts (Fig. 8.21).

CONCLUSION

From these few examples it is apparent that the diagenesis of any sandstone formation must be studied with reference to all available information on facies, pore fluids, burial history and composition. Utilization of SEM techniques can greatly aid explanation of porosity and permeability parameters, relating them to depositional and diagenetic factors by examination of pore morphologies and fillings.

8.11 LIMESTONES AND DOLOMITES

8.11.1 General

Utilization of the SEM for the study of limestones is most valuable in the examination of fine-grained porous limestones such as chalk (Scholle, 1977) and for tracing the evolution of cement and replacement fabrics in Recent to sub-Recent carbonates (Bathurst, 1975; Folk, 1974a; James *et al.*, 1976; Wilkinson *et al.*, 1982). In studies of modern fine-grained carbonate sediments the morphologies and origins of grains can be determined (Hay *et al.*, 1970) and detail of biogenic particles more easily recognized.

Textures and pore geometry of dolomitized and dedolomitized sediments are also suitable for examination as in examples of Silurian dolomite diagenesis in the Lockport Formation, USA (Shukla & Freidman, 1983), dolomitized Cretaceous chalk of Europe (Jorgensen, 1983) and recent Australian Coorong dolomites (Von Der Borch & Lock, 1979).

For a general discussion of carbonate petrology,

Bathurst (1975) provides an excellent account and Scholle (1978) illustrates the use of the SEM in carbonate studies and provides a useful bibliography to selected topics and techniques. A variety of techniques such as thin section, XRD and cathodoluminescence and isotope analysis are all of great value in carbonate studies and the SEM can make a useful contribution in many cases and provides excellent illustrative material.

Only a few examples can be quoted in the space available and these are chosen to illustrate the wide range of use of the SEM in carbonate studies, without attempting to review theories on the various examples presented.

8.11.2 **Examples**

CARBONATE GRAINS

Unconsolidated carbonate sediments can be examined as scattered grain mounts for the identification of fine carbonate grains. This technique is useful for the identification of biogenic particles as shown experimentally by Hay *et al.* (1970) who crushed examples of known invertebrate material and were able to recognize distinctive skeletal morphologies in fine sands and even in some grains as small as 4 μm. Mounting the grains in resin and examining polished and etched surfaces enabled them to recognize distinctive skeletal internal structures.

The contributions which the fragmentation products of organisms make to carbonate muds, such as aragonite needles derived from the breakdown of the calcified algae *Halimeda* and *Penicillus*, and the contribution of coccoliths, discoasters and other planktonic organisms to deep sea oozes can be ascertained (illustrations in Scholle, 1978). Alexandersson (1979) illustrates the contribution that the breakdown products of mussel (*Mytilus*) shells make to sediments of the Skagerrak, North Sea. A similar process operates in muds of the Ythan Estuary, North Scotland, where characteristic calcite needles and plates are released by the breakdown within the sediment of mussel shell fragments originally contributed to the sediment by the predatory activities of eider ducks feeding on the local mussel beds (Trewin & Welsh, 1976; Fig. 8.19). Calcite needles originating from the breakdown of mussel shells also occur in suspended sediment on the north-eastern USA Continental shelf (Fitzgerald, Parmenter & Milliman, 1979).

Sediments may also be examined for grains of non-biogenic origin such as the 2–20 μm high-Mg calcite grains precipitated in Lake Manitoba (Canada) illustrated by Last (1982) and the numerous studies of ooids and their structure such as that of Land, Behrens & Frishman (1979) on the ooids of Baffin Bay, Texas, and Sandberg (1975) and Halley (1977) on the ooids of the Great Salt Lake. Sandberg's study utilizes SEM examination of etched surfaces to show that coarse radial aragonite is of depositional rather than recrystallization origin. The SEM provides evidence not available from microscope petrography.

Evidence of aragonite replacement by calcite can also be revealed by SEM preparations such as the relic aragonite structures in calcitized Jurassic bivalves illustrated by Sandberg & Hudson (1983).

These and many other studies illustrate the general use of the SEM in studies involving carbonate grains.

CEMENTATION OF CARBONATES

There are many cases in which early cementation of carbonates occurs in marine subtidal and intertidal conditions and in freshwater phreatic and vadose situations. The morphology of the cements produced can ideally be examined using the SEM and much more detail obtained than is possible in thin section. In freshwater solutions with low Mg^{++} concentrations there is a tendency for simple rhombs to grow, but in solutions with a high Mg^{++} concentration sideways growth of the crystals may be poi-

Fig. 8.19. Bivalve (*Mytilus*) shell fragment breaking up to release individual calcite laths into the sediment. From recent intertidal mudflat, Ythan Estuary, North Scotland.

soned by Mg and fibrous crystals of Mg-calcite or aragonite form (Folk, 1974a); however, some uncertainties exist concerning this mechanism. Bathurst (1975), Folk (1974a), Friedman (1975) and Moore (1979) provided summaries of early cementation in marine and freshwater environments, and only a few examples can be mentioned here.

Oolite cementation by calcite in freshwater conditions on Joulters Cay, Bahamas described by Halley & Harris (1979) results in cementation around grain contacts with blocky calcite in the vadose zone producing a form of meniscus cement. A similar cement is found in sub-Recent oolite dune sands around Hamelin Pool, Shark Bay, West Australia (Logan *et al.*, 1974) as illustrated here in Fig. 8.20(a). Below the water table at Joulters Cay the calcite cement is not concentrated at grain contacts, but consists of small dog-tooth crystals of 20–40 μm in isopachous grain coats and in deeper samples scattered rhombohedra of 20–30 μm which decrease in size to 5–10 μm some 4 m below the water table. Using the SEM evidence Halley & Harris were able to calculate aragonite dissolution rates and calcite cementation rates and show that local rainfall is sufficient to account for the observed cementation. The result of continued aragonite solution and calcite deposition is well displayed by cemented Pleistocene oolite dune sands around Shark Bay where comoldic porosity is developed (Fig. 8.20b).

The value of SEM work is well illustrated by the work of Steinen (1978) on the diagenesis of lime mud using subsurface material from Barbados. He recognized microspar deposition in voids created or enlarged by dissolution and showed that muds cement early and in many stages. In thin section the textures resembled aggrading neomorphism, but the SEM allowed true crystal shapes and relations to be recognized.

Submarine cementation by aragonite and calcite has been widely reported from different environments [James *et al.* (1976) from tropical reefs in Belize; Adams & Schofield (1983) from gravel at Islay, Scotland]. Alexandersson (1974) illustrated a variety of aragonite and Mg-calcite cements related to biochemical activity of living red algae in calcium carbonate undersaturated waters of the Skagerrak (North Sea) and showed that the cement and coralline algae undergo dissolution following the death of the algae.

Characteristic aragonite cements comprise delicate needles coating grains as shown in an example of an oolitic hardground from Shark Bay (Fig. 8.20c, d). Submarine calcite cements frequently consist of high-Mg calcite occurring as micrite in fine pores and on grain surfaces and followed by bladed spar as reported by James *et al.* (1976) from Belize reefs. Longman (1980) provided an SEM-illustrated account of carbonate cementation features in marine and freshwater environments and related processes to the numerous cementation products both aragonitic and calcitic.

'Beach rock' formed by cementation in the intertidal to supratidal zone is common in the tropics but also occurs in more temperate climates. Cement is frequently acicular aragonite but may also be of high-Mg calcite.

Cementation and structures in calcretes such as the variety of calcified filaments of soil fungi, algae, actinomycetes and root hairs described and illustrated by Klappa (1979, 1980) from Mediterranean Quaternary calcretes can be ideally studied with the SEM, as also demonstrated by Watts (1980) in a study of calcretes from the Kalahari, southern Africa, where both high- and low-Mg calcite are deposited in passive, displacive and replacive modes and are associated with authigenic palygorskite, sepiolite and minor dolomite.

DIAGENESIS OF LIMESTONE — CHALK

Whilst many limestones can be examined advantageously using SEM techniques the examination of chalks has been particularly instructive. Scholle (1977) summarized the diagenetic modifications of true nannofossil chalks from the North Sea and surrounding European outcrops as well as North American Gulf Coast and Scotia Shelf examples.

In early diagenesis compactional dewatering of the highly porous muds leads to a grain supported frame; some selective dissolution may take place at this stage, particularly in deep water examples. Depth of burial is the single most important factor in chalk diagenesis and the progressive development of an interlocking calcite cement which overgrows coccoliths is admirably displayed by fig. 8 of Scholle (1977).

The mechanism of cementation in low-permeability chalks is by solution transfer. Cement is introduced into a load bearing frame and prevents mechanical compaction as burial loading increases. The source of the cement is internal to the formation and results partly from selective dissolution of cal-

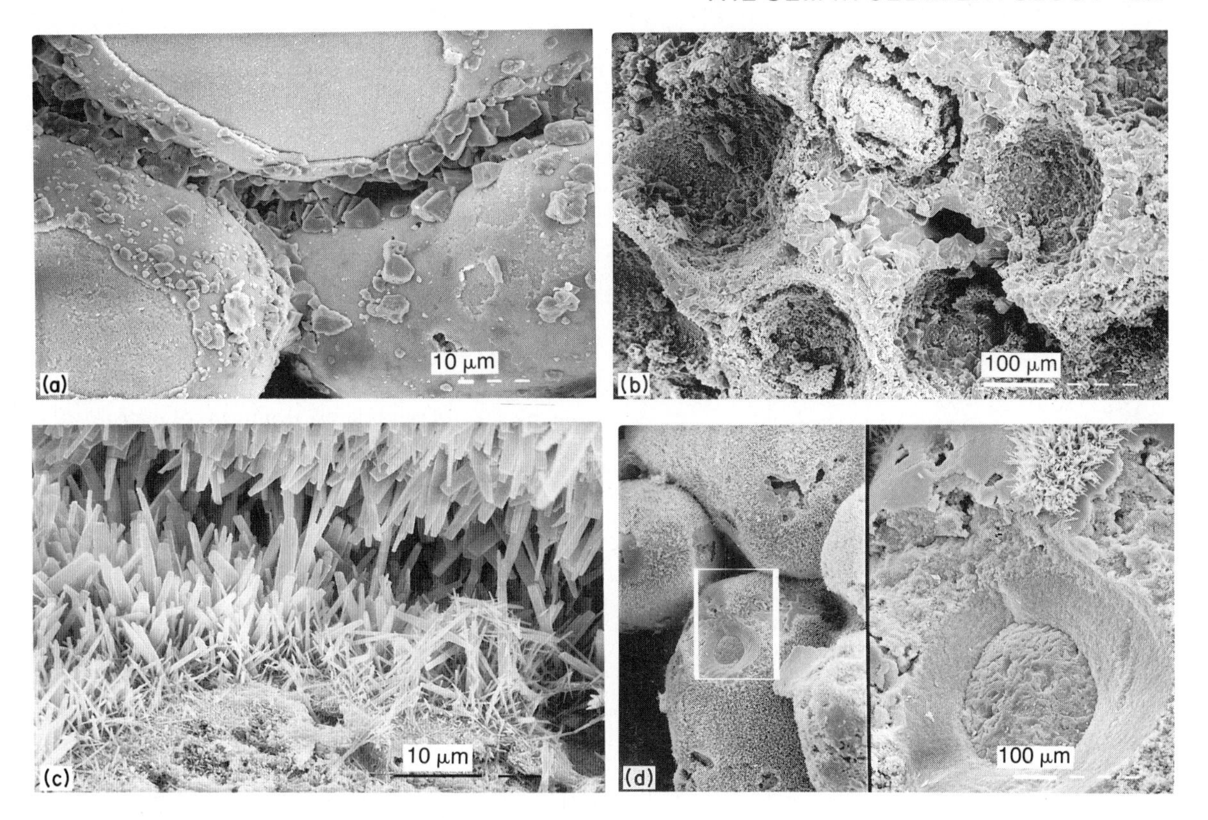

Fig. 8.20. Examples of cementation and dissolution in Recent–Pleistocene oolitic rocks and sediments from Shark Bay, West Australia.
(a) Blocky clacite cement concentrated at grain contacts as a meniscus cement in a oolite cemented in the vadose zone. Lithified sub-Recent oolitic aeolian dune deposit. Carbla Point, Hamelin Pool.
(b) Oolite with aragonitic ooids largely dissolved to leave secondary oomoldic porosity, some primary intergranular porosity remains within blocky calcite cement. Pleistocene oolitic aeolian dune sand below calcrete crust. Hutchison Embayment.
(c) Bladed aragonite cement. Recent submarine cementation of oolitic hardround in Hamelin Pool.
(d) Aragonite cement evenly coating grains in a recent oolitic hardground. Enlargement (marked scale × 5) shows etched quartz grain surface beneath an ooid rim. Some borings present in ooid surfaces. Hamelin Pool.

citic or aragonitic organisms but originates mainly from solution seams and stylolites. Over-pressured chalks retain high porosities since stress is reduced at grain contacts, and chalk in oil reservoirs may retain high porosity due to the influx of oil preventing further cementation. In fine-grained chalk reservoirs the SEM is the only practical means of examining details of pore geometry and rock framework.

DIAGENESIS — CARBONATE POROSITY

A useful example of SEM use on ancient reefs (Upper Miocene of southern Spain) is that of Armstrong, Snavely & Addicott (1980) in which thin section and SEM information is neatly combined to illustrate the modifications of primary porosity due to aragonite solution and dolomitization of lime mud.

Pore geometry in carbonates has been investigated by producing resin pore casts of chalk (Walker, 1978) to reveal intersecting laminar pores 1–2 μm × 0.1–0.2 μm resulting in non-laminar pores which connect larger pore spaces such as those within foraminifera. The small size of pore throats results in the low (less than 10 md) permeabilities of most chalks. Patsoules & Cripps (1983) have illustrated a variety of pore types in chalk by use of resin casts.

Wardlaw & Cassan (1978) illustrated a wide variety of limestone and dolomite porosity by use of pore casts and related porosity type to recovery efficiency in reservoir rocks; their illustrations of pore casts compared with thin section are most instructive. Bimodal porosity in oolites described by Keith & Pittman (1983) is due to micropores within ooids which are water filled whereas gas is confined to the larger primary pores between ooids, thus the SEM can aid in the assessment reservoir productivity. Contrasting pore casts of both primary and secondary limestone porosity are illustrated in Fig. 8.21.

DOLOMITIZATION

Dolomitization of limestones frequently results in the creation of useful interconnected porosity as discussed and illustrated by Wardlaw (1976), Wardlaw & Cassan (1978) and Davies (1979). Dolomite formation has been attributed to a variety of processes in early diagenesis (summaries in Bathurst, 1975 and Zenger, Dunham & Ethington, 1980).

Dolomitization frequently affects only specific components of a limestone, as for example when bioclasts are dissolved and cementing calcite dolomitized. Figure 8.22 shows an example from the subsurface Devonian of West Australia (geology of area summarized by Playford, 1980) where ooids have been dissolved and the cement dolomitized, and also an example with diagenetic illite and kaolinite developed in pore space of the dolomite during freshwater invasion of a marine limestone following its dolomitization. In these examples dolomitization probably took place by mixing of fresh and marine water.

Dolomitization of chalk from the North Sea (Jorgensen, 1983) results in the production of dolomite rhombs of 10–30 μm size which clearly overgrow original texture and fossils. Dolomite forms only 2–8% of the rock and appears to have formed early in diagenesis. More extensive dolomitization of limestones is described by Shukla & Friedman (1983) from the Lockport Formation (Middle Silurian) of New York State where early incipient dolomitization of micrite took place in a supratidal environment. Successive stages of dolomitization resulted in dolomitization of all the groundmass, dolomitization of groundmass and allochems and finally totally obliterative dolomitization. SEM examination of dolomites clearly reveals the problems of low permeability dolomite reservoirs which are usually caused by narrow, poorly connected pore throats in intercrystalline porosity, and the presence of non-effective porosity in small vugs.

8.12 ENDOLOTHIC MICROBORINGS

8.12.1 General

The SEM has played an important role in the study of microborings in carbonate substrates, since it enables the three-dimensional forms of micron-sized borings to be examined in detail.

Bioerosion is now recognized as a most important factor in the reduction of carbonate grains and substrates, particularly in the shallow marine environment. Golubic, Perkins & Lukas (1975) reviewed the history of the study of microborings and provided an extensive reference list. The larger borings produced by echinoids, gastropods, bivalves, polychaetes and sipunculids (summary in Warme, 1975) are not studied using the SEM, but detailed features of the walls of the borings can provide useful information. The smaller borings due to sponges (particularly *Cliona*) are of suitable size for SEM examination of fine detail and of sediment produced, but the greatest contribution of the SEM is in the study of algal, fungal and possible bacterial borings in grains and rock surfaces. Other small borings such as some of those due to cirripedes, bryozoa and foraminifera can also be usefully examined with the SEM.

8.12.2 Methods

Preparation methods are described in Section 8.5. Study of recent borings should involve identification of the organism responsible, but in dead and fossil material all that remains is a cavity or filled boring which must be prepared by a suitable impregnation or etching technique. Considerable taxonomic problems exist with respect to endolithic algae and fungi which are beyond the scope of this work.

8.12.3 Sedimentological factors

Bioerosion is of the greatest importance in carbonate environments and on coastlines with exposed carbonate-rich rocks, but is by no means confined

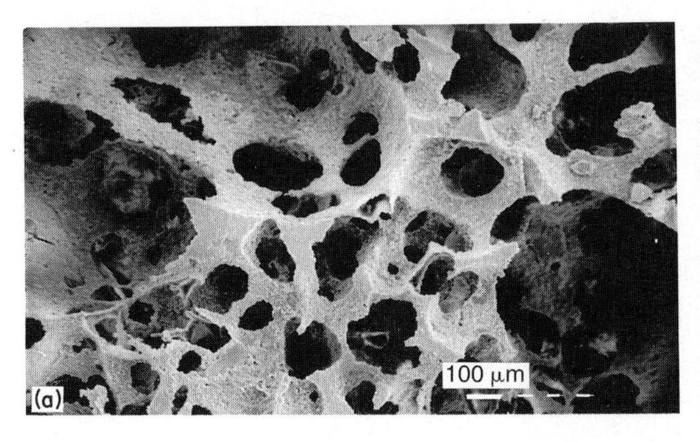

Fig. 8.21. (a) Resin pore-cast of interparticle porosity; carbonate aeolian dune sand, Pleistocene, Shark Bay, West Australia.
(b) Resin pore cast of secondary oomoldic porosity in the same rock as in Fig. 8.20(b). Calcite has been dissolved, but a dolomite rhomb remains within the pore cast at lower centre of picture.

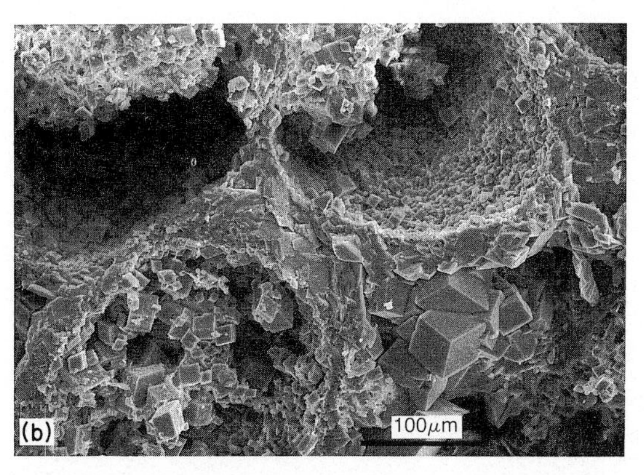

Fig. 8.22. (a) Coarse dolomite with intercrystalline porosity containing books of kaolinite; small illite plates and ribbons are also present on dolomite surfaces.
(b) Dolomite with oomoldic porosity due to dissolution of ooids, some intergranular porosity remains.
Both from Yellow Drum Formation, Upper Devonian, Canning Basin, West Australia.

to such situations. The organisms responsible for microborings have specific ecological requirements and thus a zonation of such organisms can be recognized with respect to factors such as water depth, tidal range, light penetration and climate. Thus the identification of borings can lead to interpretations (Fig. 8.23) on the depth ranges of boring algae and fungi (Golubic *et al.*, 1975; Lukas, 1979). Limestone coasts are generally subject to intense bioerosion, with distinctive zones of bioerosion leading to development of features such as a biogenic notch where destruction is particularly great. Schneider (1976), in a detailed study of an Adriatic limestone coast, illustrates the borings found in the various zones of the rocky shore. Such studies are of great value from the ecological standpoint, and increase our knowledge of the depth range of specific types of borings.

Endolithic algae (Fig. 8.24) which require very little light occur as deep as 370 m (Lukas, 1979) but are more characteristic of shallower depths to 100 m. Fungi do not require light and are found to much greater depths. It appears (e.g. Fig. 8.23) that there

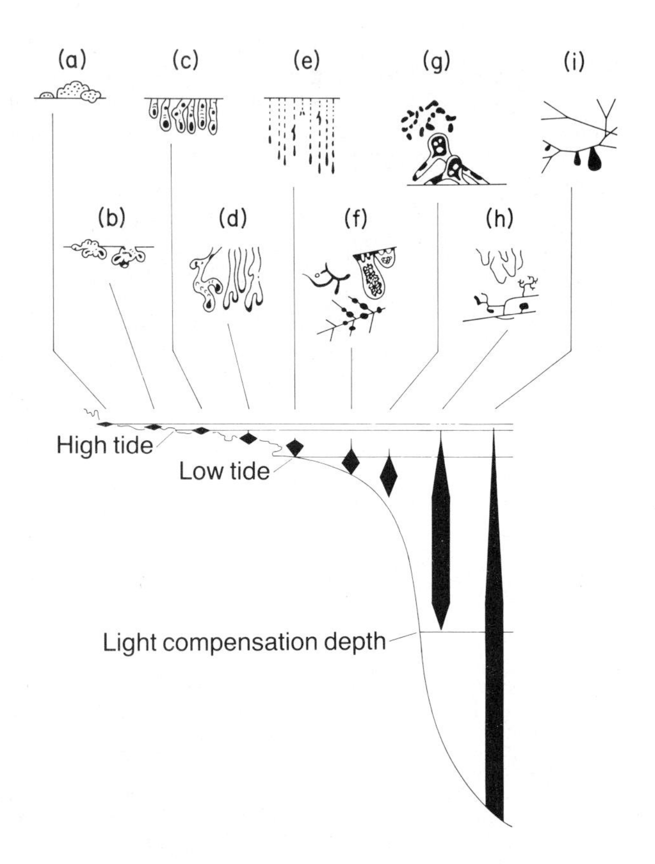

Fig. 8.23. Relative vertical distribution of common marine microboring algae superimposed on an idealized coastal profile. The upper limits above high-tide level are controlled by water supply; the lower limits by light penetration into the water column.
(a) Coccoid eiplithic cyanophytes, (b) *Hormathonema luteobrunneum*, *H. violaceo-nigrum*, (c) *Hormathonema paulocellulare*, (d) *Solentia foveolarum*, *Kyrtuthrix dalmatica*, (e) *Hyella tenuior*, (f) *Mastigocoleus testarum*, *Gomontia polyrhiza*, *Phaeophila dendroides*, *Conchocelis*-stages of various rhodophytes, (g) *Hyella caespitosa*, *Eugomontia sacculata*, (h) *Plectonema terebrans*, *Ostreobium quekettii*, (i) Fungi. Modified from Golubic *et al*. (1975) and Lukas (1979).

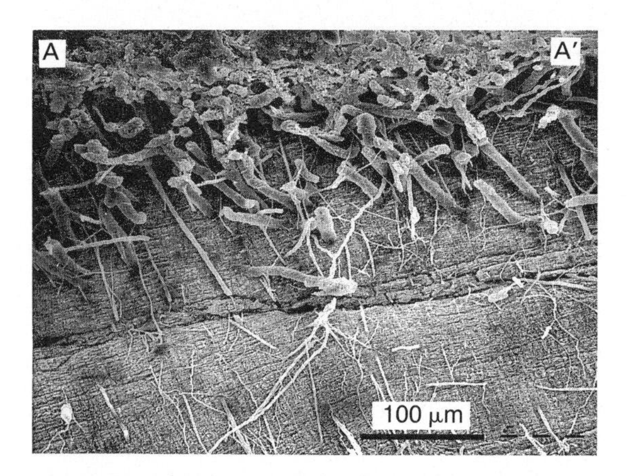

Fig. 8.24. Endolothic algal borings in mollusc shell, several different morphologies and depths of boring extend into the shell from its outer surface A–A (top of picture). The finest borings are possibly of fungal origin. Borings prepared by impregnation with resin and etching of a polished surface.

is greater variety and abundance of microborings in shallow waters and a more precise depth zonation developed than in deeper water. As indicated in Table 8.2, algae, lichens and fungi bore by corrosion with a chemical mechanism and thus do not directly produce sediment. The clionid sponges, however, release carbonate particles which contribute significantly to sediment production.

SPONGES

The mechanism of boring of *Cliona* has been described and well illustrated with SEM photos (Rutzler & Rieger, 1973) to show the method of excavation which removes 15–100 μm size chips of substrate of characteristic shape with etched surfaces. Mean dimensions of *Cliona*-produced chips are 56 × 47 × 32 μm. The chips are excavated by the production of crevices only 0.2 μm wide etched by cellular activity, thus only 2–3% of the substrate is dissolved, the rest contributing to sediment production. Typical sediment chips and wall features of the boring are illustrated in Fig. 8.25. In early colonization stages Neumann (1966) found that *Cliona* destroyed 5–7 kg limestone m^2 in 100 days, but since *Cliona* ceases to bore after reaching a particular depth this figure is well in excess of the normal rate which Rutzler (1975) calculated to be 250 mg of sediment per m^2 per year in Bermuda, which can account for up to 41% of sediment in mud pockets within a coral framework. Futterer (1974) found sponge excavated particles to constitute 30% of Fanning Island (Pacific Atoll) lagoonal sediments, and 2–3% of samples from the Arabian Gulf and the Adriatic. He also illustrated typical grains and showed how the edges become rounded by abrasion. Acker & Risk (1985) calculated rates of bioerosion by *Cliona caribbaea* in the shallow terrace zones off Grand Cayman (B.W.I.) to be 8 kg limestone m^2 per annum, most of which is rapidly transported downslope. SEM photographs of fine fractions (3–4 ϕ) were used for estimation of abundance of sponge-produced chips.

ENDOLITHIC ALGAE

Endolithic algae have been reported to bore at rates varying from 0.3 to 36 μm/day (data summary in Lukas, 1979) but most can only reach a depth of a few millimetres due to light requirements. Their activity makes the rock or grain surface weak and porous and thus more subject to mechanical erosion, but it is the activities of grazing animals such as sea urchins, gastropods, chitons and fish which rasp away the algae infested rock which result in most sediment production. Grazing activity removes the surface layer and allows the algae to bore deeper

Table 8.2. Examples of processes, mechanisms, traces produced, and habitats of some organisms responsible for bioerosion (modified from Schneider, 1976)

Microboring organisms	Destruction process	Mechanisms	Traces	Habitat
Algae, blue-green, green, red	Boring, internal corrosion	Chemical solution?	Network of fine borings with specific patterns. Up to 800 μm deep. Tubes 1 μm to 100 μm but most frequently 2–10 μm. Various rock surfaces	Littoral to lower limit of photic zone, mostly to only 100 m depth, extreme 370 m +
Lichens	Surface corrosion and boring	Chemical solution by organic acids	Various rock surfaces, disintegration of surface into small particles	Intertidal and supratidal
Fungi	Boring, internal corrosion	Chemical solution	Very fine borings up to 2 mm deep. Frequent 1–3 μm hyphae with sporangea 2–50 μm but can be 8–12 μm diameter tubes	Intertidal and below to 1000 m +
Sponges, e.g. *Cliona*	Boring	Chemical loosening of small chips of 15–100 μm size by acids and enzymes	Regular chamber system in rock with openings to surface	From biological notch near low tide downwards

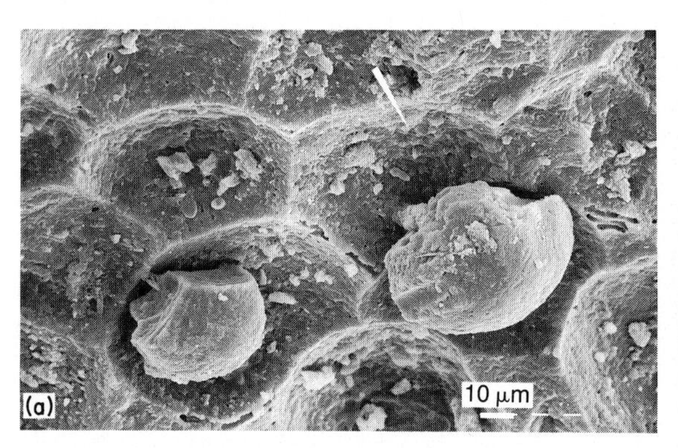

Fig. 8.25. *Cliona* borings in the bivalve *Tridacna*. Heron Island lagoon, Great Barrier Reef, Queensland, Australia.
(a) Typical sculpture of wall of boring, with two chips excavated by *Cliona* lying on the surface and showing the characteristic convex outer surface (wrt the sponge).
(b) Typical concave etched facets of the inner surface of a chip.

and so continue the bioerosion process. Farrow & Clokie (1979) illustrated typical effects of grazing by limpets and chitons on algal-infested shells and the consequent production of sediment. The rate of destruction of carbonate rock surfaces in various environments is of the order of 1 mm yr^{-1} (Schneider, 1976). Algal endoliths have a long history extending at least to the late Precambrian (Campbell, 1982). Typical algal endolithic borings are illustrated in Fig. 8.24. The rate of carbonate dissolution by micro-boring organisms was investigated in the lagoon of Davies Reef, Australia, by Tudhope & Risk (1985). Using the SEM, and observing impregnated bored grains, the percentage of borings could be point counted within grains. Examination of the surface uncoated and under a weak vacuum produced high contrast between resin-filled borings and carbonate, and simplified point-counting.

Infestation of carbonates by endoliths also occurs beneath the sediment/water interface, particularly in organic-rich mud. May & Perkins (1979) illustrate a restricted assemblage of four forms found up to 1.6 m below the surface in fine-grained reducing sediments. Two of the boring cell-like forms are considered to represent unicellular prokaryotic blue-green algae or bacteria. Such algae probably function as anaerobic heterotrophs and can form chlorophyll in the dark if a suitable organic carbon source is available.

FUNGI

Fungal borings are difficult to distinguish from algal borings, there being no simple criteria for their separation and a distinct overlap in sizes of borings. Zeff & Perkins (1979) described five distinct types of fungal borings in deep water sediments (210–1450 m) from the Bahamas area. Some are identical to shallow water forms but three are considered characteristic of deep aphotic environments. As pointed out by Zeff & Perkins the study of such borings has sedimentological significance in the recognition of aphotic zone sediments, source areas of turbidites, and deep water carbonates.

MICRITE ENVELOPES

Boring by endolithic algae and fungi on loose carbonate grains is responsible for the formation of micrite envelopes (Bathurst, 1966, 1975). Abandoned borings are filled by fine aragonite or high-Mg calcite. Margolis & Rex (1971) illustrated the relation between endolithic algae and micrite envelope formation in Bahamian oolites. Endolithic algae can extensively colonize new substrates in weeks or months (Lukas, 1979 for summary) and thus repeated abrasion and grain-size reduciton must occur due to the activities of sediment ingesting organisms in carbonate environments (e.g. fish, holothurians). There is a great deal of scope for use of the SEM in examination of carbonate grain surface textures due to endolithic borers and their destruction by organic and physical processes. The production of micrite envelopes by boring is frequently taken to be a shallow water phenomenom, but the deep water fungal borings described by Zeff & Perkins (1979) may also lead to the production of micritized grain surfaces.

ACKNOWLEDGMENTS

The author wishes to thank Dr I.J. Fairchild for Fig. 8.8(a–c), A. Hogg, E. Sellier and Total CFP for Fig. 8.8(d–f), A.T. Kearsley for Fig. 8.9(a–c), and R.L. Gawthorpe for Fig. 8.9(d–h). Figs 8.17(c, d) and 8.22(a, b) were taken by the author at the Western Australia Institute of Technology whose technical assistance is acknowledged. All other figures were photographed at Aberdeen University's Department of Geology and Mineralogy using equipment purchased through generous research grants to the author from Occidental Petroleum. Iain S.C. Spark and Robert A. Downie assisted greatly with machine operation. Barry Fulton assisted with drafting and the text was patiently word-processed and edited by Sue Castle.

9 Chemical analysis of sedimentary rocks

IAN FAIRCHILD, GRAHAM HENDRY,
MARTIN QUEST and MAURICE TUCKER

9.1 INTRODUCTION

This chapter is written for sedimentologists new to chemical analysis, or those seeking new ways of tackling rock-based chemical sedimentological problems. Since several textbooks would be needed to cover the details of the geochemical theory and analytical techniques reviewed here, the approach has necessarily been selective. Where good descriptions of experimental details exist in readily-available literature, then reference is made to those papers and books. This chapter is not simply a recipelist, but it aims to instil a philosophy of approach to analysis, its objectives and the fundamental chemical controls on sediment composition. Some major aspects of the chemistry of sedimentary rocks are omitted, notably organic geochemistry and the analysis of radioactive isotopes, but these are not usually studied by sedimentologists. Although the solution-based techniques of chemical analysis described in this chapter are common to the analysis of pore water and depositional waters, the specialized tools and procedures of the aqueous analyst and the experimentalists in crystal growth and dissolution are not to be found here (but see Whitfield, 1975; Riley, 1975; Manheim, 1976; Pamplin, 1980).

There are two key features of the chemistry of the sedimentary cycle. First, there is the role of water as a solvent, a medium of transport and source for precipitating minerals. Second, thermodynamic equilibrium is often not obtained at the low temperatures which concern us. Geology graduates often have a poor background in low-temperature aqueous geochemistry; hence the need for Section 9.2 which outlines the most relevant concepts and discusses their usefulness in practice. Those embarking on the acquisition and interpretation of chemical data are strongly advised to pursue a proper understanding of these concepts by further study. Sometimes interpretation may appear to be a matter of pattern recognition by comparison of data sets, but this is only on a superficial level. Good geochemical research involves a clear understanding of the limitations of the data due to the sampling procedure and analytical techniques used and an ability to assess the feasibility of both chemical and geological processes.

9.2 OBJECTIVES

Whereas basic chemical data on sedimentary rocks have been available for some time (Clarke, 1924) and correct conclusions drawn about the origin of many common sedimentary minerals, it is in the fields of economic mineralization, and igneous and metamorphic geology, that the results of chemical analysis were first fully exploited. The diverse origins of sedimentary rocks in general, and of the components of individual specimens in particular, make the formulation of objectives, prior to planning the nature and extent of a programme of chemical analysis, of the utmost importance. Since this is not normally explicitly treated in sedimentary texts, an attempt at a general scheme of objectives is given below.

Successful approaches to problem-solving by chemical analysis of sedimentary rocks are illustrated in Fig. 9.1. The potential presence of four *genetic components* is illustrated: unmodified terrigenous detritus, leached or otherwise chemically-weathered terrigenous detritus, chemical or biological precipitates in the sedimentary environment, and authigenic (diagenetic) phases. A rock with all four components is shown as a circle, otherwise the appropriate quarter-circles are depicted.

In the tabular part of Fig. 9.1 the analysed portion of each rock is represented by a blacked-in region. Columns 1 and 2 refer respectively to whole-rock analysis (e.g. by X-ray fluorescence spectrometry) and selective analysis (e.g. by microprobe or by analysis of the solute after dissolution of a particular mineral). It is assumed in these cases that the minerals present can be obviously assigned to one of the four petrographic components of the rock and that the elements of interest are obviously sited in par-

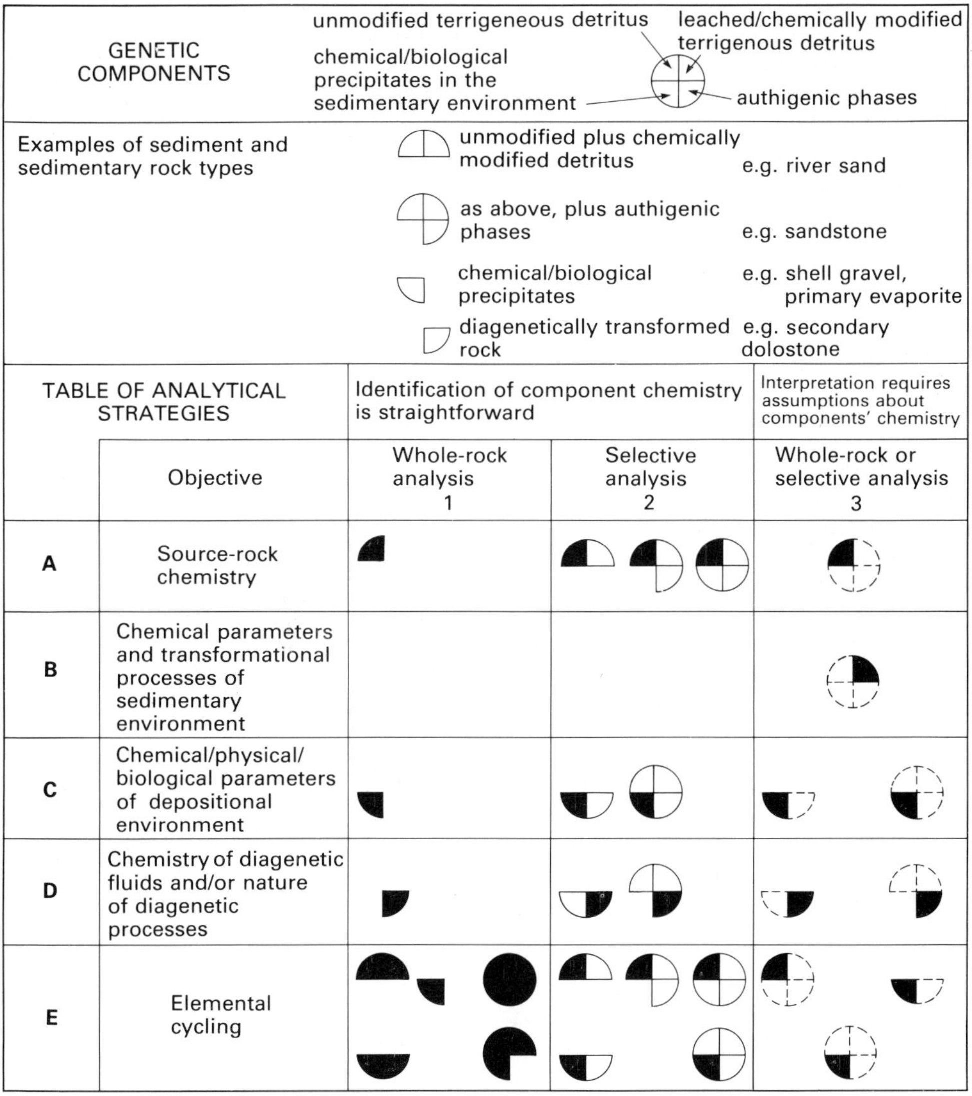

GENETIC COMPONENTS

unmodified terrigeneous detritus

leached/chemically modified terrigenous detritus

chemical/biological precipitates in the sedimentary environment

authigenic phases

Examples of sediment and sedimentary rock types

unmodified plus chemically modified detritus	e.g. river sand
as above, plus authigenic phases	e.g. sandstone
chemical/biological precipitates	e.g. shell gravel, primary evaporite
diagenetically transformed rock	e.g. secondary dolostone

TABLE OF ANALYTICAL STRATEGIES		Identification of component chemistry is straightforward		Interpretation requires assumptions about components' chemistry
	Objective	Whole-rock analysis 1	Selective analysis 2	Whole-rock or selective analysis 3
A	Source-rock chemistry			
B	Chemical parameters and transformational processes of sedimentary environment			
C	Chemical/physical/ biological parameters of depositional environment			
D	Chemistry of diagenetic fluids and/or nature of diagenetic processes			
E	Elemental cycling			

Fig. 9.1. Objectives of chemical analysis. See text for explanation.

ticular minerals. Quite often, however, one has to assume that this is probably, or largely so, in order to interpret the results. This is the situation represented by column 3. For example one might have to assume that diagenetic alteration of feldspars did not occur in order to use the feldspar composition as a guide to provenance. In column 3 the dashed lines enclose components of the rock whose contribution to the analysed chemistry is hoped to be either obvious or else insignificant. These components are not blacked-in on the diagram since they may or may not have been analysed. The column 3 approach is of course a slippery slope, but as long as one component dominates the aspects of the chemistry that are being studied then it is a perfectly reasonable strategy.

The five rows in Fig. 9.1 denote five different kinds of problems that may be tackled. Strategy A (row A) assumes that terrigenous sediments have inherited some aspects of source-rock chemistry unaltered (A1/A2) or still preserved as a clear signal (A3) in order to make deductions about the source area from which the sediments were derived.

Strategy B involves looking at the alteration processes in the sedimentary environment such as characterizing the intensity of chemical weathering by whole-rock analysis, or the salinity of the depositional environment from analysis of trace elements in clay minerals. B1 and B2 are not shown because it is extremely unlikely that some inheritance of original chemistry, formation of new precipitates (such as neoformed clays) in the depositional environments, or diagenetic alteration can be categorically ruled out in such studies.

There is a wide variety of precipitates which could be used to characterize parameters of the depositional environment (strategy C). Examples include the analysis of minor elements or oxygen isotopes in calcareous fossils to estimate the chemical composition and palaeotemperatures of ancient oceans; characterization of unusual sea water compositions near spreading centres by study of metalliferous sediments; elucidation of environmental setting or stage of evaporation of brines responsible for evaporite deposits.

Strategy D involves the analysis of diagenetically transformed rocks or the diagenetic components of sediments to ascertain parameters such as the degree of secondary alteration, or its timing, or the chemistry of the diagenetic fluids, or the evolution of porosity.

The understanding of the cycling of the elements now and at various times in the Earth's history, is the objective of strategy E. Recent successful developments have concerned Sr analyses in calcareous fossils as a monitor of Sr content of sea water, strontium isotope studies, which can be used to assess changing patterns of tectonism and sea-level, and carbon and sulphur isotope studies which have major implications about changing depositional environments with time. Bulk-rock analyses (E1) can be used (Garrels & MacKenzie, 1971) even if the origin of the rocks is not known. However, generating geochemical data in ignorance of the rock petrology is not generally a good idea!

Data derived from strategies E2 and E3 (and sometimes (E1) can also be used for strategies A and C and vice-versa. Strategies in column 2 give information vital to mineralogists. Whole-rock analyses are essential for many industrial uses of sedimentary materials but, in many cases, selective analysis or sedimentological expertise would be of great benefit in optimizing extraction or treatment procedures or predicting the extent of resources. Good geochemical data, with accompanying information on field relationships and petrography, often do prove multifarious in their application.

The purpose of this section has been to stimulate thought about the aims of the research (although Fig. 9.1 is necessarily generalized and incomplete). Before asking how to get the data, be sure you know why you want them!

9.3 CHEMICAL PRINCIPLES

Since only a skeletal treatment is possible here, particularly of the more standard physical chemistry, the reader is referred to the following texts for further enlightenment: Krauskopf (1979) or Brownlow (1979) for general geochemical background and more specific data on sedimentary geochemistry; Raiswell *et al.* (1980) and Open University (1981) for an introduction to surface geochemical processes; Garrels & Christ (1965) for an extensive introduction to the practical application of thermodynamic relations in aqueous systems; Berner (1980) for a most readable introduction to chemical processes occurring in early diagenesis together with a more advanced mathematical treatment; Berner (1971) for a wider but more concise overview of chemical sedimentology (excluding most sedimentary petrological aspects). The chemistry of natural waters are well reviewed by Holland (1978), Drever (1982), and the somewhat less readable advanced text of Stumm & Morgan (1981), while the past state of the oceans and atmosphere is the subject of a masterly book by Holland (1984). The practical prediction of mineral stability in a given solution can now be undertaken most easily and precisely by using interactive computer programs such as PHREEQE (PH-REDOX-EQUILIBRIUM-EQUATIONS) which was developed at the United States Geological Survey and is freely available (Parkhurst, Thorstenson & Plummer, 1980; Fleming & Plummer, 1983; Plummer & Parkhurst, 1985).

A basic understanding of the behaviour of elements can be achieved by considering their *ionic potential* (Fig. 9.2). This is the charge (z) of the appropriate positive ion of the element divided by its radius (r). Where z/r is less than 3, there is

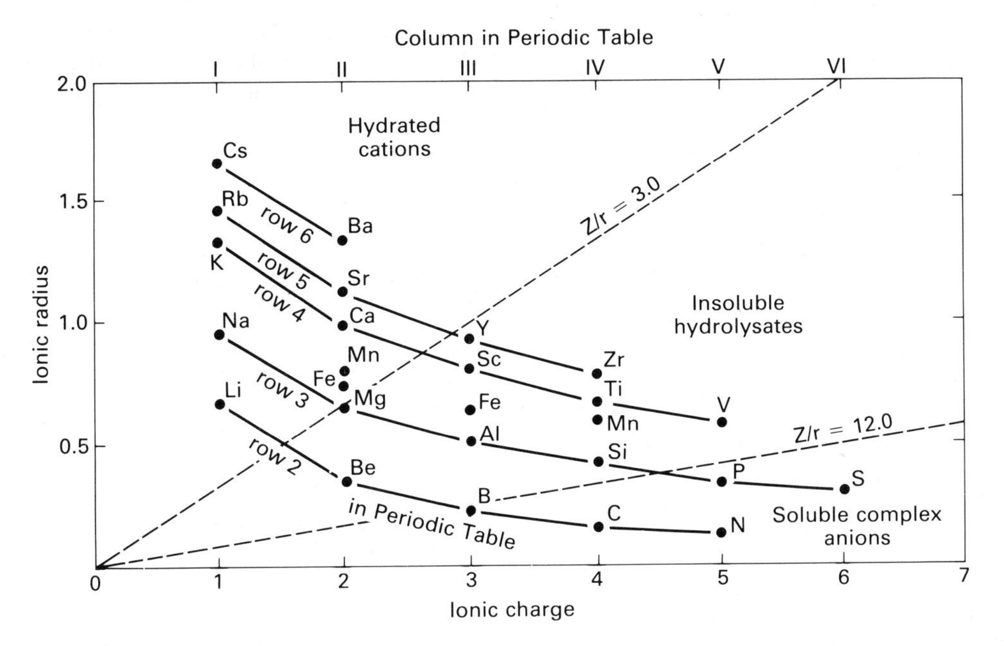

Fig. 9.2. Ionic potential of elements (modified from Blatt *et al.*, 1980).

relatively little affinity of the element for oxygen in water molecules: the element forms simple cations in solution surrounded by loosely-bound water molecules (water of hydration). With z/r between 3 and 12, the affinity for oxygen is greater and hydroxides of the element readily form. When z/r is greater than 12, the affinity for oxygen is so great that oxyanions result. Both the electropositive elements, forming hydrated cations, and the electronegative elements forming anions, tend to remain in solution at relatively high concentrations, whereas intermediate elements tend to be readily removed from solution as hydroxides or more usually by adsorption on to solids (Li, 1981). Whereas the elements with very high or very low ionic potential tend to occur in sediments as relatively soluble salts and intermediate elements as unaltered or chemically-modified detritus, all also occur as reactive metastable phases and bound on the surface of solids. The behaviour of each element is therefore highly variable, depending on precisely which minerals and dissolved species are present.

9.3.1 Concentrations and activities

Chemical analysis generally yields results in terms of weight concentrations of chemical species. In a rock sample, results are expressed as a percentage of the total sample weight (wt %) for major elements or parts per million (ppm) for minor components. Expression in terms of the equivalent oxide is normal for silicate rocks. Conversion of an analysis from a weight basis (e.g. wt %) to a molecular basis (e.g. mole %) involves division of the analysis of each species by its formula weight and normalizing (recalculating the total to 100%). Mineral analyses may then be expressed in terms of an equivalent formula, e.g. $K_{1.5}$ $(Si_{7.3}Al_{0.7})$ $(Al_{1.8}Fe^{3+}{}_{1.3}Fe^{2+}{}_{0.4}Mg_{0.5})$ O_{20} $(OH)_4$, a glauconite. For clay minerals this will normally involve an assumption about oxygen and hydroxyl stoichiometry since these components are often calculated by difference, being difficult to analyse.

Analysis is usually in terms of parts per thousand of the solution (‰ or g kg^{-1}) or, for trace components, ppm (mg kg^{-1}), or even parts per billion (ppb, μg kg^{-1}). At standard temperatures and pressures g kg^{-1} is equivalent to g l^{-1} of dilute solutions. The *molarity* (M) of a solution is the concentration of a chemical in moles l^{-1} (i.e. g l^{-1} divided by the formula weight of the species concerned). For theoretical calculations the *molality* (m) is used: it is the concentration in moles per kilogram of solvent (water). Unlike molarity, molality is independent of temperature and pressure, but at low temperatures the two scales are virtually identical except in concentrated brines. The conversion equation is:

$$m = M\frac{W}{(W - w)\,\Phi} \qquad (1)$$

where W = weight of the solution, w = weight of the dissolved chemicals and Φ = solution density. For gases, the partial pressure (P) is used to express concentration.

Except for reactions in dilute solutions the parameter of concentration is inadequate to account for the behaviour of chemical species because of interaction between chemicals in solution. Oppositely-charged ions will attract one another which reduces their ability to participate in chemical reations. Also, a certain proportion will combine to form ion-pairs (both neutral, e.g. $MgSO_4^0$, and charged, e.g. $CaHCO_3^-$) and ion-complexes dissolved in the solution. The *activity* of a species i (a_i) is the effective amount of i which is available to take part in a reaction,

$$a_i = \gamma_i m_i \qquad (2)$$

where m_i has units of moles l^{-1} and γ_i is the *activity coefficient* with units chosen to be l mole^{-1} so that activity is dimensionless.

An essential concept involved with activity is that of standard states. The *standard state* of a substance is when it has an activity of 1 at 25°C and one atmosphere pressure. For a solid, liquid or ideal gas, $a = 1$ when it is pure and at this temperature and pressure. For dissolved species the standard state that has been chosen assumes that $a_i = \gamma_i = m_i = 1$. It is impossible to make such a solution since it requires the ions to behave ideally (i.e. not to interact with one another) yet be present at the very high concentration of 1 mole l^{-1}! The convenience of this definition is that in dilute solutions a is numerically equal to m, and hence concentrations can be used in calculations.

In dilute solutions, values of γ_i allowing for ion interactions (but not ion-complex formation) are readily calculated by the Debye-Huckel equation (e.g. Berner, 1980, pp. 15–18) which quantifies the deceasing γ_i with increasing charge of ions and increasing concentration of the solution. This concentration is expressed as the *ionic strength* (I):

$$I = \tfrac{1}{2}\Sigma\, m_i(z_i)^2 \qquad (3)$$

where i refers to each ionic species in solution and z_i is the charge of the ion. In more concentrated solutions, ion complexes must be allowed for as well (to give 'total' activity coefficients), but this is difficult since these depend not only on I but also on the relative abundance of particular ions as the different ion complexes vary greatly in stability. The elucidation of total ion activity coefficients in sea water has been an obvious target for research (Garrels & Thompson, 1962) and refinements continue to be made, albeit with disagreement in some important details (Millero & Schreiber, 1982; Plummer & Sundquist, 1982; Nesbitt, 1984). More concentrated brines have proved rather intractable although it has long been clear that activity coefficients rise to greater than one (thus activities are numerically greater than concentrations — a good reason for not assigning units of concentration to activity as is done in some introductory texts). A generalized explanation for this phenomenon is that at high ionic strengths cations lose their attached water molecules and so are more free to take part in chemical reactions than in the hypothetical standard state condition (where the activity coefficient is one, but cations are assumed to be hydrated). Recently, following theoretical advances by the chemist Pitzer, considerable progress has been made in understanding the chemistry of brines and hence predicting the precipitation of minerals from them (e.g. Harvie, Moller & Weare, 1984).

9.3.2 Equilibrium

Chemical equilibrium refers to a state of dynamic balance between abundances of chemical species. It is rapidly attained for reactions involving only *dissolved* species, but at low temperatures *solids* tend to remain out of equilibrium with the solution with which they are in contact (except where this contact is extremely prolonged, e.g. Nesbitt, 1985). Nevertheless, the concept of equilibrium is extremely useful in that reactions will move in the direction of equilibrium. Therefore the sense of change in a system (e.g. dissolution or precipitation of a mineral) can be predicted.

The stoichiometry of an equilibrium reaction can be written in a general way as:

$$b\mathrm{B} + c\mathrm{C} \rightleftharpoons d\mathrm{D} + e\mathrm{E}, \qquad (4)$$

i.e. b molecules of species B react with c molecules of C to form d of D and e of E (and vice-versa). The Law of Mass Action shows that:

$$\frac{a_D{}^d a_E{}^e}{a_B{}^b a_C{}^c} = K \quad (5)$$

where $a_D{}^d$ = activity of D raised to the power d etc., and K is the *thermodynamic equilibrium constant* for the reaction. K varies only with temperature and pressure. Sometimes the relationship is formulated in terms of concentrations (K becomes K_c), but K_c will vary with solution composition as well as temperature and pressure.

Equation (5) expresses the relative stability of the chemical species: the degree to which reaction (4) goes to the right or left. A second way of expressing relative stabilities of chemicals is by the change in free energy accompanying the reaction. The *free energy* of a substance is the energy it possesses to do work. For any reaction:

$$\Delta G° = -RT \ln K \quad (6)$$

where $\Delta G°$ is the change in (Gibbs) free energy accompanying the reaction (the superscript ° denotes that this is the standard free energy change corresponding to the chemicals being in their standard state), R is the gas constant and T is the absolute temperature. At 25°C:

$$\Delta G = \Delta G° = -1.364 \log_{10} K \quad (7)$$

where ΔG is expressed in kcal mole^{-1}. For reaction (4):

$$\Delta G° = \Delta G°_{fD} + \Delta G°_{fE} - \Delta G°_{fB} - \Delta G°_{fC} \quad (8)$$

whre $\Delta G°_{fD}$ is the free energy involved in forming D from its elements in the standard state (standard free energy of formation) and likewise for E, B and C.

Values of $\Delta G°_f$ are tabulated in many texts, although constantly subject to revision. Values for K can thus be readily calculated for any reaction of interest. A negative value of ΔG indicates that a system starting with proportions of chemicals as written in equation (4) will go to the right to reach equilibrium. Conversely a positive value indicates a leftward movement to reach equilibrium.

In the special case of the dissolution-precipitation reaction of a salt BC:

$$B_b C_c \rightleftharpoons bB^+ + cC^+ \quad (9)$$

$$K = \frac{(a_B{}^+)(a_C{}^-)}{a_{BC}}. \quad (10)$$

Since the activity of a pure solid is one by definition, then:

$$K = (a_B{}^+)(a_C{}^-) = K_s \quad (11)$$

where K_s is the (activity) *solubility product.*

In a given solution, the *Ionic Activity Product* (IAP) can be determined:

$$\text{IAP} = (a_B{}^+)(a_C{}^-) \quad (12)$$

and compared with the solubility product by the *saturation index* (Ω):

$$\Omega = \frac{\text{IAP}}{K_s} \quad (13)$$

or the % *saturation*:

$$\%\ \text{saturation} = \frac{\text{IAP}}{K_s}. \quad (14)$$

NB some authors define Ω as $\log_{10}(\text{IAP}/K_s)$. If $\Omega = 1$, the solution is at equilibrium (the salt tending neither to dissolve nor to precipitate) and is said to be just (100%) saturated. With $\Omega > 1$, then the solution is supersaturated and ought to be precipitating the salt, conversely an undersaturated solution ($\Omega < 1$) should be dissolving the salt.

The correct formulation of an equilibrium constant for solid solutions has been under debate with particular reference to magnesian calcites and clay minerals. Three different approaches have been proposed. Using magnesian calcites as an example, fractional exponents can be used (e.g. Thorstenson & Plummer, 1977):

$$K_{(Mg^x Ca^{1-x} CO_3)} = (a_{Ca}{}^{2+})^{1-x}(a_{Mg}{}^{2+})^x(a_{CO_3}{}^{2-}). \quad (15)$$

Alternatively it could be assumed that equilibrium is always reached for the $CaCO_3$ component (e.g. Wollast, Garrels & MacKenzie, 1980):

$$K_{(Ca,Mg)CO_3} = (a_{Ca}{}^{2+})(a_{CO_3}{}^{2-}) = K_{s(CaCO_3)}.$$

Finally, activities of substituting ions could be added (Lippmann, 1977; Gresens, 1981):

$$K_{(Ca,Mg)CO_3} = (a_{Ca}{}^{2+} + a_{Mg}{}^{2+})a_{CO_3}{}^{2-}. \quad (17)$$

Despite the theoretical objections to the use of fractional exponents (Lippman, 1977; Gresens, 1981), the formulation of equation (15) was the only one to fit the data in the careful experiments of Walter & Morse (1982). Tardy & Fritz (1981) also continued this approach in calculating clay mineral solubilities.

For solid solutions a distinction (well reviewed by MacKenzie *et al.*, 1983) must be made between true thermodynamic equilibrium of solid and contacting solution, and *stoichiometric saturation* (Thorstenson & Plummer, 1977). In the former case, crystals equilibrating with a large reservoir of fluid have their composition determined by the fluid, which usually necessitates recrystallization. Since this is often very slow in nature, stoichiometric saturation describes the common case of dissolution ceasing in the absence of recrystallization (i.e. congruent dissolution). The converse is incongruent dissolution: the release of chemicals to solution in different proportions to the solid composition. This occurs not only with solid solutions but can also lead to the transformation of one mineral to another as in the chemical weathering of silicates.

The dissociation of water is an equilibrium of special importance:

$$H_2O \rightleftharpoons H^+ + OH^-. \quad (18)$$

At 25°C and in dilute solutions:

$$10^{-14} = K_w = \frac{(a_{H^+})(a_{OH^-})}{a_{H_2O}} = (a_{H^+})(a_{OH^-}). \quad (19)$$

In more concentrated solutions, a_{H_2O} is less than one and is equivalent to the water vapour pressure above the solution at equilibrium relative to the vapour pressure above pure water. The varying *relative* abundances of H^+ and OH^- are extremely important in nature for life processes and stability of minerals, notably carbonates. When $a_{H^+} = a_{OH^-}$ ($= 10^{-7}$ at 25°C), the solution is *neutral*. When $a_{H^+} > a_{OH^-}$, the solution is *acid* whereas when $a_{OH^-} > a_{H^+}$ the solution is *alkaline*. For convenience, a logarithmic derivative (pH) of H^+ is generally used to express this property:

$$pH = -\log_{10}(a_{H^+}). \quad (20)$$

Therefore at 25°C the pH of a neutral solution is 7, lower values corresponding to acid solutions, higher pHs to alkaline ones.

The relevance to carbonates arises because the carbonate ion is more abundant at higher pH. It is derived by the double dissociation of the weak acid H_2CO_3 (carbonic acid).

$$H_2CO_3 \rightleftharpoons H^+ + HCO_3^- \quad (21)$$

$$HCO_3^- \rightleftharpoons H^+ + CO_3^{2-}. \quad (22)$$

Low amounts of H^+ (high pH) drive these equilibria to the right. K for (21) is $10^{-6.4}$ so that the pH has to rise to greater than 6.4 for $a_{HCO_3^-}$ to be greater than $a_{H_2CO_3}$. K for (22) is $10^{-10.3}$, therefore even in an alkaline solution like sea water (around pH 8), $a_{HCO_3^-} = (10^{2.3})(a_{CO_3^{2-}}) \gg a_{CO_3^{2-}}$. Thus it is clear that the amounts of carbonate ion are often a limiting factor in carbonate mineral formation.

The amounts of CO_3^{2-} cannot be measured directly in solution but, by carrying out an acid titration, the amount of chemicals (*bases*) that will react with acids is easily computed. This is the *alkalinity* of the solution. In most cases HCO_3^- and CO_3^{2-} are by far the dominant weak bases present so that the alkalinity approximates to the *carbonate alkalinity* (A_c) where:

$$A_c = m_{HCO_3^-} + 2m_{CO_3^{2-}}. \quad (23)$$

The figure two arises in equation (23) since CO_3^{2-} can react with two H^+ ions. Knowing A_c, $a_{CO_3^{2-}}$ can be calculated from a knowledge of K from equation (22). The alkaline (high pH) nature of a particular solution should not be confused with alkalinity: the two parameters are independent variables. High values of carbonate alkalinity mean that the solution can receive much acid input without markedly changing its pH, since the excess H^+ ions are removed by reacting with the weak bases. Any solution which resists a change in pH by reacting with H^+ or OH^- ions is said to be *buffered* and the reactions are termed buffering reactions.

Although the relationship of pH to mineral stability is normally clear, complications arise when two minerals are in competition. Take for example sedimentary calcium phosphate and calcium carbonate. Although both are less soluble in alkaline solutions, calcium carbonate shows this more intensely, and so tends to win a competition for calcium ions in alkaline solutions. In early marine diagenesis, calcium phosphate would thus be expected to form in neutral to weakly acid pore waters (Nathan & Sass, 1981).

In chemical reactions involving exchange of electrons there is a third measure of equilibrium (other than ΔG and K) which is the oxidation-reduction potential (Eh). Each such reaction can be broken down into two half-reactions involving an oxidized chemical reacting with electrons to form a reduced chemical:

$$B \rightleftharpoons B^{n+} + ne \quad \text{(half-reaction)} \quad (24)$$

$$ne + C^{n+} \rightleftharpoons C \quad \text{(half-reaction)} \quad (25)$$

$$B + C^{n+} \rightleftharpoons C + B^{n+} \quad \text{(overall reaction)} \quad (26)$$

where n is an integer and e stands for an electron. Since each half reaction involves electron transfer it is associated with a putative EMF (voltage), the magnitude of which depends on the free energy for each half-reaction. Under standard conditions (25°C,

1 atmosphere pressure, unit activities of chemicals):

$$\Delta G^\circ = nFE^\circ \qquad (27)$$

where F is Faraday's number (the charge of a mole of electrons) and E° is the standard EMF for the reaction. Under non-standard conditions the voltage produced is the oxidation-reduction potential (Eh):

$$\begin{aligned} \text{Eh} &= E^\circ + \frac{RT}{nF} \ln \frac{a_{(\text{oxidized chemical species})}}{a_{(\text{reduced chemical species})}} \\ &= E^\circ + \frac{0.06(a_{\text{ox}})}{a_{\text{red}}}. \end{aligned} \qquad (28)$$

In practice, voltage can only be measured for *pairs* of half-reactions so that one half-reaction needs to be arbitrarily defined as zero to enable values for each half-reaction to be given. Reaction (29) is thus defined with $E^\circ = 0$ V.

$$H_2 \rightleftharpoons 2H^+ + 2e. \qquad (29)$$

This is equivalent to saying that the standard free energy of formation of H^+ is zero.

For any solution in nature there will be a mixture of various oxidizing and reducing agents which will react and tend towards equilibrium. By comparing this solution with a standard electrode of known Eh, the Eh of the test solution can be determined. A positive value for Eh indicates that it is relatively oxidizing ('electron-grabbing'), likewise reducing for negative Eh. Eh is one of the less satisfactory chemical concepts to apply to real solutions because it is difficult to measure, and in sedimentary systems many redox reactions are irreversible and bacterially-mediated (Hostettler, 1984).

A complicating factor when considering equilibrium in nature is the physical condition of solid materials. For example, very small particles have excess free energy (surface free energy) arising from unsatisfied bonds on their surface. This leads to the phenomenon of Ostwald's ripening (reviewed by Baronnet, 1982) whereby larger crystals may grow in solution whilst very small ones dissolve. This is a useful phenomenon in experimental studies of crystal growth carried out at low growth rates (Lorens, 1981). Excessive surface area is also a property of organic tests and complex crystal aggregates, and Williams, Parks & Crerar (1985) illustrated how the resulting increased solubility plays an important role in silica diagenesis.

Crystals that have been stressed or which contain a larger proportion of dislocations than other crystals of the same phase will dissolve preferentially. Even different surfaces of single crystals differ in susceptibilities to dissolution, as discussed for quartz by Hurst (1981). Crystal defects affect the incorporation of some trace elements into minerals (see Section 9.3.5) and probably limit the size of clay mineral crystals.

Diagrams showing the stability of different mineral phases and aqueous species are commonly used in chemical sedimentology and have a powerful visual impact as an apparently precise guide to the conditions under which minerals have formed. Garrels & Christ (1965) gave an excellent guide to their construction and use. Figure 9.3 has been deliberately badly presented to illustrate possible pitfalls in the use of stability diagrams. The lines themselves have been constructed from equations in terms of Eh and/or pH resulting from substituting free energy data into equation (8), together with (28), or (6) and (5). Values for ΔG are known with greater or lesser precision and are constantly being revised (Helgeson *et al.*, 1978; Robie, Hemingway & Fisher, 1978; Weast, 1983; Woods & Garrels, 1987). Another source of imprecision is the measurement of Eh in natural waters which, as previously mentioned, is difficult and not always meaningful (Hostettler, 1984). The stability fields refer to regions where one phase is more stable than another. When a field is labelled for a dissolved phase, it is saying 'no solid phase is stable here'. However, dissolved phases have variable activities, so that field boundaries with

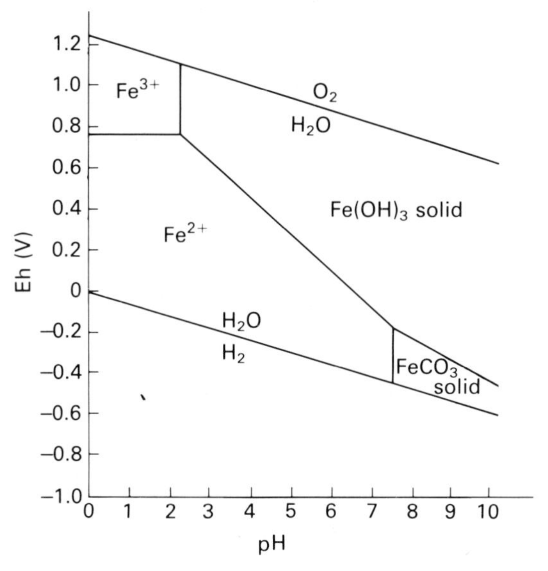

Fig. 9.3. Thermodynamic stability relations of iron species at 25°C and 1 atmosphere total pressure. See text for omissions!

solids depend on the choice of activity. In Fig. 9.3, the boundary between the Fe^{2+} and $Fe(OH)_3$ fields is actually drawn in assuming $a_{Fe}{}^{2+} = 10^{-3}$ at the boundary, lower values for the activity would correspond to moving the line parallel to itself to the upper right. The caption to Fig. 9.3 should state: activities of Fe^{2+} and Fe^{3+} at field boundaries are 10^{-3}. Choosing an appropriate value for the limiting activity of a dissolved phase requires some geological judgement. For example, Curtis & Spears (1968) decided that since $a_{Fe}{}^{2+} < 10^{-6}$ in sea water then $a_{Fe}{}^{2+} = 10^{-6}$ would be an appropriate boundary condition for the Fe^{2+} field in a diagram expressing relationships in sea water, whereas in pore waters $a_{Fe}{}^{2+} = 10^{-3}$ would be apposite since activities of ferrous iron range up to this value there. Another problem with Fig. 9.3 is that in order to construct a field boundary for $FeCO_3$ you need to know the total activity of the dissolved carboxy species (H_2CO_3, HCO_3^- and CO_3^{2-}); the caption for Fig. 9.3 should have stated that their total activity was 10^{-2} in this case. Some fields are missing in Fig. 9.3: for example $Fe(OH)^{2+}$ has a stability field between Fe^{3+} and $Fe(OH)_3$ (e.g. Garrels & Christ, 1965, fig. 7–14). Thus one must be certain that relevant equilibria have not been forgotten. The diagram also implicitly assumes that activities of HS^- are negligible, otherwise pyrite would have a stability field of substantial size. Silicates such as glauconite and berthierine must also not be forgotten (Maynard, 1986). Finally, it would be more informative to plot Eh against activity of HCO_3^- or HS^- because these species vary much more than pH in practice (Curtis & Spears, 1968). A series of diagrams is needed to show the complete picture of the system.

Good examples of the utility of stability diagrams in diagenetic studies are given by Davies *et al.* (1979), Hutcheon (1981) and Curtis (1983).

When considering systems far removed from their standard state, for example during burial diagenesis, the effects of temperature, pressure and salinity variations must be carefully considered. Take for example the relative stability of carbonate minerals and silica. Apparent alternating silica-carbonate replacements have been explained simply in terms of pH variations (Walker, 1962), but the solubility of silica shows little variation with pH until pHs of 9 or more. Actually during steady burial, temperature (T) pressure (P), pH and salinity (S) will probably all rise. For calcite the effects of T and pH are to decrease solubility, whereas P (with constant P_{CO_2}) and S increase solubility. For quartz, T, P and pH all increase solubility, but S reduces it because of the reduction in a_{H_2O} which destabilizes hydrated silica in solution (Fournier, 1983). These concepts can be applied to permeable rocks in which convective flow is occurring (Wood, 1986) and indicate that quartz and carbonates can be transferred from one part of the rock mass to another, often moving in opposite directions and hence giving rise to replacive relationships.

9.3.3 Departures from equilibrium

Where sediments are out of equilibrium with their contained fluids, this relates either to kinetic controls (the slowness of chemical reactions or movement of chemicals) or to biological interference.

The most important kinetic control concerns the behaviour of crystal surfaces involved in dissolution or growth, especially the latter. The presence of abundant early diagenetic minerals such as glauconite or carbonate fluorapatite only in *slowly*-deposited marine calcareous sediments illustrates the kinetic control on crystal growth. Conversely the fact that slowly-deposited detrital sediments exist at all attests to the incomplete dissolution of mineral grains in undersaturated waters.

Crystal growth in a clean, supersaturated solution is initially limited by a stage of *nucleation*, since nuclei below a certain critical size have excess free energy and are thus unstable. However, in the sedimentary environment there will normally already be nuclei in the system which can be overgrown, even if nucleus and overgrowth are different minerals (epitaxial overgrowth). Where supersaturation is low [Ω (equation 13) close to 1] the critical nucleus size is larger and the availability of nuclei could then be rate-limiting, but such conditions also lead to slow crystal growth anyway.

If *surface reactions* rather than movement of chemicals limit the growth rate of crystals, then the mode of growth would be expected to be by addition to growth spirals around screw dislocations occurring on the crystal surface, rather than by the birth and spreading of two-dimensional layers (Sunagawa, 1982). If Ω is close to 1, then:

$$R = K(\Omega - 1)^2 \quad (30)$$

where R is the growth rate, frequently expressed in terms of a radial increase in crystal size, and K is a constant. Often the real growth-limiting factor is adsorption (see Section 9.2.4) of foreign ions on the surface of the crystal. In this case the above equation would probably become (Berner, 1980, p. 105):

$$R = K(\Omega - 1)^n \quad (31)$$

where $n > 1$. For example, Mg-adsorption strongly inhibits calcite growth, and adsorbed humic and fulvic acids and PO_4^{3-} inhibit the growth of aragonite (Berner *et al.*, 1978). The importance of Mg as an inhibitor lies in its abundance in natural waters. Experimental studies (Meyer, 1984) on calcite growth have shown that numerous inorganic and organic species are more effective inhibitors, with Fe^{2+} being so by four orders of magnitude. Adsorption phenomena are also important in controlling crystal habit (Boistelle, 1982).

For dissolution, we have:

$$R = K(1 - \Omega)^n \text{ (cf. 31).} \quad (32)$$

where R refers to the dissolution rate. Very slow dissolution rates caused by ion-adsorption are common: for example the dissolution of calcite and aragonite shells in undersaturated sea water where $n = 4-5$ (Keir, 1980), the reaction being limited by adsorption of PO_4^{3-}. However, inhibition of dissolution of calcium carbonate by organic materials is generally related to physical covering, rather than adsorption phenomena (Morse, 1983).

Berner (1978a, 1980) has shown that where dissolution is controlled by surface reactions, the crystals have angular outlines with crystallographically-controlled etch pits: such surface-reaction control is normal in soils and nearly all surface waters.

Under some circumstances the rate of crystal growth or dissolution is controlled by the rate of movement of chemicals in solution rather than by reactions taking place on the crystal surface. This only seems feasible within pore waters and where precipitation or dissolution is unusually rapid (implying great supersaturation or undersaturation). Movement of chemicals in general may be by *advection* (movement under forces, e.g. groundwater flow or the relative movement of grains and pore fluid due to compaction), *dispersal* by burrowing organisms or turbulence, or by *molecular diffusion*. The existence of appropriate concentration gradients in solutions away from sites of, for example, manganese oxide or iron sulphide precipitation in Recent sediments demonstrates the importance of the transport-control mechanism. Growing crystals in these circumstances may be restricted to a very fine crystal size, or show dendritic or spherulitic morphologies (Rodriguez-Clemente, 1982; Sunagawa, 1982); dissolving crystals will exhibit smoothly-rounded outlines (Berner, 1978a, 1980). The existence of specifically *diffusion*-controlled growth in sediments and sedimentary rocks is demonstrated by Liesegang ring phenomena, such as iron oxide bands formed during weathering, or concentric mineralogical banding in concretions (De Celles & Gutschick, 1983). This arises where inter-diffusion of two species has led to precipitation of material at episodically-shifting sites.

The magnitude of a diffusional flux is given by Fick's first law of diffusion. It relates the mass of chemical component i transported per unit area per unit time (J) to its concentration gradient (dC/dx) across distance x:

$$J = -D_i \frac{dC}{dx} \quad (33)$$

where D_i is the diffusion coefficient, specific to the chemical component i (typically around 10^{-5} cm^2 s^{-1} for salts in water at 25°C). In a sediment, diffusion is slowed because of the effective increase in path length caused by the obstructions represented by the sediment grains. Thus the effective diffusion coefficient is smaller by a factor relating to the sediment porosity (e.g. ϕ^2 where ϕ is the fractional porosity, Lerman, 1979). There is also some variation of D_i with salinity of a solution (e.g. Lerman, 1979, p. 81). Sometimes corrections are necessary to allow for ion-pairing effects (Mangelsdorf & Sayles, 1982). The behaviour of the diffusion gradient with time is covered by Fick's second law (e.g. Berner, 1980, pp. 32–33); diffusion acts to reduce the concentration gradients but, if it is allowed to do so, the flux necessarily diminishes.

Enhanced movement of chemicals, because of the presence of open burrows in a sediment, can be modelled in a similar way, using a biodiffusion coefficient or, if complete bioturbation occurs, then a 'box modelling' approach can be used (Berner, 1980, pp. 42–53).

An advective flux can be calculated as the product of the concentration of the component (C) and the velocity of flow (U),

$$J = CU. \quad (34)$$

It is often useful to compare the effectiveness of diffusion with that of advective flow. If we imagine a situation where a diffusive flux in one direction is balanced by an advective flux in the other direction, then:

$$-D_i \frac{dC}{dx} + CU = 0. \quad (35)$$

If the concentration difference (dC) is of the same order as the concentration (C) then:

$$Udx \simeq D_i. \qquad (36)$$

For a given diffusion coefficient, this equation shows over what distance an advective flow of velocity U is equally effective as diffusion. For example, for a typical value of D_i in a sediment at 25°C of 100 $cm^2\ yr^{-1}$ then an advective flux would need to exceed 1000 $cm\ yr^{-1}$ in order to balance diffusion across a 1 mm gradient, or only 1 $cm\ yr^{-1}$ over a distance of 1 m (Berner, 1980, pp. 117–118). Unless the concentration gradient is fixed by reactions occurring at either end (such as rapid dissolution, precipitation or dispersal at an interface), diffusion will diminish with time by Fick's second law. A way of conceiving this (Lerman, 1979, pp. 58–60), again assuming $dC \approx C$, is by use of the criterion that advection will transport material in a time t over a distance (L_a) of $L_a = Ut$ whereas diffusion will transport material over a distance (L_d) = $\sqrt{(D_i t)}$. If the concentration gradient is not fixed then $L_a = L_d$ is a criterion for the time ($t = D_i/(U^2)$) after which advection starts to become more effective than diffusional transport (e.g. 14 days with $U = 1\ m\ yr^{-1}$ and $D_i = 10^{-5}\ cm^2\ s^{-1}$). In this way the most effective transport process may be determined for a given situation.

It is often stated (e.g. Pingitore, 1982) that diffusion becomes more important at higher temperatures. Certainly D_i increases steadily with increasing temperature: data (Weast, 1983) on equivalent conductance of ions (directly proportional to diffusion coefficients, Li & Gregory, 1974) yield a rough relationship to 150°C:

$$D_i = D_{oi}\left(1 + \frac{T_c}{25}\right) \qquad (37)$$

where D_{oi} is the diffusion coefficient for the ion at 0°C and T_c is temperature in degrees Celsius. The relationship with temperature is also definable in terms of fluid viscosity (η) and absolute temperature (T) such that:

$$\frac{D_i \eta}{T} = \text{constant} \qquad (38)$$

(Li & Gregory, 1974; Lerman, 1979, pp. 86–89). Rates of diffusion will only increase with temperature as long as the increase in D_i outweighs the restriction due to decreasing porosity. For example, Baker, Gieskes & Elderfield (1982) indicated that this is marginally so for Sr^{2+} and Ca^{2+} in a 500 m vertical section of deep-sea ooze with a 20°C temperature increase from top to bottom. It is more difficult to assess diffusion coefficients in extremely small pores, but the effectiveness of diffusion along intergranular boundaries during diagenesis is well known from the pressure dissolution and mineralogical replacement phenomena abundant in sedimentary rocks.

Apart from kinetic effects, organisms have a profound influence on the sedimentary environment, causing widespread departures from equilibrium. For example, the secretion of calcareous shells in undersatured waters is well known. Siliceous organisms can reduce silica concentrations in solution to vanishingly small amounts. The decomposition of organic matter during early diagenesis is immensely speeded-up by the action of bacteria and this in turn leads to greatly accelerated mineral reactions, particularly pyrite formation, and reactions involving carbonates. Such activities, not forgetting the deposition of immense quantities of reduced carbon, have obviously had a major effect on the sedimentary cycling of elements. Vital effects on carbonate trace element chemistry and isotopic ratios are mentioned in later sections.

9.3.4 **Adsorption**

Adsorption is the process of accumulation of chemical species at an interface: in our context at the surface of a solid. It is a complex phenomenon (Parks, 1975; van Olphen, 1977; Yariv & Cross, 1979) which is sometimes rendered confusing by oversimplification. Parks (1975) provided an excellent review of the chemical concepts involved with detailed examples, and Cody (1971) discussed its application to palaeoenvironmental studies of shales. The relevance of adsorption may be judged from the statement of Li (1981) that it is the most important removal mechanism for most elements into sediments from sea water. The importance for crystal growth and dissolution kinetics was mentioned in Section 9.3.3 and its relevance in coprecipitation studies is dealt with in Section 9.3.5.

Figure 9.4(a) illustrates a solid whose surface has developed a surface charge. The charge is neutralized by an electrical field in the solution featuring an excess of (in this case) positive ions close to the solid. The whole system makes up an *electrical double layer* consisting of the *fixed layer* on the solid and the diffuse or *Gouy layer* of counter-ions and co-ions which have respectively the opposite and the same charge as the fixed layer. In a more concentrated solution the Gouy layer contracts (Fig. 9.4b). The ions in the Gouy layer are described as adsorbed, but are readily exchanged for other ions if the composition of the bulk solution changes. In Fig. 9.4(c), the fixed layer now comprises the solid surface together with strongly- ('specifically-') adsorbed species (in this case positively-charged ions of two different kinds). The overall charge of the fixed layer in this instance has been reversed so that the Gouy layer now has an excess of negative ions. The change from Fig. 9.4(b) to Fig. 9.4(c) could be simply a function of time, allowing two kinds of process to occur. First, weakly-adsorbed species in the Gouy layer may develop a stronger bond with the solid (e.g. by losing water of hydration). Second, the solid may progressively scavenge any rare species in the solution for which it has a great affinity, particularly if solid and solution are in differential motion.

A thermodynamic approach (Parks, 1975) is helpful in clarifying why adsorption occurs. A particular species will adsorb if this will entail a loss of free energy (ΔG_{ads} is negative). ΔG_{ads} is the sum of a series of possible processes. In many cases electrostatic attaction of ions to a charged surface (with associated ΔG_{elec}) is the most important process, but in other cases ΔG_{elec} is smaller in magnitude than ΔG_{chem} (the formation of a specific chemical bond of the adsorbent with the solid) or ΔG_{hyd} (changes in the size or shape of the hydration envelope around ions) or ΔG_{rxh} (displacement of ions from the solid by reaction with the adsorbent), or some combination of these.

When ΔG_{elec} is dominant, the adsorption process is relatively *non-specific*, that is, it depends more on the charge of the ions being adsorbed than on any of their other properties. Here the Gouy layer concept applies, for which there are various detailed mathematical models (Parks, 1975; van Olphen, 1977).

When ΔG_{elec} is subordinate, the adsorption is *specific* to the adsorbent and the solid concerned. For example, some large organic molecules are adsorbed because the process of Van der Waal's bonding contributes much loss of free energy. Boron is known to adsorb strongly and specifically to illite, presumably by chemical bonding of $B(OH)_4^-$ groups to sites on the mineral surface (Couch & Grim, 1968). In fact any solute species which is capable of forming particularly strong complexes in solution, or insoluble compounds with some component of the solid, will potentially show strong adsorption. It should be emphasized that all gradations exist between specific and non-specific adsorption.

The sign and magnitude of the surface charge on most solids depends on the solution chemistry. For example, considering hydroxides, they will either become positively charge by reactions such as:

$$MOH + H^+ \leftharpoondown MOH_2^+$$

or negatively charged:

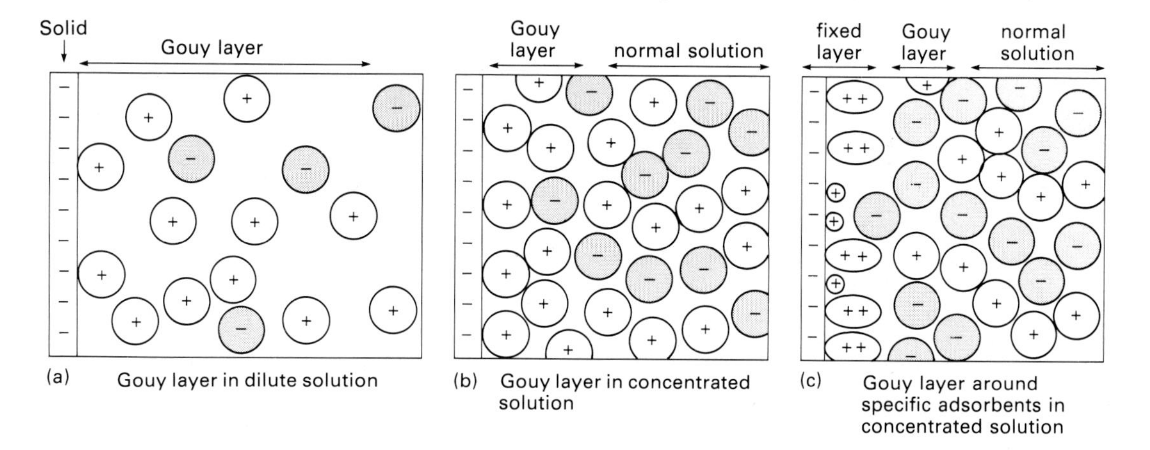

Fig. 9.4. Schematic illustration of adsorption processes. For simplicity, the surface charge of the solid is shown as if owing to lattice substitutions (as in clays) rather than because of adsorption of potential-determining ions (see text).

$$MOH \leftharpoondown MO^- + H^+.$$

Excess of H^+ ions, as in acid solutions, promotes a positive charge which decreases with increasing pH until a *point of zero charge* (pzc), specific for each substance, is reached, followed by the development of a negative charge at higher pH. Oxides behave in a similar way, since their surface oxygens become hydroxyl groups adjacent to aqueous solutions. Thus for oxides and hydroxides, H^+ determines the charge and hence the electric potential of the surface: it is the *potential-determining ion* (PDI). PDIs for salts such as $CaCO_3$ are its constituent ions, positive charge being expected when $A_{Ca^{2+}}$ minus $a_{CO_3{}^{2-}}$ in the solution exceeds a critical level, but pH is relevant too since $a_{CO_3{}^{2-}}$ depends on it.

As indicated on Fig. 9.4(c), the surface charge on a solid may be nullified or reversed by specific adsorbents. One important example of this is the observation that a variety of natural solids, regardless of their charge in artificial sea water, show a negative charge in natural sea water, probably due to adsorption of organic molecules (Neihof & Loeb, 1972).

Clay minerals are negatively-charged, but this arises out of substitution of lower- for higher-valent cations in the lattice and so is an intrinsic feature. The charge is balanced by the adsorption of cations between lattice layers and on the flat basal surfaces of clay particles. Another effect present on the edge of clay grains, and over the whole surface of some clays (e.g. kaolinite), which display little altervalent substitution, is a positive or negative charge developed by reaction of H^+ with SiOH or AlOH sites. As with other fine-grained materials, clays behave as *colloids* in aqueous solutions. A colloidal system has particles of colloidal dimensions (between 10^{-9} and 10^{-6} m in at least one direction) dispersed in a continuous phase of different composition (van Olphen, 1977). Colloidal physical properties (e.g. settling behaviour) are dominated by surface electrostatic phenomena.

The relationship between concentration of adsorbent in solution and amount of adsorption can take different forms, but normally the main feature is that adsorption reaches a maximum level corresponding to a saturation of adsorption sites. Assuming that the adsorbent molecules form a monolayer and do not interact with each other, then at equilibrium the rate of adsorption (proportional to concentration of adsorbent in bulk solution and number of available sites) will equal the rate of desorption (proportional to the number of sites occupied by the adsorbent). This logic (e.g. Yariv & Cross, 1979, pp. 120–123) leads to a form of equation known as the *Langmuir isotherm:*

$$C_{ad} = \frac{abC_{sol}}{1 + (aC_{sol})} \qquad (39)$$

where C_{ad} is the concentration of adsorbed species (per unit mass of solid), C_{sol} is the concentration in solution and a and b are constants, b being a measure of the total number of adsorption sites. The assumptions involved are probably not valid in concentrated solution where, for example, new adsorption sites on clays may become available (Cody, 1971).

Where adsorption sites are not saturated, an empirical equation (the *Freundlich isotherm*) often fits:

$$C_{ad} = K(C_{sol})^{1/n} \qquad (40)$$

where $n > 1$ and K is a constant. At low concentrations, a simplified form is often applicable:

$$C_{ad} = KC_{sol}. \qquad (41)$$

Adsorbed ions are more or less readily exchangeable. The total amount of exchangeable cations (*cation-exchange capacity*, CEC) is measured in terms of milliequivalents (millimoles times charge) of exchangeable cations per 100 g of dry substance, and is highly dependent on its specific surface area. Typical values range from 100 to 300 for organic matter and zeolites, through 70–80 for smectites, 10–40 for illites and 3–15 for kaolinite (Lerman, 1979).

Ion-exchange takes place readily when ΔG_{elec} forms a large proportion of ΔG_{ads}. With Gouy-layer type adsorption all ions of the same charge are competing for the adsorption sites. If ΔG_{elec} were *overridingly* important then the relative concentrations (strictly activities) of the adsorbed ions would be equal to their relative concentrations (or activities) in the bulk solution. In practice this is not so and, using clays as an example, adsorption affinity decreases with decreasing hydrated ion radius, for example from bottom to top of columns I and II of the periodic table. This can be expressed in terms of an exchange reaction (between equally-charged ions, as the simplest example):

$$A^+ + B_{ad} \rightleftharpoons B^+ + A_{ad} \qquad (42)$$

for which the equilibrium constant (K not equal to one) is:

$$K = \frac{(a_B{}^+)(a_{Aad})}{(a_A{}^+)(a_{Bad})}. \quad (43)$$

Since activity coefficients for adsorbed ions are not readily estimated and often appear to be close to unity, mole fractions (x) are frequently used instead of activities of adsorbed species. If the system is noticeably non-ideal (for example when large variations in compositions are being considered) then mole fractions may still be used with a different (empirical) form of equation:

$$\frac{x_A}{x_B} = K' \left(\frac{a_A{}^+}{a_B{}^+}\right)^p \quad (44)$$

where K' and p are constants for the pair of adsorbents considered (Lerman, 1979, p. 346).

Altervalent ion-exchange has been found to be an important mechanism in experiments on freshwater clays exposed to sea water (Sayles & Mangelsdorf, 1977, 1979). Equations (42) and (43) can be modified to model this process (e.g. Berner, 1980, p. 70) and carry the important implication that the less highly-charged ion will be preferentially adsorbed in solutions of progressively higher ionic strength because its activity coefficient in solution will fall more slowly. Experiments on exposure of clays to sea water also show that CEC diminishes with time as K^+ becomes specifically sorbed on to degraded illites (Russell, 1970). Ion-exchange reactions and specific adsorption processes are also very important during weathering and burial diagenesis.

Li (1981) considered adsorption as a mechanism for controlling the ratios of concentrations of elements in oceanic pelagic sediments and in sea water by plotting these ratios against equilibrium constants for reactions involved in adsorption on OH sites on solids. (These equilibrium constants correlate fairly well with ionic potential.) Most elements fall within a band on such plots, consistent with adsorption being a major control on their behaviour.

The complexing of trace metals to organic material can be extremely important as a means of introducing elements to sediments even if later released by organic decay and fixed in other ways. Unfortunately the mechanisms of complexation (e.g. by specific adsorption, or alternatively by *chelation*, i.e. bonded to O, N or S and lying centrally in ring structures) is not often known (Jackson, Jonasson & Skippen, 1978), especially in sea water. Chelation is well-known for example for V and Ni in petroleum and oil shales (e.g. Riley & Saxby, 1982). A number of elements are found to correlate with organic carbon in black shales (e.g. Mo, Ni, Cu, V, Co, U, Zn, Hg and As in the study of Leventhal & Hosterman, 1982) which is empirical evidence for the importance of the association of organic molecules and metals in sea water.

Nyffeler, Li & Santschi (1984) performed experiments on the rates of adsorption processes to test whether particles remained long enough in sea water to reach a state of equilibrium adsorption. Where particles only reside for a few days in the water column it seems that equilibration may not be complete and will certainly not be so for certain elements (Be, Mn, Co and Fe). However, for several of these elements, removal of adsorption on to manganese nodules will be continuously available. In principle, however, sedimentation rate and aspects of sedimentary environment (e.g. water depth), even with constant water chemistry, are liable to affect the minor element (adsorbed species) composition of sediments. Variations in salinity or organic carbon flux to sediments will be *major* controls on trace element content.

When considering use of adsorbed trace elements to characterize the chemistry of depositional environments, specific adsorbents are required that will not readily be diagenetically-altered. Possible problems are inheritance of trace elements in detrital phases, control of uptake by *several* chemical parameters (e.g. pH, temperature and salinity), and variations in mineralogical abundances and surface properties even of individual minerals (Cody, 1971). Success only seems likely when making comparative studies, within a geological unit, using selective analyses, on a specific mineral species (see Section 9.8).

9.3.5 Lattice incorporation of trace elements

The object of studying trace elements in minerals is normally to constrain the chemistry of the precipitating solution, or to assess the degree of chemical exchange during diagenetic alterations (Brand & Veizer, 1980; Veizer, 1983). Usually interpretations are only possible when coprecipitation of the trace element by substitution for a host (carrier) ion of similar radius and electronegativity has occurred (McIntyre, 1963). Ions occupying interstitial lattice positions may also be amenable to analysis and interpretation (Ishikawa & Ichikuni, 1984), although the abundance of these sites is related to the occurrence of lattice defects (Busenberg & Plummer, 1985). In assuming that lattice substitution has occurred, care should be taken that inclusions of other phases are absent (Angus, Raynor & Robson, 1979). Bulk analysis will be useless if a trace element occurs in significant amounts within solid or liquid inclusions in the host mineral. Even if lattice substitution has occurred, sometimes there may be two possible lattice sites which can be occupied: it is often assumed that an element is present in one site only: for example in dolomite, Fe^{2+} and Mn^{2+} are thought to occupy dominantly Mg^{2+} sites rather than Ca^{2+} sites because of their small ionic radius (like Mg^{2+}). Lumsden & Lloyd (1984) showed that variations in Mn site-partitioning do occur in different dolomite samples: in the future such variations could be useful in interpreting mineral genesis.

If a very small volume of solid is considered, in which a trace element (Tr) is substituting for a carrier element (Cr), precipitating from a large reservoir of solution then at equilibrium:

$$\left[\frac{m_{\mathrm{Tr}}}{m_{\mathrm{Cr}}}\right]_S = K\left[\frac{m_{\mathrm{Tr}}}{m_{\mathrm{Cr}}}\right]_L \tag{45}$$

where S refers to the solid phase and L the liquid phase. K is the *partition coefficient* or distribution coefficient which is constant provided that: (1) temperature and pressure and constant, (2) the solutions are dilute and (3) the ratio $m_{\mathrm{Tr}}/m_{\mathrm{Cr}}$ is low in the solution and the solid.

So far we have considered a small volume of solid with high surface area to volume ratio such that all parts of the crystal are effectively in contact with the fluid and the solution reservoir is sufficiently large that $m_{\mathrm{Tr}}/m_{\mathrm{Cr}}$ in the solution has remained constant. In the more general case, crystals grow from solutions whose compositions change either because the system is open, or because the crystallizing solids in a closed system are removing chemicals in a different ratio from their initial ratio in the solution. In either case, two end-member situations are possible.

First, the crystals could become zoned during growth so that equation (45) only applies to the outermost layer of the crystals. Theoretically this '*Doerner-Hoskins*' behaviour applies when diffusion in the solid is slow and precipitation is either slow, with constant degree of supersaturation, or else rapid but with no recrystallization occurring (McIntyre, 1963). The constant K is often symbolized λ in this case and is known as the logarithmic distribution coefficient because it also appears in equation (46) which relates initial (i) and final (f) fluid compositions of closed systems:

$$\ln\left[\frac{(m_{\mathrm{Tr}})^{\mathrm{i}}}{(m_{\mathrm{Tr}})^{\mathrm{f}}}\right] = \lambda \ln\left[\frac{(m_{\mathrm{Cr}})^{\mathrm{i}}}{(m_{\mathrm{Cr}})^{\mathrm{f}}}\right]. \tag{46}$$

The Doerner-Hoskins behaviour can of course occur in open systems too, but equation (46) is not relevant in that case.

Second, the crystals could remain homogeneous in composition as the solution composition changes. This could be due to repeated crystallization of fine-grained crystals, but is also the expected result of slow relief of supersaturation of a solution (McIntyre, 1963). In this '*Berthelot-Nernst*' behaviour, K is referred to as D (McIntyre, 1963).

Experimental results (e.g. Katz, 1973; Kushnir, 1980) indicate that in seidmentary systems one would normally expect Doerner-Hoskins behaviour, certainly whenever euhedral crystals were being formed from the outset. Although this is generally agreed, authors (e.g. Veizer, 1983) commonly use D when referring to partition coefficients. This usage arises because of the convergence of Berthelot-Nernst and Doerner-Hoskins behaviour in the situation initially described (equation 45). It would seem more logical, however, to use neither D nor λ for situations covered by this convergent behaviour, to avoid confusion (Dickson, 1985). Thus K is used here.

The thermodynamic quantities contributing to K can be seen if we consider an exchange reaction of the type:

$$\mathrm{CrM(ss)} + \mathrm{Tr}^{m+} \rightleftharpoons \mathrm{TrM(ss)} + \mathrm{Cr}^{m+}$$

where M refers to the part of the formula unaffected by solid solution, CrM(ss) and TrM(ss) refer to the components of, respectively, the carrier element and the trace element in the solid solution and Tr^{m+} and Cr^{m+} are the ions in the fluid phase. It can be shown (McIntyre, 1963) that:

$$K = \frac{K_{\mathrm{CrM}}\,(\gamma_{\mathrm{Tr}}{}^{m+})}{K_{\mathrm{TrM}}\,(\gamma_{\mathrm{Cr}}{}^{m+})}\exp(-\Delta\mu/RT) \tag{47}$$

where K_{CrM} and K_{TrM} are the solubility products of the end-members of the solid solution, $\gamma_{\mathrm{Tr}}{}^{m+}$ and $\gamma_{\mathrm{Cr}}{}^{m+}$ are the activity coefficients in aqueous solution, R is the gas constant, T the absolute temperature and $\Delta\mu$ is a measure of the departure (in terms of free energy gain) of the *real* solid solution from an *ideal* one.

If one could ignore the last two terms in equation

(47), as has been done for example by Garrels & Christ (1965, pp. 89–91) and Eriksson, McCarthy & Truswell (1975) then K would be readily calculable. Unfortunately, although the second term usually approximates to one, the solid solution will normally be far from ideal: K must be determined experimentally. Sverjensky (1984), however, has shown that it is possible to calculate relative values of K for two trace ions of similar ionic radius, at least in carbonates.

Experimental work on determination of K for evaporite minerals is summarized by Holser (1979) to which should be added new work on gypsum (Kushnir, 1980), anhydrite (Kushnir, 1982) and halite (Herrmann, 1980; McCaffrey, Lazar & Holland, 1987). For carbonates, Veizer (1983) gave an extensive bibliography which can be supplemented by the experiments of Füchtbauer (1980), Füchtbauer & Hardie (1980), Scherer & Seitz (1980), Mucci & Morse (1983), Ishikawa & Ichikuni (1984), Pingitore & Eastman (1984, 1986), Takano (1985) and Okurama & Kitano (1986). In order to obtain some estimates of K for dolomite, given the absence of experimental work except for Sr, Veizer (1983) utilized the suggestion of Kretz (1982), that the partitioning of trace elements between dolomite and calcite could be estimated solely by considering ionic radii. Kretz' model makes no allowance for temperature. Although temperature effects were thought theoretically not important enough to allow for geothermometry of calcitedolomite mineral pairs (Jacobsen & Usdowski, 1976), this is contradicted by the work of, e.g. Bodine, Holland & Borcsik (1965), Katz (1973) and Powell, Condliffe & Condliffe (1984). Therefore the use of Kretz' (1982) model to calculate K, even in the general way suggested by Veizer (1983), should be regarded very cautiously.

The role of kinetic factors in controlling minor element chemistry is a source of disagreement. Lorens (1981) demonstrated a clear kinetic effect for the coprecipitation of Sr, Mn, Co and Cd with calcite as did Busenberg & Plummer (1985) for Na and SO_4 in calcite, but no kinetic effects were found for Sr in magnesian calcite by Mucci & Morse (1983) or Ba in calcite by Pingitore & Eastman (1984). One problem in evaluating experimental work is that authors often do not give potentially important information such as the degree of supersaturation (Ω), precipitation rate in terms of lattice layers per unit time, absolute concentrations rather than just ratios of concentrations of reactants, and growth form of the precipitates. In looking at sedimentary rocks one should not expect any reliable information to come from spherulitic or dendritic crystals since these have grown relatively quickly. Lorens' (1981) data can be recast in a petrographically-meaningful way: his slowest growth rate (actually rate of recrystallization by Ostwald's ripening) would roughly correspond to the growth of a 1 mm rhombic crystal from a substrate in about 3000 years. K_{Mn} halved at growth rates ten times larger than this, whereas K_{Sr} showed no change until still higher growth rates were reached.

From the work of, for example, Herrmann (1980), Kushnir (1980), Lorens (1981) and Busenberg & Plummer (1985), it seems that one can generalize in stating that K is nearer to one at high precipitation rates than at low rates. Under conditions close to equilibrium, the rates of ion adsorption and desorption (which control K) are much quicker than crystal growth rate. At high growth rates, however (Kushnir, 1980), adsorbed ions are accidentally incorporated because they do not desorb quickly enough to avoid being incorporated into the crystal lattice: thus the mineral Tr/Cr ratio more nearly resembles the solution Tr/Cr ratio.

The influence of solution chemistry on K can be significant, particularly where the trace and carrier ions show different tendencies to form complexes (as, e.g. Br^- versus Cl^- in Mg-Cl brines, Herrmann, 1980). This effect can be minimized if K is formulated using activities rather than concentrations. Effects of varying *solid* composition should generally be significant, for example, the increase in K for Sr^{2+} in calcite with increasing content of Mg (Mucci & Morse, 1983) and Mn (Takano, 1985).

When the 'trace' component becomes a major element in solution one would not expect K to remain constant. Studies by Fuchtbauer & Hardie (1980), Fuchtbauer (1980) and Mucci & Morse (1983) for Fe and Mg in calcite, are illustrative. For Mg in calcite, as the Mg/Ca ratio in solution rises, the partition coefficient falls because of an increasing saturation of adsorption sites for Mg, adsorption being a necessary precursor to lattice incorporation. K falls by much less than an order of magnitude, however, because of an opposing tendency to increase Mg adsorption because of the affinity of Mg ions in solution for Mg already adsorbed (Mucci & Morse, 1983).

The incorporation of Na and K into calcite is now known to be an example of *interstitial solid solution*.

The experimental work of Ishikawa & Ichikuni (1984) showed that the levels of these elements was independent of the Ca^{2+} activity of the solution at fixed Na^+ and K^+ activities, but actually follow a Freundlich-type isotherm (Busenberg & Plummer, 1985). Ishikawa & Ichikuni (1984) attempted to calibrate this as a palaeosalinity indicator, but Busenberg & Plummer (1985) clearly showed that Na incorporation was related to the speed of crystal growth, fast crystal growth leading to the formation of more lattice defects which are selected by Na. Absolute levels of Na are not therefore useful as salinity indicators, although Na variation within a sample suite of similar origin may reflect salinity variations (Zhao & Fairchild, 1987). Okurama & Kitano (1986) demonstrated that the presence of Mg in the solution increased the amounts of alkali metals (Li, Na, K, Rb) incorporated in calcite.

Pingitore & Eastman (1986) provided perhaps the best example of the complexity of real chemical systems by considering Sr incorporation in calcite (Fig. 9.5). They showed that Sr displays complex behaviour best reconciled with a model in which some Sr could be incorporated in defect sites as well as substituting for Ca^{2+}. Hence the partition coefficient (K) is higher at very fast growth rates when more lattice defects would be present. At slower growth rates, there would be a fixed number of defect sites which, when filled, would cause a lowering in K since Sr would then only be entering Ca^{2+} sites. Thus K is lowered by increasing Na or Ba concentrations in the fluid (these elements preferentially fill defect sites), or high absolute concentration of Sr^{2+} (equivalent to concentrations of Sr in the solid of more than a few hundred ppm).

Organisms precipitating calcareous shells may or may not adhere to trace element concentrations appropriate for inorganic systems. Great variations, particularly in Sr and Mg, are well known (Milliman, 1974). Sometimes growth rate, often coupled to temperature changes, has an effect (Moberly, 1968; Kolesar, 1978) but in other cases K remains constant and different from the inorganic case regardless of growth rate (Lorens & Bender, 1980). A different composition of body fluids from the ambient fluid explains some, but not all, of these relationships (Lorens & Bender, 1980). Interpretation of ancient water chemistry from shell chemistry obviously requires as a prerequisite a detailed knowledge of the behaviour of the group being considered from studies of modern organisms.

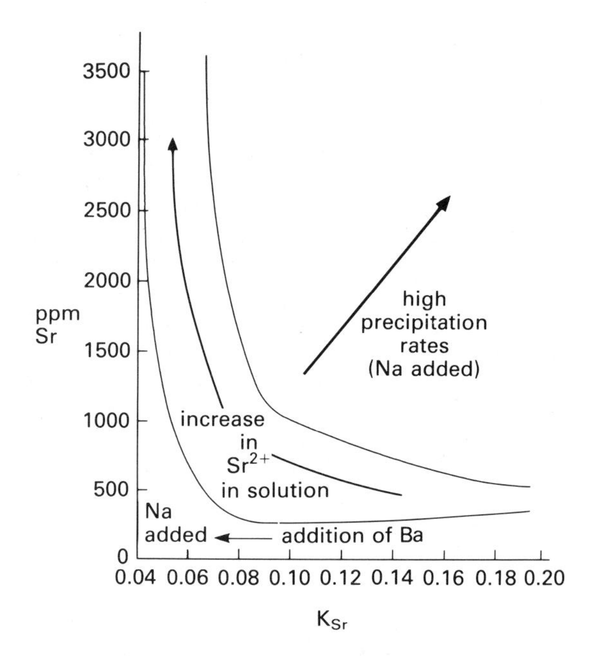

Fig. 9.5. Summary of the results of Pingitore & Eastman (1986) concerning Sr incorporation in calcite. Low values of partition coefficient K result from high Sr or addition of Ba or Na to the solution (see paper for quantitative aspects). Even with Na present in solution, a high value for K can be restored by increasing growth rate.

9.3.6 Stable isotope fractionation

The partitioning of the isotopes of a given element between different phases during chemical or physical processes is known as *fractionation*. Specific environmental or diagenetic processes are characterized by particular fractionations which are often identifiable in the rock record. Hoefs (1980), Anderson & Arthur (1983), Kaplan (1983), Longstaffe (1983) and Veizer (1983) provided readable accounts of the theory and application of isotope analysis.

In principle, both equilibrium and kinetically-controlled isotope fractionations can be recognized which are explicable in terms of quantum theory (in particular the vibrational frequencies of the molecules or lattices concerned). Differences in isotopic composition between two phases at equilibrium is inevitable whenever the element bonds differ in the two phases: the lighter element, which tends to form weaker bonds, concentrates in the phase where bonding is weaker (Coleman, 1977). Kinetic fractionations occur additionally when there is insufficient time for equilibrium to be obtained and in general it is the lightest isotope that participates in a chemical reaction the most readily.

The isotopic fractionation between two phases A and B is expressed as a *fractionation factor* (α)

$$\alpha_{A-B} = \frac{R_A}{R_B}, \tag{4a}$$

where R_A and R_B represent the ratio of abundance of a heavier, rarer isotope to the most abundant isotope (for our purposes D/H, $^{13}C/^{12}C$, $^{18}O/^{16}O$ and $^{34}S/^{32}S$). For equilibrium processes, α corresponds to the equilibrium constant for an isotope exchange reaction between the two substances, written so that only one atom is exchanged (Javoy, 1977, p. 612).

In practice, analysis of absolute abundances is imprecise so that results are normally expressed as delta (δ) values in ‰ where

$$\delta(x) = 1000\frac{(R_x - R_{std})}{R_{std}} \tag{50}$$

where R_x is the isotopic ratio of the sample and R_{std} is the same ratio in a standard for which R is known. Compared to the standard, samples with positive δ values are enriched in the heavier isotope (hence, 'heavy') whereas samples with negative δ values are described as isotopically 'light'.

Since values of α are very close to unity it is often convenient to express equilibrium fractionation in a different way, as

$$1000(\alpha_{A-B} - 1) \text{ or} \tag{51}$$

$$1000 \ln\alpha_{A-B} \text{ or} \tag{52}$$

$$\delta_A - \delta_B \tag{53}$$

These expressions yield very similar numbers (essentially identical when less than 10). The enrichment factor (E_{A-B}) is defined as the quantity in equation (51) whilst each of the these three expressions has been used by different authors as a definition of Δ_{A-B} (Hoefs, 1980; Schwarcz, 1981; Anderson & Arthur, 1983). The argument for considering 1000 $\ln\alpha_{A-B}$ (Schwarcz, 1981) is that this quantity has a special significance in that it expresses the temperature-dependence of the reaction (at geologically-relevant temperatures smooth curves are obtained when 1000 $\ln\alpha_{A-B}$ is plotted against $1/T^2$ where T is the absolute temperature). Friedman & O'Neil (1977) presented a compilation of fractionation factors of geochemical interest and their variation with temperature.

A special type of equilibrium fractionation process involves the repeated removal of one phase and was originally treated mathematically by Rayleigh (see Hoefs, 1980, pp. 10–12). Partial distillation and condensation of water vapour is the most familiar of these 'Rayleigh processes'.

The following discussion outlines the specific fractionation mechanisms of importance in sedimentary geology. Table 9.1 summarizes the isotopes, the reference standards used and the ranges of isotopic variation. Nitrogen isotopes are not discussed here but see Kaplan (1983).

CARBON

Bicarbonate dissolved in sea water (with $\delta^{13}C$ around O) provides a convenient baseline, especially since there is only a very small fractionation on $CaCO_3$ precipitation at earth surface temperatures, although calcified organisms show varying taxon-related departures from inorganic equilibrium (Dodd & Stanton, 1981; Veizer, 1983) and there are inorganic kinetic effects too (Turner, 1982). Equilibrium is approximately maintained between sea water bicarbonate and atmospheric CO_2 with $\delta^{13}C$ around −7‰.

The most pronounced fractionations are related to the fixation of carbon in organic matter by photosynthesis (summarized by Deines, 1980 and Anderson & Arthur, 1983). The organic carbon of plants has $\delta^{13}C$ in the range −10 to −30‰ with variations according to taxa, and, in the case of marine plankton, temperature. Kerogen derived by maturation of marine organic matter may show slight enrichment or depletion of ^{13}C compared to the original organic matter. Release of organic decay products into pore waters during early diagenesis causes pronounced shifts in $\delta^{13}C$ and increases in carbonate alkalinity leading to the formation of authigenic carbonates and phosphates with distinctive isotope signatures (e.g. Irwin, Curtis & Coleman, 1977; Pisciotto & Mahoney, 1981; Benmore, Coleman & McArthur, 1983; Nelson & Lawrence, 1984). Important in this respect are bacterial sulphate-, iron- and manganese-reduction which are thought to produce bicarbonate with little carbon isotope fractionation. At slightly greater depths occurs bacterial fermentation which leads to fractionation between simultaneously-generated CO_2 ($\delta^{13}C$ say +10 to +15‰) and CH_4 ($\delta^{13}C$ say −55 to −70‰). Sometimes carbon from such sources mixes with heavier bicarbonate derived from dissolution of unstable carbonates (Coleman & Raiswell, 1981).

In freshwater conditions, $\delta^{13}C$ of bicarbonate (and hence that of carbonate precipitates) typically is negative, but shows great variation depending on the size of input from organic carbon, atmospheric CO_2 and rock carbonate. Heavier values accompany equilibration with atmospheric CO_2 in shallow, agitated waters. An additional possible mechanism of fractionation arises when carbonate precipitation in an enclosed marine or continental water body is induced by CO_2 loss. The residual bicarbonate is heavy (Katz, Kolodny & Nissenbaum, 1977). More extreme enrichment is possible in cave carbonates, and in brines, where rapid CO_2 loss can lead to an additional, kinetic fractionation (Hendy, 1971; Dreybrodt, 1982; Stiller, Rounick & Shasha, 1985).

OXYGEN

Exchange of oxygen between water and dissolved inorganic species is rapid, apart from oxygen in the sulphate radical (Chiba & Sakai, 1985). Hence, since water molecules are by far the most abundant species in aqueous solutions, the $\delta^{18}O$ composition of the water will determine the oxygen isotope composition of the precipitated minerals. An exception may be some examples of relatively rapid dissolution-precipitation reactions in thin film micro-environments (Veizer, 1983). Mineral-water fractionation factors are strongly temperature-dependent (Friedman & O'Neil, 1977) which, although holding promise of palaeotemperature determinations, in practice leads to ambiguity of interpretation of isotope values since $\delta^{18}O$ values of natural waters show great variation. Low temperature minerals are enriched in ^{18}O compared to water by considerable amounts; about 20 ‰ for phosphate, low 20s ‰ for clays, high 20s ‰ for carbonates and 35 ‰ for silica.

At 25°C water is enriched in ^{18}O by about 0.8 ‰ compared to co-existing vapour because the greater velocity of the $H_2{}^{16}O$ molecule allows it to evaporate more easily. Application of a Rayleigh distillation model predicts the observed progressive lightening in isotopic composition of precipitation at higher latitudes (Fig. 9.6) because of the selective removal of ^{18}O in initial condensates and progressive lowering of temperature, although the real system is actually much more complex than this model admits (Anderson & Arthur, 1983). Evaporation into unsaturated air leads to an additional kinetic fractionation of oxygen isotopes. However, the positive excursions of $\delta^{18}O$ of evaporated water bodies are limited: first, by an oppositely-directed isotope exchange effect which causes $\delta^{18}O$ values to peak (at around 6‰ SMOW in humid coastal areas) when the liquid is 15−25% of its original mass and, second, by the possibility of addition of freshwater during evaporation (Lloyd, 1966).

Pore waters of buried carbonate sequences may undergo $\delta^{18}O$ enrichment in response to recrystallization of sedimentary carbonates at significantly greater temperatures than those of sedimentation. This process awaits detailed modelling. In terrigenous sediments enrichment of ^{18}O can occur due to clay membrane filtration (see next section), although this will be opposed by any dissolution of high-temperature silicate minerals.

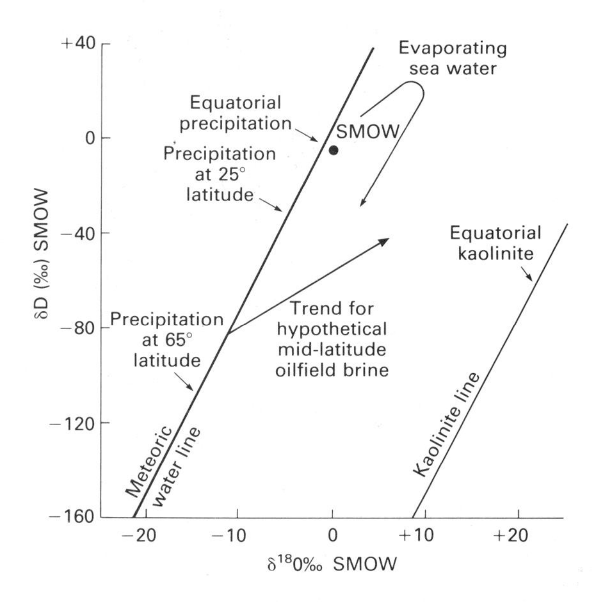

Fig. 9.6. Variations in δD and $\delta^{18}O$. Compiled from various sources. The kaolinite line indicates the potential range of equilibrium compositions of kaolinite forming from meteoric waters.

HYDROGEN

As for oxygen, the water molecule is the dominant reservoir of hydrogen. The large relative difference in mass between H and deuterium is the cause of the large isotopic variations in nature. The main fractionation mechanism is that of distillation-condensation which, as for oxygen isotopes, causes considerable isotopic lightening in the more extreme fractionation products. Meteoric waters show linear covariation on a $\delta D-\delta^{18}O$ plot (meteoric water line of Fig. 9.6). Figure 9.6 also shows the departure of waters in oilfields from the meteoric water line perhaps because of the effects of adsorption on to clay membranes (Coplen & Hanshaw, 1973) during pore fluid flow. Inter-layer water in clay minerals shows ready isotope exchange, but good estimates of the isotopic composition of the mineral-forming fluids can be obtained by isotopic analysis of structural hydrogen and oxygen.

Isotopic equilibrium can be demonstrated for the formation of clays during weathering. Figure 9.6 illustrates the fractionation involved in forming kaolinite from meteoric water at equatorial latitudes.

Knauth & Beeunas (1986) emphasized that a combination of decreased $\delta^{18}O$ and δD in highly evaporated sea water (following the hook-like trajectory of Fig. 9.6) and mixing with meteoric water may lead to isotope signatures typical of many formation waters, hitherto interpreted as flushed free of connate brines. Isotope data are thus equivocal here. On the other hand, providing appropriate extraction techniques are employed, isotopic studies on fluid inclusions can provide invaluable information (Buchbinder, Magaritz & Goldberg, 1984; Knauth & Beeunas, 1986).

SULPHUR

Sulphur isotope variation has recently been reviewed by Coleman (1977), Nielsen (1979) and Kaplan (1983). The mechanism responsible for much of the natural variation is biological sulphate reduction which yields light H_2S. This kinetic effect can usually be seen to follow one of two pathways, each producing characteristic fractionations: (a) Dissimilatory sulphate reduction occurs in anaerobic environments where organisms use sulphur in place of oxygen during respiration which results in fractionations of up to 60‰. (b) Assimilatory reduction occurs when cells utilize sulphur in biosynthesis; the effectiveness of this fractionation being governed by the speed of sulphur uptake and complexing. Reduction of this sort, however, usually yields only modest fractionation (i.e. ~5‰).

One of the main effects of bacterial sulphate reduction is that ocean water is enriched in ^{34}S relative to primordial sulphur (e.g. meteorites) since light sulphur is extracted in sediments through the production of iron sulphides. In practice isotope values for iron sulphides in particular cases are variable depending, for example, on the degree of closure of the system for sulphur.

Other sulphur fractionation mechanisms include a fractionation of up to 6‰ during the clay membrane filtration process discussed in the hydrogen section (Nriagu, 1974).

ISOTOPE REFERENCE STANDARDS

Isotope laboratories normally use their own working standard for routine isotope measurements, but quote their results relative to internationally-accepted standards (summarized in Table 9.1), which (until exhausted) are used for inter-laboratory calibrations. Standards should of course be isotopically homogeneous and easy to prepare and handle, but should also lie in the mid-range of natural variations (Hoefs, 1980), unless two standards are used. Susceptibility to isotopic re-equilibration can cause problems as in the case of the international standard Solenhofen Limestone (NBS-20) which, because of its fine crystal size, exchanges oxygen with atmospheric moisture (Friedman & O'Neil, 1977). To eliminate such exchange, samples and standards must be kept dry. If wetted by a chemical treatment they should be washed with acetone rather than being left to dry (Barbera & Savin, 1987).

All carbon isotope data are reported relative to the Chicago CO_2 standard PDB, an Upper Cretaceous belemnite. Since this is exhausted, secondary or tertiary carbonate and graphite standards are distributed by the National Bureau of Standards.

The SMOW standard (Craig, 1961a, b) is used for $\delta^{18}O$ results of waters and silicates. Defined as a hypothetical water with $\delta^{18}O$ close to average ocean water, it has now been supplemented by Vienna SMOW (V-SMOW) which has the same zero point. Hydrogen isotopes are also referable to SMOW, but V-SMOW is about -1‰ δD relative to SMOW (Anderson & Arthur, 1983).

For carbonate oxygen, PDB is more commonly used than SMOW. The conversion equations be-

Table 9.1. Characteristics of stable isotopes important in sedimentary systems (data from Hoefs, 1980)

Element	Isotope abundance	Ratio used	International reference materials ('standards')	Range of variation in sedimentary systems
H	^{1}H 99.9844% $D = {}^{2}H$ 0.0156%	D/H	V-SMOW (Vienna Standard Mean Ocean Water)	$\delta D = -430$ to $+50‰$
C	^{12}C 98.89% ^{13}C 1.11%	$^{13}C/^{12}C$	PDB Cretaceous Belemnite	$\delta^{13}C = -90$ to $+20‰$
O	^{16}O 99.763% ^{17}O 0.0375% ^{18}O 0.1995%	$^{18}O/^{16}O$	SMOW or V-SMOW (Vienna) Standard Mean Ocean Water. PDB often used for carbonates	$\delta^{18}O = -45$ to $+40‰$ SMOW
S	^{32}S 95.02% ^{33}S 0.75% ^{34}S 4.21% ^{36}S 0.02%	$^{34}S/^{32}S$	CD troilite (troilite from the Cañon Diablo meteorite)	$\delta^{34}S = -40$ to $+50‰$

tween the two scales have recently (Coplen, Kendall & Hopple, 1983) been revised as:

$$\delta^{18}O_{V\text{-}SMOW} = 1.03091\ \delta^{18}O_{PDB} + 30.91$$

$$\delta^{18}O_{PDB} = 0.97002\ \delta^{18}O_{V\text{-}SMOW} - 29.98.$$

Hudson (1977b) provided a useful review of the subtleties of the interconversions between PDB and SMOW scales.

Sulphur data are referred to sulphur from troilite of the Canon Diablo meteorite, which is used because of its great isotopic homogeneity compared to terrestrial sulphur. Recently the IAEA has distributed alternative standards of native sulphur, and $BaSO_4$ precipitated from sea water.

9.4 GEOLOGICAL SAMPLE COLLECTION

The starting point for the chemical analysis of any geological material is the collection of the sample. If this collection is not carefully planned and executed any geological conclusions drawn from the analysis are suspect and the errors introduced at the sampling stage cannot be subsequently rectified, no matter how skilled the analyst or sophisicated the equipment. The old computer aphorism of 'garbage in — garbage out' applies equally well to geochemistry.

From the point of view of sampling strategy, there are two types of investigation: large scale and small scale.

(1) Large-scale investigations in which chemical variations are sought over an area or volume that is extremely large in relation to the sample size, e.g. bulk-rock chemical variations in a major depositional unit. Knowledge of sampling statistics is particularly vital in this case.

(2) Small-scale investigations are those where chemical variations are studied on a scale comparable to that of the samples, e.g. studies of carbonate diagenesis where complete fossils and whole regions of cements are analysed.

The overall aim of a geochemical research programme should, as far as possible, be defined before any samples are collected. This allows the investigator to identify the hypotheses to be tested, to choose appropriate statistical tests and to select the sampling schemes most suited to the tests. Regretably, perhaps the majority of geological programmes proceed on an *ad hoc* basis with the investigators becoming aware of predictable problems only at a late stage. The ideal is that the material should be collected bearing in mind all the investigation techniques that **may** subsequently be applied. For example a research programme into lateral variations within one sedimentary unit may initially be thought to require only optical microscopy and bulk major element analyses. These results may show a need for trace element analysis followed by mineral separation and perhaps isotopic analysis. Forethought during collection not only avoids the need for expensive re-collection but also ensures that all techniques of analysis are applied to essentially the same samples.

The sample scheme and sample collection should always be the responsibility of the field geologist who ought to have the greatest knowledge of the material and its variability. In practice it is rarely possible to follow a rigorous scheme, especially as all the questions to be answered cannot be defined at the start of the programme. However a clear plan can help to minimize the more gross errors.

Sampling schemes have been mainly developed by statisticians working in the social sciences or in biometrics. In these subjects the collection of an adequate statistical sample from the target population is relatively simple, unlike the geological sciences where standard schemes can only be applied in the few cases where the objectives are clearly defined, for example in the geochemical exploration for minerals. For the majority of geological applications the problems are too varied for general guidelines. Krumbein & Graybill (1965), Koch & Link (1970) and Garrett (1983) discussed the problems and give a number of specific examples, but it is frequently necessary for a researcher to return to first principles using a text such as Cochran (1977). Because the topic is so important a relatively simple example is given to illustrate the questions that must be answered before geological samples are collected. The problem is presented as a simplified version of that attempted by Hickman & Wright (1983). The authors wished to establish criteria to distinguish quartzite units found in Upper Proterozoic metasediments in the Appin area of the West Highlands of Scotland, and to use the criteria to identify the units in other areas.

Considering only two units, the *TARGET* population is here made up of two populations, the two units, and is the total population. The *SAMPLED* population is that which we succeed in sampling and is made up of a number of (large ~1 kg) hand specimens, of which the *TARGET* population contains an essentially infinite number. The first task is to define the the *TARGET* population from a number of possible alternatives, for example: (1) The whole of the two units as deposited. (2) The whole of the two units remaining after erosion. (3) The units not covered by younger rocks. (4) The units currently exposed.

Clearly (1) is no longer possible, while (4) is the most likely and simplest alternative, but (2) is also a possibility if the cost of drilling is justified.

Having established the *TARGET* population, the *SAMPLED* population must then be defined with considerable thought being given to geological as well as statistical considerations. For example: (1) Are the outcrops (mainly on the tops of hills) present because these parts of the units are more resistant to erosion? (2) Do changes in metamorphic grade affect the chemical composition? (3) Is the geological interest only in the total variance of each unit, or is more detail of vertical and lateral variation required?

Finally, before a sampling design can be produced the scale and shape of expected variations must be defined.

At this stage it may become apparent that answers are not available to some of the questions, and that it is necessary to conduct an orientation survey to establish the final form of the *SAMPLED* population.

9.4.1 Search techniques

Ideally geological samples should be collected according to some pattern, with the spacing of individual points related to the scale and shape of the overall chemical variations. In the example given it might be expected that there would be vertical changes in composition on a scale of a few centimetres while lateral variations would be over many hundreds of metres. Thus the collection programme should contain a relatively widely spaced areal design with closely spaced points along vertical sections through the two units.

The simplest design is a systematic collection at equally spaced points along vertical traverses located at the intersections of square, rectangular, triangular or hexagonal grids (Fig. 9.7a). Even allowing for the fact that outcrops do not generally coincide with a regular grid this is an inefficient design.

A better plan is to collect at points randomly spaced along the traverses which are randomly located areally (Fig. 9.7b). The random numbers may be taken from published tables or more conveniently from the random number generator found in some scientific pocket calculators (Cheeney, 1983, p. 136, described a very simple technique involving poker dice). Even this method can be improved by introducing stratification where the sites for traverses are randomly located within grid areas (cells) related to the scale of variation. Points along the traverses are also stratified (Fig. 9.7c).

Further refinement can be added by using a nested hierarchical system where the main cells are subdivided and both sub-cells and the position of traverse points within sub-cells occupied randomly (Fig. 9.7d). Frequently a balanced sampling scheme is adopted where sites are sampled in duplicate, often with each sample then being analysed in duplicate. This leads to an excessive load both in collecting and in analysis. An unbalanced nested hierarchical design (Fig. 9.7e) is more efficient and can result, in favourable circumstances, in a saving of up to 50% in samples analysed. For a large-scale study this method is only viable when a computer program is available to plan the design (Garrett & Gross, 1980).

Having decided on the collection plan it only remains to identify the size of geological sample required. This must obviously be large enough to provide material for all envisaged procedures, plus a safety margin, **after** contaminated and altered areas have been removed. The appropriate mass which is required to provide a given statistical error in sampling mineral grains or fragments can be calculated by the methods covered in detail in Section 9.5.2. A reasonable approximation is given by Edelman (1962) who showed that for a rock with an even grain size of 1.2 mm, a grab sample of 1 kg is needed to represent adequately the rock for chemical analysis. If the grain size is doubled the required sample weight is increased by eight.

Such large samples may be impossible to collect logistically, and for consolidated sediments, clearly require a sledge hammer to remove from the outcrop. If a decision is taken to reduce the sample mass the level of error introduced should be quantified and allowed for in subsequent data analysis.

9.5 PREPARATION FOR CHEMICAL ANALYSIS

From the time a geological sample is chosen for collection considerable care must be taken to minimize contamination. This can be present locally from industrial activity, agricultural chemicals, as part of natural processes, for example from salt water spray, from the tools used to remove the sample or from the materials used for storage.

The containers used to transport and store the material may add or remove elements, especially when wet unconsolidated sediment is kept for a long period. The cheaper varieties of highly coloured domestic plastic goods are especially prone to add high concentrations of exotic elements. Polythene or polypropylene jars are preferable for field use, and ideally should be soaked overnight in 50% nitric or hydrochloric acid and washed in distilled water to remove contaminants, particularly zinc.

Rock specimens are frequently labelled in the field using paint or marker pens which again may lead to contamination. Where large numbers of samples are collected, according to the type of statistical scheme discussed above, laboratory numbers are often randomly allocated (see Section 9.7.1) and here the containers will have been labelled already with the grid co-ordinates and laboratory number of the sample before going into the field. The policy of using pre-numbered bags or jars can also reduce confusion during less extensive sample collection. Table 9.2 lists the more common contaminants of containers.

Unconsolidated sediments will generally require drying. In geochemical exploration surveys this is often done in the field by sun drying in high wet-strength paper bags. Most samples, however, will first be returned to the laboratory and must be dried at temperatures no higher than 65°C otherwise clays will bake hard and volatile components, for example mercury or carbon compounds, may be lost. Once dry the sample should be disaggregated by gentle pounding in an agate pestle and mortar.

9.5.1 Crushing

In any subsequent processing of sediment or rock samples, contamination is inevitable although it may be possible to reduce the effect to insignificant levels. The method(s) chosen should be dictated by the type of chemical analysis and elements required, and by any further investigations which **may** follow.

ROCK SAMPLES

Different approaches are required depending on analytical strategy. Where one spatial component of the sample is to be analysed, sawing and thin sectioning should precede analysis. Whilst microbeam techniques make use of polished sections, other methods require a powdered sample which can be chipped or drilled from sawn surfaces and then ground in an

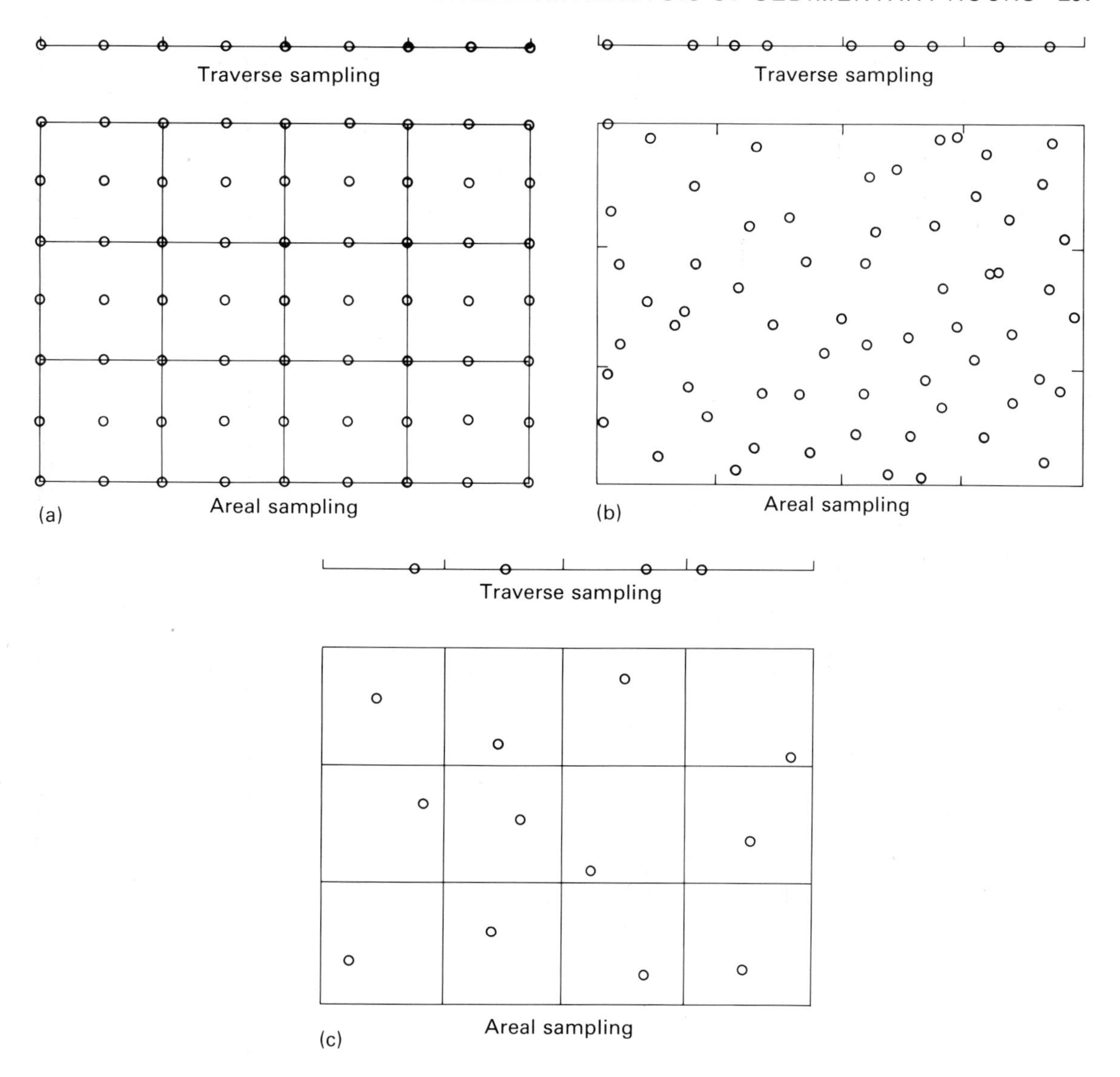

Fig. 9.7.(a) Example of traverse and areal sampling using a regular grid. The spacing is determined by the expected size of the target.
(b) Example of simple random sampling along a traverse and over an area. There may be a clustering of sample sites in parts of the area and gaps left elsewhere. Note that it is possible that one site will be sampled more than once.
(c) Stratified random sampling where a grid of pre-determined size is randomly placed over an area and samples randomly drawn from each grid cell. This reduces the clustering effect of simple random sampling.
(d) Balanced sampling using a nested hierarchical technique with duplicates collected from each site. Sample site and analytical variability can both be studied but it is not necessary for every level to be fully replicated. In this example 72 samples would be analysed in duplicate.
(e) Unbalanced sampling using a nested hierarchical design. This produces a considerable saving in analysis with only 36 samples collected. Only one of the field duplicates need be analysed in duplicate.

(d)

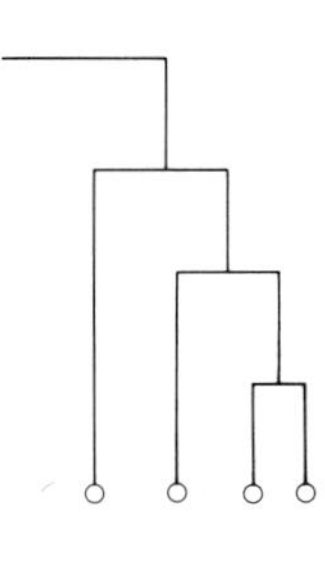

(e)

Table 9.2. Potential contamination introduced during various stages of sample preparation. This list is far from comprehensive and all new materials or new supplies should ideally be checked before use

Containers:		
	Polypropylene	Ti
	Polythene	Ti, Zn, Ba, Cd
	PVC	Ti, Zn, Na, Cd
	Brown paper	Si
	White paper	Ti, Ba
	Rubber	Zn
Crushing:		
	Steel, iron, hard alloys	Fe, Co, Cr, Mn, Ni, V, Cu, Mo, Zr
	Ceramic, alumina	Al, Ga, B, Ba, Co, Cu, Fe, Li, Mn, Zn, Zr
	W carbide	W, Co, Ti, REE, Hf, Ta, Cu
	Agate	Si, Pb
Saws:		
	Diamond	Cu, Fe, Cr, Mn, Ni
	Abrasive	Al, Fe
Sieves:		Cu, Zn, Pb, Sb, Sn
		Fe, Ni, Cr, Pb, Sn, Sb
Coloured plastic:		Various
Oils:		Various, Mo

agate pestle and mortar. When the sampled components have small areas and the analytical techniques require small sample weights (e.g. isotopes and ICP), use of a fine drill creates a powder which requires little or no further preparation. Contamination will be similar to that described below. Thought must of course be given to a sampling scheme to ensure that the samples are representative of all similar areas in the slice.

In the case where whole samples or large parts of samples need to be prepared a more complex process is required.

This will initially involve crushing the sample to an appropriate grain size after sufficient material has been retained for a hand specimen and for thin sectioning. Weathered material not trimmed in the field should first be removed and the sample then reduced to roughly 50 mm cubes using a hydraulic splitter with hardened steel knife edges. Valuable samples may be sliced with a diamond saw and weathering removed by grinding although care should be taken to prevent metal smearing from the blade. The fragments should then be washed in distilled water (unless there is a danger of leaching some components), preferably with ultrasonic cleaning.

To reduce the grain size of the sample further the most popular device is a jaw crusher where fragments are fed between moving steel jaws. Roller crushers employing rotating eccentric ribbed hard steel rollers and cone grinders where the fragments are fed between two ribbed steel cones, one within the other, are also used.

All these are efficient in quickly reducing fragment size, the last two down to approximately 60 BS mesh (250 μm), but in the process slivers of metal are removed from grinding surfaces into the sample powder. In addition they are difficult to clean and it is generally necessary to crush and then discard some sample to prevent cross contamination. Some crushers also have sieve trays to screen the sample, providing a further source of contamination.

A simpler and cheaper method uses an industrial 'fly press' or hydraulic press fitted with hardened steel or tungsten carbide plates. The straight crushing action introduces very little contamination, the sample is visible at all times and the apparatus is simple to clean. It is, however, comparatively time consuming to use.

All methods will produce a sample with a maximum size of 5 mm down to 250 μm depending on the machine settings, but with 'fines' down to a few microns. **Appropriate action is needed to avoid health**

hazards associated with silicate dust and with rock splinters. Good analytical practice also dictates that fine dust should not be lost from the sample as this will introduce bias into the analysis. Contamination of the sample at this stage will be from the elements present in hardened steel, and possibly from lubricating and hydraulic oil.

If the initial sample was relatively coarse grained a mass of 1 kg or more should have been crushed, in which case some form of blending is appropriate. This may involve simple repeated 'cone and quartering' followed by re-combination, or the use of more sophisticated mechanical blenders. The blended 'coarse' powder may be required for a number of purposes, for example mineral separation, radiometric dating or ferrous iron determination, all requiring different sample preparation, hence an appropriate number of 'splits' should be taken, again either by 'cone and quartering' or by using a proprietary sample splitter.

Further reduction in grain size to 'fine powder' is almost universally performed in a laboratory disc mill of the Tema® or Shatterbox® type. Providing the initial powder is not too coarse and that the manufacturer's instructions regarding minimum and maximum volumes are followed (maximum ~60 cm^3 for a 100 cm^3 mill) these mills will rapidly reduce samples to a median grain size of around 40 μm. The major exceptions are platey minerals such as micas which are little affected by crushing. If the mill is overloaded some coarse rock fragments will remain no matter how long the material is crushed.

Agate and tungsten carbide are the most commonly used barrels. Agate is relatively fragile and must be run at slow speeds, extending crushing times to 2–5 min for a 60 cm^3 load, but reducing the welding of softer minerals to the mill. Silica is introduced as a contaminant but equally significant is the introduction of lead from vugs of galena generally contained in the agate. Tungsten carbide usually contains approximately 6% cobalt as a binder and both elements will contaminate the sample. In addition rare earth elements may be used in the binder and the tungsten may contain Hf and Ta impurities. The heat generated in a tungsten carbide mill often causes aggregation of the sample grains and may also encourage oxidation of ferrous iron to ferric iron and possibly lead to the loss of volatile organic components and elements such as mercury and cadmium. There is also a slight possibility of altering the mineralogical composition; Burns & Breding (1956) reported the change of calcite to aragonite with prolonged grinding. In general the agate mill is to be preferred for general crushing, but it may be necessary to crush separate splits of coarse powder in each mill if all elements are to be analysed. Loss of very fine dust must obviously be avoided by maintaining the barrel seals in good condition.

Maximum contamination levels should be checked by grinding a hard, pure material such as clear quartz crystal or pure silica sand, remembering that levels may change with time as the barrels are eroded thus exposing pockets of ore in the agate or, for example, the brazing at the edge of the carbide inserts. Table 9.2 shows common contaminants and Table 9.3 the results of a routine contamination tests.

Cross contamination of samples is always possible and careful cleaning, and possibly crushing and discarding some material, is essential. The level of such cross contamination is generally low, for example if 0.1 g of sample containing 1000 ppm of an element is carried over to the next 60 g load in a swing mill only 2 ppm of that element will be added to the next sample. It is, however, advisable to record the sequence of sample crushing to isolate sources of error. The crushing history of both 'coarse' and 'fine' powders should also be recorded both in laboratory notes and on the sample containers. The international reference sample T-1 which was contaminated with cassiterite during crushing should serve as a salutary reminder.

9.5.2 The statistics of sampling

Under ideal circumstances the sample presented for chemical analysis would be totally representative of the target but in the majority of cases this is an unattainable ideal because of the essential inhomogeneity of geological materials. The magnitude of the error can, however, be quantified using very simple statistical methods and either reduced to an acceptable level or allowed for when the chemical data are interpreted.

When a bulk sample is to be analysed a sufficiently large amount should be collected in the field to allow for variations in mineral or clast composition. The 'rule of thumb' (due to Edelman, 1962, and referred to previously) shows that the majority of samples collected by geologist are too small to provide a 1% sampling error as many rocks have minerals or clasts exceeding 2.4 mm, and 8 kg samples are the

Table 9.3. Test of the contamination introduced during the crushing of an industrial silica sand in three separate Tema® swing mill barrels (ten 50 g samples in each). Milling time 1 min at high speed for W carbide and steel, 3 min for agate. Analysis by XRF of duplicate 15 g powder pellets. Oxides as percentages, elements as ppm. The low value for Ni crushed in carbide is caused by a W background interference. The difference in mean Zr values for carbide and agate can be explained by sampling errors

	W carbide			Agate			Steel		
	Range	$\bar{x}$	s	Range	$\bar{x}$	s	Range	$\bar{x}$	s
Fe_2O_3*	0.259–0.284	0.267	0.01	0.233–0.994	0.268	0.02	0.290–0.344	0.311	0.02
MnO	0.002–0.004	0.003	0.001	0.002–0.003	0.003	0.001	0.003–0.006	0.005	0.001
CaO	0.156–0.209	0.193	0.03	0.161–0.206	0.195	0.02	0.176–0.202	0.189	0.01
K_2O	0.107–0.137	0.125	0.01	0.107–0.159	0.133	0.02	0.119–0.146	0.133	0.02
Ni	1–2	1.5	0.6	4–7	4.8	1.3	4375–7163	5588	1194
Cr	14–23	18	3.8	8–24	15	6.3	629–962	775	143
Zr	48–107	70	25.6	50–137	88	33.6	84–126	102	18
Sr	12–13	12.6	0.6	11–13	12.0	0.7	11–14	12	1.3
Pb	2–4	3.0	0.7	48–134	90	37	2–4	2.8	0.8
W	751–1855	1430	457	<2	—	—	<2	—	—

exception rather than the rule! Even when an adequate amount of sample has been collected, crushed and blended for analysis it is necessary to ensure that the small sub-samples taken for analysis are representative of the whole, particularly when accessory mineral phases containing the majority of one element (e.g. zircon) are present.

A moderately extensive literature is available which discusses the whole range of sample size from mountain to milligram, but it is largely ignored except in commercial areas such as the assaying of bulk ores (where payment depends on an accurate knowledge of total composition), in the assay of very low concentrations of precious metals present in large quantities of 'gangue', and in the supply and evaluation of rock and mineral reference materials.

A general treatment is given by Wilson (1964) who considered the real cases where elements are distributed in major and minor proportions between several minerals of differing density and size range. All these approaches work on the principle that the standard deviation of the amount present can be calculated using a multi-nomial distribution, and the confidence limits assigned. Where there are a large number of particles of each mineral or clast species, a Gaussian distribution is appropriate but when minor or accessory minerals are considered the sample is most likely to be represented by a skewed Poisson distribution. An extreme example taken from Thompson (1983) is illustrated in Fig. 9.8.

The method below is taken from Moore (1979) who used the Poisson distribution to develop equations to calculate the standard deviation for minor mineral species. The simplified approach considers a particulate material containing p ppm of uniformly sized grains of an accessory mineral of weight x g. If a random sample of weight W g is taken it will contain an average number of accessory mineral grains Wp/x. As the particles are discrete they may be expected to follow a Poisson distribution with mean Wp/x. The sample may be considered as a large number of sub-samples for which there is a small probability of including an accessory mineral particle. The coefficient of variation (CV) of this mineral content is:

$$\mathrm{CV} = 100\sqrt{x/Wp}\ \%$$

for particles of similar shape $x = A\varsigma D^3$
where D is the particle diameter (μm),
ς is the particle density (g cm^{-3}),

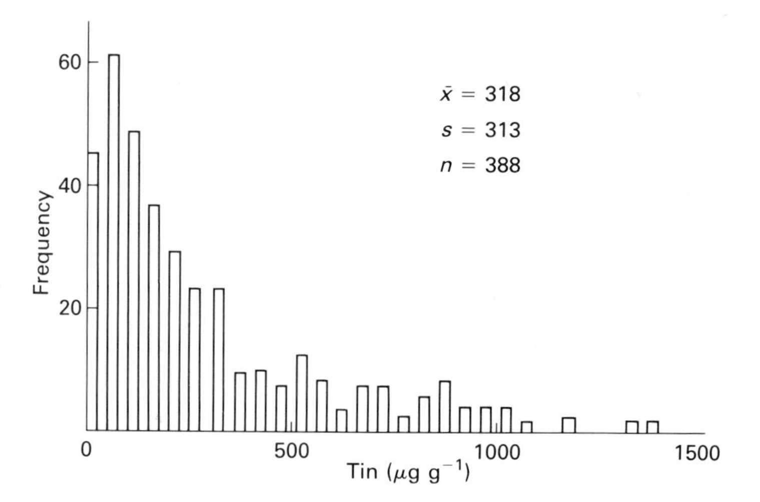

Fig. 9.8. Tin results obtained by optical emission spectroscopy of a reference sample of stream sediment in 388 successive batches. There is extreme deviation from the Normal curve because of the small sample size (15 mg), the particle size (100 μm) and segregation of tin in cassiterite (from Thompson, 1983).

A is a constant = $\pi/(6 \times 10^6)$ for spherical particles,

i.e. $$CV = \sqrt{\frac{\pi \varsigma D^3}{600Wp}}\%.$$

Hence the variability of sampling the accessory mineral increases with increasing density and diameter and with decreasing sample size and amount of mineral present. Table 9.4 illustrates this for a range of mineral species. Gold is chosen because of its extreme density but also because it has been the subject of many papers relating to its obvious commercial importance. Nichol (1986) gave an extended account of the problems and showed that for a content of Au at 64 ppb with particles at 62 μm diameter an 800 g sample must be analysed to give a precision of ±50% at a 95% confidence level.

Generally an improvement in sampling precision will be made either by decreasing the grain size, or by increasing the analysed sample weight. In the first case this may not always be practicable, for example where minerals resistant to crushing such as micas, or malleable materials like gold are present.

When the analytical method involves dissolving the sample it is usually possible to increase the mass taken but with some solid sampling techniques, such as X-ray fluorescence analysis of light elements, the **effective** mass analysed may be extremely small.

The above treatment can be extended to cover ranges of particle sizes, or the situation where the accessory mineral is present with two size distributions, for example as discrete large grains, and as small grains within another mineral. In the latter case if replicate analyses were made with small sample weights one would obtain low but relatively consistent results but with occasional high values. If the sample weight is increased the replicates may be less consistent but will have a mean close to the 'true' value. Thus with a knowledge of the sample it is possible to predict errors and to avoid mistakes in the interpretation of chemical results.

It is surprising that in spite of the available information sampling errors are often not considered even where highly accurate work is involved. Benedetti *et al.* (1987), for example, found discrepancies between published values obtained by different laboratories when analysing international rock and mineral reference samples for gold at the ppb level (e.g. 1.6–6.7 ppb for the sample Mica-Fe). For some results under 1 g of material had been used!

The principle of applying confidence limits is extremely valuable and should be attempted wherever possible if only to avoid false conclusions. An interesting example is provided by the correspondence concerning the occurrence of rare detrital zircons from Mt Narryer, Australia, which was claimed, on the basis of ion-probe analysis by Froude *et al.* (1983), to be older than 4000 Myr. The claim is questioned by Scharer & Allegre (1985) who found, after analysis of 32 grains by mass spectrometry, a younger age. Compston *et al.* (1985), however, showed that as the original analysis found only 5 'old' grains in 260 there is a strong statistical probability that Scharer & Allegre would not have found 'old' zircons in their small sample.

Simple statistical methods can also be used to

Table 9.4. Errors introduced by sampling different masses and grain sizes of minerals of differing densities (ς) using the method of Moore (1979) (see Section 9.5.2). The 95% confidence limit is approximated to 2s. Note that the ranges are as ppm of the *mineral*

Samples of 40 μm diameter containing 20 ppm of mineral

	Gold Au ς19.3			Cassiterite SnO_2 ς7			Zircon $ZrSiO_4$ ς4.7			Apatite $Ca_5F(PO_4)_5$ ς3.2		
Mass (g)	C of V %	Range at 95% confidence ppm		C of V %	Range at 95% confidence ppm		C of V %	Range at 95% confidence ppm		C of V %	Range at 95% confidence ppm	
10	5.7	17.7	22.3	3.4	18.6	21.4	2.8	18.9	21.1	2.3	19.1	20.9
5	8.0	16.8	23.2	4.8	18.1	21.9	4.0	18.4	21.6	3.3	18.7	21.3
1	18.0	12.8	27.2	10.8	15.7	24.3	8.9	16.4	23.6	7.3	17.1	22.9
0.5	25.4	9.8	30.2	15.3	13.9	26.1	12.6	15.0	25.0	10.4	15.9	24.1
0.1	56.8	—	42.8	34.3	6.3	33.7	28.1	8.8	31.2	23.2	10.7	29.3

Samples of 1 g containing 20 ppm of mineral

Grain size BS mesh	μm	Gold C of V %	Gold range		Cassiterite C of V %	Cassiterite range		Zircon C of V %	Zircon range		Apatite C of V %	Apatite range	
16	1000	2248	—	919	1354	—	561	1109	—	464	915	—	203
60	250	280	—	132	169	—	88	139	—	76	114	—	66
100	150	130	—	72	79	—	52	64	—	46	53	—	41
150	106	78	—	51	47	1.3	48.7	38	4.7	35.3	32	7.4	32.6
200	75	46	1.5	38.5	27.8	8.9	31.1	22.8	10.9	29.1	18.8	12.5	27.5
300	53	27.4	9.0	31.0	16.5	13.4	26.6	13.5	17.3	22.7	11.2	15.5	24.5
—	20	6.4	17.5	22.5	3.8	18.5	21.5	3.1	18.7	21.3	2.6	19.0	21.0

consider the errors in sampling inhomogeneous or zoned minerals.

9.5.3 Sample decomposition

Almost all methods of chemical and isotopic analysis require the initial decomposition of the sample, either directly, for example by an electrical arc or spark as in optical emission spectroscopy, or chemically by acid attack or fusion. Only the methods using the excitation of characteristic X-rays (XRF, electron probe) and radiation techniques (such as instrumental neutron activation analysis, INAA) can on a routine basis employ the finely ground sample powder alone.

In general the samples are presented for analysis as solutions (in the special case of major element determination by XRF as a solid solution in the form of a glass disc), either after 'total' decomposition by acid attack and/or fusion or after partial and selective decomposition or extraction potentially involving a wide range of reagents. Total decomposition is extremely difficult to achieve, particularly where minor elements are present in resistant accessary minerals. Geologists are frequently guilty of following general recipes without thinking of the chemical characteristics of the minerals in the sample. Discrepancies between values produced by different methods may result from incomplete decomposition, and the mineralogy of any residue should be checked microscopically or by X-ray diffraction.

Decomposition methods will also be determined by the final method of analysis and such factors as matrix effects, the solid content of flames and spectral interferences — the latter being especially important in multi-element analysis and where the same solution is used both for chemical and isotopic analysis.

Selective extraction and decomposition are important in the determination of the organic components of a sediment, in the analysis of the readily soluble carbonate fraction of samples and in analysing the geochemically 'mobile' components of unconsolidated sediments and soils.

There is an extensively literature on available methods for the dissolution of silicates and related materials, hence the methods will only be summarized. For up-to-date reviews see Jeffery & Hutchison (1981) and Potts (1987) and for exploration-related samples see Fletcher (1981).

Two types of decomposition are generally used: acid decomposition and fusion; both methods are greatly assisted by a finely ground sample.

DECOMPOSITION BY MINERAL ACIDS

Acid decomposition has been widely used for more than 30 years, both for major and minor element analysis. It is now especially important for trace analysis by atomic absorption and inductively coupled plasma methods where the relatively low salt contents of flame and plasma is a considerable advantage. A further major advantage is the purity of mineral acids. The maximum levels of each element of interest in all analytical reagents should of course be carefully checked. For trace analysis Analar® acids are the minimum purity, and it may be necessary to use the Aristar® grade or in extreme cases to re-distill the most pure commercially available material. Potts (1987) described the distillation method and lists impurities in nitric, hydrochloric and hydrofluoric acids (Potts, 1987, tables 1.21 and 1.22).

Decomposition vessels are possible sources of contamination in trace level analysis and these should be monitored. Some workers for example have been tempted to use soda glass when digesting carbonates in cold dilute HCl only to find that alkali elements are leached from the glass! Even borosilicate glass may be unsatisfactory and ideally PTFE or, more cheaply, polyethylene or polycarbonate should be used (after rigorous cleaning). Adsorption of some elements on the container walls may be a problem especially in the more extreme decomposition procedures discussed below.

Hydrochloric acid Dilute hydrochloric acid, either cold or heated, will dissolve all carbonate minerals (other than the scapolites) and is generally used for the decomposition of carbonate rocks and the separation of any silicate or oxide minerals (see section on selective dissolution). Some sulphides and calc-silicate minerals are also attacked at elevated temperatures.

The chlorides of As, B, Ge, Hg, Sb and Sn are volatile and may be lost from the solution.

Nitric acid Concentrated nitric acid will decompose carbonates and most sulphide minerals which are oxidized to sulphates and is widely used in mineral exploration geochemistry. A mixture of HNO_3 and HCl (3:1, aqua regia) is also used as a powerful solvent for the noble metals and oxides as well as sulphides. Carbon compounds may not be fully oxidized and may preclude the use of nitric acid dissolution where organic-rich material is to be analysed by ICP.

Hydrofluoric acid This acid, combined with nitric or perchloric acids, is very widely used and is able to decompose the majority of silicates as well as carbonates but, where this is done in open vessels, acid-resistant minerals such as zircon, rutile and tourmaline may not be attacked. Disadvantages are that decomposition must take place in PTFE (below 260°C) or platinum crucibles and the highly hazardous nature of both hydrofluoric and perchloric acids.

A typical procedure is to warm a 0.5 g sample for several hours in a covered vessel with 15 ml HF and 4 ml $HClO_4$, the latter to ensure oxidizing conditions. The solution is evaporated to incipient dryness, removing the Si as the volatile silicon tetrafluoride together with other fluorides which may interfere with some determinations. It may be necessary to repeat this evaporation a number of times with additions of perchloric acid. This acid is a very strong oxidizing agent which, with organic carbon and in particular oils in the sample, may explode violently at the evaporation stage. It is better with such samples to add HNO_3 to the $HClO_4$ (at least 4:1) allowing a slow oxidation of the organics. The cooled residue is then taken up in HCl (4 ml) and diluted to an appropriate volume with distilled water.

An alternative procedure, which has the advantage of retaining silicon, is to digest the sample with hydrofluoric acid plus nitric acid or aqua regia in a sealed PTFE lined 'bomb' as originally described by

Bernas (1968). The attack at temperatures of up to 180°C for 1 hour will decompose most resistant minerals but there is the disadvantage that elements may be absorbed at pressure by the PTFE. Care should be taken to monitor the possible release of such elements in subsequent samples. Excess boric acid is then added to complex the fluoride and it is preferable to re-heat in the 'bomb' for ten minutes at pressure to prevent the formation of insoluble fluorides. Even here some resistant minerals may remain and the solution should be checked and if necessary the residue fused. This method has the disadvantage of producing a solution high in solids which makes analysis by ICP spectrometry difficult.

It should be noted that the borate complexed solution will still attack glassware!

SAFETY NOTE

Hydrofluoric acid produces serious burns requiring hospital treatment with even minor contact. Users must be familiar with all necessary precautions and should have their method of work approved by a Safety Officer.

Perchloric acid is an extremely powerful oxidizing agent and should not be allowed to come into contact with organic material. In addition, it forms explosive metal perchlorates. A scheme of work must again be followed and the acid only used in an approved fume hood with wash-down facilities.

DECOMPOSITION BY FUSION

Classical gravimetric analysis initiated in the nineteenth century and early colorimetric schemes of major element analysis decompose the sample by fusing with sodium carbonate. More recently lithium borates have been widely employed both for solution-based analysis and especially for X-ray fluorescence using fused cast beads. Some specific examples of methods are discussed below.

A major problem of fusion is the relatively high ratio of sample to flux (generally more than 1:3) which produces a high solid content and also potentially introduces contaminants (see section on preparation for XRF). For these reasons decomposition by fusion is more widely employed in major element analysis or to attack any residue after acid decomposition.

Fluxes are generally extremely reactive and fusion must take place in appropriate resistant vessels, normally platinum, silver or nickel. These are themselves attacked and have a limited life, the life being further reduced by alloys forming between the crucible and metals within the sample. These elements may be taken up into subsequent fusions, hence the use of crucibles should be carefully logged.

Sodium carbonate This material in the anhydrous form will decompose all silicate rocks with fusion times of about one hour at 1000°C, although some accessory minerals may require an additional ten minutes at 1200°C. The ratio of sample to flux now recommended is 1:5 in contrast to earlier reliance on larger amounts of flux. Some analysts have further reduced the ratio to 1:1 by sintering at 1200°C. The fusion is normally performed in platinum crucibles, and care must be taken to avoid reducing conditions, which could cause alloying and so destroy the crucible! A small amount of nitrate or chlorate can be added to the flux, or where sulphide or carbon is present the sample should be pre-roasted in air. A small amount of iron from the sample normally alloys with the crucible and this may be released if a low iron sample is subsequently fused.

Alkali hydroxide Sodium and potassium hydroxide are widely used for decomposition, often with the addition of peroxide, and appear in a number of colorimetric analysis schemes. The fusion temperatures are less than for sodium carbonate but again an hour is needed to ensure total solution. Fusion is performed in either silver or nickel and care must be taken to ensure that the melt does not 'creep' out of the crucible.

Prolonged fusion of samples containing fluoride may result in loss of silica.

Alkali borates Sodium borate has been used as an efficient flux for aluminium-rich materials and also for slags from the ferrous and non-ferrous industries (West, Hendry & Bailey 1974). However, much more widely used are lithium metaborate ($LiBO_2$) and lithium metaborate-tetraborate mixtures as they allow the determination of sodium. Thompson & Walsh (1983) recommended the metaborate as a flux for major element determination by ICP at a sample to flux ratio of 1:4 which ensures that the final solid content does not exceed 2%. The sample and flux are heated, with swirling at 1000°C for 30–45 min in a Pt or Pt/Au crucible and, when cool, the melt taken up in nitric acid. Most silicates are readily dissolved if they are finely ground (<60 μm) but high concentrations of a number of accessory minerals may leave a residue. Potts (1987, table 2.4) listed the minerals and the proportion remaining after fusion.

The greatest use of borate fluxes is in X-ray fluorescence, details of which are given in the next section. Many of the problems associated with the use of such fluxes are also common to atomic absorption and ICP techniques and it is also possible to dissolve XRF beads after analysis for use in solution methods.

Lithium borate fusions for X-ray fluorescence analysis The intensity of characteristic X-ray emission lines for the lighter elements (Na-Fe) is not only dependent on the concentration of the element but also on the grain size and mineralogical composition of the sample. It is therefore essential to remove grain effects and the most efficient method is to present all samples to the spectrometer as a solid solution in the form of a glass disc.

The flux(es) used must satisfy the following criteria: (a) a wide range of sample compositions must ideally be taken into solution; (b) fusion temperatures must be sufficiently low to produce a fluid melt using gas burners or normal laboratory furnaces (1000–1200°C); (c) the cooled melt must form a stable glass disc without devitrification; (d) the glass must be relatively resistant to attack by moisture and atmospheric gases; (e) the flux should ideally not contain elements which are to be analysed in the sample.

(a) Bennett & Oliver (1976) considered the efficiency of Li metaborate and Li tetraborate for the decomposition of a wide range of silicate, carbonate and ceramic materials. They showed that the former flux is most effective for silica-rich samples (100% silica–80% alumina) and the latter for materials rich in alkali and alkaline earth oxides and aluminous samples (100% alumina–95% silica), and proposed a fluxed composed of a 1 part Li tetraborate and 4 parts Li metaborate (Johnson Matthey Spectroflux® 100B) to cover the widest possible range of materials. This flux is extremely effective at a ratio of 1 part sample to 5 parts flux and will decompose most silicates with 20 min fusion at temperatures between 1000 and 1200°C. Bennett & Oliver (1976) also considered the fluxes and sample to flux ratios needed to produce stable beads for a wide range of materials. Of particular importance to the analysis of sedimentary rocks is their recommendation of lithium metaborate (sample to flux 1:5) to decompose limestones and dolomites. Their major conclusion is that there is no universal flux and that the chemistry of the sample to be dissolved must be considered when choosing flux and sample dilution. A shortened version of their findings is given in Table 9.5.

Thomas & Haukka (1978) recommended a 1:2 fusion with lithium metaborate to allow the determination of both major and minor elements in the same bead. Their results are excellent but there must be some doubt about the ability of this low dilution to dissolve all minerals.

One flux which has been widely used is lithium metaborate with the addition of 16% lanthanum oxide. It was made popular by Norrish in a multitude of papers and was designed to reduce X-ray matrix effects by the addition of LaO as a heavy absorber. This gives the added bonus of increasing the ability to dissolve 'basic oxides' and also acting as a glass-former to produce stable beads. Recently there has been a decline in its popularity as the characteristic La lines interfere with some determinations, its purity is comparatively poor and it is more expensive than the normal lithium borate fluxes.

(b) Lithium metaborate melts at 849°C hence a burner or muffle furnace at 1000°C will produce a fluid melt with sufficient thermal agitation to aid

Table 9.5. Recommended sample-to-flux ratios for different sample types and the two common fluxes Li metaborate and Li tetraborate. * indicates minimum amount to give a clear stable bead. From Bennett & Oliver (1976)

	Parts of flux to 1 part ignited sample	
	Li metaborate	Li tetraborate
Silica-alumina range	4	1
Apatite	0	3*
Zircon	4	1
Limestone	0	5*
Dolomite	0	5*
Magnesite	0	10

solution without the need for continuous swirling. Lithium tetraborate melts at 917°C placing fluid melts beyond the range of normal burners, while the 4:1 eutectic mixture of the two fluxes melts at 832°C combining the advantages of a fluid melt and decomposition of a wide range of materials.

For most laboratories fusion is a labour intensive operation using standard burners or muffle furnaces, with limited automation, e.g. a four sample swirling system as described by Bennett & Oliver (1976). Expensive automated fusion devices are available based on either gas burners or on RF furnaces which are mainly used in industrial process control where the sample throughput justifies the cost. These may operate either to produce XRF beads or samples for ICP or atomic absorption.

In XRF, where a constant sample to flux ratio must be maintained, temperature control of the fusion is critical, especially at temperatures above 1000°C where a significant proportion of the flux will be lost (0.005% min^{-1} at 1200°C). The alkali elements are also lost, approximately 0.002% of the K_2O present per minute at 1200°C. A further problem when using gas burners is that at high temperatures platinum is porous to hydrocarbons in the flame which may produce reducing conditions in the melt, which in turn may allow metals to alloy with the crucible.

(c) When fusion is used as a precursor to acid dissolution the only requirement is total dissolution of the sample. In analysis by XRF the bead is presented to the spectrometer with a flat, polished surface. Devitrification and shattering before, and particularly during, analysis is most undesirable. If the bead is well annealed this will generally not occur unless undissolved mineral grains remain in the glass to act as nucleating centres. Some samples may have insufficient glass-forming elements in which case pure silica may be added (and subtracted from the analysis result) or BeO added. This is an especially good method for 'blank' beads but requires considerable care as beryllium is a highly toxic element.

(d) Atmospheric moisture rapidly attacks sodium borate glass, changing the composition of the bead surface and producing large errors in the lighter element analyses. Lithium borate glasses are affected more slowly but the deterioration is detectable after a few hours as the moist surface allows attack by other gases, especially the sulphur oxides. This can be a particular problem in industrial atmospheres where beads may sit in automatic sample loaders for many hours. The increase in sulphur content of the bead surface is detectable by XRF after two hours. It follows that beads should be stored in a desiccator and exposed to the atmosphere for as short a time as possible, in particular for reference samples which may be re-analysed a number of times. Changes in the chemical composition of borate glass under the X-ray beam, in particular of Si and Al, have been noted by Le Maitre & Haukka (1973) and other authors; hence prolonged and repeated exposure is to be avoided.

(e) At ratios of sample to flux of 1:3 and greater a significant blank can be introduced for many elements. Table 9.6 lists typical maximum levels of impurity specified by a number of manufacturers for the X-ray fluorescence grade of both lithium meta- and tetraborate fluxes. Higher purity is available with levels about an order of magnitude less, but these grades are extremely expensive. It should be stressed that impurity levels will changes from batch to batch, and that frequent checking of blanks is essential. The level of Ca can be taken to illustrate potential errors. If it is at the maximum specified (0.01%), the apparent CaO content of a blank sample fused at a 1:5 ratio will be 0.07%.

SELECTIVE DISSOLUTION OF CARBONATES

Analysis of the carbonate fraction of impure carbonates usually involves a compromise between an effective dissolution agent and one that does not simultaneously leach the non-carbonate components. P. Robinson (1980) reviewed previous work and found that for some Palaeozoic carbonates 1M HCl caused no more leaching of Mn, Sr or Na from the insoluble residue than did 0.1M acetic acid, yet was much more effective in attacking dolomite. Iron, however, was more strongly leached. Boyle (1981) found that finely-divided Fe–Mn oxides were leached even by

Table 9.6. Typical maximum levels of impurity (in %) in Li metaborate and Li tetraborate fluxes used in X-ray fluorescence

Si	0.0050	Mg	0.0020	Cl	0.0020
Ti	0.0020	Ca	0.0100	SO_3	0.0050
Al	0.0050	Na	0.0040	F	0.0020
Fe	0.0010	K	0.0020	As	0.0020
Mn	0.0020	P	0.0030	Pb	0.0020

an acetic acid solution buffered at pH 5.5 by excess acetate. For his analyses of very low concentrations of Cd, Zn and Ba in foraminiferal tests, dissolution in distilled water in contact with a CO_2 atmosphere was required to avoid leaching the non-carbonate fraction. Where detection limits allow, electron microbeam techniques are preferable (e.g. for Fe in impure carbonates). Alternatively by plotting chemical data against percentage of insoluble residue, some estimate can be made of the degree of leaching (Veizer *et al.*, 1977, 1978). Veizer *et al.* (1978) found a clear relationship of increasing solute (8% v/v HCl) potassium with insoluble residue content, but for sodium in the same samples Veizer *et al.* (1977) found that that positive correlation disappeared when *a priori* hypersaline and normal marine samples were considered separately.

Where analysis of the insoluble fraction is the aim, acidic ion exchange resins can be used (French, Warne & Sheedy 1984), although smectites show some leaching, and zeolites and simple salts tend to dissolve.

Phosphoric acid residues from isotopic analysis can be used for cation analysis: this method is evaluated in Section 9.6.5.

EXTRACTION OF ORGANIC MATERIAL AND DETERMINATION OF ORGANIC CARBON

The details of the separation and analysis of organic material is beyond the scope of this chapter. Extractants include methanol, acetone, benzene and chloroform, which will remove the 'bituminous fraction' for analysis by gas chromatography or infrared spectrometry. **NB** *most of these solvents present a safety hazard*!

A second 'humic acid' fraction may be recovered using sodium hydroxide solution. The remaining insoluble carbonaceous material is the kerogen fraction.

It is common practice to determine organic carbon in sedimentary rocks, and there are CHN (carbon-hydrogen-nitrogen) analysers available for this. Organic carbon contents can be readily determined by loss on ignition, although the results may not be very accurate. If there is no 'carbonate carbon' present in the rock, then the dried, weighed powdered sample can be heated in an oven at 800–1000°C for 2–3 hours to oxidize the carbonaceous material. The weight loss will give the organic carbon content. If much pyrite is present, the oxidation of this will introduce an error. For sandstones and mudrocks, carbonate carbon can be removed first by acid digestion. Before the oxidation of the organic carbon, samples should be dried at 105–110°C to drive off any water, and then weighed. With limestones and dolomites, the carbonate can be dissolved out, the residue dried (and weighed) and then combusted at 800–1000°C. Frequently, the weight loss is determined on the whole powdered limestone sample (after drying), but then the temperature is critical. If it is too high, the carbonate carbon will decompose. A temperature of 550°C is generally considered sufficient to combust most of the organic carbon, but little of the carbonate carbon. If the limestone sample is then heated again, to 1000°C for a further 2–3 hours, the second weight loss will give the carbonate carbon content.

9.6 ANALYTICAL TECHNIQUES

9.6.1 Introduction

In this section, the popular techniques for analysing sedimentary rocks are discussed, notably those involving instruments rather than the 'classic', mostly 'wet' methods of rock analysis. Analysis by electron microprobe, X-ray fluorescence (XRF), atomic absorption spectrophotometry (AAS), inductively-coupled plasma spectrometry (ICP), instrument neutron activation (INAA) and stable isotope mass spectrometry are described, with the emphasis on how sedimentary rocks are treated, rather than on the details of the technique or instrument itself. There are numerous textbooks giving full explanations of all the various techniques (see Potts, 1987, for one of the most recent), although these are often written from the point of view of the hard-rock geochemist, mostly concerned with silicate rocks. Techniques which are not covered here include flame photometry colorimetric methods, and those of organic geochemistry, notably gas chromatography, mass spectrometry and pyrolysis.

9.6.2 Electron beam microanalysis

Following a period of rapid development in the 1960s and early 1970s, reliable and fairly standardized methods of quantitative analysis on the micron scale have become widely available. The electron microprobe is an instrument specifically designed for such a purpose, but quantitative results are also obtainable from a scanning electron microscope (SEM) or a scanning transmission electron microscope (STEM) fitted with a suitable detection sytem. In all cases, a focused beam of electrons in impinged upon a specimen (Fig. 9.9) causing the generation of an X-ray spectrum containing lines, characteristic of each element, whose intensity is related to the concentration of the element in the specimen.

There are several useful references for microprobe work including the lucid book by Reed (1975) and the chapter by Long (1977). Goldstein *et al.*'s (1981) book covered nearly all aspects of SEM/microprobe work, and Heinrich (1981) provided a thorough, but rather drier survey of the field. The ion microprobe is not covered here, but could have some important applications to carbonates in the future (Mason, 1987; Veizer *et al.*, 1987).

MICROPROBE ANALYSIS: GENERAL ASPECTS

Samples (usually polished thin sections) and standards are inserted into a specimen chamber which is evacuated prior to analysis. Electron microprobes and some of the new generation of SEMs incorporate an optical microscope to allow the ready location of the areas to be analysed.

The electron beam is generated by heating a filament, usually of tungsten, within a triode electron gun. The beam is focused by two sets of magnetic lenses. The first (the condenser) determines the beam diameter whilst the second (the objective) sharply focuses the beam. Accelerating voltages are 10–30 kV corresponding to electron energies within the range 10–30 keV. The beam can be kept stationary on a spot or traverse along a line or scan (raster) an area to derive a point analysis, line scan or areal distribution of an element as required.

A proportion of the electrons do not penetrate the sample, but are backscattered. Since this effect increases with average atomic number of the area, irradiated backscattered electrons can be used to produce 'atomic number contrast' images (Chapter 8). Electrons penetrating the specimen show complex trajectories (Fig. 9.10) due to various forms of interaction. They spread out over a diffuse region typically about 1 μm deep (size depending on accelerating voltage and atomic number). Although the incident electron beam can be made much narrower than 1 μm, this spreading effect determines the spatial limit of resolution in the microprobe. The

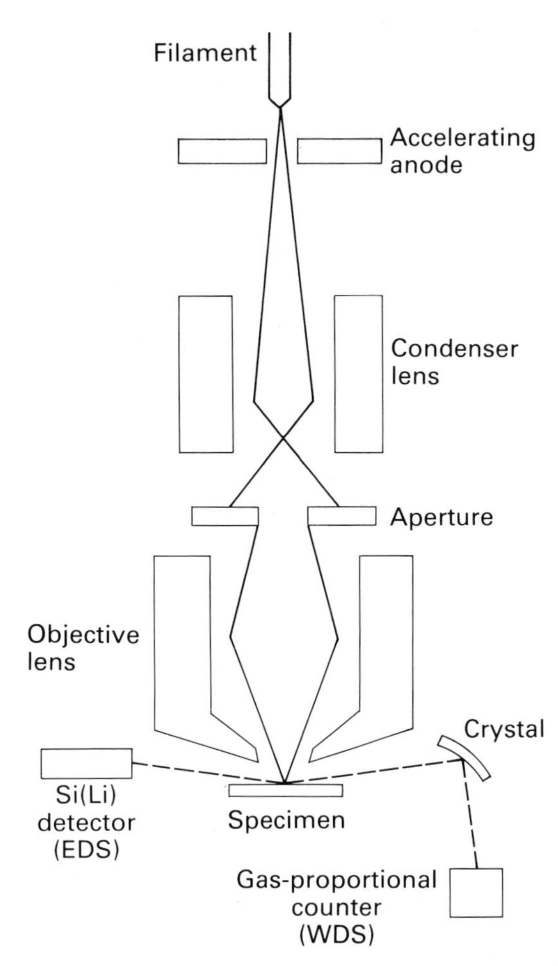

Fig. 9.9. Schematic illustration of the elements of an electron microprobe.

interactions of electrons can be divided into elastic scattering, which involves a change in direction with minimal loss of energy, and inelastic scattering. The latter covers a variety or processes leading to energy loss. For example, excitation of lattice oscillations leads to heat production which is important because of the thermal instability of certain sedimentary minerals. The ejection of low-energy (<50 eV) secondary electrons from the outer parts of electrons is another inelastic scattering process, obviously important in forming specimen images in the SEM. Another process, important in some materials, is the release of long-wavelength photons: cathodoluminescence (Chapter 6). Also there is the deceleration of electrons by charge interaction with atoms, leading to the production of X-ray photons whose

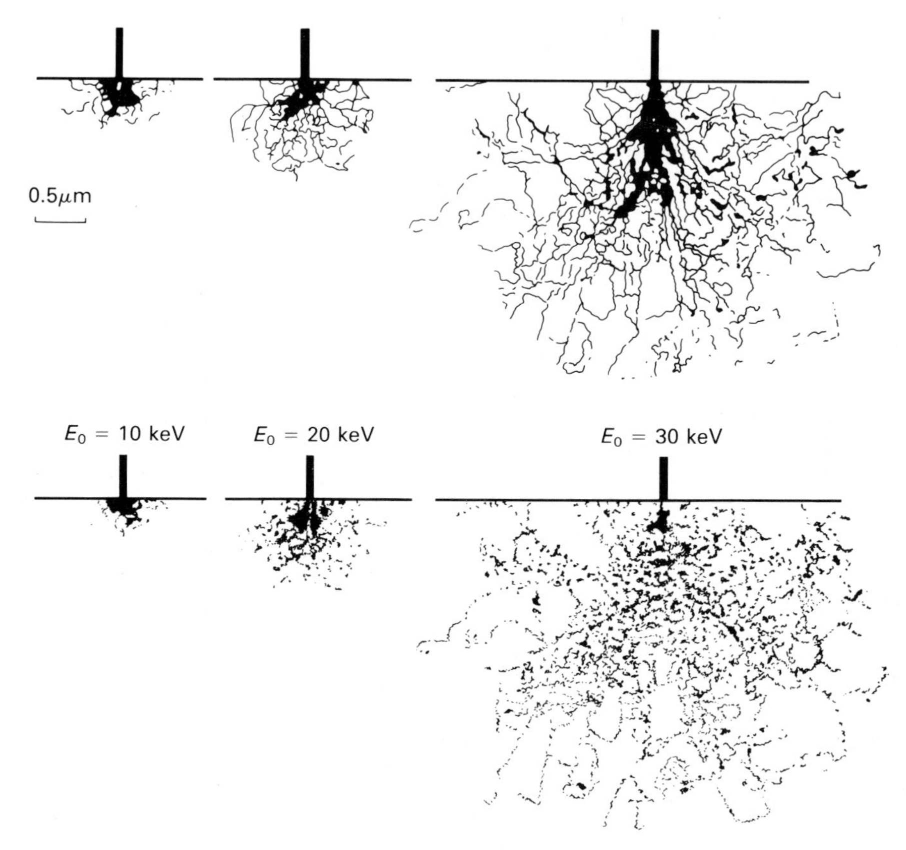

Fig. 9.10. Simulation (by a Monte Carlo procedure) of electron paths (above) and K_α X-ray photons (below) in copper. Note the emergence of backscattered electrons. The effects of atomic number are illustrated by the fact that equivalent diagrams (at 20 keV) for gold and aluminium resemble the left and right diagrams respectively. From Heinrich (1981).

energy varies up to a maximum corresponding to the energy of the incident electrons. A 'background' continuous X-ray spectrum is thus produced in which low energy X-rays (<1 keV) are missing due to absorption by the specimen. On this spectrum are superimposed peaks corresponding to specific electronic transitions in the specimen. These peaks form the characteristic spectrum and arise from decay of electrons from an outer to an inner energy shell following an initial excitation by an incident electron (Fig. 9.11). In the energy range of interest (1–10 keV) elements of higher atomic number are analysed successively by study of K, L and M lines (Fig. 9.12). Quantitative analysis is normally restricted to elements with atomic number >11 unless special techniques are used. Energy-dispersive systems (see below) have been particularly limited in this respect, but analysis down to $Z = 5$ (boron) is now available (Statham, 1982) on commercial systems.

DETECTOR SYSTEMS

Two alternative methods of analysing the X-ray spectrum are available: wavelength-dispersive systems (WDS) and energy-dispersive systems (EDS).

The original, and most analytically-sensitive arrangement (WDS) is to make use of a Bragg spectrometer in which the X-rays are diffracted by a crystal and hence separated by wavelength (see also Chapter 7). The crystal has a curved surface to improve intensities of X-rays. Bragg's law applies:

$$2d \sin \theta = n\lambda$$

where d = interplanar spacing of the diffracting

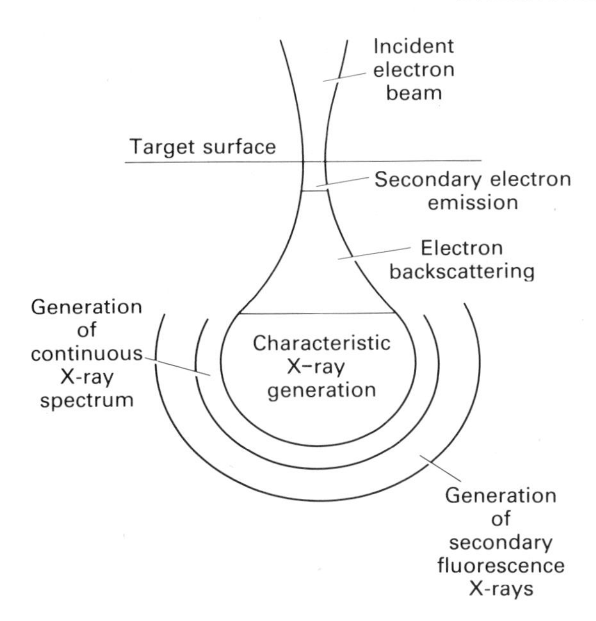

Fig. 9.11. Schematic illustration of the regions in the specimen within which electrons are emitted or reflected and X-rays generated. Scale depends on accelerating voltage, and average atomic number of specimen (see Fig. 9.10).

crystal, θ = angle of incidence of the X-rays, n = a positive integer and λ = wavelength of the X-rays. Rays satisfying this equation are rotated 2θ degrees in their angle of propagation. A detector is placed at the 2θ angle of interest for analysing the largest first-order reflection available for a given element, or can be scanned through various 2θ angles to obtain the whole spectrum. (The latter process is much more efficiently carried out by EDS, however.) X-ray photons of short wavelength can be analysed by a solid-state detector as described below for EDS, but normally in WDS the Bragg spectrometer is coupled to a gas-proportional counter. Here each X-ray photon causes ionization of the gas leading to the production of free electrons which are attracted to a wire and give rise to a pulse of charge. A sealed detector is used for shorter photon wavelengths, but otherwise a continuously-flowing gas is used. For qualitative analysis only the total number of pulses at a given 2θ value need be counted. However, in order to eliminate higher-order reflections and pulses produced by inner-shell ionization of the counter gas, the instrument is set up only to count pulses within a certain voltage range (determined empirically in each case) corresponding to photons of a given energy (= given wavelength). A 2θ scan in the area of interest reveals the form of the background continuum. Background counts are made on either side of the peak for each analysis. When two spectrometers are fitted, two elements can be determined simultaneously.

The most widely used detection system is that of EDS in which all the X-ray photons are collected in the same place. The detector used is a solid-state instrument containing a single-crystal slice of p-type silicon whose resistivity has been increased by addition of lithium: hence lithium-drifted silicon (Si(Li)). As in the gas-proportional counter, each X-ray photon causes the generation of charged particles which create an electrical pulse which, after amplification, is classified according to its amplitude. A multi-channel analyser is used so that a histogram of pulse intensities is built up which largely corresponds to the distribution of energies of the incoming photons. Counts are accumulated over a period of typically 2 min (Reed & Ware, 1975) corresponding to a 'live-time' (time when the counter is not processing a photon and thus able to accept another photon) of 100 s. Thus the overall chemistry of the analysed area is rapidly appreciated, but the histogram contains spurious peaks and overlapping peaks which complicate analysis.

In practice the energy-dispersive system is normally used for routine analyses because of its low cost and the speed with which data can be obtained. A wavelength-dispersive system is suitable when only a few elements are to be analysed or where trace elements are to be determined since the detection limits are significantly lower.

STANDARDS

The intensities of characteristic lines in the specimen's X-ray spectrum are compared with line intensities from material of known composition in order to obtain a quantitative analysis. Ideally the reference materials (standards) would be of very similar composition to the mineral analysed so that electron beam-mineral interactions would be the same, eliminating the need for a correction procedure. This approach is impractical where various minerals are to be analysed. Normally geological microprobe laboratories utilize a set of reference materials (elements, oxides and silicates) which are used in conjunction with a correction procedure for silicate and oxide analysis, and often carbonate and sulphide analysis as well. Examples of suitable materials are

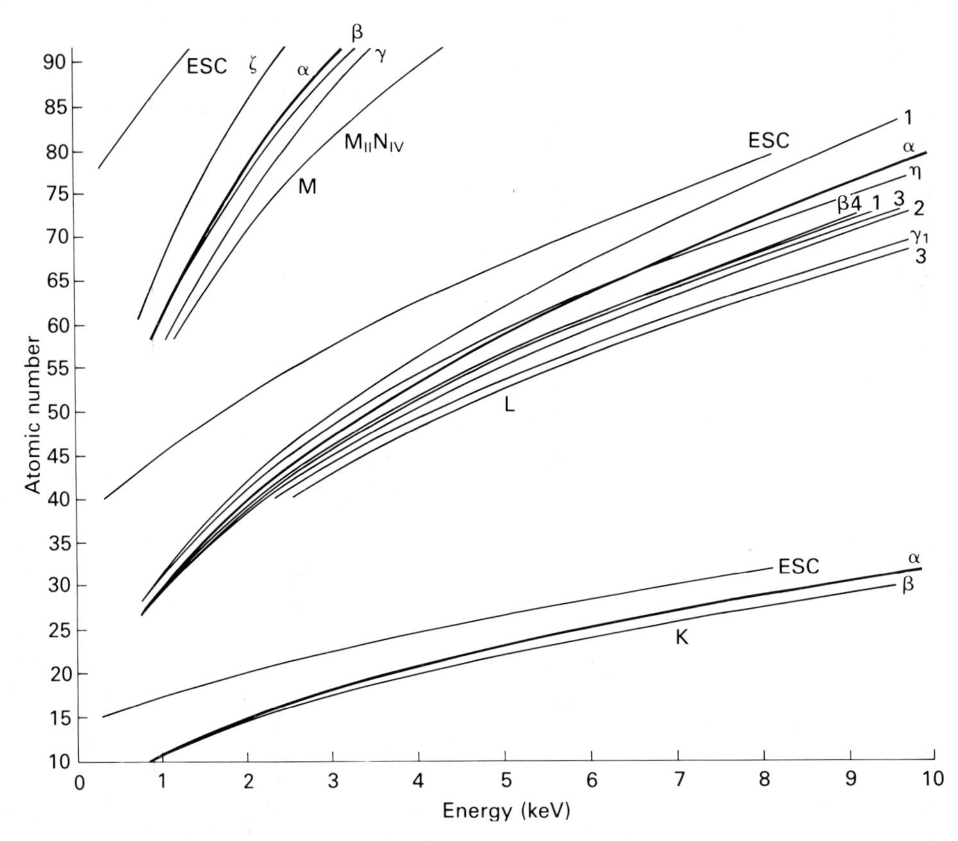

Fig. 9.12. Plot of the energy of the X-ray emission lines observed in the range 0.75–10 keV by energy-dispersive X-ray spectrometry (simplified from Fiori & Newbury, 1978). ESC indicates escape peak. An example of peak overlap would be the Kβ peak for Mn (element 25) and the Kα peak for Fe (element 26): both have the same energy. If Mn abundance ≫ Fe, then Mn Kβ will have a similar strength to Fe Kα.

periclase (for Mg), quartz (Si), jadeite (Na), wollastonite (Ca) and metals for first series transition elements. Synthetic glasses are also used and are particularly useful for rarer elements with which the glass can be 'doped', but care has to be taken because of decomposition under the electron beam. More accurate analysis should in principle be achievable by using end-members of solid-solution series, e.g. pure siderite for Fe in calcite (Moberly, 1968; Jarosewich & MacIntyre, 1983). Use of other than the normal laboratory reference materials should be considered for study of, for example, minor elements in carbonates or minerals with unusually high concentrations of uncommon elements. It is obviously vital to ensure the reference material is homogeneous on all scales for the elements of interest.

SPECIMEN PREPARATION

The normal practice is to analyse highly-polished thin sections of standard (30 μm) thickness. Successively finer abrasives are used, finishing on $\frac{1}{4}$ μm alumina. A final hand polish using a cloth with 0.05–0.1 μm alumina is recommended by Taylor & Radtke (1965). A more rapid method is described by Allen (1984) using only 0.3 μm alumina. Exceptionally, diamond may be required rather than alumina if contamination of soft or porous materials by alumina occurs and Al-analysis is required. The necessity for a good polish arises since an uneven surface increases the statistical uncertainties of X-ray analysis, whilst sharp steps or grooves can lead to widely inaccurate results. The quality of the surface polish can be judged on the microprobe by specimen current imaging.

An electrically-conducting coating, 200–400 Å thick, usually of carbon, is applied to avoid the build up of charge on the surface of non-conducting materials. Standards must have the same thickness of coating as the specimens: this is most easily obtained by coating both at the same time.

OPERATING CONDITIONS

Routinely, an accelerating voltage within the range 15–20 kV is used for analysis. Higher voltages may be used to increase the peak to background ratio for trace element analysis, whilst voltages of 10 kV may be required for very unstable (glassy) materials. The incident probe current is an important variable and needs to be monitored in some way to check constancy of analytical conditions. Sometimes authors report the specimen current (the current from specimen to earth). However, this current depends on the amount of backscattering and secondary electron generation and so varies with different materials. Energy-dispersive analysis requires a probe current of only 1 nA whereas 10–100 nA is more usual in WDS. Since specimen heating is directly proportional to current (as well as voltage), specimens liable to thermal decomposition should preferably be analysed on an EDS: this includes hydrous minerals such as most clay minerals, halides, phosphates, carbonates and phases containing alkali metals (e.g. alkali feldspar). Alternatively, using a WDS, decomposition may be reduced by defocusing the beam to 20 μm or more in diameter. Since spatial resolution is important for clay mineral analysis, Velde (1984) recommended use of a 5 μm spot on an EDS system with its associated low current and low counting times for a complete analysis. Another strategy (Fairchild, 1980b), designed particularly for characterizing minor elements in zoned carbonates, is to allow sample decomposition to occur under a probe current, but to use chemically-similar standards since these decompose to a comparable degree. Totals are normalized to allow for differential decomposition (Moberly, 1968). This ensures a high peak to background ratio. Gold coating and use of continuous scans rather than spot analyses were found to improve precision. The optimal coating material would appear to be silver, as its high thermal conductivity minimizes sample decomposition at high values of probe current (Smith, 1986).

The rapid procedures of the EDS meet most requirements, but with a detection limit of the order of 1000 ppm (0.1 wt %), it is not appropriate for trace element analysis (e.g. Na or Sr in carbonates). Detection limits using WDS, although variable, are of the order of 100 ppm.

CORRECTION PROCEDURES

The simplest case is a WDS where standards are similar in composition to specimens. Following deduction of the background counts, the ratio of X-ray intensities are presumed equal to the ratio of concentrations of specimen and standard. More generally in WD analysis a series of corrections have to be applied (Long, 1977, p. 313). First, a 'dead-time' correction is made which allows for photons which were not counted because the proportional counter was processing the previous photon: this is only important at high count rates. Then there are corrections for atomic number, absorption and fluorescence differences between specimen and standard: the ZAF corrections.

The atomic number correction is the resultant of two opposing tendencies: backscattering which increases with increasing atomic number, and electron-stopping power which decreases with increasing atomic number. The absorption correction arises because the X-rays are generated below the surface of the specimen and particular wavelengths are then preferentially absorbed by different atoms. The fluorescence correction allows for the generation of 'secondary' X-rays by interaction of higher energy X-ray photons with atoms. These corrections are applied iteratively until satisfactory convergence has been reached.

A more complex method is required for an EDS (Statham, 1981) because of the presence of spurious peaks and common overlapping peaks in the spectrum: these must be removed before carrying out the ZAF corrections. 'Escape' peaks (caused by ionization of Si inner shell electrons in the detector, and displaced by the energy of the SiK_α transition from the parent peak) and sum peaks (caused by simultaneous arrival of two photons) must clearly be removed. In order to separate peaks from background, either a model for the shape of the background can be generated or the peaks can be removed in stages until none remain. Overlapping peaks provide the major problem in EDS analysis. A major element will generate several peaks, some of which may coincide with the main peak of a minor element. To separate peaks, an accurate idea of their shape, either from a theoretical model, or a 'library' of standard shapes, is required. Potential problems with peak overlap can be judged from Fig. 9.12. In severe cases (e.g. the analysis of small amounts of Fe in an Mn mineral), quantitative analysis with an EDS may not be possible. It is thus important to be aware of the particular problems of ED correction procedures in order to eliminate spurious results although many modern systems automatically advise on possible overlaps.

As an alternative to the ZAF procedure, many authors correct using empirical 'α-factors' (Bence & Albee, 1968) which summarize the effect of one element on the others present. This process involves less computation than the ZAF method and can obtain near-equivalent results, but is specific to one accelerating voltage and take-off angle (angle between electron beam and specimen surface).

Generally, results for unusual materials should, wherever possible, be checked against reference materials of similar composition in case of deficiencies in the specific programme used.

ANALYSIS USING AN SEM

In the past there has been very little use of SEMs by sedimentologists for quantitative analysis. The problem is that the pre-eminent design aim has been spatial resolution rather than analytical convenience. Low take-off angles have led to high absorption corrections being necessary and the lack of optical microscope attachments have led to difficulties in specifying the area to be analysed. This situation has now changed with the introduction of combined SEM-microprobes and the increasing use of back-scattered electron imagery to relate chemical and petrographic features (Huggett, 1984a). If specimens are polished as for normal microprobe work then the situation is the same as that described in the previous sections.

Use of the SEM in qualitative analysis (and hence identification) is widespread in sedimentological studies and is striaghtforward if the alternatives are known and chemically distinct. Otherwise more care is needed: an excellent summary of guidelines for qualitative analysis is given in Goldstein *et al.* (1981, chapter 6). Often analysis is undertaken on fracture surfaces of rock chips: here one must be particularly aware of possible stray X-rays generated from adjacent minerals or the sample holder.

ANALYSIS USING A STEM (SCANNING TRANSMISSION ELECTRON MICROSCOPE)

This instrument enables analysis of areas as small as 5–10 nm in diameter, since the electron-transparent specimens required are so thin that there is no spreading effect of the electron beam. Also, given a negligible specimen thickness, there are no absorption or fluorescence effects so the correction procedure is simplified. High accelerating voltages, of the order of 100 kV, are used usually in conjunction with a Si(Li) detector (see review article by Goodhew & Chesco, 1980).

Polished thin sections are mounted on a metal grid and thinned by ion-beam milling until holes start to appear (Phakey, Curtis & Oertel, 1972). To be certain that specimen thickness is negligible, areas next to holes are analysed (Ireland, Curtis & Whiteman, 1983). The beam size must be above a certain minimum for each mineral to avoid loss of volatiles such as alkali metals. The correction procedure involves a knowledge of the relative efficiencies of X-ray generation of the element in question relative to, for example silicon, determined by analysis of thinned standards (Cliff & Lorimer, 1975). The use of a STEM is powerful, not only because of the spatial resolution of the analysis, but also because TEM images can be obtained to illustrate textures, and selected-area diffraction patterns may be obtained from grains analysed: hence texture, mineralogy and crystal structure can be determined on submicron-sized grains. The STEM can also be used in other kinds of ways: for example in the analysis of fine suspended sediment, STEM analysis of individual particles can allow the interpretation of 'bulk' (100 μm area) analyses in terms of mineralogical percentages (Bryant & Williams, 1982).

9.6.3 X-ray fluorescence

XRF analysis is a standard technique in hard-rock petrology and in soft-rock circles it is frequently used for the whole-rock analysis of mudrocks, less often for sandstones and rarely for carbonates and evaporites. The principle behind this technique is that when a sample is bombarded with high energy X-rays, secondary radiation is emitted, with the wavelengths and intensities dependent on the elements present. Measurement of the intensity of the characteristic radiation for a particular element gives a value reflecting its concentration in the sample.

The emission from standards is measured first to produce a calibration curve, against which the unknown samples an be compared. Primary X-rays are produced in an X-ray tube, commonly with a rhodium target, and the secondary radiation emitted is passed through a collimator system. The various wavelengths are then resolved by diffraction off LiF or PET crystals. The radiation passes to a counter system, with a flow proportional counter being used for light elements (<22) and a scintillation counter for heavy elements. Two types of XRF system are currently available: wavelength dispersive (WD) and energy dispersive (ED). ED systems are a more recent development (early 1970s) than the WD units, and have some advantages in certain applications. For example, ED is cheaper and measures all elements, however, in general the detection limit is not as low as with WD, and there are many instances of spectral line overlap.

Many textbooks give details of the XRF technique; see, e.g. Norrish & Chappell (1977), Jenkins & de Vries (1970), Bertin (1975), Johnson & Maxwell (1981), Tertian & Claisse (1982) and Potts (1987).

Samples for XRF analysis mostly consist of powdered rocks made up into pressed pellets (briquettes) or fused discs. For the pellets, the powder is mixed with a commercially mixed binder, such as Mowoil, and pressed to 20,000 psi (1400 kg cm^{-2}) or more to form a briquette. Standard rocks and samples need to be prepared in exactly the same manner to achieve the same packing density. Fused discs involve the fusion of powdered rock with lithium tetraborate or metaborate and the casting of the melt into a mould. With this technique, the sample is homogenized, and synthetic standards can be prepared more readily. Also, discs are durable and can be many times. One disadvantage is that trace constituents are diluted further by the addition of the flux. The preparation of fusion discs is fully discussed in Section 9.5.3. In recent years, pressed pellets have become more popular than the disc although discs are still considered better for major elements.

XRF is ideal for the determination of major and minor elements, such as Si, Al, Mg, Ca, Fe, K, Na, Ti, S and P in siliciclastic rocks and also for trace elements, such as metals Pb, Zn, Cd, Cr and Mn. Limestones, on the other hand, are rarely analysed with XRF since a powdered sample will include the clay fraction, and it is the elements in the carbonate fraction which are mostly being sought. XRF has been very successfully used for the analysis of bromine (and other elements) in halite.

9.6.4 Atomic absorption analysis

Atomic absorption spectrophotometry (AAS) has been widely used as a technique for elemental analysis of rocks and minerals, and also waters, since the 1960s. More than 50 useful elements can be detected and the technique has been popular since it is a relatively simple procedure, the instruments are generally sensitive and reliable, and a basic AA machines is not too expensive. Information on AAS is presented by Angino & Billings (1972), Johnson & Maxwell (1981), Potts (1987), and others, and most instruction manuals supplied with AA instruments give much useful background material, as well as guidance. However, it is possible that in the future AAS will become less widespread in its use as ICP (Section 9.6.5) becomes more readily available.

The principle of AAS is the absorption of radiant energy by ground state atoms. When a substance is dispersed as an atomic vapour, it possesses the property of absorbing particular radiations, identical in wavelength to those which the substance can emit, as when it is heated for example. If a parallel beam of radiation of intensity I_0 is incident on an atomic vapour, I_v is the intensity of transmitted radiation and v is the frequency, then

$$I_v = I_0 \exp(-K_v l)$$

where K_v is the absorption coefficient and l is the atomic vapour thickness. The absorption coefficient is also proportional to the concentration of the free atoms in the vapour (Beer's law):

$$\int K_v dv = \frac{\pi e^2}{mc} N v f$$

where f is the oscillator strength (i.e. average number of electrons per atom which can be excited by the incident radiation), N_v is the number of atoms per cm^3 which are capable of absorbing in the frequently v to dv, c is the velocity of light, m is the electronic mass and e the electronic charge. For absorption to take place, the atoms must be in the ground state, not excited. The fraction of total atoms available which exist in the excited stated becomes significant only at high temperatures, or for atoms which have low ionization potentials. The spectral lines absorbed by atoms in their ground state are referred to as resonance lines.

AA analysis involves passing the characteristic spectrum of an element through a flame in which atoms are present. If the atoms are of the same element, then there is a reduction in intensity, due to absorption, of a particular wavelength (Fig. 9.13). The amount of light absorbed depends upon the concentration of the element in the vapour (Beer's law, above). Generally, other metal atoms in the flame do not interfere, since there is no light of suitable wavelength for them to absorb.

The AA instrument requires a light source, a flame, an atomizer and a photomultiplier to measure the transmitted raditation (Fig. 9.14). A monochromator is inserted between the flame and the detection device so that any interfering light, such as generated by emission from the flame itself, is excluded. Hollow-cathode lamps supply the radiation, with the cathode made of the element whose spectrum is required. Multi-element lamps are popular, and save on time. Resonance lines are emitted from the lamps when a current is passed and the cathode is bombarded by ions from the filler gas, usually argon or neon. The characteristic wavelengths of the element are emitted as excited-state atoms return to the ground state. Many lines are produced for one element, and several prominent ones, not absorbed or interfered with by the atoms of the filler gas, are usually available for use in the analysis. The flame (or furnace) is the heart of the AA machine where the sample is atomized and through which the light from the spectral source is passed. The temperature of the flame is important; it must be one at which dissociation of all molecules in the sample occurs, but at which a minimum of ionization takes place. The gases used in the flame are usually mixtures of two gases from acetylene, air, nitrous oxide, oxygen and hydrogen. Air-acetylene is used for much work, set to give a temperature of around 2350 K. The gases and sample are generally mixed in a burner chamber or nebulizer before entering the flame. For analysing elements in very low concentration, a graphite furnace, consisting of a tube of graphite heated by electrodes, can be used instead of a flame. Higher concentrations of the atoms are obtained and there is a more precise control on temperature.

After transmission through the flame, the spectral line of interest is selected at the monochromator and then its intensity measured at the photomultiplier. The reduction in light intensity before and while the sample is in the flame, i.e. the absorbance, is displayed as a meter reading, chart record or digital printout. While measurements are being made the instrument must be stable, i.e. the temperature of the flame must not change, nor the output from the lamp. Rates of atomization of samples and standards must also be similar. Modern instruments are quite complicated in order to overcome these sorts of problems. For example, many instruments have a double-beam system, whereby the light is passed alternately through the flame and round the flame by a system of mirrors, and the ratio of the intensities is measured (Fig. 9.15).

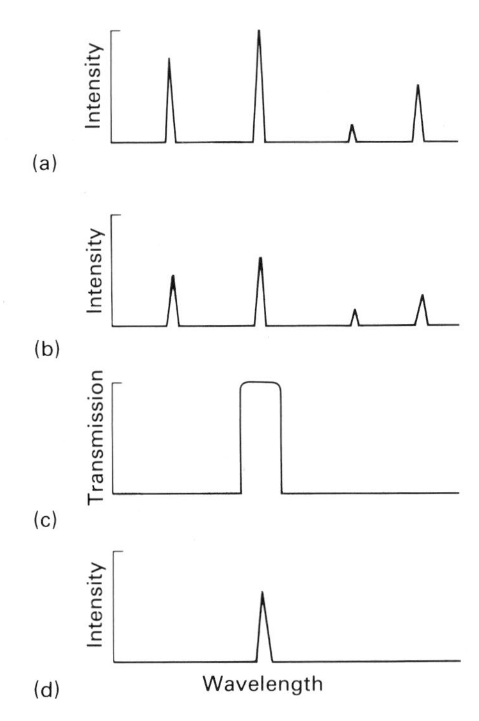

Fig. 9.13. Sketches to illustrate the principle of atomic absorption analysis. (a) In the hollow-cathode lamp, a spectrum of the element to be determined is produced. (b) In the flame, if the element in question is present in the atomic vapour of the sample then absorption takes place, so that the intensity of the spectrum is 'reduced'. (c) At the monochromator, only a narrow band of wavelength is allowed to pass, that including the resonance line of the element to be determined. (d) At the detector, the reduced intensity of the resonance line is measured.

Samples are presented to the AA machine in solution and the instrument is calibrated by running standards, either pure, off-the-shelf solutions with known concentrations of the element, or solutions of rocks of known composition. A range of standard solutions is made up with different concentrations; absorbances are measured and a calibration line is constructed. This should be a straight line, as predicted by Beer's law, but at higher concentrations it may be a curve.

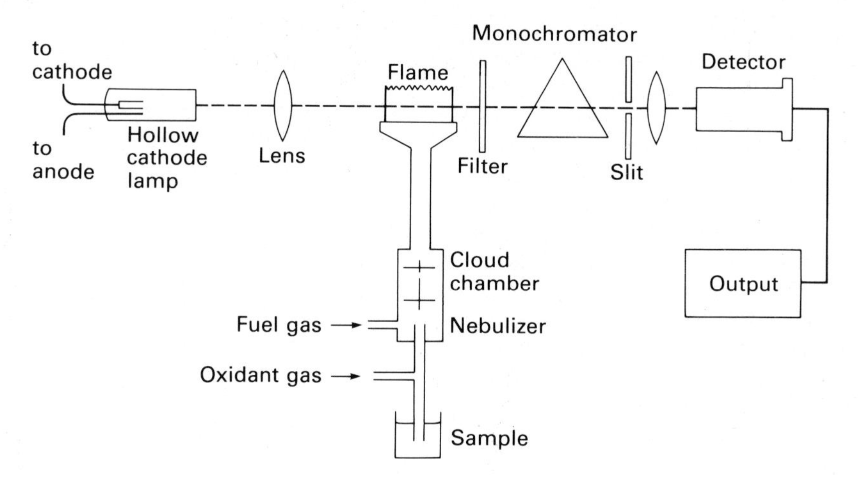

Fig. 9.14. Basic features of an atomic absorption spectrometer.

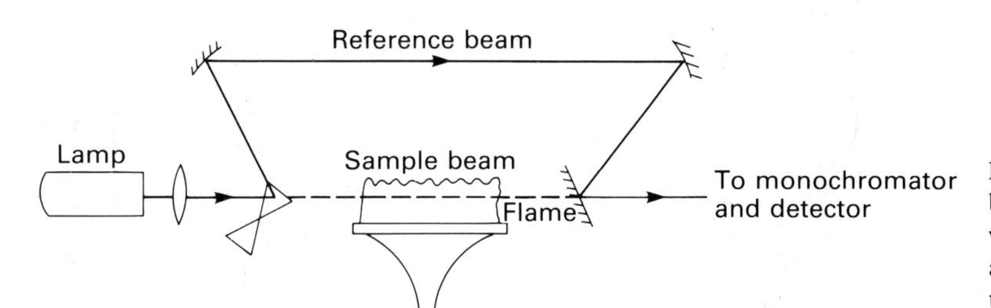

Fig. 9.15. Basic features of a double-beam atomic absorption instrument whereby light from the lamp is alternately passed through and past the flame.

INTERFERENCE

In atomic absorption analysis, interference is not generally a major problem. Spectral interference can arise where two elements have resonance lines at a similar wavelength, resulting in a positive error as the two signals are added together. An alternative line to measure can usually be found. Some metals have a low ionization potential so that a smaller number of atoms remain in the ground state where they can absorb their characteristic radiation. This ionization interference can be overcome by adding certain elements to the solution to suppress the ionization. Chemical interference results from elements combining to form stable compounds which do not break down in the flame to form ground state atoms. Negative errors arise, or even no absorption at all occurs. To overcome this, an excess of a metal is added which can compete with the metal being determined for combination with the interfering element. Alkaline earth metals are prone to interference. Ca^{2+} for example, shows the effect of interference in the presence of SO_4^{2-}, PO_4^{3-}, Al and Si, when complexes like Ca–Al or Ca–Si are formed. The addition of large amounts of strontium (in the form of $SrCl_2$) or lanthanum ($LaCl_3.6H_2O$) can overcome this source of error. Manganese can be interfered with by Si, and then Ca^{2+} is added. With strontium, there is a problem of ionization interference, especially in the presence of Na^+ and K^+, and so lanthanum or potassium (KCl) is added. The problem of interference is dealt with at length in Angino & Billings (1972), Johnson & Maxwell (1981) and Potts (1987).

SAMPLE PREPARATION

AAS has been widely used in the analysis of limestones and dolomites since the interests here are mostly in the elements contained within the carbonate lattice. To obtain the solutions, powdered samples (0.2 g is a suitable weight) are dissolved in acid which is sufficiently strong to take up the carbonate, but not so strong that it completely dissolves or leaches the insoluble residue (clays, opaques, etc.). For limestones, 3% v/v cold acetic acid (25 ml on 0.2 g) overnight is sufficiently strong, and then solutions can be made up to 50 ml. Brand & Veizer (1980) used 3% (8% v/v) HCl for 5½ hours; after this length of time the insoluble residue was extensively leached. With dolomites, stronger acid is required and 10% HCl at 60°C for 4 hours is widely used. With dolomite-calcite mixtures some trial and error may be required since dolomite solubility does depend on its ordering and stoichiometry. Burns & Baker (1987) first leached a mixed sample with acetic acid buffered with ammonium acetate (pH=5) for 30 minutes to remove the calcite. After filtering and rinsing, the remaining solid was leached in 1M HCl for 15 minutes to take up the dolomite.

Once solutions of powdered rock have been prepared then aliquots can be removed and diluted as appropriate for the machine. Several dilutions may be necessary to bring the unknown sample within the range for the element in question in the instrument, but with automatic diluters this is neither time consuming nor likely to introduce errors.

The common elements determined for carbonates are Ca, Mg, Sr, Na, Fe and Mn. For some of these it will be necessary to add solutions to prevent interference, as noted earlier. The quantity of rock frequently used in AAS (0.2 g) is not great, but much smaller weights can be used if judicious use of the solution is made. This is particularly useful for carbonates, where grains and different cement generations can be extracted with a dental drill or scalpel from thin sections. If a graphite furnace is available, then sample weights as low as 10 mg can be used.

AAS can also be used for other sedimentary rocks, sandstones, and mudrocks for example, although X-ray fluorescence can be easier. A whole-rock take-up is obviously necessary so that much stronger (and more dangerous) acids are required. A mixture of HF and $HClO_4$ (perchloric), 2 ml of each on 0.2 g powder, heated on a hot plate, will break down the rock. A mixture of HF, H_2SO_4 and HCl (aqua regia) is also popular. Blanks are also made in the same take-up procedure, as are standard rocks, if these are being used.

9.6.5 Inductively-coupled plasma spectrometry

A plasma is a luminous volume of gas with atoms and molecules in an ionized state. In ICP analysis the gas used is usually argon. The plasma is formed by passing the gas through a torch made of quartz glass. A temperature of around 10 000 K is produced by a radiofrequency generator connected to copper work coils surrounding the torch (Fig. 9.16). The plasma is constrained by the nature of the orifice of the gas injector tube in the torch and the gas flow rate so as to produce a toroidal or doughnut shaped fireball. Nitrogen, or more argon, is used as a coolant gas to stabilize the plasma centrally in the torch and to prevent the outer glass jacket of the torch from fusing or distorting. With the plasma established, the sample, in acid-solution, is passed through a nebulizer to form an aerosol and then mixed with the argon injector gas. In the fireball of the torch the sample is completely atomized at the very high temperatures there.

Now that the sample has been atomized and ionized, there are two completely different techniques for measuring the concentrations of the various

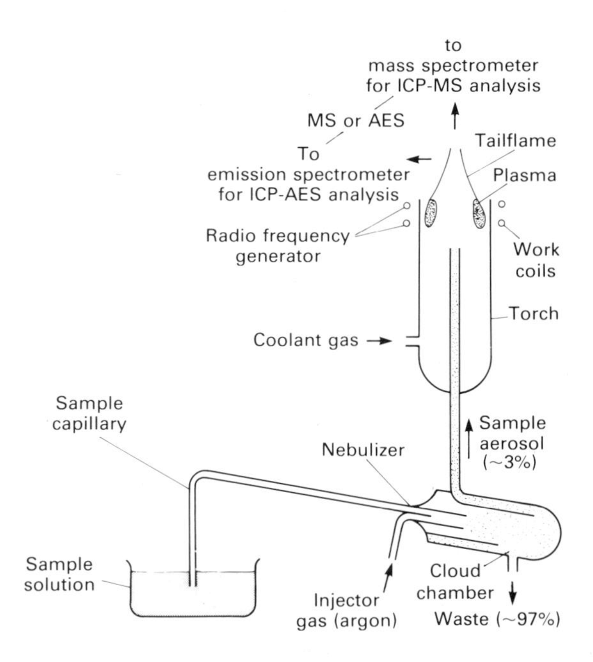

Fig. 9.16. Schematic representation of the inductively-coupled plasma instrument.

elements present: atomic (or optical) emission spectrometry (ICP-AES or ICP-OES) and mass spectrometry (ICP-MS). The instrumentation for ICP-AES has been available since the late 1970s, whereas ICP-MS is an early 1980s development. ICP-AES is related to AAS, whereas ICP-MS is more related to stable isotope analysis.

Inductively-coupled plasma atomic emission spectrometry (ICP-AES) works on the principle that when atoms and ions are excited then light is emitted, and the wavelengths and intensities of the light reflect the elements present in the sample. The emission spectra from atomization in the plasma are analysed with a high resolution spectrometer within the wavelength range of 170–780 nm. The general arrangement of an ICP-AES machine is shown in Fig. 9.17. One of the major advantages of ICP-AES over AAS is that many elements can be analysed simultaneously; some instruments have a polychromator spectrometer capable of accommodating more than 50 spectral lines. One of the problems with ICP-AES has been spectral interferences from overlaps between emission lines of different elements, so that line selection and the identification and correction of overlap interferences are important considerations. Another attraction of ICP-AES is that for many elements the detection limit is much lower than for AAS, although for many metallic elements the detection limits are about the same and, for Zn, Na and K, AAS is better. Another advantage is that calibrations are linear over 4 or 5 orders of magnitude so that dilutions are generally unnecessary and major, minor and trace elements can be determined in one run, lasting only a few minutes. Only a small amount of solution is also necessary for ICP-AES, 0.5–2.0 ml typically, so that very small quantities of sample, 1–10 mg, can be used. Powdered rocks are dissolved in acid and presented to the machine in a very dilute form. For carbonates, where ICP analysis is increasing in popularity, a scheme of solution preparation is presented later in this ICP section. Standard solutions can be prepared with a similar matrix to the samples (see Potts, 1987, p. 180 for recipes) or they can be bought off the shelf. Sample solutions are frequently spiked with an internal standard to

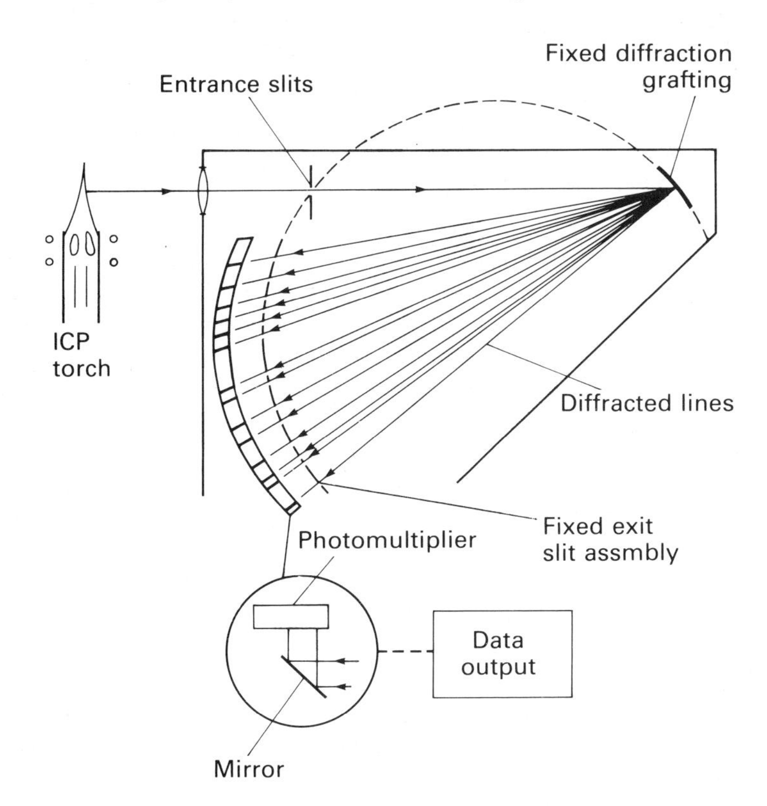

Fig. 9.17. Schematic representation of inductively-coupled plasma atomic emission spectrometer (ICP-AES).

compensate for variations in instrument sensitivity, matrix effects, electronic drift, etc. Identical concentrations of an element not present in the samples are added to all solutions and then the ratio of each measurement to the intensity of the internal standard will permit determination of any error.

Inductively-coupled plasma mass spectrometry (ICP-MS) is perhaps the most exciting development on the analytical front in decades. The technique has all the advantages of ICP-AES: rapid, multi-element trace analysis, but with much lower detection limits (ppb-level), including the determination of individual isotope ratios. In ICP-MS, the inductively-coupled argon plasma is used as a source of ions and their mass spectrum is measured using a quadrupole mass spectrometer. The argon plasma is produced in the same way as in ICP-AES (Fig. 9.16) and the sample in solution is nebulized into the plasma, although the torch is now in an horizontal position. At the very high temperatures of the ICP, the sample is very efficiently atomized and ionized, and there are few interferences. The ion beam emerging from the torch is collimated by cones and skimmers and focussed by a number of plates set at particular potentials (see Fig. 9.18). The beam is then transmitted through the quadrupole mass filter where the mass peaks of the ions are measured in an electron channel multiplier ion detector, on the basis of their mass to charge ratio. The pulses are amplified and stored in a multi-channel analyser and quickly processed by microcomputer.

ICP-MS could well be the routine instrument for geochemical analysis in the years to come, with its detection limits lower than most other techniques, the rapidity of multi-element analyses and relatively simple sample preparation. Much ICP-MS data will be forthcoming in the next few years and it is possible that particular elements will be found to have environmental and diagenetic significance.

Two potentially exciting developments with ICPS are in the use of lasers on thin sections of rock and in the study of fluid inclusions. The energy of a laser beam can be used to vaporize material so that it is possible to zap a rock slice and have the volatilized material carried into an ICP machine for analysis (Thompson, Goulter & Sieper, 1981). Fluid inclusions can be analysed by heating the crystals containing them until the inclusions break and then having the contents carried in a stream of argon into the ICP instrument for analysis (Thompson *et al.*, 1980).

Inductively-coupled plasma techniques are treated at length in Thompson & Walsh (1983) and Potts (1987) and a review is given by Thompson (1986).

ANALYSIS OF CARBONATES USING ICP-AES

Carbonate workers are likely to want to use ICP-AES in preference to other solution-based techniques whenever it is available: both because of its rapid sample throughput and capability for simultaneous analysis of many elements, and because of the small sample size (1–10 mg) required. Microsampled powders can thus be analysed for both cations and isotopes. Since there is little published on techniques specific to carbonates, a detailed account is given here. It also serves to illustrate more generally an approach to standardization of chemical analysis and correction of results.

Although data can be obtained using general-purpose standards designed for silicate rocks, considerable errors are likely to result because of the

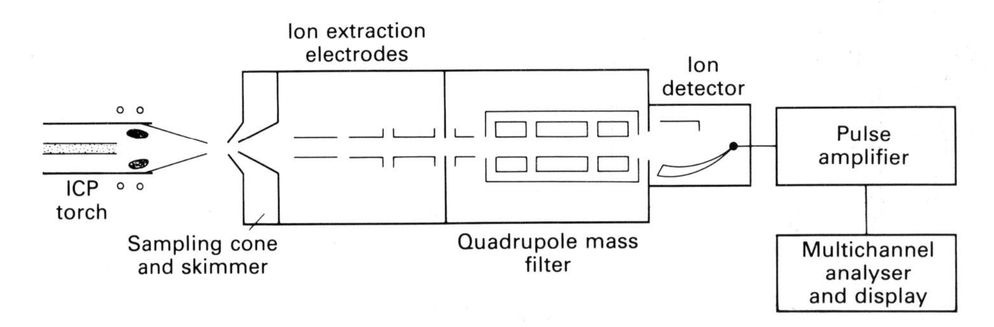

Fig. 9.18. Schematic representation of the inductively-coupled plasma mass spectrometer (ICP-MS).

Table 9.7. Equivalent concentrations of standards solutions for calcite and dolomite analysis by ICP-AES. The figures represent the model concentrations of an element expressed as if it were a solid carbonate. For example, calcite contains 400,000 ppm Ca and is modelled by standards 1C, and C1 to C4. Stoichiometric dolomite contains 200,000 ppm Ca and 120,000 ppm Mg and is modelled by standards 1CM, and D1 to D5. When the samples are dissolved, there will be a dilution factor, for example 10 mg solid in 10 g (about 10 ml) solution corresponds to a dilution factor of 1000. In this case the actual concentrations of Ca and Mg in standard solutions would be 1/1000 of those given in the table.

In making up standards in practice it proves convenient to use commercially available cation solutions wherever possible, otherwise solid salts. The quoted levels of impurities for elements to be analysed must be carefully checked to see if they intrude significantly on background levels. For Ca, only Specpure® grade solid $CaCO_3$ is sufficiently free of Sr for this purpose

Name of standard	Ca	Mg	Sr	Fe	Na	Mn
			ppm ($\mu g\ g^{-1}$)			
Acid-blank	—	—	—	—	—	—
2C	800,000	—	—	—	—	—
1C	400,000	—	—	—	—	—
0.5C	200,000	—	—	—	—	—
0.1C	40,000	—	—	—	—	—
1.5CM	300,000	120,000	—	—	—	—
1CM	200,000	80,000	—	—	—	—
0.5CM	100,000	40,000	—	—	—	—
0.1CM	20,000	8,000	—	—	—	—
C1	400,000	200	50	200	200	40
C2	400,000	400	200	400	400	200
C3	400,000	1,000	500	1,000	1,000	500
C4	400,000	5,000	4,000	5,000	5,000	2,000
D1	200,000	120,000	50	200	200	40
D2	200,000	120,000	200	1,000	400	200
D3	200,000	120,000	500	5,000	1,000	2,000
D4	200,000	120,000	2,000	20,000	2,000	10,000
D5	200,000	120,000	2,000	100,000	2,000	10,000

different acid matrix and major cation compositions of samples and standards. Instead, synthetic standards, closely matching the samples in terms of major-ion composition and acid type and strength should be prepared. Table 9.7 illustrates a possible range of standards. The acid-blank is a solution containing impurity-free acid of the same concentration as in samples and all standards and is used to monitor the baseline for background levels. The series 2C, 1C, 0.5C, 0.1C provides a Ca calibration, the top standard having twice the Ca concentration expected in a pure carbonate sample. Although a linear calibration between concentration and output voltage from the instrument is expected this should be directly tested; deviations from a straight-line relationship are likely at high concentrations (Fig. 9.19a). This series of standards additionally allows the effect of major element (Ca) interference on the background of a minor element to be quantified (line C on Fig. 9.19b). The CM series are for Ca and Mg calibrations for dolomites, 1CM being the appropriate strength for a stoichiometric dolomite sample. The effects on backgrounds are shown as line CM in Fig. 9.19b; combining data from lines C and CM allows the separate effects of Ca and Mg on background to be quantified. Solutions C1 to C4 are mixed standards designed to calibrate minor elements across their likely range in calcites; likewise D1 to D5 for dolomites, D5 being included for ankerites. More extensive Fe-rich standards would be required for siderites. Other minor elements such as Ba, Pb, Cu and Zn might well need to be included in other investigations.

Dolomite samples and calcite samples should be run separately to simplify the process of calibration and possible problems of machine drift with time. Dolomite samples should be preceded by the full set of dolomite-matching standards and every tenth sample followed by a mixed dolomite-matching standard (e.g. D4) and an acid-blank to monitor drift in background values or sensitivity (Fig. 9.19c).

A correction procedure could be written into the instrumental output, but the carbonate worker can do it personally with the advantage of a full understanding of the corrections made and potential sources of error. It is not particularly difficult given the availability of computer spreadsheet software on which the calculations can be laid out. Also there are only a few major elements for whose interference effects minor element concentrations have to be corrected.

A spreadsheet is a table or matrix of material typically consisting partly of text, partly of input data and partly of the results of calculations. The instructions for carrying out calculations are 'hidden' in matrix cells. It is much easier to learn how to write a spreadsheet to carry out a series of corrections to results than to write a computer program, and intermediate steps are more easily seen. The examples given here were written using Microsoft Corporation's Multiplan® speadsheet software.

A data spreadsheet is created of all the output voltages for the elements of interest for each sample and standard. A second spreadsheet file (e.g. Table 9.8) contains correction factors. Calculations for each element in turn are performed on separate spreadsheets (e.g. Table 9.9) which copy in data and correction factors as required from the other files. Another spreadsheet then copies the results of these calculations, totals them, normalizes them, and calculates molar percentages.

A comparative study of the analysis of HCl and H_3PO_4 solutes was made on 50 samples of calcite and dolomite. A table of results is given in Fairchild & Spiro (1987).

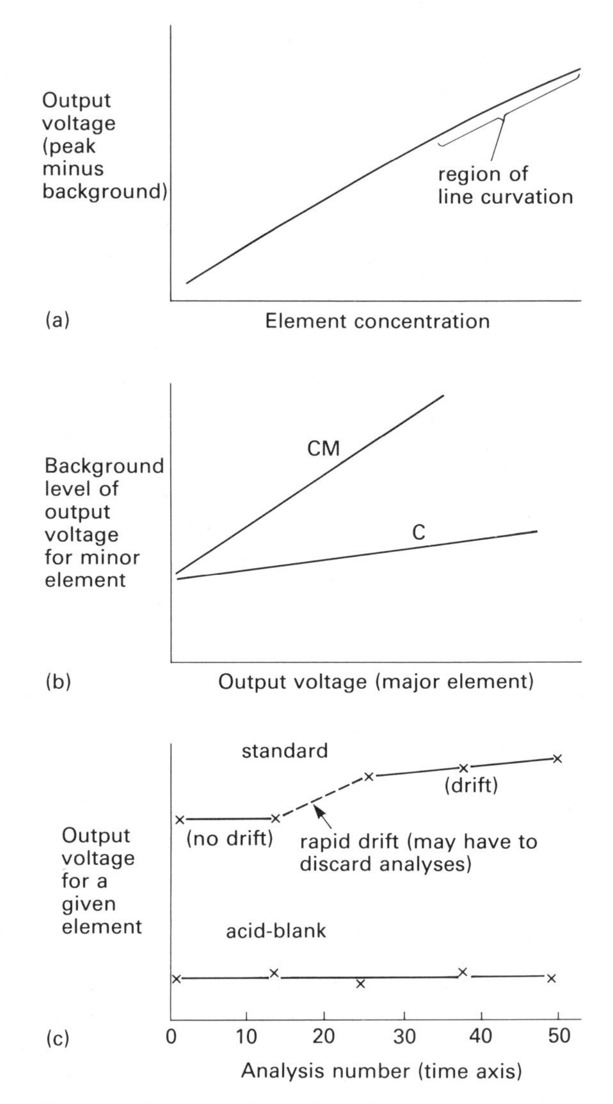

Fig. 9.19. Schematic illustrations of plots used to calculate correction factors for ICP-AES analysis of carbonates. CM and C refer to series of standards (Table 9.7).

The HCl residues were obtained by dissolving a dried and accurately-weighed sample (in the range 5–40 mg) in 8 ml 10% v/v Aristar®-grade HCl. Each sample was decanted into a new vial whenever there was a discernable insoluble residue. The insoluble residue was weighed after twice dilution and decantation of the acid and evaporation on a hotplate at 100°C. The results show detection limits of 3.5 ppm (Sr) to 70 ppm (Fe) calculated by the methods of

Table 9.8. Example of a spreadsheet with factors used to correct data to concentrations. The analysis numbers identify which block of samples have similar correction factors (as identified on plots like that of Fig. 9.19c) and so can be corrected together by the factors in this table. Stdwt is the standard weight (e.g. 10 mg of carbonate) assumed to be dissolved. Cabkd is the Ca background in volts, Cacorr is the inverse slope of the line in a plot like that of Fig. 9.19(a), Cadrift is the rate of change of Cacorr per analysis derived from a plot like that of Fig. 9.19(c). For Mg similar factors apply except that there is a correction to apply to the Mg background because of the presence of major Ca (CacorrMgbkd is equal to the slope of plots like that of C in Fig. 9.19(b). For the other elements similar factors apply except that for dolomites there is also a magnesium correction for the background representing the difference in slope per unit Ca of lines CM and C on Fig. 9.19(b)

Date	30/10/86
User	ijf
Analysis nos	17–44
Stdwt	10
Cabkd	70
Cacorr	49.8
Cadrift	0
Mgbkd	82
CacorrMgbkd	0.0068
Mgcorr	4.6
Mgdrift	0
Febkd	86
CacorrFebkd	0
MgcorrFebkd	0.0005
Fecorr	5.1
Fedrift	0
Mnbkd	103
CacorrMnbkd	0
MgcorrMnbkd	0.0005
Mncorr	1.17
Mndrift	−0.0028
Srbkd	115
CacorrSrbkd	0.023
MgcorrSrbkd	0
Srcorr	0.0354
Srdrift	−0.000056
Nabkd	175
CacorrNabkd	0.0050
MgcorrNabkd	0.0025
Nacorr	1.36
Nadrift	−0.002

Thompson & Walsh (1983) or 1–5 ppm calculated using the less rigorous definition that detection limits are three times the standard deviation of the background.

The phosphoric acid solutes resulted from sample dissolution for isotopic analysis. Because of the high viscosity of 100% H_3PO_4, dilution to 12% H_3PO_4 by weight was required, even though this gives a high dilution factor (w/w of the original *c.* 10 mg carbonate solid in final solution) of about 1:6000. The proportion of insoluble residue was taken to be the same for each powder as determined by HCl dissolution. Detection limits were of the order of 2000 ppm for Fe, 100 ppm for Mn, and greater than 40 ppm for Sr. Na analysis was essentially impossible because of high Na impurity from P_2O_5, used to increase the concentration of H_3PO_4, and difficult to buy in sufficiently pure form. However, the accuracy of major element results was as good as for the HCl residues. It should be concluded that as long as a few milligrams of sample can be spared for ICPES analysis by HCl dissolution, this will be preferable to analysing phosphoric acid residues, especially given the physical difficulties in handling H_3PO_4, but for minute samples the latter might be appropriate. Where minor element and isotopic variation occurs on a very fine scale the phosphoric acid method could also come into its own (M. Coleman, 1986, pers. comm.), although this would depend on the degree of crushing of the sample. The deliberate repeated analysis of an inhomogeneous sample crush, with an intimate mixture of two components differing in minor element and isotopic composition, could be used to find the composition of one end member if the composition of the other is known by other means.

The handling of small samples does create difficulties as is shown by analytical totals for the HCl residues of Fairchild & Spiro (1987) which do not closely cluster around 100% (Fig. 9.20). Normalized analyses of reference materials in the same sample batch are closely comparable with results using other analytical methods, suggesting that the correction factors used are generally valid. An exception may be the consistently low totals obtained on dolomites with high sample weights (Fig. 9.20) which could be due to inadequate allowance for line curvature of the type shown in Fig. 9.19(a). Other contributory factors to totals deviating from 100% are errors in weighing such small samples (variable water content if insufficiently dried); errors in measuring insoluble

Table 9.9. Example of a spreadsheet used to correct output voltages (in this case for Fe) to concentrations. The first four columns and column 15 are automatically input from a data file. Columns 5 to 7, 10, 11 and 15 are automatically input from the file of correction factors (Table 9.8). The first steps are to compute the background Fe value allowing for the interference of Ca and Mg; the final background (column 8) is then subtracted from the Fe value output to give peak minus background in column 9. The Fe correction (column 10) is then modified for any drift (column 11, see also Table 9.8) to give a final correction (column 12) by which column 9 is multiplied to give the Fe value in column 13. This value must then be corrected by the ratio of the actual weight of carbonate minerals dissolved (column 14: total sample weight minus insoluble residue) to the standard sample weight (column 15) to yield Fe ppm (column 16)

	1	2	3	4	5	6	7
1	Sample	Ca	Mg	Fe	Febkd	CacorrFebkd	MgcorrFebkd
2	bcs368v	3731	23300	305	86	0	0.0005
3	f6962a	4112	22250	5924	86	0	0.0005
4	f6975a	4858	31330	5999	86	0	0.0005
5	f6989b	2865	16141	10533	86	0	0.0005
6	f6890a	1287	6768	2556	86	0	0.0005
7	m4208b	5917	33200	3172	86	0	0.0005
8	m3852	4102	20770	5488	86	0	0.0005
9	f6890b	4910	26130	6720	86	0	0.0005
10	f6963b	4314	23070	2370	86	0	0.0005
11	m3855	2282	11191	6186	86	0	0.0005

8	9	10	11	12	13	14
Fin bkd	p−b	Fecorr	Fedrift	Fin corr	Fe value	Weight
98	207	5.1	0	5.1	1057	9.7
97	5827	5.1	0	5.1	29717	10.9
102	5897	5.1	0	5.1	30076	11.3
94	10439	5.1	0	5.1	53239	7.7
89	2467	5.1	0	5.1	12580	3
103	3069	5.1	0	5.1	15654	13.1
96	5392	5.1	0	5.1	27497	8.8
99	6621	5.1	0	5.1	33767	10.9
98	2272	5.1	0	5.1	11590	9.5
92	6094	5.1	0	5.1	31081	5.7

15	16	17
Std wt	Feppm	FeCO3
10	1090	0.23
10	27263	5.66
10	26616	5.52
10	69141	14.34
10	41932	8.70
10	11950	2.48
10	31247	6.48
10	30979	6.43
10	12200	2.53
10	54529	11.31

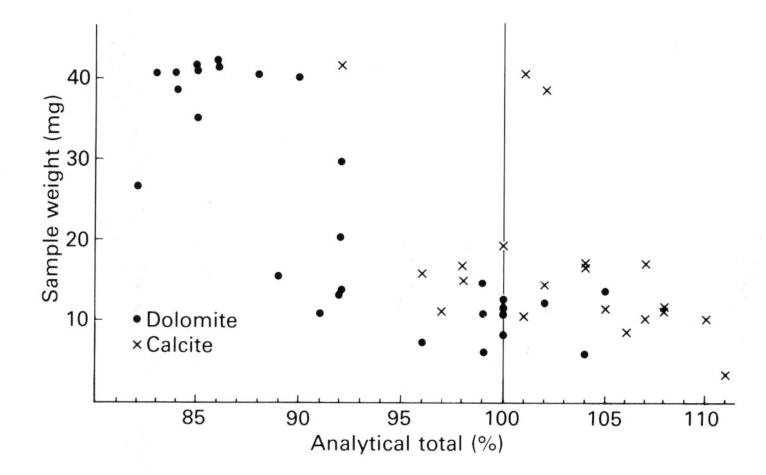

Fig. 9.20. Plot of sample weights against analytical totals for HCl residues of Fairchild & Spiro (1987).

residues; minor dilution or evaporation of samples. These factors will not affect the accuracy of normalized results. Even so, the calculation of analytical totals (necessarily involving weighing insoluble residues) is an important part of the analytical procedure and will allow errors to be readily spotted.

9.6.6 Instrument neutron activation analysis

This technique is mostly used to determine the contents of rare earth elements (La, Ce, Nd, Sm, Eu, Tb, Yb and Lu), and uranium, thorium, hafnium and tantalum (and some other trace elements) in a variety of rock-types. The basis of NAA is that when material, in this case powdered rock, is irradiated by nuclear particles in a reactor, some of the atoms in the sample interact with bombarding particles and are converted into radioactive isotopes. The induced radioactivity is separated, identified and measured relative to known standards. The elements are identified by the characteristic energy (gamma rays) emitted during the radioactive decay. A gamma-ray spectrometer is used as a detector, measuring the energies and intensities of various γ-ray peaks above background, which emanate from an activated sample. Complicated mathematical processing of the data is required to separate the complex spectrum into components that correspond to individual radioactivities.

NAA is mostly used by hard-rock petrologists for rare earth and other elements, which are generally present at the level of tens of ppb. The technique has been used to determine REE in sandstones and mudrocks, and data from Archean and Proterozoic sediments have been used to discuss the origin of continental crust (see Taylor & McLennan, 1985) and the amount of recycling which has taken place. There are few studies of REE in carbonate rocks, but Tlig & M'Rabet (1985) have recently shown that typical REE distribution patterns are preserved during dolomitization, although total amounts are reduced. The decrease correlates with Sr and $\delta^{18}O$ and is related to lower salinity (meteoric-marine mixed) dolomitizing fluids. REE determinations can also be made by ICP.

9.6.7 Stable isotopes

The main variations in technique arise from the mineralogical siting of the isotope concerned and so the text is divided up accordingly. Covered here are sulphur and oxygen isotopes in sulphates, sulphate isotopes and sulphides, carbon and oxygen isotopes in carbonates and oxygen isotopes in silica. The reader is referred to the following references for analyses of other minerals. For the preparation of organic carbon in its various forms: see Dunbar & Wilson (1983) for coal, Jackson, Fritz & Drimmie (1978) and Schopf (1983) for kerogen and extractable organic matter, and Schoell (1980) for methane. The extraction of oxygen from phosphatic shell material for $\delta^{18}O$ determinations is covered by Tudge (1960), while Carothers & Kharaka (1980) deal with the measurement of $\delta^{13}C$ on dissolved HCO_3^-. Finally, the standard reference for $\delta^{18}O$ work on waters is the CO_2 equilibration method of Epstein & Mayeda (1953).

MASS SPECTROMETRY

In simple terms a mass spectrometer may be thought of as an instrument to measure relative differences in the abundance of certain isotopes of a given element. In all cases the sample is introduced as a gas (e.g. CO_2, SO_2) via an inlet system designed to allow the rapid comparison of the sample with a reference gas prepared at the same time (Fig. 9.21). Once inside the instrument the gas is ionized by electrical bombardment in the 'source region', the ions are accelerated in an electrical field and collimated to emerge as an ion beam. The ions within this beam are then separated according to their mass by passage through a magnetic field, emerging as a series of beams, each with a given mass and a charge. Collectors for each mass, usually arranged in the form of a metal cup or Faraday cage, are placed at the appropriate spot to collect particular ions, which then discharge on to them. The strength of the discharge, proportional to the number of ions of each mass present, is then registered electronically and displayed as a ratio or, normally, transferred automatically for storage in a microcomputer.

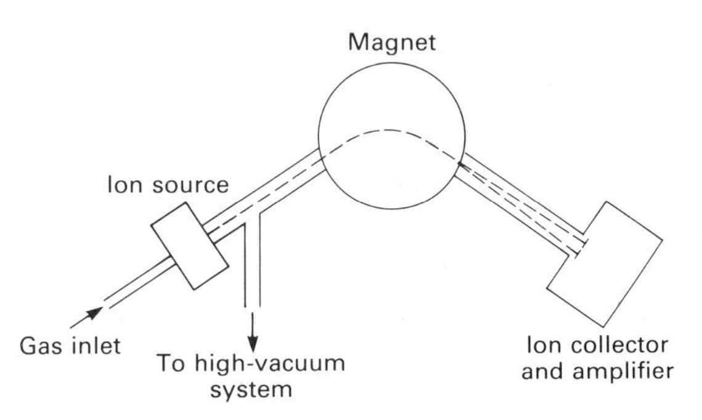

Fig. 9.21. Schematic representation of mass spectrometer (from Dodd & Stanton, 1981).

SULPHATES AND SULPHIDES

Generally the preparation techniques for measuring sulphur isotope ratios seek to yield pure sulphur dioxide (SO_2), although some authors prefer SF_6 as an end-product. Sulphides are converted to SO_2 by reaction with a suitable oxidizing agent such as Cu_2O or V_2O_5, while sulphates have been traditionally converted first to the sulphide by a variety of chemical means, and then oxidized in a similar manner. Since about 1970, however, a variety of thermal decomposition methods have been developed which allow the *direct* reduction of sulphates to SO_2.

PURIFICATION AND REDUCTION PROCEDURES FOR SULPHATES

The treatment of sulphates may be conveniently divided into three steps: (a) purifying the sample sulphate, (b) reduction to sulphide-sulphur and (c) oxidation to SO_2. The last step is essentially the same as that for sulphides and will be outlined under that heading.

(a) Isolation of pure sulphate is usually achieved by dissolution of solid sulphates (in HCl or NaCl solutions for calcium sulphate) and precipitation as the insoluble salt $BaSO_4$ by addition of $BaCl_2$ solution (Longinelli & Craig, 1967; Claypool *et al.*, 1980; Sakai *et al.*, 1980; Cortecci *et al.*, 1981). Where impure sulphates are to be analysed for oxygen isotopes, the additional precaution of passing the solute through an ion-exchanger prior to addition of $BaCl_2$ is desirable to avoid coprecipitation of metal oxides with the $BaSO_4$ (Claypool *et al.*, 1980).

(b) The rapid quantitative reduction of sulphate requires powerful reducing agents, a variety of which have been found satisfactory. Usually Ag_2S is designed to be the end-product of the process. Details are given in Thode, Monster & Dunford (1961), Gavelin, Parwel & Ryhage (1960), Sasaki, Arikawa & Folinsbee (1979), Sakai *et al.* (1980) and Kiyosu (1980).

Direct-reduction of sulphate to SO_2 (thus eliminating the sulphide stage altogether) was advocated by Holt & Engelkemeir (1970). They headed $BaSO_4$ coated in pulverized quartz to 1400°C *in vacuo*:

$$BaSO_4 \rightarrow BaO + \tfrac{1}{2}O_2 + SO_2.$$

The BaO fused with the silica and pure SO_2 was collected in a nitrogen cold trap. Yields were reported as 100 ± 1%. There was a little contamination by CO_2 from carbon residues on the vacuum line which was removed by a 1:1 solution of HF. Sample weights ranged from 20 to 50 mg. Bailey & Smith (1972) modified the method by introducing copper metal to the system and reducing the temperature of reaction to 800°C which allowed a better control over the generation of SO_2, keeping production of SO_3 to a minimum. They also pointed out that powdered silica need not be added since there was sufficient on the walls of the vacuum line.

Coleman & Moore (1978) outlined a further improvement by mixing their sulphate sample (15 mg) with cuprous oxide (Cu_2O, 200 mg) and pure quartz sand (600 mg, <75 μm) and heating the mixture under vacuum to 1120°C. Although they were unsure of the exact reaction involved, repeated use proved yields of 99.8 ± 1.3% and gave $\delta^{34}S$ values on a laboratory reference sample, virtually identical to those obtained via the technique of reduction and subsequent oxidation.

ANALYSIS OF SULPHIDE-SULPHUR

The preparation procedure is relatively straightforward for a pure sample, requiring only the oxidation of the sample to yield SO_2. Use of oxygen as the oxidizing agent (Thode *et al.*, 1961) required extremely high temperatures (1350°C) to hinder fractionation by SO_3 production, and to avoid formation of $AgSO_4$ from Ag_2S (Robinson & Kusakabe, 1975). Neither problem arises if a solid oxidant is used, as is now normal.

A range of solid materials are utilized including CuO (Ripley & Nicol, 1981), Cu_2O (Robinson & Kusakabe, 1975; Kiyosu, 1980) and V_2O_5 (Makela & Vartiainen, 1978; Gavelin *et al.*, 1960) and at significantly reduced temperatures; generally in the range 600–850°C. Robinson & Kusakabe (1975) carried out their reactions at 800°C producing yields of 99–100% in approximately 10 minutes with contaminating CO_2 being removed via an *n*-pentane-liquid nitrogen trap.

A pre-concentration step is needed for the analysis of disseminated sulphide in a sample. Ripley & Nicol (1981) dissolved their sample in concentrated acid and used Thode solution (a boiling mixture of HI, H_3PO_4 and HCl, Thode *et al.*, 1961) to convert the sulphur to H_2S, and subsequently Ag_2S. Sasaki *et al.* (1979) stressed that Kiba reagent (a mixture of tin II and H_3PO_4) was far more rapid than Thode solution in extracting sulphide from pyrite. Cameron (1983) used Kiba reagent, but found that since this also extracted organic S and S in $BaSO_4$ a sulphide-specific dissolution stage was needed first (Dinur, Spiro & Aizenshtat, 1980).

SO_2 VERSUS SF_6

Despite the chemical stability of SO_2, it does cause some problems (Rees, 1978). Inter-laboratory correlations are made difficult by the tendency of SO_2 to stick in the inlet system, causing memory effects. Also, corrections need to be made for mass interferences in the spectrometer due to the presence of several isotopes of oxygen. The alternative is SF_6: chemically inert, insensitive to moisture, not prone to adsorb on to the vacuum line and needing no corrections for interference effects since ^{19}F is the only stable isotope. The intensity of the SF_5^+ beam can be measured readily as it occurs in a mass region free of instrument background (Puchelt, Sables & Hoering, 1971).

Rees *et al.* (1978) produced SF_6 from Ag_2S by reaction with BrF_5, while Puchelt *et al.* (1971) preferred BrF_3 because it was less dangerous; although it is still very toxic, fumes in moist air and reacts violently with water and organic matter! Despite these problems, they maintained that elemental sulphur and many metallic sulphides react rapidly and completely with BrF_3 according to the reaction:

$$2FeS_2 + 10BrF_3 \rightarrow 4SF_6 + 2FeF_3 + 5Br_2.$$

The reaction products are sufficiently distinct chemically to allow the easy separation and purification of SF_6.

The apparent lack of any widespread acceptance of SF_6 in sulphur isotope work in the face of its obvious advantages may be related to the toxicity of BrF_5^-–BrF_3 and their high cost. Moreover, sulphates still require reduction to metal sulphide before reaction to SF_6 so that the direct reduction method to SO_2 is still more convenient.

OXYGEN ISOTOPES IN SULPHATES

There are two possible objectives: the analysis of $^{18}O/^{16}O$ in the sulphate or that associated with interstitial water of crystallization.

In the former case, the most widely used preparation technique is the graphite reduction method of Rafter (1967). Earlier work by Rafter (1957) had proved the general feasibility of producing CO_2 by reduction with carbon according to the reaction:

$$BaSO_4 + 2C \rightarrow BaS + 2CO_2$$

but at the temperatures involved (900–1000°C) a CO_2-graphite reduction reaction also occurred which converted some of the CO_2 to CO. He therefore developed a modified technique in which the first appearance of CO caused a pressure change in the vacuum line which induced a high voltage electrical discharge across two elements; converting CO to

CO_2 by the reaction:

$$2CO \rightarrow CO_2 + C.$$

According to Sakai (1977) this reaction proceeds to the right if the CO_2 formed is continuously condensed on to a glass wall cooled by liquid air. Rafter (1967) also showed that the intimate mixing and fine grinding of sample ($BaSO_4$) and graphite allowed the reaction temperature to be lowered, diminishing the formation of CO. Sakai & Krouse (1971) modified this technique slightly by using internally heated reaction cells with cooled quartz walls rather than the externally heated quartz tube of Rafter (1967). This eliminated memory effects caused by oxygen isotope exchange between the hot quartz walls and CO. Other examples of the use of graphite reduction include Longinelli & Craig (1967), Claypool *et al.* (1980) and Cortecci *et al.* (1981).

The measurement of $^{18}O/^{16}O$ associated with the water of crystallization of sulphates is a relatively simple affair (Sofer, 1978; Halas & Krouse, 1982). The basic technique involves heating the sample *in vacuo* and collecting any expelled water in a cold trap. The oxygen isotopic composition of this water is then determined using a standard CO_2 equilibration method (Epstein & Mayeda, 1953).

CARBONATES

Both $^{18}O/^{16}O$ and $^{13}C/^{12}C$ ratios can be determined in one operation by the evolution of CO_2 from carbonates. McCrea (1950) laid the basis for the technique now used. Measurements of the ratios of CO_2 with mass 44 ($^{12}C^{16}O_2$) 45 ($^{13}C^{16}O_2$) and 46 ($^{12}C^{18}O^{16}O$) allows the two ratios to be determined since ^{12}C and ^{16}O are by far the most abundant isotopes. Successful mass spectrometry of CO_2 thus required a quantitative and reproducible extraction procedure, lacking impurities in the mass range 44–46, as well as a lack of opportunity for oxygen isotope exchange. McCrea (1950) found that thermal decomposition did not give reproducible results, and experimented with various acids. The use of 100% phosphoric acid (H_3PO_4) at 25°C was found to be a reliable technique for $CaCO_3$. The reaction takes place *in vacuo* and the gas should be passed through a cold trap of dry ice and methanol (safer than dry ice/actone) to remove impurities (particularly H_2O) before CO_2 is frozen in a collection vessel immersed in liquid nitrogen. The sample size required will typically be 10 mg or less. Some laboratories routinely process milligram-sized samples (e.g. 0.3–0.5 mg at the University of Michigan, Given & Lohmann, 1986) if very small petrographic components (Given & Lohmann, 1986) or hand-picked microfossils (Shackleton, Hall & Boersma, 1984; Fig. 9.22) are to be analysed. When the volume of evolved CO_2 is very small it may be necessary to use 'cold finger' attachment which allows *all* the gas to be introduced to the machine. For pure calcium carbonate the time allowed to elapse before collection of the CO_2 will depend very much on the degree of automation incorporated within the preparation line-mass spectrometer system; that is, crudely, on the age of the assembly. Some of the older mass spectrometers are only semi-automatic and require 'setting-up' by the operator prior to analysing each sample and hence the number of gases that can be measured in a day is limited. In this case the sample and acid mixture is usually prepared, left overnight in a water bath at 25°C and the CO_2 collected and measured the following day. However, because the reaction itself will go to completion in a matter of tens of minutes, modern fully automated systems allow batches of samples to be prepared and measured more or less continuously. In this latter case some laboratories react their carbonates with H_3PO_4 at 50°C to hasten the evolution of CO_2 (e.g. Fig. 9.22).

Because of the relatively small quantities of phosphoric acid consumed annually, to our knowledge it is not available commercially in the desired purity. Individual laboratories therefore usually prepare their own according to their needs; adding phosphorous pentoxide (P_2O_5) to orthophosphoric acid (85–88%) to produce 100% H_3PO_4.

ISOTOPIC CORRECTIONS

No attempt will be made here to provide a thorough account of the various corrections that may be applied to 'raw' isotopic data, since a good number are concerned with machine errors (e.g. inlet valve leakage) or interference effects which are, strictly speaking, outside the scope of this discussion; but see Craig (1957), Deines (1970) and Blattner & Hulston (1978). Since the processing of raw data is usually done by computer, all necessary corrections may be incorporated within the program and applied automatically to yield the adjusted delta value. It is recommended that all correction factors should be quoted when publishing isotopic data together with an approximation of the *total* analytical error (i.e. the preparation and machine error combined).

Fig. 9.22. Reaction system used in the laboratory of N.J. Shackleton to generate CO_2 from foraminifera for isotopic analysis. The whole region illustrated is kept at 50°C. Orthophosphoric acid is dropped on to the foraminifera which have previously been cleaned and vacuum-roasted in the sample thimble within container A. The acid is replenished monthly and kept pumped so that it does not take up moisture (from Shackleton *et al.*, 1984, with permission of the Ocean Drilling Program, Texas A & M University).

One correction factor that needs particular judgement is that in respect of isotopic fractionation during phosphoric acid digestion. Although all the sample may be consumed, only about 66% of the total oxygen finds its way into the CO_2; the remainder reacting with H^+ ions to form water. This imparts a fractionation factor (α) of the form:

$$\alpha = \frac{(^{18}O/^{16}O)_{CO_2}}{(^{18}O/^{16}O)_{carbonate}}$$

which results in the uncorrected $\delta^{18}O$ value being approximately 10‰ too heavy. This fractionation mechanism is temperature-dependent, hence it is important always to perform the acid decomposition reaction at the same temperature (e.g. 25°C *or* 50°C). Sharma & Clayton (1965) provided an account of this particular correction, while Tarutani, Clayton & Mageda (1969) and Rubinson & Clayton (1969) outlined additional corrections for Mg calcites and aragonite. Rosenbaum & Sheppard (1986) provided the best compilation of data on Fe- and Mg-bearing carbonates at varying temperatures. Dolomite poses particular problems since there is no agreement on its fractionation factor with phosphoric acid. Most authors after 1965 followed Sharma & Clayton (1965) in applying a correction that approximates to −0.8‰ with respect to calcite. Land (1980) noted that corrections applied were varied, and often not stated. Both he and Rosenbaum & Sheppard (1986) found a substantially larger correction most likely in the light of their experimental evidence, although Land (1980) recommended treating all carbonates effectively as calcites in this respect until the matter is resolved. Clearly it is very important to state the phosphoric acid correction factor used in published reports.

PREPARATION OF CARBONATES OTHER THAN $CaCO_3$ AND MIXED CARBONATE SAMPLES

The standard preparation method of McCrea (1950) outlined above was designed specifically to cope with calcium carbonate, either as calcite or aragonite which react readily with phosphoric acid at room temperature. With other carbonate species, however, the reaction at 25°C is very much slower and can considerably lengthen the time to process each sample. Becker & Clayton (1972) prepared samples of calcite, dolomite, ankerite and siderite at 25°C and found that while the calcite reaction was complete in 24 hours, periods of one week, two weeks and two to three months respectively were required for the remainder. Not only are such extended periods of reaction inconvenient in routine work but there is always a danger that the reaction vessel may leak while not being actively pumped, leading to contamination with atmospheric CO_2. Fine grinding of the sample to increase the total surface area can be used to speed up the reaction but such treatment is time-consuming and may cause isotopic exchange with the atmosphere. Much more convenient is to increase the reaction temperature. Gould & Smith (1979), for example, working with siderite, initially mixed sample and acid at 80°C and then allowed the mixture to react at 25°C for 72 hours. Standard calcites treated in the same manner showed no change in $\delta^{13}C$ although $\delta^{18}O$ values were found to be too light by 1.2‰ and hence all oxygens were corrected by this amount (see also Gautier, 1982). Raising the reaction temperature is particularly useful when working with dolomites, which at 50°C react to completion overnight. Bearing in mind the temperature-related changes in the phosphoric acid correction factor, the standards should be run at the same temperature.

Mixtures of carbonates, particularly calcite and dolomite, often need to be analysed. If the composition of only one of the minerals is needed, it may be possible to remove it physically (see microsampling section), otherwise a chemical pre-treatment is required. Videtich (1981) ground the sample to <3 μm and selectively leached the calcite with EDTA (Glover, 1961) in order to analyse the dolomite, but warned that this method was not suitable for calcian dolomites which dissolve more readily in EDTA. Wada & Suzuki (1983), seeking to analyse calcite in dolomitic marbles, ground samples to <50 μm and used a heavy-liquid method to separate the calcite. Other techniques include that of Magaritz & Kafri (1981) who removed calcite by dissolution in 3% HCl and Land (1973) who used a dilute acetic acid digestion, both sets of authors using XRD techniques to prove the purity of the remaining dolomite. When it is necessary to know the isotopic composition of both minerals present, and physical separation is not possible, chemical methods are again necessary. The most favoured method of separation (although not without its critics) makes use of the variable reaction rates of calcite and dolomite with phosphoric acid; a technique originally pioneered by Epstein, Graf & Degens (1964) and Degens & Epstein (1964). In their experiments, mixtures of calcite and dolomite having first been ground to <50 μm were reacted with phosphoric acid at 25°C and the CO_2 produced in the first hour (assumed to be principally from the more reactive calcite phase) analysed on the mass spectrometer. The reaction was then allowed to proceed for a further three hours during which time the CO_2 formed was pumped away, while the remainder of the gas produced from the fourth hour up to a maximum of 72 hours was assumed to represent the dolomitic fraction. Hence, they believed that by using this double collection procedure it was possible to determine the isotopic composition of both minerals with minimal cross contamination; although they suggested it was unsuitable for samples in which the two phases differed greatly in isotopic composition or where the ratio of one mineral to the other was large. More recently Walters, Claypool & Choquette (1972) have pointed out that unless close bracketing of grain size is attempted by sieving, the large range of grain sizes will introduce errors into this technique because of the variable solubility in H_3PO_4. Despite these problems many workers have made use of the double collection procedure and a number of modifications have been made to meet specific needs. Clayton *et al.* (1968) altered their reaction times to cope with very fine-grained samples while Becker & Clayton (1972) used the method to separate ankerite and siderite.

REMOVAL OF ORGANIC MATTER

The presence of organic carbon in samples of carbonate leads to the introduction of volatile substances into the mass spectrometer which form a variety of reaction products in the ionization chamber and interfere with the weak mass 45 and 46 ion beams.

Although pre-treatments for the removal of organic matter are most relevant to Recent skeletal aragonite, which has a very high organic content, it repays to give some thought to the effect on ancient carbonates. Charef & Sheppard (1984) noted that sulphur-rich organic material is particularly troublesome.

By far the simplest technique for removing organic carbon involves soaking the sample in a solution of sodium hypochlorite ($NaClO_3$, clorox) or hydrogen peroxide (H_2O_2) to oxidize the organic material (e.g. Wefer & Berger, 1981; Cummings & McCarty, 1982). A more elaborate procedure was developed by Epstein *et al.* (1953) and involved 'roasting' the sample in a continuous stream of helium which swept away the decomposition products of the heated organic materials and preserved the sample in an inert atmosphere. Weber *et al.* (1976), Land, Long & Barnes (1977), Durazzi (1977) and Weil *et al.* (1981) are among those that have attempted experimental work to quantify the effectiveness of each of these techniques and any isotopic fractionation effects they may have. The general consensus is that neither the clorox/H_2O_2 treatments nor high temperature roasting are capable of removing *all* organic carbon and there is considerable evidence that the former will impart some error to the measured isotopic value; although provided all samples are prepared in the same manner any errors will be systematic. Because of the possibility of some organic carbon remaining even after thorough pretreatment, Weber *et al.* (1976) devised a rejection procedure based on a careful examination of the reference and sample chart recorder traces that allowed an excessive input from organic matter to be recognized. It should also be noted that roasting is not suited to aragonite samples since some conversion of the unstable phase to calcite is likely under the high temperature conditions.

MICROSAMPLING

In common with chemical analyses discussed elsewhere, the effectiveness of isotope studies on carbonates is often greatly enhanced by being able to resolve differences between individual rock components such as grain and cement types (Hudson, 1977a).

One of the most popular methods of microsampling carbonate rocks is to use a dental drill modified to accept very small drill bits which ideally need to be <0.5 mm in diameter. Prezbindowski (1980) outlined some of the requirements of such a sampling procedure and described a miniature vertical milling machine developed for such a role and capable of both vertical and horizontal (furrow) cutting movements.

A precise, but more time-consuming alternative is to use a scalpel or razor blade to cut out particular components of interest from a thin-section. Dickson & Coleman (1980) used multiple 40 μm thin sections cut from a single hand specimen which had been mounted in Lakeside 70 and stained. Samples of >5 mg were cut from each thin section using a scalpel and binocular microscope and they were able to prove that neither the mounting glue nor the stain affected the measured isotopic composition of the carbonate.

SILICATES

Techniques designed for crystalline rocks have been adapted for sedimentary silicates. $^{18}O/^{16}O$ and D/H ratios are the only important ones; see Douthitt (1982) for silicon isotopes.

OXYGEN IN SILICATES

Oxygen isotope ratio measurements are performed exclusively on CO_2 in preference to O_2 because it lies in a mass region relatively free of background interference. Moreover, CO_2 is less reactive than O_2 and thus less likely to be involved in chemical reactions with substances within the preparation line-mass spectrometer assembly which could introduce an isotope fractionation effect. Hence, the method of sample preparation is usually a two-step process involving the extraction of O_2 from the sample followed by its quantitative conversion to CO_2 for measurement on the mass spectrometer.

The initial liberation of O_2 from the silicate structure requires oxidation with a suitable reagent, the two most commonly in use being F_2 (Taylor & Epstein, 1962) and BrF_5 (Clayton & Mayeda, 1963); which have largely superseded the high temperature carbon reduction method of Baertschi & Schwander (1952), which is prone to contamination.

Bromine pentafluoride (BrF_5) at room temperature is highly corrosive to glass although it may be handled safely when cooled by liquid nitrogen; while at typical reaction temperatures (400–600°C) nickel is the only suitable material for handling the substance. According to Clayton & Mayeda (1963) a five-fold excess of BrF_5 over stoichiometric requirements was desirable, sample size ranging from 5 to 30 mg which yielded 100–400 μmol of O_2. On completion of the reaction, O_2 was collected via a liquid nitrogen cold-trap (Fig. 9.23) which removed unwanted product gases (e.g. BrF_3, Br_2, SiF_4 and HF). The extraction of O_2 using F_2 follows similar lines to the BrF_5 technique with most published accounts quoting Taylor & Epstein (1962) as the definitive treatment. A minor modification to this method was suggested by Savin & Epstein (1970) who used hot mercury in addition to a KBr trap to remove excess F_2 and other contaminants from the oxygen.

Once pure O_2 has been isolated from other product gases the next step is to convert the O_2 to CO_2 quantitatively, and in both the F_2 and BrF_5 examples cited above this was achieved by passing the gas over an electrically heated graphite rod. The CO_2 produced is then measured in the conventional way with standard corrections (Craig, 1957), while Taylor & Epstein (1962) pointed out that O_2 introduced with fluorine in their method required a correction of 0.1–0.2‰. Pisciotto (1981) provided an account of how to calculate the $\delta^{18}O$ value for opal CT from an opal CT/quartz mixture.

The only substantial modification to either of the above techniques to date involves the method of sample entry into the reaction vessel. While Clayton & Mayeda (1963) opened their reaction vessels in a P_2O_5 dry-box to prevent entry of water vapour, Friedman & Gleason (1973) and Labeyrie & Juillet (1982) suggested loading the sample directly under a positive pressure of dry nitrogen. This procedure is less complex and providing the reaction vessel is open for no more than one minute the amount of water vapour that enters is negligible and has no apparent effect on the measured $^{18}O/^{16}O$ ratio.

In recent years there has been some attempt to find a cheaper alternative to BrF_5, which has become prohibitively expensive for routine O_2 isotope work as supplies have dwindled. Borthwick & Harmon (1982) examined the suitability of a variety of potential oxidation reagents and concluded that ClF_3 had the most desirable characteristics, being easily flushed from the preparation line while its ability to freeze completely in liquid nitrogen prevented contamination of evolved O_2. They also provided a short description of their analytical procedure to highlight the small differences compared with conventional methods.

Sedimentary rocks often require pre-treatment to remove carbonates, organic carbon and oxides which provide a source of non-silicate oxygen. McMurty, Chong-Ho Wang & Yeh (1983) removed calcium carbonate with an acetic acid solution buffered with sodium acetate to pH 5. Organic matter can be oxidized with 30% hydrogen peroxide (H_2O_2) or sodium hypochlorite 'clorox' ($NaClO_3$) adjusted to pH 9.5 with 1N HCl. Iron and possibly manganese oxides can be removed using the sodium citrate-dithionite solution buffered with sodium bicarbonate developed by Mehra & Jackson (1960). In some cases it may also be necessary to analyse specific grain-size fractions. Le Roux, Clayton & Jackson (1980), for example, isolating the 1–10 μm portion of their samples by a combination of sedimentation and centrifuge techniques (see also Jackson, 1979). They also isolated quartz from the remainder of the sediment using a modified form of the sodium pyrosulphate fusion-hexafluorosilicic acid technique of Syers, Chapman & Jackson (1968) developed by Sridhar, Jackson & Clayton (1975) and Jackson, Sayin & Clayton (1976). Pisciotto (1981) used this technique to remove clays and feldspars from siliceous mudrocks.

Also of crucial importance is the removal of interlayer or adsorbed water from clays without affecting H/D or $^{18}O/^{16}O$ ratios of structural H and O. Savin & Epstein (1970) suggested that clays for $\delta^{18}O$ analysis should be kept in a P_2O_5 dry-box for at least 24 hours prior to oxidation, while Yeh & Savin (1977) replaced the atmosphere in their dry-box with dry

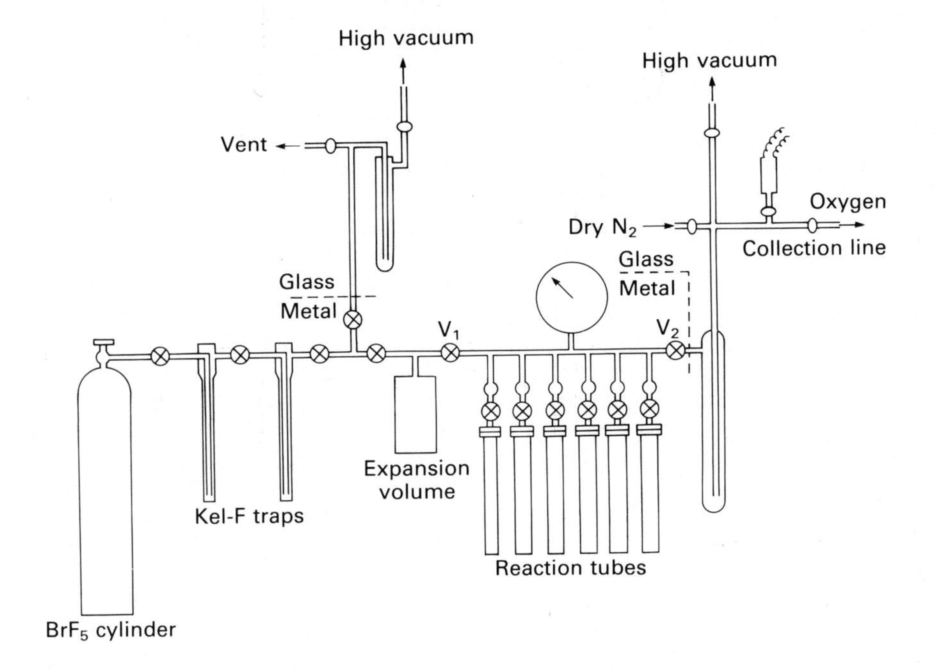

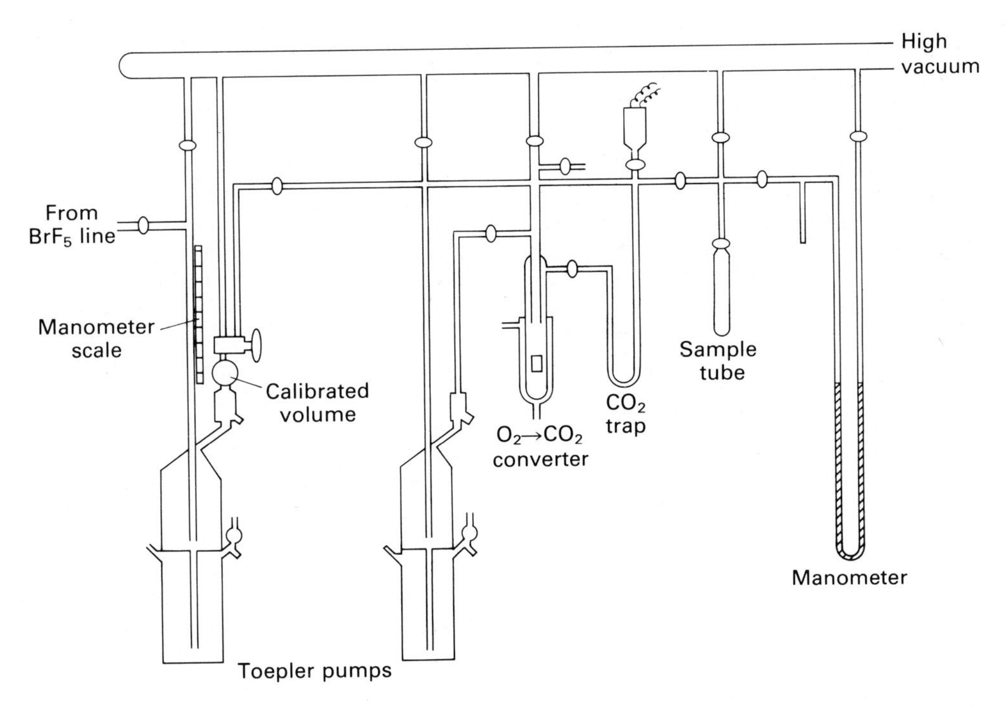

Fig. 9.23. Top: apparatus for reaction of oxygen compounds with bromine pentafluoride. Bottom: apparatus for collection of oxygen and conversion to carbon dioxide (from Clayton & Mayeda, 1963).

nitrogen two hours after the sample had been introduced. The small amount of water vapour remaining after 24 hours was not expected to effect the measured $\delta^{18}O$ value by more than a few tenths per mil. Particular care should be taken when preparing smectitic clays which have the ability to adsorb relatively large amounts of water and may require more thorough drying.

For H/D measurements preliminary out-gassing needs to be much more intensive. Savin & Epstein (1970) recommended heating the sample under vacuum at 100–250°C for 24 hours or more, while Yeh (1980) dried his samples at between 250 and 300°C for not less that three hours followed by storage in a nitrogen filled P_2O_5 dry-box. Both methods should be capable of removing virtually all adsorbed and inter-layer water.

D/H RATIO MEASUREMENTS ON SILICATES

The measurement is performed on hydrogen gas which is prepared by a fairly standard technique. The initial step after out-gassing is to extract hydrogen from the silicate structure as molecular water by heating the sample under vacuum at high temperature, anywhere between 900°C (Savin & Epstein, 1970) and 1500–1700°C (Godfrey, 1962). Since some hydrogen is also liberated in the molecular form this is converted to water by reaction with CuO. This water may then be quantitatively converted to H_2 by reaction with hot uranium (400–700°C) or hot zinc after which it is collected via a liquid nitrogen cold trap (Bigeleisen, Perlman & Prosser, 1952; Craig, 1961a, b; Godfrey, 1962; Coleman *et al.*, 1983). The evolved H_2 is measured isotopically against a suitable reference with a correction being made for the H^{3+} ion interference effect (Friedman, 1953). Because of the frequently large deviation from the standard, δD values may be quoted in per cent rather than parts per mil when necessary.

9.7 ANALYTICAL QUALITY

9.7.1 Accuracy and precision

When chemical concentrations have been obtained it is desirable to know both their internal consistency and their relation to the 'true' concentration.

If a single technique is used to analyse repeatedly a perfectly homogeneous geological material then a range of results will be obtained for the elemental concentration(s). The variation arises because of the small errors introduced at all stages of the preparation and measurement of the sample, and if sufficient replicates are plotted as a frequency diagram they will generally follow a Gaussian distribution. The results can thus be described by the mean ($\bar{x}$) and the standard deviation (s), the latter term being known as the repeatability (reproducibility is often used synonymously but it may also be used for the specific meaning of the variation between laboratories which have analysed that sample by the same method). Relative repeatability is more convenient than standard deviation with the terms *relative standard deviation* (rsd $= s/\bar{x}$), *coefficient of variation* ($C = s/\bar{x}.100\%$) and *precision* ($P = 2s/\bar{x}.100\%$) being used. Precision is often used loosely instead of repeatability.

Accuracy is the extent to which the mean approaches the 'true' concentration and bias the difference between the mean or median and the 'true' value. For geological materials the true concentration is unknowable and is approximated by a consensus 'usable value' (see Section 9.7.2).

The properties of the Normal distribution allow an analyst to predict the proportion of the total number of measurements which lie between the given ranges (for a Gaussian distribution 68.3% of observations lie in the range $x \pm s$). This allows criteria for the rejection of results which may be in error, the rejection limit usually being set at 95% confidence (strictly this is $\pm 1.96s$ but is usually approximated to $\pm 2s$). It should be stressed that 5% of valid results will be rejected and possibly some erroneous results accepted. The observations in error are usually termed 'fliers' and arise from unusual situations, for example the misreading of a balance or transient electrical 'noise'. They can usually be detected where there are a large number of replicates but when, for example, only triplicate measurements are available the rejection limits in terms of s become very large ($\bar{x} \pm 10s$) (Harvey, 1974). Where fliers are present the median may be a better estimator of central tendency than the mean. The repeatability can be illustrated by frequency diagrams or, more usefully, by control charts where the individual observations in order of analysis are plotted against the deviation from the mean.

The precision of a measurement varies as a hyperbolic function with the concentration reaching a level, the detection limit, below which a value cannot be detected. This theoretical limit is usually defined as being greater than two standard deviations of the measurements taken at zero concentration. It should be stressed that this is a theoretical limit with the practical limit of determination being 3–5 times the detection limit. It follows that when precisions are being quoted they should be given for the range of concentrations being studied.

9.7.2 Practical quality control

There are three broad areas of an analysis where it is desirable to have information regarding repeatability.

(a) The repeated analysis of a single *prepared* sample as in analyses of a solution by ICP or of a powder pellet by XRF. These, assuming the sample stays unaffected, give an estimate of the instrumental performance over the period of the test (short- or long-term drift). The overall coefficient of variation is a combination of errors introduced by individual components, for example a series of measurements by XRF could include generator and tube variation (1), sample changer reset (2) and goniometer reset (3), all contributing small errors to the total where:

$$C_{\text{total}} = \sqrt{C_1^2 + C_2^2 + C_3^2}.$$

This type of experiment is especially valuable where an instrument malfunction is suspected as individual functions can be tested and their contribution to the total error isolated.

The use of mini and microcomputers attached to automated analytical instruments allows such information to be readily collected and stored, and statistical evaluation software is usually provided by the manufacturer. Stable samples such as pressed powder pellets for XRF may be re-measured for several years, giving a very comprehensive measure of instrument performance, whereas solutions used, for example, in ICP and AAS have a relatively limited life. The sophistication of instrumental hardware and software varies from the ability to measure one or more references at fixed intervals to a program where the samples are selected within a random sequence. Computer control also allows automatic repetition of samples until a required precision of measurement is reached.

The above information is also desirable where less automated equipment is used although it is much more time consuming to produce. It should be stressed that only the instrument repeatability is being tested and that the precision figure represents the *best possible case*.

(b) The second and more important measurement is the repeatability of sample preparation which should be tested for each analytical method and sample type. Unfortunately this is time consuming and frequently ignored. Ideally it should be established by preparing a relatively large number (10 or more) of replicates from the same homogeneous bulk sample, either from the analysts collection or an 'in house' reference material. The factors concerning sample homogeneity discussed in Section 9.5.2 should be very carefully considered. It is likely that the bulk samples, and especially 'in house' references, will have a smaller grain size than routine samples and consequently will be more readily decomposed during sample preparation. This will tend to give an optimistic view of the repeatability of sample preparation. The value for routine samples can be obtained from duplicates included in the batches. The variation measured will of course be a combination of the total instrumental error and the sample preparation variation. It should be stressed that the physical and chemical characteristics of rocks and minerals can strongly influence the efficiency of sample preparation; hence each major sample type should be tested.

(c) Finally, it is rare for a large collection of samples to be prepared and measured at the same time. Normally they are split into batches and hence may be prepared with different reagents or analysed with different instrumental calibrations. It follows that some check must be made of the overall consistency of results during the period of the project.

First, the within batch variation should be tested, the precise method depending on the analytical technique used. Instrumental drift may be checked and corrected with monitor samples or standard solutions preferably introduced randomly. In addition, 'in house' reference samples should be prepared and analysed, again randomly, and preferably 'blind'. This last requirement is obviously more simple to arrange where analysis is performed by an external laboratory. An average of one reference per ten unknowns inserted at random in each batch is ideal with *at least* two reference samples with chemical compositions near the top and bottom of the range of the unknowns plus one blank. The choice of such reference samples is especially difficult where multi-element analytical methods are used as it is unlikely that the range of elemental concentrations will be covered in just two or three samples.

A further test of analytical reproducibility can be provided by analysing a number of unknown samples in duplicate, the duplicates again being spread randomly within the batch. Between batch variation can be tested by following the same procedure with the same reference samples.

In large-scale surveys, where the unknown samples' numbers are randomized to avoid systematic analytical errors being interpreted as geochemical trends, numbers are also allocated to duplicates and to reference samples (Plant *et al.*, 1975; Howarth, 1977).

The schemes described obviously increase effort and cost by adding to the number of samples but this is offset by the improved knowledge of the quality of results and by early indications of systematic bias. An absolute minimum is to analyse 'in house' reference samples whenever reagents or calibrations are changed.

Accuracy is finally judged by comparison with international reference materials which are discussed in Section 9.7.2. Because of their very high value they should not be used as replicates in the above schemes, particularly in analytical methods which destroy the sample. Instead, the 'in house' references should have been calibrated against international samples, and the latter on rare occasions treated as unknowns within a batch.

9.7.2 Standardization

All the analytical techniques previously described require calibration, that is comparison of their instrumental response with that of known concentrations of the element(s) of interest. Broadly three methods can be used for calibration: (1) Standardization against known weights of extremely pure elements or compounds. (2) Standard addition or spiking. (3) Comparison with geological reference materials.

The distinction should be clearly made between *standards* and *reference materials*. The former are usually simple compounds of known stoichiometry and chemical composition, whereas the latter are natural materials for which consensus concentration values are available. Despite the common practice of referring to them as 'geostandards' or 'international standards' they are not primary standards and strictly their true composition can never be known!

(1) This method is generally used where the sample is presented for analysis as a solution. Potts (1987, table 5.4) listed suitable standard materials for all elements in the periodic table. Such compounds must be available in a very pure form (generally 99.999% or better), be stoichiometric and generally be stable in air to allow for accurate weighing. Finally, they must be readily soluble in water or dilute acids, and these solutions should be stable over a reasonable period of time. Particular care is needed in drying the substances before weighing as too high a temperature may cause some to decompose partially while others may retain water or carbon dioxide even at high temperatures (e.g. La_2O_3 may retain some 20% CO_2 below 700°C). Such standard solutions are used for initial calibration but also to monitor drift of calibration lines. In techniques such as ICP multi-element standard solutions are needed and it is necessary to establish that individual elements do not precipitate or polymerize in the presence of others. Potts (1987) identified Nb, Ta, Mo and W as particularly difficult. Standard solutions are available commercially (usually at concentrations of 1000 $\mu g\ ml^{-1}$) and similar precautions should be taken if these are mixed.

Finally it is usually necessary to 'matrix-match' standards to unknowns in terms of major element levels and salt or acid content. This may cause problems if acids or fluxes are not sufficiently pure (Table 9.6) or if the matrix elements are not added as pure compounds. For example, in the analysis of Sr in carbonates the matrix Ca must be added as Specpure® or equivalent carbonate as Analar® material can contain in excess of 0.06% Sr. Such considerations are also important where the multi-element standards are used to determine interference effects (Table 9.7).

The same method can also be used to standardize the fused bead technique in XRF analysis. Here oxides are generally used and the possible loss of elements at high fusion temperatures must be considered (Oliver, 1979).

In general the set of standards with concentrations covering the whole range of the calibration should be individually prepared from the 'master' solution to avoid propagation of errors.

(2) Standard addition is used mainly in areas

where the matrix of the unknown must be exactly matched with that of the calibration standards, for example in flame photometry. The unknown solution is divided into a minimum of six aliquots, five of which are spiked with increasing amounts of the element to be analysed. A graph is then made of added concentration against instrument response, the negative concentration intercept giving the value of the unspiked sample. A similar method can be employed to calibrate powder samples in XRF and INAA. Here standards can be added for a range of elements, as powdered oxides, as aqueous solutions or as organo-metal solutions. Great care is needed to ensure homogeneity, usually involving further grinding and mixing of the samples. Although good (linear) calibration lines are generally produced they may not provide true concentrations because of differing analytical response between the element in a mineral and the same element in a simple laboratory chemical. This is particularly so for the analysis of lighter elements by XRF.

(3) Geological reference materials are unfortunately increasingly being used for standardization. This practise should be avoided wherever practicable and the reference materials used only as a means of comparing sets of samples analysed in different laboratories, or by different methods which have been calibrated by *primary* methods. One great danger in using *comparative* standardization is the propagation of errors in each new reference material whose 'consensus' concentrations may relate to a very few original 'standards' analysed by primary methods.

9.7.3 Geological reference materials

Over the past 25 years a large number of geologically related reference materials have been produced. The most comprehensive list of those available is that of Flanagan (1986) who detailed more than 830 samples for use as chemical, isotopic or microprobe references. Unfortunately the greater proportion of these are of igneous materials, and even where natural sedimentary samples are available the quality of information is generally inferior to that for igneous samples.

The most recent compilation of values is due to Govindaraju (1984) in a special issue of *Geostandards Newsletter* where 'working values' for a wide range of elements in 170 reference samples are listed. This journal is the major source of information on GRS with periodic updates of values for existing material plus the announcement of new samples.

Working values are generally obtained by distributing samples for analysis to interested laboratories and the large amounts of data independently returned are statistically treated to remove outliers before arriving, with a greater or lesser degree of subjectivity, at an accepted 'recommended', 'certified', 'best' or 'working' value. If the original results are studied there is usually a considerable spread, even after the removal of outliers, and the user of GRMs is advised to study these before making use of compiled 'recommended' values. It should be obvious that analytical methods whose calibration relies on these consensus values should not be used to provide values for new reference samples, or those values should be treated with caution. There are many examples in the literature where there has been extensive extrapolation of calibrations, and where many analysts use the same method and comparative calibration samples a consistent but biassed 'working value' may be obtained. A good example is the analysis of Zr where most values are obtained by XRF.

Most reference materials are distributed free of charge by their producers, usually national geological surveys or similar organizations. However, they are extremely costly to produce and analyse and so should be used sparingly. Indeed the supplies of a number of early reference samples were exhausted before reliable working values were reached.

The main purpose of a GRM is to allow users (and suppliers) of analytical data to judge its reliability, and hence the results of geochemical analyses should be accompanied by equivalent data on reference materials of similar composition and analysed by the same methods. This particularly allows the efficiency of sample preparation to be tested.

The aim should be to employ a reference only when an analytical technique is essentially proven and for each laboratory to provide its own 'in-house' references for day to day quality control and for method development. This is particularly important for all methods which destroy the original material. Users of GRMs and in particular those provided free of charge should be prepared to provide the issuing organization with reliable data on the samples within a reasonable time.

9.7.4 Reporting and documentation

Geochemical data from sedimentary rocks are reported in several forms, and there are many ways in which the data can be presented to illustrate trends, patterns and variations in distribution. In addition, data are frequently examined statistically to determine means, standard deviations, correlation coefficients and other parameters.

Analyses of major and some minor elements in *siliciclastic* rocks are usually quoted in percentages as the oxides. This is the convention adopted in hard-rock petrology, and it is useful since it does give an indication of the completeness of the analysis. Many of the less common metals and other elements present in only trace quantities are expressed as the elemental form in ppm or ppb. With *carbonates* it is usual to give the results in elemental form, as % or ppm, although for limestones the acid-soluble Mg content is often quoted as the mole % $MgCO_3$ present in the calcite lattice. With dolomites, the Ca and Mg values are often recalculated to give the mole % of each carbonate in 100% dolomite, so that the stoichiometry is obvious.

Where limestones in an analysed suite have a variable insoluble residue, then direct comparison of trace element (and major element) values between samples is difficult. To get round this problem, analyses of the acid-soluble fraction of limestones and dolomites are frequently recalculated to give insoluble residue-free values (i.e. 100% soluble carbonate).

Insoluble Residues (IR) are frequently determined for carbonate rocks, as they give a value for the purity of the rock. In addition, one can see if there is any positive correlation between IR and trace elements in the acid-soluble fraction (notably Fe^{2+}, Mg^{2+}, Al^{3+} and Mn^{2+}), which would indicate leaching from the insoluble residue. After acid digestion, the IR is removed by filtering and then it is weighed after ashing the filter paper at, say, 900°C for 2 hours. If organic carbon is present, then this would be oxidized and removed along with the paper on ashing.

The validity of the interpretation of geochemical data depends on the quality of those data, and it is therefore necessary for an author to provide the reader with sufficient information to make a judgement. It follows that such information should also be fully documented in the laboratory before publication.

A brief description of collection and crushing procedures should be given together with sample pretreatment (for example the drying temperature before analysis). The method of analysis should be decribed, and if this is modified from a published technique any deviations or improvements should be indicated. Sample weights and the method of calibration used are especially important. The aim should be to allow the reader to repeat the method of analysis exactly. Reference materials quoted should have been analysed by the same method.

Steele (1978) has listed the essential information which an analyst should provide when reporting on values of reference materials; this list is an excellent model for the reporting of all analytical data. Where extensive use is made of unpublished primary data (for example through lack of journal space) the full set and analytical information should be available on request.

The precision and accuracy of analyses are often given in methods sections of papers and here it is normal to quote the reference materials used and a figure for accuracy and instrumental and analytical precision in relative per cent. Where no indiation of the uncertainty is given then, in terms of the number of significant figures for an element in an analysis, it is generally held that the last figure is correct to ±2 or 3; that is, a figure of 35.2% means the actual value is probably between 35.5 and 34.9%; with 35.24, it would lie between 35.21 and 35.27.

Geochemical data are frequently displayed as scatter plots of one element against another. With limestones, Na v. Sr, Na or Sr v. Mn, Mn v. Fe, $\delta^{13}C$ v. $\delta^{18}O$ and Na or Sr v. $\delta^{18}O$ are frequently shown (see e.g. Al-Aasm & Veizer, 1982; Tucker, 1986). In some instances two elements are added together and plotted against a third; e.g. Fe + Mn v. Sr. Fe and Mn are normally low in marine carbonate precipitates (Sr is high) and they are picked up during diagenesis (Sr is lost), so that Fe + Mn v. Sr will reflect the degree of diagenetic alteration. Ratios of elements are also used, e.g. 1000 Sr/Ca, which allows for variations in Ca content between samples. Brand & Veizer (1980) used a plot of 1000 Sr/Ca v. Mn to show the amount of alteration of fossils during diagenesis.

In limestones, as most rocks, the trace element contents are log-normally distributed. To clarify chemical trends in carbonates, data are frequently plotted on log scales.

Geochemical data can be treated at length statistically, especially now that data manipulation can be undertaken by computer. Necessary values, such as mean, standard deviation, precision and accuracy are easily calculated. It is common to go further than this and undertake tests of significance, analysis of variance and calculations of correlation and regression. Correlation coefficients are useful and allow one to find meaningful relationships between the different parameters. Significant correlations can help in the interpretation of the data. For example, with dolomites a positive correlation between $\delta^{13}C$ and $\delta^{18}O$ or Sr and $\delta^{13}O$ can support a mixing-zone origin for the dolomitization. As noted earlier, negative correlations of Sr and Mn, or Na and Mn, reflect degrees of diagenetic change. Mn and Fe normally co-vary, as do Na and Sr, and Ca and Mg.

More elaborate statistical techniques can be used and include factor analysis, to identify the reasons for variations in the measured elements and discriminant analysis, to distinguish between different suites of samples. Factor analysis was used by Brand & Veizer (1980) and Al-Aasm & Veizer (1982) to account for the variations of trace elements in various components of Silurian and Mississippian limestones, and in rudist bivalves. Walters *et al.* (1987) used discriminant analysis to separate Jurassic–Cretaceous shales of marine and non-marine origin. Details of these and other methods of statistical analysis are contained in the many textbooks on this subject.

9.8 EXAMPLES

In this section, a few of the applications of chemical analysis of sedimentary rocks are presented, roughly in the order of the objectives A to E outlined in Section 9.1.

9.8.1 Provenance and weathering

These two topics (objective A and part of objective B) are treated together because of the difficulty in establishing that weathering effects were minimal in provenance determinations. Chemical analysis has much ground to make up on purely mineralogical studies in these areas. For example, the precise interpretation of provenance from heavy mineral studies has been long established, as are interpretations of the tectonic significance of the source and depositional areas deduced from the modal percentages of the main minerals; the climatic significance of the clay mineralogy of detrital sediments is also well known. In many cases, therefore, the direct route to the objective is by purely mineralogical rather than chemical study, but some notable exceptions are given below.

Blatt, Middleton & Murray (1980) reviewed examples where knowledge of mineral chemistry, particularly of heavy minerals, can be used to reconstruct provenance. Garnet is a notable example because of its chemical variability in the source area and chemical stability (Morton, 1985). Among major constituents, feldspar has particular promise. Trevena & Nash (1981) argued that microprobe analysis of detrital feldspars should be more widely used in provenance determinations, and they spelt out the significance of feldspar chemistry in terms of source rocks. In contrast to whole-rock work, only fresh, unweathered grains are chosen for analysis. Although weathering is not a problem in this approach (providing some fresh grains are present), diagenetic alteration must be carefully evaluated.

If one considers the whole-rock analytical approach, only in glacially-transported sediments can it be realistically supposed that minimal chemical weathering has occurred. Unless inert chemical components are studied, such analyses are likely to reveal more about chemical weathering processes than source rock chemistry. The study of Nesbitt, Markovics & Price (1980) on granodiorite weathering in an 'average' climate illustrates the varying susceptibility of cations to the combination of leaching and adsorption processes operating here. Their results are consistent with adsorption theory and the leachate has very similar element ratios to world-average river water. Applying this approach to the Early Proterozoic Huronian Supergroup, Nesbitt & Young (1982) reasoned that mudrocks would give a good index of possible changes in intensity of chemical weathering with time. Using a chemical alteration index (Fig. 9.24) they were apparently able to demonstrate climatic changes between (i) a regime of active chemical weathering at the top and bottom of the studied section and (ii) the glacial Gowganda Formation in the central part of the section in which chemical weathering effects were small, particularly in the matrix of diamictites. This approach is worthy of further development, which will be helped by theoretical and experimental studies (e.g. Nesbitt & Young, 1984).

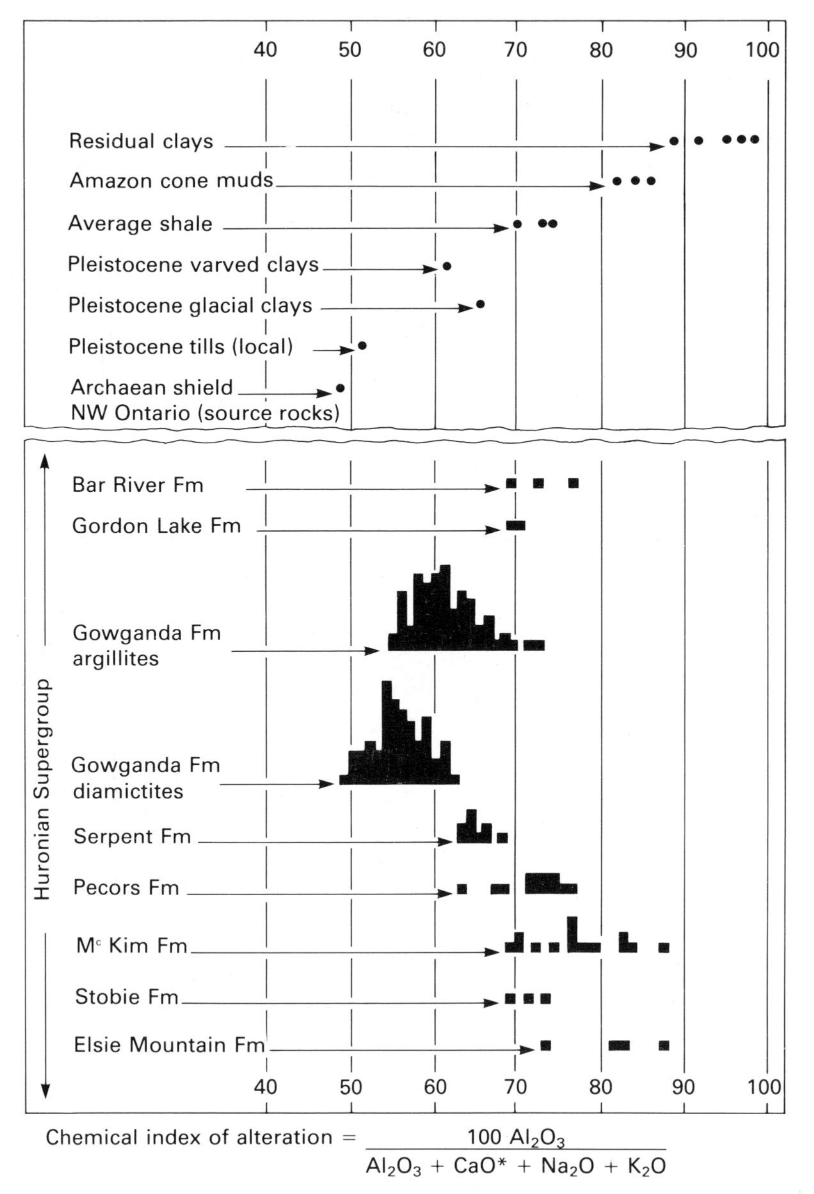

Fig. 9.24. Plot of 'chemical index of alteration' of sedimentary rocks from the Huronian Supergroup compared with Quaternary clays and the Archaean source rocks for the Huronian Supergroup. Note that CaO* refers to silicate Ca, i.e. the data are whole-rock data on a carbonate-free basis. For comparison, CIAs for basalts are 30–45, and for granites 45–55. High CIAs denote greater chemical weathering (after Nesbitt & Young, 1982).

The device of using inert chemical components to monitor provenance has been explored by various authors. Strontium has been advocated as an evaluator of the geographic siting and type of terrigenous input by Parra, Puechmaille & Carrkesco (1981). However, their use of whole-rock analyses causes problems since there is abundant Sr sited in biogenic carbonates in their studied sediments. They estimated the Sr content of the non-carbonate fraction as the intercept on plots of Sr versus $CaCO_3$ content. It would have been more straightforward to have analysed carbonate-free insoluble residues.

Rare-earth elements (REE) and certain other trace elements such as Th are regarded as the most reliable elemental inert components in the sedimentary cycle (Piper, 1974; Bhatia & Taylor, 1981; McLennan, 1982; Pacey, 1984). Relatively little work has been done, however, on the effects of weathering processes on these elements. It will not usually be possible to tackle the problem in the elegant fashion of Tieh, Ledger & Rowe (1980), who used fission-track images to demonstrate that U was lost only from *intergranular* sites during weathering of granites. Nesbitt (1979) showed in a study of granodiorite weathering that REE elements were leached in acid, upper portions of soils and precipitated or adsorbed, in accordance with theoretical predictions, as pH increased to neutral down the soil profile. He argued that Ti will be the most inert element here. In a similar study, but on a more reactive volcanogenic sandstone, Duddy (1980) came to similar conclusions, but by using microprobe analyses was able to show that vermiculite was a major repository of rare earths, in quantities sufficiently large (up to 10%!) to indicate siting within the lattice rather than by adsorption. Thus little of the REE find their way into solution, and are there in similar proportions to the host rocks anyway (Piper, 1974). Taylor, McLennan & McCulloch (1983), in testing loess as a sampling agent for the composition of the upper crust, found REE patterns (but not other elemental abundances) very constant. This testifies to the inert nature of REE, but also illustrates their imprecision as source indicators. Basu, Blauchard & Brannon (1982) found that weathering climate was irrelevant to REE patterns (but not absolute abundances), but indicated that some variation with rock type may occur. Nevertheless, REE may be more useful in assessing the terrigenous contribution to a sediment rather than specifying provenance. The study of Tlig & Steinberg (1982) is instructive in illustrating that this will not be possible in sediments with complex mixtures of chemical components. In their sediments, a bland whole-rock REE pattern, similar to that of typical shales, is seen to consist of a mixture of two patterns: one with a negative Ce-anomaly in the coarser, biogenic size fractions (similar to the REE pattern of sea water) and a pattern with a positive Ce-anomaly, associated with authigenic smectite in the finer size fractions (Fig. 9.25).

The possibilities of using geochemical parameters to estimate the size of the terrigenous component in sediments has been exploited to good effect by

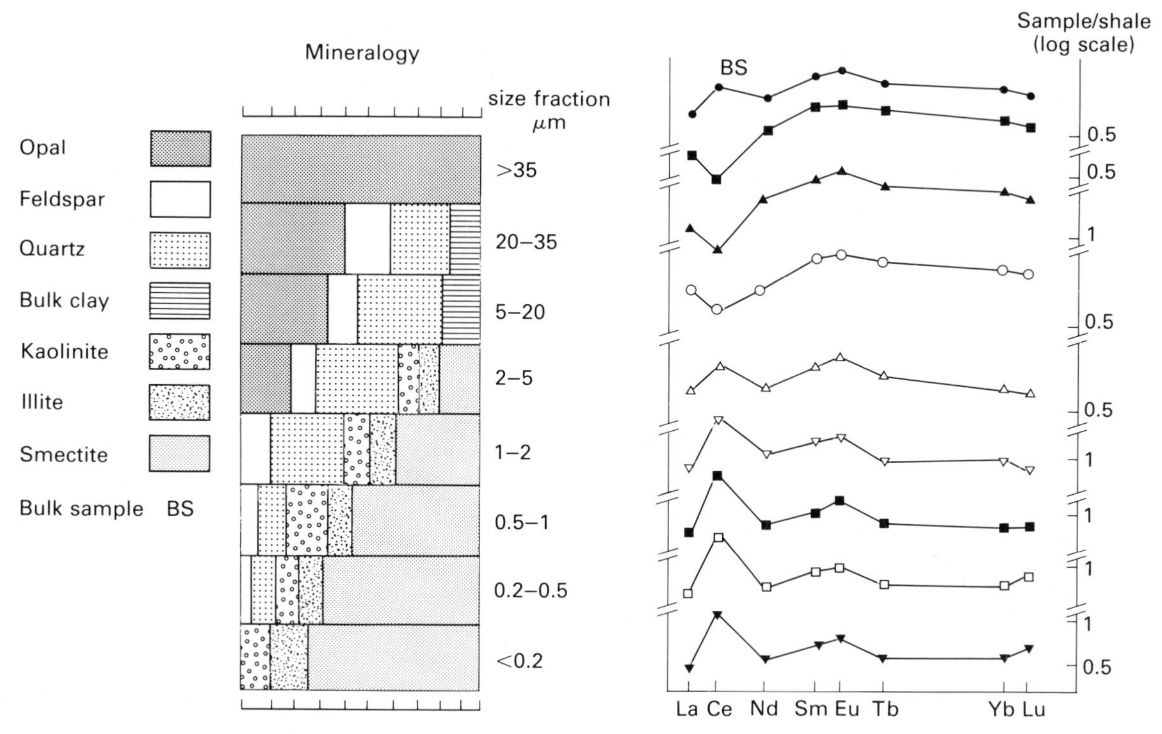

Fig. 9.25. Rare earth element patterns and mineralogical variation for different size fractions of a Recent sediment from a site in the Indian Ocean. BS = bulk sample (after Tlig & Steinberg, 1982).

Sugisaki (1984) who used ratios of Mn, Co and Ni to Ti as an expression of the relative abundance of authigenic oxide precipitates to terrigenous sediment. These ratios correlate well with sedimentation rate which in turn correlates with distance from source of terrigenous detritus. This allows distinction of pelagic from arc sediments in ophiolite suites.

Hickman & Wright (1983), in an empirical study, have shown that whole-rock geochemical data can allow stratigraphic correlations to be made in deformed terrains for slates, carbonates, and even some quartzites. When combined with selective analysis and petrographic data, this could prove to be a powerful tool.

9.8.2 Environmental parameters

These are the objectives of strategy C and sometimes strategy B (Fig. 9.1). The range of material studied is very considerable but only two topics are covered here, both classics of sedimentary geochemistry: boron palaeosalinity studies on clays, and geochemical studies on calcareous shells.

The following are useful reviews and recent notable papers on other applications: the geochemistry of metalliferous oceanic deposits and manganese nodules (e.g. Cronan & Moorby, 1981; Moorby, Cronon & Glasby, 1984; and a special issue of *Geochimica cosmochimica Acta*, May 1984); the geochemistry of phosphorites (e.g. McArthur *et al.*, 1986); isotopic methods for measuring sedimentation rates (e.g. *Special Issue of Chemical Geology*, Volume 44, parts 1/3, 1984); the geochemistry of marine-authigenic silicates (e.g. articles in Burns, 1979; Harder, 1980; Cole & Shaw, 1983; Berg-Madsen, 1983); the origin of brines responsible for evaporite deposits (Hardie, 1984).

BORON PALAEOSALINITY STUDIES

Environmental discrimination based on trace element analysis is founded on the assumption that there is some link between the measured concentration of an element in a sediment and its concentration in solution at the time of deposition. In other words, 'the chemical composition of detrital clay minerals ... is affected by the chemistry of their aqueous depositional environment' (Cody, 1971). This definition is equally applicable to sandstones, limestones and shales since it is in the clay fraction that elements of use in palaeosalinity work are encountered.

It follows, therefore, that for an element to be of use as an index of salinity it must meet certain criteria: (a) It must be widespread in common sediments and abundant enough to be detected and measured with a reasonable degree of precision. (b) Its concentration should depend as far as possible on salinity and not on any other factors. (c) Equilibrium should exist between its concentration in the sediment and the solution from which the sediment was deposited. (d) Diagenetic, metamorphic and weathering processes should have no effect on the concentration of the element.

Those elements which broadly satisfy these requirements include V, Cr, Ga, Ni, Rb, B and Li; although with the exception of boron most of these are treated with scepticism today. Degens, Williams & Keith (1957) and Potter, Shimp & Witters (1963) are among the few authors who have successfully applied such a bundle of elements to specific palaeoenvironmental problems.

More commonly it is to boron that most geologists have turned in the hope of arriving at credible salinity interpretations. However, even with boron relatively few attempts have been made to apply the technique to real problems directly. Many researchers have concentrated instead on validating the boron-salinity relationship using as guides sediments with good faunal-salinity data. This is in large part due to the wide range of factors that can affect its measured concentration (i.e. it does not strictly conform with (b) above). In this respect boron is a good example of how an apparently simple geochemical relationship becomes complicated when attempting to use it to solve geological problems (Walker, 1975).

The concentration of boron in natural waters is primarily linked to salinity via the ability of certain dissolved salts to promote the dissociation of boric acid (H_3BO_3). Hence in sea water the presence of Ca, Na, K and Mg salts increases the number of boron anion groups (e.g. $B(OH)_4^-$) being liberated into solution and thus available for adsorption by clays. Conversely the reduced concentration of these salts in river water inhibits the dissociation process and explains the low values for dissolved boron in such solutions (i.e. 0.013 ppm compared to 4.8 ppm). From this relationship it may be reasoned that marine clays will contain more boron than freshwater clays, and indeed this is what the earliest research found (e.g. Goldschmidt & Peters, 1932; Landergen, 1945; Frederickson & Reynolds, 1960).

However, in the intervening years it has become clear that this simple boron-salinity relationship is complicated by a range of other natural phenomena.

Not all clay minerals fix boron at an equal rate, most research showing that the element has a strong affinity for illite (Hingston, 1964; Fleet, 1965). Hence it is the concentration of boron in illite that is usually determined, while since grain size also affects the amount of boron adsorbed it is usual to confine all measurements to a single particle size or range (e.g. <2 μm). This also removes the possibility of introducing boron as tourmaline. To complicate matters, some research has shown that smectitic clays are capable of adsorbing as much or more boron than illite (Tourtelot, Schultz & Huffman, 1961; Lerman, 1966), while successful palaeosalinity interpretations have been made using kaolinite (Couch, 1971).

Poorly crystallized illites are capable of adsorbing more boron than well crystallized varieties (Porrenga, 1967), a phenomenon apparently reflecting the division of each particle into a 'structurally coherent silicate core' with little or no vacant sites, surrounded by an 'incoherent rind' or 'frayed edge' with abundant adsorbing sites (Jackson, 1963; Gaudette, Grim & Metzger, 1966; Fig. 9.26). Hence, as crystallinity improves, the silicate core occupies a larger proportion of the illite particle and less boron can be fixed. The reverse is true for poorly crystallized particles.

Even when apparently pure illite is isolated, compositional variations can affect the uptake of boron. C.T. Walker (1962) used the concentration of K_2O in illite to determine the number of muscovite-type layers present, deriving what he termed 'adjusted'

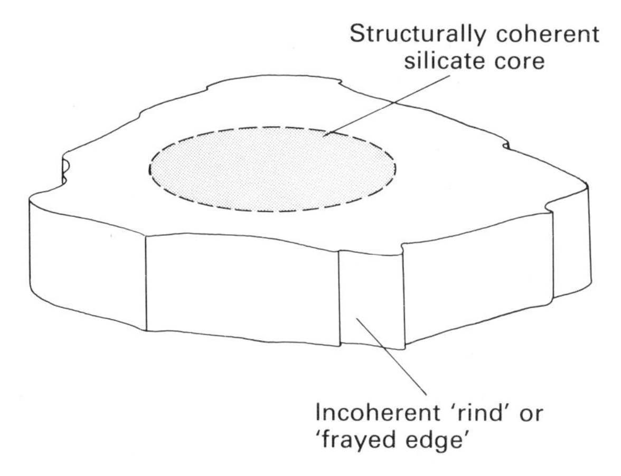

Fig. 9.26. The 'core-rind' model of boron incorporation in illite.

boron. This correction was refined by Walker & Price (1963) because 'adjusted' boron was found to be dependent on the potassium content. Their 'equivalent' boron they suggested was capable of resolving small salinity variations.

Spears (1965), while accepting the validity of 'adjusted' boron, rejected 'equivalent' boron because it was based on unique empirical data which could not be expected to be applicable to all sedimentary basins. Spears went further, however, and appeared to reject both corrections; believing that the boron/K_2O ratio was not solely influenced by the chemistry of the depositional solution but was greatly affected by weathering of source rocks, boron being more stable than potassium.

The concentration of organic matter in solution can also affect the fixation process, Bader (1962) showing that clays have the potential to adsorb 350% of their own mass of colloidal organic carbon from solution, hence forming a 'protective skin' (Landergren & Carvajal, 1969) around each clay particle, thus retarding or even stopping the adsorption of boron.

Purely detrital organic matter has a simple dilutant effect, as does $CaCO_3$ and SiO_2, such that the fluctuating concentration of these substances through a sedimentary sequence will produce changes in boron unrelated to salinity, unless corrected for.

Sedimentation rate is also important since the second step in the fixation mechanism after initial adsorption is the diffusion of boron into the clay's interior (Couch & Grim, 1968) which is slow at surface temperatures and pressures. Hence alternations of limestone (slowly deposited) and sandstone (rapidly deposited) may also produce spurious boron variations.

This tenaceous bonding of boron to illite via structural incorporation has profound consequences for the use of boron as a salinity tool because it introduces the problem of element re-cycling. It is this problem of 'inherited' boron which has dogged the technique and which tempered much of the early enthusiasm for the method.

Landergren & Carvajal (1969) defined inherited boron in terms of 'relic (detrital)' boron which did not reflect the existing set of environmental conditions but nevertheless belonged to the same depositional episode and 'recycled or redeposited' boron which had undergone one or more sedimentary cycles.

Experimental work by Fleet (1965) and Couch & Grim (1968) revealed that illites with an inherited boron concentration of up to 500 ppm were still capable of fixing fresh boron from solution, while Spears (1965) convincingly showed that much of the boron work on Coal Measure sediments from the north of England was invalid because of the strong source control. This problem of recycled boron clearly hinges on the acceptance that illites cannot be degraded by normal weathering processes and forced to lose their boron. However, Bohor & Gluskoter (1973), while accepting the great resilience of illite to chemical attack, nevertheless believed that surface processes were capable of achieving this aim and that as a consequence, 'our rivers are full of degraded illites'. Couch (1971) supported this view, pointing out that under moderate to humid climates appreciable leaching of boron occurs during soil formation.

Recent work by one of us (Quest, 1985) on the Purbeck Beds (Upper Jurassic–Lower Cretaceous) of Dorset appears to confirm that syn-sedimentary leaching of boron is possible. Illite in these beds is thought to have been derived from erosion of the underlying marine Kimmeridge Clay and thus should give the whole sequence a marine signature, while faunal studies (Clements, 1973) record a fluctuating salinity regime from hypersaline to near fresh water. From an examination of approximately 30 beds chosen on the basis of their fauna as belonging to one of four salinity groups (hypersaline, marine, brackish and freshwater), Quest was able to show that boron in illite was a realistic measure of palaeosalinity and that boron recycling was unimportant (Fig. 9.27). Leaching of boron from the marine Kimmeridge source rocks must have occurred prior to deposition. Smith & Briden (1976) suggested a palaeolatitude of about 36°N for the Dorset Purbeck and thus appreciable chemical weathering in the warm humid climate would be expected.

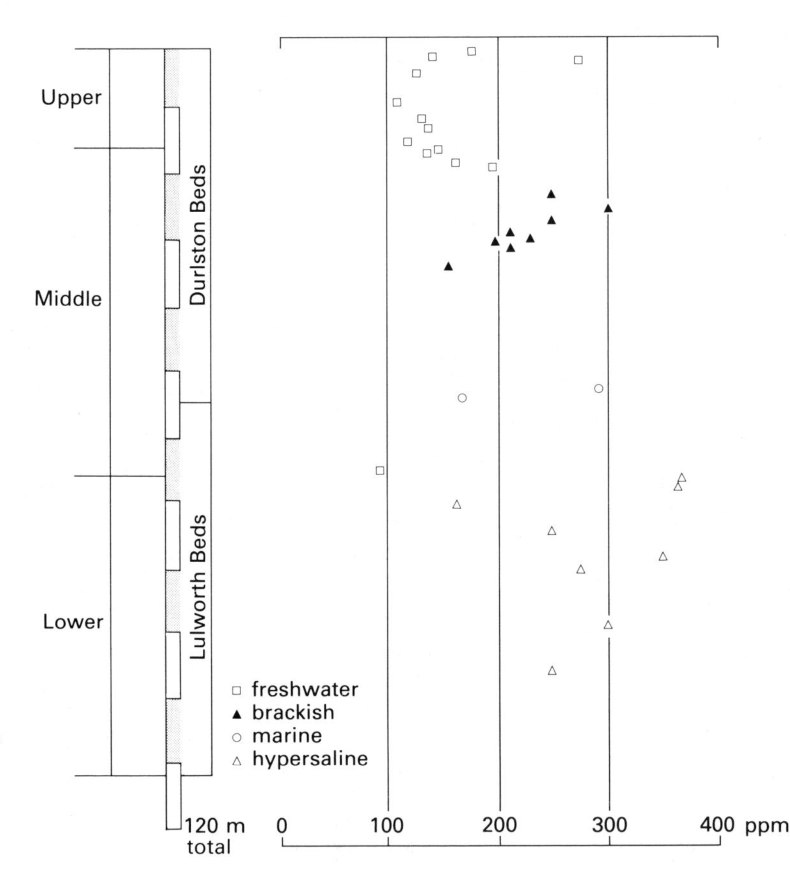

Fig. 9.27. Concentration of boron in <2 μm illite for the Purbeck at Durlston Bay (data of M. Quest). Faunal/salinity classification scheme after Clements (1973).

SKELETAL GEOCHEMISTRY

A good introduction to this topic was provided in the text by Dodd & Stanton (1981) who detailed the chemical properties of fossils as tools in palaeoenvironmental reconstruction, concentrating on skeletal mineralogy, trace chemistry and isotopic techniques. They suggested that the main factors controlling skeletal chemistry (physical-chemical, environmental, physiological and diagenetic) can be regarded as the four corners of a tetrahedron (Fig. 9.28), with the actual trace element or isotopic composition of the shell plotting as a point whose position is determined by the relative importance of each factor. Clearly shells with much diagenetic alteration need to be avoided (see Dodd & Stanton, 1981, and other works reviewed in Section 9.8.3). They believed that the study of biogeochemistry can help in the solution of four major types of geological problem: (a) general carbonate geochemistry, (b) carbonate diagenesis, (c) element recycling within the crust and (d) palaeoenvironmental interpretation. Numerous examples of such applications are provided including an examination of oxygen isotope palaeothermometry.

A particularly good example of the combined use of trace element and isotope geochemistry was provided by the same authors (Stanton & Dodd, 1970) in their study of Plio–Pleistocene invertebrates from the Kettleman Hills, central California. The purpose of this work, among other things, was to compare and contrast faunal and geochemical methods for reconstructing palaeoenvironments and, in particular, temperature and salinity. The sediments consisted of sandstones, siltstones and claystones with local conglomerates deposited in a marine embayment which became gradually isolated until lacustrine and fluvial environments were established. Hence relatively large-scale salinity fluctuations could be expected while there was evidence for a gradual cooling towards the early Pleistocene. Fossil invertebrates were abundant and well preserved with extant varieties for comparison.

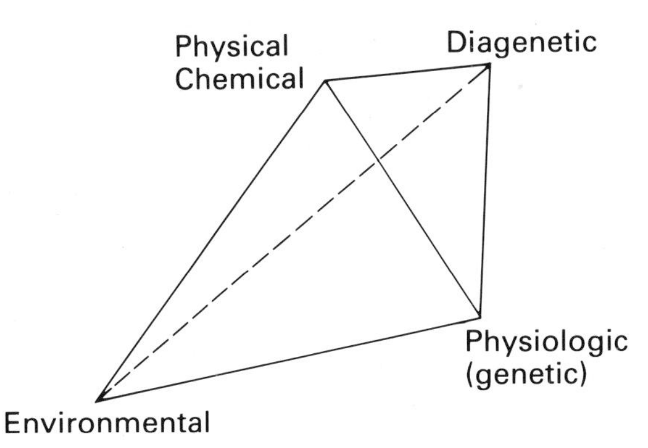

Fig. 9.28. Schematic representation of the factors controlling the chemistry of skeletal components (from Dodd & Stanton, 1981).

Palaeotemperature estimates were made using the concentration of Sr in the outer calcitic layer of *Mytilus* shells. All analyses were performed on the same species of bivalve to avoid phylogenetic effects. Estimating the Sr/Ca ratio in the depositional waters proved difficult because of the restricted nature of the palaeoenvironment and a variable freshwater input. Extensive diagenetic modification was, however, ruled out because ontogenetic Sr variations could be recognized.

The Sr palaeotemperature data they produced were found to be in broad agreement with faunal interpretations with no systematic variations. Unfortunately the authors failed to include an explanation of how the Sr concentrations were converted to palaeotemperatures, this information being provided by Dodd (1966). He found that the concentration of Sr expressed as mol % $SrCO_3$ in the calcite layer of modern *Mytilus* shells was related to temperature over the range 12–20°C by the equation:

$$T = 347\text{Sr} - 32.9.$$

When attempting to apply oxygen isotope data to an understanding of palaeosalinity, the authors had to concede that only crude trends in salinity could be deduced because the $^{18}O/^{16}O$ ratio for the Kettleman environment was not known due to the uncertain influence of inflowing fresh water. This also ruled out palaeotemperature work, although they suggested that if the Sr palaeotemperature and the oxygen isotopic composition of the shell were put into the palaeotemperature equation of Epstein *et al.* (1953), then it could be solved to give the $^{18}O/^{16}O$ ratio of the water in which the shell grew. Diagenetic alteration was once again found to be unimportant due to the lack of any significant lightening of the measured $\delta^{18}O$ values. Oxygen isotope palaeosalinities also corresponded well to those deduced faunally.

The authors concluded that the interpretative value of a combined faunal and geochemical study was very strong because the two methods complemented each other. Geochemistry allowed the quantification of temperature and salinity while the structure and composition of the fauna provided an estimate of the degree of isolation from the sea. In addition the agreement between geochemistry and palaeontology was thought to be a good indication of a lack of diagenetic alteration although independent petrographic verification of lack of alteration is preferable.

In a later paper, Dodd & Stanton (1975) were able to provide palaeosalinity contours for the Kettleman region in two stratigraphic zones. They assumed that during the deposition of each zone the temperature at which calcification occurred was more or less the same such that the $^{18}O/^{16}O$ ratio in the water (and hence in the shell) was a function of salinity alone. By analysing known marine and freshwater bivalves they were able to establish end-members for their salinity spectrum. Thus a shell with a $\delta^{18}O$ intermediate between these points would correspond to a given amount of mixing of marine and fresh waters. Using this technique they were able to identify the mouth of the embayment and the points of inflow of river water.

Most theoretical discussions of oxygen isotope palaeothermometry assume that the skeletal material being analysed is derived from fully mixed open marine conditions. As the above example shows, however, this may not be the case since many invertebrates inhabit marginal near-shore regions with an input of light ^{16}O that is difficult to quantify. Mook (1971) overcame this problem in his study of molluscs from four Dutch estuaries. He found that when the isotopic composition of the shell carbonate from these estuaries was plotted on a graph of $\delta^{18}O$ v. $\delta^{13}C$ they each described a straight line (Fig. 9.29). This relationship arose because the isotopic composition of the brackish water in which the organisms calcified was related linearly to the $\delta^{18}O$ and $\delta^{13}C$ content of the mixed marine and fresh waters and their dissolved bicarbonate. The fact that molluscs from each estuary plotted on separate lines was because the freshwater source in each case was isotopically distinct. Since all four estuaries contained some marine water of common isotopic composition, Mook found that the data for each converged on a point equivalent to the $\delta^{18}O$ and $\delta^{13}C$ content of marine carbonate. Since the $^{18}O/^{16}O$ ratio of sea water is known, a palaeotemperature could be calculated for the estuaries.

These studies are precise because of the younger age of the studied material. In older rocks, material which has suffered minimal alteration is harder to find. Cathodoluminescence can help to select unaltered portions of shells (Popp, Anderson & Saud-

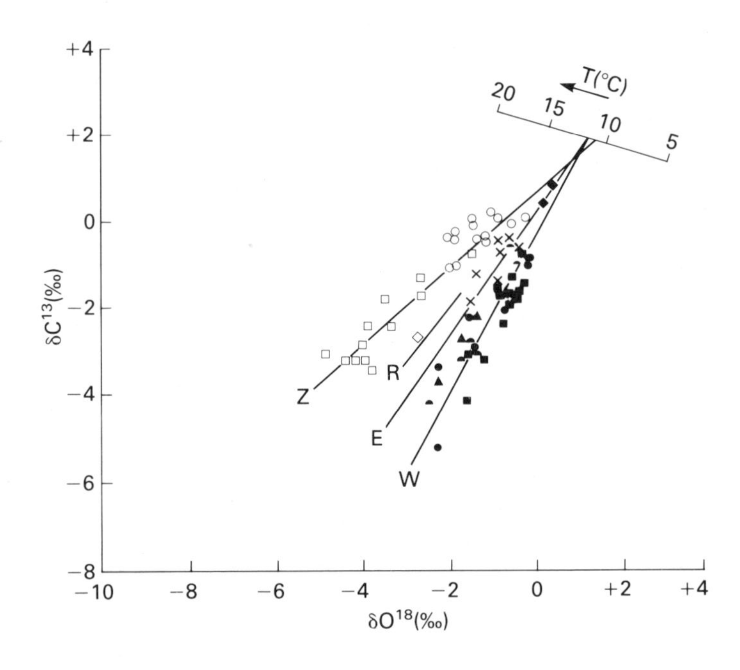

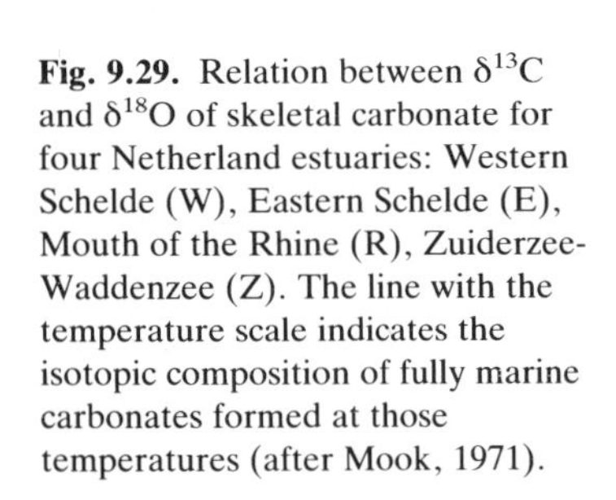
Fig. 9.29. Relation between $\delta^{13}C$ and $\delta^{18}O$ of skeletal carbonate for four Netherland estuaries: Western Schelde (W), Eastern Schelde (E), Mouth of the Rhine (R), Zuiderzee-Waddenzee (Z). The line with the temperature scale indicates the isotopic composition of fully marine carbonates formed at those temperatures (after Mook, 1971).

berg, 1986); this will be particularly useful when combined with analytical techniques (such as ICP) that use minimal sample. Alternatively, preserved aragonite can yield valuable information, even using the mobile element Na (Brand, 1986). Brand & Morrison (1987) provided a good review of fossil skeletal geochemistry.

9.8.3 **Diagenesis**

Chemical analysis has played a full part in the revolutionary advances in the understanding of diagenesis that have taken place in recent years, yet we are far short of the goal of predictive, rather than deductive, diagenesis. Most space will be devoted here to carbonate rocks because of the special impact of chemical rather than merely mineralogical analysis.

Whilst most carbonate workers have carried out integrated field-petrographic-geochemical studies on individual stratigraphic units, an alternative approach, spearheaded by Veizer and co-workers in numerous publications, seeks to characterize chemical diagenetic processes in a more general way by the empirical geochemical study of more widely-drawn samples, or by focussing on diagenetic alteration of particular depositional components. This alternative approach is reviewed first.

The study of Veizer (1974) on the diagenesis of certain Jurassic belemnites is illustrative and particularly apposite given the choice of a belemnite as the isotopic (PDB) standard, presumably in the hope that it was both homogeneous and unaltered in composition. Veizer (1974) found that although the Sr (Fig. 9.30a) and Mg contents of belemnite rostra were fairly constant, Mn (Fig. 9.30b) and Fe were much more variable. The chemical composition of the phragmacone septa (originally aragonitic) was nearly identical to the carbonate in the host rock. Veizer inferred that the rostra must have been partly recrystallized or had had internal pores cemented (or both) and deduced the degree of this diagenetic alteration from the approach of the Mn content to the host rock value (Fig. 9.30b). The secondary calcite must have contained much more Fe and Mn than the original rostra, but differed little in Sr and Mg. Those rostra with slightly more negative $\delta^{18}O$ values also have high Mn contents, invalidating previous conclusions of palaeotemperature variations in the Jurassic seas supposedly evidenced by $\delta^{18}O$ values. Some points from this study that have been found to be generally valid are (1) carbonate rocks are chemically heterogeneous with original mineralogy having an important control on chemical diagenetic changes, (2) diagenetic alteration may not be detectable by a single chemical parameter and (3) the direction of diagenetic change of minor elements can be predicted from partition coefficients (e.g. Sr decreases, Mn and Fe increase).

Empirical observations suggest that solid-state diffusion and consequent enhanced exchange of ions with pore fluid is insignificant in carbonate rocks even for oxygen isotopes (Hudson, 1977a; Brand, 1982). Neither does *repeated* recrystallization by dissolution-reprecipitation occur (Veizer, 1977). A contrary opinion regarding isotope data is sometimes expressed by isotope geochemists working on burial diagenesis (e.g. Longstaffe, 1983) and indeed little empirical evidence is available to check the experimental studies of Anderson (1969) which quantified the increasing amount of oxygen isotope exchange in calcite with increasing temperature. Carbon isotopes and minor elements would be less susceptible to exchange; Fairchild (1980b, 1985) showed that fine chemical zones in dolomite (but not calcite) survived regional metamorphism with temperatures attaining 450°C, and Tucker (unpublished data) has found oxygen isotope data from massive dolomites in the same geological setting to show a similar range as unmetamorphosed ones of similar age elsewhere.

The evidence for different geochemical pathways for different original carbonate minerals has been summarized by Veizer (1977, 1983). Brand & Veizer (1980, 1981) carried out a particularly sophisticated study of this type in which they recognize three different diagenetic trends on Sr–Mn plots, corresponding to original calcite, aragonite and Mg-calcite, allowing original mineralogy to be determined as long as diagenetic alteration is in the lesser half of the studied samples. Al-Aasm & Veizer (1982) and Brand (1981a, b, 1982) confirmed the small degree of recrystallization apparently normal for calcitic bioclasts and illustrated the preservation in special circumstances of arguably original chemistry of preserved metastable carbonates, although Brand (1983) recognized the importance of later cementation of pores within bioclasts. Veizer (e.g. 1977) and Pingitore (1976, 1978, 1982) have stressed the importance in neomorphism of a 'thin' or 'messenger' aqueous film, between dissolving and precipitating crystals, of different chemistry to the bulk pore fluid. Thus,

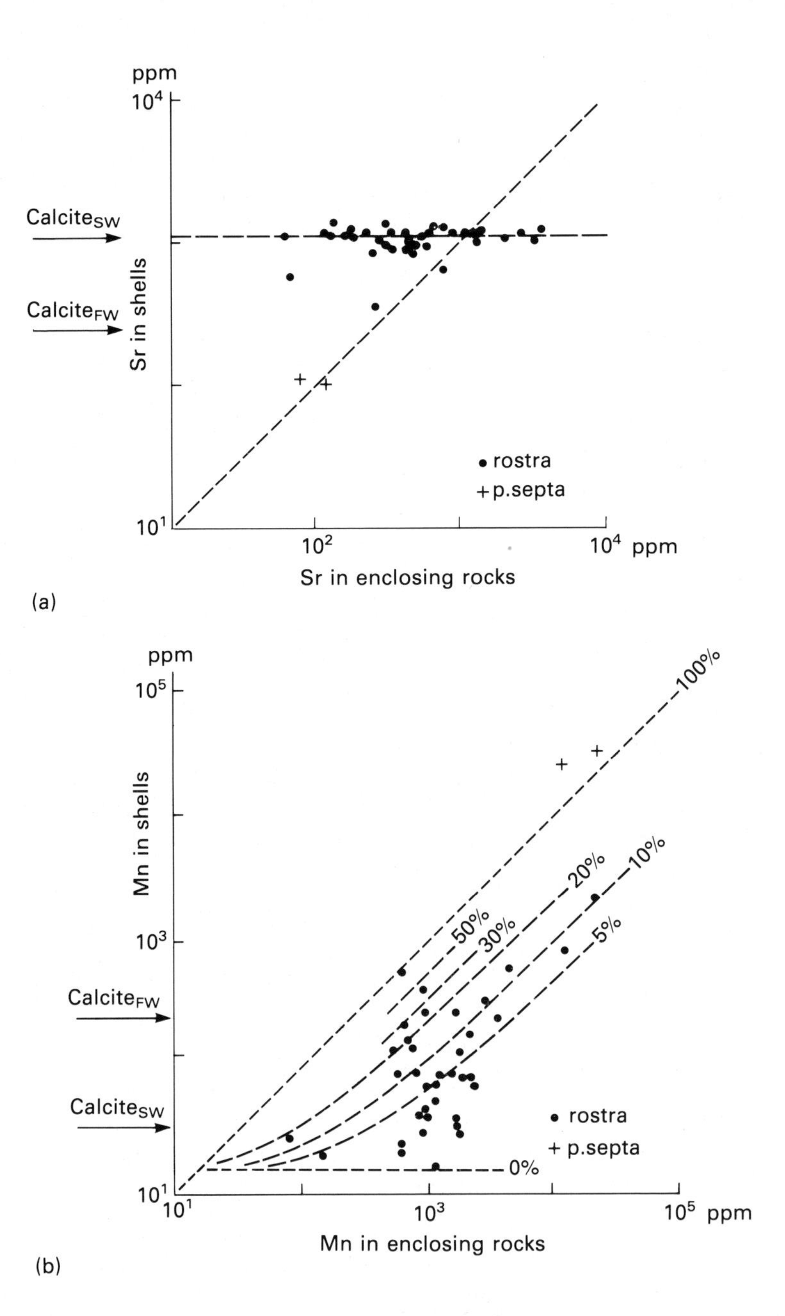

Fig. 9.30. (a) Sr and (b) Mn analyses of Jurassic belemnites from SW Germany (from Veizer, 1974). Calcite$_{SW}$ and Calcite$_{FW}$ indicate compositions in equilibrium with sea water and average fresh water respectively. In (b) the percentage figures indicate calculated degree of recrystallization from an original content of 15 ppm in rostra.

neomorphic or replacive carbonates can retain some memory of the chemistry of their precursor, the more so the more 'closed' the system. Veizer *et al.* (1978) expressed this concept in terms of the percentage of ions in the thin film derived from the dissolving solid phase; this illustrates the important point that the system will be more 'open' to some components (particularly oxygen isotopes) than others.

Veizer (1983) reviewed future developments in the approach to chemical diagenesis of carbonates. More physical separation of components, or availability or use of microanalytical techniques, together with use of conventional and luminescence petrography, will clearly help interpretations. When components are physically mixed, factor analysis of data (e.g. Brand & Veizer, 1980, 1981) can be employed, but is clearly a less direct method. There is also the danger of over-interpreting the results: of assuming that every factor represents a single physical phenomenon (such as salinity variations). This need not be so (see Gould, 1980, for a lucid explanation of this). Interpretations of Na data (e.g. Veizer *et al.*, 1977, 1978; Rao, 1981) need to be rethought in the light of the work of Busenberg & Plummer (1985). Finally the explicit assumption that stabilization only occurs in meteoric waters (Veizer 1977, 1983) needs evaluation. Possible stabilization in marine-derived fluids should be considered and modelled.

Chemical characteristics of marine cements in modern and ancient carbonates are not discussed here, but see Milliman (1974), Bathurst (1975) and James & Choquette (1983).

Following earlier controversy about the extent to which raw carbon isotope data can tell us about the prevalence of meteoric water diagenesis (Hudson, 1975; Allan & Matthews, 1977; Bathurst, 1980), Allan & Matthews (1982) have provided useful examples of the characteristics of isotope profiles across ancient exposure surfaces, interpreting their results by analogy with Quaternary sections on Barbados. Among other points made are the observations that the variability in oxygen isotope ratios is much less than that of carbon isotopes when diagenesis is by a single fresh groundwater system, that abrupt decreases in $\delta^{13}C$ are likely across exposure surfaces and that covariance of C and O is to be expected when diagenesis in the freshwater–marine water mixing zone occurs. Figure 9.31 (from Allan & Matthews, 1982) illustrates a Mississippian profile which can be interpreted in this way. It ought to be said, however, that since so little of the stratigraphic column has been studied by detailed isotopic profiles, chemical data alone are insufficient to be conclusive.

Elemental characters of meteoric carbonate cements are variable. Vadose cements will be iron-free; also Benson (1974) demonstrated, by microprobe analysis, a zonation with successive Mg and Sr peaks corresponding to the successive stabilization of Mg-calcite and of aragonite in the aquifer. Since meteoric lenses change in response to seasonal and longer-term climatic changes, phreatic cements should be expected to show complex zonation (Steinen, Matthews & Sealy, 1978), as is apparently

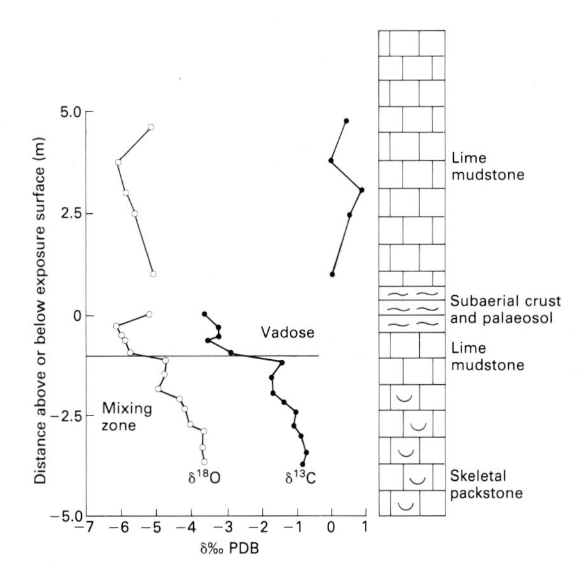

Fig. 9.31. Isotope profiles across a Mississippian emersion surface (Newman Limestone, Kentucky). Top five samples below exposure surface are depleted in ^{18}O and ^{13}C. Deeper samples show covariant increase in $\delta^{13}C$ and $\delta^{18}O$ thought to indicate diagenesis in marine-meteoric mixing zone. After Allan & Matthews (1982).

demonstrated in Mississippian limestones of New Mexico (Meyers, 1978). Complexly zoned cements and replacements have, however, also been argued to originate by burial diagenetic processes (Fairchild, 1980b; Wong & Oldershaw, 1981).

Many studies have demonstrated geochemical differences between depositional components, earlier and later cements, allowing assignment of the later cements to burial diagenesis. Dickson & Coleman (1980) provided a particularly convincing example on Dinantian limestones from the Isle of Man, England where a sequence of zoned calcite cements followed by dolomite and kaolinite cements display increasingly negative $\delta^{18}O$ values corresponding to progressively higher temperatures (coupled with an unknown degree of isotopic evolution of the pore fluid).

Chemical analyses of calcareous oozes have provided important evidence supporting the prevalence of pressure solution-reprecipitation as the process of lithification. Thus Scholle (1977) showed that $\delta^{18}O$ values of Cretaceous Chalks systematically lighten in more deeply buried, less porous Chalks since the reprecipitated calcite crystallized at elevated temperatures. Chalks which were cemented on the seafloor (hardgrounds) show a weaker, similar trend with higher $\delta^{18}O$ values as their residual porosity is diminished. Killingley (1983) proposed that burial recrystallization could account for the apparent climatic cooling trend deduced from isotopic study of Tertiary oozes, but this suggestion has not yet been rigorously assessed. Minor element changes, notably loss of Sr and gain of Mg (Fig. 9.32), accompanying the burial diagenesis of oozes have been well documented by Baker *et al.* (1982) who studied pore water and sediment chemical data. Particularly significant is their demonstration that, when diffusion is allowed for, there can be a major export of Sr from sediments (particularly when sedimentation rates are slow, see Fig. 9.32); meteoric diagenesis need not be invoked for Sr-depleted limestones! Baker *et al.* (1982) further considered that their data constrain K_{Sr} for calcite to be around 0.04, consistent with the experimental work of Katz, Sass & Starinsky (1972) whereas K_{Mg} appeared to be 8.1×10^{-4}, 60 times smaller than experimentally determined (Katz, 1973). In fact this conclusion depends on the assumption of an open diagenetic system and it is also inconsistent with the data from site 305. Modelling (by IJF) indicates that the Sr and Mg data from site 305 is satisfied by a number of combinations of *K* values and percentage closure of the system with respect to the minor elements. The importance of explicitly considering the degree of closure of a system has often been stressed (Pingitore, 1978, 1982; Veizer *et al.*, 1978). Chalks with low porosity (20%) do not show depth-related $\delta^{18}O$ changes. This arises not from a diagenetic microenvironment isolated from bulk pore fluid ('closed' system referred to above), but as a cumulative result of dissolution-precipitation reactions: the pore fluid becomes so enriched in $\delta^{18}O$ that no further reduction in rock $\delta^{18}O$ occurs despite increased temperatures (Jorgensen, 1987).

Geochemical interpretation of dolomitization is particularly complicated by major uncertainties in partition coefficients as well as the question of the degree of isotopic and minor element interference from precursor carbonates (if any). Veizer *et al.* (1977, 1978) showed for some Palaeozoic examples how high Na and Sr concentrations appear to correlate with early dolomitization from more continental waters. Generally, absolute amounts of Na and Sr are liable to vary particularly given the complexly variable chemistries of potential dolomitizing

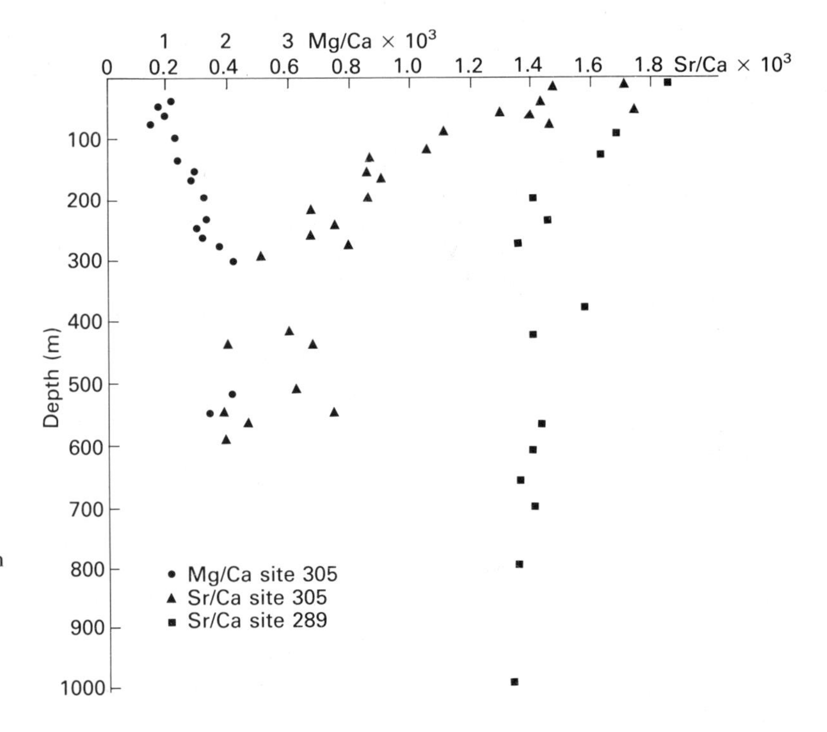

Fig. 9.32. Variations in composition with depth of pelagic carbonates from two DSDP drill sites. Data replotted from Baker *et al.* (1982). Site 305 has the slower rate of sedimentation.

fluids. Land (1980) has critically reviewed these topics. Bein & Land (1983) stressed that a simplistic interpretation of one geochemical parameter (e.g. Sr in terms of salinity) is certainly to be avoided (see also Sass & Katz, 1982 and Zhao & Fairchild, 1987).

Whereas in pure carbonate sediments the effects of non-carbonate components are insignificant, reactive silicates or organic matter, where present, have a major impact. Even small amounts of alteration of reactive volcanic material can be detected by $^{87}Sr/^{86}Sr$ analysis of recrystallized carbonate (Elderfield & Gieskes, 1982). More widely recognized now in carbonate minerals in terrigenous rocks are carbon isotopic signatures indicative of carbon derivation from particular bacterial reactions (e.g. Irwin *et al.*, 1977; Pisciotto & Mahoney, 1981; Marshall, 1981). Another development has been the recognition that carbonates of extremely variable chemical composition are very common (e.g. Curtis, Pearson & Somogyi, 1975; Matsumoto & Iijima, 1981; Tasse & Hesse, 1984). Matsumoto & Iijima's (1981) study is instructive in correlating the chemistry of authigenic carbonates with pore fluid chemistry as controlled by depositional setting, whilst Tasse & Hesse (1984) have utilized a particularly effective mode of data presentation (Fig. 9.33). Matsumoto & Matsuhisa (1985) not only describe the elemental variation of Neogene authigenic carbonates from the Japan trench, but elegantly place them in burial context by oxygen isotope analysis. Dolomite formation in sulphate-bearing pore waters is just one of the many interesting inferences that can be made from their data.

As for carbonates, the true chemical variability of clay minerals is now becoming apparent, aided particularly by microbeam analysis. Glauconitic clays are amenable to wet chemical analyses because pellets can be physically separated: Odin & Matter (1981) summarized these data. Although they distinguish two separate families: high-Fe glauconitic minerals and low-Fe illites, Berg-Madsen (1983), on the basis of microprobe results, showed that intermediate-Fe, high-Al glauconites occur and proposed a cool-water origin for them. In contrast, Ireland *et al.* (1983), in presenting STEM analyses of clays, noted the association (also present in Berg-Madsen's clays) of high-Al glauconite with pyrite and proposed a diagenetic origin for the clay chemistry. Clearly, development of our understanding of the nature and origin of clay-mineral chemical varia-

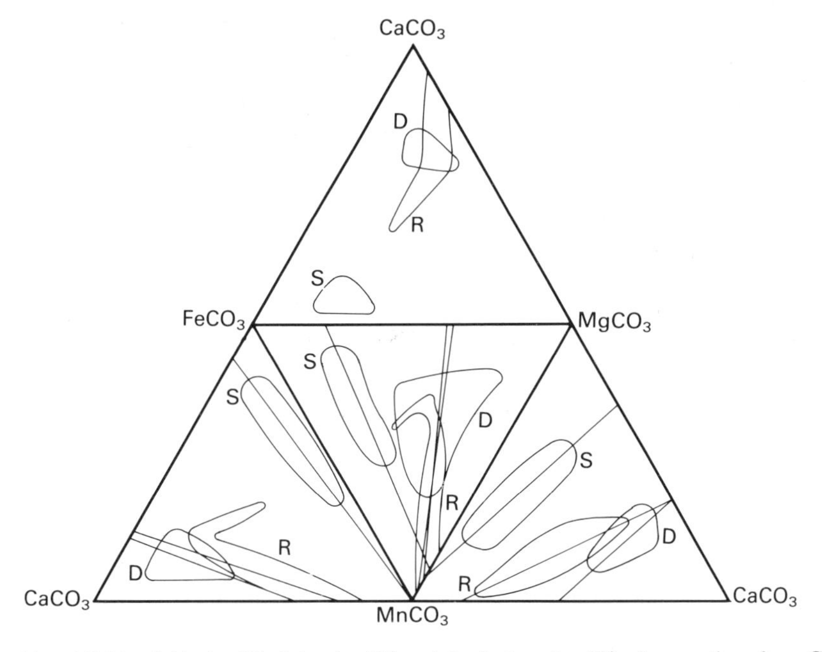

Fig. 9.33. Compositional fields of siderite (S), dolomite (D) and rhodochrosites (R) of concretions from Cretaceous black shales of the western Alps projected on to each face of the $CaCO_3$-$FeCO_3$-$MnCO_3$-$MgCO_3$ tetrahedron. Straight lines are regression lines calculated for the most elongated fields (simplified from Tasse & Hesse, 1984).

tions is most likely to be rapidly advanced by combined microchemical-microstructural studies.

Isotopic studies on silicates have been relatively few because of the complexity of the technique, but have yielded some extremely useful results (reviewed by Longstaffe, 1983). Particularly appealing is the prospect of the possibility of using co-existing minerals to tie down palaeotemperatures, irrespective of isotopic values of the pore fluid (Elsinger, Savin & Yeh, 1979).

9.8.4 Elemental cycling

The geological cycle involves transfer of sediments between several reservoirs which, on the simplest level, are the atmosphere, continental waters, sea water and sediments and rocks. Models of the kinetics of these processes (e.g. Garrels & MacKenzie, 1971; Veizer & Jansen, 1979; Berner, Lasaga & Garrels, 1983) depend on the existence of reliable chemical and mineralogical data from rocks of various ages. Possible secular changes in the composition of these reservoirs are of interest from many points of view.

Analyses of terrigenous material should provide evidence of the composition of the upper crust. Obtaining average estimates of the chemical composition of sedimentary sequences of various ages is, of course, replete with pitfalls. Schwab (1978) made an extensive compilation of 'terrigenous' compositions and correctly excluded the carbonate fraction, but he did not disregard other chemical sediments. Another criticism (McLennan, 1982) is that he placed too much emphasis on glacially transported, chemically unweathered sediments in early Proterozoic sections (Fig. 9.24). When this is allowed for it is clear that the Archaean–Proterozoic boundary marks the most significant change in the sedimentary record with changes in Na_2O/K_2O, Th, light REE/heavy REE (and emergence of a negative Eu anomaly in REE patterns normalized to chondrites) corresponding to a change from a more mafic to a more felsic upper crust (Veizer & Jansen, 1979; McLennan, Taylor & Eriksson, 1983; McLennan, 1982).

Considerable attention continues to be given to possible changes in oceanic chemistry with time. Being for most purposes and at most times well-mixed, sea water provides a sensitive indicator of changes in patterns of cycling between other reservoirs and itself. Perhaps the most important putative change was an increase in oxidation state accompanied by increased oxygen accumulation in the atmosphere, largely during the Proterozoic. It is very difficult to provide chemical data from marine sediments to constrain this. For example, it is clear that the Fe and Mn content of dolostones decreased during the Precambrian (whole rock data of Veizer, 1978; for microprobe data and a review of whole rock data for pure finely crystalline dolostones see Fairchild, 1980a), which could be explained by an increasing Eh of surface environments. This cannot be readily proven, however, because of the great variation in chemistry at any one time due to palaeogeographic considerations, and potential variability in the timing of formation of the dolomite. Less equivocal evidence comes from chemical analyses of palaeosols (and studies on the survivability of easily oxidized detrital minerals in placers) which constrain atmospheric CO_2/O_2 ratios (Holland, 1984).

The proposition that seawater chemistry can be treated in equilibrium terms (Sillen, 1967) has given way to the view that kinetic factors (fluxes in and out of sea water) predominate, although Holland (1972, 1984) showed that the mineralogy of preserved sediments constrained the possible past excursions in oceanic chemistry within half an order of magnitude limits of abundance of many individual elements, at least in the Phanerozoic. Establishing the major element composition of sea water at different times within these limits is extremely difficult. In some circumstances, minor elements are tractable. For example, Graham *et al.* (1982) demonstrated small, but significant variations in oceanic Sr/Ca by analysing well-preserved Tertiary planktonic foraminifera and utilizing their well-known partition coefficient for Sr.

Even if aspects of the composition of sea water were known, establishing the nature of the temporally-changing kinetic processes is a more important, yet more remote objective. For example, the model of Berner *et al.* (1983) of the control of the carbonate-silicate geochemical cycle for the last 100 million years illustrates the complexity of the controls on oceanic pH and abundance of Ca^{2+} and Mg^{2+}, and conversely the difficulty in finding geochemical parameters that give information about few (preferably only two) kinetic processes. There are certain parameters that meet this requirement, however: notably Sr/Ca, $^{34}S/^{32}S$, $^{13}C/^{12}C$ and $^{87}Sr/^{86}Sr$. For example, Graham *et al.*'s (1982) results cited above can be interpreted in terms of two processes: seafloor spreading rate (affecting Ca removed by hydrothermal action) and aragonite precipitation rate (removing material with high Sr/Ca).

Strontium isotope variation reflects the balance of input of Sr in solution with relatively high $^{87}Sr/^{86}Sr$ (derived from the upper continental crust) and input by interaction with basaltic oceanic crust (with low $^{87}Sr/^{86}Sr$). The composition of carbonates in equilibrium with sea water readily yields information on Sr isotopes in sea water since there is no fractionation during carbonate precipitation. By analysis of the carbonate fraction of carbonate-rich sediments, major data sets have been compiled by Veizer & Compston (1974, 1976), Burke *et al.* (1982) and Veizer *et al.* (1983). Samples with low Sr abundance, or which may have been contaminated by abundant ^{87}Sr in a high insoluble residue (Burke *et al.*, 1982), or low Ca/Sr or high Mn indicative of diagenetic alteration (Veizer *et al.*, 1983) are rejected. The curve of Burke *et al.* (1982) (Fig. 9.34) is the best available data set covering the whole of the Phanerozoic for any chemical parameter, because of the precision both of the analyses and of the stratigraphic age of the samples. Realization of the full potential of this approach requires more careful screening of samples, using both the valuable geochemical approach of Veizer *et al.* (1983), and more integration with petrography.

Carbon and sulphur isotope variations in carbonates and sulphate evaporites respectively have been well documented. Despite the cruder age-grouping, particularly of the carbon isotopic data, antipathetic variation is clearly shown (Veizer, Holser & Wilgus, 1980) illustrating a coupling of C and S cycles by an *overall* reaction such as:

$$15CH_2O + 8CaSO_4 + 2Fe_2O_3 + 7MgSiO_3 = 4FeS_2 + 8CaCO_3 + 7MgCO_3 + 7SiO_2 + 15H_2O.$$

The geological consequences are very interesting (Berner & Raiswell, 1983) as are the short-term violations of this relationship (Anderson & Arthur, 1983). Because of the heterogeneity of carbon isotope values in carbonate rocks at any one time, there is much scope for refinement of the sampling procedure, particularly for examining shorter-term variations. Magaritz *et al.* (1983) have demonstrated an extraordinarily rapid rise in $\delta^{13}C$ of 8.5‰ in only 4000 years in the Upper Permian of Texas that correlates with other rapid rises world-wide. Likewise, Claypool *et al.* (1980) showed that some $\delta^{34}S$ rises are very rapid and postulate that sudden de-

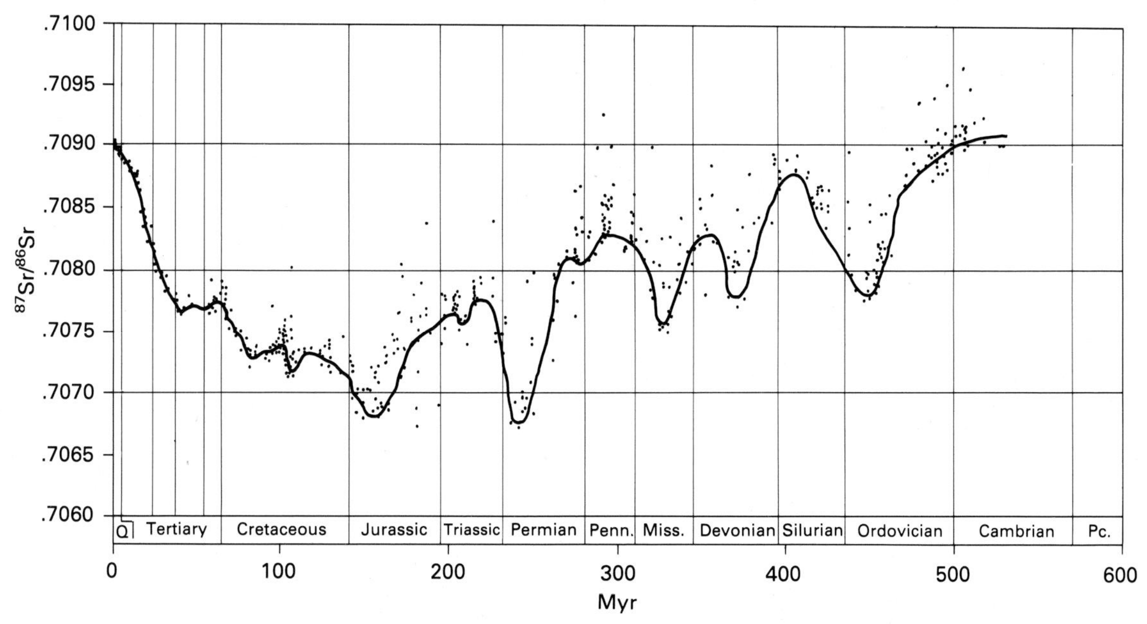

Fig. 9.34. Plot of Sr isotope variation with age of 744 samples of marine carbonates, evaporites and phosphorites (from Burke *et al.*, 1982). Data points scatter above the line because of diagenetic alteration and short-lived positive excursions (e.g. Cretaceous–Tertiary boundary. Hess, Bender & Schilling, 1986).

stratification of an ocean basin is the only mechanism that can explain them.

Secular variation in the oxygen isotope composition of carbonates, cherts and phosphorites takes the form of a progressive decrease in average $\delta^{18}O$ in older rocks (Veizer & Hoefs, 1976; Knauth & Lowe, 1978; Anderson & Arthur, 1983). The cause of this variation, whether increased total recrystallization with age, temperature change, or secular variation in seawater isotopic composition, is hotly disputed, although the strong buffering of sea water $\delta^{18}O$ by interaction with new ocean crust is increasingly evident (Holland, 1984). The careful analysis of little-altered portions of Palaeozoic brachiopod shells (Popp *et al.*, 1986; Veizer, Fritz & Jones, 1986) demonstrated that Palaeozoic sea water was not as isotopically depleted as previously thought. Similarly, Tucker (1986) and Fairchild & Spiro (1987) argued that the average pronounced isotopic depletion in late Proterozoic carbonates relates to mineralogical stabilization rather than changes in oxygen isotopic composition of sea water. The alternative approach of using chert-phosphate pairs to derive both palaeotemperatures and isotopic compositions of pore fluids (Karhu & Epstein, 1986) has yielded startlingly high palaeotemperatures in Precambrian samples. As so often with the chemical analysis of sedimentary rocks, such interpretations cannot be accepted until the mineralogical and petrological history of the analysed samples is well understood.

References

Acker, K.L. & Risk, M.J. (1985) Substrate destruction and sediment production by the boring sponge *Cliona caribbaea* on Grand Cayman Island. *J. sedim. Petrol.* **55**, 705–711.

Adams, A.E., MacKenzie, W.S. & Guildford, C. (1984) *Atlas of Sedimentary Rocks under the Microscope*. Longmans, London.

Adams, A.E. & Schofield, K. (1983) Recent submarine aragonite, magnesian calcite and hematite cements in a gravel from Islay, Scotland. *J. sedim. Petrol.* **53**, 417–421.

Al-Aasm, I.S. & Veizer, J. (1982) Chemical stabilization of low-Mg calcite: an example of brachiopods. *J. sedim. Petrol.* **52**, 1101–1109

Alexander, L.E. & Klug, H.P. (1948) Basic aspects of X-ray absorption in quantitative diffraction analysis of powder mixtures. *Anal. Chem.* **20**, 886–889

Alexander-Marrack, P.D., Friend, P.F. & Yeats, A.K. (1970) Mark sensing for recording and analysis of sedimentological data. In: *Data Processing in Biology and Geology* (Ed. by J. Cutbill), pp. 1–16. Systematics Association Special Vol. 3. Academic Press, London.

Alexandersson, E.T. (1974) Carbonate cementation in coralline algal nodules in the Skagerrak, North Sea: biochemical precipitation in undersaturated waters. *J. sedim. Petrol.* **44**, 7–26

Alexandersson, E.T. (1979) Marine maceration of skeletal carbonates in the Skagerrak, North Sea. *Sedimentology*, **26**, 845–852.

Al-Hashimi, W. & Hemingway, J.E. (1974) Recent dolomitization and the origin of the rusty crusts of Northumberland: A reply. *J. sedim. Petrol.* **44**, 271–274.

Ali, A.D & Turner, P. (1982) Authigenic K-feldspar in the Bromsgrove Sandstone Formation (Triassic) of Central England. *J. sedim. Petrol.* **52**, 187–197.

Allan, J. R. & Matthews, R.K. (1977) Carbon and oxygen isotopes as diagenetic and stratigraphic tools: surface and subsurface data, Barbados, West Indies. *Geology*, **5**, 16–20.

Allan, J.R. & Matthews, R.K. (1982) Isotope signatures associated with early meteoric diagenesis. *Sedimentology*, **29**, 791–817.

Allen, D. (1984) A one-stage precision polishing technique for geological specimens. *Miner. Mag.* **48**, 298–300.

Allen, G. P. (1971) Relationship between grain size distribution and current patterns in the Gironde estuary (France). *J. sedim. Petrol.* **41**, 74–88.

Allen, J.R.L. (1966) On bed forms and palaeocurrents. *Sedimentology*, **6**, 153–190.

Allen, J.R.L. (1967) Notes on some fundamentals of palaeocurrent analysis, with reference to preservation potential and sources of variance. *Sedimentology*, **9**, 75–88.

Allen, J.R.L. (1970) Studies in fluviatile sediment: a comparison of fining upwards cyclothems, with special reference to coarse member composition and interpretation. *J. sedim. Petrol.* **40**, 298–324.

Allen, J.R.L. (1971) Transverse erosional marks of mud and rock; their physical basis and geological significance. *Sediment. Geol.* **5**, 167–385.

Allen, J.R.L. (1982) *Sedimentary Structures: their Character and Physical Basics*. Vols 1 & 2. Elsevier, Amsterdam.

Allen, J.R.L. (1983) Studies in fluviatile sedimentation: bars, bar complexes and sandstone sheets (low sinuousity braided streams) in the Brownstones (L. Devonian), Welsh Borders. *Sediment. Geol.* **33**, 237–293.

Allen, P.A. (1981) Sediments and processes on a small stream-flow dominated, Devonian alluvial fan, Shetland Islands. *Sediment. Geol.* **29**, 31–66.

Allen, P.A. & Matter, A. (1982) Oligocene meandering stream sedimentation in the eastern Ebro Basin, Spain. *Eclog. geol. Helv.* **75**, 33–49.

Allen, P.A., Cabrera, L., Colombo, F. & Matter, A. (1983) Variations in fluvial style on the Eocene/Oligocene alluvial fan of the Scala Dei Group, SE Ebro Basin, Spain. *J. geol. Soc. London*, **140**, 133–146.

Allman, M. & Lawrence, D.F. (1972) *Geological Laboratory Techniques*. Arco, New York.

Amieux, P. (1982) La cathodoluminescence: methode d'étude sedimentologique des carbonates. *Bull. Centres Réch. Explor.-Elf-Aqu.* **6**, 437–483.

Anderson, T.F. (1969) Self diffusion of carbon and oxygen in calcite by isotope exchange with carbon dioxide. *J. geophys. Res.* **74**, 3918–3932.

Anderson, T.F. & Arthur, M.A. (1983) Stable isotopes of oxygen and carbon and their application to sedimentologic and palaeoenvironmental problems. In: *Stable Isotopes in Sedimentary Geology*, 1-1–1-151. Soc. econ. Palent. Miner. Short Course 10.

Angino, E.E. & Billings, G.K. (1972) *Atomic Absorption Spectrometry in Geology*. Elsevier, Amsterdam.

Angus, J.G., Raynor, J.B. & Robson, M. (1979) Reliability of experimental partition coefficients in carbonate systems: evidence for inhomogeneous distribution of impurity cations. *Chem. Geol.* **27**, 181–205.

Armstrong, A.K., Snavely, P.D. Jr & Addicott, W.O. (1980) Porosity evolution of Upper Miocene Reefs, Almeria Province Southern Spain. *Bull. Am. Ass. Petrol. Geol.* **64**, 188–208.

Assereto, R. & Folk, R.L. (1980) Diagenetic fabrics of aragonite, calcite and dolomite in ancient peritidal-spelean environment: Triassic Calcare Rosso, Lombardia, Italy. *J. sedim. Petrol.* **50**, 457–474.

Baba, J. & Komar, P. D. (1981) Measurements and analysis of settling velocities of natural quartz sand grains. *J. sedim. Petrol.* **51**, 631–640.

Bader, R.G. (1962) Some experimental studies with organic compounds and minerals. *Occ. Publ.* No. 1. Graduate School of Oceanography, University of Rhode Island.

Baertschi, P. & Schwander, H. (1952) A new method for measuring differences in 0^{18} content of silicate rocks. *Helv. Chim. Acta*, **35**, 1748–1751.

Bagnold, R. A. (1941) *The Physics of Blown Sands and Desert Dunes.* Chapman & Hall, London.

Bagnold, R. A. (1968) Deposition in the process of hydraulic transport. *Sedimentology*, **10**, 45–56.

Bagnold, R. A. & Barndorff-Nielsen, O. (1980) The pattern of natural size distributions. *Sedimentology*, **27**, 199–207.

Bailey, S.A. & Smith, J.W. (1972) Improved method for the preparation of sulphur dioxide from barium sulphate for isotope ratio studies. *Anal. Chem.* **44**, 1542–1543.

Baker, E.T. (1973) Distribution and composition of suspended sediment in the bottom waters of the Washington Continental Shelf and Slope. *J. sedim. Petrol.* **43**, 812–821.

Baker, P.A., Gieskes, J.M. & Elderfield, H. (1982) Diagenesis of carbonates in deep-sea sediments — evidence from Sr/Ca ratios and interstitial dissolved Sr^{2+} data. *J. sedim. Petrol.* **52**, 71–82.

Barbera, E. & Savin, S.M. (1987) Effect of sample preparation on the $\delta^{18}O$ value of fine-grained calcite. *Chem. Geol.* **66**, 301–305.

Barndorff-Nielsen, O. (1977) Exponentially decreasing distributions for the logarithm of particle size. *Proc. R. Soc. Lond., A*, **353**, 401–419.

Barnes, J. (1981) *Basic Geological Mapping.* Open University Press, Milton Keynes.

Baronnet, A. (1982) Ostwald ripening in solution. The case of calcite and mica. *Estud. geol.* **38**, 185–198.

Barrett, P.J. (1980) The shape of rock particles, a critical review. *Sedimentology*, **27**, 291–304.

Basan, P.B. (1978) *Trace Fossil Concepts.* Soc. econ. Paleont. Mineral. Short Course Notes 5.

Basu, A., Blanchard, D.P. & Brannon, J.C. (1982) Rare earth elements in the sedimentary cycle: a pilot study of the first leg. *Sedimentology*, **29**, 737–742.

Bathurst, R.G.C. (1966) Boring algae, micrite envelopes and lithification of molluscan biosparites. *Geol. J.* **5**, 15–32.

Bathurst, R.G.C. (1975) *Carbonate Sediments and their Diagenesis.* Developments in Sedimentology, 12. Elsevier, Amsterdam.

Bathurst, R.G.C. (1980) Lithification of carbonate sediments. *Sci. Prog.* **66**, 451–471.

Batschelet, E. (1981) *Circular Statistics in Biology.* Academic Press, London.

Baynes, F.J. & Dearman, W.R. (1978) Scanning electron microscope studies of weathered rocks: a review of nomenclature and methods. *Bull. Int. Ass. Eng. Geol.* **18**, 199–204.

Beall, A.O. & Fischer, A.G. (1969) Sedimentology. In: *Initial Reports of the Deep Sea Drilling Project*, vol. 1, pp. 521–593, US Government Printing Office, Washington DC.

Becker, R.H. & Clayton, R.N. (1972) Carbon isotopic evidence for the origin of a banded iron-formation in Western Australia. *Geochim. cosmochim. Acta*, **36**, 577–595.

Bein, A. & Land, L.S. (1983) Carbonate sedimentation and diagenesis associated with Mg-Ca-Chloride brines: the Permian San Andres Formation in the Texas Panhandle. *J. sedim. Petrol.* **53**, 243–260.

Bence, A.E. & Albee, A.L. (1968) Empirical correction factors for the electron microanalysis of silicates and oxides. *J. Geol.* **76**, 382–403.

Benedetti, M.F., de Kersabiec, A.M. & Boulègue, J. (1987) Determination of gold in twenty geochemical reference samples by flameless atomic absorption spectrometry. *Geostand Newsl.* **11**, 127–129.

Benmore, R.A., Coleman. M.L. & McArthur, J. M. (1983) Origin of sedimentary francolite from its sulphur and carbon isotope composition. *Nature*, **302**, 516–8.

Bennett, H. & Oliver, G.J. (1976) Development of fluxes for the analysis of ceramic materials by X-ray fluorescence spectrometry. *Analyst (London)*, **101**, 803–807.

Benson, L.V. (1974) Transformations of a polyphase sedimentary assemblage into a single phase rock: a chemical approach. *J. sedim. Petrol.* **44**, 123–135.

Berg-Madsen, V. (1983) High-alumina glaucony from the Middle Cambrian of Oland and Bornholm, southern Baltoscandia. *J. sedim. Petrol.* **53**, 875–893.

Bernas, B. (1968) A new method for decomposition and comprehensive analysis of silicates by atomic absorption spectrometry. *Anal. Chem.* **42**, 1682–1686.

Berner, R.A. (1971) *Principles of Chemical Sedimentology.* McGraw-Hill, New York.

Berner, R.A. (1978) Rate control of mineral dissolution under earth surface conditions. *Am. J. Sci.* **278,** 1235–1252.

Berner, R.A. (1980) *Early Diagenesis: a Theoretical Approach.* Princeton University Press.

Berner, R.A., Lasaga, A.C. & Garrels, R.M. (1983) The carbonate-silicate geochemical cycle and its effect on atmospheric carbon dioxide over the past 100 million years. *Am. J. Sci.* **283**, 641–683.

Berner, R.A. & Raiswell, R. (1983) Burial of organic carbon and pyrite sulfur in sediments over Phanerozoic time: a new theory. *Geochim. cosmochim. Acta*, **47**, 855–862.

Berner, R. A., Westrich, J.T., Graber, R., Smith, J & Martens, C.S. (1978) Inhibition of aragonite precipitation from supersaturated seawater: a laboratory and field study. *Am. J. Sci.* **278**, 816–837.

Bertin, E.P. (1975) *Principles and Practice of X-ray Spectrometric Analysis.* Plenum Press, New York.

Bhatia, M.R. & Taylor, S.R. (1981) Trace-element geochemistry and sedimentary provinces: a study from the Tasman Geosyncline, Australia. *Chem. Geol.* **33**, 115–125.

Bigeleisen, J., Perlman, M.L. & Prosser, H.C. (1952) conversion of hydrogenic materials to hydrogen for isotopic analysis. *Anal. Chem.* **24**, 1356–1357.

Biscaye, P.E. (1965) Mineralogy and sedimentation of recent deep sea clay in the Atlantic Ocean and adjacent seas and oceans. *Geol. Soc. Am. Bull.* **76**, 803–832.

Bjørlykke, K. (1983) Diagenetic reactions in sandstones. In: *Sediment Diagenesis* (Ed. by A. Parker and B.W. Sellwood), pp. 169–214. NATO ASI Series C: vol. 115. Reidel, Dordrecht.

Blanche, J.B. & Whittaker, J.H.McD. (1978) Diagenesis of part of the Brent Formation (Middle Jurassic) of the northern North Sea Basin. *J. geol. Soc. London,* **135**, 73–82.

Blatt, H. (1982) *Sedimentary Petrology.* W.H. Freeman, San Francisco.

Blatt, H., Middleton, G. & Murray, R. (1980) *Origin of Sedimentary Rocks.* Prentice-Hall, New Jersey.

Blattner, P. & Hulston, J.R. (1978) Proportional variations of geochemical $\delta^{18}O$ scales — an interlaboratory comparison. *Geochim. cosmochim. Acta*, **42**, 59–62.

Bluck, B.J. (1967) Deposition of some Upper Old Red Sandstone conglomerates in the Clyde area, a study in the significance of bedding. *Scott. J. Geol.* **3**, 139–167.

Bluck, B.J. (1974) Structure and directional properties of some coarse Icelandic valley sandurs. *Sedimentology*, **21**, 533–554.

Bluck, B.J. (1978) Sedimentation in a late orogenic basin: the Old Red Sandstone of the Midland Valley of Scotland. In: *Crustal Evolution in Northwestern Britain and Adjacent Regions* (Ed. by D.R. Bowes and B.E. Leake), pp. 249–278.

Bluck, B.J. (1981) Scottish alluvium: modern and ancient. In: *Field Guides to Modern and Ancient Fluvial Systems in Britain and Spain* (Ed. by T. Elliot). International Fluvial Conference, Keele.

Bluck, B.J. (1984) Pre-Carboniferous history of the Midland Valley of Scotland. *Trans. R. Soc. Edinb., Earth Sci.* **75**, 275–295.

Bockelie, T.G. (1973) A method for displaying sedimentary structures in micritic limestones. *J. sedim. Petrol.* **43**, 537–539.

Bodine, M.W., Holland, H.D. & Borcsik, M. (1965) Co-precipitation of manganese and strontium with calcite. In: *Problems of Post-Magmatic Ore Deposits*, pp. 401–406, Proc. Symp., Prague, II.

Bohor, B.F. & Gluskoter, H.J. (1973) Boron in illite as an indicator of paleosalinity of Illinois coals. *J. sedim. Petrol.* **43**, 945–956.

Bohor, B.F. & Hughes, R.E. (1971) Scanning electron microscopy of clays and clay minerals. *Clays Clay Miner.* **19**, 49–54.

Boistelle, R. (1982) Mineral crystallization from solution. *Estud. geol.* **38**, 135–153.

Boles, J.R. (1984) Secondary porosity reactions in the Stevens Sandstone, San Joaquin valley, California. In: *Clastic Diagenesis* (Ed. by D.A. McDonald and R.C. Surdam), pp. 217–224. *Mem. Am. Ass. Petrol. Geol.* 37.

Bomback, J.L. (1973) Stereoscopic techniques for improved X-ray analysis of rough SEM specimens. In: *Scanning Electron Microscopy 1973* (Ed. by O. Johari and I. Corvin), pp. 97–104. Chicago, Illinois.

Bonham, L.C. & Spotts, J.H. (1971) Measurement of grain orientation. In: *Procedures in Sedimentary Petrology* (Ed. by R.E. Carver), pp. 285–312. Wiley-Interscience, New York.

Borthwick, J. & Harmon, R.S. (1982) A note regarding ClF_3 as an alternative to BrF_5 for oxygen isotope analysis. *Geochim. cosmochim. Acta*, **46**, 1665–1668.

Bosellini, A. (1984) Progradation geometries of carbonate platforms: examples from the Triassic of the Dolomites, northern Italy. *Sedimentology*, **31**, 1–24.

Bouma, A.H. (1962) *Sedimentology of some Flysch Deposits: a Graphic Approach to Facies Interpretation.* Elsevier, Amsterdam.

Bouma, A.H. (1969) *Methods for the Study of Sedimentary Structures.* Wiley-Interscience, New York.

Bourgeois, J. (1980) A transgressive shelf sequence exhibiting hummocky stratification: the Cape Sebastian Sandstone (Upper Cretaceous), Southwestern Oregon. *J. sedim. Petrol.* **50**, 681–702.

Boyle, E.A. (1981) Cadmium, zinc, copper and barium in foraminifera tests. *Earth planet. Sci. Lett.* **53**, 111–135.

Bradley, W.F. & Grim, R.E. (1961) Mica clay minerals. In: *The X-ray Identification and Crystal Structures of Clay Minerals* (Ed. by G. Brown), pp. 208–241. Mineralogical Society, London.

Brand, U. (1981a) Mineralogy and chemistry of the Lower Pennsylvanian Kendrick fauna, Eastern Kentucky. 1. Trace elements. *Chem. Geol.* **32**, 1–16.

Brand, U. (1981b) Mineralogy and chemistry of the Lower Pennsylvanian Kendrick fauna, Eastern Kentucky. 2. Stable isotopes. *Chem. Geol.* **32**, 17–28.

Brand, U. (1982) The oxygen and carbon isotope composition of Carboniferous fossil components: sea-water effects. *Sedimentology*, **29**, 139–147.

Brand, U. (1983) Mineralogy and chemistry of the Lower Pennsylvanian Kendrick fauna, Eastern Kentucky,' U.S.A. 3. Diagenetic and paleoenvironmental analysis. *Chem. Geol.* **40**, 167–181.

Brand, U. (1986) Paleoenvironmental analysis of Middle Jurassic (Callovian) ammonoids from Poland: trace elements and stable isotopes. *J. Paleont.* **60**, 293–301.

Brand, U. & Morrison, J.O. (1987) Biogeochemistry of fossil marine invertebrates. *Geosci. Can.* **14**, 85–107.

Brand, U. & Veizer, J. (1980) Chemical diagenesis of a multicomponent carbonate system — 1: trace elements. *J. sedim. Petrol.* **50**, 1219–1236.

Brand, U. & Veizer, J. (1981) Chemical diagenesis of a multicomponent carbonate system — 2: stable isotopes. *J. sedim. Petrol.* **51**, 987–997.

Brenchley, P.J. (1969) Origin of matrix in Ordovician greywackes, Berwyn Hills, North Wales. *J. sedim. Petrol.* **39**, 1297–1301.

Briggs, D. (1977) Sediments. In: *Sources and Methods in Geography*, Butterworths London.

Brindley, G.W. (1980) Quantitative X-ray mineral analysis of clays. In: *Crystal Structures of Clay Minerals and their X-ray Identification* (Ed. by G. W. Brindley and G. Brown), pp. 411–438. Mineralogical Society, London.

British Standards Institution (1973) *BS 882*. Specification for aggregates from natural sources for concrete (including granolithic).

British Standards Institution (1975) *BS 812 Part 1*. Methods for sampling and testing mineral aggregate sands and filters.

British Standards Institution (1975) *BS 1377*. Methods of test for soils for civil engineering purposes.

British Standards Institution (1976) *BS 1199*.

Bromberger, S.H. & Hayes, J.B. (1965) Quantitative determination of calcite–dolomite–apatite mixtures by X-ray diffraction. *J. sedim. Petrol.* **36**, 358–361.

Bromley, R.G. (1981) Enhancement of visibility of structures in marly chalk: modification of the Bushinsky oil technique. *Bull. geol. Soc. Denmark*, **29**, 111–118.

Brown, G. & Brindley, G.W. (1980) X-ray diffraction procedures for clay mineral identification. In: *Crystal Structures of Clay Minerals and their X-ray Identification* (Ed. by G.W. Brindley and G. Brown), pp. 305–360, Mineralogical Society, London.

Brown, M.B. (1974) Identification of sources of significance in two-way contingency tables. *Appl. Statis.* **23**, 405–413.

Brown, P.R. (1969) Compaction of fine-grained terrigenous and carbonate sediments — a review. *Bull. Can. Petrol. Geol.* **17**, 486–495.

Brownlow, A.H. (1979) *Geochemistry*. Prentice-Hall, New Jersey.

Brunton, G. (1955) Vapour pressure glycolation of oriented clay minerals. *Am. Miner.* **40**, 124–126.

Bryant, R & Williams, D.J.A. (1982) Mineralogical analysis of suspended cohesive sediments using an analytical electron microscope. *J. sedim. Petrol.* **52**, 299–306.

Buchbinder, L.G., Magaritz, M. & Goldbery, M. (1984) Stable isotope study of karstic-related dolomitization: Jurassic rocks from the coastal plain, Israel. *J. sedim. Petrol.* **54**, 236–256.

Bull, P.A. (1978) A quantitative approach to scanning electron microscope analysis of cave sediments. In: *Scanning Electron Microscopy in the Study of Sediments* (Ed. by W.B. Whalley), pp. 201–226. Geo Abstracts, Norwich.

Bull, P.A. (1981) Environmental reconstruction by electron microscopy. *Prog. Phys. Geogr.* **5**, 368–397.

Buller, A.T. & McManus, J. (1972) Simple metric sedimentary statistics used to recognise different environments. *Sedimentology*, **18**, 1–21.

Buller, A.T. & McManus, J. (1973a) Modes of turbidite deposition deduced from grain size analyses. *Geol. Mag.* **109**, 491–500.

Buller, A.T. & McManus, J. (1973b) The quartile deviation — median diameter relationship of glacial deposits. *Sediment. Geol.* **10**, 135–146.

Buller, A.T. & McManus, J. (1973c) Distinction among pyroclastic deposits from their grain size frequency distributions. *J. Geol.* **81**, 97–106.

Buller, A.T. & McManus, J. (1979) Sediment sampling and analysis. In: *Estuarine Hydrography and Sedimentation* (Ed. by K.R. Dyer), pp. 87–130. Cambridge University Press.

Burke, W.H., Demson, R.E., Hetherington, E.A., Koepnick, R.B., Nelson, H.F. & Oto, J.B. (1982) Variation of seawater $^{87}Sr/^{86}Sr$ throughout Phanerozoic time. *Geology*, **10**, 516–519.

Burley, S.D. (1984) Patterns of diagenesis in the Sherwood Sandstone Group (Triassic), United Kingdom. *Clay Miner.* **19**, 403–440.

Burley, S.D. (1986) The development and destruction of porosity within Upper Jurassic reservoir sandstones of the Piper and Tartan Fields, Outer Moray Firth, North Sea. *Clay Miner.* **21**, 649–694.

Burley, S.D. & Kantorowicz, J.D. (1986) Thin section and S.E.M. textural criteria for the recognition of cement-dissolution porosity in sandstones. *Sedimentology*, **33**, 587–604.

Burley, S.D., Kantorowicz, J.D. & Waugh, B. (1985) Clastic diagenesis. In: *Sedimentology: Recent Developments and Applied Aspects* (Ed. by P.J. Brenchley and B.P.J. Williams), pp. 189–226. Spec. Publ. geol. Soc. London, No. 18. Blackwell Scientific Publications, Oxford.

Burns, J.H. & Breding, M.A. (1956) Transformation of calcite to aragonite by grinding. *J. chem. Phys.* **25**, 1281.

Burns, R.G. (1979) *Marine Minerals*. Min. Soc. Am. Rev. Mineral. 6.

Burns, S.J. & Baker, P.A. (1987) A geochemical study of dolomite in the Monterey Formation, California. *J. sedim. Petrol.* **57**, 128–139.

Burruss, R.C., Cercone, K.R. & Harris, P.M. (1985) Timing of hydrocarbon migration: evidenced from fluid inclusions in calcite cements, tectonics and burial history.

In: *Carbonate Cememts* (Ed. by N. Schneidermann and P.M. Harris), pp. 277–289. Soc. econ. Paleont. Miner. Spec. Publ. 36.

Busenberg, E. & Plummer, L.N. (1985) Kinetic and thermodynamic factors controlling the distribution of SO_4^{2-} and Na^+ in calcites and selected aragonites. *Geochim. cosmochim. Acta*, **49**, 713–725.

Cameron, E.M. (1983) Genesis of Proterozoic iron-formations: sulphur isotope evidence. *Geochim. cosmochim. Acta*, **47**, 1069–1074.

Campbell, C.V. (1967) Lamina, laminaset, bed and bedset. *Sedimentology*, **8**, 7–26.

Campbell, S.E. (1982) Precambrian endoliths discovered. *Nature*, **299**, 429–431.

Cant, D.J. & Walker, R.G. (1976) Development of a braided-fluvial facies model for the Devonian Battery Point Sandstone, Quebec. *Can. J. Earth Sci.* **13**, 102–119.

Carothers, W.W. & Kharaka, Y.K. (1980) Stable carbon isotopes of HCO_3^- in oil field waters — implications for the origin of CO_2. *Geochim. cosmochim. Acta*, **44**, 323–332.

Carozzi, A.V. (1960) *Microscopic Sedimentary Petrography*. Wiley, New York.

Carr, T.R. (1982) Log-linear models. Markov chains and cyclic sedimentation. *J. sedim. Petrol.* **52**, 905–912.

Carroll, D. (1970) *Clay minerals: A guide to their X-ray identification*. Geol. Soc. Am. Spec. Pap. 126. Geological Society of America.

Carruthers, A. (1985) The Upper Palaeozoic geology of the Galtee Mountains and adjacent areas, Ireland. *PhD thesis*, University of Dublin.

Cassyhap, S.M. (1968) Huronian stratigraphy and paleocurrent analysis in Ontario, Canada. *J. sedim. Petrol.* **38**, 920–942.

Cater, J.M.L. (1984) An application of scanning electron microscopy of quartz sand surface textures to the environmental diagnosis of Neogene carbonate sediments, Finestrat Basin, south-east Spain. *Sedimentology*, **31**, 717–731.

Channon, R.D. (1971) The Bristol fall column for coarse sediment grading. *J. sedim Petrol.* **41**, 867–870.

Charef, A. & Sheppard, S.M.F. (1984) Carbon and oxygen isotope analyses of calcite or dolomite associated with organic matter. *Chem. Geol.* **46**, 325–333.

Chayes, F. (1956) *Petrographic Modal Analysis — an Elementary Statistical Appraisal.* Wiley, New York.

Cheeney, R.F. (1983) *Statistical Methods in Geology*. Allen & Unwin, London.

Chiba, H. & Sakai, H. (1985) Oxygen isotope exchange rate between dissolved sulphate and water at hydrothermal temperatures. *Geochim. cosmochim. Acta*, **49**, 993–1000.

Chilingar, G.V. (1982) Graph for determining the size of sedimentary particles. *AGI Data Sheet*, 16. American Geological Institute.

Chisholm, J.I. & Dean, J.M. (1974) The Upper Old Red Sandstone of Fife and Kinross: a fluviatile sequence with evidence of marine incursion. *Scott. J. Geol.* **10**, 1–30.

Choquette, P.W. & Pray, L.C. (1970) Geologic nomenclature and classification of porosity in sedimentary carbonates. *Bull. Am. Ass. Petrol. Geol.* **54**, 207–250.

Choquette, P.W. & Trusell, F.C. (1978) A procedure for making the titan-yellow stain for Mg-calcite permanent. *J. sedim. Petrol.* **48**, 639–641.

Christiansen, C. (1984) *A comparison of sediment parameters from log-probability plots and log-log plots of the same sediments.* University of Aarhus, Geoskrifter No. 20.

Clarke, F.W. (1924) *The Data of Geochemistry* (5th edn). U.S. Geol. Surv. Bull. 770.

Clarke, I.C. (1975) In: *Principles and Techniques of Scanning Electron Microscopy, vol. 3, Biological Applications* (Ed. by M.A. Hayat), pp. 154–194. Van Nostrand, Reinhold.

Claypool, G.E., Holser, W.T., Kaplan, I.R., Sakai, H. & Zak, I. (1980) The age curves of sulfur and oxygen isotopes in marine sulfate and their mutual interpretation. *Chem. Geol.* **28**, 199–260.

Clayton, R.N. & Mayeda, T.K. (1963) The use of bromine pentafluoride in the extraction of oxygen from oxides and silicates for isotopic analysis. *Geochim. cosmochim. Acta*, **27**, 43–52.

Clayton, R.N., Jones, B.F. & Berner, R.A. (1968) Isotope studies of dolomite formation under sedimentary conditions. *Geochim. cosmochim. Acta*, **32**, 415–432.

Clayton, R.N., Skinner, H.C.W., Berner, R.A. & Rubinson, M. (1968) Isotopic compositions of recent South Australian lagoonal carbonates. *Geochim. cosmochim. Acta*, **32**, 983–988.

Clements, R.G. (1973) A study of certain non-marine Gastropods from the Purbeck Beds of England. *Unpublished PhD thesis*, University of Hull.

Cliff, G. & Lorimer, G.W. (1975) The quantitative analysis of thin specimens. *J. Microsc.* **103**, 203–207.

Cochran, W.G. (1977) *Sampling Techniques.* Wiley, New York.

Cody, R.D. (1971) Adsorption and the reliability of trace elements as environmental indicators for shales. *J. sedim. Petrol.* **41**, 461–471.

Cole, T.G. & Shaw, H.F. (1983) The nature and origin of authigenic smectites in some recent marine sediments. *Clay Miner.* **18**, 239–252.

Coleman, M.L. (1977) Sulphur isotopes in petrology. *J. geol. Soc. London*, **133**, 593–608.

Coleman, M.L. & Moore, M. P. (1978) Direct reduction of sulphate to sulphur dioxide for isotopic analysis. *Anal. Chem.* **50**, 1594–1595.

Coleman, M.L. & Raiswell, R. (1981) Carbon, oxygen and sulphur isotope variations in concretions from the Upper

Lias of N.E. England. *Geochim. cosmochim. Acta*, **45**, 329–340.

Coleman, M.L., Shepherd, T.J., Durham, J.J., Rouse, J.E. & Moore, G.R. (1983) Reduction of water with zinc for hydrogen isotope analysis. *Anal. Chem.* **54**, 993–995.

Collinson, J.D. (1968) Deltaic sedimentation units in the Upper Carboniferous of northern England. *Sedimentology*, **10**, 233–254.

Collinson, J.D. (1971) Current vector dispersion in a river of fluctuating discharge. *Geologie Mijnb.* **50**, 671–678.

Collinson, J.D. & Thompson, D.B. (1982) *Sedimentary Structures*. Allen & Unwin, London.

Compston *et al.* (1985) The age of (a tiny part of) the Australian continent. *Nature*, **317**, 559–560.

Compton, R.R. (1962) *Manual of Field Geology*. Wiley, London.

Connor, C C. (1953) Some effects of the grading of sand on masonry mortar. *Proc. Am. Soc. test. Materials*, **52**, 933–948.

Conybeare, C.E.B. & Crook, K.A.W. (1968) *Manual of Sedimentary Structures*. Aust. Bur. Miner. Res., Geol. Geophys. Bull. 102.

Cook, P.J. & Mayo, W. (1977) Sedimentology and Holocene history of a tropical estuary (Broad Sound, Queensland). *Bur. Min. Res. Geol. Geophys. Aust. Bull.* 170.

Cooper, M.A. & Marshall, J.D. (1981) ORIENT: a computer program for the resolution and rotation of palaeocurrent data. *Computers Geosci.* **7**, 153–165.

Coplen, T. B. & Hanshaw, B.B. (1973) Ultrafiltration by a compacted clay membrane I. Oxygen and hydrogen isotopic fractionation. *Geochim. cosmochim. Acta*, **37**, 2295–2310.

Coplen, T.B., Kendall, C. & Hopple, J. (1983) Comparison of stable isotope reference samples. *Nature*, **302**, 236–238.

Cortecci, G., Reyues, E., Berti, G. & Casati, P. (1981) Sulphur and oxygen isotopes in Italian marine sulphates of Permian and Triassic ages. *Chem. Geol.* **34**, 65–79.

Couch, E.L. (1971) Calculation of paleosalinities from boron and clay mineral data. *Bull. Am. Ass. Petrol. Geol.* **55**, 1829–1837.

Couch, E.L. & Grim, R.E. (1968) Boron fixation by illites. *Clays Clay Miner.* **16**, 249–256.

Coy-yll, R. (1970) Quelques aspects de la cathodoluminescence des mineraux. *Chem. Geol.* **5**, 243–254.

Craig, H. (1957) Isotopic standards for carbon and oxygen and correction factors for mass-spectrometric analysis of carbon dioxide. *Geochim. cosmochim. Acta*, **12**, 133–149.

Craig, H. (1961a) Isotopic variations in meteoric waters. *Science*, **133**, 1702.

Craig, H. (1961b) Standard for reporting concentrations of deuterium and oxygen-18 in natural waters. *Science*, **133**, 1833.

Crimes, T.P. & Harper, J.C. (1970) *Trace fossils*. Geol. J. Spec. Issue No. 3.

Cronan, D.S. & Moorby, S. A. (1981) Manganese nodules and other ferro-manganese oxide deposits from the Indian Ocean. *J. geol. Soc. London*, **138**, 527–539.

Crooks, W. (1880) Magnetic deflection of molecular trajectory — laws of magnetic rotation in high and low vacua — phosphorogenic properties of molecular discharge. *Phil. Trans. R. Soc. London*, **170**, part II, 641–662.

Culver, S.J., Bull, P.A., Campbell, S., Shakesby, R.A. & Whalley, W.B. (1983) Environmental discrimination based on quartz grain surface textures: a statistical investigation. *Sedimentology*, **30**, 129–136.

Cummings, C.E. & McCarty, H.B. (1982) Stable carbon isotope ratios in *Astrangia danae*: evidence for algal modification of carbon pools used in calcification. *Geochim. cosmochim. Acta*, **46**, 1125–1129.

Cummins, W.A. (1962) The greywacke problem. *Geol. J.* **3**, 51–72.

Curray, J.R. (1956) The analysis of two dimensional orientation data. *J. Geol.* **64**, 117–131.

Curray, J.R. (1960) Tracing sediment masses by grain-size modes. *Proc. 21st Geol Congress*, Norden, pp. 119–130.

Curry, A., Grayson, R.F. & Hosey, G.R. (1982) *Under the Microscope*. Blandford Press, Poole.

Curtis, C.D. (1983) Link between aluminium mobility and destruction of secondary porosity. *Bull. Am. Ass. Petrol. Geol.* **67**, 380–393.

Curtis, C.D., Pearson, M.J. & Somogyi, V.A. (1975) Mineralogy, chemistry and origin of a concretionary siderite sheet (clay-ironstone band) in the Westphalian of Yorkshire. *Min. Mag.* **40**, 385–393.

Curtis, C.D. & Spears, D.A. (1968) The formation of sedimentary iron minerals. *Econ. Geol.* **63**, 258–270.

Davies, D.K., Almon, W.R., Bonis, S.B. & Hunter, B.E. (1979) Deposition and diagenesis of Tertiary–Holocene volcaniclastics, Guatemala. In: *Aspects of Diagenesis* (Ed. by P.A. Scholle and P.R. Schluger), pp. 281–306. Spec. Publ. Soc. econ. Paleont. Miner. 26.

Davies, G.R. (1979) Dolomite reservoir rocks: processes, controls, porosity development. In: *Geology of Carbonate Porosity*, pp. C1–17. Am. Ass. Petrol. Geol., Course Notes, 11.

Davies, I.C. & Walker, R.G. (1974) Transport and deposition of resedimented conglomerates, the Gap Enragé Formation, Cambro–Ordovician, Gaspe, Quebec. *J. sedim. Petrol.* **44**, 1200–1216.

Davies, P.J. & Till, R. (1968) Stained dry cellulose peels of ancient and Recent impregnated carbonate sediments. *J. sedim. Petrol.* **38**, 234–237.

De Celles, P.G. & Gutschick, R.C. (1983) Mississippian wood–grained chert and its significance in the Western Interior, United States. *J. sedim. Petrol.* **53**, 1175–1191.

De Celles, P.G., Longford, R.P. & Schwartz, R.K. (1983) Two new methods of paleocurrent determination from trough cross stratification. *J. sedim. Petrol.* **53**, 629–642.

Degens, E.T. & Epstein, S. (1964) Oxygen and carbon

isotope ratios in coexisting calcites and dolomites from Recent and ancient sediments. *Geochim. cosmochim. Acta*, **28**, 23–44.

Degens, E.T., Williams, E.G. & Keith, M.L. (1957) Environmental studies of Carboniferous sediments, 1. Geochemical criteria for differentiating marine and freshwater shales. *Bull. Am. Ass. Petrol. Geol.* **41**, 2427–2455.

Deines, P. (1970) Mass spectrometer correction factors for the determination of small isotopic composition variations of carbon and oxygen. *Int. J. Mass Spectrom., Ion Phys.* **4**, 283.

Deines, P. (1980) The isotopic composition of reduced organic carbon. In: *Handbook of Environmental Isotope Geochemistry*, Vol. 1 (Ed. by P. Fritz and J.C. Fontes), pp. 329–406. Elsevier, Amsterdam.

Delgado, F. (1977) Primary textures in dolostones and recrystallized limestones: a technique for their microscopic study. *J. sedim. Petrol.* **47**, 1339–1341.

De Raaf, J.F.M., Boersma, R. & Van Gelder, A. (1977) Wave-generated structures and sequences from a shallow marine succession, Lower Carboniferous, County Cork, Ireland. *Sedimentology*, **24**, 451–484.

De Raaf, J.F.M., Reading, H.G. & Walker, R.G. (1965) Cyclic sedimentation in the Lower Westphalian of North Devon, England. *Sedimentology*, **4**, 1–52.

Dickinson, W.R. (1985) Interpreting provenance relations for detrital modes of sandstones. In: *Provenance of Arenites* (Ed. by G.G. Zuffa), pp. 333–361. NATO ASI Series C: vol. 148. Reidel, Dordrecht.

Dickson, J.A.D. (1965) A modified staining technique for carbonates in thin section. *Nature*, **205**, 587

Dickson, J.A.D. (1966) Carbonate identification and genesis as revealed by staining. *J. sedim. Petrol.* **36**, 491–505.

Dickson, J.A.D. (1980) Artificial colouration of fluorite by electron bombardment. *Min. Mag.* **43**, 820–822.

Dickson, J.A.D. (1983) Graphical modelling of crystal aggregates and its relevance to cement diagnosis. *Phil. Trans. R. Soc. London, A*, **309**, 465–502.

Dickson, J.A.D. (1985) Diagenesis of shallow-marine carbonates. In: *Sedimentology: Recent Developments and Applied Aspects* (Ed. by P.J. Brenchley and B.J.P. Williams), pp. 173–188. Spec. Publ. geol. Soc. London, No. 18. Blackwell Scientific Publications, Oxford.

Dickson, J.A.D. & Coleman, M.L. (1980) Changes in carbon and oxygen isotope composition during limestone diagenesis. *Sedimentology*, **27**, 107–118.

Dilks, A. & Graham, S.C. (1985) Quantitative mineralogical characterization of sandstones by back-scattered electron image analysis. *J. sedim. Petrol.* **55**, 347–355.

Dinur, D., Spiro, B. & Aizenshtat, L. (1980) The distribution and isotopic composition of sulphur in organic-rich sedimentary rocks. *Chem. Geol.* **31**, 37–51.

Dodd, J.R. (1966) Diagenetic stability of temperature-sensitive skeletal properties in *Mytilus* from the Pleistocene of California. *Bull. geol. Soc. Am.* **77**, 1213–1224.

Dodd, J.R. & Stanton, R.J. (1975) Paleosalinities within a Pliocene Bay, Kettleman Hills, California: A study of the resolving power of isotopic and faunal techniques. *Bull. geol. Soc. Am.* **86**, 51–64.

Dodd, J.R. & Stanton, R.J. (1981) *Paleoecology, Concepts and Applications*. Wiley, London.

Dott, R.H. (1973) Paleocurrent analysis of trough cross stratification. *J. sedim. Petrol.* **43**, 779–783.

Dott, R.H. (1974) Paleocurrent analysis of severely deformed flysch-type strata — a case study from South Georgia Island. *J. sedim. Petrol.* **44**, 1166–1173.

Douthitt, C.B. (1982) The geochemistry of the stable isotopes of silicon. *Geochim. cosmochim. Acta*, **46**, 1449–1458.

Doveton, J.H. (1971) An application of Markov chain analysis to the Ayrshire Coal Measures succession. *Scott. J. Geol.* **7**, 11–27.

Dravis, J.D. & Yurewicz, D.A. (1985) Enhanced carbonate petrography using fluorescence microscopy. *J. sedim. Petrol.* **55**, 795–804.

Dreimanis, A. (1984) Discussion. Lithofacies types and vertical profile models; an alternative approach to the description and environmental interpretation of glacial diamict and diamictite sequences. *Sedimentology*, **31**, 885–886.

Drever, J.I. (1982) *The Geochemistry of Natural Waters*. Prentice-Hall, New Jersey.

Dreybrodt, W. (1982) A possible mechanism for growth of calcite speleothems without participlation of biogenic carbon dioxide. *Earth planet. Sci. Lett.* **58**, 293–299.

Duck, R.W. (1983) Settling velocity analysis of fine grained, freshwater sediments: a cautionary note. *J. Wat. Resour.* **2**, 23–29.

Duddy, I.R. (1980) Redistribution and fractionation of rare-earth and other elements in a weathering profile. *Chem. Geol.* **30**, 363–381.

Dunbar, J. & Wilson, A.T. (1983) The use of O^{18}/O^{16} ratios to study the formation and chemical origin of coal. *Geochim. cosmochim. Acta*, **47**, 1541–1543.

Dunham, R.J. (1962) Classification of carbonate rocks according to depositional texture. In: *Classification of Carbonate Rocks*, Vol. 1 (Ed. by W.E. Ham), pp. 108–121. American Association of Petroleum Geologists, Tulsa.

Dunnington, H.V. (1967) Aspects of diagenesis and shape change in stylolitic limestones reservoirs. *Proc. 7th World Petroleum Congr.* Mexico, **2**, 339–352.

Dunoyer de Segonzac, G. (1970) The transformation of clay minerals during diagenesis and low-grade metamorphism. *Sedimentology*, **15**, 281–348.

Durazzi, J.T. (1977) Stable isotopes in the ostracod shell: a preliminary study. *Geochim. cosmochim. Acta*, **41**, 1168–1170.

Eberl, D.D. (1984) Clay mineral formation and transformation in rocks and soils. *Phil. Trans. R. Soc. London A*, **311**, 241–257.

Edelman, N. (1962) Mathematics and geology. *Geol. För. Stockh. Förh.* **84**, 343–350.

Ednam, J.D. & Surdam, R.C. (1986) Organic-inorganic interactions as a mechanism for porosity enhancement in the Upper Cretaceous Ericson Sandstone, Green River Basin, Wyoming. In: *Roles of Organic Matter in Sediment Diagenesis* (Ed. by D.L. Gautier), pp. 85–110. Spec. Publ. Soc. econ. Paleont. Miner. 38.

Ekdale, A., Bromley, R.G. & Pemberton, S.G. (1984) *The use of trace fossils in sedimentology and stratigraphy*. Soc. econ. Paleont. Mineral. Short Course Notes 15.

Elderfield, H. & Gieskes, J.M. (1982) Sr isotopes in interstitial waters of marine sediments from Deep Sea Drilling Project cores. *Nature*, **300**, 493–497.

Elsinger, E.V., Savin, S.M. & Yeh, H.W. (1979) Oxygen isotope geothermometry of diagenetically altered shales. In: *Aspects of Diagenesis* (Ed. by P.A. Scholle and P.R. Schluger) Spec. Publ. Soc. econ. Paleont. Miner., 26.

Emery, K. O. (1983) Rapid method of mechanical analysis of sands. *J. sedim. Petrol.* **8**, 105–111.

Epstein, S., Buchsbaum, R., Lowenstam, H.A. & Urey, H.C. (1953) Revised carbonate-water isotopic temperature scale. *Geol. Soc. Am. Bull.* **64**, 1315–1326.

Epstein, S., Graf, D.L. & Degens, E.T. (1964) Oxygen isotope studies on the origin of dolomites. In: *Isotope and Cosmic Chemistry* (Ed. by H. Craig, S.L. Miller and C.J. Wasserburg), pp. 169–180. North Holland, Amsterdam.

Epstein, S. & Mayeda, T. (1953) Variation of O^{18} content of waters from natural sources. *Geochim. cosmochim. Acta*, **4**, 213–224.

Eriksson, K.A., McCarthy, T.S. & Truswell, J.F. (1975) Limestone formation and dolomitization in a Lower Proterozoic succession from South Africa. *J. sedim. Petrol.* **45**, 604–614.

Erlich, R. (1983) Size analysis wears no clothes, or have moments come and gone? *J. sedim. Petrol.* **53**, 1.

Erol, O., Lohnes, R.A. & Demirel, T. (1976) Preparation of clay-type, moisture-containing samples for scanning electron microscopy. In: *Proc. Workshop on Techniques for Particulate Matter Studies*, pp. 769–776. Illinois Institute of Technology Press, Chicago.

Ethier, V.G. (1975) Application of Markov analysis to the Banff Formation (Mississippian), Alberta. *Math. Geol.* **7**, 47–61.

Eyles, N., Eyles, C.H. & Miall, A.D. (1983) Lithofacies types and vertical profile analysis; an alternative approach to the description and environmental interpretation of glacial diamict and diamictite sequences. *Sedimentology*, **30**, 393–410.

Fairchild, I.J. (1980a) Sedimentation and origin of a Late Precambrian 'dolomite' from Scotland. *J. sedim. Petrol.* **50**, 423–446.

Fairchild, I.J. (1980b) Stages in a Precambrian dolomitization, Scotland: cementing versus replacement textures. *Sedimentology*, **27**, 631–650.

Fairchild, I.J. (1983) Chemical controls of cathodoluminescence of natural dolomites and calcites: new data and review. *Sedimentology*, **30**, 579–583.

Fairchild, I.J. (1985) Petrography and carbonate chemistry of some Dalradian dolomitic metasediments: preservation of diagenetic textures. *J. geol. Soc. London*, **142**, 167–185.

Fairchild, I.J. & Spiro, B. (1987) Petrological and isotopic implications of some contrasting Precambrian carbonates. *Sedimentology*, **34**, 973–990.

Fang, J.H. & Bloss, D.F. (1966) *X-ray Diffraction Tables*. Southern Illinois University Press, Carbondale, Illinois.

Fang, J.H. & Zevin, L. (1985) Quantitative X-ray diffractometry of carbonate rocks. *J. sedim. Petrol.* **55**, 611–613.

Farrow, G.E. & Clokie, J. (1979) Molluscan grazing of sublittoral algal-bored shells and the production of carbonate mud in the Firth of Clyde, Scotland. *Trans. R. Soc. Edinb.* **70**, 139–148.

Fiegl, F. (1937) *Qualitative Analysis by Spot Tests*. Nordemann, New York.

Field, M.E., Nelson, C.H., Cacchione, D.A. & Drake, D.E. (1981) Sand waves on an epicontinental shelf: northern Bering Sea. *Mar. Geol.* **42**, 233–258.

Fiori, C.E. & Newbury, D.E. (1978) *SEM/1978/I*, pp. 401–410. SEM Inc., AMF O'Hare, Illinois.

Fitzgerald, M.G., Parmenter, C.M. & Milliman, J.D. (1979) Particulate calcium carbonate in New England shelf waters: result of shell degradation and resuspension. *Sedimentology*, **26**, 853–857.

Flanagan, F.J. (1986) Reference samples in geology and geochemistry. *U.S. geol. Surv. Bull. 1582*.

Fleet, M.E.L. (1965) Preliminary investigations into the sorption of boron by clay minerals. *Clay Miner. Bull.* **6**, 3–16.

Fleming, G.W. & Plummer, L.N. (1983) PHRQINPUT — an interactive computer program for constructing input data sets to the geochemical simulation program PHREEQE. *U.S. Geol. Surv. Wat. Resour. Invest.* 83–4236.

Fletcher, W.K. (1981) *Analytical Methods in Geochemical Prospecting*. Elsevier, Amsterdam.

Folk, R.L. (1955) Student operator error in determination of roundness, sphericity and grain size. *J. sedim. Petrol.* **25**, 297–301.

Folk, R.L. (1962) Spectral subdivision of limestone types. In: *Classification of Carbonate Rocks* (Ed. by W.E. Ham), pp. 62–84. Am. Ass. Petrol. Geol. Mem. 1.

Folk, R.L. (1966) A review of grain-size parameters. *Sedimentology*, **6**, 73–93.

Folk, R.L. (1974a) The natural history of crystalline calcium carbonate: effects of magnesium content and salinity. *J. sedim. Petrol.* **40**, 40–53.

Folk, R.L. (1974b) *Petrology of Sedimentary Rocks*. Hemphill, Austin.

Folk, R.L., Andrews, P.B. & Lewis, D.W. (1970) Detrital sedimentary rock classification and nomenclature for use in New Zealand. *N.Z. J. Geol. Geophys.* **13**, 937–968.

Folk, R.L. & Ward, W.C. (1957) Brazos River bar: a study

in the significance of grain size parameters. *J. sedim. Petrol.* **27**, 3–26.

Fournier, R.O. (1983) A method of calculating quartz solubilities in aqueous sodium chloride solutions. *Geochim. cosmochim. Acta*, **47**, 579–586.

Frederickson, A.F. & Reynolds, R.C. (1960) How measuring palaeosalinity aids exploration. *Oil Gas J.* **1**, 154–158.

Freeman, T. & Pierce, K. (1979) Field statistical assessment of cross-bed data. *J. sedim. Petrol.* **49**, 624–625.

French, D.H., Warne, S.St. J. & Sheedy, M.T. (1984) The use of ion-exchange resins for the dissolution of carbonates. *J. sedim. Petrol.* **54**, 641–642.

Frey, M. (1970) The step from diagenesis to metamorphism in pelitic rocks during Alpine orogenesis. *Sedimentology*, **15**, 261–279.

Frey, R.W. (Ed.) (1975) *The Study of Trace Fossils*. Springer-Verlag, Berlin.

Friedman, G.M. (1958) Determination of sieve-size distribution from thin section data for sedimentary studies. *J. Geol.* **66**, 394–416.

Friedman, G.M. (1959) Identification of carbonate minerals by staining methods. *J. sedim. Petrol.* **29**, 87–97.

Friedman, G.M. (1961) Distinction between dune, beach and river sands from their textural characteristics. *J. sedim. Petrol.* **31**, 514–529.

Friedman, G.M. (1962) Comparison of moment measures for sieving and thin-section data for sedimentary petrological studies. *J. sedim. Petrol.* **32**, 15–25.

Friedman, G.M. (1967) By name processes and statistical parameters compared for size frequency distribution of beach and river sands. *J. sedim. Petrol.* **37**, 327–354.

Friedman, G.M. (1971) Staining. In: *Procedures in Sedimentary Petrology* (Ed. by R.E. Carver), pp. 511–530. Wiley-Interscience, New York.

Friedman, G.M. (1975) The making and unmaking of limestones or the downs and ups of porosity. *J. sedim. Petrol.* **45**, 379–398.

Friedman, G.M. & Johnson, K.G. (1982) *Exercises in Sedimentology*. Wiley, New York.

Friedman, G.M. & Sanders, J.E. (1978) *Principles of Sedimentology*. Wiley, New York.

Friedman, I. (1953) Deuterium content of natural waters. *Geochim. cosmochim. Acta*, **4**, 89–103.

Friedman, I. & Gleason, J.D. (1973) Notes on the bromine pentafluoride technique of oxygen extraction. *J. Res. U.S. geol. Surv.* **1**, (6), 679–680.

Friedman, I. & O'Neil, J.R. (1977) Compilation of stable isotope fractionation factors of geochemical interest. *U.S. geol. Surv. Prof. Pap.* 440–KK.

Friend, P.F., Alexander-Marrack, P.D., Nicholson, J. & Yeats, A.K. (1976) Devonian sediments of east Greenland. 1. Introduction, classification, sequences, petrographic notes. *Medd Gronland*, **206**, 1–56.

Friend, P.F., Slater, M.J. & Williams, R.C. (1979) Vertical and lateral building of river sandstone bodies, Ebro Basin, Spain. *J. geol. Soc. London*, **136**, 34–46.

Froude, D.O. *et al.* (1983) Ion microprobe identification of 4100–4200 Myr-old terrestrial zircons. *Nature*, **304**, 616–618.

Fryberger, S.G. & Dean, B.G. (1979) Dune forms and wind regime. In: *A Study of Global Sand Seas* (Ed. by E.D. McKee), pp. 137–169. U.S. geol. Surv. Prof. Pap. 1052.

Füchtbauer, H. (1980) Experimental precipitation of ferroan calcites (abstract), pp. 170–171. *IAS 1st European Meeting,* Bochum.

Füchtbauer, J. & Hardie, L.A. (1980) Comparison of experimental and natural magnesian calcites (abstract), pp. 167–169. *IAS 1st European Meeting*, Bochum.

Fütterer, D.K. (1974) Significance of the boring sponge *Cliona* for the origin of fine grained material of carbonate sediments. *J. sedim. Petrol.* **44**, 79–84.

Galehouse, J.S. (1971a) Point counting. In: *Procedures in Sedimentary Petrology* (Ed. by R.E. Carver), pp. 385–407. Wiley-Interscience, New York.

Galehouse, J.S. (1971b) Sedimentation analysis. In: *Procedures in Sedimentary Petrology* (Ed. by R.E. Carver), pp. 65–94. Wiley-Interscience, New York.

Garrels, R.M. & Christ, C.L. (1965) *Solutions, Minerals and Equilibria*. Harper & Row, New York.

Garrels, R.M. & MacKenzie, F.T. (1971) *Evolution of Sedimentary Rocks*. Norton, New York.

Garrels, R.M. & Thompson, M.E. (1962) A chemical model for sea water at 25°C and one atmosphere total pressure. *Am. J. Sci.* **260**, 57–66.

Garrett, R.G. (1980) HANOVA — a FORTRAN IV program for unbalanced analysis of variance. *Comp. Geosci.* **6**, 35–60.

Garrett, R.G. (1983) Sampling methodology. In: *Statistics and Data Analysis in Geochemical Prospecting* (Ed. by R.J. Howarth), pp. 83–107. Handbook of Exploration Geochemistry, 2, Elsevier, Amsterdam.

Garrett, R.G. & Gross, T.I. (1980) HANOVA — a FORTRAN IV program for unbalanced regional geochemical surveys. In: *Geochemical Exploration 1978* (Ed. by J.B. Watterson and P.K. Theobald), pp. 371–383. Association of Exploration Geochemists, Rexdale, Ontario.

Garrison, R.E., Douglas, R.G. *et al.* (eds) (1981) The Monterey Formation and related siliceous rocks of Carlifornia. *Soc. econ. Paleont. Miner. Pacific Sec. Spec. Publ.* Los Angeles, California.

Gaudette, H.E., Grim, R.E. & Metzger, C.F. (1966) Illite: a model based on the sorption behaviour of cesium. *Am. Miner.* **51**, 1649–1656.

Gautier, D.C. (1982) Siderite concretions: Indications of early diagenesis in the Gammon Shale (Cretaceous). *J. sedim. Petrol.* **52**, 859–872.

Gavelin, S., Parwel, A. & Ryhage, R. (1960) Sulphur isotope fractionation in sulphide mineralisation. *Econ. Geol.* **55**, 510–530.

Gavish, E. & Friedman, G.M. (1973) Quantititive analysis of calcite and Mg-calcite by X-ray diffraction: effect of grinding on peak height and peak area. *Sedimentology*, **20**, 437–444.

Gawthorpe, R.L. (1987) Burial dolomitization and porosity development in a mixed carbonate-clastic sequence: an example from the Bowland Basin, northern England. *Sedimentology*, **34**, 533–558.

Geake, J.E., Walker, G., Telfer, D.J. & Mills, A.A. (1977) The cause and significance of luminescence in lunar plagioclase. *Phil. Trans. R. Soc. Lond. A*, **285**, 403–408.

Gensmer, R.P. & Weiss, M.P. (1980) Accuracy of calcite/dolomite ratios by X-ray diffraction and comparison with results from staining techniques. *J. sedim. Petrol.* **50**, 626–629.

Gibbons, G.S. (1972) Sandstone imbrication study in planar sections: dispersion, biases and measuring methods. *J. sedim. Petrol.* **42**, 966–972.

Gibbs, R.J. (1967) Quantitative X-ray diffraction analysis using clay mineral standards extracted from the samples to be analysed. *Clay Miner.* **7**, 79–90.

Gibbs, R.J. (1977) Clay mineral segregation in the marine environment. *J. sedim. Petrol.* **47**, 237–243.

Gibbs, R.J., Mathews, M.D. & Link, D.A. (1971) The relationship between sphere size and settling velocity. *J. sedim. Petrol.* **41**, 7–18.

Gill, W.D., Khalaf, F.E. & Massoud, M.S. (1977) Clay minerals as an index of the degree of metamorphism of the carbonate and terrigenous rocks in the South Wales coalfield. *Sedimentology*, **24**, 675–691.

Gillett, S.L. (1983) Major, through-going stylolites in the Lower Ordovician Goodwin Limestone, southern Nevada: petrography with dating from paleomagnetism. *J. sedim. Petrol.* **53**, 209–291.

Gingerich, P.D. (1969) Markov analysis of cyclic alluvial sediments. *J. sedim. Petrol.* **39**, 330–332.

Gipson, M. (1963) Ultrasonic disaggregation of clays. *J. sedim. Petrol.* **33**, 955–958.

Given, R.K. & Lohmann, K.C. (1986) Isotopic evidence for the early meteoric diagenesis of the reef facies, Permian Reef Complex of West Texas and New Mexico. *J. sedim. Petrol.* **56**, 183–93.

Given, R.K. & Wilkinson, B.H. (1985) Kinetic control of morphology, composition and mineralogy of abiotic sedimentary carbonates. *J. sedim. Petrol.* **55**, 109–119.

Glennie, K.W. (1970) *Desert Sedimentary Environments.* Elsevier, Amsterdam.

Glennie, K.W., Mudd, G.C. & Nagtegaal, P.J.C. (1978) Depositional environment and diagenesis of Permian Rotliegendes sandstones in Leman Bank and Sole Pit areas of the UK southern North Sea. *J. geol. Soc. London,* **135**, 25–34.

Glover, E.D. (1961) Method of solution of calcareous materials using the complexing agent, EDTA. *J. sedim. Petrol.* **31**, 622–626.

Goddard, E.N., Trask, P.D., De Ford, R.K., Rove, O.N., Singlewald, J.T. & Overbeck, R.M. (1975) *Rock Color Chart.* Geological Society of America.

Godfrey, J.D. (1962) The deuterium content of hydrous minerals from the East-Central Sierra Nevada and Yosemite National Park. *Geochim. cosmochim. Acta*, **26**, 1215–1245.

Goldschmidt, V.M. & Peters, C. (1932) Zur Geochemie des Bors, I, II. *Nachr. Ges. Wiss. Gottingen, Math. Physik. Kl.* III: 402–407; IV: 528–545.

Goldsmith, J.R. & Graf, D.L. (1958a) Relations between lattice constraints and composition of the Ca-Mg carbonates. *Am. Miner.* **43**, 84–101.

Goldsmith, J.R. & Graf, D.L. (1958b) Structural and compositional variations in some natural dolomites. *J. Geol.* **66**, 678–693.

Goldsmith, J.R., Graf, D.L. & Heard, H.C. (1961) Lattice constants of the calcium-magnesium carbonates. *Am. Miner.* **46**, 453–457.

Goldstein, J.I., Newbury, D.E., Echlin, P., Joy, D.C., Fiori, C. & Lifshin, E. (1981) *Scanning Electron Microscopy and X-ray Microanalysis.* Plenum Press, New York.

Golubic, S., Brent, G. & Lecampion, T. (1974) Scanning electron microcopy of endolithic algae and fungi using a multipurpose casting-embedding technique. *Lethaia*, **3**, 203–209.

Golubic, S., Perkins, R.D. & Lukas, K.J. (1975) Boring microorganisms and microborings in carbonate substrates. In: *The Study of Trace Fossils* (Ed. by R.W. Frey), pp. 229–259. Springer-Verlag, Berlin.

Goodhew, P.J. & Chesco, D. (1980) Microanalysis in the transmission electron microscope. *Micron*, **11**, 153–181.

Goodman, L.A. (1968) The analysis of cross-classified data: independence, quasi-independence and interactions in contingency tables with or without missing entries. *J. Am, statist. Ass.* **63**, 1091–1131.

Gould, K.W. & Smith, J.W. (1979) The genesis and isotopic composition of carbonates associated with some Permian Australian coals. *Chem. Geol.* **24**, 137–150.

Gould, S.J. (1980) *The Mismeasure of Man.* Norton, New York.

Govindaraju, K. (1984) 1984 compilation of working values and sample description for 170 international reference samples of mainly silicate rocks and minerals. *Geostand. Newsl.* **8**.

Graham, D.W., Bender, M.L. Williams, D.F. & Keigwin, L.D. (1982) Strontium-calcium ratios in Cenozoic planktonic foraminifera. *Geochim. cosmochim. Acta*, **46**, 1281–1292.

Graham, J.R. (1982) Transition from basin-plain to shelf deposits in the Carboniferous flysch of southern Morocco. *Sediment. Geol.* **33**, 173–944.

Graham, J.R. (1983) Analysis of the Upper Devonian Munster Basin, an example of a fluvial distributary system. In: *Modern and Ancient Fluvial Systems* (Ed. by J.D. Collinson and J. Lewin), pp. 473–483. Spec. Publ. int. Ass. Sediment. 6. Blackwell Scientific Publications, Oxford.

Grant, P. (1978) The role of the scanning electron microscope in cathodoluminescence petrology. In: *Scanning Electron Microscopy in the Study of Sediments* (Ed. by W.B. Whalley), pp. 1–12. Geo Abstracts, Norwich.

Gray, D.I. & Benton, M.J. (1982) Multidirectional palaeocurrents as indicators of shelf storm beds. In: *Cyclic and Event Stratification* (Ed. by G. Einsele and A. Seilacher), pp. 350–353. Springer-Verlag, Berlin.

Greene-Kelly, R. (1973) The preparation of clay soils for determination of structure. *J. Soil Sci.* **24**, 277–283.

Gresens, R.L. (1981) The aqueous solubility product of solid solutions. 1. Stoichiometric saturation; partial and total solubility product. *Chem. Geol.* **32**, 59–72.

Griffin, J.J., Windom, H. & Goldberg, E.D. (1968) The distribution of clay minerals in the World Ocean. *Deep-Sea Res.* **15**, 433–459.

Griffin, O.G. (1954) A new internal standard for the quantitative X-ray analysis of shales and mine dust. *Res. Rep. No. 101, S.Af. Mines Res. Est. Ministry of Fuel and Power*, 1–25.

Griffiths, J.C. (1967) *Scientific Method in Analysis of Sedimentary*. McGraw-Hill, New York.

Grover, G. Hr & Read, J.E. (1983) Paleoaquifer and deep burial related cements defined by regional cathodoluminescent patterns, Middle Ordovician carbonates, Virginia. *Bull. Am. Ass. Petrol. Geol.* **67**, 1275–1303.

Gunatilaka, H.A. & Till, R. (1971) A precise and accurate method for the quantitative dermination of carbonate minerals by X-ray diffraction using a spiking technique. *Mineralog. Mag.* **38**, 481–487.

Halas, S. & Krouse, H.R. (1982) Isotopic abundance of water of crystallisation of gypsum from the Miocene evaporite formation, Carpathian Foredeep, Poland. *Geochim. cosmochim. Acta*, **46**, 293–296.

Hall, M.G. & Lloyd, G.E. (1981) The SEM examination of geological samples with a semi-conductor backscattered electron detector. *Am. Miner.* **66**, 362–368.

Halley, R.B. (1977) Ooid fabric and fracture in the Great Salt Lake and the geologic record. *J. sedim. Petrol.* **47**, 1099–1120.

Halley, R.B. (1978) Estimating pore and cement volumes in thin section. *J. sedim. Petrol.* **48**, 642–650.

Halley, R.B. & Harris, P.M. (1979) Fresh-water cementation of a 1,000-year-old oolite. *J. sedim. Petrol.* **49**, 969–988.

Hamblin, W.K. (1971) X-ray photography. In: *Procedures in Sedimentary Petrology* (Ed. by R.E. Carver), pp. 251–284. Wiley-Interscience, New York.

Hancock, N.J. (1978a) An application of scanning electron microscopy in pilot water injection studies for oilfield development. In: *Scanning Electron Microscopy in the Study of Sediments* (Ed. by W.B. Whalley), pp. 61–70. Geo Abstracts, Norwich.

Hancock, N.J. (1978b) Possible causes of Rotliegend sandstone diagenesis in northern West Germany. *J. geol. Soc. London,* **135**, 35–40.

Hancock, N.J. & Taylor, A.M. (1978) Clay mineral diagenesis and oil migration in the Middle Jurassic Brent Sand Formation. *J. geol. Soc. London,* **135**, 69–72.

Hansen, W.R. (1960) Improved Jacob Staff for measuring inclined stratigraphic intervals. *Bull. Am. Ass. Petrol. Geol.* **44**, 252–254.

Hansley, P.L. (1987) Petrologic and experimental evidence for the etching of garnets by organic acids in the Upper Jurassic Morrison Formation, northwestern New Mexico. *J. sedim. Petrol.* **57**, 666–681.

Hantzschel, W. (1975) Trace fossils and problematica. Part W. In: *Treatise on Invertebrate Paleontology*. University of Kansas Press.

Harder, H. (1980) Syntheses of glauconite at surface temperatures. *Clays Clay Min.* **28**, 217–222.

Hardie, L.A. (1984) Evaporites: marine or non-marine. *Am. J. Sci.* **284**, 193–240.

Harms, J.C., Southard, J.B. & Walker, R.G. (1982) *Structures and sequences in clastic rocks*. Soc. econ. Paleont. Mineral. Short Course Notes 9.

Harper, D.W. (1984) Improved methods of facies sequence analysis. In: *Facies Models* (2nd edn) (Ed. by R.G. Walker), pp. 11–13. Geoscience, Canada.

Harrell, J. (1984) A visual comparator for degree of sorting in thin and plane sections. *J. sedim. Petrol.* **54**, 646–650.

Harrell, J. & Eriksson, K.A. (1979) Empirical conversion equations for thin-section and sieve derived size distribution parameters. *J. sedim. Petrol.* **49**, 273–280.

Harris, D.C. & Kendall, A.C. (1986) Secondary porosity in Upper Lisburne Group carbonates (Wahoo Limestone: Lower Pennsylvanian); North Slope, Alaska, (abstract). *SEPM Annual Midyear Meeting Abstracts*, **3**, 50.

Harris, P.M., Kendall, C.G.St.C. & Lerche, I. (1985) Carbonate cementation — a brief review. In: *Carbonate Cements* (Ed. by N. Schneidermann and P.M. Harris), pp. 79–96. Soc. econ. Paleont. Miner. Spec. Publ. 36.

Harvey, P.K. (1974) The detection and correction of outlying determinations that may occur during geochemical analysis. *Geochim. cosmochim. Acta*, **38**, 435–451.

Harvey, P.K. & Ferguson, C.C. (1976) On testing orientation data for goodness of fit to a von Mises distribution. *Computers Geosci.* **2**, 261–268.

Harvie, C.E., Moller, N & Weare, J.H. (1984) The prediction of mineral solubilities in natural water: the Na-K-Mg-Ca-H-SO_4-OH-HCO_3-CO_3-CO_2-H_2O system to high ionic strengths at 25°C. *Geochim. cosomochim. Acta*, **48**, 723–725.

Harwood, G.M. (1980) Calcitized anhydrite and associated sulphides in the English Zechstein First Cycle Carbonate (EZ1 Ca). In: *The Zechstein Basin with Emphasis on Carbonate Sequences* (Ed. by H. Fuchtbauer and T.M. Peryt), pp. 61–72. Contr. Sedimentology, 9.

Harwood, G.M. (1986) The diagenetic history of the

Cadeby Formation (EZ1 Ca), Upper Permian, eastern England. In: *The English Zechstein and Related Topics* (Ed. by G.M. Harwood and D.B. Smith), pp. 75–86. Spec. Publ. Geol. Soc. Lond. 22. Blackwell Scientific Publications, Oxford.

Harwood, G.M. & Moore, C.H. Jr (1984) Comparative sedimentology and diagenesis of Upper Jurassic ooid grainstone sequences, East Texas Basin. In: *Carbonate Sands: a Core Workshop* (Ed. by P.M. Harris), pp. 176–232. Soc. econ. Paleont. Miner. Core Workshop No. 5.

Hattori, I. (1976) Entropy in Markov chains and discrimination of cyclic patterns in lithologic successions. *Math. Geol.* **8**, 477–497.

Hawkins, P.J. (1978) Relationship between diagenesis, porosity reduction, and oil emplacement in late Carboniferous sandstone reservoirs, Bothamsall Oilfield, E. Midlands. *J. geol. Soc. London*, **135**, 7–24.

Hay, W.W., Wise, S.W. & Stieglitz, R.D. (1970) Scanning electron microscope study of fine grain size biogenic carbonate particles. *Trans. Gulf Coast Ass. Geol. Soc*, **20**, 287–302.

Hayat, M.A. (ed.) (1974–78) *Principles and Techniques of Scanning Electron Microscopy*, Vols 1–6. Van Nostrand Reinhold, New York.

Hayes, J.B. (1979) Sandstone diagenesis — the hole truth. In: *Aspects of Diagenesis* (Ed. by P.A. Scholle and P.R. Schlager), pp. 127–140. Soc. Econ. Paleont. Min. Spec. Publ. No. 26.

Heald, M.T. (1955) Stylolites in sandstones. *J. Geol.* **63**, 101–114.

Heald, M.T. (1956) Cementation of Triassic arkoses in Connecticut and Massachusetts. *Bull. geol. Soc. Am.* **67**, 1133–1154.

Heald, M.T. (1959) Significance of stylolites in permeable sandstones. *J. sedim. Petrol.* **29**, 251–253.

Hein, J.R., Scholl, D.W. & Gutmacher, C.E. (1976) Neogene clay minerals of the far northwest Pacific and southern Bering Sea: sedimentation and diagenesis. In: *Proc. Int. Clay Conf.*, 1975, Mexico City (Ed. by S.W. Bailey), pp. 71–80.

Heinrich, K.F.J. (1981) *Electron Beam X-ray Microanalysis*. Van Nostrand Reinhold, New York.

Helgeson, H.C., Delany, J.M., Nesbitt, H.W. & Bird, D.K. (1978) Summary and critique of the thermodynamic properties of rock-forming minerals. *Am. J. Sci.* **278A**.

Hendy, C.H. (1971) The isotopic geochemistry of speleothems — I. The calculation of the effects of different modes of formation on the isotopic composition of speleothems and their applicability as palaeoclimatic indicators. *Geochim. cosmochim. Acta*, **35**, 801–824.

Herrmann, A.G. (1980) Bromide distribution between halite and NaCl-saturated seawater. *Chem. Geol.* **28**, 171–177.

Hess, J., Bender, M.L. & Schilling, J.-G. (1986) Evolution of Strontium-87 to Strontium-86 from Cretaceous to present. *Science*, **231**, 979–984.

Heward, A.P. (1978) Alluvial fan and lacustrine sediments from the Stephanian A and B (La Magdalena, Cinera-Matallane and Sabero) coalfields, northern Spain. *Sedimentology*, **25**, 451–488.

Hickman, A.H. & Wright, A.E. (1983) Geochemistry and chemostratigraphical correlation of slates, marbles and quartzites of the Appin Group, Argyll, Scotland. *Trans. R. Soc. Edinb.* **73**, 251–278.

Higgs, R. (1979) Quartz grain surface features of Mesozoic-Cenozoic sands from the Labrador and Western Greenland continental margins. *J. sedim. Petrol.* **49**, 599–610.

Hill, P.R. & Nadeau, O.C. (1984) Grain-surface textures of late Wisconsinan sands from the Canadian Beaufort Shelf. *J. sedim. Petrol.* **54**, 1349–1357.

Hinckley, D.N. (1963) Variability in "crystallinity" values among the kaolin deposits of the coastal plain of Georgia and South Carolina. In: *Clays, Clay Min., Proc. 11th Conf.* (Ed. by E. Ingerson), pp. 229–235. Pergamon Press, Oxford.

Hingston, F.J. (1964) Reactions between boron and clays. *Aust. J. Soil Res.* **2**, 83–95.

Hird, K. (1986) Petrography and geochemistry of some Carboniferous and Precambrian dolomites. *Unpublished Thesis*. University of Durham.

Hoefs, J. (1980) *Stable Isotope Geochemistry* (2nd edn). Springer-Verlag, Berlin.

Holland, H.D. (1972) The geologic history of sea-water — an attempt to solve the problem. *Geochim. cosmochim. Acta*, **36**, 637–651.

Holland, H.D. (1978) *The Chemistry of the Atmosphere and Oceans*. Wiley-Interscience, New York.

Holland, H.D. (1984) *The Chemical Evolution of the Atmosphere and Oceans*. Princeton University Press.

Holser, W.T. (1979) Trace elements and isotopes in evaporites. In: *Marine Minerals* (Ed. by R.G. Burns), pp. 295–346. Rev. Mineral. Min. Soc. Am. 6.

Holt, B.D. & Engelkemeir, A.G. (1970) Thermal decomposition of barium sulphate to sulphur dioxide for mass spectrometric analysis. *Anal. Chem.* **42**, 1451–1453.

Hooton. D.H. & Giorgetta, N.E. (1977) Quantitative X-ray diffraction analysis by a direct calculation method. *X-ray Spectrom.* **6**, 2–5.

Horne, J.C., Ferm, J.C., Caruccio, F.T. & Baganz, B.P. (1978) Depositional models in coal exploration and mine planning in Appalachian region. *Bull. Am. Ass. Petrol. Geol.* **62**, 2379–2411.

Hostettler, J.D. (1984) Electrode electrons, aqueous electrons and redox potentials in natural waters. *Am. J. Sci.* **284**, 734–759.

Houghton, H. F. (1980) Refined techniques for staining plagioclase and alkali feldspars in thin section. *J. sedim. Petrol.* **50**, 629–631.

Hounslow, A.W. (1979) Modified gypsum/anhydrite stain. *J. sedim. Petrol.* **49**, 636–637.

Howarth, R.J. (1977) Automatic generation of randomised sample submittal schemes for laboratory analysis. *Comp. Geosci.* **3**, 327–334.

Hower, J., Eslinger, E.V., Hower, M.E. & Perry, E.A. (1976) Mechanism of burial metamorphism of argillaceous sediment. *Bull. geol. Soc. Am.* **87**, 725–737.

Hsü, K.J. & Jenkyns, H.C (eds) (1974) *Pelagic Sediments: on Land and under the Sea.* Spec. Publ. Int. Ass. Sediment. 1. Blackwell Scientific Publications, Oxford.

Huang, W.L., Bishop, A.M. & Brown, R.W. (1986) The effect of fluid/rock ratio on feldspar dissolution and illite formation under reservoir conditions. *Clay Miner.* **21**, 585–601.

Hubert, J.F. (1971) Analysis of heavy-mineral assemblages. In: *Procedures in Sedimentary Petrology* (Ed. by R.E. Carver), pp. 452–478. Wiley-Interscience, New York.

Hubert, J.F. & Hyde, M.G. (1982) Sheet flow deposits of graded beds and mudstones on an alluvial sandflat-playa system: Upper Triassic Blomiden red beds, St. Mary's Bay, Nova Scotia. *Sedimentology*, **29**, 457–474.

Hubert, J.F. & Mertz, K.A. (1984) Eolian sandstones in Upper Triassic — Lower Jurassic red beds of the Fundy Basin, Nova Scotia. *J. sedim. Petrol.* **54**, 788–810.

Hudson, J.D. (1975) Carbon isotopes and limestone cements. *Geology*, **3**, 19–22.

Hudson, J.D. (1977a) Stable isotopes and limestone lithification. *J. geol. Soc. London*, **133**, 637–660.

Hudson, J.D. (1977b) Oxygen isotope studies of Cenozoic temperatures, oceans and ice accumulation. *Scott. J. Geol.* **13**, 313–325.

Huggett, J.M. (1984a) An SEM study of phyllosilicates in a Westphalian Coal Measures sandstone using backscattered electron imagery and wavelength dispersive spectral analysis. *Sediment. Geol.* **40**, 233–247.

Huggett, J.M. (1984b) Controls on mineral authigenesis in Coal Measures sandstones of the East Midlands, UK. *Clay Miner.* **19**, 343–357.

Huggett, J.M. (1986) An SEM study of phyllosilicate diagenesis in sandstones and mudstones in the Westphalian coal measures using back-scattered electron microscopy. *Clay Miner.* **21**, 603–616.

Hurst, A.R. (1981) A scale of dissolution for quartz and its implications for diagenetic processes in sandstones. *Sedimentology*, **28**, 451–459.

Hurst, A.R. & Irwin, H. (1982) Geological modelling of clay diagenesis in sandstones. *Clay Miner.* **17**, 5–22.

Hurst, J.M. & Surlyk, F. (1984) Tectonic control of Silurian carbonate-shelf margin morphology and facies, North Greenland. *Bull. Am. Ass. Petrol. Geol.* **68**, 1–17.

Hutcheon, I, (1981) Applications of thermodynamics to clay minerals and authigenic mineral equilibria. In: *Short Course in Clays and the Resource Geologist* (Ed. by F.J. Longstaffe), pp. 169–193. Min. Ass. Can. Short Course Handbook 7.

Hutcheon, I. (1984) A review of artificial diagenesis during thermally enhanced recovery. In: *Clastic Diagenesis* (Ed. by D.A. McDonald and R.C. Surdam), pp. 413–429. Mem. Am. Ass. Petrol. Geol. 37.

Hutchison, C.S. (1971) *Laboratory Handbook of Petrographic Techniques*. Wiley-Interscience, New York.

Iman, M.B. & Shaw, H.F. (1985) The diagenesis of Neogene clastic sediments from the Bengal Basin, Bangladesh. *J. sedim. Petrol.* **55**, 665–671.

Ingersoll, R.V., Bullard, T.F., Ford, R.L., Grim, J.P., Puckle, J.D. & Sares, S.W. (1984) The effect of grain size on detrital modes: a test of the Gazzi-Dickinson point-counting method. *J. sedim. Petrol.* **54**, 103–116.

Ingram, R.L. (1953) Fissility of mudrocks. *Bull. geol. Soc. Am.* **64**, 869–878.

Ingram, R.L. (1954) Terminology for the thickness of stratification and parting units in sedimentary rocks. *Bull. geol. Soc. Am.* **65**, 935–938.

Inmam, D.L. (1952) Measures for describing the size distribution of sediments. *J. sedim. Petrol.* **22**, 125–45.

Ireland, B.J., Curtis, C.D. & Whiteman, J.A. (1983) Compositional variation within some glauconites and illites and implications for their stability and origins. *Sedimentology*, **30**, 769–786.

Ireland, H.A. (1973) Xeroxing for reproduction of rock sections. *Bull. geol. Soc. Am.* **57**, 2280–2281.

Irwin, H., Curtis, C.D. & Coleman, M. (1977) Isotopic evidence for source of diagenetic carbonates in organic-rich sediments. *Nature*, **269**, 209–213.

Ishikawa, M. & Ichikuni, M. (1984) Uptake of sodium and potassium by calcite. *Chem. Geol.* **42**, 137–146.

Ixer, R.A., Turner, P. & Waugh, B. (1979) Authigenic iron and titanium oxides in Triassic red beds: (St. Bees Sandstone), Cumbria, Northern England. *Geol. J.* **14**, 179–192.

Jackson, K.S., Jonasson, I.R. & Skippen, G.B. (1978) The nature of metals-sediment-water interactions in freshwater bodies, with emphasis on the role of organic matter. *Earth Sci. Rev.* **14**, 97–146.

Jackson, M.L. (1958) *Soil Chemical Analysis.* Prentice-Hall, Englewood Cliffs, New Jersey.

Jackson, M.L. (1963) Interlayering of expansible layer silicates in soils by chemical weathering. *Proc. 11th Nat. Conf. Clays and Clay Min.*, Ottawa, Ontario (1962), pp. 29–46.

Jackson, M.L. (1979) *Soil Chemical Analysis — Advanced Course* (2nd edn). Published by the author.

Jackson, M.L., Sayin, M. & Clayton, R.N. (1976) Hexafluorosilicic acid reagent modification for quartz isolation. *Soil Sci. Soc. Am. J.* **40**, 958–960.

Jackson, M.L., Whitting, L.D. & Pennington, R.P. (1950)

Segregation procedures for mineralogical analysis of soils. *Proc. Soil. Sci. Soc. Am.* **14**, 77–81.

Jackson, T.A., Fritz, P. & Drimmie, R. (1978) Stable carbon isotope ratios and chemical properties of kerogen and extractable organic matter in Pre-Phanerozoic and Phanerozoic sediments — their inter-relations and possible paleobiological significance. *Chem. Geol.* **21**, 335–350.

Jacobs, M.B. (1974) Clay mineral changes in Antarctic deep-sea sediments and Cenozoic climatic events. *J. sedim. Petrol.* **44**, 1079–1056.

Jacobsen, R.L. & Usdowski, H.E. (1976) Partitioning of strontium between calcite, dolomite and liquids: an experimental study under high temperature diagenetic conditions and a model for the prediction of mineral pairs for geothermometry. *Contr. Miner. Petrol.* **59**, 171–185.

James, N.P. & Choquette, P.W. (1983) Diagenesis 6. Limestones — the sea floor diagenetic environment. *Geosci. Can.* **10**, 162–179.

James, N.P. & Choquette, P.W. (1984) Diagenesis 9. Limestone — the meteoric diagenetic environment. *Geosci. Can.* **11**, 161–194.

James, N.P., Ginsburg, R.N., Marszalek, D.S. & Choquette, P.W. (1976) Facies and fabric specificity of early subsea cements in shallow Belize (British Honduras) reefs. *J. sedim. Petrol.* **46**, 523–544.

James, W.C., Wilmar, G.C. & Davidson, B.G. (1986) Roles of quartz type and grain size in silica diagenesis, Nuggett sandstone, south-central Wyoming. *J. sedim. Petrol.* **56**, 657–662.

Jarosewich, E. & Macintyre, I.G. (1983) Carbonate reference samples for electron microprobe and scanning electron microscope analyses. *J. sedim. Petrol.* **53**, 677–678.

Javoy, M. (1977) Stable isotopes and geothermometry. *J. geol. Soc. London*, **133**, 609–636.

JCPDS (1974) Joint Committee for the Powder Diffraction Standards: *Selected Powder Diffraction Data for Minerals.* JCPDS, Pennsylvania, USA.

Jeans, C.V. (1978) The origin of the Triassic clay assemblages of Europe with special reference to the Keuper Marl and Rhaetic of parts of England. *Phil. Trans. R. Soc. Lond. A*, **289**, 549–639.

Jeffery, P.G. & Hutchison, D. (1981) *Chemical Methods of Rock Analysis* (3rd edn). Pergamon Press, Oxford.

Jenkins, R. & de Vries, J.L. (1970) *Practical X-ray Spectrometry.* MacMillan, London.

Jennings, S. & Thompson, G.R. (1986) Diagenesis of Plio–Pleistocene sediments of the Colorado River Delta, Southern California. *J. sedim. Petrol.* **56**, 89–98.

Jipa, D. (1966) Relationship between longitudinal and transversal currents in the Paleogene of the Tarcau Valley (Eastern Carpathians). *Sedimentology*, **7**, 299–305.

Johansson, C.E. (1965) Structural studies of sedimentary deposits. *Geol. För. Stockh. Förh.*, **87**, 3–61.

Johns, W.D., Grim, R.E. & Bradley, W.F. (1954) Quantitative estimations of clay minerals by diffraction methods. *J. sedim. Petrol.* **24**, 242–251.

Johnson, H.D. (1975) Tide- and wave-dominated inshore and shoreline sequences from the late Precambrian, Finnmark, North Norway. *Sedimentology*, **22**, 45–74.

Johnson, W.M. & Maxwell, J.A. (1981) *Rock and Mineral Analysis*. Wiley, New York.

Jones, J.B & Segnit, E.R. (1971) The nature of opal. *J. geol. Soc. Aust.* **18**, 57–68.

Jorgensen, N.O. (1983) Dolomitization in chalk from the North Sea Central Graben. *J. sedim. Petrol.* **53**, 557–564.

Jorgensen, N.O. (1987) Oxygen and carbon isotope compositions of Upper Cretaceous chalk from the Danish sub-basin and the North Sea Graben. *Sedimentology*, **34**, 559–570.

Joslin, I. (1977) A tool to collect rock samples for scanning electron microscopy. *J. sedim. Petrol.* **47**, 1363–1364.

Jourdan, A., Thomas, M., Brevart, O., Robson, P., Sommer, F. & Sullivan, M. (1987) Diagenesis as the control of the Brent sandstone reservoir properties in the Greater Alwyn area (East Shetland Basin). In: *Petroleum Geology of north-east Europe* (Ed. by J. Brooks and K.W. Glennie), pp. 951–961. Graham & Trotman, London.

Kaldi, J., Krinsley, D.J. & Lawson, D. (1978) Experimentally produced aeolian surface textures on quartz sand grains from various environments. In: *Scanning Electron Microscopy in the Study of Sediments* (Ed. by W.B. Whalley), pp. 261–277. Geo. Abstracts, Norwich.

Kantorowicz, J.D. (1985) The petrology and diagenesis of Middle Jurassic clastic sediments, Ravenscar Group, Yorkshire. *Sedimentology*, **32**, 833–853.

Kaplan, I.R. (1983) Stable isotopes of sulfur, nitrogen and deuterium in Recent marine environments. In: *Stable Isotopes in Sedimentary Geology,* pp. 2–1 to 2–108. Soc. econ. Paleont. Miner. Short Course No. 10.

Karhu, J. & Epstein, S. (1986) The implication of the oxygen isotope records in coexisting cherts and phosphates. *Geochim. cosmochim. Acta*, **50**, 1745–1756.

Kastner, M. (1971) Authigenic feldspars in carbonate rocks. *Am. Miner.* **56**, 1403–1442.

Katz, A. (1973) The interaction of magnesium with calcite during crystal growth at 25–90°C and one atmosphere pressure. *Geochim. cosmochim. Acta*, **37**, 1563–1586.

Katz, A & Friedman, G.M. (1965) The preparation of stained acetate peels for the study of carbonate rocks. *J. sedim. Petrol.* **35**, 248–249.

Katz, A., Kolodny, Y, & Nissenbaum, A. (1977) The geochemical evolution of the Pleistocene Lake Lisan — Dead Sea system. *Geochim. cosmochim. Acta*, **41**, 1609–1626.

Katz, A., Sass, E. & Starinsky, A. (1972) Strontium behaviour in the aragonite-calcite transformation: an experimental study at 40–98°C. *Geochim. cosmochim. Acta*, **36**, 481–496.

Keir, R.S. (1980) The dissolution kinetics of biogenic calcium carbonate in seawater. *Geochim. cosmochim. Acta*, **44**, 241–252.

Keith, B.D. & Pittman, E.D. (1983) Bimodal porosity in oolitic reservoir-effect on productivity and log response, Rodessa limestone (Lower Cretaceous) East Texas Basin. *Bull. Am. Ass. Petrol. Geol.* **67**, 1391–1399.

Kelling, G. (1969) The environmental significance of cross-stratification parameters in an Upper Carboniferous fluvial basin. *J. sedim. Petrol.* **39**, 857–875.

Kemmis, T.J. & Hallberg, G.R. (1984) Discussion. Lithofacies types and vertical profile analysis; an alternative approach to the description and environmental interpretation of glacial diamict and diamictite sequences. *Sedimentology*, **31**, 886–890.

Kendall, A.C. (1984) Evaporites. In: *Facies Models* (2nd edn) (Ed. by R.G. Walker), pp. 259–296. Geoscience Canada.

Kennedy, S.K., Meloy, T.P. & Durney, T.E. (1985) Sieve data — size and shape information. *J. sedim. Petrol.* **55**, 356–360.

Kerr, P.F. (1959) *Optical Mineralogy*. McGraw-Hill, London.

Kiersch, G.A. (1950) Small scale structures and other features of Navajo Sandstone, northern part of San Rafael Swell, Utah. *Bull. Am. Ass. Petrol. Geol.* **34**, 923–942.

Killingley, J.S. (1983) Effects of diagenetic recrystallisation on $^{18}O/^{16}O$ values of deep-sea sediments. *Nature*, **301**, 594–597.

Kittleman, L.R. (1964) Application of Rosin's Distribution to size-frequency analysis of clastic rocks. *J. sedim. Petrol.* **34**, 483–502.

Kiyosu, Y. (1980) Chemical reduction and sulphur isotope effects of sulphate. *Chem. Geol.* **30**, 47–56.

Klappa, C.F. (1979) Calcified filaments in Quaternary calcretes: organo-mineral interactions in the subaerial vadose environment. *J. sedim. Petrol.* **49**, 955–968.

Klappa, C.F. (1980) Rhizoliths in terrestrial carbonates: classification, recognition, genesis and significance. *Sedimentology*, **27**, 613–629.

Klein, G. de V. (1967) Paleocurrent analysis in relation to modern marine sediment dispersal patterns. *Bull. Am. Ass. Petrol. Geol.* **51**, 366–382.

Klovan, J.E. (1966) The use of factor analysis in determining depositional environments from grain-size distributions. *J. sedim. Petrol.* **36**, 115–125.

Klug, H.P. & Alexander, L.E. (1974) *X-ray Diffraction Procedures for Polycrystalline and Amorphous Material*. Wiley, New York.

Knauth, L.P. & Beeunas, M.A. (1986) Isotope geochemistry of fluid inclusions in Permian halite with implications for the isotopic history of ocean water and the origin of saline formation waters. *Geochim. cosmochim. Acta*, **50**, 419–433.

Knauth, L.P. & Epstein, S. (1976) Hydrogen and oxygen isotope ratios in nodular and bedded cherts. *Geochim. cosmochim. Acta*, **40**, 1095–1108.

Knauth, L.P. & Lowe, D.R. (1978) Oxygen isotope geochemistry of cherts from the Onverwacht Group (3–4 billion years), Transvaal, South Africa with implications for similar variations in the isotopic composition of cherts. *Earth planet. Sci. Lett.* **41**, 209–222.

Knebel, J.E., Conomos, T.J. & Commeau, J.A. (1977) Clay-mineral variability in the suspended sediments of the San Francisco Bay system, California. *J. sedim. Petrol.* **47**, 229–236.

Koch, G.S. & Link, R.F. (1970) *Statistical Analysis of Geological Data*. Wiley, New York.

Kocurek, G. (1981) Erg reconstruction: the Entrada Sandstone (Jurassic) of northern Utah & Colorado. *Palaeogeogr. Palaeoclim. Palaeoecol.* **36**, 125–153.

Kodama, H. & Oinuma, K. (1963) Identification of kaolin minerals in the presence of chlorite by X-ray diffraction and infrared absorption spectra. In: *Clays, Clay Min. Proc. 11th Conf.* (Ed. by E. Ingerson), pp. 236–249. Pergamon Press, Oxford.

Koepnick, R.B. (1985) Impact of stylolites on carbonate reservoir continuity: example from Middle East (abstracts). *Bull. Am. Ass. Petrol. Geol.* **69**, 274.

Kolesar, P.T. (1978) Magnesium in calcite from a coralline alga. *J. sedim. Petrol.* **48**, 815–820.

Komar, P.D. & Cui, B. (1984) The analysis of grain-size measurements by sieving and settling-tube techniques. *J. sedim. Petrol.* **54**, 603–614.

Koster, E.H., Rust, B.R. & Gendzwill, D.J. (1980) The ellipsoidal form of clasts with apolications to fabric and size analysis of fluvial gravels. *Can. J. Earth Sci.* **17**, 1725–1739.

Kottlowski, F.E. (1965) *Measuring Stratigraphic Sections*. Holt, Rinehart & Winston, New York.

Krauskopf, K.B. (1979) *Geochemistry* (2nd edn). McGraw-Hill, Tokyo.

Kretz, R. (1982) A model for the distribution of trace elements between dolomite and calcite. *Geochim. cosmochim. Acta*, **46**, 1979–1981.

Krinsley, D.H. & Donahue, J. (1968) Environmental interpretation of sand grain surface textures by electron microscopy. *Bull. geol. Soc. Am.* **79**, 743–748.

Krinsley, D.H. & Doornkamp, J.C. (1973) *Atlas of Quartz Sand Surface Textures*. Cambridge University Press.

Krinsley, D.H., Friend, P. & Klimentidis, R. (1976) Eolian transport textures on the surfaces of sand grains of early Triassic age. *Bull. geol. Soc. Am.* **87**, 130–132.

Krinsley, D.H. & McCoy, F.W. (1977) Significance and origin of surface textures on broken sand grains in deep-sea sediments. *Sedimentology*, **24**, 857–862.

Krinsley, D.H., Pye, K. & Kearsley, A.T. (1983) Application of backscattered electron microscopy in shale petrology. *Geol. Mag.* **120**, 109–114.

Krinsley, D.H. & Tovey, N.K. (1978) Cathodoluminescence in quartz sand grains. In: *Scanning Electron Microscopy 1978*, Vol. 1 (Ed. by O. Johari), pp. 887–894. SEM Inc., O'Hare, Illinois.

Krinsley, D.H. & Wellendorf, W. (1980) Wind velocities determined from the surface textures of sand grains. *Nature*, **283**, 372–373.

Krumbein, W.C. (1934) Size frequency distributions of sediments. *J. sedim. Petrol.* **4**, 65–77.

Krumbein, W.C. (1964) Some remarks on the phi notation. *J. sedim. Petrol.* **34**, 195–196.

Krumbein, W.C. & Graybill, F.A. (1965) *An Introduction to Statistical Models in Geology*. Wiley, New York.

Krumbein, W.C. & Pettijohn, F.J. (1961) *Manual of Sedimentary Petrography*. Appleton-Century-Crofts, New York.

Krumbein, W.C. & Sloss, L.L. (1963) *Stratigraphy and Sedimentation* (2nd edn). W.H. Freeman, San Francisco.

Krynine, P.D. (1940) Petrology and genesis of the Third Bradford Sand. *Bull. Penn. State College Mineral Industries Expt Sta.* **29**, 134.

Krynine, P.D. (1946) Microscopic morphology of quartz types. *Anal. 2nd Cong. Panam. Ing. Min. Geol.* **3**, 35–49.

Kubler, B. (1968) Evaluation quantitative du metamorphisme par la cristallinité de l'illite. *Bull. Centre Rech. Pau.* **2**, 385–397.

Kuenen, Ph. H. (1968) Settling convection and grain size analysis. *J. sedim. Petrol.* **38**, 817–831.

Kuijpers, E.P. (1972) Upper Devonian tidal deposits and associated sediments south and south-west of Kinsale (Southern Ireland). *PhD thesis*, University of Utrecht.

Kushnir, J. (1980) The coprecipitation of strontium, magnesium, sodium, potassium and chloride with gypsum, an experimental study. *Geochim. cosmochim. Acta*, **44**, 1471–1482.

Kushnir, J. (1982) The partitioning of seawater cations during the transformation of gypsum to anhydrite. *Geochim. cosmochim. Acta*, **46**, 433–446.

Labeyrie, L.D. & Juillet, A. (1982) Oxygen isotopic exchangeability of diatom valve silica; interpretation and consequences for paleoclimatic studies. *Geochim. cosmochim. Acta*, **46**, 967–975.

Lajoie, J. (1984) Volcaniclastic rocks. In: *Facies Models* (2nd edn) (Ed. by R.G. Walker), pp. 39–52. Geoscience Canada.

Laming, D. (1966) Imbrication, palaeocurrents and other sedimentary features in the lower New Red Sandstone, Devonshire. *J. sedim. Petrol.* **36**, 940–959.

Land, L.S. (1973) Holocene meteoric dolomitization of Pleistocene limestones, N. Jamaica. *Sedimentology*, **20**, 411–424.

Land, L.S. (1980) The isotopic and trace element geochemistry of dolomite: the state of the art. In: *Concepts and Models of Dolomitization* (Ed. by D.A. Zenger, J.B. Dunham and R.L. Ethington), pp. 87–110. Soc. econ. Paleont. Miner. Spec. Publ. 28.

Land, L.S., Behrens, E.W. & Frishman, S.A. (1979) The ooids of Baffin Bay, Texas. *J. sedim. Petrol.* **49**, 1269–1278.

Land, L.S., Long, J.C. & Barnes, D.J. (1977) On the stable carbon and oxygen isotopic composition of some shallow water, ahermatypic, scleractinian coral skeletons. *Geochim. cosmochim. Acta*, **37**, 169–172.

Landergen, S. (1945) Contribution to the geochemistry of boron. *Arkiv. Kemi.* **25**, 1–7; **26**, 1–31.

Landergren, S. & Carvajal, M.C. (1969) Geochemistry of boron. III The relationship between boron concentration in marine clay sediments and the salinity of the depositional environments expressed as an adsorption isotherm. *Arkiv. Mineral. Geol.* **5**, 13–22.

Landim, P.M.B. & Frakes, L.A. (1968) Distinction between tills and other diamictons based on textural characteristics. *J. sedim. Petrol.* **38**, 1213–1223.

Lanesky, D.E., Logan, B.W., Brown, R.G. & Hine, A.C. (1979) A new approach to portable vibracoring underwater and on land. *J. sedim. Petrol.* **49**, 654–657.

Last, W.M. (1982) Holocene carbonate sedimentation in Lake Manitoba, Canada. *Sedimentology*, **29**, 691–704.

Lawrence, G.P. (1977) Measurment of pore sizes in fine-textured soils: A review of existing techniques. *J. Soil Sci.* **28**, 527–540.

Leeder, M.R. (1973) Sedimentology and palaeogeography of the Upper Old Red Sandstone in the Scottish Border Basin. *Scott. J. Geol.* **9**, 117–144.

Lees, A. (1958) Etching technique for use on thin sections of limestones. *J. sedim. Petrol.* **28**, 200–202.

Lees, A. (1962) Sedimentology finds a new use for a standard industrial measuring projector. *Res. Develop. Indust.* July, 2 pp.

Le Maitre, R.W. & Haukka, M.T. (1973) The effect of prolonged X-ray irradiation on lithium tetraborate glass discs as used in XRF analysis. *Geochim. cosmochim. Acta*, **37**, 708–710.

Le Ribault, L. (1978) The exoscopy of quartz sand grains. In: *Scanning Electron Microscopy in the Study of Sediments* (Ed. by W.B. Whalley), pp. 319–328. Geo Abstracts, Norwich.

Lerman, A. (1966) Boron in clays and estimation of paleosalinity. *Sedimentology*, **6**, 267–286.

Lerman, A. (1979) *Geochemical Processes. Water and Sediment Environments*. Wiley-Interscience, New York.

Le Roux, J., Clayton, R.N. & Jackson, M.L. (1980) Oxygen isotopic ratios in fine quartz silt from sediments and soils of southern Africa. *Geochim. cosmochim. Acta*, **44**, 533–538.

Leventhal, J.S. & Hosterman, J.W. (1982) Chemical and mineralogical analysis of Devonian black-shale samples from Martin County, Kentucky; Caroll and Washington Counties, Ohio; Wise County, Virginia; and Overton County, Tennessee, U.S.A. *Chem. Geol.* **37**, 239–264.

Lewis, D.W. (1984) *Practical Sedimentology*. Hutchinson & Ross, Stroudsburg.

Li, Y-H. (1981) Ultimate removal mechanisms of elements from the oceans. *Geochim. cosmochim. Acta*, **45**, 1659–1664.

Li, Y-H. & Gregory, S. (1974) Diffusion of ions in seawater and deep-sea sediments. *Geochim. cosmochim. Acta*, **38**, 703–714.

Lindholm, R.C. & Dean, D.A. (1973) Ultra-thin thin sections in carbonate petrology: a valuable tool. *J. sedim. Petrol.* **43**, 295–297.

Link, M.H., Squires, R.L. & Colburn, I.P. (1984) Slope and deep sea fan facies and palaeogeography of Upper Cretaceous Chatsworth Formation, Simi Hills, California. *Bull. Am. Ass. Petrol. Geol.* **68**, 850–873.

Lippmann, F. (1977) The solubility products of complex minerals, mixed crystals and three layer clay minerals. *Neues J. Miner. Abh.* **130**, 243–263.

Lister, B. (1978) The preparation of polished sections. *Rep. Inst. Geol. Sci.* No. 78/27.

Lloyd, R.M. (1966) Oxygen isotope enrichment of seawater by evaporation. *Geochim. cosmochim. Acta*, **30**, 801–814.

Logan, B.W., Read, J.F., Hagan, G.M., Hoffman, P., Brown, R.G., Woods, P.J. & Gebelein, C.D. (1974) Evolution and diagenesis of Quaternary carbonate sequences, Shark Bay, Western Australia. *Mem. Am. Ass. Petrol. Geol.* **22**.

Logan, B.W. & Semeniuk, V. (1976) Dynamic metamorphism; processes and products in Devonian carbonate rocks, Canning Basin, Western Australia. *Geol. Soc. Aust. Spec. Publ.* 16.

Long, D.G.F. & Young, G.M. (1978) Dispersion of cross-stratification as a potential tool in the interpretation of Proterozoic arenites. *J. sedim. Petrol.* **48**, 857–862.

Long, J.V.P. (1977) Electron probe microanalysis. In: *Physical Methods in Determinative Mineralogy* (2nd edn) (Ed. by J. Zussman), pp. 273–341. Academic Press, London.

Long, J.V. & Agrell, S.O. (1965) The cathodoluminescence of minerals in thin-section. *Min. Mag.* **34**, 318–326.

Longinelli, A. & Craig, H. (1967) Oxygen 18 variations in sulphate ions in seawater and saline lakes. *Science*, **156**, 56–58.

Longman, M.W. (1980) Carbonate diagenetic textures from near-surface diagenetic environments. *Bull. Am. Ass. Petrol. Geol.* **64**, 461–486.

Longstaffe, F.J. (1983) Diagenesis 4. Stable isotope studies of diagenesis in clastic rocks. *Geosci. Can.* **10**, 43–57.

Lorens, R.B. (1981) Sr, Cd, Mn and Co distribution coefficients in calcite as a function of calcite precipitation rate. *Geochim. cosmochim. Acta*, **45**, 553–562.

Lorens, R.B. & Bender, M.L. (1980) The impact of solution chemistry on *Mytilus edulis* calcite and aragonite. *Geochim. cosmochim. Acta*, **44**, 1265–1278.

Loucks, R.G., Dodge, M.M. & Galloway, W.E. (1984) Regional controls on diagenesis and reservoir quality in Lower Tertiary Sandstones along the Texas Gulf Coast. In: *Clastic Diagenesis* (Ed. by D.A. McDonald and R.C. Surdam), pp. 15–45. Mem. Am. Ass. Petrol. Geol. 37.

Loughman, D.L. (1984) Phosphate authigenesis in the Aramachay Formation (Lower Jurassic) of Peru. *J. sedim. Petrol.* **54**, 1147–1156.

Love, L.G., Al-Kaisy, A.T.H. & Brockley, H. (1984) Mineral and organic material in matrices and coatings of framboidal pyrite from Pennsylvanian sediments, England. *J. sedim. Petrol.* **54**, 869–876.

Ludwick, J.C. & Henderson, P.L. (1968) Particles shape and inference of size from sieving. *Sedimentology*, **11**, 197–235.

Lukas, K.J. (1979) The effects of marine microphytes on carbonate substrata. In: *Scanning Electron Microscopy*, **11**, 447–455.

Lumsden, D.N. (1971) Markov chain analysis of carbonate rocks: applications, limitations and implications as exemplified by the Pennsylvanian system in southern Nevada. *Bull. geol. Soc. Am.* **82**, 447–462.

Lumsden, D.N. (1979) Discrepancy between thin section and X-ray estimates of dolomite in limestone. *J. sedim. Petrol.* **49**, 429–436.

Lumsden, D.N. & Chimahusky, J.S. (1980) Relationship between dolomite non-stoichiometry and carbonate facies parameters. *Spec. Publ. Soc. econ. Paleont. Miner., Tulsa*, **28**, 123–137.

Lumsden, D.L. & Lloyd, R.V. (1984) Mn (II) partitioning between calcium and magnesium sites in studies of dolomite origin. *Geochim. cosmochim. Acta*, **48**, 1861–1865.

Machel, H. (1985) Cathodoluminescence in calcite and dolomite and its chemical interpretation. *Geosci. Can.* **12**, 139–147.

MacKenzie, F.T., Bischoff, W.D., Bishop, F.C., Loijens, M., Schoonmaker, J. & Wollast, R. (1983) Magnesian calcites: low temperature occurrences, solubility and solid solution behaviour. In: *Carbonates: Mineralogy and Chemistry* (Ed. by R.J. Reeder), pp. 97–144. Rev. Mineral. Min. Soc. Am. 11.

Magaritz, M., Anderson, R.Y., Holser, W.T., Saltzman, G.S. & Garber, J. (1983) Isotope shifts in the Late Permian of the Delaware Basin, Texas, precisely timed by varved sediments. *Earth planet. Sci. Lett.* **66**, 111–124.

Magaritz, M. & Kafri, U. (1981) Stable isotope and Sr/Ca

evidence of diagenetic dedolomitization in a schizohaline environment: Cenomanian of northern Israel. *Sediment. Geol.* **28**, 29–41.

Makela, M. & Vartiainen, H. (1978) A study of sulphur isotopes in the Sokli multi-stage carbonate (Finland). *Chem. Geol.* **21**, 257–265.

Mangelsdorf, P.C. & Sayles, F.L. (1982) Multicomponent electrolyte diffusion corrections for North Atlantic sediment. *Am. J. Sci.* **282**, 1655–1665.

Manheim, F.T. (1976) Interstitial waters of marine sediments. In: *Chemical Oceanography*, Vol. 6 (Ed. by J.P. Riley and G. Skirrow), pp. 115–186. Academic Press, London.

Manickam, S. & Barbaroux, L. (1987) Variations in the surface texture of suspended quartz grains in the Loire River: an SEM study. *Sedimentology*, **34**, 495–510.

Manker, J.P. & Ponder, R.D. (1978) Quartz grain surface features from fluvial environments of northeastern Georgia. *J. sedim. Petrol.* **48**, 1227–1232.

Mardia, K.V. (1972) *Statistics of Directional Data*. Academic Press, London.

Marfunin, A.S. (1979) *Spectroscopy, Luminescence and Radiation Centres in Minerals*. Springer-Verlag, Berlin.

Margolis, S.V. & Krinsley, D.H. (1974) Processes of formation and environmental occurrence of microfeatures on detrital quartz grains. *Am. J. Sci.* **274**, 449–464.

Margolis, S.V. & Rex, R.W. (1971) Endoliothic algae and micrite envelope formation in Bahamian oolites as revealed by scanning electron microscopy. *Bull. Geol. Soc. Am.* **82**, 843–852.

Mariano, A.N. & Ring, P.J. (1975) Europium-activated cathodoluminescence in minerals. *Geochim. cosmochim. Acta*, **39**, 649–660.

Mariano, A.N., Ito, J. & Ring, P.J. (1973) Cathodoluminescence of plagioclase. *Geol. Soc. Am. Abstr. Progr.* **5**, 726.

Marshall, D.J. (1978) Suggested standards for the reporting of cathodoluminescence results. *J. sedim. Petrol.* **48**, 651–653.

Marshall, D.J. (1980) The influence of microscope performance on cathodoluminescence observations. *Nuclide Corp. Publs* 1019/0281, Acton, Massachussetts.

Marshall, J.D. (1981) Zoned calcites in Jurassic ammonite chambers: trace elements, isotopes and neomorphic origin. *Sedimentology*, **28**, 867–887.

Marshall, J.D. & Ashton, M. (1980) Isotopic and trace element evidence for submarine lithification of hardgrounds in the Jurassic of eastern England. *Sedimentology*, **27**, 271–289.

Martinez, B. & Plana, F. (1987) Quantitative X-ray diffraction of carbonate sediments: mineralogical analysis through fitting of Lorentzian profiles to diffraction peaks. *Sedimentology*, **34**, 169–174.

Martinsson, A. (1970) Toponomy of trace fossils. In: *Trace Fossils* (Ed. by T.P. Crimes and J.C. Harper), pp. 323–330. Seel House Press, Liverpool.

Mason, R.A. (1987) Ion microprobe analysis of trace elements in calcite with an application to the cathodoluminescence zonation of limestone cements from the Lower Carboniferous of South Wales. *Chem. Geol.* **64**, 209–264.

Mathisen, M.E. (1984) Diagenesis of Plio-Pleistocene nonmarine sandstones, Cagayan Basin, Philippines: early development of secondary porosity in volcanic sandstones. In: *Clastic Diagenesis* (Ed. by D.A. McDonald and R.C. Surdam), pp. 177–193. Mem. Am. Ass. Petrol. Geol. 37.

Matsumoto, R. & Iijima, A. (1981) Origin and diagenetic evolution of Ca-Mg-Fe carbonates in some coalfields of Japan. *Sedimentology*, **28**, 239–259.

Matsumoto, R. & Matsuhisa, Y. (1985) Chemistry, carbon and oxygen isotope ratios, and origin of deep-sea carbonates at sites 438, 439, and 584: inner slope of the Japan Trench. In: *Initial Reports of the Deep Sea Drilling Project*, **87**, 669–678. US Government Printing Office, Washington DC.

Matter, A. & Ramseyer, K. (1985) Cathodoluminescence petrography as a tool for provenance studies of sandstones. In: *Provenance of Arenites* (Ed. by G.G. Zuffa), pp. 191–211. Proc. Cetraro, Cosenza, 1984, NATO ASI Ser C148. Reidel, Dordrecht.

May, J.A. & Perkins, R.D. (1979) Endolithic infestation of carbonate substrates below the sediment-water interface. *J. sedim. Petrol.* **49**, 357–378.

Maynard, J.B. (1986) Geochemistry of oolitic iron ores, an electron microprobe study. *Econ. Geol.* **81**, 1473–1483.

Mazzullo, J.M. & Ehrlich, R. (1983) Grain shape variation in the St. Peter Sandstone: A record of eolian and fluvial sedimentation of an early Palaeozoic cratonic sheet sand. *J. sedim. Petrol.* **53**, 105–119.

McArthur, J.M., Benmore, R.A., Coleman, M.L., Soldi, C., Yeh, H.-W. & O'Brien, G.W. (1986) Stable isotope characterization of francolite formation. *Earth planet. Sci. Lett.* **77**, 20–34.

McBride, E.F. (1980) Importance of secondary porosity in sandstones to hydrocarbon exploration. *Bull. Am. Ass. Petrol. Geol.* **64**, 742–757.

McBride, E.F. (1985) Diagenetic processes that affect provenance determinations in sandstone. In: *Provenance of Arenites* (Ed. by G.G. Zuffa), pp. 95–113. NATO ASI Series C: vol. 148. Reidel, Dordrecht.

McBride, E.F. (1986) Influence of diagenesis on provenance interpretations in sandstones (abstract). *SEPM Ann. Midyear Meeting Abstr.* **3**, 74.

McCabe, A.M., Dardis, G.F. & Hanvey, P.M. (1984) Sedimentology of a Late Pleistocene submarine-moraine complex, County Down, Northern Ireland. *J. sedim. Petrol.* **54**, 716–730.

McCaffrey, M.A., Lazar, B. & Holland, H.D. (1987) The evaporation path of seawater and the coprecipitation of Br^- and K^+ with halite. *J. sedim. Petrol.* **57**, 928–937.

McCave, I.N. (1979a) Suspended sediment. In: *Estuarine*

Hydrography and Sedimentation (Ed. by K.R. Dyer), pp. 131–185. Cambridge University Press.

McCave, I.N. (1979b) Diagnosis of turbidites at sites 386 & 387 by particle-counter size analysis of the silt (2–40 μm) fraction. In: *Initial Reports of the Deep Sea Drilling Project*, Vol. 43. US Government Printing Office, Washington DC.

McCave, I.N. & Jarvis, J. (1973) Use of the Model T Coulter Counter in size analysis of fine to coarse sand. *Sedimentology*, **20**, 305–315.

McCrea, J.M. (1950) On the isotopic chemistry of carbonates and a paleotemperature scale. *J. Chem. Phys.* **18**, 849–857.

McDonald, D.A. & Surdam, R.C. (eds) (1984) Clastic diagenesis. *Mem. Am. Ass. Petrol. Geol.* **37**.

McDonald, D.I.M. & Tanner, P.G.W. (1983) Sediment dispersal patterns in part of a deformed Mesozoic back-arc basin on South Georgia, South Atlantic. *J. sedim. Petrol.* **53**, 83–104.

McHardy, W.J. & Birnie, A.C. (1987) Scanning electron microscopy. In: *A Handbook of Determinative Methods in Clay Mineralogy* (Ed. by M.J. Wilson), pp. 173–208. Blackie, Glasgow.

McHardy, W.J., Wilson, M.J. & Tait, J.M. (1982) Electron microscope and X-ray diffraction studies of filamentous illitic clay from sandstones of the Magnus Field. *Clay Miner.* **17**, 23–39.

McIntyre, W.L. (1963) Trace element partition coefficients — a review of theory and applications to geology. *Geochim. cosmochim. Acta*, **27**, 1209–1264.

McKee, E.D. (1966) Structures of dunes at White Sands National Monument (New Mexico) (and a comparison with structures of dunes from other selected areas). *Sedimentology*, **7**, 1–70.

McKee, E.D. & Weir, G.W. (1953) Terminology for stratification and cross-stratification in sedimentary rocks. *Bull. geol. Soc. Am.* **64**, 381–389.

McLennan, S.M. (1982) On the geochemical evolution of sedimentary rocks. *Chem. Geol.* **37**, 335–350.

McLennan, S.M., Taylor, S.R. & Eriksson, K.A. (1983) Geochemistry of Archaean shales from the Pilbara Supergroup, Western Australia. *Geochim. cosmochim. Acta*, **47**, 1211–1222.

McManus, D.A. (1963) A criticism of certain usage of the phi notation. *J. sedim. Petrol.* **33**, 670–674.

McManus, D.A. (1965) A study of maximum load for small diameter sieves. *J. sedim. Petrol.* **35**, 792–796.

McManus, J., Buller, A.T. & Green, C.D. (1980) Sediments of the Tay Estuary VI, Sediments of the lower and outer reaches. *Proc. R. Soc. Edinb. B*, **78**, 133–154.

McMurty, G.M., Chung-Ho Wang & Yeh, H-W. (1983) Chemical and isotopic investigations into the origin of clay minerals from the Galapagos hydrothermal mounds field. *Geochim. cosmochim. Acta*, **47**, 475–489.

Meckel, L.D. (1967) Tabular and trough cross bedding: comparison of dip azimuth variability. *J. sedim. Petrol.* **37**, 80–86.

Meshri, I.D. (1986) On the reactivity of carbonic and organic acids and generation of secondary porosity. In: *Roles of Organic Matter in Sediment Diagenesis* (Ed. by D.L. Gautier), pp. 123–128. Spec. Publ. Soc. econ. Paleont. Miner. 38.

Metz, R. (1985) The importance of maintining horizontal sieve screens when using a Ro-tap. *Sedimentology*, **32**, 613–614.

Meyer, H.J. (1984) The influence of impurities on the growth rate of calcite. *J. Crystal Growth*, **66**, 639–646.

Meyer, R. & Pena dos Rois, R.B. (1985) Paleosols and alunite silcretes in continental Cenozoic of Western Portugal. *J. sedim. Petrol*, **55**, 76–85.

Meyers, W.J. (1974) Carbonate cement stratigraphy of the Lake Valley Formation (Mississippian), Sacramento Mts., New Mexico. *J. sedim. Petrol.* **44**, 837–861.

Meyers, W.J. (1978) Carbonate cements: their regional distribution and interpretation in Mississippian limestones of southwestern New Mexico. *Sedimentology*, **25**, 371–399.

Meyers, W.J. (1980) Compaction in Mississippian skeletal limestones, southwestern New Mexico. *J. sedim. Petrol.* **50**, 457–474.

Meyers, W.J. & Hill, B.E. (1983) Quantitative studies of compaction in Mississippian skeletal limestones, New Mexico. *J. sedim. Petrol.* **53**, 231–242.

Meyers, W.J. & Lohmann, K.C. (1978) Microdolomite-rich syntaxial cements: proposed meteoric-marine mixing zone phreatic cements from Mississippian limestones, New Mexico. *J. sedim. Petrol.* **48**, 475–488.

Miall, A.D. (1973) Markov chain analysis applied to an ancient alluvial plain succession. *Sedimentology*, **20**, 347–364.

Miall, A.D. (1974) Paleocurrent analysis of alluvial sediments, discussion of directional variance and vector magnitude. *J. sedim. Petrol.* **44**, 1174–1185.

Miall, A.D. (1976) Paleocurrent and palaeohydrologic analysis of some vertical profiles through a Cretaceous braided stream deposit, Banks Island, Arctic Canada. *Sedimentology*, **23**, 459–483.

Miall, A.D. (1977) A review of the braided river depositional environment. *Earth Sci. Rev.* **13**, 1–62.

Miall, A.D. (1978) Lithofacies types and vertical profile models in braided river deposits: a summary. In: *Fluvial Sedimentology* (Ed. by A.D. Miall), pp. 597–604. Can. Soc. Petrol. Geol. Mem. 5.

Michelson, P.C. & Dott, R.H. (1973) Orientation analysis of trough cross stratification in Upper Cambrian sandstones of western Wisconsin. *J. sedim. Petrol.* **43**, 784–794.

Middleton, G.V. (1973) Johannes Walther's Law of correlation of facies. *Bull. geol. Soc. Am.* **84**, 979–988.

Middleton, G.V. & Kassera, C.A. (1987) Variations in density of V-shaped impact pits on quartz grains with size of grains, intertidal sands, Bay of Fundy. *J. sedim. Petrol.* **57**, 88–93.

Miller, J. & Clarkson, E.N.K.C. (1980) The post-ecdysial

development of the cuticle and the eye of the Devonian trilobite *Phacops rana milleri* Stewart 1927. *Phil. Trans. R. Soc. Edinb. B*, **288**, 461–480.

Miller, J. & Gillies, D. In press. Cathodoluminescence and cement stratigraphy. *Sedimentology*.

Millero, F.J. & Schreiber, D.R. (1982) Use of ion pairing model to estimate activity coefficients of the ionic components of natural waters. *Am. J. Sci.* **282**, 1508–1540.

Milliman, J.D. (1974) *Marine Carbonates*. Springer-Verlag, Berlin.

Milliman, J.D. & Bornhold, B.D. (1973) Peak height versus peak intensity analysis of X-ray diffraction data. *Sedimentology*, **20**, 445–448.

Millot, G. (1970) *Geology of Clays*. Springer-Verlag, New York.

Milner, H.B. (1962a) *Sedimentary Petrography, Volume I: Methods in Sedimentary Petrography*. Allen & Unwin, London.

Milner, H.B. (1962b) *Sedimentary Petrography, Volume II: Principles and Applications*. Allen & Unwin, London.

Mimran, Y. (1977) Chalk deformation and large-scale migration of calcium carbonate. *Sedimentology*, **24**, 333–360.

Mizutani, S. (1963) A theoretical and experimental consideration on the accuracy of sieving analysis. *J. Earth Sci.* **11**, 1–27, Magoya, Japan.

Mizutani, S. (1977) Progressive ordering of cristobalite in the early stages of diagenesis. *Contr. Miner. Petrol.* **61**, 129–140.

Moberly, R. (1968) Composition of magnesian calcites of algae and pelecypods by electron microprobe analysis. *Sedimentology*, **11**, 61–82.

Moilola, R.J. & Weiser, D. (1968) Textural parameters: an evaluation. *J. sedim. Petrol.* **38**, 45–53.

Mook, W.G. (1971) Paleotemperatures and chlorinities from stable carbon and oxygen isotopes in shell carbonate. *Palaeogeogr. Palaeoclim. Palaeoecol.* **9**, 245–263.

Moorby, S.A., Cronan, D.S. & Glasby, G.P. (1984) Geochemistry of hydrothermal Mn-oxide deposits from the S.W. Pacific island arc. *Geochim. cosmochim. Acta*, **48**, 433–441.

Moore, C.H. (1979) Porosity in carbonate rock sequences. In: *Geology of Carbonate Porosity*, pp. A1–124. Am. Ass. Petrol. Geol. Course Notes 11, Houston.

Moore, C.H. (1985) Upper Jurassic subsurface cements: a case history. In: *Carbonate Cements* (Ed. by N. Schneidermann and P.M. Harris), pp. 291–308. Spec. Publ. Soc. econ. Paleont. Miner. 36.

Moore, C.H. & Druckman, Y. (1981) Burial diagenesis and porosity evolution: Upper Jurassic Smaskover, Arkansas and Louisana. *Bull. Am. Ass. Petrol. Geol.* **65**, 597–628.

Moore, F. (1979) Some statistical calculations concerning the determination of trace constituents. *Geostand. Newsl.* **3**, 105–108.

Morod, S. (1984) Diagenetic matrix in Proterozoic greywack from Sweden. *J. sedim. Petrol.* **54**, 1157–1168.

Morow, D.W. (1978) The influence of the Mg/Ca ratio and salinity on dolomitization in evaporite basins. *Can. Petrol. Geol. Bull.* **26**, 389–392.

Morow, D.W. (1982) Diagenesis 2. Dolomite — part 2. Dolomitization models and ancient dolostones. *Geosci. Can.* **9**, 95–107.

Morse, J.W. (1983) The kinetics of calcium carbonate dissolution and precipitation. In: *Carbonates : Mineralogy and Chemistry* (Ed. by R.J. Reeder), pp. 227–264. Rev. Mineral. Miner. Soc. Am. 11.

Morton, A.C. (1985a) A new approach to provenance studies: electron microprobe analysis of detrital garnets from Middle Jurassic sandstones of the northern North Sea. *Sedimentology*, **32**, 553–566.

Morton, A.C. (1985b) Heavy minerals in provenance studies. In: *Provenance of Arenites* (Ed. by G.G. Zuffa), pp. 249–278. NATO ASI Series C, vol. 148. Reidel, Dordrecht.

Moseley, F. (1981) *Methods in Field Geology*. Freeman, San Francisco.

Moss, A.J. (1962) The physical nature of common sandy and pebbly deposits Part I. *Am. J. Sci.* **260**, 337–373.

Moss, A.J. (1963) The physical nature of common sandy and pebbly deposits part II. *Am. J. Sci.* **261**, 297–343.

Mossop, G.D. (1972) Origin of the peripheral rim, Redwater Reef, Alberta. *Bull. Can. Petrol. Geol.* **20**, 238–280.

Mount, J.F. (1985) Mixed siliclastic and carbonate sediments: a proposed first order textural and compositional classification. *Sedimentology*, **32**, 435–442.

Moussa, M.T. (1976) Plastic spray thin section cover. *J. sedim. Petrol.* **46**, 252–253.

Moussa, M.T. (1978) Reply: plastic thin section cover. *J. sedim. Petrol.* **48**, 672–674.

Mucci, A. & Morse, J.W. (1983) The incorporation of Mg^{2+} and Sr^{2+} into calcite overgrowths: influences of growth rate and solution composition. *Geochim. cosmochim. Acta*, **476**, 217–233.

Muir, M.D. & Grant, P.R. (1974) Cathodoluminescence. In: *Quantitative Scanning Electron Microscopy* (Ed. by D.B. Holt, M.D. Muir, P.R. Grant and I.M. Boswarva), pp. 287–334. Academic Press, London.

Mukherjee, B. (1984) Cathodoluminescence spectra of Indian calcites, limestone, dolomites and aragonites. *Ind. J. Phys.* **22**, 305–310.

Murray, R.S. & Quirk, J.P. (1980) Clay-water interactions and the mechanism of soil swelling. *Colloids Surfaces*, **1**, 17–32.

Nagtegaal, P.J.C. (1979) Relationship of facies and reservoir quality in Rotliegendes desert sandstones, Southern North Sea region. *J. Petrol. Geol.* **2**, 145–158.

Nathan, Y. & Sass, E. (1981) Stability relations of apatites and calcium carbonates. *Chem Geol.* **34**, 103–111.

Neihof, R.A. & Loeb, G.I. (1972) The surface charge of particulate matter in seawater. *Limnol. Oceanogr.* **17**, 7–16.

Nelson, C.S. & Lawrence, M.F. (1984) Methane-derived high-Mg calcite submarine cement in Holocene nodules from the Fraser Delta, British Columbia, Canada. *Sedimentology*, **31**, 645–654.

Nemec, W. & Steel, R.J. (1984) Alluvial and coastal conglomerates: their significant features and some comments on gravelley mass-flow deposits. In: *Sedimentology of Gravels and Conglomerates* (Ed. by E.H. Koster and R.J. Steel), pp. 1–31. Can. Soc. Petrol. Geol. Mem. 10.

Nesbitt, H.W. (1979) Mobility and fractionation of rare earth elements during weathering of a granodiorite. *Nature*, **279**, 206–210.

Nesbitt, H.W. (1984) Activity coefficients of ions in alkali and alkaline-earth chloride dominated waters including seawater. *Chem. Geol.* **43**, 127–142.

Nesbitt, H.W. (1985) A chemical equilibrium model for the Illinois basin formation waters. *Am. J. Sci.* **285**, 436–458.

Nesbitt, H.W., Markovics, G. & Price, R.C. (1980) Chemical processes affecting alkalis and alkaline earths during continental weathering. *Geochim. cosmochim. Acta*, **44**, 1659–1666.

Nesbitt, H.W. & Young, G.M. (1982) Early Proterozoic climates and plate motions inferred from major element chemistry of lutites. *Nature*, **299**, 715–717.

Nesbitt, H.W. & Young, G.M. (1984) Prediction of some weathering trends of plutonic and volcanic rocks based on thermodynamic and kinetic considerations. *Geochim. cosmochim. Acta*, **48**, 1523–1534.

Neumann, A.C. (1966) Observations on coastal erosion in Bermuda and measurements of the boring rate of the sponge, *Cliona lampa*. *Limn. Oceanogr.* **11**, 92–108.

Newell, N.D., Rigby, J.K., Fischer, A.G., Whiteman, A.J., Hickox, J.E. & Bradley, J.S. (1953) *The Permian Reef Complex of the Guadalupe Mountains Region, Texas and New Mexico — a Study in Palaeoecology*. Freeman, San Francisco.

Nichol, I. (1986) Geochemical exploration for gold. In: *Applied Geochemistry in the 1980's* (Ed. by I. Thornton and R.J. Howarth), pp. 60–85. Graham & Trotman, London.

Nickel, E. (1978) The present status of cathode luminescence as a tool in sedimentology. *Miner. Sci. Engng*, **10**, 73–100.

Nielsen, H. (1979) Sulfur isotopes. In: *Lectures in Isotopes Geology* (Ed. by E. Jüger and J.C. Hunziker), pp. 283–312. Springer-Verlag, Berlin.

Nilsen, T.H. (1968) The relationship of sedimentation to tectonics in the Solund Devonian distinct of southwestern Norway. *Norges geol. Unders*. **259**.

Norrish, K. & Chappell, B.W. (1977) X-ray fluorescence spectrometry. In: *Physical Methods in Determinative Mineralogy* (Ed. by J. Zussman), pp. 201–272. Academic Press, London.

Nriagu, J.O. (1974) Fractionation of sulfur isotopes by sediment adsorption of sulphate. *Earth planet. Sci. Lett.* **22**, 366–370.

Nyffeler, V.P., Li, Y-H. & Santschi, P.H. (1984) A kinetic approach to describe trace-element distribution between particles and solution in natural aquatic systems. *Geochim. cosmochim. Acta*, **48**, 1513–1522.

O'Brien, N.R. (1987) The effects of bioturbation on the fabric of shale. *J. sedim. Petrol.* **57**, 449–455.

O'Brien, N.R., Nakazawa, K. & Tokuhashi, S. (1980) Use of clay fabric to distinguish turbiditic and hemipelagic siltstones and silts. *Sedimentology*, **27**, 47–61.

Odin, G.S. & Matter, A. (1981) De glauconiarum origine. *Sedimentology*, **28**, 611–641.

Okurama, M. & Kitano, Y. (1986) Coprecipitation of alkali metal ions with calcium carbonate. *Geochim. cosmochim. Acta*, **50**, 49–58.

Olaussen, S., Dallan, A., Gloppen, T.G. & Johannessen, E. (1984) Depositional environment and diagenesis of Jurassic reservoir sandstones in the eastern part of Troms I area. In: *Petroleum Geology of the North European Margin*, pp. 61–79. Norwegian Petroleum Society, Graham & Trotman, London.

Oldershaw, A.E. & Scoffin, T.P. (1967) The source of ferroan and non-ferroan calcite cements in the Halkin and Wenlock limestones. *Geol. J.* **5**, 309–320.

Oliver, G.J. (1979) XRF analysis of ceramic materials at the British Ceramic Research Association. *Brit. Ceramic Res. Ass. Spec. Publ.* **98**, 75–100.

Olson, J.S. & Potter, P.E. (1954) Variance components of cross-bedding direction in some basal Pennsylvanian sandstones of the Eastern Interior Basin: statistical methods. *J. Geol.* **62**, 26–49.

Open University (1981) *The Earth: Structure, Composition and Evolution, Block 5. Surface Processes, Weathering to Diagenesis*. Open University Press, Milton Keynes.

Open University (1987) *Basin Analysis Techniques*. S338, Block 3. Open University Press, Milton Keynes.

Osipov, V.I. & Sokolov, V.N. (1978) Microstructure of recent clay sediments examined by scanning electron microscopy. In: *Scanning Electron Microscopy in the Study of Sediments* (Ed. by W.B. Whalley), pp. 29–40. Geo Abstracts, Norwich.

Otto, G.H. (1938) The sedimentation unit and its use in field sampling. *J. Geol.* **46**, 569–582.

Pacey, N.R. (1984) Bentonites in the Chalk of central eastern England and their relation to the opening of the northeast Atlantic. *Earth planet. Sci. Lett.* **67**, 48–60.

Page, H.G. (1955) Phi-millimeter conversion tables. *J. sedim. Petrol.* **25**, 285–292.

Pamplin, B.R. (ed.) (1980) *Crystal Growth* (2nd edn). Pergamon Press, Oxford.

Park, W.C. & Schot, E.H. (1968) Stylolites: their nature and origin. *J. sedim. Petrol.* **38**, 175–191.

Parkhurst, D.L., Thorstenson, D.C. & Plummer, L.N. (1980) PHREEQE — a computer program for geochemical calculations. *U.S. geol. Surv. Water Resour. Invest.* 80–96.

Parks, G.A. (1975) Adsorption in the marine environment. In: *Chemical Oceanography*, 1 (Ed. by J.P. Riley and G. Skirrow), pp. 241–308. Academic Press, London.

Parra, M., Puechmaille, C. & Carrkesco, C. (1981) Strontium as marker of the origin of biogenic and terrigenous materials and as a hydrodynamic tracer in the deep-sea North Atlantic area. *Chem. Geol.* **34**, 91–102.

Parrish, W. & Mack, M. (1963) *Charts for the solution of Bragg's equation*, Vols 1–3. Phillips Technical Library Centre, Eindhoven.

Passega, R. (1957) Textures as characteristic of clastic deposition. *Bull. Am. Ass. Petrol. Geol.* **41**, 1952–1984.

Passega, R. (1964) Grain-size representation by C M patterns as a geological tool. *J. sedim. Petrol.* **34**, 840–847.

Patsoules, M.G. & Cripps, J.C. (1983) A quantitative analysis of chalk pore geochemistry using resin casts. *Energy Sources*, **7**, 15–31.

Pauling, L. (1970) *General Chemistry*. Freeman, San Francisco.

Peirce, T.J. & Williams, D.J.A. (1966) Experiments on certain aspects of sedimentation of estuarine muds. *Proc. Inst. civ. Engrs*, **54**, 391–402.

Pelletier, B.R. (1958) Pocono paleocurrents in Pennsylvania and Maryland. *Bull. geol. Soc. Am.* **69**, 1033–1064.

Pelletier, B.R. (1973) A re-examination of the use of the silt/clay ratios as indicators of sedimentary environments: a study for students. *Mar. Sedim.* **9**, 1–12.

Perillo, G.M.E., Albero, M.C., Angiolini, F. & Codignotto, J.O. (1984) An inexpensive portable coring device for intertidal sediments. *J. sedim. Petrol.* **54**, 654–655.

Perry, E.F. & Hower, J. (1970) Burial diagenesis in Gulf Coast pelitic sediments. *Clays Clay Miner.* **18**, 165–177.

Pettijohn, F.J. (1975) *Sedimentary Rocks* (3rd edn). Harper & Row, London.

Pettijohn, F.J. & Potter, P.E. (1964) *Atlas and Glossary of Primary Sedimentary Structures*. Springer-Verlag, Berlin.

Phakey, P.P., Curtis, C.D. & Oertel, G. (1972) Transmission electron microscopy of fine-grained phyllosilicates in ultra-thin rock sections. *Clays Clay Miner.* **20**, 193–197.

Phillips, W.J. & Phillips, N. (1980) *Introduction to Mineralogy for Geologists.* Wiley, London.

Picard, M.D. & High, L.J. (1968) Shallow marine currents on the early Triassic Wyoming shelf. *J. sedim. Petrol.* **38**, 411–423.

Pickering, K.T. (1981) The Kongsfjord Formation — a Late Precambrian submarine fan in North-east Finnmark, North Norway. *Norges geol. Unders.* **367**, 77–104.

Pierson, B.J. (1981) The control of cathodoluminescence in dolomite by iron and manganese. *Sedimentology*, **28**, 601–610.

Pingitore, N.E. (1976) Vadose and phreatic diagenesis: processes, products and their recognition in corals. *J. sedim. Petrol.* **46**, 985–1006.

Pingitore, N.E. (1978) The behaviour of Zn^{2+} and Mn^{2+} during carbonate diagenesis: theory and applications. *J. sedim. Petrol.* **41**, 799–814.

Pingitore, N.E. (1982) The role of diffusion during carbonate diagenesis. *J. sedim. Petrol.* **52**, 27–39.

Pingitore, N.E. & Eastman, M.P. (1984) The experimental partitioning of Ba^{2+} into calcite. *Chem. Geol.* **45**, 113–120.

Pingitore, N.E. & Eastman, M.P. (1986) The coprecipitation of Sr^{2+} with calcite at 25°C and 1 atm. *Geochim. cosmochim. Acta*, **50**, 2195–2203.

Piper, D.J.W., Harland, W.B. & Cutbill, J.L. (1970) Recording of geological data in the field using forms for input to the IBM handwriting reader. In: *Data Processing in Biology and Geology* (Ed. by J.L. Cutbill), pp. 17–38. Systematics Association Spec. Vol. 3. Academic Press, London.

Piper, D.Z. (1974) Rare earth elements in the sedimentary cycle: a summary. *Chem. Geol.* **14**, 285–304.

Pisciotto, K.A. (1981) Diagenetic trends in the siliceous facies of the Monterey Shale in the Santa Maria region, California. *Sedimentology*, **28**, 547–571.

Pisciotto, K.A. & Mahoney, J.J. (1981) Isotopic survey of diagenetic carbonate, DSDP Leg 63. In: *Initial Reports of the Deep Sea Drilling Projects*, **63**, 595–609. US Government Printing Office, Washington DC.

Pittman, E.D. (1972) Diagenesis of quartz in sandstone as revealed by the scanning electron microscope. *J. sedim. Petrol.* **42**, 507–519.

Pittman, E.D. & Duschatko, R.W. (1970) Use of pore casts and scanning electron microscopy to study pore geometry. *J. sedim. Petrol.* **40**, 1153–1157.

Playford, P.E. (1980) Devonian "Great Barrier Reef" of the Canning Basin, Western Australia. *Bull. Am. Ass. Petrol Geol.* **64**, 814–840.

Plint, A.G. (1983) Facies, environments and sedimentary cycles in the Middle Eocene, Bracklesham Formation of the Hampshire Basin: evidence for global sea-level changes? *Sedimentology*, **30**, 625–654.

Plummer, L.N. & Parkhurst, D.L. (1985) PHREEQE: status and applications. In: *Proceedings of Conference on the Application of Geochemical Models to High-Level Nuclear Waste Repository Assessment* (Ed. by G.K. Jacobs and S.K. Whatley), pp. 37–45. Oak Ridge National Laboratory, TN73831.

Plummer, L.N. & Sundquist, E.T. (1982) Total individual ion activity coefficients of calcium and carbonate in seawater at 25°C and 35‰ salinity and implications as to the agreement between apparent and thermodynamic constants of calcite and aragonite. *Geochim. cosmochim. Acta*, **46**, 247–258.

Popp, B.N., Anderson, T.F. & Sandberg, P.A. (1986) Textural, elemental and isotopic variations among constituents in Middle Devonian limestones, North America. *J. sedim. Petrol.* **56**, 715–727.

Porrenga, D.H. (1967) Influence of grinding and heating of layer silicates on boron sorption. *Geochim. cosmochim. Acta*, **31**, 309–312.

Portnov, A.M. & Gorobets, B.S. (1969) Luminescence of apatite from different rock types. *Dokl. Akad. Nauk. SSSR*, **184**, 110–114.

Potter, P.E. & Blakey, R.F. (1968) Random processes and lithologic transitions. *J. Geol.* **76**, 154–170.

Potter, P.E. & Olson, J.S. (1954) Variance components of crossbedding direction in some basal Pennsylvanian sandstones of the Eastern Interior Basin: geological application. *J. Geol.* **62**, 50–73.

Potter, P.E. & Pettijohn, F.J. (1977) *Paleocurrents and Basin Analysis* (2nd edn). Springer-Verlag, Berlin.

Potter, P.E. & Scheidegger, A.E. (1965) Bed thickness and grain size: graded beds. *Sedimentology*, **7**, 233–240.

Potter, P.E., Shimp, N.F. & Witters, J. (1963) Trace elements in marine and fresh-water argillaceous sediments. *Geochim. cosmochim. Acta*, **27**, 669–694.

Potts, P.J. (1987) *A Handbook of Silicate Rock Analysis*. Blackie, Glasgow.

Powell, R., Condliffe, D.M. & Condliffe, E. (1984) Calcite-dolomite geothermometry in the system $CaCO_3$-$MgCO_3$-$FeCO_3$: an experimental study. *J. Metamorph. Geol.* **2**, 33–41.

Powers, M.C. (1982) Comparison chart for estimating roundness and sphericity. *AGI Data Sheet 18*. American Geological Institute.

Powers, D.W. & Easterling, R.G. (1982) Improved methodology for using embedded Markov chains to describe cyclical sediments, *J. sedim. Petrol.* **52**, 913–923.

Prezbindowski, D. (1980) Microsampling technique for stable isotopic analysis of carbonates. *J. sedim. Petrol.* **50**, 643–645.

Prezbindowski, D.R. (1985) Burial cementation — is it important? A case study, Stuart City trend, south central Texas. In: *Carbonate Cements* (Ed. by N. Schneidermann and P.M. Harris), pp. 241–264. Soc. econ. Paleont. Miner. Spec. Publ. 36.

Price, I, (1975) Acetate peel techniques applied to cherts. *J. sedim. Petrol.* **45**, 215–216.

Puchelt, H., Sables, B.R. & Hoering, T.C. (1971) Preparation of sulfur hexafluoride for isotope geochemical analysis. *Geochim. cosmochim. Acta,* **35**, 625–628.

Pye, K. & Krinsley, D.H. (1984) Petrographic examination of sedimentary rocks in the SEM using backscattered electron detectors. *J. sedim. Petrol.* **54**, 877–888.

Pye, K. & Krinsley, D.H. (1986a) Diagenetic carbonate and evaporite minerals in Rotliegend aeolian sandstones of the southern North Sea: their nature and relationship to secondary porosity development. *Clay Miner.* **21**, 443–457.

Pye, K. & Krinsley, D.H. (1986b) Microfabric, mineralogy and early diagenetic history of the Whitby Mudstone Formation (Toarcian), Cleveland Basin, U.K. *Geol. Mag.* **123**, 191–203.

Pye, K. & Sperling, C.H.B. (1983) Experimental investigation of silt formation by static breakage processes: the effect of temperature, moisture and salt on quartz dune sand and granite regolith. *Sedimentology*, 30, 49–62.

Pye, K. & Windsor-Martin, J. (1983) SEM analysis of shales and other geological materials using the Philips Multi-Function Detector System. *The Edax Editor*, **13**, 4–6.

Quest, M. (1985) Petrographical and geochemical studies of the Portland and Purbeck strata of Dorset. *Unpublished PhD Thesis*. University of Birmingham.

Rafter, T.A. (1957) Sulphur isotopic variations in nature, part 2. A quantitative study of the reduction of barium sulphate by graphite for recovery of sulphide-sulphur for sulphur isotopic measurements. *N.Z. J. Sci. Tech. B*, **38**, 955–968.

Rafter, T.A. (1967) Oxygen isotopic composition of sulphates — Part I. A method for the extraction of oxygen and its quantitative conversion to carbon dioxide for isotope radiation measurements. *N.Z. J. Sci.* **10**, 493–510.

Ragan, D.M. (1973) *Structural Geology. An Introduction to Geometrical Techniques* (2nd edn). Wiley, New York.

Raiswell, R.W., Brimblecombe, P., Dent, D.L. & Liss, P.S. (1980) *Environmental Chemistry — the Earth — Water Factory*. Edward Arnold, London.

Ramsay, J.G. (1961) The effects of folding upon the orientation of sedimentary structures. *J. Geol.* **69**, 84–100.

Ramsay, J.G. (1967) *Folding and Fracturing of Rocks*. McGraw-Hill, New York.

Rao, C.P. (1981) Geochemical difference between tropical (Ordovician) and subpolar (Permian) carbonates, Tasmania, Australia. *Geology*, **9**, 205–209.

Read, W.A. (1969) Analysis and simulation of Namurian rocks of Central Scotland using a Markov-process model. *Math. Geol.* **1**, 199–219.

Reading, H.G. (ed.) (1978a) *Sedimentary Environments and Facies*. Blackwell Scientific Publication, Oxford.

Reading, H.G. (1978b) Facies. In: *Sedimentary Environments and Facies* (Ed. by H.G. Reading), pp. 4–14. Blackwell Scientific Publications, Oxford.

Reed, S.J.B. (1975) *Electron Microprobe Analysis*. Cambridge University Press.

Reed, S.J.B. & Ware, N.G. (1975) Quantitative electron microprobe analysis of silicates using energy-dispersive X-ray spectrometry. *J. Petrol.* **16**, 499–519.

Rees, C.E. (1978) Sulpur isotope measurements using SO_2 and SF_6. *Geochim. cosmochim. Acta*, **42**, 383–389.

Rees, C.E., Jenkins, W.J. & Monster, J. (1978) The sulphur isotopic composition of ocean water sulphate. *Geochim. cosmochim. Acta*, **42**, 377–381.

Rehmer, J.A. & Hepburn, J.C. (1974) Quartz surface textural evidence for a glacial origin of the Squantum 'Tillite', Boston Basin, Massachusetts. *Geology*, **2**, 413–415.

Reiche, P. (1938) An analysis of cross-lamination from the Coconino Sandstone. *J. Geol.* **46**, 905–932.

Reineck, H.E. & Singh, I.B. (1975) *Depositional Sedimentary Environments*. Springer-Verlag, Berlin.

Remond, G, Le Gressus, C. & Okuzumi, H. (1979) Electron beam effects observed in cathodoluminescence and auger electron spectroscopy in natural materials: evidence for ionic diffusion. In: *Scanning Electron Microscopy*, Vol. 1 (Ed. by D. O'Hare), pp. 237–244.

Retallack, G.J. (1986) Fossil soils as grounds for interpreting long-term controls on ancient rivers. *J. sedim. Petrol.* **56**, 1–18.

Reyment, R.A. (1971) *Introduction to Quantitative Palaeoecology*. Elsevier, Amsterdam.

Reynolds, R.C. (1980) Interstratified clay minerals. In: *Crystal Structures of Clay Minerals and their X-ray Identification* (Ed. by G.W. Brindley and G. Brown), pp. 249–303. Mineralogical Society, London.

Reynolds, R.C. Jr. & Hower, J. (1970) The nature of interlayering in mixed-layer illite-montmorillonites. *Clays Clay Miner.* **18**, 25–36.

Ricci Lucchi, F. (1975a) Depositional cycles in two turbidite formations for Northern Appenines (Italy). *J. sedim. Petrol.* **45**, 3–43.

Ricci Lucchi, F. (1975b) Sediment dispersal in turbidite basins: examples from the Miocene of the northern Appenines. *Int. Congr. Sed. Nice, 1975*, **5**, 345–352.

Ricci Lucchi, F. (1981) The Miocene Marnoso–Arenacea turbidites, Romagna & Umbria Appenines. In: *Excursion Guidebook* (Ed. by F. Ricci Lucchi), pp. 231–303. 2nd European Meeting of the IAS.

Richardson, J.F. & Zaki, W.N. (1954) Sedimentation and fluidisation. *Trans. Inst. Chem. Engl.* **32**, 35–51.

Richter, D.K. & Zinkernagel, U. (1975) Petrographie des "Permoskyth" der Jaggl-Plawers-Einheit (Sudtirol) und Diskussion der Detrituserkunft mit Hiffe von Kathoden-Lumineszenze-Untersuchungen. *Geol. Rdsch.* **64**, 783–807.

Richter, D.K. & Zinkernagel, U. (1981) Zur Anwendung der Kathodolumineszenz in der Karbonatpetrographie. *Geol. Rdsch.* **70**, 1276–1302.

Riley, J.P. (1975) Analytical chemistry of sea water. In: *Chemical Oceanography*, vol. 3 (Ed. by J.P. Riley and G. Skirrow), pp. 193–514. Academic Press, London.

Riley, K.W. & Saxby, J.D. (1982) Association of organic matter and vanadium in oil shale from the Toolebu Formation of the Eromanga Bassin, Australia. *Chem. Geol.* **37**, 265–275.

Ripley, E.M. & Nicol, D.L. (1981) Sulphur isotopic studies of Archean slate and graywacke from northern Minnesota: evidence for the existence of sulphate reducing bacteria. *Geochim. cosmochim. Acta*, **45**, 839–846.

Risk, M.J. & Szczuczko, R.B. (1977) A method for staining trace fossils. *J. sedim. Petrol.* **47**, 855–859.

Rittenhouse, G. (1943) Relation of shape to the passage of grains through sieves. *Ind. Eng. Chem. (Anal. Ed.)* **15**, 153–155.

Robie, R.A., Hemingway, B.S. & Fisher, J.R. (1978) Thermodynamic properties of minerals and related substances at 298.15 K and 1 bar (105 Pascals) pressure and at higher temperatures. *U.S. geol. Surv. Bull.* 1452.

Robinson, B.W. (1980) The backscattered-electron/low vacuum SEM technique: A user's evaluation. *Micron*, **11**, 333–334.

Robinson, B.W. & Kusakabe, M. (1975) Quantitative preparation of sulphur dioxide for $^{34}S/^{32}S$ analysis from sulphides by combustion with cuprous oxide. *Anal. Chem.* **47**, 1179–1181.

Robinson, B.W. & Nickel, E.H. (1979) A useful new technique for mineralogy: the backscattered-electron/low vacuum mode of SEM operation. *Am. Miner.* **64**, 1322–1328.

Robinson, P. (1980) Determination of calcium, magnesium, manganese, strontium, sodium and iron in the carbonate fraction of limestones and dolomites. *Chem. Geol.* **28**, 135–146.

Robinson, V.N.E. (1975) Backscattered electron imaging. In: *Scanning Electron Microscopy* (Ed. by O. Johari and I. Corwin), pp. 52–60. IITRI, Chicago.

Robinson, V.N.E. (1980) Imaging with backscattered electrons in a scanning electron microscope. *Scanning*, **3**, 15–26.

Rodriguez-Clemente, R. (1982) The crystal morphology as geological indicator. *Estud. geol.* **38**, 155–172.

Rosenbaum, J. & Sheppard, S.M.F. (1986) An isotopic study of siderites, dolomites and ankerites at high temperatures. *Geochim. cosmochim. Acta*, **50**, 1147–1150.

Rosenfield, M.A. & Griffiths, J.C. (1953) An experimental test of visual comparison technique in estimating two dimensional sphericity and roundness of quartz grains. *Am. J. Sci.* **251,** 553–585.

Rosenfeld, M.A., Jacobsen, L. & Ferm, J.C. (1953) A comparison of sieve and thin-section technique for size analysis. *J. Geol.* **61**, 114–132.

Rosin, P. E. & Rammler, E. (1934) Die Kornzusammensetzung des Mahlgutes im Lichte der Wahrscheinlickeitslehre. *Kolloid zeitschr.* **67**, 16–26.

Rossel, N.C. (1982) Clay mineral diagenesis in Rotliegend aeolian sandstones of the southern North Sea. *Clay Miner.* **17**, 69–77.

Royse, C.F., Waddell, J.S. & Petersen, L.E. (1971) X-ray determination of calcite-dolomite: an evaluation. *J. sedim. Petrol.* **41**, 483–488.

Rubinson, M & Clayton, R.N. (1969) Carbon-13 fractionation between aragonite and calcite. *Geochim. cosmochim. Acta*, **33**, 997–1002.

Runnells, D.D. (1970) Errors in X-ray analysis of carbonates due to solid solution variation in the composition of component minerals. *J. sedim. Petrol.* **40**, 1158–1166.

Rupke, N.A. (1977) Growth of an ancient deep sea fan. *J. Geol.* **85**, 725–744.

Ruppert, L.F., Blaine Cecil, C., Stanton, R.W. & Christian, R.P. (1985) Authigenic quartz in the Upper Freeport coal bed, west-central Pennsylvania. *J. sedim. Petrol.* **55**, 334–339.

Russell, K.J. (1984) The sedimentology and palaeogeography of some Devonian sedimentary rocks in southwest Ireland. *PhD thesis*, Council for National Academic Awards.

Russell, K.L. (1970) Geochemistry and halmyrolysis of clay minerals, Rio Ameca, Mexico. *Geochim. cosmochim. Acta*, **34**, 893–907.

Rust, B.R. (1972) Pebble orientation in fluvial sediments. *J. sedim. Petrol.* **42**, 384–388.

Rust, B.R. (1975) Fabric and structure in glaciofluvial gravels. In: *Glaciofluvial and Glaciolacustrine Sedimentation* (Ed. by A.V. Jopling and B.C. McDonald), pp. 238–245. Soc. econ. Paleont. Mineral. Spec. Publ. 23.

Rust, B.R. (1978) A classification of alluvial channel systems. In: *Fluvial Sedimentology* (Ed. by A.D. Miall), pp. 187–198. Can. Soc. Petrol. Geol. Mem. 5.

Rutzler, K. (1975) The role of burrowing sponges in bioerosion. *Oecologia*, **19**, 203–216.

Rutzler, K. & Rieger, G. (1973) Sponge burrowing: Fine structure of *Cliona lampa* penetrating calcareous substrata. *Mar. Biol.* **21**, 144–162.

Ryan, D.E. & Szabo, J.P. (1981) Cathodoluminescence of detrital sands: a technique for rapid determination of the light minerals of detrital sands. *J. sedim. Petrol.* **51**, 669–670.

Ryan, J.J. & Goodell, H.G. (1972) Marine geology and estuarine history of Mobile Bay, Alabama. Part I. Contemporary sediments. *Geol. Soc. Am. Mem.* **133**, 517–554.

Ryer, T.A. (1981) Deltaic coals of Ferron Sandstone Member of Mancos Shale: a predictive model for Cretaceous coal-bearing strata of Western Interior. *Bull. Am. Ass. Petrol. Geol.* **65**, 2323–2340.

Sakai, H. (1977) Sulphate-water isotope thermometry applied to geothermal systems. *Geothermics*, **5**, 67–74.

Sakai, H., Gunnlaugssen, E., Tomasson, J. & Rouse, J.E. (1980) Sulpur isotope systematics in Icelandic geothermal systems and influence of sea water circulation at Reykjanes. *Geochim. cosmochim. Acta*, **44**, 1223–1231.

Sakai, H. & Krouse, H.R. (1971) Elimination of memory effects in $^{18}O/^{16}O$ determinations in sulphates. *Earth planet. Sci. Lett.* **11**, 369–373.

Sandberg, P.A. (1975) New interpretations of Great Salt Lake ooids and of ancient non-skeletal carbonate mineralogy. *Sedimentology*, **22**, 497–538.

Sandberg, P.A. (1985) Aragonite cements and their occurrence in ancient limestone. In: *Carbonate Cements* (Ed. by N. Schneidermann and P.M. Harris), pp. 33–58. Soc. econ. Paleont. Min. Spec. Publ. 36.

Sandberg. P.A. & Hudson, J.D. (1983) Aragonite relic preservation in Jurassic calcite-replaced bivalves. *Sedimentology*, **30**, 879–892.

Sanderson, I.D. (1984) Recognition and significance of inherited quartz overgrowths. *J. sedim. Petrol.* **54**, 646–650.

Sanford, R.B. & Swift, D.J.P. (1971) Comparison of sieving and settling techniques for size analysis, using a Benthos rapid sediment analyser. *Sedimentology*, **7**, 257–264.

Sasaki, A., Arikawa, Y. & Folinsbee, R.E. (1979) Kiba reagent method of sulphur extraction applied to isotopic work. *Bull. geol. Surv. Japan*, **30**, 241–245.

Sass, E. & Katz, A. (1982) The origin of platform dolomites: new evidence. *Am. J. Sci.* **282**, 1184–1213.

Savin, S.M. & Epstein, S. (1970) The oxygen and hydrogen isotope geochemistry of clay minerals. *Geochim. cosmochim. Acta*, **34**, 25–42.

Sayles, F.L. & Mangelsdorf, P.C. (1977) The equilibration of clay minerals with seawater: exchange reactions. *Geochim. cosmochim. Acta*, **41**, 951–960.

Sayles, F.L. & Mangelsdorf, P.C. (1979) Cation-exchange characteristics of Amazon river suspended sediment and its reaction with seawater. *Geochim. cosmochim. Acta*, **43**, 767–779.

Scharer, U. & Allecre, C.J. (1985) Determination of the age of the Australian continent by single-grain zircon analysis of Mt Narryer metaquartzite. *Nature*, **315**, 52–55.

Scherer, M. & Seitz, H. (1980) Rare-earth element distribution in Holocene and Pleistocene corals and their redistribution during diagenesis. *Chem. Geol.* **28**, 279–289.

Schlager, W. & James, N.P. (1978) Low magnesian calcite limestones forming at the deep sea floor, Tongue of the Ocean, Bahamas. *Sedimentology*, **25**, 675–702.

Schlee, J. (1966) A modified Woods Hole rapid sediment analyser. *J. sedim. Petrol.* **36**, 404–413.

Schlee, J., Uchupi, E. & Trumbull, J.V.A. (1965) Statistical parameters of Cape Cod Beach and Eolian sands. *U.S. geol. Surv. Prof. Pap.* 501–D, 118–122.

Schmidt, V. & McDonald, D.A. (1979a) The role of secondary porosity in the course of sandstone diagenesis. In: *Aspects of Diagenesis* (Ed. by P.A. Scholle and

P.R. Schluger), pp. 175–207. Spec. Publ. Soc. econ. Paleont. Miner. 26, Tulsa.

Schmidt, V. & McDonald, D.A. (1979b) Texture and recognition of secondary porosity in sandstones. In: *Aspects of Diagenesis* (Ed. by P.A. Scholle and P.R. Schluger), pp. 209–225. Spec. Publ. Soc. econ. Paleont. Miner. 26, Tulsa.

Schmidt, V., McDonald, D.A. & Platt, R.L. (1977) Pore geometry and reservoir aspects of secondary porosity in sandstones. *Bull. Can. Petrol. Geol.* **25**, 271–290:

Schneider, J. (1976) *Biological and Inorganic Factors in the Destruction of Limestone Coasts.* Contributions to Sedimentology, 6. Elsevier, Amsterdam.

Schoell, M. (1980) The hydrogen and carbon isotopic composition of methane from natural gases of various origins. *Geochim. cosmochim. Acta*, **44**, 649–661.

Scholle, P.A. (1971) Diagenesis of deep-water carbonate turbidites, Upper Cretaceous Monte Antola Flysch, northern Appenines, Italy. *J. sedim. Petrol.* **41**, 233–250.

Scholle, P.A. (1977) Chalk diagenesis and its relation to petroleum exploration: oil from chalks: a modern miracle? *Bull. Am. Ass. Petrol. Geol.* **61**, 982–1009.

Scholle, P.A. (1978) A color illustrated guide to carbonate rock constituents, textures, cements and porosities. *Am. Ass. Petrol. Geol. Mem.* 27.

Scholle, P.A. (1979) A color illustrated guide to constituents, textures, cements and porosities of sandstones and associated rocks. *Am. Ass. Petrol. Geol. Mem*, 28.

Scholle, P.A. & Halley, R.B. (1985) Burial diagenesis: out of sight, out of mind! In: *Carbonate Cements* (Ed. by N. Schneidermann and P.M. Harris), pp. 309–334. Spec. Publ. Soc. econ. Paleont. Miner. 36.

Schopf, J.W. (ed.) (1983) *Earth's Earliest Biosphere. Its Origin and Evolution.* Princeton University Press, New Jersey.

Schreiber, B.C. (1978) Environments of subaqueous gypsum deposition, In: *Marine Evaporites* (Ed. by W.E. Dean and B.C. Schreiber), pp. 43–74. Soc. econ. Paleont. Miner. Short Course No. 4.

Schulman, J.H., Evans, L.W., Ginther, R.J. & Murata, K.J. (1947) The sensitized luminescence of manganese-activated calcite. *J. appl. Phys.* **18**, 732–739.

Schultz, L.G. (1960) Quantitative X-ray determination of some aluminous clay minerals in rocks. In: *Clays, Clay Minerals, Proc. 7th Nat. Conf.* (Ed. by E. Ingerson), pp. 216–224. Pergamon Press, Oxford.

Schultz, L.G. (1964) Quantitative interpretation of mineralogical composition from X-ray and chemical data for the Pierre Shale. *U.S. geol. Surv. Prof. Pap.* 391–C.

Schwab, F.L. (1978) Secular trends in the composition of sedimentary rock assemblages — Archaean through Phanerozoic time. *Geology*, **6**, 532–536.

Schwarcz, H.P. (1981) Book review of 'stable isotope geochemistry'. *Geochim. cosmochim. Acta*, **45**, 2295.

Schwarzacher, W. (1969) The use of Markov chains in the study of sedimentary cycles. *Math. Geol.* **1**, 17–39.

Schwarzacher, W. (1975) *Sedimentation Models and Quantitative Stratigraphy*. Elsevier, Amsterdam.

Sears, S.O. (1984) Porcelaneous cement and microporosity in California Miocene turbidites — origin and effect on reservoir properties. *J. sedim. Petrol.* **54**, 159–169.

Sedimentation Seminar (1981) Comparison of methods of size analysis for sands of the Amazon-Solimoes Rivers of Brazil and Peru. *Sedimentology*, **28**, 123–128.

Selley, R.C. (1968) A classification of palaeocurrent models. *J. Geol.* **76**, 99–110.

Selley, R.C. (1969) Studies of sequences in sediments using a simple mathematical device. *Q. J. geol. Soc. London*, **125**, 557–581.

Selley, R.C. (1985) *Elements of Petroleum Geology*. Freeman, New York.

Sellwood, B.W. & Parker, A. (1978) Observations on diagenesis in North Sea reservoir sandstones. *J. geol. Soc. London*, **135**, 133–135.

Sengupta, S. & Rao, J.S. (1966) Statistical analysis of cross bedding azimuths from the Kamthi Formation around Bheenmaram, Pranhita — Godavari valley, Sankhya. *Ind. J. Stat.* **288**, 165–174.

Sengupta, S. & Veenstra, H.J. (1968) On sieving and settling techniques for sand analysis. *Sedimentology*, **11**, 83–98.

Shackleton, N.J., Hall, M.A. & Boersma, A. (1984) Oxygen and carbon isotope data from Leg 74 Foraminifers. *Initial Reports of the Deep Sea Drilling Project*, **74**, 599–612. US Government Printing Office, Washington DC.

Shanmugam, G. (1980) Rhythms in deep sea, fine-grained turbidite and debris-flow sequences, Middle Ordovician, eastern Tennessee. *Sedimentology*, **27**, 419–432.

Shanmugan, G. (1985) Types of porosity in sandstones and their significance in interpreting provenance. In: *Provenance of Arenites* (Ed. by G.G. Zuffa), pp. 115–137. NATO ASI Series C: vol. 148. Reidel, Dordrecht.

Sharma, T. & Clayton, R.N. (1965) Measurement of $0^{18}/0^{16}$ ratios of total oxygen of carbonates. *Geochim. cosmochim. Acta*, **29**, 1347–1353.

Sheldon, R.W. & Parsons, T.R. (1967) *A Practical Manual on the Use of the Coulter Counter in Marine Science.* Coulter Electronics Sales, Toronto.

Shinn, E.A. & Robbin, D.M. (1983) Mechanical and chemical compaction in fine-grained shallow-water limestones. *J. sedim. Petrol.* **53**, 595–618.

Shukla, V. & Friedman, G.M. (1983) Dolomitization and diagenesis in a shallowing-upward sequence: the Lockport Formation (Middle Silurian). New York State. *J. sedim. Petrol.* **53**, 703–717.

Shultz, A.W. (1984) Subaerial debris flow deposition in the Upper Paleozoic Cutler Formation, western Colorado. *J. sedim. Petrol.* **54**, 759–772.

Sibley, D.F. & Blatt, H. (1976) Intergranular pressure solution and cementation of the Tuscarora orthoquartzite. *J. sedim. Petrol.* **46**, 881–896.

Siebert, R.M., Moncure, G.K. & Lahann, R.W. (1984) A theory of framework grain dissolution in sandstones. In: *Clastic Diagenesis* (Ed. by D.A. McDonald and R.C.

Surdam), pp. 163–175. Mem. Am. Ass. Petrol Geol. 37.

Sillen, L.G. (1967) The ocean as a chemical system. *Science*, **156**, 1189–1197.

Simpson, J. (1985) Stylolite-controlled layering in a homogenous limestone: pseudo-bedding produced by burial diagenesis. *Sedimentology*, **32**, 495–506.

Sippel, R.F. (1968) Sandstone petrology, evidence from luminescence petrography. *J. sedim. Petrol.* **38**, 530–554.

Sippel, R.F. (1971) Quartz grain orientations — 1 (the photometric method). *J. sedim. Petrol.* **41**, 38–59.

Sippel, R.F. & Glover, E.D. (1965) Structures in carbonate rocks made visible by luminescence petrography. *Science*, **150**, 1283–1287.

Slingerland, R.L. & Williams, E.G. (1979) Paleocurrent analysis in light of trough cross-stratification geometry. *J. Geol.* **87**, 724–732.

Sly, P.G. (1977) Sedimentary environments in the Great Lakes. In: *Proceedings of the SIL-UNESCO Symposium on Interactions Between Sediments and Freshwater* (Ed. by H.L. Golterman), pp. 76–82. Amsterdam, Junk.

Sly, P.G. (1978) Sedimentary processes in lakes. In: *Lakes — Chemistry, Geology, Physics* (Ed. by A. Lerman), pp. 65–89. Springer-Verlag, Berlin.

Smart, P. & Tovey, N.K. (1982) *Electron Microscopy of Soils and Sediments: Techniques.* Clarendon Press, Oxford.

Smith, A.G. & Briden, J.C. (1976) *Mesozoic and Cenozoic Paleocontinental Maps*. Cambridge University Press.

Smith, D.G. (1984) Vibracoring fluvial and deltaic sediments: tips on improving penetration and recovery. *J. sedim. Petrol.* **54**, 660–663.

Smith, M.P. (1986) Silver coating inhibits electron microprobe beam damage of carbonates. *J. sedim. Petrol.* **56**, 560–561.

Smith, J.V. & Stenstrom, R.C. (1965) Electron-excited luminescence as a petrologic tool. *Geology*, **73**, 627–635.

Sofer, Z. (1978) The isotopic composition of hydration water in gypsum. *Geochim. cosmochim. Acta*, **42**, 1141–1150.

Soloman, M. & Green, R. (1966) A chart for designing modal analysis by point counting. *Geol. Rdsch*, **55**, 844–848.

Sommer, F. (1978) Diagenesis of Jurassic sandstones in the Viking Graben. *J. geol. Soc. London*, **135**, 63–67.

Sommer, S.E. (1972a) Catholuminescence of carbonates. I. Characterization of cathodoluminescence from carbonate solid solutions. *Chem. Geol.* **9**, 257–273.

Sommer, S.E. (1972b) Cathodoluminescence of carbonates. II. Geological applications. *Chem. Geol.* **9**, 275–284.

Sommer, S.E. (1975) Effect of staining on microprobe determination of iron in carbonates. *J. sedim. Petrol.* **45**, 541–542.

Sorby, H.C. (1851) On the microscopical structure of the Calcareous Grit of the Yorkshire coast. *Q. J. geol. Soc. London*, **7**, 1–6.

Spark, I.S.C. & Trewin, N.H. (1986) Facies related diagenesis in the Main Claymore Oilfield sandstone. *Clay Miner.* **21**, 479–496.

Spears, D.A. (1965) Boron in some British Carboniferous sedimentary rocks. *Geochim. cosmochim. Acta*, **29**, 315–328.

Spears, D.A. & Sezgin, H.I. (1985) Mineralogy and geochemistry of the Subcrenatum Marine Band and associated coal-bearing sediments, Langsett, South Yorkshire. *J. sedim. Petrol.* **55**, 570–578.

Sperber, C.M., Wilkinson, B.H. & Peacor, D.R. (1984) Rock composition, dolomite stoichiometry and rock/water reactions in dolomitic carbonate rocks. *J. Geol.* **92**, 609–622.

Sridhar, K., Jackson, M.L. & Clayton, R.N. (1975) Quartz oxygen isotopic stability in relation to isolation from sediments and diversity of source. *Proc. Soil Sci. Soc. Am.* **39**, 1209–1213.

Stablein III, N.K. & Dapples, E.C. (1977) Feldspars of the Tunnel City Group (Cambrian), Western Wisconsin. *J. sedim. Petrol.* **47**, 1512–1538.

Stalder, P.J. (1979) Organic and inorganic metamorphism in the Taveyannaz Sandstone of Swiss Alps and equivalent sandstones in France and Italy. *J. sedim. Petrol.* **49**, 463–482.

Stanton, R.J. & Dodd, J.R. (1970) Paleoecologic techniques — comparison of faunal and geochemical analyses of Pliocene paleoenvironments, Kettleman Hills, California. *J. Paleont.* **44**, 1092–1121.

Starkey, H.C., Blackmon, P.D. & Hauff, P.L. (1984) The routine mineralogical analysis of clay-bearing samples. *U.S. geol. Surv., Bull.* 1563.

Statham, P.J. (1981) X-ray microanalysis with Si (Li) detectors. *J. Microsc.* **123**, 1–23.

Statham, P.J. (1982) Prospects for improvement in EDX micro analysis. *J. Microsc.* **130**, 165–176.

Steel, R.J. (1974) New Red Sandstone floodplain and piedmont sedimention in the Hebridean province, Scotland. *J. sedim. Petrol.* **44**, 336–357.

Steel, R.J., Maehle, S., Nilsen, H., Roe, S.L. & Spinnangr, A. (1977) Coarsening-upward cycles in the alluvium of Hornelen Basin (Devonian) Norway, sedimentary response to tectonic events. *Bull. geol. Soc. Am.* **88**, 1124–1134.

Steel, R.J. & Thompson, D.B. (1983) Structures and textures in Triassic braided stream conglomerates ('Bunter' Pebble Beds) in the Sherwood Sandstone Group, North Staffordshire, England. *Sedimentology*, **32**, 495–506.

Steele, T.W. (1978) A guide to the reporting of analytical results relating to the certification of geological reference materials. *Geostand. Newsl.* **2**, 31–33.

Steinen, R.P. (1978) On the diagenesis of lime mud: scanning electron microscope observations of subsurface material from Barbados, W.I. *J. sedim. Petrol.* **48**, 1139–1148.

Steinen, R.P., Matthews, R.K. & Sealy, H.A. (1978)

Temporal variation in geometry and chemistry of the freshwater phreatic lens: the coastal carbonate aquifer of Christchurch, Barbados, West Indies. *J. sedim. Petrol.* **48**, 733–742.

Stewart, D.J. (1981) A meander-belt sandstone of the Lower Cretaceous of Southern England. *Sedimentology*, **28**, 1–20.

Stewart, H.B. Jr (1958) Sedimentary reflections on depositional environments in San Migue Lagoon, Baja, California, Mexico. *Bull. Am. Ass. Petrol. Geol.* **42**, 2567–2618.

Stiller, M., Rounick, J.S. & Shasha, S. (1985) Extreme carbon-isotope enrichments in evaporating brines. *Nature*, **316**, 434–435.

Stockdale, P.B. (1926) The stratigraphic significance of solution in rocks. *J. Geol.* **34**, 399–414.

Stow, D.A.V. & Miller, J. (1984) Mineralogy, pertrology and diagenesis of sediments at Site 530, southeast Angola Basin. In: *Initial Reports of the Deep Sea Drilling Project*, **75**, 857–873. US Government Printing Office, Washington DC.

Stumm, W. & Morgan, J.J. (1981) *Aquatic Chemistry — an Introduction Emphasizing Chemical Equilibria in Natural Waters* (2nd edn). Wiley, New York.

Sugisaki, R. (1984) Relation between chemical composition and sedimentation rate of Pacific ocean-floor sediments deposited since the Middle Cretaceous: basic evidence for chemical constraints on depositional environments of ancient sediments. *J. Geol.* **92**, 235–259.

Sunagawa, I. (1982) Morphology of crystals in relation to growth conditions. *Estud. geol.* **38**, 127–134.

Surdam, R.C., Boese, S.W. & Crossey, L.J. (1984) The chemistry of secondary proosity. In: *Clastic Diagenesis* (Ed. by D.A. McDonald and R.C. Surdam), pp. 127–149. Mem. Am. Ass. Petrol. Geol. 37.

Surlyk, F. (1978) submarine fan sedimentation along fault scarps on tilted fault blocks (Jurassic–Cretaceous boundary, East Greenland). *Grnl. Geol. Unders.* **128**.

Sverjensky, D.A. (1984) Prediction of Gibbs free energies of calcite-type carbonates and the equilibrium distribution of trace elements between carbonates and aqueous solutions. *Geochim. cosmochim. Acta*, **48**, 1127–1134.

Swallow, H.T.S. (1964) Building mortars and bricks. *Brit. Ceramics Res. Ass. Tech. Note* 57.

Syers, J.K., Chapman, S.L. & Jackson, M.L. (1968) Quartz isolation from rocks, sediments and soils for determination of oxygen isotope composition. *Geochim. cosmochim. Acta*, **32**, 1022–1025.

Takano, B. (1985) Geochemical implications of sulfate in sedimentary carbonates. *Chem. Geol.* **49**, 393–403.

Tanner, W.F. (1959) The importance of modes in cross-bedding data. *J. sedim. Petrol.* **29**, 221–226.

Tanner, W.F. (1969) The particle size scale. *J. sedim. Petrol.* **39**, 509–512.

Tardy, Y. & Fritz, B. (1981) An ideal solid solution model for calculating solubility of clay minerals. *Clay Miner.* **16**, 361–373.

Tarutani, T., Clayton, R.N. & Mayeda, Y.K. (1969) The effect of polymorphism and magnesium substitution on oxygen isotope fractionation between calcium carbonate and water. *Geochim. cosmochim. Acta*, **33**, 987–996.

Tasse, N. & Hesse, R. (1984) Origin and significance of complex authigenic carbonates in Cretaceous black shales of the western Alps. *J. sedim. Petrol.* **54**, 1012–1027.

Taylor, C.M. & Radtke, A.S. (1965) Preparation and polishing of ores and mill products for microscopic examination and electron microprobe analysis. *Econ. Geol.* **60**, 1306–1319.

Taylor, H.P. & Epstein, S. (1962) Relationship between $0^{18}/0^{16}$ ratios in coexisting minerals of igneous and metamorphic rocks. *Bull. geol. Soc. Am.* **73**, 461–480.

Taylor, S.R. & McLennan, S.M. (1985) *The Continental Crust: its Composition and Evolution.* Blackwell Scientific Publications, Oxford.

Taylor, S.R., McLennan, S.M. & McCulloch, M.T. (1983) Geochemistry of loess, continental crustal composition and crustal model ages. *Geochim. cosmochim. Acta*, **47**, 1897–1905.

Ten Have, A. & Heynen, W. (1985) Cathodoluminescence activation and zonation in carbonate rocks: an experimental approach. *Geol. Mijnb.* **64**, 297–310.

Tennant, C.B. & Berger, R.W. (1957) X-ray determination of dolomite calcite ratio of a carbonate rock. *Am. Miner.* **42**, 23–29.

Terry, R.D. & Chilingar, G.V. (1955) Summary of "Concerning some additional aids in studying sedimentary formations" by M.S. Shretsor. *J. sedim. Petrol.* **25**, 229–234.

Tertian, R. & Claisse, F. (1982) *Principles of Quantitative X-ray Fluorescence Analysis*. Heyden, London.

Theide, J., Chriss, T., Clauson, M. & Swift, S.A. (1976) *Settling tubes for size analysis of fine and coarse fractions of oceanic sediments.* Ref. 76–78 Oregon State University, College of Oceanography.

Thode, H.G., Monster, J. & Dunford, H.B. (1961) Sulphur isotope geochemistry. *Geochim. cosmochim. Acta*, **25**, 159–174.

Thomas, I. L. & Haukka, M.T. (1978) XRF determination of trace and major elements using a single fused disc. *Chem. Geol.* **21**, 39–50.

Thompson, A. (1959) Pressure solution and porosity. In: *Silica in Sediments* (Ed. by H.A. Ireland), pp. 92–110. Soc. econ. Paleont. Miner. Spec. Publ. 7.

Thompson, M. (1983) Control procedures in geochemical analysis. In: *Statistics and Data Analysis in Geochemical Prospecting* (Ed. by R.J. Howarth), pp. 39–58. Elsevier, Amsterdam.

Thompson, M. (1986) The future role of inductively-coupled plasma atomic emission spectrometry in applied geochemistry. In: *Applied Geochemistry in the 1980's* (Ed. by I. Thornton and P.J. Howarth), pp. 191–211. Graham & Trotman, London.

Thompson, M., Goulter, J.F. & Sieper, F. (1981) Laser ablation for the introduction of solid samples into an inductively-coupled plasma for atomic emission spectrometry. *Analyst (London)*, **106**, 32–39.

Thompson, M., Rankin, A.M., Waltom, S.J., Halls, C. & Foo, R.N. (1980) The analysis of fluid inclusion descrepitate by inductively-coupled plasma atomic emission spectroscopy: an exploratory study. *Chem. Geol.* **30**, 121–133.

Thompson, M. & Walsh, J.N. (1983) *A Handbook of Inductively Coupled Plasma Spectrometry*, Blackie, Glasgow.

Thorez, J. (1974) *Phyllosilicates and Clay Minerals: a Laboratory Handbook for their X-ray Diffraction Analysis*. Editions G. Lelotte, Dison, Belgium.

Thorez, J. (1976) *Practical Identification of Clay Minerals*. Editions G. Lelotte, Dison, Belgium.

Thornton, P.R. (1968) *Scanning Electron Microscopy*. Chapman & Hall, London.

Thorstenson, D.C. & Plummer, L.N. (1977) Equilibrium criteria for two component solids reacting with fixed composition in an aqueous phase — example: the magnesian calcites. *Am. J. Sci.* **277**, 1203–1223.

Tieh, T.T., Ledger, E.B. & Rowe, M.W. (1980) Release of uranium from granitic rocks during *in-situ* weathering and initial erosion (Central Texas). *Chem. Geol.* **29**, 227–248.

Tillman, R.W. & Almon, W.R. (1979) Diagenesis of Frontier Formation offshore bar sandstones, spearhead Ranch field, Wyaming. In: *Aspects of Diagenesis* (Ed by P. A. Scholle and P. R. Schluger), pp. 337–378. Spec. Publ. Soc. econ. Miner. Palaeont.

Ting, F.T.C. (1977) Microscopical investigations of the transformation (diagenesis) from peat to lignite. *J. Microsc.* **109**, 75–83.

Tlig, S. & M'Rabet, A. (1985) A comparative study of the rare earth element (REE) distributions within the Lower Cretaceous dolomites and limestones of Central Tunisia. *Sedimentology*, **32**, 897–907.

Tlig, S. & Steinberg, M. (1982) Distribution of rare-earth elements (REE) in size fractions of Recent sediments of the Indian Ocean. *Chem. Geol.* **37**, 317–333.

Tourtelot, H.A., Schultz, L.G. & Huffman, C. (1961) Boron in bentonites and shales from the Pierre shale, South Dakota, Wyoming, and Montana. *U.S. geol. Surv. Prof. Pap.* 424C, 288–292.

Tovey, N.K. (1978) Potential developments in stereoscopic scanning electron microscope studies of sediments. In: *Scanning Electron Microscopy in the Study of Sediments* (Ed. by W.B. Whalley), pp. 105–117. Geo Abstracts, Norwich.

Tovey, N.K. & Wong, K. Y. (1978) Preparation, selection and interpretation problems in scanning electron microscope studies of sediments. In: *Scanning Electron Microscopy in the Study of Sediments* (Ed. by W.B. Whally), pp. 181–199. Geo Abstracts, Norwich.

Towe, K.M. (1974) Quantitative clay petrology: The trees but not the forest? *Clays Clay Miner.* **22**, 375–378.

Trevena, A.S. & Nash, W.P. (1981) An electron microprobe study of detrital feldspar. *J. sedim. Petrol.* **51**, 137–150.

Trewin, N.H. & Welsh, W. (1976) Formation of a graded estuarine shell bed. *Palaeogeogr. Palaeoclim. Palaeoecol.* **19**, 219–230.

Trurnit, P. (1968) Analysis of pressure solution contacts and classificatioin of pressure solution phenomena. In: *Recent Developments in Carbonate Sedimentology in Central Europe* (Ed. by G. Müller and G.M. Friedman), pp. 75–54. Springer-Verlag, Berlin.

Tucker, M.E. (1981) *Sedimentary Petrology, an Introduction*. Blackwell Scientific Publications, Oxford.

Tucker, M.E. (1982) *The Field Description of Sedimentary Rocks*. Open University Press, Milton Keynes.

Tucker, M.E. (1986) Formerly aragonitic limestones associated with tillites in the Late Proterozoic of Death Valley, California. *J. sedim. Petrol.* **56**, 818–830.

Tudge, A.P. (1960) A method of analysis of oxygen isotopes in orthophosphate — its use in the measurement of paleotemperature. *Geochim. cosmochim. Acta*, **18**, 81–93.

Tudhope, A.W. & Risk, M.J. (1985) Rate of dissolution of carbonate sediments by microboring organisms, Davies Reef, Australia. *J. sedim. Petrol.* **55**, 440–447.

Turk, G. (1979) Transition analysis of structural sequences, disscussion. *Bull. geol. Soc. Am.* **90**, 989–991.

Turner, J.V. (1982) Kinetic fractionation of carbon-13 during calcium carbonate precipitation. *Geochim. cosmochim. Acta*, **46**, 1183–1191.

Turner, P. (1974) Lithostratigraphy and facies analysis of the Ringerike Group of the Oslo Region. *Norges geol. Unders.* **314**, 101–132.

Valloni, R. (1985) Reading provenance from modern marine sands. In: *Provenance of Arenites* (Ed. by G.G. Zuffa), pp. 309–332. NATO ASI Series C: vol. 148. Reidel, Dordrecht.

Valloni, R. & Maynard, J.B. (1981) Detrital modes of recent deep-sea sands and their relation to tectonic setting: a first approximation. *Sedimentology*, **28**, 75–83.

Van der Plas, L. & Tobi, A.C. (1965) A chart for judging the reliability of point counting results. *Am. J. Sci.* **263**, 87–90.

Van Gelder, A. (1974) Sedimentation in the marine margin of the Old Red Sandstone continent, south of Cork, Ireland. *PhD thesis*. University of Utrecht.

Van Houten, F.B. (1974) Northern Alpine Molasse and similar Cenozoic sequences of Southern Europe. In: *Modern and Ancient Geosynclinal Sedimentation* (Ed. by R.H. Dott and R.H. Shaver), pp. 260–273. Soc. econ. Paleont. Mineral. Spec. Publ. 19.

Van Olphen, H. (1977) *An Introduction to Clay Colloid Chemistry* (2nd edn). Wiley, New York.

Van Olphen, H. & Fripiat, J.J. (1979) *Data Handbook for Clay Materials and other Non-Metallic Minerals*. Pergamon Press, Oxford.

Van Olphen, H. & Veniale, F. (1982) *International Clay Conference 1981*. Elsevier, Amsterdam.

Veizer, J. (1974) Chemical diagenesis of belemnite shells and possible consequences for paleotemperature determination. *Neues Jb. Geol. Paleont. Abh.* **147**, 91–111.

Veizer, J. (1977) Diagenesis of pre-Quaternary carbonates as indicated by tracer studies. *J. sedim. Petrol.* **47**, 565–581.

Veizer, J. (1978) Secular variations in the composition of sedimentary carbonate rocks, II. Fe, Mn, Ca, Mg, Sr and minor constituents. *Precamb. Res.* **6**, 381–413.

Veizer, J. (1983) Chemical diagenesis of carbonates: theory and application of trace element techniques. In: *Stable Isotopes in Sedimentary Geology*, pp. 3–1 to 3–100. Soc. econ. Paleont. Miner. Short Course 10.

Veizer, J. & Compston, W. (1974) $^{87}Sr/^{86}Sr$ composition of seawater during the Phanerozoic. *Geochim. cosmochim. Acta*, **38**, 1461–1484.

Veizer, J., Compston, W. (1976) $^{87}Sr/^{86}Sr$ in Precambrian carbonates as an index of crustal evolution. *Geochim. cosmochim. Acta*, **40**, 905–915.

Veizer, J., Compston, W., Clauer, N. & Schidlowski, M. (1983) $^{87}Sr/^{86}Sr$ in Late Proterozoic Carbonates: evidence for a "mantle event" at ~900 Ma ago. *Geochim. cosmochim. Acta*, **47**, 295–302.

Veizer, J., Fritz, P. & Jones, B. (1986) Geochemistry of brachiopods: oxygen and carbon isotopic records of Paleozoic oceans. *Geochim. cosmochim. Acta*, **50**, 1679–1696.

Veizer, J., Hinton, R.W., Clayton, R.N. & Lerman, A. (1987) Chemical diagenesis of carbonates in thin-sections: ion microprobe as a trace element tool. *Chem. Geol.* **64**, 225–237.

Veizer, J. & Hoefs, J. (1976) The nature of O^{18}/O^{16} and C^{13}/C^{12} secular trends in sedimentary carbonate rocks. *Geochim. cosmochim. Acta*, **40**, 1387–1395.

Veizer, J., Holser, W.T. & Wilgus, C.K. (1980) Correlations of $^{13}C/^{12}C$ and $^{34}S/^{32}S$ secular variations. *Geochim. cosmochim. Acta*, **44**, 579–581.

Veizer, J. & Jansen, S.L. (1979) Basement and sedimentary recycling and continental evolution. *J. Geol.* **87**, 341–370.

Veizer, J., Lemieux, J., Jones, B., Gibling, M.R. & Savelle, J. (1977) Sodium: palaeosalinity indicator in ancient carbonate rocks. *Geology*, **5**, 177–179.

Veizer, J., Lemieux, J., Jones, B., Gibling, M.R. & Savelle, J. (1978) Paleosalinity and dolomitization of a Lower Paleozoic carbonate sequence, Somerset and Prince of Wales Islands, Arctic Canada. *Can. J. Earth Sci.* **15**, 1448–1461.

Velde, B. (1977) *Clays and Clay Minerals in Natural and Synthetic Systems*. Elsevier, Amsterdam.

Velde, B. (1984) Electron microprobe analysis of clay minerals. *Clay Miner.* **19**, 243–247.

Videtich, P.E. (1981) A method for analysing dolomite for stable isotopic composition. *J. sedim. Petrol.* **50**, (2), 661–662.

Vinopal, R.J. & Coogan, A.H. (1978) Effect of particle shape on the packing of carbonate and sands and gravels. *J. sedim. Petrol.* **48**, 7–24.

Visher, G.S. (1969) Grain size distributions and depositional processes. *J. sedim. Petrol.* **39**, 1074–1106.

Von Der Borch, C.C. & Lock, D. (1979) Geological significance of Coorong dolomites. *Sedimentology*, **26**, 813–824.

Von Rad, U. & Rosch, H. (1974) Petrography and diagenesis of deep-sea cherts from the central Atlantic. In: *Pelagic Sediments: on Land and under the Sea* (Ed. by K.J. Hsü and H.C. Jenkyns), pp. 327–347. Spec. Publ. Int. Ass. Sediment. 1. Blackwell Scientific Publications, Oxford.

Wachs, D. & Hein, J.R. (1974) Petrography and diagenesis of Franciscan limestones. *J. sedim. Petrol.* **44**, 1217–1231.

Wada, H. & Suzuki, K. (1983) Carbon isotopic thermometry calibrated by dolomite-calcite solvus temperatures. *Geochim. cosmochim. Acta*, **47**, 697–706.

Walkden, G.M. & Berry, J.R. (1984) Syntaxial overgrowths in muddy crinoidal limestones: cathodoluminescence sheds new light on an old problem. *Sedimentology*, **31**, 251–267.

Walker, B.M. (1978) Chalk pore geometry using resin pore casts. In: *Scanning Electron Microscopy in the Study of Sediments* (Ed. by W.B. Whalley), pp. 17–27. Geo Abstracts, Norwich.

Walker, C.T. (1962) Separation techniques in sedimentary geochemistry, illustrated by studies of boron. *Nature*, **194**, 1073–1074.

Walker, C.T. (ed.) (1975) *Geochemistry of Boron. Benchmark Papers in Geology. V23.* Dowden, Hutchinson & Ross, Stroudsburg.

Walker, C.T. & Price, N.B. (1963) Departure curves for computing paleosalinity from boron in illites and shales. *Bull. Am. Ass. Petrol. Geol.* **47**, 833–841.

Walker, G. (1983) Luminescence centres in minerals. *Chem. Britain*, October, pp. 824–826.

Walker, G. (1985) Mineralogical applications of luminescence techniques. In: *Chemical Bonding and Spectroscopy in Mineral Chemistry*, pp. 103–140. Chapman & Hall, London.

Walker, R.G. (1975) Generalized facies models for re-sedimented conglomerates of turbidite association. *Bull. geol. Soc. Am.* **86**, 737–748.

Walker, R.G. (1984) Facies and facies models. 1. General Introduction. In: *Facies Models* (2nd edn) (Ed. by R.G. Walker), pp. 1–9. Geoscience, Canada.

Walker, T.E. (1962) Reversible nature of chert-carbonate replacement in sedimentary rocks. *Bull. geol. Soc. Am.* **73**, 237–242.

Walker, T.R. (1967) Formation of red beds in modern and ancient deserts. *Bull. geol. Soc. Am.* **78**, 353–368.

Walker, T.R., Waugh, B. & Crone, A.J. (1978) Diagenesis in first cycle desert alluvium of Cenozoic age, south-western United States and northwestern Mexico. *Bull. geol. Soc. Am.* **89**, 19–32.

Walter, L.M. & Morse, J.W. (1984) Magnesian calcite stabilities: a reevaluation. *Geochim. cosmochim. Acta*, **48**, 1059–1069.

Walters, L.J., Claypool, G.E. & Choquette, P.W. (1972) Reaction rates and δO^{18} variation or the carbonate-phosphoric acid preparation method. *Geochim. cosmochim. Acta*, **36**, 129–140.

Walters, L.J., Owen, D.E., Henley, A.L., Winsten, M.S. & Valek, K.W. (1987) Depositional environments of the Dakota Sandstone and adjacent units in the San Juan Basin utilizing discriminant analysis of trace elements in shales. *J. sedim. Petrol.* **57**, 265–277.

Wang, Y., Piper, D.J.W. & Vilks, G. (1982) Surface textures of turbidite sand grains, Laurentian Fan and Sohm Abyssal Plain. *Sedimentology*, **29**, 727–736.

Wanless, H.R. (1979) Limestone response to stress: pressure solution and dolomitization. *J. sedim. Petrol.* **49**, 437–462.

Wanless, H.R. (1983) Burial diagenesis in limestones. In: *Sediment Diagenesis* (Ed. by A. Parker and B.W. Sellwood), pp. 379–417. NATO ASI Series C: vol. 115. Reidel, Dordrecht.

Ward, W.C. & Halley, R.B. (1985) Dolomitization in a mixing zone of near-seawater composition, Later Pleistocene, Northeastern Yucatan Peninsula. *J. sedim. Petrol.* **55**, 407–420.

Wardlaw, N.C. (1976) Pore geometry of carbonate rocks as revealed by pore casts and capillary pressure. *Bull. Am. Ass. Petrol. Geol.* **60**, 245–257.

Wardlaw, N.C. & Cassan, J.P. (1978) Estimation of recovery efficiency by visual observation of pore systems in reservoir rocks. *Bull. Can. Petrol. Geol.* **27**, 117–138.

Warme, J.E. (1975) Borings as trace fossils, and the processes of marine bioerosion. In: *The Study of Trace Fossils* (Ed. by R.W. Frey), pp. 181–227. Springer-Verlag, Berlin.

Watts, N.L. (1980) Quaternary pedogenic calcretes from the Kalahari (southern Africa): mineralogy, genesis and diagenesis. *Sedimentology*, **27**, 661–686.

Waugh, B. (1970) Formation of quartz overgrowths in the Penrith Sandstone (Lower Permian of northwest England as revealed by scanning electron microscopy. *Sedimentology*, **14**, 309–320.

Waugh, B. (1978a) Diagenesis in continental red beds as revealed by scanning electron microscopy: a review. In: *Scanning Electron Microscopy in the Study of Sediments* (Ed. by W.B. Whalley), pp. 329–346. Geo Abstracts, Norwich.

Waugh, B. (1978b) Authigenic K-feldspar in British Permo-Triassic sandstones. *J. geol. Soc. London*, **135**, 51–56.

Weast, R.C. (ed.) (1983) *CRC Handbook of Chemistry and Physics*. CRC Press, Baton Rouge, Florida.

Weaver, C.E. (1960) Possible uses of clay minerals in the search for oil. *Bull. Am. Ass. Petrol. Geol.* **44**, 1505–1518.

Weber, J.N., Deines, P., Weber, P.H. & Baker, P.A. (1976) Depth related changes in the 13C/12C ratio of skeletal carbonate deposited by the Caribbean reef-frame building coral *Montastrea annularis*: further implications of a model for stable isotope fractionation by scleractinian corals. *Geochim. cosmochim. Acta*, **40**, 31–39.

Weber, J.N. & Smith, F.G. (1961) Rapid determination of calcite-dolomite ratios in sedimentary rocks. *J. sedim. Petrol.* **31**, 130–131.

Wefer, G. & Berger, W.H. (1981) Stable isotope composition of benthic calcareous algae from Bermuda. *J. sedim. Petrol.* **51**, 459–465.

Weil, S.M., Buddemeir, R.W., Smith, S.V. & Kroopnick, P.M. (1981) The stable isotopic composition of coral skeletons: control by environmental variables. *Geochim. cosmochim. Acta*, **45**, 1147–1153.

Weir, A.H., Ormerod, E.C. & El-Mansey, M.I. (1975) Clay mineralogy of sediments of the western Nile Delta. *Clay Miner.* **10**, 369–386.

Wellendorf, W. & Krinsley, D. (1980) The relation between crystallography of quartz and upturned aeolian cleavage plates. *Sedimentology*, **27**, 447–453.

Weller, J.M. (1959) Compaction of sediments. *Bull. Am. Ass. Petrol. Geol.* **43**, 273–319.

Welton, J.E. (1984) *SEM Petrology Atlas*. Am. Ass. Petrol. Geol., Methods in Exploration series no. 4.

Wentworth, C.K. (1922) A scale of grade and class terms for clastic sediments. *J. Geol.* **30**, 377–392.

West, N.G., Hendry, G.L. & Bailey, N.T. (1974) The analysis of slags from primary and secondary copper smelting processes by X-ray fluorescence. *X-ray Spectrom.* **3**, 78–87.

Weyl, P.K. (1959) Pressure solution and the force of crystallization — phenomenological theory. *J. geophys. Res.* **64**, 2001–2025.

Whalley, W.B. (ed.) (1978) *Scanning Electron Microscopy in the Study of Sediments*. Geo Abstracts, Norwich.

Whitaker, J.H.McD. (1978) Diagensis of the Brent Sand Formation: a scanning electron microscope study. In: *Scanning Electron Microscopy in the Study of Sediments*. (Ed. by W.B. Whalley), pp. 363–380. Geo Abstracts, Norwich.

White, S.H., Shaw, H.F. & Huggett, J.M. (1984) The use of back-scattered electron imaging for the petrographic study of sandstones and shales. *J. sedim. Petrol.* **54**, 487–94.

Whitfield, M. (1975) The electroanalytical chemistry of seawater. In: *Chemical Oceanography*, vol. 4 (2nd edn) (Ed. by J.P. Riley and G. Skirrow), pp. 111–154. Academic Press, London.

Wilkinson, B.H., Janecke, S.U. & Brett, C.E. (1982) Low-magnesian calcite marine cement in Middle Ordovician hardgrounds from Kirkfield, Ontario. *J. sedim. Petrol.* **52**, 47–57.

Williams, L.A., Parks, G.A. & Crerar, D.A. (1985) Silica diagenesis. I. Solubility controls. *J. sedim. Petrol.* **55**, 301–311.

Wilson, A.D. (1964) The sampling of silicate rock powders for chemical analysis. *Analyst (London)*, **89**, 18–30.

Wilson, M.D. & Pittman, E.D. (1977) Authigenic clays in sandstones: Recognition and influence on reservoir properties and palaeoenvironmental analysis. *J. sedim. Petrol.* **47**, 3–31.

Wilson, P. (1978) A scanning electron microscope examination of quartz grain surface textures from the weathered Millstone Grit (Carboniferous) of the southern Pennines, England. A preliminary report. In: *Scanning Electron Microscopy in the Study of Sediments* (Ed. by W.B. Whalley), pp. 307–318. Geo Abstracts, Norwich.

Wollast, R., Garrels, R.M. & MacKenzie, F.T. (1980) Calcite-seawater reactions in ocean waters. *Am. J. Sci.* **280**, 831–848.

Wong, P.K. & Oldershaw, A. (1981) Burial cementation in the Devonian, Kaybob Reef Complex, Alberta, Canada. *J. sedim. Petrol.* **51**, 507–520.

Wood, J.R. (1986) Advective diagenesis: thermal mass transfer in systems containing quartz and calcite. In: *Roles of Organic Matter in Sediment Diagenesis* (Ed. by D.L. Gautier). Spec. Publ. Soc. econ. Paleont Mineral. **38**, 169–180.

Woodcock, N.H. (1976a) Structural style in slump sheets, Ludlow Series, Powys, Wales. *J. geol. Soc. London*, **132**, 399–416.

Woodcock, N.H. (1976b) Ludlow Series slumps and turbidites and the form of the Montgomery Trough, Powys, Wales. *Proc. geol. Ass.* **87**, 169–182.

Woodcock, N.H. (1979) The use of slump folds as palaeoslope orientation estimators. *Sedimentology*, **26**, 83–99.

Woods, T.L. & Garrels, R.M. (1987) *Thermodynamic Values at Low Temperatures: an Uncritical Survey*. Oxford University Press, New York.

Wyrwoll, K.-H. & Smyth, G.K. (1985) On using the log-hyperbolic distribution to describe the textural characteristics of Eolian sediments. *J. sedim. Petrol.* **55**, 471–478.

Yariv, S. & Cross, H. (1979) *Geochemistry of Colloid Systems for Earth Scientists*. Springer-Verlag, Berlin.

Yeh, H.-W. (1980) D/H ratios and late-stage dehydration of shales during burial. *Geochim. cosmochim. Acta*, **44**, 341–352.

Yeh, H.-W. & Savin, S.M. (1977) Mechanism of burial metamorphism of argillaceous sediments: O-isotope evidence. *Bull. geol. Soc. Am.* **88**, 1321–1330.

Zeff, M.L. & Perkins, R.D. (1979) Microbial alteration of Bahamian deep-sea carbonates. *Sedimentology*, **26**, 175–201.

Zeigler, J.M. & Gill, A. (1959) *Tables and graphs for the settling velocity of quartz in water, above the range of Stoke's Law.* Refs No. 59–36, Woods Hole Oceanographic Institution.

Zeigler, J.M., Whitney, G.G. & Hayes, C.R. (1960) Woods Hole Rapid Sediment Analyser. *J. sedim. Petrol.* **30**, 490–495.

Zenger, D.H. (1979) Primary textures in dolostones and recrystallized limestones: a technique for their microscopie study. *J. sedim. Petrol.* **49**, 677–678.

Zenger, D.H., Dunham, J.B. & Ethington, R.L. (eds) (1980) Concepts and models of dolomitization. *Soc. econ. Paleont. Miner. Spec. Publ.* 28.

Zhao Xun & Fairchild, I. (1987) Mixing zone dolomitization of Devonian carbonates, Guangxi, South China. In: *Diagenesis of Sedimentary Sequences* (Ed. by J.D. Marshall), pp. 157–170. Spec. Publ. geol. Soc. London No. 36. Blackwell Scientific Publications, Oxford.

Zingg, T. (1935) Beitrage zur Schotteranalyse. *Min. Petrog. Mitt. Schweiz.* **15**, 39–140.

Zinkernagel, U. (1978) Cathodoluminescence of quartz and its application to sandstone petrography. *Contrib. Sediment.* 8. 69 pp.

Zuffa, G.G. (1985) Optical analyses of arenites: influence of methodology on compositional results. In: *Provenance of Arenites* (Ed. by G.G. Zuffa), pp. 165–189. NATO ASI Series C: vol. 148. Reidel, Dordrecht.

Zuffa, G.G. (1986) Carbonate particles in reading provenance from arenites: examples from the Mediterranean area (abstract). *Soc. econ. Paleont. Miner. Ann. Midyear Meeting Abstr.* **3**, 121.

Index

Pages on which figures appear are printed in *italic*, and those with tables in **bold**

Table 7.5. Flow sheet for clay mineral identification (from Starkey *et al.*, 1984)

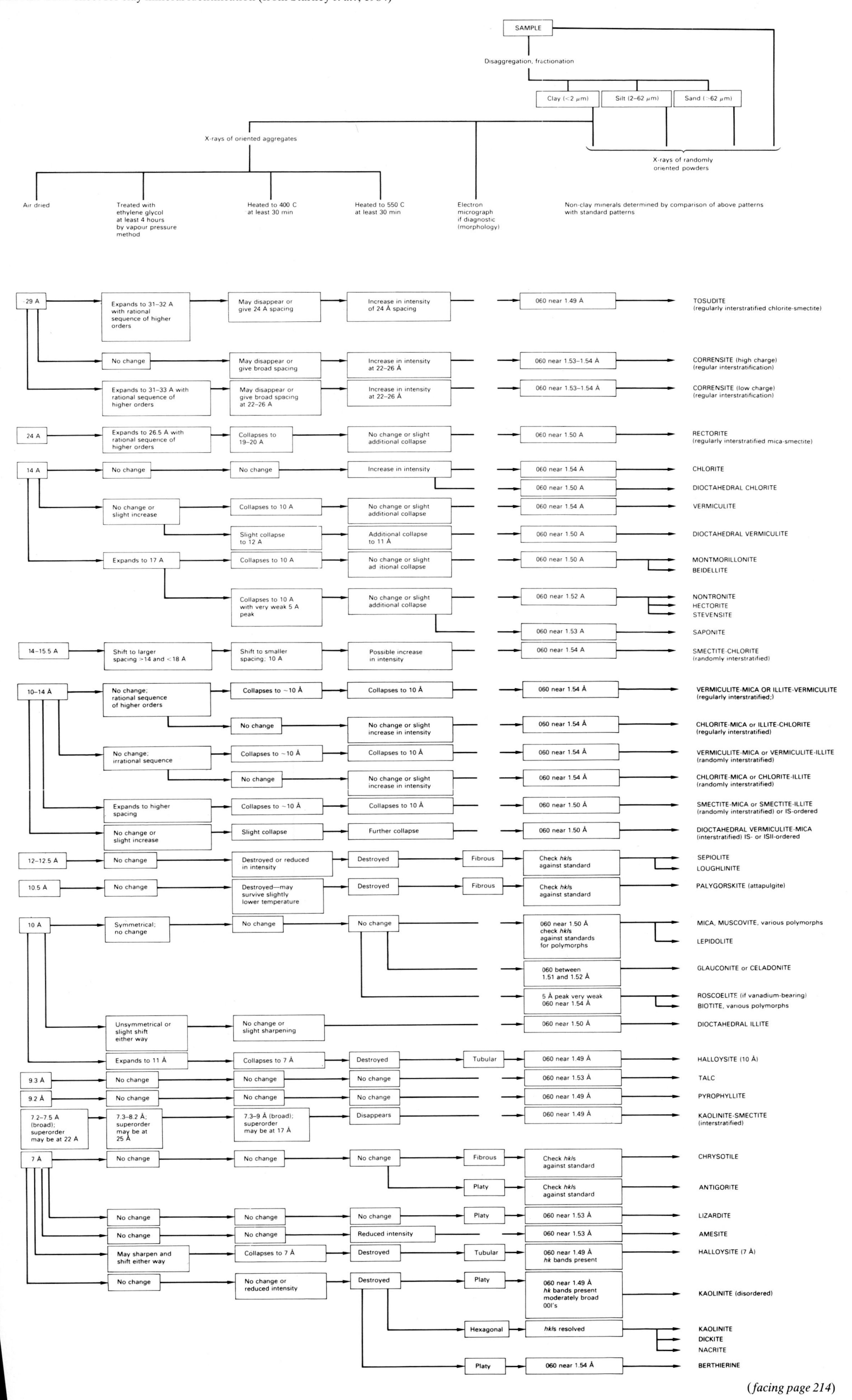